# Statistical Machine Learning

**CHAPMAN & HALL/CRC**
**Texts in Statistical Science Series**



**Recently Published Titles**

**Sampling**
Design and Analysis, Second Edition
*Sharon L. Lohr*

**The Analysis of Time Series**
An Introduction with R, Seventh Edition
*Chris Chatfield and Haipeng Xing*

**Time Series**
A Data Analysis Approach Using R
*Robert H. Shumway and David S. Stoffer*

**Practical Multivariate Analysis, Sixth Edition**
*Abdelmonem Afifi, Susanne May, Robin A. Donatello, and Virginia A. Clark*

**Time Series: A First Course with Bootstrap Starter**
*Tucker S. McElroy and Dimitris N. Politis*

**Probability and Bayesian Modeling**
*Jim Albert and Jingchen Hu*

**Surrogates**
Gaussian Process Modeling, Design, and Optimization for the Applied Sciences
*Robert B. Gramacy*

**Statistical Analysis of Financial Data**
With Examples in R
*James Gentle*

**Statistical Rethinking**
A Bayesian Course with Examples in R and STAN, Second Edition
*Richard McElreath*

**Randomization, Bootstrap and Monte Carlo Methods in Biology**
Fourth Edition
*Bryan F. J. Manly, Jorje A. Navarro Alberto*

**Statistical Machine Learning**
A Unified Framework
*Richard M. Golden*

**For more information about this series, please visit: https://www.crcpress.com/Chapman--HallCRC-Texts-in-Statistical-Science/book-series/CHTEXSTASCI**

# Statistical Machine Learning

## A Unified Framework

Richard M. Golden

CRC Press is an imprint of the
Taylor & Francis Group, an **informa** business
A CHAPMAN & HALL BOOK

First edition published 2020
by CRC Press
6000 Broken Sound Parkway NW, Suite 300, Boca Raton, FL 33487-2742

and by CRC Press
2 Park Square, Milton Park, Abingdon, Oxon, OX14 4RN



**Library of Congress Cataloging-in-Publication Data**

Names: Golden, Richard M., author.
Title: Statistical machine learning : a unified framework / Richard M. Golden.
Description: First edition. | Boca Raton, FL : CRC Press, 2020. | Includes bibliographical references and index.
Identifiers: LCCN 2020004853 | ISBN 9781138484696 (hardback) | ISBN 9780367494223 (paperback) | ISBN 9781351051507 (ebook)
Subjects: LCSH: Machine learning--Statistical methods. | Computer algorithms.
Classification: LCC Q325.5 .G65 2020 | DDC 006.3/10727--dc23
LC record available at https://lccn.loc.gov/2020004853

ISBN: 978-1-138-48469-6 (hbk)
ISBN: 978-1-351-05150-7 (ebk)

Typeset in CMR
by Nova Techset Private Limited, Bengaluru & Chennai, India

**Visit the Taylor & Francis Web site at**
**http://www.taylorandfrancis.com**

**and the CRC Press Web site at**
**http://www.crcpress.com**

# Contents

# IV Generalization Performance 359

# *Symbols*

## Symbol Description

## Linear Algebra and Matrix Notation

| | |
|---|---|
| $f : D \rightarrow R$ | Function on domain D with range R |
| $a$ or $A$ | Scalar variable |
| $\mathbf{a}$ | Column vector (lowercase boldface) |
| $\mathbf{a}^T$ | Transpose of column vector $\mathbf{a}$ |
| $\mathbf{A}$ | Matrix (uppercase boldface) |
| $\ddot{\mathbf{a}}$ | $\ddot{\mathbf{a}} : D \rightarrow R$ is an approximation for $\mathbf{a} : D \rightarrow R$ |
| $d\mathbf{f}(\mathbf{x}_0)/d\mathbf{x}$ | Jacobian of function $\mathbf{f}$ evaluated at $\mathbf{x}_0$ |
| $\exp(x)$ | Exponential function evaluated at $x$ |
| $\mathbf{exp}(\mathbf{x})$ | Vector whose $i$th element is $\exp(x_i)$ |
| $\log(x)$ | Natural logarithm function evaluated at $x$ |
| $\mathbf{log}(\mathbf{x})$ | Vector whose $i$th element is $\log(x_i)$ |
| $\lvert\mathbf{x}\rvert$ | Square root of scalar $\mathbf{x}^T\mathbf{x}$ |
| $\mathbf{I}_d$ | Identity matrix with dimension $d$ |
| $\det(\mathbf{M})$ | Determinant of matrix $\mathbf{M}$ |
| $\mathrm{tr}(\mathbf{M})$ | Trace of matrix $\mathbf{M}$ |
| $\mathbf{1}_d$ | Column vector of ones with dimension $d$ |
| $\mathbf{0}_d$ | Column vector of zeros with dimension $d$ |
| $A \times B$ | Cartesian product of set A and set B |
| $\mathcal{R}^d$ | Set of $d$-dimensional real column vectors |
| $\mathcal{R}^{m \times n}$ | Set of $m$ by $n$ real matrices |
| $\mathbf{A} \odot \mathbf{B}$ | Hadamard product (element-by-element multiplication) |
| $\mathbf{A} \otimes \mathbf{B}$ | Kronecker tensor product |

## Random Variables

| | |
|---|---|
| $\tilde{a}$ | Scalar random variable |
| $\tilde{\mathbf{a}}$ | Random column vector |
| $\tilde{\mathbf{A}}$ | Random matrix |
| $\hat{a}$ | Random scalar-valued function |
| $\hat{\mathbf{a}}$ | Random vector-valued function |
| $\hat{\mathbf{A}}$ | Random matrix-valued function |
| $\mathcal{B}^d$ | Borel sigma-field generated by open sets in $\mathcal{R}^d$ |

## Probability Theory and Information Theory

| | |
|---|---|
| $p_x(\mathbf{x})$ or $p(\mathbf{x})$ | Probability density for $\tilde{\mathbf{x}}$ |
| $p(\mathbf{x};\boldsymbol{\theta})$ | Density of $\tilde{\mathbf{x}}$ with parameter vector $\boldsymbol{\theta}$ |
| $p(\mathbf{x}\vert\boldsymbol{\theta})$ | Conditional density of $\tilde{\mathbf{x}}$ given $\boldsymbol{\theta}$ |
| $E\{\mathbf{f}(\tilde{\mathbf{x}})\}$ | Expectation of $f(\mathbf{x})$ with respect to density of $\tilde{\mathbf{x}}$ |
| $\mathcal{M}$ | Probability model |
| $\mathcal{H}(\tilde{\mathbf{x}})$ or $\mathcal{H}(p)$ | Entropy of random vector $\tilde{\mathbf{x}}$ with density $p$ |
| $\mathcal{H}(p_e, p)$ | Cross entropy of density $p$ with respect to $p_e$ |
| $D(p_e \Vert p)$ | Kullback-Leibler divergence of density $p$ from $p_e$ |

## Special Functions and Symbols

| | |
|---|---|
| $\mathcal{S}(x)$ | Logistic sigmoid $\mathcal{S}(x) = 1/(1+\exp(-x))$ |
| $\mathcal{J}(x)$ | Softplus sigmoid $\mathcal{J}(x) = \log(1+\exp(x))$ |
| $\boldsymbol{\theta}$ | Parameters of a learning machine |
| $\mathcal{D}_n$ | Data sample of $n$ pattern vectors |
| $\hat{\ell}_n$ | Empirical risk function for $\mathcal{D}_n$ |
| $\ell$ | Risk function estimated by $\hat{\ell}_n$ |
| $p_e$ | Environmental (data generating process) probability density |

# Preface

## Objectives

Statistical machine learning is a multidisciplinary field that integrates topics from the fields of machine learning, mathematical statistics, and numerical optimization theory. It is concerned with the problem of the development and evaluation of machines capable of inference and learning within an environment characterized by statistical uncertainty. The recent rapid growth in the variety and complexity of new machine learning architectures requires the development of improved methods for analyzing, designing, evaluating, and communicating machine learning technologies. The main objective of this textbook is to provide students, engineers, and scientists with practical established tools from mathematical statistics and nonlinear optimization theory to support the analysis and design of old, state-of-the-art, new, and not-yet-invented machine learning algorithms.

It is important to emphasize that this is a mathematics textbook intended for readers interested in a concise, mathematically rigorous introduction to the statistical machine learning literature. For readers interested in non-mathematical introductions to the machine learning literature, many alternative options are available. For example, there are many useful software-oriented machine learning textbooks which support the rapid development and evaluation of a wide range of machine learning architectures (Geron 2019; James et al. 2013; Muller and Guido 2017; Raschka and Mirjalili 2019). A student can use these software tools to rapidly create and evaluate a bewilderingly wide range of machine learning architectures. After an initial exposure to such tools, the student will want to obtain a deeper understanding of such systems in order to properly apply and properly evaluate such tools. To address this issue, there are now many excellent textbooks providing comprehensive and excellent descriptions of a wide variety of important machine learning algorithms (e.g., Duda et al. 2001; Hastie el al. 2001; Bishop 2006; Murphy 2012; Goodfellow et al. 2016). Such textbooks specifically omit particular technical mathematical details under the assumption that students without the relevant technical background should not be distracted, while students with graduate-level training in optimization theory and mathematical statistics can obtain such details elsewhere.

However, such mathematical and technical details are essential for providing a principled methodology for supporting the communication, analysis, design, and evaluation of novel nonlinear machine learning architectures. Thus, it is desirable to explicitly incorporate such details into self-contained, concise discussions of machine learning applications. Technical and mathematical details support improved methods for machine learning algorithm specification, validation, classification, and understanding. In addition, such methods can provide important support for rapid machine learning algorithm development and deployment, as well as novel insights into reusable modular software design architectures.

## Book Overview

A unique feature of this textbook is a statistical machine learning framework based upon the premise that machine learning algorithms learn a best-approximating probability distribution for representing the true data generating process (DGP) probability distribution. This statistical machine learning framework is defined by a collection of core theorems that support a unified framework for analyzing the asymptotic behavior of many commonly encountered machine learning algorithms. Explicit examples from the machine learning literature are provided to show students how to properly interpret the assumptions and conclusions of the framework's core theorems.

Part 1 is concerned with introducing the concept of machine learning algorithms through examples and providing mathematical tools for specifying such algorithms. Chapter 1 informally shows, by example, that many supervised, unsupervised, and reinforcement learning algorithms may be viewed as empirical risk function optimization algorithms. Chapter 3 provides a formal discussion of how optimization algorithms that minimize risk functions may be semantically interpreted as rational decision-making machines.

Part 2 is concerned with characterizing the asymptotic behavior of deterministic learning machines. Chapter 6 provides sufficient conditions for characterizing the asymptotic behavior of discrete-time and continuous-time time-invariant dynamical systems. Chapter 7 provides sufficient conditions for ensuring a large class of deterministic batch learning algorithms converge to the set of critical points of the objective function for learning.

Part 3 is concerned with characterizing the asymptotic behavior of stochastic inference and stochastic learning machines. Chapter 11 develops the asymptotic convergence theory for Monte Carlo Markov Chains for the special case where the Markov chain is defined on a finite state space. Chapter 12 provides relevant asymptotic convergence analyses of adaptive learning algorithms for both passive and reactive learning environments. Example applications include: the Gibbs sampler, the Metropolis-Hastings algorithm, minibatch stochastic gradient descent, stochastic approximation expectation maximization, and policy gradient reinforcement adaptive learning.

Part 4 is concerned with the problem of characterizing the generalization performance of a machine learning algorithm even in situations where the machine learning algorithm's probabilistic model of its environment is only approximately correct. Chapter 13 discusses the analysis and design of semantically interpretable objective functions. Chapters 14, 15, and 16 show how both bootstrap simulation methods (Chapter 14) and asymptotic formulas (Chapters 15, 16) can be used to characterize the generalization performance of the class of machine learning algorithms considered here.

In addition, the book features a comprehensive discussion of advanced matrix calculus tools to support machine learning applications (Chapter 5). The book also includes self-contained relevant introductions to real analysis (Chapter 2), linear algebra (Chapter 4), measure theory (Chapter 8), and stochastic sequences (Chapter 9) specifically designed to reduce the required mathematical prerequisites for the core material covered in Chapters 7, 10, 11, 12, 13, 14, 15, and 16.

## Targeted Audience

The material in this book is suitable for first- or second-year graduate students or advanced, highly-motivated undergraduate students in the areas of statistics, computer science, electrical engineering, or applied mathematics.

In addition, practicing professional engineers and scientists will find the material in this book to be a useful reference for verifying sufficient conditions for: (1) ensuring convergence of many commonly used deterministic and stochastic machine learning optimization algorithms, and (2) ensuring correct usage of commonly used statistical tools for characterizing sampling error and generalization performance.

The book does not make strong assumptions about the student's math background. Therefore, it is also appropriate for the professional engineer or the multidisciplinary scientist. The only prerequisites are lower-division linear algebra and an upper-division calculus-based probability theory course. Students possessing only these minimal mathematical prerequisites will find this text challenging yet accessible.

## Notation

Scalars will not be written in boldface notation and typically will be lowercase letters (e.g., $a$ might be used to denote a scalar variable). Following conventional notation in the engineering and optimization sciences, an uppercase boldface letter denotes a matrix (e.g., $\mathbf{A}$ might be used to denote a matrix), while a lowercase boldface letter denotes a vector (e.g., $\mathbf{a}$). The notation $\tilde{\mathbf{A}}$ indicates a matrix consisting of random variables. The notation $\hat{\mathbf{A}}$ typically indicates a matrix consisting of random functions. Thus, $\mathbf{A}$ may be a realization of $\hat{\mathbf{A}}$ or $\tilde{\mathbf{A}}$. This notation has the advantage of clearly distinguishing between random scalar variables, random vectors, and random matrices as well as their realizations.

However, this latter notation is non-standard in mathematical statistics where uppercase letters typically denote random variables and lowercase letters their realizations. Unfortunately, mathematical statistics notation typically does not explicitly distinguish between matrices and vectors. Because of the complexity of the engineering problems encountered in this text, a notation that can distinguish between matrices, vectors, and scalars is necessary, so standard statistics notation is not adequate. The standard engineering and optimization theory notation cannot be used without modification, however, because engineering and optimization notation does not explicitly distinguish between random vectors and their realizations. These notation difficulties have been recognized by econometricians (Abadir and Magnus 2002) and statisticians (Harville 2018).

The notation developed in this book corresponds to one possible resolution of this problem. In particular, the notation introduced here is an enhanced version of the typical matrix-vector notation used in areas of engineering and optimization theory that carefully distinguishes between random variables and their realizations.

## Instructional Strategies

The first course in an instructional sequence, which could be titled *Machine Learning Math*, covers Chapters 1, 2, 3, 4, 5, 8, 9, and 10 and moves quickly through Chapter 3. This first course is a required prerequisite for the second course in the sequence, *Statistical Machine Learning*, which covers Chapters 6, 7, 11, 12, 13, 14, 15, and 16 and moves quickly through Chapter 6. In both courses, proofs of the theorems should be de-emphasized or omitted. The instruction should focus on why the theorems are important and how the theorems are relevant for practical machine learning analysis and design problems. Figure 1 provides other recommendations regarding strategies for selecting chapter sequences for instruction or self-study.

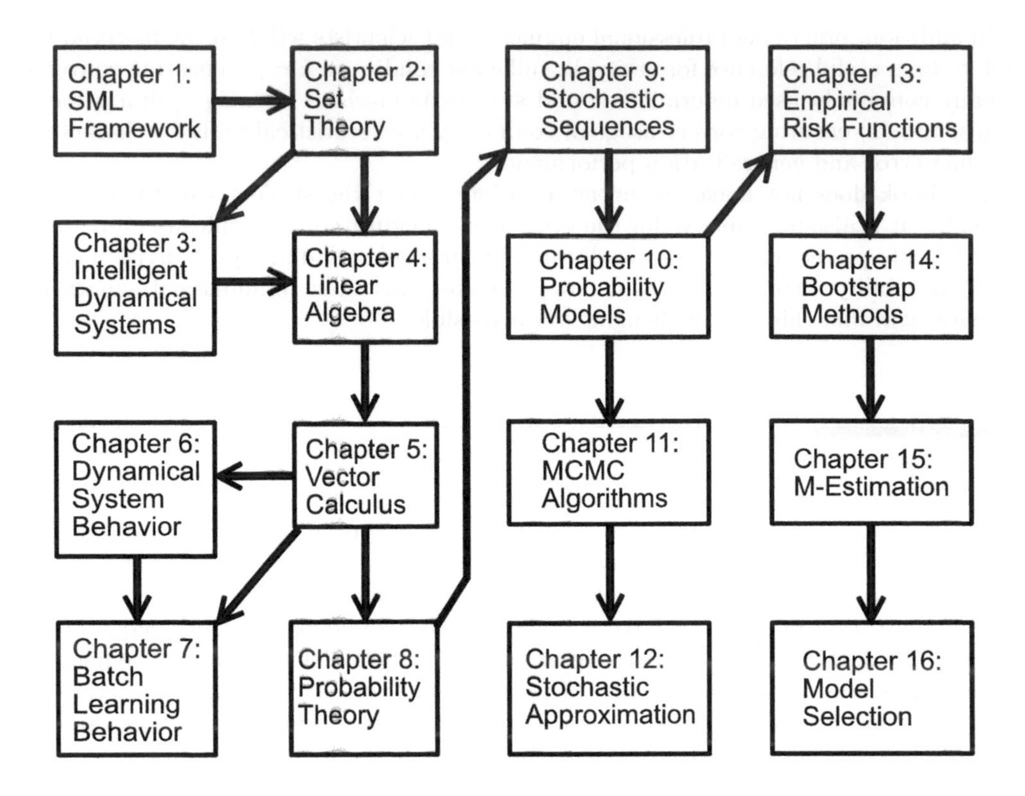


**FIGURE 1**
**Alternative suggested instructional paths for teaching or self-study.** Each pathway through the flowchart corresponds to a suggested instructional sequence.

The material in this book can also be used for a third advanced graduate seminar course titled *Advanced Statistical Machine Learning Math* which covers the proofs of the theorems in this text. The prerequisites for this advanced third graduate course would be completion of the second course in the three-course sequence. Only after the students understand the importance of the theorems and demonstrate the ability to apply the theorems in practice should technical details of proofs be discussed. It is recommended that the basic idea of each proof is first presented, followed by a more detailed discussion of each step of the proof.

At the end of each chapter, there is a section entitled *Further Reading* to enable the student to pursue the concepts in the chapter in greater depth. The majority of the problems are designed to help students master the key theorems and apply them in the analysis and design of a wide range of machine learning algorithms. In addition, algorithm design problems are included to help students understand the practical relevance of the theory.

Example course syllabi, sample exams, supplementary instructional materials, software problems, and book revisions are available at the author's website:

`www.statisticalmachinelearning.com`

## Acknowledgments

I thank my doctoral student Athul Sudheesh for the hypercube book cover concept.

I thank my doctoral student James Ryland for discussions related to learning algorithms and matrix calculus. I also thank my former doctoral student Shaurabh Nandy for discussions related to the empirical risk framework, convergence issues associated with MCMC sampling and adaptive learning, as well as model selection. My discussions with both James and Shaurabh were very helpful in identifying specific issues in earlier versions of this text which ultimately benefited from further elaboration.

My wife, Karen, has been ridiculously patient and understanding during this multi-year book-writing effort and, more generally, throughout my academic career. I can't thank her enough.

I dedicate this text to my parents, Sandy and Ralph, for their unconditional encouragement and support.

And finally, I am eternally grateful to the legions of students who struggled through earlier versions of this text over the past decade as the ideas and examples presented here evolved and self-organized. My students have contributed to the development of the text in profound ways by posing a simple question or displaying a quick puzzled expression!

## For the Student: Some Advice and Encouragement

### Studying Strategies

At the beginning of each chapter, specific learning objectives are outlined in a *Learning Objectives Box*. Please read those learning objectives carefully and use those learning objectives to develop strategies for reading the material in the chapter. In addition, *Recipe Boxes* are provided within many of the chapters for the purpose of providing straightforward recipes for applying key theorems described in the main text; for example, Recipe Box 0.1 provides some suggested study strategies.

### Encouragement and Guidance

Here's some additional useful advice for students preparing to read this book. I have personally found this advice valuable in my work throughout the years.

- **"Genius is 1% inspiration and 99% perspiration."**
  *Thomas Alva Edison*, twentieth-century American inventor
- **"From one thing, know ten thousand things."**
  *Miyamoto Musashi*, sixteenth-century samurai warrior
- **"There are many paths to enlightenment. Be sure to take one with a heart."**
  Unknown ancient Chinese philosopher

And, last but not least, *both* of the following important insights:

- **"The good God is in the details."**
  *Gustave Flaubert*, nineteenth-century French novelist
- **"The devil is in the details."**
  *Friedrich Wilhelm Nietzsche*, nineteenth-century German philosopher and poet

**Recipe Box 0.1 Suggestions for Reading This Book**

- **Step 1: Read entire Chapter 1 once.**
  Do not worry if you are puzzled by specific details. Revisit this chapter as necessary while reading Chapters 3 through 16.
- **Step 2: Read entire Chapter 2 once.**
  Do not worry about memorizing specific definitions. Carefully study all definitions and examples. Revisit this chapter as necessary while reading Chapters 3 through 16.
- **Step 3: Read entire Chapter 3 for the first time.**
  Try to understand how the theorems in the chapter relate to the chapter learning objectives. Do not worry about the assumptions of the theorems. Skim the examples with the goal of understanding their importance. Skip the proofs of the theorems. Focus attention on the recipe boxes in the chapter.
- **Step 4: Read entire Chapter 3 for the second time.**
  Focus attention on the statement of each theorem. Study carefully the assumptions and conclusion of each theorem. Carefully study every example in the entire chapter. Skip the proofs of the theorems. Try working the problems.
- **Step 5: Read the rest of the book.**
  Repeat Steps 3 and 4 for the remaining chapters of the book.
- **Step 6: Study basic idea of proofs.**
  When you are ready, study the *basic idea* of the proofs of the theorems, but only after you have understood the theorem statements and why those theorem statements are relevant in engineering applications. Understanding the proofs will give you more confidence in applying the theorems and help you apply the theorems correctly in practice. Eventually, you will want to verify all technical details of every proof.

# Part I

# Inference and Learning Machines

# 1

# *A Statistical Machine Learning Framework*

**Learning Objectives**

- Explain benefits of the empirical risk minimization framework.
- Design supervised learning gradient descent algorithms.
- Design unsupervised learning gradient descent algorithms.
- Design reinforcement learning gradient descent algorithms.

## 1.1 Statistical Machine Learning: A Brief Overview

Machine learning algorithms are now widely used throughout society. Example machine learning applications include: weather prediction, handwriting recognition, translation of speech to text, document search and clustering, missing data analysis, stock market predictions, detecting fraudulent financial transactions, voice recognition systems, detecting if an email message is spam, identifying computer viruses, identifying the identity of faces and places in an on-line photo library, identifying medical disorders, analyzing DNA sequences, solving complex sorting and scheduling tasks, predicting customer purchasing preferences, detecting and classifying signals, supporting intelligent tutoring systems, controlling robots in manufacturing and vehicles, and supporting bionic implants.

Statistical machine learning algorithms are fundamentally inductive learning machines which learn by example. Inductive inference machines are fundamentally different from deductive inference machines. Finite state automata, production systems, relational databases, Boolean logic, and Turing machines are important examples of deductive inference machines. A deductive inference machine generates logical inferences from input patterns using a knowledge base which may be interpreted as a collection of rules. In contrast, inductive inference machines generate plausible inferences from input patterns using a knowledge base which may be interpreted as a collection of beliefs. Probabilistic inductive logic can be shown, under general conditions, to be the only inductive logic which is consistent with the Boolean logic in the special case where one's belief that an event will occur is certain (Cox 1946).

Statistical machine learning is built on the framework of probabilistic inductive logic. It is assumed that the training data is generated from the environment by sampling from a probability distribution called the *environmental distribution*. The process which generates the training data is called the *data generating process*. A statistical learning machine's knowledge base is a set of probability distributions which is called the learning machine's

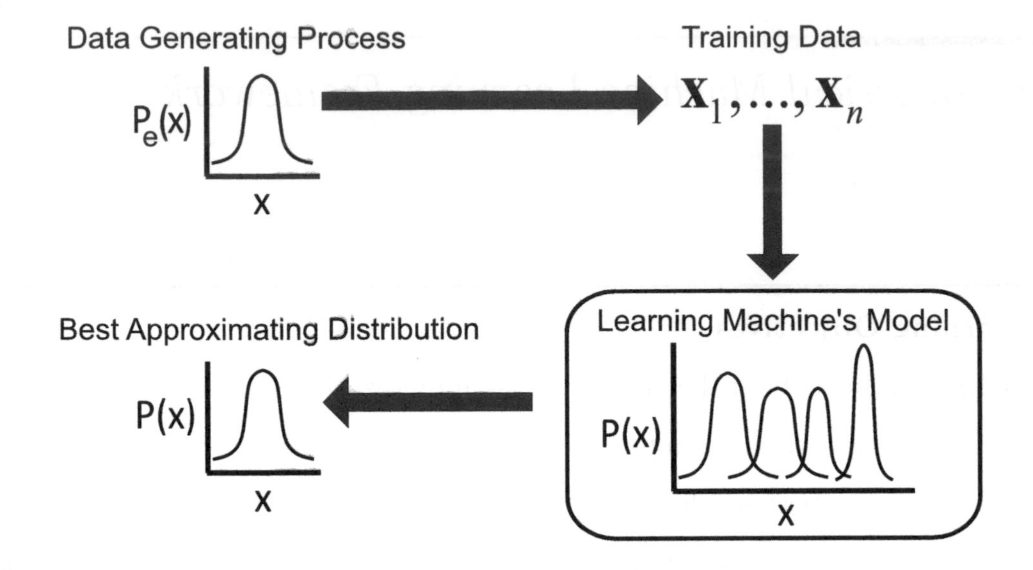


**FIGURE 1.1**
**The statistical machine learning framework.** The Data Generating Process (DGP) generates observable training data from the unobservable environmental probability distribution $P_e$. The learning machine observes the training data and uses its beliefs about the structure of the environmental probability distribution in order to construct a best-approximating distribution $P$ of the environmental distribution $P_e$. The resulting best-approximating distribution of the environmental probability distribution supports decisions and behaviors by the learning machine for the purpose of improving its success when interacting with an environment characterized by uncertainty.

*probability model.* The learning machine additionally may have specific beliefs regarding the likelihood that a probability distribution in its probability model is relevant for approximating the environmental distribution.

In the statistical machine learning framework, the goal of the learning process is not to memorize the training data. Rather, the statistical machine's goal is to learn the data generating process by searching for the probability distribution in the learning machine's probability model which best approximates the environmental distribution. This best-approximating distribution allows the learning machine to estimate the probability of specific environmental events. The learning machine, in turn, can use this best-approximating distribution to make appropriate probabilistic inductive inferences (see Figure 1.1). These issues are discussed in further detail in Section 10.1.

## 1.2 Machine Learning Environments

### 1.2.1 Feature Vectors

#### 1.2.1.1 Feature Vector Representations

Because the experiences of a learning machine will influence its behavior, it is useful to discuss the environment within which a learning machine lives before discussing specific types of learning machines. An "environmental event" corresponds to a snapshot of the learning machine's environment over a particular time period in a particular location. Examples of "events" include: the pattern of light rays reflecting off a photographic image, air pressure vibrations generated by a human voice, an imaginary situation, the sensory and physical experience associated with landing a lunar module on the moon, a document and its associated search terms, a consumer's decision-making choices on a website, and the behavior of the stock market.

A *feature map* is a function that maps an environmental event into a *feature vector* whose elements are real numbers. Note that a critical consequence of this representation is that some information is suppressed or destroyed while other information is emphasized or exaggerated in the feature vector representation. Figure 1.2 illustrates this key idea. A variety of different coding schemes which correspond to different choices of feature maps are commonly used in machine learning applications. The choice of an appropriate coding scheme is crucial. An inappropriate coding scheme can make an easy machine learning problem computationally impossible to solve. Conversely, a clever appropriate scheme can make an extremely difficult machine learning problem very easy to solve.

A *numerical coding scheme* is typically used to model the output of a sensor or a collection of sensors. For example, a photographic image might be represented by an array of pixels where each pixel's value is represented by a number indicating the gray level of that pixel. Other examples of numerical coding schemes might be the output voltage of a microphone at a particular time or the closing price of a security on a particular day.

A *binary coding scheme* is used to model the presence or attribute of a feature. There are two major types of binary coding schemes. A *present-absent binary coding scheme* corresponds to a feature map that returns the value of one if some environmental event is present and returns the value of zero if that environmental event is absent. A *present-unobservable binary coding scheme* corresponds to a feature map that returns the value of one if some environmental event is present and returns the value of zero if that environmental event is

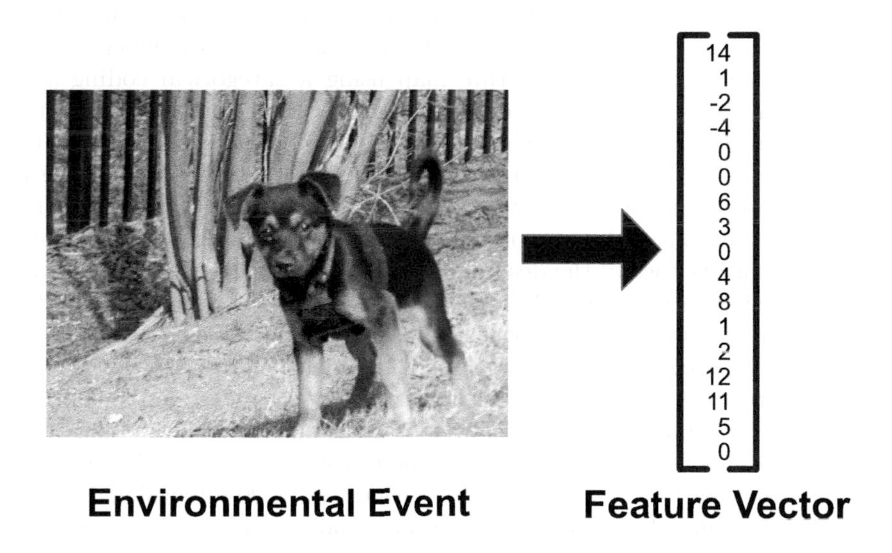


FIGURE 1.2
**Representation of real-world events as feature vectors.** An event in the environment of a learning machine is represented as a feature vector whose elements are real numbers. The feature vector suppresses irrelevant information and exaggerates relevant information in order to provide an appropriate abstract representation of the event.

unobservable. An image processing example of a binary coding scheme might be a feature map which looks at a particular region of an image and attempts to determine whether or not an edge is present in that region. A knowledge-based example of a binary coding scheme might be a feature map which returns the value of one when a proposition is true and the value of zero when that proposition is false. A feature vector consisting of $d$ binary features can take on $2^d$ possible values.

A *categorical coding scheme* models an $M$-valued categorical variable such that any pair of the $M$ values are semantically equally dissimilar. An example of such a categorical coding scheme would be the color of a traffic light which can take on the values of `red`, `green`, and `yellow`. Another example might be a word variable which can take on one of $M$ possible values corresponding to possible words from a vocabulary consisting of $M$ possible words. A third example might be an action variable used to specify the behavior of a robot which could take on values such as `move-forward`, `move-backward`, `pick-up-block`, or `do-nothing`.

There are two major types of categorical coding schemes which are commonly used in the machine learning literature. The first coding scheme is called a *one-hot encoding scheme* where the categorical variable can take on $M$ possible values such that the $k$th value of the categorical variable is an $M$-dimensional vector with a one in the $k$th position and zeros in the remaining $M - 1$ elements of the vector. A second type of categorical variable coding scheme is called *reference cell coding.* In a reference cell coding scheme, a categorical variable that takes on $M$ possible values is represented by an $M - 1$-dimensional vector such that $M - 1$ values of the $M$-valued categorical variable are represented as the columns of an $M - 1$-dimensional identity matrix and the remaining value of the categorical variable is represented as a column vector of zeros.

Larger feature maps are often constructed as a grouping of smaller feature maps. For example, suppose the goal of a learning machine is to associate an image with a particular label. Assume there are 200 possible labels for the image. The 300 by 500 array of image pixels is represented by an array of 300 by 500 feature maps where each feature map returns a number indicating the grey scale level of a particular pixel in the image using a numerical coding scheme. The output of the 150,000-pixel feature maps is a large 150,000-dimensional feature vector. The desired response of the learning machine is a label which could be represented as a single categorical feature map using a categorical coding scheme. The single categorical feature map generates a 200-dimensional vector which has the number one in its $k$th element and zeros elsewhere to specify that the $k$th label should be assigned to the image. Thus, in this example, the environment presents to the learning machine a large feature vector consisting of 150,000 + 200 = 150,200 elements where the first 150,000 elements specify the image which is presented as input to the learning machine and the remaining 200 elements specify the desired response of the learning machine. The learning machine is trained with examples of labeled images and then its labeling performance is ultimately evaluated using a new set of images which are not labeled.

In this example, notice that the large 150,200-dimensional feature vector is naturally partitioned into two subvectors. The first subvector with dimension 150,000 is called the *input pattern vector* while the second subvector with dimensionality 200 is called either the *desired response pattern vector* or *reinforcement feedback vector.* The learning machine processes the input pattern vector and then generates a 200-dimensional *output pattern vector* which is typically also coded using the same coding scheme as the desired response pattern vector.

#### 1.2.1.2 The Curse of Feature Vector Dimensionality

At first glance, one might think that increasing the number of features in a feature vector would make the learning process for a learning machine easier because it could simply learn to use the features that it requires. However, increasing the number of features results in a situation which Bellman (1961) refers to as the "curse of dimensionality". As the number of features in a feature vector increases, the number of possible mappings from input to output increases at an exponential rate. To see this, suppose that each input feature in a learning machine can take on $M$ possible values. And suppose that the output of the learning machine is a binary response which generates either a zero or a one. The number of possible input patterns is $M^d$ and the number of possible responses is 2 so the number of possible input-to-output mappings is $2M^d$. Therefore, the number of input-to-output mappings increases exponentially as the dimensionality $d$ of the feature vector increases linearly.

A learning machine which can represent a large number of possible input-to-output mappings has the advantage that because it can represent more functions it should, in principle, have a better chance of approximating the input-to-output mapping that optimizes its performance within its environment. However, this advantage is counterbalanced by the disadvantage that the learning machine now has a more difficult learning problem. It is more challenging to find a good input-to-output mapping when the number of possible input-to-output mappings is extremely large.

In order to reduce feature vector dimensionality, the goal is to select a subset of all possible input features that will improve the performance of the classification machine or regression model. Choosing a good subset of the set of all possible input features can dramatically improve the performance of a classification machine or regression model. This problem of selecting a best subset of features is called "best subsets regression" in the statistical literature (Beale et al. 1967; Miller 2002). One approach to solving the best subsets regression problem is to simply try all possible subsets of input features. If the number of input features is less than about 20 and if the data set is not too large, then a supercomputer can approach the best subsets regression problem by exhaustively considering all possible $2^d$ subsets of set of $d$ possible input features. This is called "exhaustive search". For situations, however, involving either larger data sets or more than about 20 input features, then the method of exhaustive search is not a viable alternative because it is too computationally intensive.

#### 1.2.1.3 Feature Vector Engineering Methods

The importance of choosing an appropriate feature space representation of a problem can be traced back to the early days of machine learning (e.g., Rosenblatt 1962; Nilsson 1965; Minsky and Papert 1969; Andrews 1972; Duda and Hart 1973) and continues to play a critical role in machine learning engineering practice. Dong and Liu (2018) provide a modern review of a wide range of feature engineering methods for machine learning. One strategy for selecting a feature representation is to design some method of *automatically learning* useful feature vector representations. This approach was initially explored by Rosenblatt (1962) in his studies of the Perceptron (a linear classification machine that learned randomly sampled feature vector recoding transformations) and is a central component of "deep learning" methods (see Goodfellow et al. 2016 for a review).

### 1.2.2 Stationary Statistical Environments

The assumption that a statistical environment is "stationary" means that environmental statistics cannot change as a function of time. An important type of stationary statistical environment can be defined by specifying a feature transformation that maps sequences of events in the real world into a sequence of feature vectors (see Chapter 9). In order for learning to be effective, some assumptions about the characteristics of the statistical learning environment are required. It is typically assumed that the statistical environment is stationary. Stationarity implies, for example, that if a learning machine is trained with a particular data set, then the particular time when this training occurs is irrelevant. Stationarity also implies that there exist statistical regularities in training data which will be useful for processing novel test data.

One can imagine many potential adverse consequences of nonstationary statistical environments. For example, a learning machine might be trained to classify a given pattern vector as a member of a specific category of patterns. This classification is deemed correct, but then the statistical environment changes its characteristics. After that change, the learning machine makes an error when classifying a previously learned pattern because now the previously learned response is no longer correct. In many practical applications, the stationarity assumption does not automatically hold. For example, for weather prediction problems, data collected over the summer months may not be predictive over the winter months. Or, for stock market prediction problems, data collected during a time period in the economy where inflation is present may not be predictive over a time period where inflation is absent. An understanding of these issues often leads to improved methods for sampling or recoding the statistical environment so that the statistical environment experienced by the learning machine is approximately stationary.

In many important machine learning applications, machine learning environments are represented as either a stationary environment, a slowly changing statistical environment, or a sequence of statistical environments that converges to a stationary environment (e.g., Konidaris and Barto 2006; Pan and Yang 2010). Another approach assumes that the machine learning environment consists of a collection of stationary environments and the probability distribution of observing a particular sequence of such stationary environments is also stationary.

### 1.2.3 Strategies for Teaching Machine Learning Algorithms

In *adaptive learning problems*, the learning machine observes a sequence of events and adaptively revises its current state of knowledge each time it experiences a new event. For *batch learning problems* (also known as "off-line" learning problems), the learning machine observes a collection of events and updates its current state of knowledge simultaneously with the entire collection of events. For *minibatch learning problems*, the learning machine observes a subset of events at a time and revises its current state of knowledge each time a subset of events is experienced. In situations where all of the training data is available, batch learning methods are usually preferred because they are often more effective than adaptive learning methods. However, in many real-world situations only adaptive learning methods are applicable because either processing the entire batch of training data is too computationally intensive or parts of the data set are only incrementally available over time.

Another important class of learning problems are the *reactive environment reinforcement learning problems*. For this class of problems, the machine learning algorithm interacts with its environment during the learning process such that the actions of the learning machine influence the behavior of the machine learning algorithm's environment. For example, suppose that the goal of the learning process is to teach the machine learning algorithm to land a helicopter. After experiencing a helicopter flying episode in its statistical environment, the learning machine updates its parameters. This parameter update causes the learning machine to behave differently and generate different actions in response to its inputs. These different actions, in turn, will influence the future learning experiences available for the learning process. That is, an inexperienced machine learning algorithm will continually crash the helicopter when practicing landings and will never have the opportunity to experience the perfect helicopter landing! However, if the inexperienced machine learning algorithm is provided some initial instruction in the form of specialized training in artificial environments or prior knowledge in the form of architectural constraints, the machine learning algorithm may have enough knowledge to keep the helicopter in the air long enough so that relevant learning experiences can be acquired. This is an example of how learning experiences in reactive learning environments are determined not only by the environment but additionally how the learning machine interacts with its environment.

Now consider the more difficult problem of teaching a machine learning algorithm to control the landing of a helicopter in the presence of violent wind gusts. It might be impossible for such a learning machine to acquire the relevant statistical regularities through exposure to this type of complex environment. On the other hand, a teacher might first ensure that the learning machine practices landing the helicopter without wind gusts. When this skill is mastered, then the teacher might then make the statistical environment more realistic and include wind gusts. The parameters estimated in the first statistical environment correspond to hints about what parameters should be estimated in the second more realistic statistical environment. This strategy of training a learning machine using sequences of distinct statistical environments is extremely important and provides a practical way for teaching nonlinear learning machines to learn complex problems. An important research topic of interest is the development of new methods of teaching learning machines through the construction of a carefully crafted sequence of stationary environments (Konidaris and Barto 2006; Bengio et al. 2009; Taylor and Stone 2009; Pan and Yang 2010).

### 1.2.4 Prior Knowledge

Although machine learning problems are often characterized by large data sets, a large data set does not necessarily imply that the data set contains sufficient information to support reliable and effective inferences. For example, performance in predicting an individual's name given their age is unlikely to improve regardless of the size of the database. Therefore, in many situations, the performance of a learning machine can be dramatically improved through the incorporation of hints and prior knowledge.

By incorporating prior environmental knowledge constraints properly into the structure of a learning machine, improvements in learning efficiency and generalization performance can often be realized. On the other hand, incorporating the wrong types of constraints will decrease generalization performance. The concept of a probability model and methods for incorporating prior knowledge into such models is discussed in Chapters 10 and 13. In addition, both simulation methods (Chapters 12, 14) and analytical methods (Chapters 15, 16) are discussed in this text for the purpose of evaluating the quality and implications of specific types of hints and constraints.

#### 1.2.4.1 Feature Maps Specify Relevant Statistical Regularities

An extremely important source of prior knowledge is the methodology for representing events in the real world as feature vectors. A poorly designed feature vector representation can transform an easy machine learning problem into an impossible problem; while a well-designed feature vector representation can make an impossible machine learning problem easy to solve. For example, suppose that a medical researcher develops a simple machine learning algorithm which generates a "health score number" $\ddot{y}$ where larger values of $\ddot{y}$ indicate increased levels of health while small values indicate poor health. Typically, the environmental event characterized by a person's body temperature is measured by a thermometer which generates a number $s$ where $s = 98.6$ indicates a normal body temperature in degrees. This measurement scale is arbitrary so define a preprocessing transformation $f$ which maps a body temperature $s$ into a health score $f(s)$. The health score $f(s)$ is a one-dimensional feature vector generated from the event in the environment corresponding to a patient's body temperature. The simple machine learning algorithm proposed by the researcher is $\ddot{y} = f(s - \beta)$ where $\beta$ is a free parameter which is estimated by the learning machine from examples of body temperature measurements from both healthy and unhealthy patients. The choice of the preprocessing transformation $f$ is crucial. If $f$ is chosen such that $f(s) = s - \beta$, then there does not exist a choice of $\beta$ such that a patient will be assigned a large health score if their temperature is much larger or much smaller than 98.6. However, if $f$ is chosen such that $f(s) = \exp\left(-(s - \beta)^2\right)$, then the learning machine can easily learn to detect abnormal body temperatures by estimating $\beta = 98.6$.

#### 1.2.4.2 Similar Inputs Predict Similar Responses

The assumption that very similar input patterns should generate similar responses is a useful prior knowledge heuristic for processing novel input patterns. That is, this prior knowledge assumption involves placing constraints on estimated conditional probability distributions such that similar input patterns have similar response probabilities. For example, given $d$ binary input variables, it is possible to represent $2^d$ possible $d$-dimensional binary input pattern vectors. Assume the goal of learning is to estimate the probability a binary response variable will take on the value of one given a particular $d$-dimensional binary input pattern vector. Given $2^d$ free parameters, it is possible to specify the probability the response variable takes on the value of one for each of the possible $2^d$ input patterns. This latter probability model will be referred to as the "contingency table model". The contingency table model does not incorporate the prior knowledge constraint that similar input patterns should generate similar response probabilities.

In a logistic regression model, however, the probability that the binary response variable takes on the value of one is a monotonically increasing function of a weighted sum of the $d$ binary input variables where the $d$ weights are the free parameters. Thus, only $d$ free parameters rather than $2^d$ free parameters are used in the probability model. This exponential reduction in the number of free parameters has multiple benefits including increased learning speed as well as the ability of the system to incorporate prior knowledge about how the learning machine should generate a response to a novel input pattern which is similar to a previously learned input pattern. In environments, however, when the assumption that similar input patterns should yield similar response patterns fails, then the logistic regression model may exhibit catastrophic difficulties in learning. Such difficulties are a consequence of the logistic regression model's inability to adequately represent the wide range of probability distributions which can be represented using a contingency table representation scheme.

#### 1.2.4.3 Only a Small Number of Model Parameters Are Relevant

To ensure that a learning machine has the flexibility to learn a wide variety of concepts, it is often helpful to include many additional free parameters in the learning machine. Using the concept of a regularization term (see Section 1.3.3) which forces the learning process to penalize models with redundant and irrelevant free parameters, it is possible to incorporate prior knowledge that informs the learning machine that many of its free parameters are redundant and irrelevant. In situations where this assumption holds, this can improve generalization performance by reducing the effects of overfitting.

#### 1.2.4.4 Different Feature Detectors Share Parameters

Another example of the incorporation of prior knowledge constraints is the recent success of Convolutional Neural Networks (CNNs) and Recurrent Neural Networks (RNNs) for successfully addressing very difficult problems in many areas of science and engineering. Goodfellow et al. (2016) provides a comprehensive introduction to CNNs and RNNs, while Schmidhuber (2015) provides a comprehensive recent review of the literature.

Although convolutional neural networks and recurrent neural networks are often described as learning machines that make minimal assumptions about the structure of their statistical environments, they incorporate considerable prior knowledge about their statistical environment. In particular, a key property of the CNN is that if a feature detector learns a statistical regularity in one region of the image then the CNN automatically learns that statistical regularity in other regions of the image as well. That is, spatial features are not location-specific within an image. This constraint is implemented by constraining groups of feature detectors in a CNN such that they share the same parameter values. Similarly, recurrent neural networks make extensive use of parameter sharing for the purpose of learning time-invariant statistical regularities.

## Exercises

1.2-1. Explain how each of the following environmental events could be represented as a feature vector: (1) a digital image consisting of thousands of colored pixels, (2) the output of a microphone which is a time-varying voltage, (3) the behavior of a collection of financial securities which is a time-varying collection of closing prices, (4) consumer purchasing decisions based upon the placement of website ads at particular space-time locations, and (5) an episode consisting of a sequence of events where a human pilot lands a helicopter by interacting with the helicopter control panel.

1.2-2. During a learning process, a learning machine is provided a collection of digital images of faces. After the learning process, the learning machine observes a digital image of a face and generates either the number 0 or 1 to indicate whether or not the digital image of the face is a familiar face. Provide three different ways of representing a digital image of a face as a feature vector. Provide one example where the feature vectors used to train this system are generated from a stationary statistical environment and one example where the feature vectors used to train this system are not generated from a stationary environment. What types of prior knowledge could one provide the learning machine to facilitate the learning process?

1.2-3. A medical doctor wishes to use a machine learning algorithm to generate a medical diagnosis regarding the presence or absence of `CANCER`. The basis of the medical diagnosis is determined by the following list of patient characteristics: `GENDER`,`BLOOD-PRESSURE`, and `HEART-RATE` and patient symptoms: `FEVER`, `MUSCLE-SORENESS`, `SORE-THROAT`, and `PURPLE-DOTS-ON-FOREHEAD`. Accordingly, the medical doctor collects data on the patients at his clinic and trains a learning machine to predict the presence or absence of one or more particular diseases given patient characteristics and symptoms. Provide three different ways to represent this event as a feature vector which has an "input pattern" component and a "response pattern" component. Provide one example where the feature vectors used to train this system are generated from a stationary statistical environment and one example where the feature vectors used to train this system are not generated from a stationary environment. What types of prior knowledge could one provide the learning machine to facilitate the learning process?

1.2-4. A machine learning company has developed a self-driving car whose current state includes information obtained from the car's image, audio, radar, and sonar sensors at time $t$. Each test of the self-driving car generates a sequence of states and at the end of the sequence of states the car is either punished for poor driving with a reinforcement signal of -10 or rewarded for excellent driving with a reinforcement signal of +10. Provide three different methods for representing the event of observing a particular test drive with such a reinforcement signal as a feature vector. Provide one example where the feature vectors used to train this system are generated from a stationary statistical environment and one example where the feature vectors used to train this system are generated from a nonstationary environment. What types of prior knowledge could one provide the learning machine to facilitate the learning process?

1.2-5. Consider one situation where a machine learning algorithm is trained to drive a car by showing the machine learning algorithm video clips of humans driving cars. Consider another situation where the machine learning algorithm actually controls the behavior of a car in a physical environment and learns from these interactive experiences. Which of these situations would correspond to learning in a reactive learning environment?

1.2-6. Suppose a researcher attempts to train a learning machine to parallel park a self-driving car with examples of parallel parking but the learning machine is unable to solve this problem. Explain how to design a nonstationary environment $E$ consisting of a sequence of different stationary environments $E_1$, $E_2$, $\ldots$ which can help the learning machine solve the parallel parking problem. More specifically, assume that the learning machine learns the stationary environment $E_1$ to obtain parameter estimates $\hat{\boldsymbol{\theta}}_1$, then uses parameter estimates $\hat{\boldsymbol{\theta}}_1$ as an initial hint for the parallel parking solution for stationary environment $E_2$ whose solution yields the parameter estimates $\hat{\boldsymbol{\theta}}_2$, and so on.

## 1.3 Empirical Risk Minimization Framework

### 1.3.1 ANN Graphical Notation

It will be convenient to discuss the architecture of certain classes of machine learning algorithms using a particular graphical notation originally developed in the Artificial Neural Network (ANN) literature (e.g., Rosenblatt 1962; Rumelhart, Hinton, and McClelland 1986) and also used in the deep learning literature (e.g., Goodfellow et al. 2016). This *ANN graphical notation* specifies a machine learning algorithm as a collection of nodes or *units* where each node has a state called its *activity level*. Typically, the activity level of a node is a real number. An output unit generates its state or activity level from the states of the input nodes which pass information in the direction of the arrows connecting the input units to the output unit. Each output unit also has a set of parameters which are specific to that unit and which may influence how the states of the input units influence computation of the output unit's state. This set of unit-specific parameters are called the *connection weights* for the unit. More technically, this type of formulation of state-updating and learning dynamics corresponds to a dynamical system specification of the machine learning algorithm. Chapter 3 provides a formal systematic approach for specifying machine learning algorithms as dynamical systems.

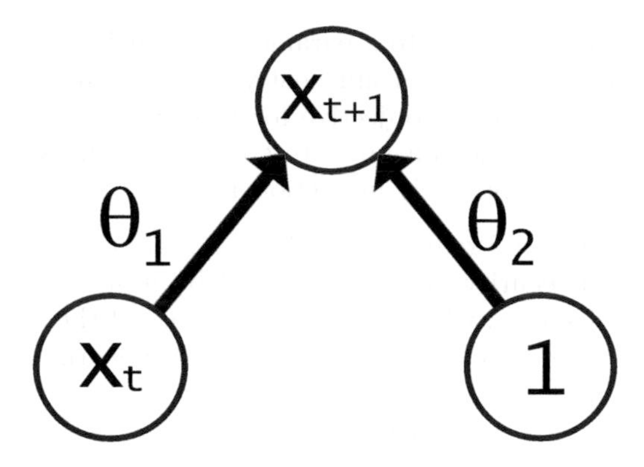


**FIGURE 1.3**
**Artificial neural network node representation of stock market linear regression predictive model.** An artificial neural network node representation of a linear regression model which predicts tomorrow's closing price $x_{t+1}$ given today's closing price $x_t$ using the formula $\hat{x}_{t+1} = \theta_1 x_t + \theta_2$.

Figure 1.3 illustrates the structure of a prediction machine using this graphical artificial neural network node representation. The input pattern consists of two nodes whose respective states are $x_t$ and 1. The output node is the prediction of state $x_{t+1}$ denoted by $\hat{x}_{t+1}(\boldsymbol{\theta})$ and which is computed as a weighted sum of states of the two input pattern nodes so that $\hat{x}_{t+1}(\boldsymbol{\theta}) = \theta_1 x_t + \theta_2$.

### 1.3.2 Risk Functions

A large class of machine learning algorithms can be interpreted as Empirical Risk Minimization (ERM) learning algorithms which are formally introduced in Chapter 13. In an ERM learning problem it is assumed that the learning machine's statistical environment generates $n$ events which are detected by the learning machine as a collection of $n$ $d$-dimensional feature vectors or "training vectors". The set of $n$ training vectors $\mathbf{x}_1, \ldots, \mathbf{x}_n$ is called the "training data set" $\mathcal{D}_n$. In addition, the learning machine's state of knowledge is specified by a *parameter vector* $\boldsymbol{\theta}$. The goal of the learning process is to choose particular values of the learning machine's state of knowledge $\boldsymbol{\theta}$.

The learning machine's actions and decisions are entirely determined by the parameter vector $\boldsymbol{\theta}$ of the learning machine. Let $\mathbf{x} \in \mathcal{D}_n$ denote a training vector corresponding to an environmental event observed by the learning machine. The quality of the learning machine's ability to make good actions and decisions when the machine observes $\mathbf{x}$ with a current parameter vector $\boldsymbol{\theta}$ is measured by a *loss function* $c$ which maps $\mathbf{x}$ and $\boldsymbol{\theta}$ into the number $c(\mathbf{x}, \boldsymbol{\theta})$. The number $c(\mathbf{x}, \boldsymbol{\theta})$ is semantically interpreted as the penalty received by the learning machine for experiencing the event vector $\mathbf{x}$ in its environment given its knowledge state parameter vector $\boldsymbol{\theta}$. In order to minimize the average received penalty, the learning machine seeks a parameter vector $\boldsymbol{\theta}$ that minimizes the empirical risk function

$$\hat{\ell}_n(\boldsymbol{\theta}) = (1/n) \sum_{i=1}^{n} c(\mathbf{x}_i, \boldsymbol{\theta}) \tag{1.1}$$

where the set $\mathcal{D}_n \equiv [\mathbf{x}_1, \ldots, \mathbf{x}_n]$ is the training data. If the sample size $n$ is larger, then the learning machine will be able to more accurately find a parameter vector $\boldsymbol{\theta}$ which minimizes the "true risk". The true risk assigned to a parameter vector is approximately equal to the empirical risk $\hat{\ell}_n$ when the sample size is sufficiently large. Minimizing the true risk as well as the empirical risk is an important objective for learning machines which must exhibit effective generalization performance on novel test data as well as familiar training data. These ideas are discussed more precisely in Section 13.1.

**Example 1.3.1** (Stock Market Prediction Model). Consider a prediction machine which uses the closing price of a security at day $t$ denoted by input pattern $x_t$ and parameter vector $\boldsymbol{\theta} = [\theta_1, \theta_2]$ to predict $x_{t+1}$, which is the closing price of the security at day $t+1$ (see Figure 1.3). In particular, the closing price predicted by the learning machine for day $t+1$ is expressed by the formula

$$\hat{x}_{t+1}(\boldsymbol{\theta}) = \theta_1 x_t + \theta_2.$$

Now assume that one has a history of closing prices for the security over the past $n$ days which is denoted by: $x_1, x_2, \ldots, x_n$. A measure of the prediction machine's performance is specified by the empirical risk function

$$\hat{\ell}_n(\boldsymbol{\theta}) = (1/(n-1)) \sum_{t=1}^{n-1} \left(x_{t+1} - \hat{x}_{t+1}(\boldsymbol{\theta})\right)^2 .$$

Notice that when $\hat{\ell}_n(\boldsymbol{\theta})$ is smaller in value, the prediction machine generates improved predictions. Thus, the goal of the learning process is to find a parameter vector $\hat{\boldsymbol{\theta}}_n$ which minimizes the objective function $\hat{\ell}_n(\boldsymbol{\theta})$. △

**Example 1.3.2** (Nonlinear Least Squares Regression Risk Function). Let $\mathbf{x}_i = [\mathbf{s}_i, y_i]$ be the $i$th training stimulus in a data set consisting of $n$ training stimuli where $y_i$ is the desired response for the learning machine when the $d$-dimensional input pattern vector $\mathbf{s}_i$ is presented to the learning machine, $i = 1, \ldots, n$. Let the prediction of a learning machine be defined as a smooth nonlinear function $\ddot{y}$ such that $\ddot{y}(\mathbf{s}_i, \boldsymbol{\theta})$ is the learning machine's generated response for a given input pattern $\mathbf{s}_i$ and the learning machine's $q$-dimensional parameter vector $\boldsymbol{\theta}$. The goal of the learning process for the learning machine is to find a parameter vector $\hat{\boldsymbol{\theta}}_n$ that minimizes the average sum-squared error between the learning machine's prediction and the desired response of the learning machine for a given input pattern given by

$$\hat{\ell}_n(\boldsymbol{\theta}) = (1/n) \sum_{i=1}^{n} (y_i - \ddot{y}(\mathbf{s}_i, \boldsymbol{\theta}))^2. \tag{1.2}$$

△

**Example 1.3.3** (Learning as Empirical Risk Function Minimization). Let $\boldsymbol{\Theta} \subseteq \mathcal{R}^q$ be the parameter space. Let the loss function $c : \mathcal{R}^d \times \Theta \rightarrow \mathcal{R}$ be defined such that $c(\mathbf{x}, \boldsymbol{\theta})$ is the loss incurred for choosing parameter vector $\boldsymbol{\theta}$ when the learning machine experiences event $\mathbf{x}$. Let $\mathbf{x}_1, \ldots, \mathbf{x}_n$ be a data set consisting of $n$ data records corresponding to a realization of the $n$ random observations $\tilde{\mathbf{x}}_1, \ldots, \tilde{\mathbf{x}}_n$. Let $\hat{\ell}_n : \Theta \rightarrow \mathcal{R}$ be an empirical risk function defined such that for all $\boldsymbol{\theta} \in \Theta$:

$$\hat{\ell}_n(\boldsymbol{\theta}) = (1/n) \sum_{i=1}^{n} c(\tilde{\mathbf{x}}_i, \boldsymbol{\theta}).$$

The goal of a machine learning algorithm is to minimize the risk function $\ell(\boldsymbol{\theta})$ which is defined as the expected value of $\hat{\ell}_n(\boldsymbol{\theta})$. Thus, the empirical risk function $\hat{\ell}_n$ is an "estimator" based upon the observable data for the true risk function $\ell$ which is defined with respect to the unobservable data generating process. △

**Example 1.3.4** (Classification as Probability of Error Minimization). Let $p(y = 1|\mathbf{s})$ denote the probability that a learning machine believes the response category $y = 1$ is correct given input pattern $\mathbf{s}$ for a binary classification task with only two response categories. The learning machine's belief that a particular response $y$ is correct is therefore given by the formula:

$$p(y|\mathbf{s}) = yp(y = 1|\mathbf{s}) + (1 - y)(1 - p(y = 1|\mathbf{s})).$$

Show that choosing $y$ to minimize the probability of error $1 - p(y|\mathbf{s})$ is equivalent to choosing the response category $y = 1$ when $p(y = 1|\mathbf{s}) > 0.5$ and choosing the response category $y = 0$ when $p(y = 1|\mathbf{s}) < 0.5$.

**Solution.** The minimum probability of error decision rule is to choose $y = 1$ when $p(y = 1|\mathbf{s}) > p(y = 0|\mathbf{s})$ and choose $y = 0$ otherwise. Equivalently, the minimum probability of error decision rule can be written as:

$$p(y = 1|\mathbf{s}) > 1 - p(y = 1|\mathbf{s})$$

which, in turn, may be rewritten as: $p(y = 1|\mathbf{s}) > 0.5$. △

In Chapters 10, 13, 14, 15, and 16 both simulation methods and analytical formulas are provided for explicitly characterizing the generalization performance of learning machines.

### 1.3.3 Regularization Terms

High-dimensional learning machines such as deep neural networks, unsupervised learning machines, and reinforcement learning situations often rely upon the use of regularization terms. A regularization term $k_n(\boldsymbol{\theta})$ associates an additional penalty for a particular choice of parameter vector $\boldsymbol{\theta}$. An empirical risk function which includes a regularization term may be constructed by modifying (1.1) so that the following penalized empirical risk function is obtained:

$$\hat{\ell}_n(\boldsymbol{\theta}) = (1/n) \sum_{i=1}^{n} c(\mathbf{x}_i, \boldsymbol{\theta}) + k_n(\boldsymbol{\theta}). \tag{1.3}$$

It is usually assumed the regularization term $k_n(\boldsymbol{\theta})$ takes on a smaller value if more elements of $\boldsymbol{\theta}$ are set equal to zero which corresponds to a "sparse solution". Such solutions have a tendency to reduce overfitting to the training data because they penalize non-sparse solutions thus reducing the model's (undesirable) flexible ability to learn statistical regularities present in the training data which are not typically present in the test data.

Let $\lambda$ be a positive number. Let $\boldsymbol{\theta} = [\theta_1, \ldots, \theta_q]^T$. A regularization term of the form

$$k_n(\boldsymbol{\theta}) = \lambda|\boldsymbol{\theta}|^2$$

where $|\boldsymbol{\theta}|^2 = \sum_{j=1}^{q} \theta_j^2$ is called an $L_2$ or *ridge regression* regularization term. A regularization term of the form

$$k_n(\boldsymbol{\theta}) = \lambda\,|\boldsymbol{\theta}|_1$$

where $|\boldsymbol{\theta}|_1 = \sum_{j=1}^{q} |\theta_j|$ is called an $L_1$ or *lasso* regularization term. A regularization term that includes both $L_1$ and $L_2$ regularization terms such as:

$$k_n(\boldsymbol{\theta}) = \lambda|\boldsymbol{\theta}|^2 + (1 - \lambda)|\boldsymbol{\theta}|_1, \quad 0 \leq \lambda \leq 1$$

is called an *elastic net* regularization term.

The design and evaluation of machine learning algorithms is often facilitated when the objective function for learning is differentiable. Unfortunately, however, the $L_1$ regularization term $|\boldsymbol{\theta}|_1$ is not differentiable. This problem may be solved by simply "smoothing" the absolute value function so that the absolute value of the $j$th element of $\boldsymbol{\theta}$, $\theta_j$, is approximated using the formula:

$$|\theta_j| \approx ((\theta_j)^2 + \epsilon^2)^{1/2}$$

when $\epsilon$ is a small positive number which is chosen such that when $|\theta_j| << \epsilon$ this corresponds to the case where $\theta_k$ is considered to be numerically equivalent to zero.

Alternatively, one can use the following differentiable formula to approximate the $L_1$ regularization term $|\theta_j|$ using the formula:

$$|\theta_j| \approx \tau \mathcal{J}(\theta_j/\tau) + \tau \mathcal{J}(-\theta_j/\tau)$$

where $\mathcal{J}(\phi) \equiv \log(1 + \exp(\phi))$ and $\tau$ is a sufficiently small positive number.

### 1.3.4 Optimization Methods

#### 1.3.4.1 Batch Gradient Descent Learning

Chapters 5, 6, and 7 discuss various batch learning methods for finding the optimal parameters $\hat{\boldsymbol{\theta}}_n$ that minimize the Empirical Risk Function $\hat{\ell}_n$. The essential idea of many of these methods is that one begins with an initial guess for the learning machine's parameters denoted by $\boldsymbol{\theta}(0)$. Then, for $t = 0, 1, 2, \ldots$, this initial guess is refined so that the refined guess has a smaller empirical risk $\hat{\ell}_n(\boldsymbol{\theta}(t+1))$ than the original guess $\hat{\ell}_n(\boldsymbol{\theta}(t))$. This process is then repeated multiple times until one obtains a set of parameter values that minimizes the empirical risk function.

Many batch machine learning algorithms may be interpreted as variations of a single type of iterative algorithm. This important iterative algorithm is called the "method of gradient descent". Let the notation

$$\frac{d\hat{\ell}_n(\boldsymbol{\theta}(t))}{d\boldsymbol{\theta}}$$

denote the vector derivative of $\hat{\ell}_n$ evaluated at $\boldsymbol{\theta}(t)$. The essential idea of the method of gradient descent is that one begins with an empirical risk function $\hat{\ell}_n$ and wishes to find a vector of parameter values $\hat{\boldsymbol{\theta}}_n$ which is a minimizer of the function $\hat{\ell}_n$. To achieve this goal, one makes an initial guess $\boldsymbol{\theta}(0)$ for $\hat{\boldsymbol{\theta}}_n$. Next, let $t = 1$ and use $\boldsymbol{\theta}(0)$ to compute $\boldsymbol{\theta}(1)$ using the formula:

$$\boldsymbol{\theta}(t+1) = \boldsymbol{\theta}(t) - \gamma_t \frac{d\hat{\ell}_n(\boldsymbol{\theta}(t))}{d\boldsymbol{\theta}} \tag{1.4}$$

where the "stepsize" or "learning rate" $\gamma_t$ is a strictly positive number (see Chapter 5 and Chapter 7). By applying Equation (1.4) to the revised estimator $\boldsymbol{\theta}(1)$, one obtains the refined estimate $\boldsymbol{\theta}(2)$. Continuing in this manner generates a sequence of parameter estimates $\boldsymbol{\theta}(0), \boldsymbol{\theta}(1), \ldots$ which can be shown to converge to a set of points which contain the minimizers of $\hat{\ell}_n$. Variations of the method of gradient descent can be developed such as the Levenberg-Marquardt and L-BFGS descent algorithm methods which may converge much more rapidly in certain situations (see Chapter 7 for further discussion).

One possible method for assessing convergence is to check if $\boldsymbol{\theta}(t+1) \approx \boldsymbol{\theta}(t)$. However, a better criterion would be to check to see if the magnitude of the derivative of $\hat{\ell}_n$ is sufficiently small since this latter criterion focuses upon the shape of the objective function $\hat{\ell}_n$ and is not dependent upon the dynamics of the learning algorithm. The problem of deciding when to stop iterating is important and is discussed throughout this text (Chapters 5, 6, 7, 11, and 12).

Chapters 6 and 7 also provide conditions ensuring that batch learning algorithms implemented by the method of gradient descent generate a sequence of states which converge in an appropriate manner to solution sets of interest.

**Example 1.3.5** (Design a Batch Gradient Descent Stock Market Prediction Algorithm). Consider the stock market prediction problem in Example 1.3.1. A gradient descent algorithm designed to estimate a minimizer $\hat{\boldsymbol{\theta}}_n$ of $\hat{\ell}_n$ involves first picking a value for the two-dimensional vector $\boldsymbol{\theta}(0)$ at random. Design a learning algorithm with learning rate $\gamma_k$ to minimize $\hat{\ell}_n$ using the gradient descent learning formula:

$$\boldsymbol{\theta}(k+1) = \boldsymbol{\theta}(k) - \gamma_k \frac{d\hat{\ell}_n(\boldsymbol{\theta}(k))}{d\boldsymbol{\theta}}.$$

**Solution.** Let $c([x_{t+1}, x_t], \boldsymbol{\theta}) = (x_{t+1} - \theta_1 x_t - \theta_2)^2$ so that

$$\hat{\ell}_n(\boldsymbol{\theta}) = \frac{1}{n-1} \sum_{t=1}^{n-1} c([x_{t+1}, x_t], \boldsymbol{\theta}).$$

Substituting the derivatives of $c([x_{t+1}, x_t], \boldsymbol{\theta}) = (x_{t+1} - \theta_1 x_t - \theta_2)^2$ given respectively by

$$dc/d\theta_1 = -2(x_{t+1} - \theta_1 x_t - \theta_2)x_t \text{ and } dc/d\theta_2 = -2(x_{t+1} - \theta_1 x_t - \theta_2)$$

into the gradient descent learning rule formula specified in (1.4) gives:

$$\theta_1(k+1) = \theta_1(k) + \frac{2\gamma_k}{n-1} \sum_{t=1}^{n-1} (x_{t+1} - \theta_1 x_t - \theta_2)x_t \tag{1.5}$$

and

$$\theta_2(k+1) = \theta_2(k) + \frac{2\gamma_k}{n-1} \sum_{t=1}^{n-1} (x_{t+1} - \theta_1 x_t - \theta_2) \tag{1.6}$$

where $\gamma_1, \gamma_2, \ldots$ is a sequence of positive learning rates. △

**Example 1.3.6** (Design a Batch Learning Rule for Nonlinear Least Squares Regression). Consider the nonlinear least squares regression model in Example 1.3.2. Design a learning rule for estimating the parameters of that model using the batch gradient descent formula.

**Solution.** The empirical risk function $\hat{\ell}_n$ is defined such that:

$$\hat{\ell}_n(\boldsymbol{\theta}) = (1/n) \sum_{i=1}^{n} \left(y_i - \ddot{y}(\mathbf{s}_i, \boldsymbol{\theta})\right)^2.$$

The goal of the learning process is to find a minimizer of $\hat{\ell}_n$. To find the minimizer of $\hat{\ell}_n$, an algorithm specified by the gradient descent update rule (1.4) is proposed. Using the vector chain rule described in Chapter 5, the derivative of $\hat{\ell}_n$ with respect to $\boldsymbol{\theta}$ is given by the formula:

$$\frac{d\hat{\ell}_n}{d\boldsymbol{\theta}} = -(1/n) \sum_{i=1}^{n} \left(y_i - \ddot{y}(\mathbf{s}_i, \boldsymbol{\theta})\right) \frac{d\ddot{y}(\mathbf{s}_i, \boldsymbol{\theta}(k))}{d\boldsymbol{\theta}}. \tag{1.7}$$

Thus, a gradient descent learning rule for computing a minimizer of $\hat{\ell}_n$ is computed using the formula:

$$\boldsymbol{\theta}(k+1) = \boldsymbol{\theta}(k) + \gamma_k (1/n) \sum_{i=1}^{n} \left(y_i - \ddot{y}(\mathbf{s}_i, \boldsymbol{\theta}(k))\right) \frac{d\ddot{y}(\mathbf{s}_i, \boldsymbol{\theta}(k))}{d\boldsymbol{\theta}} \tag{1.8}$$

where the positive number $\gamma_k$ is the learning rate. △

#### 1.3.4.2 Adaptive Gradient Descent Learning

Assume that a batch gradient descent algorithm as specified in (1.4) is used to minimize the empirical risk function

$$\hat{\ell}_n(\boldsymbol{\theta}) = (1/n) \sum_{i=1}^{n} c(\mathbf{x}_i, \boldsymbol{\theta}). \tag{1.9}$$

Note that the empirical risk function $\hat{\ell}_n$ is an estimator of the expected loss incurred by the learning machine when the number of training stimuli, $n$, is extremely large.

Batch learning typically provides a high quality refinement of the current parameter estimates at each iteration but may be computationally expensive in situations involving very large numbers of training stimuli. Furthermore, in many important situations, it is necessary to develop adaptive learning algorithms capable of operating in statistical environments where only relatively small amounts of training data become incrementally available to the learning machine during the learning process. Using the loss function $c(\mathbf{x}, \boldsymbol{\theta})$ in (1.9), an example adaptive gradient descent algorithm for minimizing an approximation to the empirical risk function in (1.9) is given by the update rule:

$$\boldsymbol{\theta}(t+1) = \boldsymbol{\theta}(t) - \gamma_t \frac{dc(\mathbf{x}(t), \boldsymbol{\theta}(t))}{d\boldsymbol{\theta}} \tag{1.10}$$

where $\mathbf{x}(t)$ is the feature vector observed by the learning machine at time $t$ and $\gamma_t$ is the positive stepsize constant at time $t$. Chapter 12 provides sufficient conditions to ensure the learning rule in (1.10) minimizes the expected value of the empirical risk function in (1.9).

## Exercises

1.3-1.Modify the empirical risk function in Example 1.2.5, so that the predicted value of the closing price of a security for tomorrow is some constant plus the weighted sum of today's closing price, the security's closing price averaged over the past five days, and the security's closing price averaged over the past 30 days. Then, derive a gradient descent algorithm for updating the four-dimensional parameter vector for this learning machine.

1.3-2.Modify the empirical risk function developed in Exercise 1.3-1 to include a differentiable regularization term which biases the learning process to seek sparse solutions. Then, derive a gradient descent algorithm for updating the four-dimensional parameter vector of this new empirical risk function which incorporates a regularization term.

1.3-3.Assume the output of a learning machine is the predicted probability that a patient has brain cancer given the patient's individual characteristics. In particular, assume the patient characteristics are: `BLOOD-PRESSURE`, `HEART-RATE`, `WHITE-BLOOD-CELL-COUNT`, `AGE`, and `GENDER`. Show how to construct an input pattern vector $\mathbf{s}$ which represents these patient characteristics. Let $p_{\mathbf{s}}$ specify a number between zero and one which indicates the probability that the patient has cancer. To actually determine if the patient has cancer, it is necessary to take a biopsy of the patient's brain which is an expensive and risky procedure. An insurance company decides that a decision error in the initial screening stage where the learning machine incorrectly decides the patient has cancer based upon the initial screening costs the insurance company $30,000, while a decision error where the learning machine incorrectly decides the patient is cancer free ultimately costs the insurance company $1,000,000. Derive a decision rule based upon the estimated probability of brain cancer from the initial screening, $p_s$, which saves the most money for the insurance company. In particular, assume a decision rule which states that the initial screening test indicates the presence of cancer if $p_s$ is greater than some threshold constant $\psi$ and show how to compute $\psi$.

## 1.4 Theory-Based System Analysis and Design

A major objective of this text is to introduce a modular theory-based design strategy for supporting the mathematical analysis and design of machine learning algorithms. Detailed technical conditions that characterize the asymptotic behavior and performance of machine learning algorithms are provided in the form of explicit definitions and mathematical theorems. The theory is then illustrated with explicit examples of a wide range of commonly used machine learning algorithms. Such a unified theoretical framework supports: (1) an understanding of the relationships among existing machine learning algorithms, (2) the design and evaluation of machine learning algorithms, and (3) the creation of novel task-appropriate machine learning algorithms. This section provides an overview and discussion of how this unified theoretical framework is applicable in real-world engineering applications.

### 1.4.1 Stage 1: System Specification

A prerequisite to theoretical analyses of an inference or learning machine is that the inference or learning machine must be formally represented. Within this framework, the inference and learning algorithm is specified as a dynamical system whose goal is to minimize an objective function. These two processes are often uncoupled so that in one phase the learning algorithm dynamical system estimates parameter values as a result of its interactions with its statistical environment. Then, in the next phase the inference algorithm dynamical system uses the estimated parameter values to make inferences.

A complete mathematical specification of the learning and inference system should be implemented and documented in a word processing document. This specification should include a careful analysis of the structural features of the environment within which the learning and inference machine is embedded. Chapters 3 and 13 provide a formal introduction to this methodology, while this chapter provides an informal overview of such specifications. The mathematical specification consists of two components: (i) specification of the machine learning optimization algorithm, and (ii) specification of a scalar-valued objective function which is minimized by the optimization algorithm. Although some important machine learning algorithms cannot be interpreted as optimization algorithms, the theoretical framework presented here focuses on an optimization approach to the analysis and design of statistical learning machines. Chapters 1, 3, and 13 of the text discuss how to develop and design dynamical system representations of inference and learning machines for minimizing an objective function.

### 1.4.2 Stage 2: Theoretical Analyses

The outcome of Stage 1 is a representation of the learning machine as a dynamical system. In addition, Stage 1 specifies an objective function which is minimized during the learning process by the learning machine. Stage 2 of the design process involves theoretical investigations of conditions which ensure that the learning machine dynamical system generates a sequence of parameter updates that converges to a desirable knowledge state. Convergence of batch learning algorithms is discussed in Chapter 7, while convergence of adaptive learning algorithms is discussed in Chapter 12.

The learning machine then uses the obtained knowledge state parameter vector to guide inferences about its statistical environment. However, the estimated parameter vector is functionally dependent only upon the training data observed by the learning machine, and a theoretical understanding of how the learning machine will behave in novel previously unencountered situations is desirable. Such analyses can be achieved by theoretically investigating the learning machine's generalization performance. Bootstrap simulation methods are covered in Chapter 14 for the purpose of characterizing generalization performance, while non-simulation methods for characterizing generalization performance are covered in Chapters 13, 15, and 16.

### 1.4.3 Stage 3: Physical Implementation

After theoretical analyses of the learning machine's performance have been completed, a physical instantiation of the system is implemented. Such a physical implementation is typically a software implementation but could also be a hardware implementation. The mathematical specification methodology developed in Stage 1 provides explicit documentation for guiding either a software or hardware implementation of the learning and inference machine. The theoretical analyses in Stage 2 provide explicit theorems which assert specific classes of learning machines will have specific types of asymptotic behaviors. These specific theoretically motivated algorithm classes and theoretically motivated evaluation measures derived from Stage 2 can be used to specify reusable software modules. Such software modules support not only the development and evaluation of known learning machines but, because of their generality, can naturally suggest appropriate novel learning machine architectures.

### 1.4.4 Stage 4: System Behavior Evaluation

The performance of the physical system developed in Stage 3 is then compared with respect to the theoretical predictions of Stage 2. If the implemented system fails to perform as predicted, then the software implementation in Stage 3 can be checked to make sure that it faithfully reflects the algorithm specifications in Stage 1. In addition, the assumptions and theoretical predictions of the theorems in Stage 2 can be re-examined in order to determine possible revisions to the algorithm specifications in Stage 1.

In practice, such comparisons provide important insights not only into which types of empirical studies are required but additionally provide important insights into the interpretation of empirical results. Empirical results, in turn, can often provide insights that complement theoretical analyses and can suggest important changes to the original mathematical systems specifications developed in Stage 1. The design engineer then returns to Stage 1 for the purpose of refining and revising the original design. In practice, this entire design cycle is often reinstated several times.

### Recipe Box 1.1 Machine Learning Algorithm Design

- **Step 1: Specify learning machine's statistical environment.**
  What are the events in the learning machine's environment? How should the events be represented as feature vectors? Is the learning machine's statistical environment stationary?

- **Step 2: Specify machine learning architecture.**
  In this step, a specific machine learning architecture is chosen. This step might involve selecting a particular parametric form for the architecture for the purpose of implicitly providing the learning machine with hints about the structural characteristics of its probabilistic environment. Begin with the simplest possible architecture powerful enough to solve the target problem and understand its strengths and limitations before considering more complicated architectures. Introduce regularization constraints if appropriate.

- **Step 3: Specify loss function for empirical risk function.**
  In this step, a differentiable loss function for the empirical risk function is chosen which incorporates prior knowledge about the machine learning inference problem. An important component of the loss function is the machine learning architecture developed in Step 2.

- **Step 4: Design inference and learning algorithms.**
  Use loss function derivatives to design inference and learning algorithms (see Chapters 6, 7, 11, and 12).

- **Step 5: Design evaluative methods for analyzing performance.**
  Use loss function derivatives to design performance evaluation methods (see Chapters 13, 14, 15, and 16).

- **Step 6: Implement algorithm and evaluation methods.**
  Implement the learning algorithm and evaluation methods in software. Verify that the software implementation correctly implements documented description of the algorithms (Steps 2 and 3) and the evaluation methods (Step 5).

- **Step 7: Evaluate algorithm's behavior.**
  Use implemented software to evaluate learning machine's behavior.

- **Step 8: Repeat the process.**
  Repeat Steps 1, 2, 3, 4, 5, 6, and 7 until satisfactory performance is achieved.

**Example 1.4.1** (An Optimal Learning Machine Which Does Not Learn). A theoretician proves a theorem which states that in a stationary environment a particular learning machine which minimizes a differentiable objective function will generate a sequence of parameter updates which converge to an optimal state of knowledge provided that the number of learning experiences is sufficiently large. An engineer implements this learning machine in software and finds that the algorithm is not converging but exhibits crazy oscillations. Discuss some possible reasons for the failure of the learning algorithm.

**Solution.** There are several reasons why the learning algorithm may fail.

First, it is possible that there is a software implementation problem which corresponds to a physical implementation problem in Stage 3. This can be addressed by checking that the software module implementation in Stage 3 correctly implements the algorithm specification in Stage 1.

Second, it is possible the specification of a stationary environment in Stage 1 is not correct. This can be checked in Stage 4 by constructing different data sets associated with different time periods from the same statistical environment and checking that the statistical regularities extracted by the learning machine are comparable across these data sets. If the environment is not stationary, it might be possible to specify in Stage 1 a recoding of the statistical characteristics of the environment using an alternative feature vector representation. Another approach to addressing the problem of a nonstationary environment would be to investigate alternative strategies for sampling data from the environment by designing an improved algorithm (Stage 1). Stage 2 is then repeated to generate revised theoretical predictions.

Third, it is possible that the objective function is not specified in Stage 1 as differentiable causing the theoretician's conclusions to be invalid. This can be checked by a theoretical examination of the objective function specification to check if it is differentiable and fixed if necessary by a revised specification of the objective function in Stage 1 and then generating new theoretical predictions in Stage 2.

Fourth, it is possible that as a result of the Stage 4 analyses additional learning experiences are required. If the data set size is sufficiently large, one would expect that learning performance should remain approximately constant as the size of the data set is increased.

Fifth, it is possible that even though the algorithm and objective function are both correctly specified and implemented, performance is poor due to algorithm choice. This may be resolved by specifying a new algorithm in Stage 1 and generating new theoretical predictions in Stage 2.

Sixth, there could be an error in the theoretical derivation in Stage 2. This can be resolved by double-checking that the theoretical derivation assumptions are consistent with the system specification in Stage 1 and that the mathematics used to derive algorithm behavior predictions is correct. △

## Exercises

1.4-1. A theoretician proves a theorem which states that in a stationary environment that a particular learning machine, which has converged to a strict local minimizer of a differentiable objective function, will make predictions which have 1% error rate on novel test data provided that the number of learning experiences is sufficiently large. An engineer implements this learning machine in software and finds that the learning machine's predictions have an error rate of 5% on the training data set and an error rate of 20% on a novel test data set. Discuss some possible reasons why these experimental results may differ from the theoretical predictions including at least one example from each of the categories: (i) specification error, (ii) theoretical analysis error, (iii) implementation error, and (iv) evaluation error. Explain how the proposed theory-based system analysis and design strategy can be helpful in possibly improving the performance of the learning machine.

1.4-2. Define the Empirical Risk Minimization (ERM) Framework in your own words. List several benefits and several limitations of the ERM Framework.

## 1.5 Supervised Learning Machines

The goal of a *supervised learning machine* is to predict an appropriate output pattern vector given an input pattern vector. The training data set for a supervised learning machine is often represented as an unordered set of pairs of input pattern vectors and response pattern vectors. For example, given an input pattern vector corresponding to an image, the learning machine is taught to produce the "desired response" or "label" of the given input pattern. In addition, supervised learning machines are often expected to generalize from their experiences. This means that if the supervised learning machine is presented with a novel input pattern vector, then it is expected that the machine will produce the appropriate desired response even though the ordered pair of novel input pattern vector and appropriate desired response were never included in the training data of the learning machine (see Figure 1.4).

### 1.5.1 Discrepancy Functions

In a supervised learning paradigm, the training vector $\mathbf{x} = [\mathbf{s}, \mathbf{y}]$ consists of an input pattern vector $\mathbf{s}$ and the desired response vector $\mathbf{y}$ which we would like the learning machine to generate when it is presented with input pattern vector $\mathbf{s}$. Typically, the loss function $c$ for the ERF in (1.1) has the parametric form:

$$c(\mathbf{x}, \boldsymbol{\theta}) = D\left(\mathbf{y}, \ddot{\mathbf{y}}(\mathbf{s}, \boldsymbol{\theta})\right) \tag{1.11}$$

where $D$ is called the *discrepancy function* which compares the desired response $\mathbf{y}$ to the predicted response $\ddot{\mathbf{y}}(\mathbf{s}, \boldsymbol{\theta})$ for a given input pattern $\mathbf{s}$ and a given parameter vector (i.e., "knowledge state") $\boldsymbol{\theta}$.

**Algorithm 1.5.1** A generic adaptive gradient descent supervised learning algorithm. Let $\mathbf{y}$ be the desired response of a supervised learning machine to a given input pattern $\mathbf{s}$ and let $\ddot{\mathbf{y}}(\mathbf{s}, \boldsymbol{\theta})$ be the predicted response of the supervised learning machine for $\mathbf{s}$ given a parameter vector $\boldsymbol{\theta}$. Let $c$ be defined as in (1.11). The small positive number $\gamma_t$ controls the rate of learning at a particular algorithm iteration $t$.

1: **procedure** SUPERVISED-LEARNING-GRADIENT-DESCENT($\boldsymbol{\theta}(0)$)
2: $\quad t \Leftarrow 0$
3: $\quad$ **repeat**
4: $\quad\quad$ Select next input pattern $\mathbf{s}(t)$
5: $\quad\quad$ Observe Desired Response $\mathbf{y}(t)$ for input pattern $\mathbf{s}(t)$
6: $\quad\quad \mathbf{g}([\mathbf{y}(t), \mathbf{s}(t)], \boldsymbol{\theta}(t)) \Leftarrow dc\left([\mathbf{y}(t), \mathbf{s}(t)]; \boldsymbol{\theta}(t)\right)/d\boldsymbol{\theta}$
7: $\quad\quad \boldsymbol{\theta}(t+1) \Leftarrow \boldsymbol{\theta}(t) - \gamma_t \mathbf{g}([\mathbf{y}(t), \mathbf{s}(t)], \boldsymbol{\theta}(t))$
8: $\quad\quad t \Leftarrow t + 1.$
9: $\quad$ **until** $\boldsymbol{\theta}(t) \approx \boldsymbol{\theta}(t-1)$
10: $\quad$ **return** $\{\boldsymbol{\theta}(t)\}$
11: **end procedure**

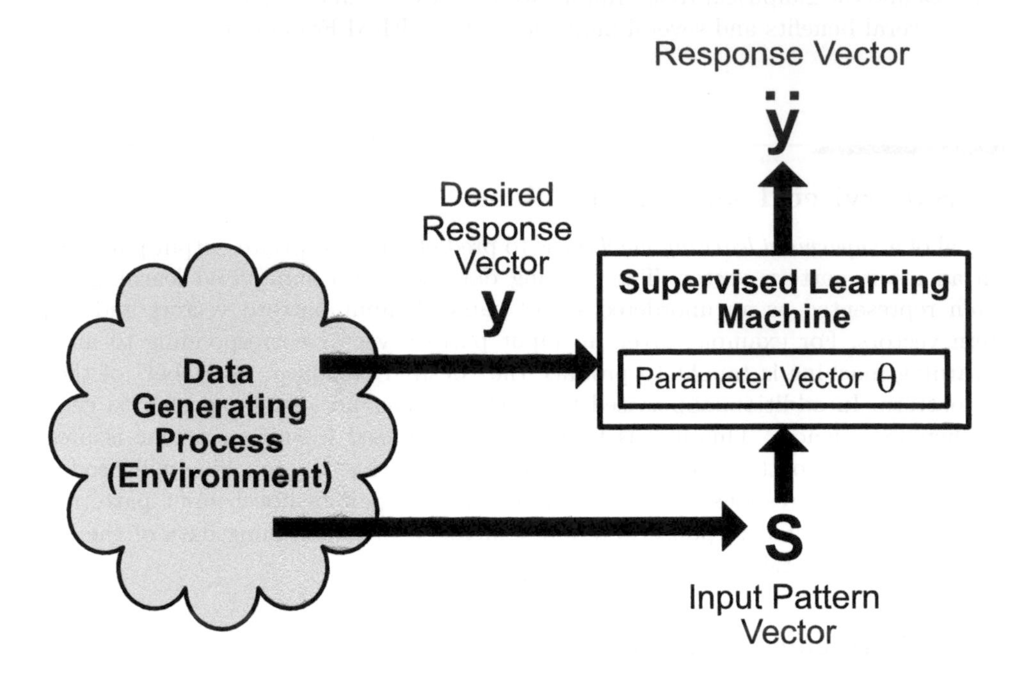


**FIGURE 1.4**
**Overview of a supervised learning machine.** The data generating process generates an event which is detected as a feature vector by the supervised learning machine. The feature vector consists of two components: An input pattern vector $\mathbf{s}$ and a desired response vector $\mathbf{y}$. The learning machine uses the input pattern and its internal knowledge state specified by a parameter vector $\boldsymbol{\theta}$ to produce a response $\ddot{\mathbf{y}}$. A discrepancy between the learning machine's produced response $\ddot{\mathbf{y}}$ and the desired response $\mathbf{y}$ results in an update to the learning machine's internal knowledge state $\boldsymbol{\theta}$.

**Example 1.5.1** (Least Squares Discrepancy Measure for Numerical Targets). A *Linear Regression* or *least squares* discrepancy measure is most appropriate in situations where the desired response $y$ of the inference machine is a continuous numerical quantity (e.g., $-\infty < y < \infty$) when the inference machine is presented with the input pattern $\mathbf{s}$. Let

$$\ddot{y}(\mathbf{s}, \boldsymbol{\theta}) = \boldsymbol{\theta}^T [\mathbf{s}^T \;\; 1]^T$$

denote the predicted response for a given input pattern $\mathbf{s}$ and a given parameter vector $\boldsymbol{\theta}$. The discrepancy function $D$ is chosen such that

$$D(y, \ddot{y}(\mathbf{s}, \boldsymbol{\theta})) = (y - \ddot{y}(\mathbf{s}, \boldsymbol{\theta}))^2 .$$

With this choice of discrepancy function $D$, the empirical risk function $\hat{\ell}_n$ is defined for the training data set $\mathcal{D}_n = \{(y_1, \mathbf{s}_1), \ldots, (y_n, \mathbf{s}_n)\}$ such that:

$$\hat{\ell}_n(\boldsymbol{\theta}) = (1/n) \sum_{i=1}^{n} (y_i - \ddot{y}(\mathbf{s}_i, \boldsymbol{\theta}))^2 . \tag{1.12}$$

The parameters of the linear regression model corresponding to a minimizer of $\hat{\ell}_n$ in (1.12) may be estimated using Algorithm 1.5.1 and the loss function

$$c([y, \mathbf{s}], \boldsymbol{\theta}) = (y - \ddot{y}(\mathbf{s}, \boldsymbol{\theta}))^2 .$$

△

**Example 1.5.2** (Logistic Regression Discrepancy Measure for Binary Targets). A *Logistic Regression discrepancy measure* is most appropriate in situations where the desired response of the inference machine $y$ is binary-valued (e.g., $y \in \{0, 1\}$). Specifically, the inference machine estimates the probability $y = 1$ for a given input pattern $\mathbf{s}$ and a given parameter vector $\boldsymbol{\theta}$ as:

$$\ddot{p}(\mathbf{s}, \boldsymbol{\theta}) = [1 + \exp(-\ddot{y}(\mathbf{s}, \boldsymbol{\theta}))]^{-1}$$

where

$$\ddot{y}(\mathbf{s}, \boldsymbol{\theta}) = \boldsymbol{\theta}^T [\mathbf{s}^T \;\; 1]^T.$$

The discrepancy function $D$ is chosen such that

$$D(\mathbf{y}, \ddot{\mathbf{y}}) = -[y \log(\ddot{p}(\mathbf{s}, \boldsymbol{\theta})) + (1 - y) \log(1 - \ddot{p}(\mathbf{s}, \boldsymbol{\theta}))].$$

With this choice of discrepancy function $D$, the empirical risk function $\hat{\ell}_n$ is defined for the training data set $\mathcal{D}_n = \{(y_1, \mathbf{s}_1), \ldots, (y_n, \mathbf{s}_n)\}$ such that:

$$\hat{\ell}_n(\boldsymbol{\theta}) = -(1/n) \sum_{i=1}^{n} [y_i \log(\ddot{p}(\mathbf{s}_i, \boldsymbol{\theta})) + (1 - y_i) \log(1 - \ddot{p}(\mathbf{s}_i, \boldsymbol{\theta}))]. \quad (1.13)$$

The parameters of the logistic regression model corresponding to a minimizer of $\hat{\ell}_n$ in (1.13) are then estimated using Algorithm 1.5.1 with the loss function

$$c([y, \mathbf{s}], \boldsymbol{\theta}) = -y \log \ddot{p}(\mathbf{s}, \boldsymbol{\theta}) - (1 - y) \log(1 - \ddot{p}(\mathbf{s}, \boldsymbol{\theta})).$$

△

**Example 1.5.3** (Multinomial (Softmax) Logistic Discrepancy for Categorical Targets). A *multinomial logistic regression machine* or *softmax output network architecture* is appropriate when the desired response of the learning machine is a nominal categorical variable. This architecture can be viewed as a generalization of the logistic regression discrepancy function presented in Example 1.5.2.

Let the $k$th column, $\mathbf{y}^k$, of the $m$-dimensional identity matrix denote the $k$th possible categorical desired response of a learning machine when the learning machine is presented a $d$-dimensional input pattern vector $\mathbf{s}$ for $k = 1, \ldots, m$. Let the parameter vector $\boldsymbol{\theta}$ of the learning machine be defined as the columns of a parameter matrix $\mathbf{W}$ with $m$ rows and $d$ columns such that the first $m - 1$ rows of $\mathbf{W}$ are free parameters and the last row of $\mathbf{W}$ is a row vector of zeros. The parameter vector $\boldsymbol{\theta}$ is used to specify the non-zero elements of $\mathbf{W}$. Let $\mathbf{1}_m$ be an $m$-dimensional column vector of ones. Let $\mathbf{exp}$ be a vector-valued function defined such that the $k$th element of $\mathbf{exp}(\mathbf{u})$ is equal to $\exp(u_k)$ where $u_k$ is the $k$th element of $\mathbf{u}$ for $k = 1, \ldots, m$.

Let $\boldsymbol{\psi}(\mathbf{s}; \mathbf{W}) = \mathbf{W}\mathbf{s}$. Let $\psi_k(\mathbf{s}, \mathbf{W})$ denote the $k$th element of $\boldsymbol{\psi}(\mathbf{s}; \mathbf{W})$ for $k = 1, \ldots, m$. Let the probability that the input pattern vector $\mathbf{s}$ is assigned to the $k$th out of $m$ possible categories be specified by the $k$th element,

$$p(\mathbf{y}_i^k | \mathbf{s}_i; \mathbf{W}) = \frac{\exp(\psi_k(\mathbf{s}_i; \mathbf{W}))}{\mathbf{1}_m^T \mathbf{exp}(\boldsymbol{\psi}(\mathbf{s}_i; \mathbf{W}))},$$

of the $m$-dimensional probability vector $\mathbf{p}_i(\boldsymbol{\theta})$.

The softmax empirical risk function is defined as

$$\hat{\ell}_n(\boldsymbol{\theta}) = (1/n) \sum_{i=1}^{n} D\left(\mathbf{y}_i, \mathbf{log}(\mathbf{p}_i(\boldsymbol{\theta}))\right)$$

where the *softmax discrepancy*

$$D\left(\mathbf{y}_i, \mathbf{log}(\mathbf{p}_i(\boldsymbol{\theta}))\right) = -\mathbf{y}_i^T \mathbf{log}(\mathbf{p}_i(\boldsymbol{\theta})).$$

△

### 1.5.2 Basis Functions and Hidden Units

It is often computationally convenient and useful to assume the predicted response function $\ddot{y}(\mathbf{s}, \boldsymbol{\theta})$ is a weighted sum of the elements of $\mathbf{s}$ so that $\ddot{y}(\mathbf{s}, \boldsymbol{\theta}) = \boldsymbol{\theta}^T [\mathbf{s}^T, 1]^T$. However, this assumption is not appropriate when the predicted response $\ddot{y}(\mathbf{s}, \boldsymbol{\theta})$ is a nonlinear function of $\mathbf{s}$ and $\boldsymbol{\theta}$. Moreover, the problem of picking a parametric formula for $\ddot{y}(\mathbf{s}, \boldsymbol{\theta})$ is further complicated by the fact that many possible nonlinear representations are possible.

Figure 1.5 illustrates a popular strategy for addressing these problems. The predicted response $\ddot{y}(\mathbf{s}, \boldsymbol{\theta})$ for a given input pattern $\mathbf{s}$ is specified by the formula:

$$\ddot{y}(\mathbf{s}, \boldsymbol{\theta}) = \sum_{j=1}^{J} w_j h_j(\mathbf{s}, \mathbf{v}_j) \tag{1.14}$$

where the parameter vector $\boldsymbol{\theta} \equiv [w_1, w_2, \ldots, w_J, \mathbf{v}_1^T, \mathbf{v}_2^T, \ldots, \mathbf{v}_J^T]^T$. The basis functions or "hidden units" $h_1, h_2, \ldots, h_j$ for a particular choice of $\mathbf{v}_1, \ldots, \mathbf{v}_J$ specify a transformation

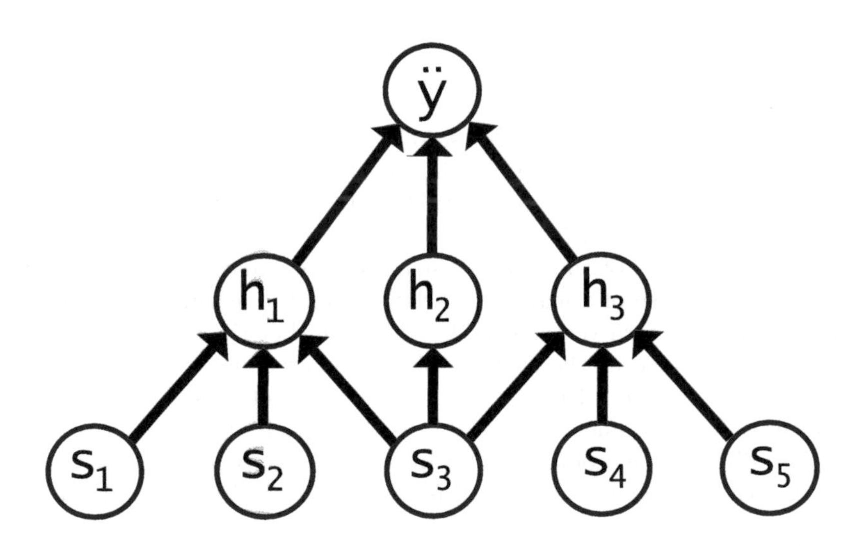


**FIGURE 1.5**
**Multilayer perceptron representation of a basis function approximation strategy.** The response of the multilayer perceptron network $\ddot{y}$ is a weighted sum of the hidden states $h_1, h_2, h_3$. Each hidden state $h_k$ is the value of the basis function computed from the basis function parameters and the elements of the input pattern vector $s_1, s_2, s_3, s_4, s_5$.

of $\mathbf{s}$ which results in the feature vector

$$\mathbf{h}^T \equiv [h_1(\mathbf{s}, \mathbf{v}_1), \ldots, h_J(\mathbf{s}, \mathbf{v}_J)]$$

designed to have the property that the predicted response $\ddot{y}(\mathbf{s}, \boldsymbol{\theta})$ in (1.14) may be represented as a weighted sum of the elements of $\mathbf{h}$. As illustrated in Figure 1.5, the $d$-dimensional input pattern vector $\mathbf{s}$ is a set of $d$ real numbers that specify the states of $d$ input units. The feature vector $\mathbf{h}$ is a set of $J$ real numbers that specifies the states of the set of $J$ hidden units. The state of the output unit is a real number that specifies the scalar predicted response $\ddot{y}(\mathbf{s}, \boldsymbol{\theta})$.

The network in Figure 1.5 consists of one layer of hidden units and is called a *multilayer perceptron* or *shallow neural network*. Additional layers of hidden units may also be introduced. In the special case where the inputs to the $k$th layer of hidden units are only functionally dependent upon the outputs from the $(k-1)$th layers of hidden units, then the resulting network architecture is called a *feedforward neural network* or *feedforward network*. A network with two or more layers of hidden units is typically referred to as a *deep learning network*. The majority of important recent empirical advances in machine learning research (e.g., Goodfellow et al., 2016) have used multilayer perceptrons with multiple layers of hidden units.

The number of hidden units $J$ in the multilayer perceptron depicted in Figure 1.5 plays a critical role in determining the perceptron's performance characteristics. If the number of hidden units $J$ is too large, then each hidden unit may learn a different input pattern so that the network essentially memorizes the training data set. Although this can result in excellent performance on the training data set, performance on a test data set often suffers in such situations because the learning machine simply "memorizes" the training data and is not forced to extract out the important statistical regularities. Choosing too few hidden units, however, results in a multilayer perceptron which is not capable of adequately representing all of the important structural statistical regularities in its environment. Conceptually, the hidden unit layer defines an *information bottleneck* that forces the system to learn the most important and most stable statistical regularities in its environment and prevents the system from learning "statistical flukes" in the training data. In practice, one can experiment with different numbers of hidden units in a layer or use regularization techniques which encourage the system to eliminate irrelevant and redundant hidden units.

A nonlinear transformation which maps an input pattern state vector specifying the states of the input units generated from an environmental event into the hidden unit state vector pattern may be interpreted as a recoding transformation of the input patterns. The recoding transformation is learnable so the learning process actually designs the recoding transformation to improve predictive performance of the learning machine.

The state activation pattern over the hidden units produced by the input unit activation pattern is called an *embedding feature representation* and the learnable nonlinear transformation from input pattern to the embedding feature representation is called an *embedding layer*.

It can be shown that when the hidden units $h_1, \ldots, h_J$ are continuous and bounded functions and the number of hidden units, $J$, is sufficiently large, that the predicted response $\ddot{y}(\mathbf{s}, \boldsymbol{\theta})$ in (1.14) can closely approximate any arbitrary continuous function (Hornik 1991; Pinkus 1999). Thus, the formula in (1.14) implements a universal approximator when the number of hidden units is sufficiently large. Moreover, recent theoretical advances have provided conditions that show multilayer perceptrons with *two layers* of hidden units are capable of approximating an arbitrary continuous function with a *finite* number of hidden units (Guliyev and Ismailov 2018).

The choice of the initial guess for the parameter values of a gradient descent algorithm that minimizes the empirical risk function for a multilayer perceptron also plays an influential role in determining the ultimate performance of the learning algorithm. Typically, the parameter values of the hidden units should be initially set equal to small mean-zero random values. Because the objective function is not convex and may have multiple local minima, maxima, and saddlepoints, a gradient descent type algorithm that seeks to minimize the prediction error of the response of a network with hidden units is at the mercy of the initial guess for the parameter values (see Section 5.3 for further discussion). By selecting an initial set of parameter values which have small random values, this biases different hidden units to learn to detect different classes of patterns. A useful review of parameter initialization strategies and their consequences for deep learning networks can be found in Sutsekevar et al. (2013).

**Example 1.5.4** (Multilayer Perceptron (MLP) ). Let $\mathbf{y}$ denote the desired response of the learning machine given input pattern $\mathbf{s}$. Let $\ddot{\mathbf{y}}(\mathbf{s}, \boldsymbol{\theta})$ denote the learning machine's predicted response to input pattern $\mathbf{s}$ for a particular parameter vector $\boldsymbol{\theta}$. Assume the goal of the learning process is to find a parameter vector $\boldsymbol{\theta}$ such that the predicted response is similar to the desired response for a given input pattern $\mathbf{s}$. Let the vector-valued function $\mathbf{h} \equiv [h_1, \ldots, h_M]$ be defined such that the $j$th *basis function* or *hidden unit* $h_j$ maps a subset of the elements of $\mathbf{s}$ and the parameter vector $\boldsymbol{\theta}$ into a number $h_j(\mathbf{s}, \boldsymbol{\theta})$. The predicted response $\ddot{\mathbf{y}}(\mathbf{s}, \boldsymbol{\theta})$ is then explicitly computed using the formula:

$$\ddot{\mathbf{y}}(\mathbf{s}, \boldsymbol{\theta}) = \mathbf{V}\mathbf{h}(\mathbf{s}, \mathbf{W})$$

where the matrices $\mathbf{V}$ and $\mathbf{W}$ are parameters of the learning machine which are adjusted during the learning process. The parameter vector $\boldsymbol{\theta}$ is defined as the concatenation of all elements of $\mathbf{W}$ and $\mathbf{V}$.

A gradient descent learning algorithm (see Algorithm 1.5.1) may be used to estimate $\hat{\boldsymbol{\theta}}_n$ by minimizing the empirical risk function:

$$\hat{\ell}_n(\boldsymbol{\theta}) = (1/n) \sum_{i=1}^{n} |\mathbf{y}_i - \ddot{\mathbf{y}}(\mathbf{s}, \boldsymbol{\theta})|^2 \tag{1.15}$$

where

$$\ddot{\mathbf{y}}(\mathbf{s}, \boldsymbol{\theta}) = \mathbf{V}\mathbf{h}(\mathbf{s}, \boldsymbol{\theta}).$$

△

One popular choice of basis function is the *Radial Basis Function* (RBF) hidden unit defined such that:

$$h_j(\mathbf{s}, \boldsymbol{\theta}) = G\left(\mathbf{s} - \boldsymbol{\theta}_j\right) \tag{1.16}$$

where

$$G(\mathbf{x}) = \exp\left(-|\mathbf{x}|^2/2\sigma^2\right)$$

is called a *Gaussian radial function* centered at $\boldsymbol{\theta}_j$ with radius (or "width") $\sigma$. The radial basis function $h_j$ takes on large values for $\mathbf{s}$ in a region centered at $\boldsymbol{\theta}_j$ of approximately radius $|\sigma|$ and small values elsewhere. Thus, as shown in Figure 1.6 an RBF models a "response bump".

By choosing $M$ and the coefficients $v_1, \ldots, v_M$ and the RBF centers $\boldsymbol{\theta}_1, \ldots, \boldsymbol{\theta}_M$ appropriately, one can approximate any arbitrary smooth function $\mathbf{f}$ with arbitrary precision as illustrated in Figure 1.7.

*Sigmoidal basis functions* (also known as *sigmoidal hidden units*) are typically defined such that

$$h_j(\mathbf{s}, \boldsymbol{\theta}) = \mathcal{S}\left(\boldsymbol{\theta}_j^T \mathbf{s}/\mathcal{T}\right)$$

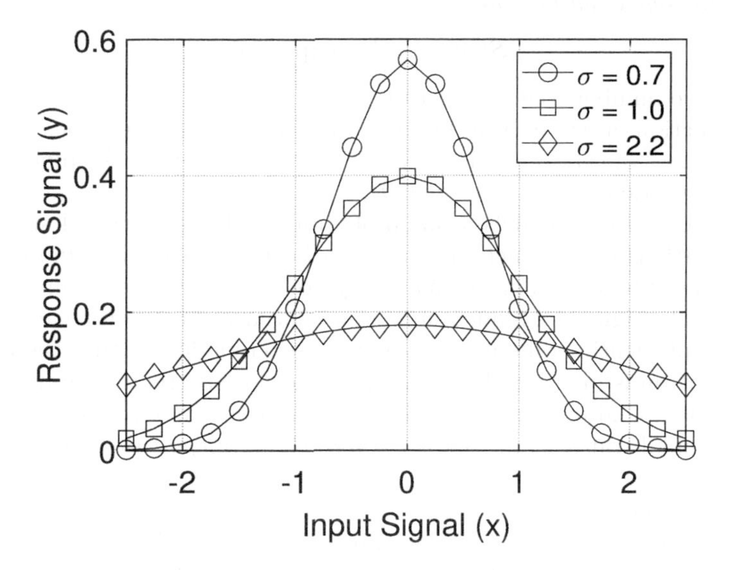


**FIGURE 1.6**
**Response characteristics of a radial basis function.** A radial basis function has a symmetric response curve which looks like a "bump" centered at mean 0 with an approximate width of $\sigma$. In this figure, the response signal $y$ is computed using the formula $y(x) = (\sigma\sqrt{2\pi})^{-1} \exp\left(-x^2/(2\sigma^2)\right)$.

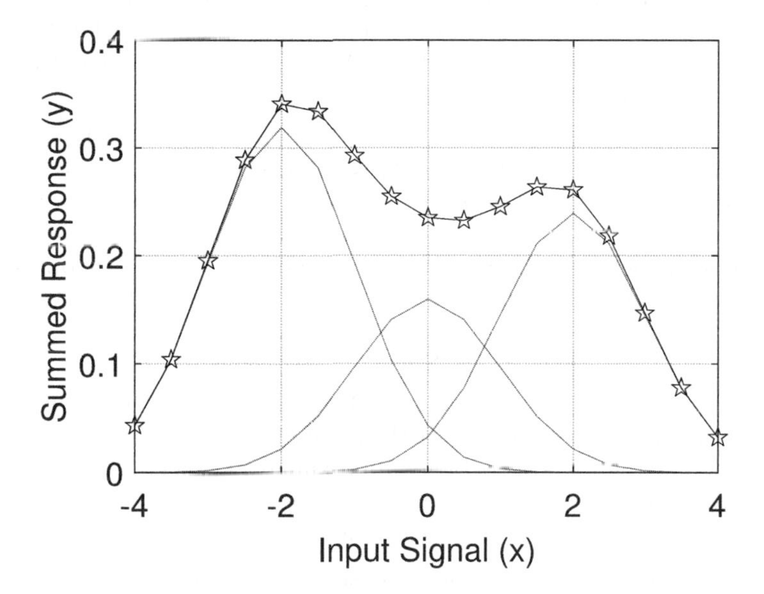


**FIGURE 1.7**
**A weighted sum of radial basis functions can approximate an arbitrary smooth function.** In this diagram, the response of an individual radial basis function is plotted using solid lines. The weighted sum of the responses of the three individual unimodal radial basis functions is plotted as a darker solid line with stars as markers. This weighted sum is bimodal rather than unimodal. More specifically, let $K = (2\pi)^{-1/2}$. Next, define the basis functions $h_1(x) = K \exp\left(-0.5(x+2)^2\right)$, $h_2(x) = K \exp\left(-0.5x^2\right)$, and $h_3(x) = K \exp\left(-0.5(x-2)^2\right)$. And then define the weighted sum of basis functions $y(x) = 0.8h_1(x) + 0.4h_2(x) + 0.6h_3(x)$.

where the *logistic sigmoidal function*

$$\mathcal{S}(\phi) = 1/(1 + \exp(-\phi)) \tag{1.17}$$

and $\mathcal{T}$ is a positive number (see Figure 1.8).

Sigmoidal functions are important because, like RBF transfer functions, they are capable of modeling "bumps" of arbitrary width and at arbitrary locations for function approximation purposes. In particular, given the definition $\mathcal{S}(\phi) = 1/(1+\exp(-\phi))$, the weighted sum $\mathcal{S}(\phi) - \mathcal{S}(\phi - b)$ essentially models a bump of width $b$. Thus, weighted sums of sigmoidal functions, like RBF transfer functions, can also be used to model bumps and therefore implement nonlinear function approximators by computing weighted sums of such bumps (see Figure 1.7). The sigmoidal hidden units in a multilayer perceptron can alternatively be interpreted as implementing different logistic regression models so that the response of the multilayer perceptron is interpretable as a mixture of logistic regression models.

The logistic sigmoidal function in (1.17) has asymptotes at zero and one. Another commonly used sigmoidal transfer function is the *hyperbolic tangent sigmoidal function* defined by the formula $2\mathcal{S}(2\phi) - 1$ where $\mathcal{S}(\phi) = 1/(1+\exp(-\phi))$. The asymptotes of the hyperbolic tangent sigmoidal function are $+1$ and $-1$.

Both logistic and hyperbolic sigmoidal functions are widely used in the construction of basis functions to support function approximation in machine learning but logistic sigmoidal functions have the additional interesting property that they can be interpreted as parameterized logic gates. That is, by selecting the parameters of a logistic sigmoidal basis function in an appropriate manner, one can transform the logistic sigmoidal basis function into an `AND` gate, `OR` gate, or `NOT` gate (see Example 1.5.5). Given a network consisting of `AND`, `OR`, and `NOT` gates it follows that any arbitrary logic function can be implemented. This critical building block of the modern digital computer was first published by the Neuroscientist/Physicist Warren McCulloch and the young genius mathematician Walter Pitts in a *Bulletin of Mathematical Physics* paper titled "A logical calculus of the ideas immanent in nervous activity" (McCulloch and Pitts 1943).

**Example 1.5.5** (Approximating Logic Gates with Logistic Sigmoidal Basis Functions). A *logical and* function $f : \{0,1\} \times \{0,1\} \rightarrow \{0,1\}$ is defined such that: $f(0,0) = 0$, $f(0,1) = 0$, $f(1,0) = 0$, and $f(1,1) = 1$. A *logical or* function $r : \{0,1\} \times \{0,1\} \rightarrow \{0,1\}$ is defined such that $r(0,0) = 0$, $r(0,1) = 1$, $r(1,0) = 1$, and $r(1,1) = 1$. A *logical not* function $n : \{0,1\} \rightarrow \{0,1\}$ is defined such that $n(0) = 1$ and $n(1) = 0$. (i) Show how to set the parameters $w_0, w_1, w_2$ of a sigmoidal basis function $q : \{0,1\} \times \{0,1\} \rightarrow (0,1)$ defined such that:

$$q(x_1, x_2) = \mathcal{S}(w_1 x_1 + w_2 x_2 + w_0) \tag{1.18}$$

where $\mathcal{S}(\phi) = (1 + \exp(-\phi))^{-1}$ so that $q(x_1, x_2)$ in (1.18) closely approximates the logical `AND` function $f(x_1, x_2)$. (ii) Show how to set the parameters $w_0, w_1, w_2$ of the sigmoidal basis function $q$ in (1.18) so that $q(x_1, x_2)$ closely approximates the logical `OR` function $r(x_1, x_2)$. (iii) Show how to set the parameters $w_0, w_1, w_2$ of the sigmoidal basis function $q$ in (1.18) so that $q(x, 0)$ closely approximates the logical `NOT` function $n(x)$. △

*Softplus basis functions* are also effective smooth function approximators and have recently become increasingly popular because of increasing evidence they exhibit superior advantages over sigmoidal basis functions in deep learning neural networks which possess multiple layers of hidden units. A *softplus function* $\mathcal{J}$ is defined such that:

$$\mathcal{J}(\phi) = \log\left(1 + \exp(\phi)\right). \tag{1.19}$$

Note that the formula $\mathcal{J}(\phi) - \mathcal{J}(\phi - b)$ approximates a "bump" and then weighted combinations of such bumps can be used to approximate a variety of nonlinear functions as illustrated in Figure 1.7.

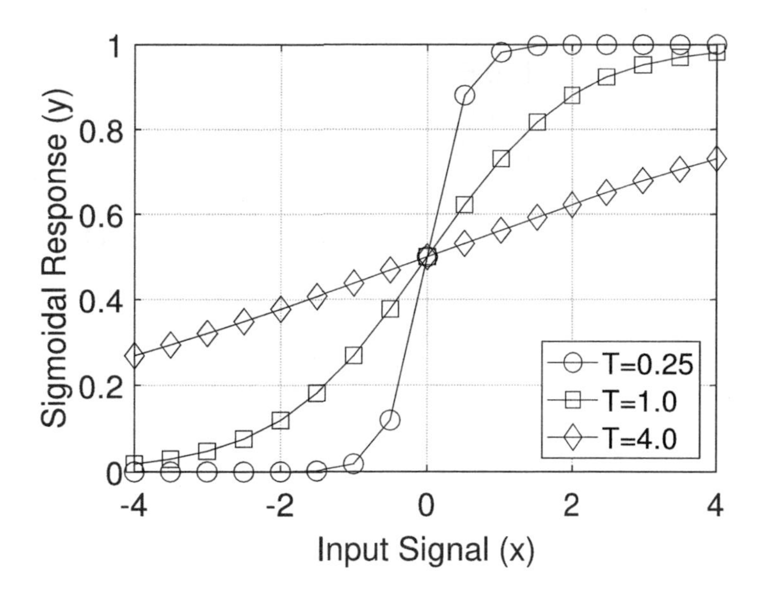


**FIGURE 1.8**
**Response characteristics of a sigmoidal basis function.** A sigmoidal function is a monotonic increasing function with lower and upper bounds. In this figure, the sigmoidal function $y(x) = \mathcal{S}(x/T)$ where $\mathcal{S}(x) = 1/(1+\exp(-x))$ is plotted for different choices of the positive parameter $T$. When $T$ is close to zero (e.g., $T = 0.25$), then the sigmoidal response function approximates a step function. When $T$ is large (e.g., $T = 4$), the sigmoidal response function approximates a linear function when $x$ is close to zero.

Let $\phi_j \equiv \boldsymbol{\theta}_j^T \mathbf{s}$ where $\boldsymbol{\theta}_j$ is the parameter vector for the $j$th softplus basis function and $\mathbf{s}$ is the input pattern vector. The formula for a softplus basis function is given by:

$$h_j(\mathbf{s}, \boldsymbol{\theta}) = \tau \mathcal{J}(\phi_j/\tau)$$

where $\tau$ is a small positive number. As $\tau$ approaches zero, the softplus basis function converges to a *rectilinear basis function* which returns $\phi_j$ when $\tau > 0$ and returns zero otherwise (see Figure 1.9).

**Example 1.5.6** (Residual Net Multilayer Perceptron with Skip Connections). Let $\mathbf{y}$ denote the desired response of the learning machine given input pattern $\mathbf{s}$. Let $\ddot{\mathbf{y}}(\mathbf{s}, \boldsymbol{\theta})$ denote the learning machine's predicted response to input pattern $\mathbf{s}$ for a particular parameter vector $\boldsymbol{\theta}$. Assume the goal of the learning process is to find a parameter vector $\boldsymbol{\theta}$ such that the predicted response is similar to the desired response for a given input pattern $\mathbf{s}$. Let the vector-valued function $\mathbf{h} \equiv [h_1, \ldots, h_M]$ be defined such that the $j$th *basis function* or *hidden unit* $h_j$ maps a subset of the elements of $\mathbf{s}$ and the parameter vector $\boldsymbol{\theta}$ into a number $h_j(\mathbf{s}, \boldsymbol{\theta})$. The predicted response $\ddot{\mathbf{y}}(\mathbf{s}, \boldsymbol{\theta})$ for a *residual net feedforward perceptron* with a *skip layer* $\mathbf{Q}$ is then explicitly computed using the formula:

$$\ddot{\mathbf{y}}(\mathbf{s}, \boldsymbol{\theta}) = \mathbf{V}\mathbf{h}(\mathbf{s}, \mathbf{W}) + \mathbf{Q}\mathbf{s}$$

where the connection matrix $\mathbf{V}$ specifies the parameters which specify the connection values from the hidden units to the output units, and the connection matrix $\mathbf{W}$ specifies the parameters from the input units to the hidden units. In addition, a third connection matrix

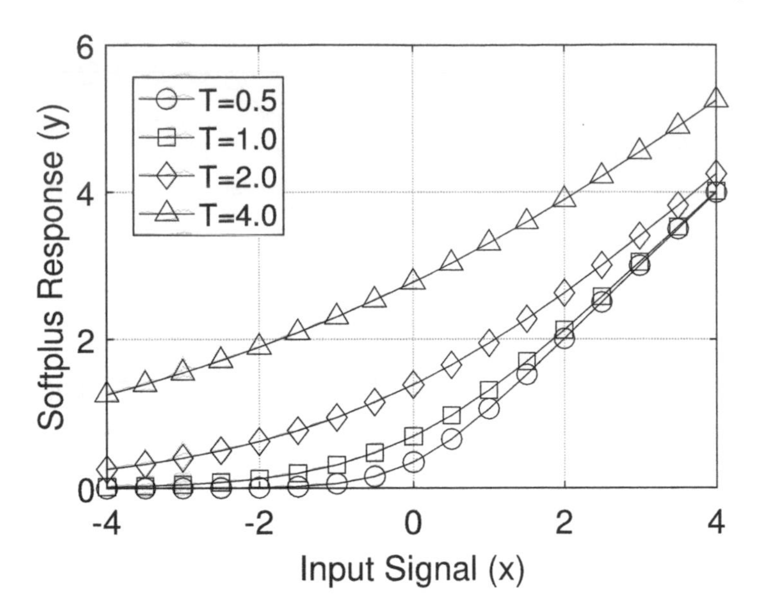


**FIGURE 1.9**
**Response characteristics of a softplus basis function.** A softplus function is a smooth approximation to a rectilinear unit whose output is equal to its input when the input is positive and whose output is zero otherwise. In particular, this figure plots the softplus response function $y(x) = T\mathcal{J}(x/T)$ where $\mathcal{J}(\phi) = \log(1 + \exp(\phi))$. When the positive constant $T$ is sufficiently small (e.g., $T = 0.5$), the softplus response function is positive for large positive $x$ and zero for large negative $x$. When the positive constant $T$ is sufficiently large, the softplus response function is approximately linear.

$\mathbf{Q}$ is used to specify how the input units are directly connected to the output units. That is, these connections "skip" the hidden layer. The parameter vector $\boldsymbol{\theta}$, which specifies all of the free parameters of the learning machine, consists of the elements of $\mathbf{W}$, $\mathbf{V}$, and $\mathbf{Q}$.

The essential idea of the residual network is that the linear mapping parameters $\mathbf{Q}$ can be quickly learned to find a linear solution. In situations where the linear solution is inadequate, then the parameters $\mathbf{W}$ and $\mathbf{V}$ play more important roles by providing the learning machine with options for using nonlinearities to compensate for the inadequate linear transformation specified by $\mathbf{Q}$. Thus, the hard nonlinear learning process only needs to focus on learning the residual nonlinear mapping which cannot be learned by the linear component of the residual network. Networks with skip connections can be interpreted as a type of regularization that introduces smoothness constraints on the objective function (Orhan and Pitkow 2018). △

### 1.5.3 Recurrent Neural Networks (RNNs)

In many situations, learning machines must extract statistical regularities over time as well as space. Temporal learning machines are trained with sequences of various lengths and then are used to make predictions of either the next item in a sequence or possibly even a prediction of a sequence of items given an initial item. One approach to modeling such situations is to assume that the statistical environment generates a sequence of independent and identically distributed "episodes". It will be convenient to model the $i$th episode $\mathbf{x}_i$ of length $T_i$ as a sequence of state vectors $(\mathbf{s}_i(1), \mathbf{y}_i(1)), \ldots, (\mathbf{s}_i(T_i), \mathbf{y}(T_i))$ for $i = 1, 2, \ldots$. Note that the $T_i$ elements of $(\mathbf{s}_i(1), \mathbf{y}_i(1)), \ldots, (\mathbf{s}_i(T_i), \mathbf{y}(T_i))$ are not presumed to be independent and are assumed to be highly correlated.

Semantically, $\mathbf{s}_i(t)$ corresponds to the $t$th pattern observed by the learning machine in the $i$th episode while $\mathbf{y}_i(t)$ is a representation of the learning machine's response which is functionally dependent upon both the currently observed input pattern $\mathbf{s}_i(t)$ and the past history of observations within the current episode $(\mathbf{s}_i(1), \mathbf{y}_i(1)), \ldots, (\mathbf{s}_i(t-1), \mathbf{y}(t-1))$. In a typical RNN, the desired response vector, $\mathbf{y}_i(t)$, for the input pattern presented at time $t$ within the $i$th episode specifies the $k$th category out of $m$ possible categories by the $k$th column of an $m$-dimensional identity matrix.

Four important applications of RNNs are: "sequence prediction", "sequence classification", "sequence generation", and "sequence translation".

In a *sequence prediction application*, the RNN is presented with the $i$th episode consisting of a temporally ordered sequence of input patterns $\mathbf{s}_i(1), \mathbf{s}_i(2), \ldots, \mathbf{s}_i(T_i - 1)$ corresponding to a recent history of temporally ordered events and the goal is to predict the next event in the sequence. That is, a sequence of $T_i - 1$ input patterns are presented to the learning machine and the learning machine is taught to predict the next pattern by making the desired response at time $T_i$, $\mathbf{y}_i(T)$, equal to the next input pattern in the sequence. This type of prediction problem would correspond to a case where the RNN is using a previously given sequence of notes to predict the next note in the sequence.

In a *sequence classification application*, the RNN is presented with the $i$th episode consisting of a temporally ordered sequence of input patterns followed by the desired response of the RNN at the end of the sequence. That is, a sequence of $T_i$ input patterns $(\mathbf{s}_i(1), \mathbf{s}_i(2), \ldots, \mathbf{s}_i(T_i)$ associated with the $i$th episode might be presented to the RNN and the desired response of the RNN would be $\mathbf{y}_i(T_i)$. Note the desired response vector $\mathbf{y}_i(t)$ is observable only at the end of the sequence. For example, this would correspond to a situation where the RNN is trained to recognize a song comprised of a sequence of notes.

In a *sequence generation application*, the RNN generates a temporally ordered sequence of responses $\mathbf{y}_i(1), \ldots, \mathbf{y}_i(T_i)$ given an input pattern $\mathbf{s}_i(1)$ for the $i$th episode. When the RNN is used for sequence generation, the input pattern sequence $\mathbf{s}_i(1), \ldots, \mathbf{s}_i(T_i)$ is a sequence of identical vectors which specify a "fixed context" (i.e., $\mathbf{s}_i(t) = \mathbf{c}_i$ for $t = 1, \ldots, T_i$ where $\mathbf{c}_i$ is the context vector for the $i$th episode. For example, this would correspond to a situation where the RNN is trained to generate a motor sequence that produces a sequence of notes given the name of a song.

In a *sequence translation application*, the RNN generates a response $\mathbf{y}_i(t)$ for a given input pattern $\mathbf{s}_i(t)$ while taking into account the temporal context in which the input pattern $\mathbf{s}_i(t)$ was generated. In particular, the temporal context for $\mathbf{s}_i(t)$ is typically comprised of both the past history of input patterns $\mathbf{s}_i(t-1), \mathbf{s}_i(t-2), \ldots$, and the past history of desired responses $\ddot{\mathbf{y}}_i(t-1), \ddot{\mathbf{y}}_i(t-2), \ldots$. For example, consider an RNN which is trained to translate a phonemic representation of a word into a semantic representation of the word. For this problem, the RNN would not only use the phonemic representation of the word to predict the word's semantic representation but also would use temporal context constraints to refine that prediction.

**Example 1.5.7** (Simple Recurrent Neural Network). The most important key ideas of RNNs are best understood by examining an early version of modern RNNs which is called a simple recurrent Elman neural network (Elman-SRN) (see Elman 1990, 1991). Consider a sequence translation application where it is desired to map a sequence of $d$-dimensional input pattern vectors $\mathbf{s}_i(1), \ldots, \mathbf{s}_i(T_i)$ into a sequence of $m$-dimensional desired response pattern vectors $\mathbf{y}_i(1), \ldots, \mathbf{y}_i(T_i)$.

The *desired response* pattern vector $\mathbf{y}_i(t)$ specifies the $k$th out of $m$ possible categories when $\mathbf{y}_i(t)$ is equal to the $k$th column of an $m$-dimensional identity matrix. Let the *predicted response* of a simplified Elman-SRN in a sequence translation application be defined by the softmax response function

$$\ddot{\mathbf{y}}_i(t) = \frac{\mathbf{exp}[\boldsymbol{\psi}_i(t)]}{\mathbf{1}_m^T \mathbf{exp}[\boldsymbol{\psi}_i(t)]}, \quad \boldsymbol{\psi}_i(t) \equiv \mathbf{W}\mathbf{h}_i(t) \tag{1.20}$$

where $\mathbf{1}_m$ is an $m$-dimensional vector of ones and the $k$th element of $\ddot{\mathbf{y}}_i(t)$ is the predicted probability that input pattern vector $\mathbf{s}_i(t)$ is labeled as belonging to category $k$. The $k$th element of the vector $\boldsymbol{\psi}_i(t)$ may be interpreted as the evidence that input pattern vector $\mathbf{s}_i(t)$ is labeled as belonging to category $k$.

The vector $\mathbf{h}_i(t)$ is the hidden unit activation pattern generated by input pattern $\mathbf{s}_i(t)$ and the previous hidden unit activation pattern $\mathbf{h}_i(t-1)$. More specifically,

$$\mathbf{h}_i(t) = \mathcal{S}\left(\mathbf{V}\left[\mathbf{h}_i(t-1)^T, \ \mathbf{s}_i(t)^T\right]^T\right) \tag{1.21}$$

where $\mathcal{S}$ is a vector-valued function defined such that the $j$th element of $\mathcal{S}$ is equal to $\mathcal{S}(\phi) = 1/(1+\exp(-\phi))$. Assume $\mathbf{h}_i(0)$ is a vector of zeros.

The empirical risk function for learning is given by the formula:

$$\hat{\ell}_n(\boldsymbol{\theta}) = -(1/n)\sum_{i=1}^{n}\sum_{t=1}^{T_i} \mathbf{y}_i(t)^T \mathbf{log}(\ddot{\mathbf{y}}_i(t)) \tag{1.22}$$

where $\mathbf{log}$ is a vector-valued log function defined such that the $k$th element of $\mathbf{log}(\mathbf{u})$ is the natural logarithm of the $k$th element of $\mathbf{u}$.

Inspection of (1.21) shows that for the $i$th episode, the hidden unit representation at time $t$ is a compressed representation of the hidden unit representation at time $t-1$ and the current input pattern $\mathbf{s}_i(t)$. Using a recursive argument, it follows that the hidden unit representation at time $t$ not only has the potential to represent a compressed version of the past history of episode $i$ from time 1 to time $t-1$, but the information in this compressed representation has been chosen such that it minimizes the episode prediction error of the Elman-SRN. Figure 1.10 provides a graphical illustration of an Elman SRN.

△

**Example 1.5.8** (Gated Recurrent Unit RNN). Although the key ideas of the Elman-SRN form the basis of modern RNNs, most real-world applications of RNNs use additional temporal-context dependent learning mechanisms which determine when the hidden unit should be turned on or off. One such modern version of an RNN is called the Gated Recurrent Unit (GRU) RNN (see Cho et al. 2014). A generic GRNN is now defined for expository reasons, but students are encouraged to consult Schmidhuber (2015) and Goodfellow et al. (2016) for more recent detailed surveys of this literature.

Essentially, the GRNN is qualitatively very similar to the Elman-SRN in Example 1.5.7. In fact, assume in this example that the predicted response of the GRNN is specified by (1.20) and the empirical risk function for the GRNN is specified by (1.22). The big important difference is how the input pattern vector $\mathbf{s}_i(t)$ is combined with the previous hidden unit pattern vector $\mathbf{h}_i(t-1)$ to generate the current hidden unit pattern vector $\mathbf{h}_i(t)$ so (1.21) is replaced with a new formula which will be called the Gated Recurrent Unit (GRU) formula.

Let the operator $\odot$ denote element-by-element matrix multiplication so that, for example,

$$[a_1, a_2, a_3] \odot [b_1, b_2, b_3] = [a_1 b_1, a_2 b_2, a_3 b_3].$$

Let

$$\mathbf{h}_i(t) = (\mathbf{1} - \mathbf{z}_i(t)) \odot \mathbf{h}_i(t-1) + \mathbf{z}_i(t) \odot \ddot{\mathbf{h}}_i(t) \tag{1.23}$$

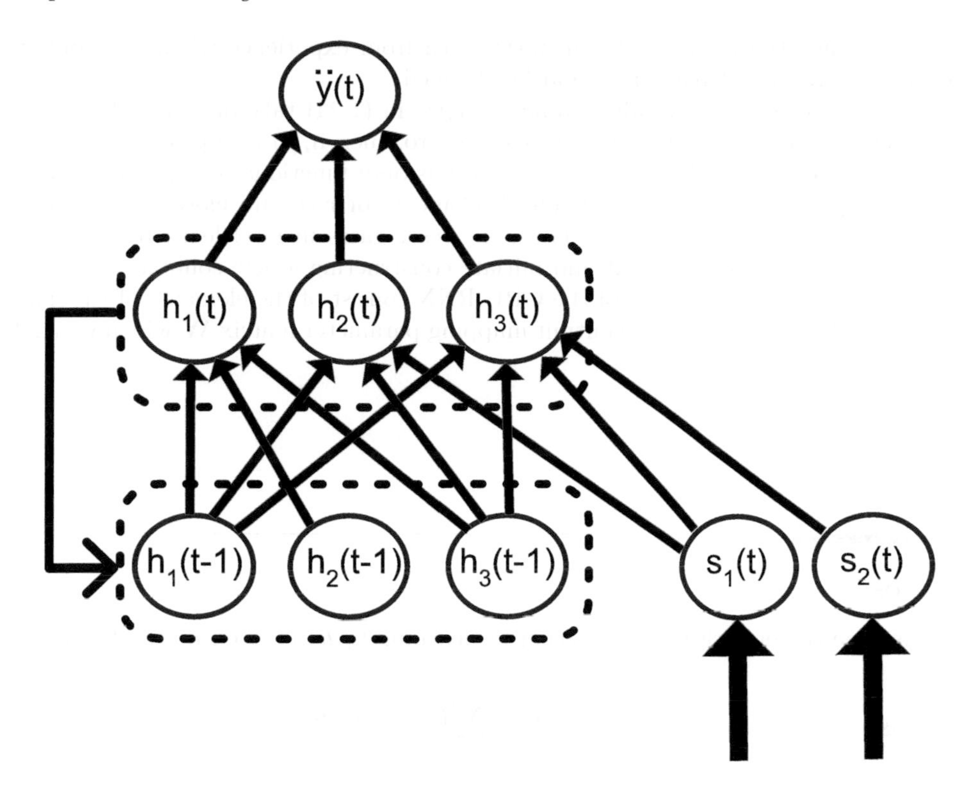


**FIGURE 1.10**
**Recurrent neural network (RNN) for learning sequences.** An example of an Elman SRN as described in Example 1.5.7 for learning sequences updates the current hidden unit responses by combining the previous responses of the hidden units and the current input pattern vector. The current hidden unit responses are then used to generate a prediction which is functionally dependent upon the past history of input patterns. Modern RNNs are based upon a similar principle but include additional free parameters which adjust the behavior of the hidden units based upon temporal context (see Example 1.5.8).

where

$$\ddot{\mathbf{h}}_i(t) = \mathcal{S}\left(\mathbf{V}_h\left[\mathbf{r}_i(t)^T \odot \mathbf{h}_i(t-1)^T,\ \ \mathbf{s}_i(t)^T\right]^T\right) \tag{1.24}$$

the *update gate function*

$$\mathbf{z}_i(t) = \mathcal{S}\left(\mathbf{V}_z\left[\mathbf{h}_i(t-1)^T,\ \ \mathbf{s}_i(t)^T\right]^T\right) \tag{1.25}$$

and the *reset gate function*

$$\mathbf{r}_i(t) = \mathcal{S}\left(\mathbf{V}_r\left[\mathbf{h}_i(t-1)^T,\ \ \mathbf{s}_i(t)^T\right]^T\right) \tag{1.26}$$

The elements of the vector-valued function $\mathbf{z}_i(t)$ take on values ranging between zero and one. When all elements of $\mathbf{z}_i(t)$ are close to zero, then $\mathbf{h}_i(t) \approx \mathbf{h}_i(t-1)$ in (1.23). That is, the context does not change when the elements of $\mathbf{z}_i(t)$ are all close to zero. On the other hand, when all elements of $\mathbf{z}_i(k)$ are close to one, then the hidden unit state vector $\mathbf{h}_i(t)$ is updated by integrating its previous value $\mathbf{h}_i(t-1)$ with the current input pattern vector $\mathbf{s}_i(t)$. Thus, the update gates in $\mathbf{z}_i(t)$ learn from experience when the context for interpreting a given input pattern should be changed.

The elements of the vector-valued function $\mathbf{r}_i(t)$ in (1.24) take on values between zero and one. When all elements of $\mathbf{r}_i(k)$ are close to zero, then $\mathbf{h}_i(t)$ is approximately a vector of zeros and the current hidden unit state vector is only functionally dependent upon the current input pattern vector $\mathbf{s}_i(t)$. When all elements of $\mathbf{r}_i(t)$ are close to one, then the updating of $\mathbf{h}_i(t)$ proceeds normally. Thus, the reset gates in $\mathbf{r}_i(t)$ learn from experience when to forget the previous context and initiate constructing a new context.

To summarize, the parameters of the GRU RNN consist of the elements of the matrices $\mathbf{V}_h$, $\mathbf{V}_z$, $\mathbf{V}_r$ and the hidden unit to output mapping parameter matrix $\mathbf{W}$ which was defined in (1.20).

△

## Exercises

1.5-1. Assume the empirical risk function for learning $\hat{\ell}_n(\boldsymbol{\theta})$ is given by the formula:

$$\hat{\ell}_n(\boldsymbol{\theta}) = (1/n)\sum_{i=1}^{n}(y_i - \ddot{y}(\mathbf{s}_i, \boldsymbol{\theta}))^2.$$

Run a computer program which evaluates the classification performance of a linear regression machine defined such that

$$\ddot{y}(\mathbf{s}_i, \boldsymbol{\theta}) = \boldsymbol{\theta}^T \mathbf{s}_i$$

on both a training data set and a test data set. Now use the same empirical risk function with a nonlinear regression model using radial basis functions defined such that

$$\ddot{y}(\mathbf{s}_i, \boldsymbol{\theta}) = \sum_{k=1}^{m} v_k \exp(-|\mathbf{s}_i - \mathbf{w}_k|^2/\tau)$$

where $\boldsymbol{\theta} \equiv [v_1, \ldots, v_m, \mathbf{w}_1, \ldots, \mathbf{w}_m^T]^T$ and $\tau$ is a positive constant. Evaluate the classification performance of the nonlinear regression machine on both a training data set and test data set and compare this performance with the linear regression machine. If $n$ is the number of training stimuli, try approximately $n$ hidden units. Next, try using $K$ hidden units where $K \equiv \log_2(n)$ is the number of binary bits required to recode $n$ distinct pattern vectors as $n$ binary $K$-dimensional vectors.

## 1.6 Unsupervised Learning Machines

The goal of an *unsupervised learning machine* is to learn statistical regularities among all components of a feature vector generated from the environment. As illustrated in Figure 1.11, the data generating process generates the training data which consists of a collection of input pattern vectors. This collection of input pattern vectors $\mathbf{x}_1, \ldots, \mathbf{x}_n$ is then used to adjust the parameters of the unsupervised learning machine until the parameter

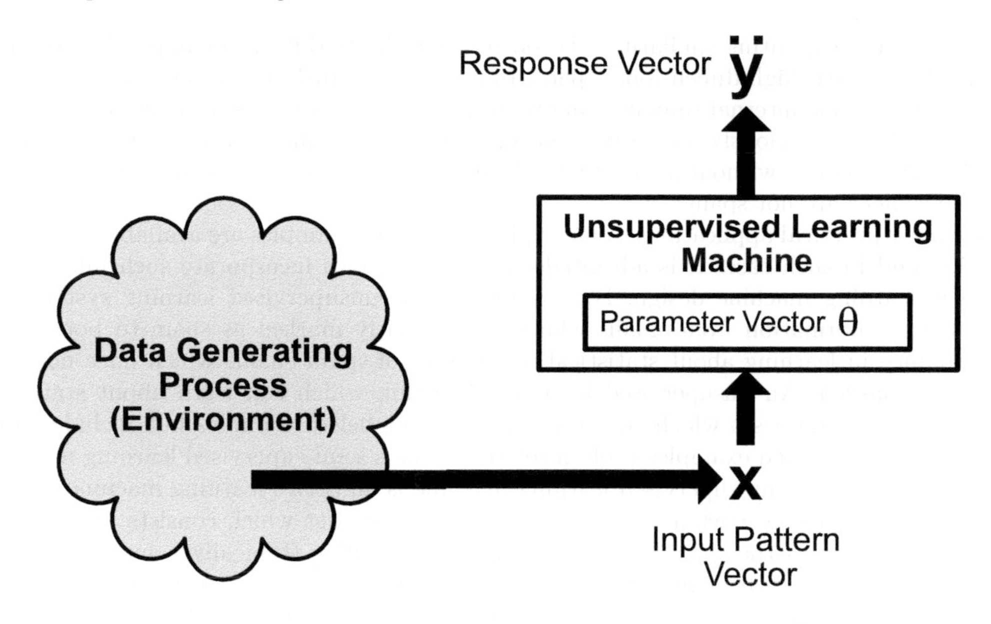


**FIGURE 1.11**
**Overview of an unsupervised learning machine.** Although the training data for an unsupervised learning machine consists only of input pattern vectors, an unsupervised learning machine can still learn to generate an appropriate response for a given input pattern.

estimate $\hat{\boldsymbol{\theta}}_n$ is obtained. After the training process is completed, the unsupervised learning machine generates a response $\ddot{y}$ in response to a given input pattern vector $\mathbf{x}$ and its current parameter vector $\hat{\boldsymbol{\theta}}_n$. For example, an unsupervised learning machine might be trained on a collection of email messages and then asked to identify unusual email messages which don't conform to the typical observed patterns of statistical regularities. This type of machine could be useful for identifying suspicious emails. The unsupervised learning machine might learn without feedback to distinguish spam email messages from non-spam email messages.

Unsupervised learning machines can be developed to provide solutions to a variety of inference tasks. These tasks may be roughly characterized as: (1) filtering, (2) reconstruction, (3) information compression, and (4) clustering. An example of a filtering task involves presenting the unsupervised learning machine with examples of images or acoustic signals which have been corrupted with noise and having the unsupervised learning machine attempt to remove the noise from the signal. In a reconstruction task, a pattern vector is presented to the unsupervised learning machine which has some missing components and the unsupervised learning machine must use its experience with its environment to reconstruct the missing components. This type of application arises, for example, in systems which represent large amounts of probabilistic knowledge and are intended to infer which probability-specific conclusions (e.g., medical diagnoses) hold given different combinations of assumptions (e.g., symptoms and patient characteristics). In an information compression task, a high-dimensional pattern vector is transformed into a low-dimensional compressed representation which preserves important characteristics of the original high-dimensional pattern vector and suppresses irrelevant characteristics. For example, algorithms which compress large audio, image, or video files without observable degradations in content quality. In addition, to providing a mechanism for generating compact representations of high-dimensional pattern vectors, compressed representations are useful for identifying and exploiting similarity relationships which are difficult or impossible to identify in the original high-dimensional space. In a clustering task, the unsupervised learning machine uses some internal similarity metric to group input patterns into different clusters. For example, as previously mentioned, an unsupervised learning machine which processes email might identify without supervision which email messages are spam email and which email messages are not spam.

In many practical applications, some explicitly labeled examples are available to support learning and in such cases it is advantageous to attempt to incorporate such information into the learning machine design. For example, some unsupervised learning systems are able to use information from emails which are explicitly marked as spam to boost their performance in learning about statistical regularities of spam emails which have not been explicitly marked. An unsupervised learning algorithm which can learn about statistical regularities in a data set which consists of mostly unlabeled examples but includes some small explicitly labeled examples is often referred to as a semi-supervised learning machine. A special case of a semi-supervised learning machine is an active learning machine.

An active learning machine typically processes a data set which consists of unlabeled examples but it has the capability to ask some other entity (typically a human) how to label specific input pattern vectors in the training data set. Active learning machines may be designed to make such requests only for the input pattern vectors which are maximally informative to the learning machine with the specific goal of reducing the number of times such questions are generated.

---

**Algorithm 1.6.1** A generic adaptive gradient descent unsupervised learning algorithm. The learning machine receives a penalty of $c(\mathbf{x};\boldsymbol{\theta})$ for observing event $\mathbf{x}$ when its parameters are set to $\boldsymbol{\theta}$. The goal of learning is to find a set of parameter values $\boldsymbol{\theta}$ which minimize the expected penalty. The small positive number $\gamma_t$ controls the rate of learning at a particular algorithm iteration $t$.

---

```
 1: procedure UNSUPERVISED-LEARNING-GRADIENT-DESCENT(θ(0))
 2:     t ⇐ 0
 3:     repeat
 4:         Observe Environmental Event x(t)
 5:         g(x(t), θ(t)) ⇐ dc(x(t); θ(t))/dθ
 6:         θ(t + 1) ⇐ θ(t) − γ_t g(x(t), θ(t))
 7:         t ⇐ t + 1.
 8:     until θ(t) ≈ θ(t − 1)
 9:     return {θ(t)}
10: end procedure
```

---

In an unsupervised learning paradigm, the training data is a collection of feature vectors $\mathbf{x}_1, \ldots, \mathbf{x}_n$ and the goal of the learning process is to discover statistical regularities characterizing the internal statistical structure of those feature vectors. As in supervised learning, the unsupervised learning machine minimizes the empirical risk function in Equation (1.1) with respect to a particular loss function $c$. Typically, the learning machine's incurred loss for choosing parameter vector $\boldsymbol{\theta}$ is given by the formula $c(\mathbf{x}, \boldsymbol{\theta})$ for a particular input pattern feature vector $\mathbf{x}$ and loss function $c$. The loss function $c$ implicitly incorporates prior knowledge regarding what are the relevant statistical regularities in the training data as well as prior knowledge regarding what types of solutions to the empirical risk function optimization problem in Equation (1.1) are preferable. Algorithm 1.6.1 provides a simplified example of an Unsupervised Learning Adaptive Gradient Descent Algorithm.

| | Document 1 | Document 2 | Document 3 | Document 4 |
|---|---|---|---|---|
| "the" | 20 | 8 | 5 | 14 |
| "dog" | 3 | 2 | 0 | 0 |
| "cat" | 0 | 4 | 0 | 0 |
| "animal" | 0 | 1 | 0 | 0 |
| "ball" | 0 | 0 | 8 | 0 |
| "paper" | 0 | 0 | 0 | 6 |

**FIGURE 1.12**
**An example of a term by document matrix.** The number of times the $j$th term occurs in document $k$ is specified by a number in row $j$ of the $k$th column of the *term by document* matrix. Note that documents 1 and 2 are considered similar because they have more terms in common. The terms "dog" and "paper" are considered less similar because they have fewer documents in common.

**Example 1.6.1** (Document Retrieval Using Latent Semantic Indexing (LSI)). A simple model for document retrieval based upon keyword matching may be interpreted as a type of unsupervised learning algorithm. Let the training data set

$$\mathcal{D}_n = \{d_1, \ldots, d_n\}$$

be a collection of $n$ documents. For example, the $i$th document, $d_i$, might be the $i$th web site in a collection of $n$ websites. Or alternatively, the $i$th document $d_i$, might be the $i$th book title in a collection of $n$ books.

Let the set $T = \{t_1, \ldots, t_m\}$ be a collection of $m$ keywords or *terms* that tend to appear in the documents in $\mathcal{D}_n$. One may choose, for example, $T$ to consist of all of the unique words in all of the documents in $\mathcal{D}_n$.

Let $\mathbf{W} \in \{0, 1\}^{m \times d}$ be called the *term by document matrix* which is defined such that the $ij$th element in $\mathbf{W}$, $w_{ij} = 1$ if term $t_i$ occurs at least once in document $d_j$. If term $t_i$ does not occur at all in document $d_j$ then $w_{ij} = 0$. Now notice that the columns of the matrix $\mathbf{W}$ can be interpreted as points in an $m$-dimensional *term space*. The columns of $\mathbf{W}$ are called *document column vectors* and have dimension $m$. These ideas are illustrated in Figure 1.12.

Let a *query* $Q$ be a subset of $T$ which is used for the purpose of finding documents that contain terms which are also contained in the query. Let $\mathbf{q}$ be an $m$-dimensional *query column vector* consisting of zeros and ones defined such that the $i$th element of $\mathbf{q}$ is equal to 1 if term $t_i \in Q$. A keyword matching algorithm can then be implemented by computing the expression $\mathbf{y}^T = \mathbf{q}^T \mathbf{W}$ where $\mathbf{y}$ is a $d$-dimensional column vector whose $j$th element is a measure of the number of times that term $t_i$ occurs in document $d_j$. If the $j$th element of $\mathbf{y}^T$ is larger in magnitude, this indicates that the document $j$ tended to include more query terms in the query $Q$. This is essentially a keyword matching algorithm.

For many document retrieval applications, keyword matching algorithms are not very effective for two major reasons. First, semantically related documents may not share terms. For example, five different documents associated with the topics `CARS`, `BICYCLES`, `AIRPLANES`, `TRAINS`, and `CRUISE-SHIPS` may not share the keyword query term *transportation*. Second, documents that are unrelated to a query may have many terms in common.

For example, a query involving the terms *Brazil* and *transportation* may retrieve documents associated with the topics `CARS`, `BICYCLES`, `AIRPLANES`, `TRAINS`, `CRUISE-SHIPS`, `COFFEE`, and `SOUTH-AMERICA`.

Latent Semantic Indexing (LSI) (Landauer et al. 2014) provides a method for addressing these issues. The procedure for implementing an LSI analysis begins by first forming a term by document matrix $\mathbf{W}$ as before, but now the $ij$th element of $\mathbf{W}$, $w_{ij}$, is a measure of the number of times, $n_{ij}$, that term $t_i$ occurs in document $d_j$. One common choice for $w_{ij}$ is TF-IDF (term frequency inverse document frequency) defined by the formula

$$w_{ij} = (\log(1 + n_{ij})) \log(1 + (d/M_i))$$

where $d$ is the number of documents and $M_i$ is the number of documents which contain term $t_i$.

Second, the "noise" in the matrix $\mathbf{W}$ is filtered out to obtain an approximation of $\mathbf{W}$ which may be denoted as $\ddot{\mathbf{W}}$. The approximation $\ddot{\mathbf{W}}$ is chosen to include the stable dominant statistical regularities in the original term by document matrix $\mathbf{W}$ while the unstable weak statistical regularities of the original term by document matrix $\mathbf{W}$ are not included in $\ddot{\mathbf{W}}$. Consequently, the "latent semantic structure" is revealed by $\ddot{\mathbf{W}}$ since elements in the original matrix $\mathbf{W}$ which were equal to zero may be non-zero in the approximation $\ddot{\mathbf{W}}$. The construction of such approximations are typically accomplished using the matrix factorization technique called "singular value decomposition" (see Chapter 4).

Finally, under the assumptions that the number of documents does not increase during the learning process and the number of terms does not increase during the learning process, then the problem of constructing the approximation $\ddot{\mathbf{W}}$ can be interpreted as minimizing an empirical risk function (e.g., see Theorem 4.3.1). △

**Example 1.6.2** (Nonlinear Denoising Autoencoder (Vincent et al. 2008)). A popular nonlinear unsupervised learning machine is the denoising autoencoder which corrupts an input pattern $\mathbf{s}$ with a noise source $\tilde{\mathbf{n}}$ and then is trained to reconstruct the original input pattern $\mathbf{s}$. The desired response for the denoising autoencoder is the input pattern $\mathbf{s}$ before it was corrupted with noise. The noise source might randomly select large regions of the input pattern and then set the values of the input pattern vector in these regions equal to zero in order to teach the denoising autoencoder to "reconstruct" input patterns. Alternatively, the noise source might add zero-mean noise to all elements of the input pattern $\mathbf{s}$ in order to teach the denoising autoencoder to "filter out noise" from the input pattern.

For example, let the parameter vector for a denoising autoencoder be the $q$-dimensional column vector $\boldsymbol{\theta}$. Define the response function

$$\ddot{\mathbf{s}}(\mathbf{s} \odot \tilde{\mathbf{n}}, \boldsymbol{\theta})) = \sum_{j=1}^{m} w_{k,j} \mathcal{J}\left(\boldsymbol{\theta}_j^T (\mathbf{s} \odot \tilde{\mathbf{n}})\right) \quad (1.27)$$

where $\tilde{\mathbf{n}}$ is a noise mask whose elements are zeros and ones chosen at random, and $\mathcal{J}(\phi) \equiv \log(1 + \exp(\phi))$ specifies a softplus basis function. The notation $\mathbf{s} \odot \tilde{\mathbf{n}}$ indicates element-by-element vector multiplication so that the $i$th element of $\mathbf{s} \odot \tilde{\mathbf{n}}$ is the $i$th element of $\mathbf{s}$ multiplied by the $i$th element of $\tilde{\mathbf{n}}$. In this example, Equation (1.27) only considers one layer of hidden units but in practice, multiple layers of hidden units are commonly used in deep learning applications. Figure 1.13 provides a network diagram of a nonlinear denoising autoencoder with one layer of hidden units.

Define the discrepancy function

$$D(\mathbf{s}, \ddot{\mathbf{s}}(\mathbf{s} \odot \tilde{\mathbf{n}}, \boldsymbol{\theta})) = |\mathbf{s} - \ddot{\mathbf{s}}(\mathbf{s} \odot \tilde{\mathbf{n}}, \boldsymbol{\theta})|^2$$

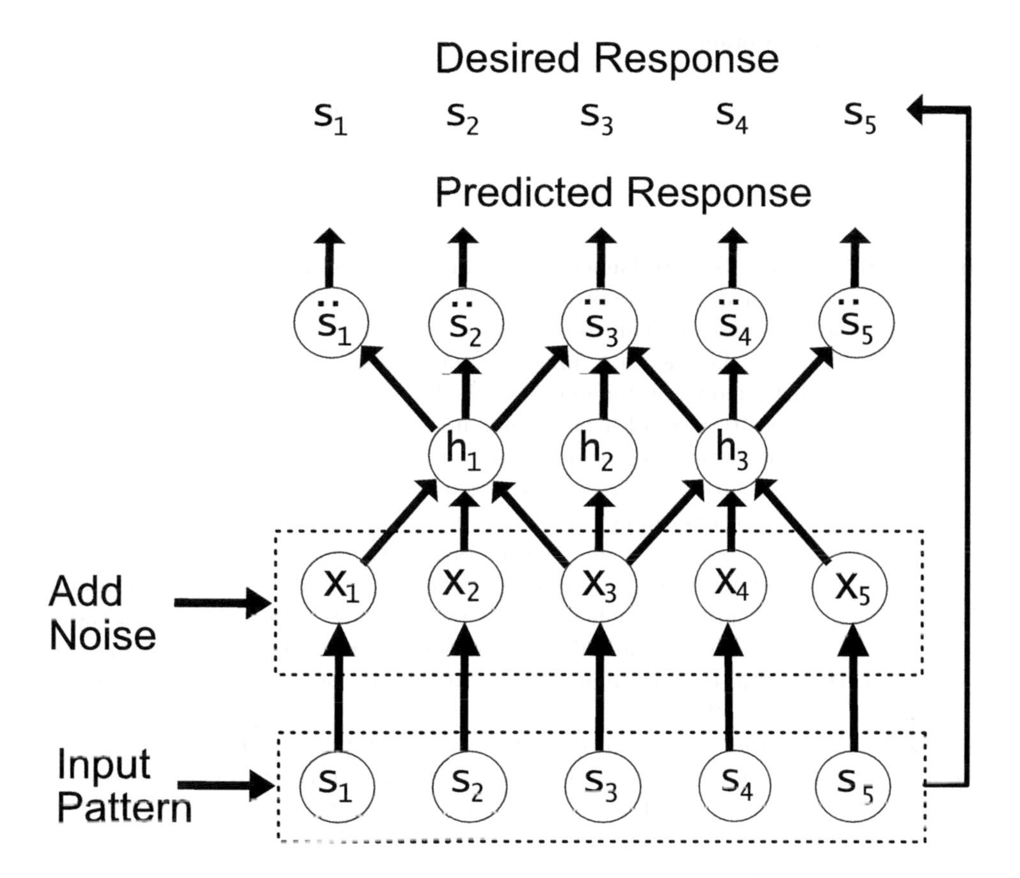


**FIGURE 1.13**
**Denoising autoencoder with one layer of hidden units.** An input pattern $s_1, s_2, s_3, s_4, s_5$ is presented to the denoising autoencoder and then corrupted by additive noise to obtain the noise corrupted pattern $x_1, x_2, x_3, x_4, x_5$. Next, the noise-corrupted input pattern is mapped into a pattern $[h_1, h_2, h_3]$ of hidden unit states using only a relatively small number of hidden units. The predicted response of the autoencoder $\ddot{s}_1, \ddot{s}_2, \ddot{s}_3, \ddot{s}_4, \ddot{s}_5$ is then computed from the hidden unit states $[h_1, h_2, h_3]$. The desired response of the autoencoder is the original input pattern before it was corrupted with noise. Thus, the denoising autoencoder learns how to filter out the noise. The small number of hidden units prevents the autoencoder from memorizing input patterns and forces the autoencoder to represent the corrupted input pattern using only statistical regularities which are dominant in the statistical environment.

which compares the predicted response of the nonlinear denoising autoencoder $\ddot{\mathbf{s}}(\mathbf{s} \odot \tilde{\mathbf{n}}, \boldsymbol{\theta})$ with the noise-free pattern vector $\mathbf{s}$.

A penalized empirical risk function, $\hat{\ell}_n$, can then be constructed using the loss function

$$c(\mathbf{s}, \boldsymbol{\theta}) = D\left(\mathbf{s}, \ddot{\mathbf{s}}(\mathbf{s} \odot \tilde{\mathbf{n}}, \boldsymbol{\theta})\right) \tag{1.28}$$

with an elastic net regularization term, $k_n(\boldsymbol{\theta})$,

$$k_n(\boldsymbol{\theta}) = \eta(1-\lambda)\sum_{j=1}^{q}(\theta_j^2 + \epsilon^2)^{1/2} + \eta\lambda\sum_{j=1}^{q}\theta_j^2 \tag{1.29}$$

where $0 \leq \lambda, \eta \leq 1$ to obtain:

$$\hat{\ell}_n(\boldsymbol{\theta}) = k_n(\boldsymbol{\theta}) + (1/n) \sum_{i=1}^{n} c(\mathbf{s}_i, \boldsymbol{\theta}). \tag{1.30}$$

△

**Example 1.6.3** (Image Texture Generation Problem (Cross and Jain 1983)). Image texture modeling is concerned with the problem of building a synthetic model of a small patch of an image. Realistic image texture models are useful in a variety of areas of image processing including: image classification, image segmentation, and image compression. In this problem, a probabilistic model of image texture is described. The probabilistic model works by predicting the intensity level of a particular image pixel based upon the intensity level values of the image pixels which are neighbors of that target image pixel. Unsupervised learning algorithms are used to learn image texture models because they extract relevant statistical regularities without supervision from image texture samples.

Let $S \equiv \{0, \ldots, G-1\}$ be a set of G numbers specifying a range of grey levels where 0 is the lightest grey level and G is the darkest grey level. A typical value of $G$ might be 256. Let $\tilde{\mathbf{X}} \in S^{d \times d}$ be a random matrix whose $ij$th element in row $i$ and column $j$ of the random image matrix $\tilde{\mathbf{X}}$, $\tilde{x}_{i,j}$, takes on one value $x_{i,j}$ which specifies one of G possible grey level values. The $mn$th pixel in the image is called a "neighbor" of the $ij$th pixel in the image if $0 < (i-m)^2 + (j-n)^2 \leq 2$.

It is assumed that the conditional probability the $ij$th pixel in the image, $\tilde{x}_{i,j}$, takes on the value of $x_{i,j}$ is functionally dependent upon the values of the pixels in the image which are physical neighbors of the $ij$th pixel. The set of physical neighbors of the $ij$th pixel in image $\mathbf{X}$ will be denoted as $\mathcal{N}_{i,j}(\mathbf{X})$.

In particular, let the binomial conditional probability mass function

$$p(\tilde{x}_{i,j} = K | \mathcal{N}_{i,j}(\mathbf{X}); \boldsymbol{\theta}) = \left( \frac{G!}{K!(G-K)!} \right) \eta_{i,j}^K (1 - \eta_{i,j})^{(G-K)}$$

where

$$\eta_{i,j} = \mathcal{S}\left(\boldsymbol{\theta}^T \boldsymbol{\phi}_{i,j}(\mathbf{X})\right), \quad \mathcal{S}\left(\boldsymbol{\theta}^T \boldsymbol{\phi}_{i,j}(\mathbf{X})\right) \equiv \left(1 + \exp(-\boldsymbol{\theta}^T \boldsymbol{\phi}_{i,j}(\mathbf{X})\right)^{-1},$$

and

$$\boldsymbol{\phi}_{i,j}(\mathbf{X}) \equiv [x_{i-1,j-1}, x_{i-1,j}, x_{i,j+1}, x_{i-1,j+1}, \ldots x_{i,j-1}, x_{i,j+1}, x_{i+1,j-1}, x_{i+1,j}, x_{i+1,j+1}, 1].$$

Assume the goal of learning is to find the parameters of the probability model such that the probability that the $ij$th pixel takes on the value of $x_{i,j}$ is given by $p\left(x_{i,j} | \mathcal{N}_{i,j}(\mathbf{X}); \boldsymbol{\theta}\right)$, then this unsupervised learning objective may be reformulated as minimizing the empirical risk function:

$$\hat{\ell}_n(\hat{\boldsymbol{\theta}}_n) = -(1/n) \sum_{i=1}^{n} \sum_{j=1}^{n} \log p\left(x_{i,j} | \mathcal{N}_{i,j}(\mathbf{X}); \boldsymbol{\theta}\right)$$

△

**Example 1.6.4** (Clustering: Within-Cluster Dissimilarity Minimization). A clustering algorithm automatically divides a set of input patterns into several clusters of input patterns. Clustering algorithms with different structural features and different parameter values will perform this task in different ways. A clustering algorithm can be interpreted as an unsupervised learning problem where one is provided with a collection of input patterns which are unlabeled and the goal of the learning machine is to assign each input pattern with an appropriate label by examining the similarities and differences among the input patterns.

Let $\mathbf{x}_1, \ldots, \mathbf{x}_n$ be a collection of $n$ $d$-dimensional feature vectors. A $d$-dimensional vector $\mathbf{x}_i$ specifies the $i$th object in some database of $n$ events, $i = 1, \ldots, n$. For example, the first element of $\mathbf{x}_i$ might be the number of exclamation points in an email message, the second element of $\mathbf{x}_i$ might be the number of characters in the email message, and the third element of $\mathbf{x}_i$ might be the number of verbs in the email message. Emails, coded in this manner, might fall naturally into different clusters where spam emails are associated with clusters of points that have more exclamation points.

A *cluster* $C_k$ is a subset of $\{\mathbf{x}_1, \ldots, \mathbf{x}_n\}$. The goal of the clustering problem is to assign each of the $n$ feature vectors to one of $K$ clusters of feature vectors subject to the constraint that each cluster contains at least one feature vector. To facilitate the solution of this problem, a measure of similarity between two feature vectors is provided which indicates the two feature vectors are either very similar or very different.

Let $x_j$ denote the $j$th element of a feature vector $\mathbf{x}$ and $y_j$ denote the $j$th element of a feature vector $\mathbf{y}$. A *dissimilarity function* $D$ is a function defined such that: (i) $D(\mathbf{x}, \mathbf{y})$ takes on larger values when the two feature vectors $\mathbf{x}$ and $\mathbf{y}$ are more dissimilar, and (ii) $D(\mathbf{x}, \mathbf{y}) = D(\mathbf{y}, \mathbf{x})$. Examples of possible choices for a dissimilarity function are:

- *Spherical Cluster Dissimilarity:* $D(\mathbf{x}, \mathbf{y}) = |\mathbf{x} - \mathbf{y}|^2$.
- *Ellipsoidal Cluster Dissimilarity:* $D(\mathbf{x}, \mathbf{y}) = |\mathbf{M}(\mathbf{x} - \mathbf{y})|^2$ where $\mathbf{M}$ is a diagonal matrix.
- *Directional Dissimilarity:* $D(\mathbf{x}, \mathbf{y}) = -\cos\psi(\mathbf{x}, \mathbf{y})$ where $\psi$ is the angle between $\mathbf{x}$ and $\mathbf{y}$.

A *selection matrix* $\mathbf{S}$ is a matrix with $n$ columns corresponding to the $n$ feature vectors to be categorized and $K$ rows corresponding to the $K$ possible clusters $C_1, \ldots, C_K$. Each column of a selection matrix is a column of a $K$-dimensional identity matrix. If the $k$th element of the $i$th column of $\mathbf{S}$ is equal to one, this means that feature vector $\mathbf{x}_i \in C_k$.

Let $\rho(\mathbf{x}_i, C_k)$ denote the *classification penalty* incurred for placing feature vector $\mathbf{x}_i$ into the $k$th cluster $C_k$ which consists of $n_k$ feature vectors. The classification penalty $\rho(\mathbf{x}_i, C_k)$ is defined as:

$$\rho(\mathbf{x}_i, C_k) = (1/n_k) \sum_{j \in C_k} D(\mathbf{x}_i, \mathbf{x}_j) \tag{1.31}$$

which basically means that the average dissimilarity of $\mathbf{x}_i$ to each element of $C_k$ is the incurred penalty.

The average classification performance associated with the assignment of $\mathbf{x}_1, \ldots, \mathbf{x}_n$ to the clusters $C_1, \ldots, C_K$ is specified by a function $V_w$ which assigns a number to a particular selection matrix $\mathbf{S}$. In particular, the clustering performance measure $V_w$ is defined such that:

$$V_w(\mathbf{S}) = (1/K) \sum_{k=1}^{K} \sum_{i \in C_k} \rho(\mathbf{x}_i, C_k). \tag{1.32}$$

The goal of the clustering algorithm is to find a selection matrix $\mathbf{S}$ such that the objective function $V_w(\mathbf{S})$ is minimized subject to the constraint that every cluster contains at least one feature vector. Exercise 6.3-7 in Chapter 6 discusses a method for designing such algorithms. The *within-cluster dissimilarity function* $V_w$ in (1.32) has the semantic interpretation that it measures the average dissimilarity of feature vectors within a particular cluster.

The widely used *K-means clustering algorithm* corresponds to the special case where $D$ is the Spherical Cluster dissimilarity defined such that $D(\mathbf{x}, \mathbf{y}) = |\mathbf{x} - \mathbf{y}|^2$:

$$V_w(\mathbf{S}) = (1/K) \sum_{k=1}^{K} \sum_{i \in C_k} (1/n_k) \sum_{j \in C_k} |\mathbf{x}_i - \mathbf{x}_j|^2. \tag{1.33}$$

Thus, *K-means clustering* is useful when the feature clusters correspond to non-overlapping hyperspheres of feature vectors. If the feature vector clusters have ellipsoidal shapes then an Ellipsoidal Cluster dissimilarity function might be considered.

Ideally, a clustering algorithm should simultaneously minimize the dissimilarity of feature vectors within a particular cluster while maximizing the dissimilarity of feature vectors which are in different clusters. The function $V_w$ in (1.32) measures the within-cluster dissimilarity. The *between-cluster dissimilarity function* $V_b$ is defined by the formula:

$$V_b(\mathbf{S}) = (1/K) \sum_{k=1}^{K} \sum_{i \notin C_k} \rho(\mathbf{x}_i, C_k). \tag{1.34}$$

Instead of minimizing $V_w$ in (1.32), the between-cluster dissimilarity measure in (1.34) can be maximized.

In order to investigate the relationship between the within-cluster dissimilarity $V_w$ and between-cluster dissimilarity $V_b$, let

$$\bar{D}_n = \frac{1}{n^2} \sum_{i=1}^{n} \sum_{j=1}^{n} D(\mathbf{x}_i, \mathbf{x}_j) = \frac{1}{n} \sum_{i=1}^{n} \frac{1}{K} \sum_{k=1}^{K} \rho(\mathbf{x}_i, C_k)$$

be the average dissimilarity between all pairs of the $n$ feature vectors. The average dissimilarity $\bar{D}_n$ is not functionally dependent upon the choice of the clusters $C_1, \ldots, C_K$ but does depend upon the $n$ feature vectors $\mathbf{x}_1, \ldots, \mathbf{x}_n$ that participate in the clustering process.

Given the definition of the average feature vector dissimilarity $\bar{D}_n$, it follows that

$$V_w(\mathbf{S}) + V_b(\mathbf{S}) = \frac{1}{K} \sum_{k=1}^{K} \left( \sum_{i \in C_k} \rho(\mathbf{x}_i, C_k) + \sum_{i \notin C_k} \rho(\mathbf{x}_i, C_k) \right) = n\bar{D}_n. \tag{1.35}$$

This means that minimizing the within-cluster dissimilarity $V_w(\mathbf{S}) = n\bar{D}_n - V_b(\mathbf{S})$ is equivalent to maximizing the between-cluster dissimilarity $V_b$. Therefore, minimizing the within-cluster dissimilarity $V_w$ in (1.32) not only minimizes the within-cluster dissimilarity but maximizes the between-cluster dissimilarity as well!

Note that simultaneous minimization of within-cluster dissimilarity and maximization of between-cluster dissimilarity depends upon the choice of the dissimilarity function $D(\mathbf{x}, \mathbf{y})$. A learning process may be defined with respect to the empirical risk function framework by assuming a *parameterized dissimilarity function* $D(\mathbf{x}, \mathbf{y}; \boldsymbol{\theta})$ that specifies a different dissimilarity function $D(\mathbf{x}, \mathbf{y}; \boldsymbol{\theta})$ given a particular choice of the parameter vector $\boldsymbol{\theta}$. For example, assume the stimuli clusters are ellipsoidal but the lengths of the principal axes of the ellipsoids must be estimated. In such a situation, one might choose $D(\mathbf{x}, \mathbf{y}; \boldsymbol{\theta}) = |\mathbf{M}(\mathbf{x} - \mathbf{y})|^2$ where the parameter vector $\boldsymbol{\theta}$ specifies the on-diagonal elements of the diagonal matrix $\mathbf{M}$. Let $\mathbf{X}_i$ denote the $i$th collection of $n$ feature vectors. Then, an empirical risk function for learning can be constructed using a loss function $c(\mathbf{X}_i, \boldsymbol{\theta})$ defined such that $c(\mathbf{X}, \boldsymbol{\theta})$ is the within-clustering performance $V_w(\mathbf{S})$ which is a function of both $\mathbf{X}$ and $\boldsymbol{\theta}$.

△

**Example 1.6.5** (Stochastic Neighborhood Embedding (Hinton and Roweis 2002)). The first panel of Figure 1.14 illustrates a situation where clusters can be identified by minimizing a within-cluster dissimilarity measure that assumes the clusters are spheres such as K-means clustering (see Example 1.6.4). The second panel of Figure 1.14 permits the clusters to be ellipsoids as well as spheres. The third panel in Figure 1.14 shows a situation where the cluster of points associated with a particular category does not have an ellipsoidal shape.

Stochastic Neighborhood Embedding (SNE) (Hinton and Roweis 2002; van der Maaten and Hinton 2008) clustering is a technique that can sometimes effectively handle clustering problems in situations where clusters are not compact ellipsoidal shapes and a point in the original feature space belongs to multiple clusters. Assume that the original high-dimensional feature space consists of $n$ $d$-dimensional feature vectors $\mathbf{x}_1, \ldots, \mathbf{x}_n$. The embedded feature space consists of $n$ $m$-dimensional feature vectors $\mathbf{y}_1, \ldots, \mathbf{y}_n$ where the dimensionality of the embedded feature space, $m$, is assumed to be much less than the dimensionality of the original feature space $d$. It is assumed that the original $n$ objects to be clustered are represented by feature vectors in the original feature space $\mathbf{x}_1, \ldots, \mathbf{x}_n$. In addition, a dissimilarity function $D_x$ is provided for the original $d$-dimensional feature space which specifies the dissimilarity between object $i$ and object $j$ as $D_x(\mathbf{x}_i, \mathbf{x}_j)$ and a

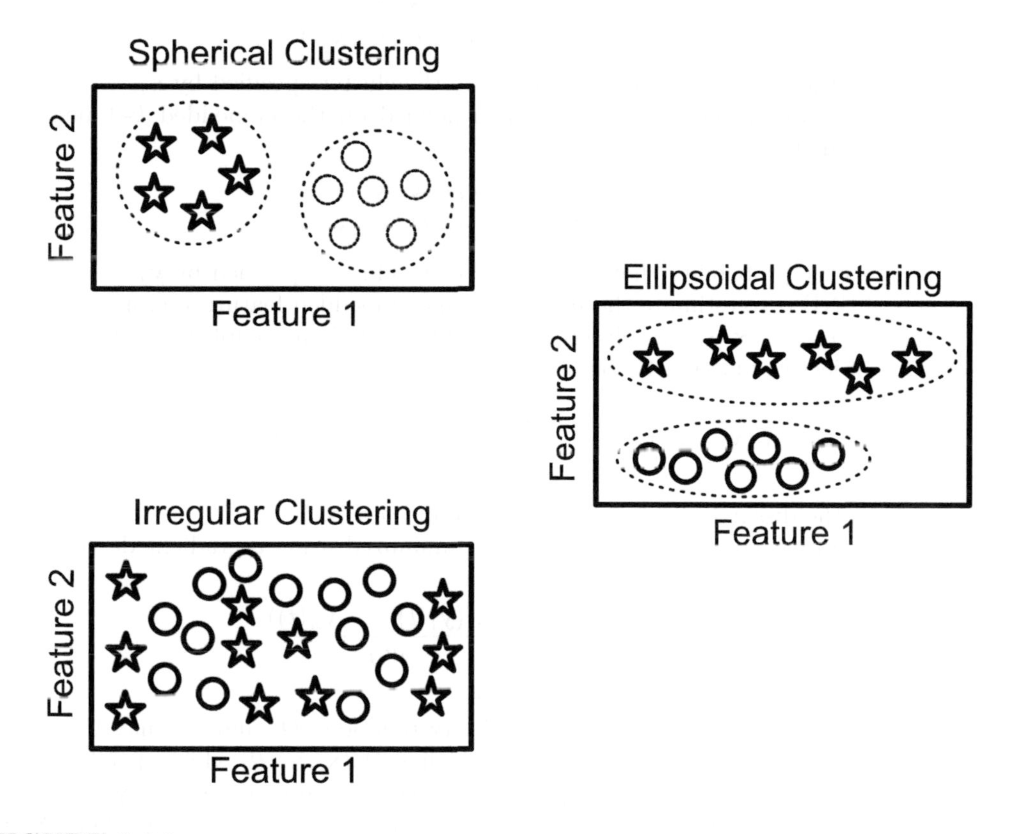


FIGURE 1.14
**Examples of clustering problems.** Dissimilarity measure within-cluster minimization problems such as K-means which assume clusters have spherical shapes. Other dissimilarity measure within-clustering algorithms allow for clusters to have ellipsoidal and other compact shapes. Stochastic neighborhood embedding clustering methods are applicable in situations where data clusters are not compact.

dissimilarity function $D_y$ is provided for the embedded $m$-dimensional feature space which specifies the dissimilarity between object $i$ and object $j$ as $D_y(\mathbf{y}_i, \mathbf{y}_j)$. The goal of SNE is to estimate $\mathbf{y}_1, \ldots, \mathbf{y}_n$ such that the dissimilarity between object $i$ and object $j$ in the embedded $m$-dimensional feature space is as similar as possible to the dissimilarity between object $i$ and object $j$ in the original $d$-dimensional feature space.

The dissimilarity functions $D_x$ and $D_y$ specify when two vectors $\mathbf{a}$ and $\mathbf{b}$ are similar and when they are different. Different choices of $D_x$ and $D_y$ will generate different SNE clustering strategies because the similarity relationships among points in the embedded feature space changes directly as a function of the choice of $D_x$ and $D_y$. Several possible choices for $D_x$ and $D_y$ are possible. Typically, the dissimilarity function $D$ is a Euclidean distance function so that: $D(\mathbf{a}, \mathbf{b}) = |\mathbf{a} - \mathbf{b}|^2$, but other choices are possible (e.g., see discussion in Example 1.6.4).

After the dissimilarity functions are defined, conditional probability distributions on the original feature space and the embedded feature space are specified according to the following procedure (see Chapter 10 for further discussion regarding conditions for the construction of such conditional distributions which correspond to Markov random fields). The conditional probability

$$p_x(\mathbf{x}_i|\mathbf{x}_j) = \frac{\exp\left(-D_x(\mathbf{x}_i, \mathbf{x}_j)\right)}{\sum_{u=1}^{n} \exp(-D_x\left(\mathbf{x}_u, \mathbf{x}_j\right))}$$

is interpreted as the probability that $\mathbf{x}_i$ belongs to the cluster specified by $\mathbf{x}_j$.

A similar conditional probability $p_y$ is then defined on the embedded $k$-dimensional feature space such that

$$p_y(\mathbf{y}_i|\mathbf{y}_j) = \frac{\exp\left(-D_y(\mathbf{y}_i, \mathbf{y}_j)\right)}{\sum_{u=1}^{n} \exp\left(-D_y(\mathbf{y}_u, \mathbf{y}_j)\right)}$$

is interpreted as the probability that $\mathbf{y}_i$ belongs to the cluster specified by $\mathbf{y}_j$.

The aim of SNE is to estimate the $m$-dimensional embedded feature vectors $\mathbf{y}_1, \ldots, \mathbf{y}_n$ so that $p_y(\mathbf{y}_i|\mathbf{y}_j) \approx p_x(\mathbf{x}_i|\mathbf{x}_j)$ as closely as possible. This is accomplished by defining a cross-entropy objective function $V(\mathbf{y}_1, \ldots, \mathbf{y}_n)$ (see Chapter 13 for details) such that:

$$V(\mathbf{y}_1, \ldots, \mathbf{y}_n) = -\sum_{i=1}^{n} \sum_{j \neq i} p(\mathbf{x}_i|\mathbf{x}_j) \log q(\mathbf{y}_i|\mathbf{y}_j)$$

which measures the dissimilarity between $q(\mathbf{y}_i|\mathbf{y}_j)$ and $p(\mathbf{x}_i|\mathbf{x}_j)$.

The objective function $V(\mathbf{y}_1, \ldots, \mathbf{y}_n)$ is then minimized using a gradient descent algorithm defined such that:

$$\mathbf{y}_k(t+1) = \mathbf{y}_k(t) - \gamma_t \frac{dV(\mathbf{y}_1(t), \ldots, \mathbf{y}_n(t))}{d\mathbf{y}_k}$$

where $k = 1, \ldots, n$ and $\gamma_t$ is the positive stepsize.

Note that one may assume that the dissimilarity functions $D_x$ and $D_y$ may be defined as parameterized dissimilarity functions. That is, define $D_x(\mathbf{x}_i, \mathbf{x}_j; \boldsymbol{\theta})$ which specifies a particular dissimilarity function for a given parameter vector $\boldsymbol{\theta}$ and define $D_y(\mathbf{y}_i, \mathbf{y}_j; \boldsymbol{\psi})$ which specifies a particular dissimilarity function for a given parameter vector $\boldsymbol{\psi}$. Let $(\mathbf{X}, \mathbf{Y})$ denote a particular collection of $n$ feature vectors $\mathbf{X} \equiv [\mathbf{x}_1, \ldots, \mathbf{x}_n]$ and a particular collection of $n$ feature vectors $\mathbf{Y} \equiv [\mathbf{y}_1, \ldots, \mathbf{y}_n]$ specifying a particular recoding of $\mathbf{x}_1, \ldots, \mathbf{x}_n$. Let $c([\mathbf{X}, \mathbf{Y}], [\boldsymbol{\theta}, \boldsymbol{\psi}])$ be the loss incurred by the learning machine when it experiences $\mathbf{X}$ recoded as $\mathbf{Y}$ for a particular parameter vector $[\boldsymbol{\theta}, \boldsymbol{\psi}]$. Using the loss function $c([\mathbf{X}, \mathbf{Y}], [\boldsymbol{\theta}, \boldsymbol{\psi}])$, one can construct an empirical risk function for the purpose of learning the parameter vector $[\boldsymbol{\theta}, \boldsymbol{\psi}]$ from the training data $[\mathbf{X}, \mathbf{Y}]$.

△

**Example 1.6.6** (Markov Logic Net (Domingos and Lowd, 2009)). A propositional calculus is a system of logic for representing statements about the world using logical formulas and then computing implications of those logical expressions. An English statement such as "*John placed the bat in a box*" is not adequate for formally representing a statement about the world because it is intrinsically ambiguous. The English statement "*John placed the bat in a box*" could mean that John placed baseball equipment in a box or it could mean that John placed a small flying animal in a box. Moreover, two distinct English statements such as "*John placed the baseball bat in a box*" and "*The baseball bat was placed in a box by John*" may represent different ways of specifying exactly the same semantic statement.

In order to unambiguously make a statement about a particular world, it is necessary to have a separate semantic language which assigns a unique semantic interpretation to each symbol. Here, we use notation such as `PLACE(AGENT:JOHN,OBJECT:BASEBALL-BAT,TO:BOX)` or `PLACE(AGENT:JOHN,OBJECT:ANIMAL-BAT,TO:BOX)` to indicate that a unique semantic interpretation has been assigned. A short-hand version of this notation would be `PLACE(JOHN,BASEBALL-BAT,BOX)` and `PLACE(JOHN,ANIMAL-BAT,BOX)`.

It is additionally assumed that it is possible to assign a truth value of either `TRUE` or `FALSE` to a particular semantic statement about a particular world. The number 1 is used to denote `TRUE` and the number 0 is used to denote `FALSE`. For example, the semantic statement `JOHN` cannot be assigned a truth value but the semantic statement `EXISTS(JOHN)` can be assigned a truth value. An unambiguous semantic statement about the world which can be assigned the value of 1 for `TRUE` or the value of 0 for `FALSE` is called an *atomic proposition*.

The logical connectives $\neg$ (`NOT`), $\wedge$ (`AND`), $\vee$ (`OR`), and $\implies$ (`IMPLIES`) may be used to generate "compound propositions" about the world from the atomic propositions. For example, the semantic statement `PLACE(JOHN,BASEBALL-BAT,BOX)` $\wedge$ `PLACE(JOHN,ANIMAL-BAT,BOX)` may be expressed as the English statement that "*John placed a baseball bat in a box and, in addition, John also placed a small flying mammal in a box*". More formally, a *compound proposition* is defined recursively as either: (1) an atomic proposition, (2) the negation of a compound proposition, or (3) the union of two compound propositions. This definition is sufficiently general to include combining atomic propositions with any of the logical connectives $\neg$, $\wedge$, $\vee$, and $\implies$.

Let the binary vector $\mathbf{x} \equiv [x_1, \ldots, x_d]$ denote a collection of $d$ compound propositions where $x_i = 1$ specifies that the $i$th compound proposition is `TRUE` and $x_i = 0$ specifies that the $i$th proposition is `FALSE`. A *possible world* $\mathbf{x} \in \{0, 1\}^d$ corresponds to an assignment of truth values to all $d$ propositions. Therefore, there are $2^d$ possible worlds.

A *rulebase* is a collection of $q$ functions $\phi_1, \ldots, \phi_q$ such that the $k$th function $\phi_k$ maps a possible world $\mathbf{x}$ into either the value of 0 or 1. The rulebase corresponds to a collection of assertions about a possible world $\mathbf{x}$. Note that some of these assertions correspond to facts specifying particular assertions about the possible world $\mathbf{x}$ which are always equal to 0 or always equal to 1. An important goal of an inference engine is to identify the set of possible worlds which are consistent with a given rulebase. This problem can be reformulated as a type of nonlinear optimization problem where the goal is to find a possible world $\mathbf{x}$ that minimizes the objective function $V(\mathbf{x}) = -\sum_{k=1}^{q} \phi_k(\mathbf{x})$.

However, for real-world applications, identification of the set of possible worlds consistent with a given rulebase can be challenging. First, in many cases one may not have a sufficient number of rules in the knowledge base to determine if a particular possible world is true or false. This implies that the solution is not a unique possible world but a collection of possible worlds. Second, one may have contradictory logical formulas in the knowledge base. And third, it is often difficult to model the physical world as a collection of logical rules. For example, birds fly unless they are ostriches.

Markov logic nets address these challenges by embedding logical constraints within a probabilistic framework. The essential idea underlying Markov logic nets is that different logical constraints are assigned different degrees of importance and the relative importance of these different logical constraints is then combined to generate specific inferences.

The weighting factor for the $k$th logical constraint $\phi_k$ is assumed to be equal to $\theta_k$ where larger values of $\theta_k$ increase the importance weighting of the $k$th logical constraint $\phi_k$ in order to support probabilistic conflict resolution. In particular, define

$$p(\mathbf{x}; \boldsymbol{\theta}) = \frac{\exp(-V(\mathbf{x}; \boldsymbol{\theta}))}{Z(\boldsymbol{\theta})} \tag{1.36}$$

where

$$V(\mathbf{x}; \boldsymbol{\theta}) = -\sum_{k=1}^{q} \theta_k \phi_k(\mathbf{x}), \tag{1.37}$$

and

$$Z(\boldsymbol{\theta}) = \sum_{\mathbf{y} \in \{0,1\}^d} \exp\left(-V(\mathbf{y}; \boldsymbol{\theta})\right)$$

for each $q$-dimensional real *parameter vector* $\boldsymbol{\theta}$ in the parameter space $\Theta \subseteq \mathcal{R}^q$. Choosing different values of $\boldsymbol{\theta}$ has the effect of readjusting the probability assigned to every possible world. Also note that since $p(\mathbf{x}; \boldsymbol{\theta})$ is a decreasing function of $V(\mathbf{x}; \boldsymbol{\theta})$ for a fixed constant $\boldsymbol{\theta}$, this means that a *Markov Logic Net Inference Algorithm* which searches for a possible world $\mathbf{x}$ that minimizes $V(\mathbf{x}; \boldsymbol{\theta})$ in (1.37) is equivalently searching for a most probable possible world $\mathbf{x}$ that maximizes $p(\mathbf{x}; \boldsymbol{\theta})$ in (1.36).

Moreover, using the method of gradient descent, one can estimate the parameter vector $\boldsymbol{\theta}$ by observing a collection of $n$ possible worlds $\mathbf{x}_1, \ldots, \mathbf{x}_n$ and finding the value of $\boldsymbol{\theta}$ that "best fits" the observed probability distribution of possible worlds in the environment using maximum likelihood estimation methods (see Chapter 10). In particular, one can develop a *Markov Logic Net Learning Algorithm* that minimizes the empirical risk function:

$$\ell_n(\boldsymbol{\theta}) = -(1/n) \sum_{i=1}^{n} \log p(\mathbf{x}_i; \boldsymbol{\theta})$$

using standard gradient descent type methods. △

## Exercises

1.6-1. *Deterministic Iterative K-Means Clustering.* The objective function for learning for a $K$-means clustering algorithm minimizes the within-cluster similarity using a Euclidean distance function. It works as follows. First, each pattern vector is arbitrarily assigned to one of $K$ clusters, $C_1, \ldots, C_K$ which are disjoint subsets of a $d$-dimensional vector space. Let $\bar{\mathbf{x}}_k$ be the average of the points currently assigned to cluster $C_k$ for $k = 1, \ldots, K$. Second, the performance of the clustering assignment is computed using the function

$$V(\mathbf{s}_1, \ldots, \mathbf{s}_K) = \sum_{k=1}^{K} \sum_{j \in C_k} |\mathbf{x}_j - \bar{\mathbf{x}}_k|^2$$

where the $j$th element of $\mathbf{s}_k$ is equal to one if $\mathbf{x}_j$ is an element of category $C_k$ and the $j$th element of $\mathbf{s}_k$ is equal to zero otherwise. Third, a point $\mathbf{x}_i$ is moved from one cluster to another and this move is accepted if $V(\mathbf{s}_1, \ldots, \mathbf{s}_K)$ decreases. Note that in order to determine if $V(\mathbf{s}_1, \ldots, \mathbf{s}_K)$ decreases only terms involving $\mathbf{x}_i$ in $V(\mathbf{s}_1, \ldots, \mathbf{s}_K)$ need to be recomputed. This process is repeated until the assignment of points to clusters does not decrease $V(\mathbf{s}_1, \ldots, \mathbf{s}_K)$. Implement this algorithm in software for the case where $d = 2$ and $K = 2$ and explore its performance for different starting conditions.

1.6-2. *Stochastic Neighborhood Embedding with Gaussian Neighborhoods.* Implement SNE for special case of Gaussian neighborhoods. In particular, define an objective function $V(\mathbf{y}_1, \ldots, \mathbf{y}_M)$ for SNE which is defined with respect to a distance function $D_x(\mathbf{x}_i, \mathbf{x}_j) = |\mathbf{x}_i - \mathbf{x}_j|^2$ which measures distance in the original high-dimensional state space and a distance function $D_y(\mathbf{y}_i, \mathbf{y}_j) = |\mathbf{y}_i - \mathbf{y}_j|^2$ which measures distance in the new low-dimensional state space.

1.6-3. *Linear Autoassociative Encoder.* Let $\mathbf{s}_1, \ldots, \mathbf{s}_n$ be a set of $n$ $d$-dimensional pattern column vectors where $n < d$. A linear (autoassociative) autoencoder can be defined as a learning machine which seeks to minimize the objective function

$$\hat{\ell}_n(\boldsymbol{\theta}) = (1/n) \sum_{i=1}^{n} |\mathbf{s}_i - \mathbf{W}\mathbf{s}_i|^2 \tag{1.38}$$

where $\boldsymbol{\theta}$ is defined as the elements of the matrix $\mathbf{W}$. A gradient descent algorithm for minimizing this objective function is given by the iterative learning rule:

$$\mathbf{W}(k+1) = \mathbf{W}(k) + \gamma(2/n) \sum_{i=1}^{n} (\mathbf{s}_i - \mathbf{W}(k)\mathbf{s}_i)\mathbf{s}_i^T.$$

Initialize $\mathbf{W}(0)$ to be a vector of zeros and then use this learning rule to estimate $\mathbf{W}(1)$. Then, use $\mathbf{W}(1)$ and the learning rule to estimate $\mathbf{W}(2)$. Let $\hat{\mathbf{W}}_n$ be a minimizer of $\hat{\ell}_n$. The response of the linear autoassociative encoder is defined as $\ddot{y} = \mathbf{W}\mathbf{s}$ for a particular input pattern $\mathbf{s}$. Interpret the objective function $\hat{\ell}_n$ is attempting to force the linear autoencoder to construct a matrix $\mathbf{W}$ such that $\mathbf{s}_i$ is an eigenvector of $\mathbf{W}$ with eigenvalue one. Use this observation to conclude the linear autoencoder is actually a linear subspace learning machine which learns a linear subspace that best represents the training stimuli it has learned.

1.6-4. *Latent Semantic Indexing Document Retrieval Performance Study.* Run a computer program which does latent semantic analysis on a collection of documents and then evaluate the performance of the program in determining between document and between term similarity.

## 1.7 Reinforcement Learning Machines

The goal of a *reinforcement learning machine* is to generate a response pattern given an input pattern and *only* vague hints about the nature of the desired response. Typically, the terminology "reinforcement learning" assumes a temporal component is present in the machine learning problem. That is, the entire feature vector consists of features representing the state of the learning machine at different points in time while other elements of the feature vector provide non-specific feedback in the form of a hint regarding whether the learning machine's behavior over time was successful.

For example, suppose a robot is learning a complex sequence of actions in order to achieve some goal such as picking up a coke can. There are many different action sequences which are equally effective. The specific details of a particular action sequence choice tend to be largely irrelevant. The important objective is that the robot has successfully picked up the coke can at the end of the action sequence.

Suppose the robot training procedure involves having the robot attempt to pick up a coke and then telling the robot whether or not it has successfully achieved the objective of picking up a coke can. The feedback received from the environment is non-specific in the sense that it does not provide explicit instructions to the robot's actuators regarding what they should do and how they should activate in sequence. Moreover, the feedback is provided to the robot after the sequence of actions has been executed so the robot must solve the "temporal credit assignment problem" in order to figure out which actions in the previously executed action sequence should be modified to improve future performance.

Learning to land a lunar spacecraft embodies the same principles of machine learning as learning to pick up a coke can. The learning machine's goal is to pilot the lunar lander by generating a sequence of signals designed to control the lunar lander's position and velocity so that the final state of the sequence corresponds to a safe lunar landing. The learning machine does not receive explicit information from the environment (e.g., an external human flight instructor) regarding how to choose an appropriate control law. Rather, the learning machine simply is provided hints such as: "the current descent is not slow and controlled" and "the current landing was not safe" and must use this feedback provided by the environment in conjunction with temporal statistical regularities to learn an acceptable control law. For some learning machines, operating in very complex environments, such hints specify an "internal model of the environment" for the learning machine whose predictions can be compared with observations from the actual environment (see Figure 1.15).

It is important to emphasize that there are two fundamentally different types of learning processes that are relevant to the discussion of reinforcement learning. The first type of learning process assumes a "passive reinforcement learning environment". In a *passive reinforcement learning environment*, the learning machine's actions do not alter the learning machine's statistical environment. Typically, the learning methods used in supervised learning can be applied in a straightforward manner for designing learning machines within the passive reinforcement learning paradigm. For example, a student might learn to fly a helicopter by learning very specific procedures for taking off and landing the helicopter based upon carefully imitating the behavior of an expert helicopter instructor. The student would not be allowed to actually fly a helicopter until the student had perfectly mastered specific procedures for taking off and landing the helicopter. Feedback in the passive reinforcement learning paradigm is continuously provided to the student and the feedback is very detailed in nature.

On the other hand, in a *reactive learning environment*, the learning machine's actions influence the nature of its statistical environment. Learning within a reactive learning environment paradigm is also a substantially more challenging learning problem. For example, suppose a learning machine without helicopter landing experience is trained to land helicopters in a simulation environment. It is likely that regardless of the number of learning trials, the learning machine would find the problem of learning to successfully land the helicopter particularly challenging because its actions would be unlikely to create the types of statistical regularities which must be learned to ensure a successful landing.

A useful strategy for assisting learning in reactive learning environments is to provide the learning machine with useful hints about the solution. For example, the initial parameter values of the learning machine could be initialized by training the learning machine using a passive reinforcement learning paradigm. Such pre-training can greatly improve the efficiency of the learning process.

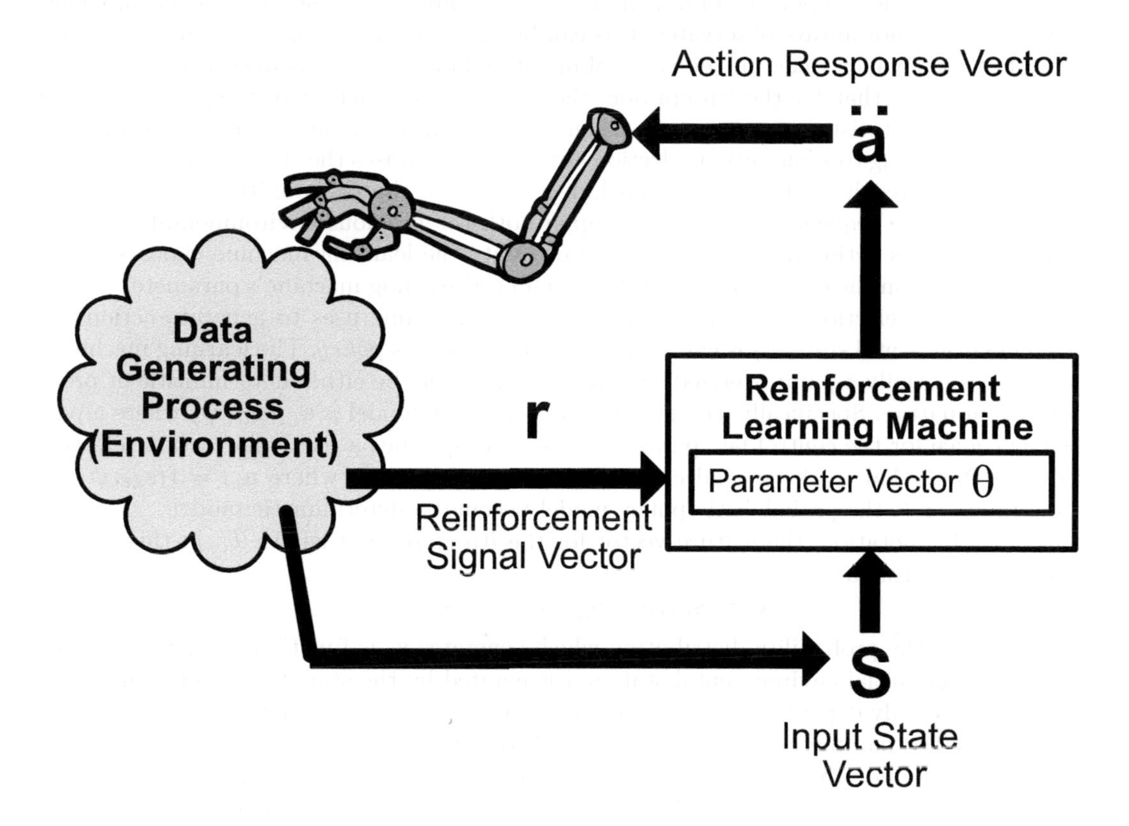


**FIGURE 1.15**
**Overview of reinforcement learning in a reactive learning environment.** In an episode-based reinforcement adaptive reactive learning environment, the statistical environment generates an initial state $\mathbf{s}$ of an episode. The learning machine then generates an action $\ddot{\mathbf{a}}$ using its current parameter vector $\boldsymbol{\theta}$ which modifies the statistical characteristics of the learning machine's environment. The modified statistical environment then generates the next state of the episode and possibly may also generate a punishment signal to the reinforcement machine represented by the reinforcement signal vector $\mathbf{r}$. The interactions of the learning machine and environment continue until the episode is concluded. The parameter values of the learning machine are then updated at the end of an episode. It is assumed that episodes are independent and identically distributed when the parameter values of the learning machine are fixed.

### 1.7.1 Reinforcement Learning Overview

It is assumed that reinforcement learning involves presenting the learning machine with *episodes* $\mathbf{x}_1, \mathbf{x}_2, \ldots$ The $i$th episode, $\mathbf{x}_i$, is defined such that $\mathbf{x}_i \equiv [\mathbf{x}_{i,1}, \ldots, \mathbf{x}_{i,T_i}, T_i]$ where $\mathbf{x}_{i,k}$ is the $k$th feature vector in trajectory $i$ and $T_i$ is called the length of episode $\mathbf{x}_i$. The feature vector $\mathbf{x}_{i,k}$, in turn, is represented as the concatenation of an environmental state vector $\mathbf{s}_{i,k}$ and a learning machine action vector $\mathbf{a}_{i,k}$ so that $\mathbf{x}_{i,k} \equiv [\mathbf{s}_{i,k}, \mathbf{a}_{i,k}]$.

Note that some components of $\mathbf{x}_{i,k}$ are not always fully observable. If the $j$th component of $\mathbf{x}_{i,k}$, $\mathbf{x}_{i,k,j}$ is not always observable, this can be modeled by adding an additional binary variable to $\mathbf{x}_{i,k}$ which takes on the value of one if and only if $\mathbf{x}_{i,k,j}$ is observable.

It is assumed that for the $i$th episode, the initial environmental state $\mathbf{s}_{i,1}$ is generated entirely by the statistical environment, then the learning machine executes action $\mathbf{a}_{i,1}$ in state $\mathbf{s}_{i,1}$ according to some action selection strategy. This causes the statistical environment to change as a result of the learning machine's actions so the probability of observing the next state in the sequence, $\mathbf{s}_{i,2}$, depends upon both the previous environmental state $\mathbf{s}_{i,1}$ and the previous action $\mathbf{a}_{i,1}$. That is, the probability the learning machine chooses action $\mathbf{a}_{i,2}$ depends upon the environmental state $\mathbf{s}_{i,2}$ and the learning machine's parameter vector.

The action selection strategy which the learning machine uses to generate action $\mathbf{a}_{i,t}$ given environmental state $\mathbf{s}_{i,t}$ is called the learning machine's *policy*. The learning machine's policy is essentially a parameterized control law which may be either deterministic or probabilistic in nature. Specifically, the policy is a probability model $p(\mathbf{a}_{i,t}|\mathbf{s}_{i,t}, \boldsymbol{\psi})$ where a particular choice for the control law parameter vector $\boldsymbol{\psi}$ specifies a particular action selection strategy. Note that in the special case where $p(\mathbf{a}_{i,t}|\mathbf{s}_{i,t}, \boldsymbol{\psi}) = 1$ where $\mathbf{a}_{i,t} = \mathbf{f}(\mathbf{s}_{i,t}, \boldsymbol{\psi})$ for some function $f$, the probabilistic policy model becomes a deterministic model.

Using this notation, the feature vector for the $i$th episode of length $T_i$, in this case, is given by the formula:

$$\mathbf{x}_i = [\mathbf{s}_{i,1}, \mathbf{a}_{i,1}, \mathbf{s}_{i,2}, \mathbf{a}_{i,2}, \ldots, \mathbf{s}_{i,T_i}, T_i]. \tag{1.39}$$

Note that the probability distribution which generates $\mathbf{x}_i$ is functionally dependent not only upon the initial environmental state $\mathbf{s}_{i,1}$ generated by the statistical environment but also is functionally dependent upon the learning machine's state of knowledge which is represented by the parameter vector $\boldsymbol{\theta}$. Let $p(\mathbf{x}_i|\boldsymbol{\theta})$ denote the probability of the $i$th episode $\mathbf{x}_i$ conditioned upon the learning machine's parameter vector $\boldsymbol{\theta}$. As a consequence of learning, the parameters of an adaptive learning machine will change. As the learning machine's parameters change during the learning process, the learning machine's behavior will change. Changes in the learning machine's behavior, in turn, modify the learning machine's environment. Thus, for reactive learning environments, the learning machine's statistical environment is generated by the behavioral interactions of the learning machine with its environment resulting in a statistical environment which evolves as the learning process evolves.

The penalty reinforcement received by the learning machine for experiencing episode $\mathbf{x}$ when its state of knowledge is the parameter vector $\boldsymbol{\theta}$ is denoted by $c(\mathbf{x}, \boldsymbol{\theta})$ and the observed episodes $\mathbf{x}_1, \mathbf{x}_2, \ldots, \mathbf{x}_n$ are assumed to be a realization of a sequence of independent and identically distributed random vectors. The goal of the ERM reinforcement learning problem is to minimize the expected value of $c(\tilde{\mathbf{x}}, \boldsymbol{\theta})$ where $\tilde{\mathbf{x}}$ denotes the process of randomly selecting an episode from the environment. To simplify notation, consider the special case where $\tilde{\mathbf{x}}$ is a discrete random vector. Then, the goal of the reinforcement learning problem for a passive statistical environment is to minimize the objective function

$$\ell(\boldsymbol{\theta}) = \sum_{\mathbf{x}} c(\mathbf{x}, \boldsymbol{\theta}) p_e(\mathbf{x})$$

where $p_e$ is the probability that the statistical environment generates the episode $\mathbf{x}$.

For a reactive statistical environment, the probability that episode $\mathbf{x}$ is observed is functionally dependent upon the current state of knowledge $\boldsymbol{\theta}$ of the learning machine so the objective function for learning is given by the formula:

$$\ell(\boldsymbol{\theta}) = \sum_{\mathbf{x}} c(\mathbf{x}, \boldsymbol{\theta}) p(\mathbf{x}|\boldsymbol{\theta})$$

where the probability of the learning machine experiencing episode $\mathbf{x}$, $p(\mathbf{x}|\boldsymbol{\theta})$, is functionally dependent upon not only the statistical environment, but the learning machine's current

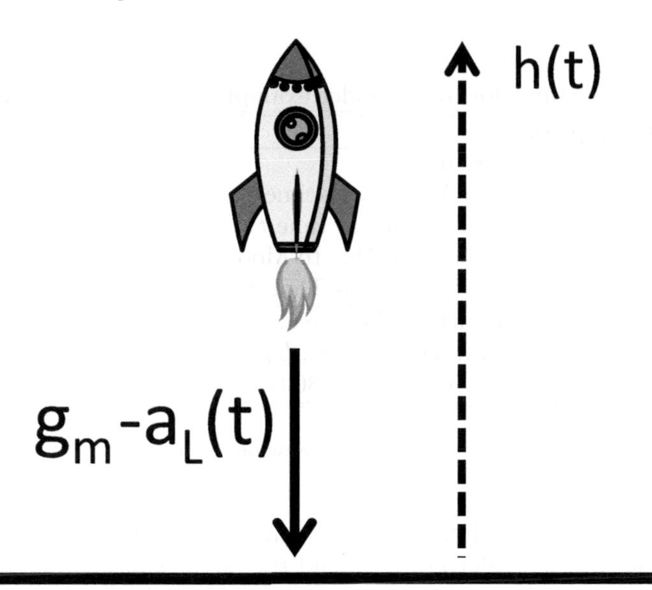


**FIGURE 1.16**
**A simplified one-dimensional lunar lander problem.** The lunar lander module begins its descent with a downward velocity at some fixed height. A gravitational accelerative force moves the lunar lander downward at a progressively faster rate. During the landing process, the lunar lander can choose to generate or choose to not generate a fixed amount of thrust to counteract the gravitational accelerative force. The binary decision to apply thrust is generated as a function of the lunar lander's physical state and current parameter values. These parameter values are adjusted throughout the lunar lander's descent.

state of knowledge $\boldsymbol{\theta}$. For example, under the assumption that the learning machine generates an action $\mathbf{a}_{i,t}$ given environmental state $\mathbf{s}_{i,t}$ and the additional assumption that the environment transitions from $\mathbf{s}_{i,t}$ given action $\mathbf{a}_{i,t}$ to the new state $\mathbf{s}_{i,t+1}$, it follows that:

$$p(\mathbf{x}_i|\boldsymbol{\theta}) = p_e(\mathbf{s}_{i,1}) \prod_{t=1}^{T_i-1} p(\mathbf{a}_{i,t}|\mathbf{s}_{i,t};\boldsymbol{\theta})p_e(\mathbf{s}_{i,t+1}|\mathbf{a}_{i,t},\mathbf{s}_{i,t}).$$

**Example 1.7.1** (A Simplified Lunar Landing Problem)**.** In this example, a simplified one-dimensional model of the lunar landing problem is discussed for the purpose of studying the reinforcement learning problem.

Figure 1.16 depicts a lunar lander whose state at time $t$ is characterized by the state vector $\mathbf{s}_i(t) = [h_i(t), v_i(t), f_i(t), E_i(t), C_i(t)]$ where for a particular time index $t$ in trajectory $i$ is defined such that: $h_i(t)$ is the distance in meters from the lander to the lunar surface, $v_i(t)$ is the downward velocity in meters per second of the lander, $f_i(t)$ is the amount of fuel in kilograms in the lander's gas tank, $E_i(t)$ is a binary indicator which equals one when trajectory terminates due to landing or vehicle is off course $C_i(t)$ is a binary indicator which equals one if the spacecraft has crashed and zero otherwise. These state variables and related constants are presented in Table 1.1.

**Environment and Spacecraft Dynamics.** The lunar lander's initial height $h(0)$ is randomly chosen with a mean value of 15000 kilometers and standard error of 20 kilometers. The lunar lander's initial velocity $v(0)$ is also randomly chosen with a mean value of 100 kilometers per second and a standard error of 40 kilometers per second. The initial fuel

**TABLE 1.1**
State variables and constants for lunar lander example.

| Variable | Description | Units [a] |
|---|---|---|
| $h(t)$ | Lander Height at time $t$ | m |
| $v(t)$ | Downward Lander Velocity at time $t$ | m/s |
| $f(t)$ | Amount of Fuel in Lander at time $t$ | kg |
| $g_M$ | Gravitation Acceleration Due to Moon ($g_M = 1.63$) | m/s |
| $\delta_T$ | Time Duration of Simulation Time Slice (e.g., $\delta_T = 1$) | s |
| $M_L$ | Mass of Lander without Fuel (e.g., $M_L = 4000$) | kg |
| $U_L$ | Maximum Upward Thrust (e.g., $U_L = 25000$) | N |
| $e_L$ | Engine Efficiency (e.g., $e_L = 2300$) | m/s |

[a] *Note:* m = meters, s=seconds, kg=kilograms, N=Newton kg/(m $s^2$)

level, $f(0)$, is chosen to be 3500 kilograms. At a given time $t$, the new height $h(t+1)$ of the lander, new velocity of the lander $v(t+1)$, and the amount of fuel currently in the lander $f(t+1)$ is computed from the previous states $h(t)$, $v(t)$, and $f(t)$ as well as the amount of rocket thrust requested by the lander's pilot.

The specific update equations for the behavior of the spacecraft in its lunar environment are given by:

$$h_i(t+1) = \mathcal{J}_0\left(h_i(t) - \delta_T v_i(t)\right) \tag{1.40}$$

where $\mathcal{J}_0(\phi) = \phi$ if $\phi > 0$ and $\mathcal{J}_0(\phi) = 0$ otherwise,

$$v_i(t+1) = v_i(t) + \delta_T\left(g_M - a_L(i,t)\right), \tag{1.41}$$

and where the lander rocket thruster acceleration at time $t$, $a_L(t)$, is defined by

$$a_L(i,t) = \frac{u_i(t)U_L}{M_L + f_i(t)}.$$

The binary thrust indicator variable $u_i(t)$ is defined such that $u_i(t) = 1$ if the pilot of the lunar lander applies thrust at time $t$ and $u_i(t) = 0$ if the pilot of the lunar lander does not apply thrust at time $t$.

The amount of fuel in the space craft at time $t$ is given by:

$$f_i(t+1) = \mathcal{J}_0(f_i(t) - \delta_T \nabla f_i(t)) \tag{1.42}$$

where the velocity of the ejected fuel relative to the lunar lander at time $t$, $\nabla f_i(t)$, is defined by

$$\nabla f_i(t) = (U_L/e_L)u_i(t).$$

The *rough ride penalty* $\Delta_i(t)$ is given by:

$$\Delta_i(t) = \sqrt{(h_i(t) - h_i(t+1))^2 + (v_i(t) - v_i(t+1))^2}.$$

At time $t$ for trajectory $i$, let the end of trajectory indicator $E_i(t) = 1$ when the lunar lander reaches the lunar surface or if the lunar lander is moving in the wrong direction; and let $E_i(t) = 0$ otherwise.

At time $t$ for trajectory $i$, let the crash indicator $C_i(t) = 1$ when the lunar lander reached the lunar surface and impacts the surface at a velocity greater than the permissible landing velocity. Let $C_i(t) = 0$ if the lunar lander has not landed on the lunar surface or has landed at a velocity which is less than the permissible landing velocity.

The reinforcement signal received for episode $(\mathbf{s}_i(t), \mathbf{a}_i(t), \mathbf{s}_i(t+1))$ is given by the formula:

$$r_i(t) = \lambda_\Delta \Delta_i(t) + L_i(t+1)\lambda_v \sqrt{(v_i(t))^2 + (v_i(t+1))^2} + \lambda_c C_i(t+1) \tag{1.43}$$

where $\lambda_\Delta$, $\lambda_v$, and $\lambda_c$ are positive constants. The first term on the right-hand side of (1.43) corresponding to the *rough ride penalty* penalizes an episode if the height or velocity of the spacecraft changes too rapidly. The second term on the right-hand side of (1.43) is equal to zero if the spacecraft has not landed and is proportional to the magnitude of the velocity when it hits the lunar surface when the spacecraft has landed. Thus, the second term on the right-hand side of (1.43) penalizes episodes involving landings where the impact velocity is large by an amount proportional to the impact velocity. The third term on the right-hand side of (1.43) also penalizes episodes involving landings but generates a fixed penalty if the landing velocity is above some threshold and no penalty if the landing velocity is below that threshold.

Notice that the formulation presented here assumes the environment is deterministic but this assumption can be relaxed. Only the initial state $\mathbf{s}_i(t)$ is randomly generated by the environment. $\triangle$

### 1.7.2 Value Function Reinforcement Passive Learning

Assume for the $i$th episode, a learning machine is learning in a passive environment by watching the behavior of some other agent demonstrating how to solve a reinforcement learning problem using a policy $\rho$ where the agent situated in environmental state $\mathbf{s}_{i,t}$ executes an action $\mathbf{a}_{i,t}$ and then finds itself in a new situation denoted by $\mathbf{s}_{i,t+1}$. The environment generates a penalty reinforcement signal, $r_{i,t}$, which is received by the agent when the agent transitions from state $\mathbf{s}_{i,t}$ to state $\mathbf{s}_{i,t+1}$. Notice that the learning machine may be trained by having it watch expert agents or novice agents since it can learn the consequences of good policies from expert agents and the consequences of poor policies from novice agents. In fact, most treatments of value function reinforcement learning assume that the agent learns an improved policy by watching how the current policy $\rho$ influences the reinforcement signals received by the environment.

The learning machine's goal is to observe the agent in order to learn the parameter vector $\boldsymbol{\theta}$ which specifies a particular *value function* $V$ that estimates the expected total *future* penalty reinforcement $\ddot{V}(\mathbf{s}_{i,t}, \rho; \boldsymbol{\theta})$ the agent will receive in the $i$th episode when the agent is in state $\mathbf{s}_{i,t}$. Since the probability that a particular episode is observed by the learning machine is not functionally dependent upon the learning machine's behavior, this is a passive learning environment.

Assume the learning machine observes that the total future penalty reinforcement received by the agent is $V(\mathbf{s}_{i,1}; \rho) = \sum_{t=1}^{\infty} r_{i,t}$, then the penalty reinforcement signal $r_{i,1}$ received by the agent for executing action $\mathbf{a}_{i,1}$ in state $\mathbf{s}_{i,1}$ is given by the formula:

$$r_{i,1} = V(\mathbf{s}_{i,1}; \rho) - V(\mathbf{s}_{i,2}; \rho). \tag{1.44}$$

A common alternative model of cumulative future penalty reinforcement received is to assume $V(\mathbf{s}_{i,1}; \rho) = \sum_{t=1}^{\infty} \mu^{t-1} r_{i,t}$ where the *discount factor* $\mu$ is a number between zero and one. This alternative model assumes that the penalties the agent expects to receive in the distant future of the episode will be discounted in value. Using this alternative model of future discounted reinforcement, the penalty reinforcement signal received by the agent for executing action $\mathbf{a}_{i,1}$ in state $\mathbf{s}_{i,1}$ is given by the formula:

$$r_{i,1} = V(\mathbf{s}_{i,1}; \rho) - \mu V(\mathbf{s}_{i,2}; \rho). \tag{1.45}$$

Equation (1.44) is a special case of (1.45) when $\mu = 1$.

Let the learning machine's expectation of the total cumulative future penalty reinforcement the agent receives for following policy $\rho$ when in state $\mathbf{s}_{i,1}$ be denoted as $\ddot{V}(\mathbf{s}_{i,1};\rho,\boldsymbol{\theta})$ where $\boldsymbol{\theta}$ is the modifiable parameter vector for the learning machine, which is adjusted as a function of the learning machine's experiences.

Given a collection of $n$ episodes $\mathbf{x}_1,\ldots,\mathbf{x}_n$, an empirical risk function $\hat{\ell}_n(\boldsymbol{\theta})$ can be constructed to estimate the parameter vector $\boldsymbol{\theta}$ by using the formula:

$$\hat{\ell}_n(\boldsymbol{\theta}) = (1/n)\sum_{i=1}^{n} c([\mathbf{s}_{i,1},\mathbf{s}_{i,2},r_{i,1}];\boldsymbol{\theta}) \tag{1.46}$$

where

$$c([\mathbf{s}_{i,1},\mathbf{s}_{i,2},r_{i,1}];\boldsymbol{\theta}) = \left(r_{i,1} - \ddot{r}_{i,1}(\mathbf{s}_{i,1},\mathbf{s}_{i,2};\boldsymbol{\theta})\right)^2 \tag{1.47}$$

and

$$\ddot{r}_{i,1}(\mathbf{s}_{i,1},\mathbf{s}_{i,2};\boldsymbol{\theta}) = \ddot{V}(\mathbf{s}_{i,1},\rho,\boldsymbol{\theta}) - \mu\ddot{V}(\mathbf{s}_{i,2},\rho,\boldsymbol{\theta}). \tag{1.48}$$

**Algorithm 1.7.1 Generic value function reinforcement learning algorithm.**

1: **procedure** VALUE-FUNCTION-REINFORCEMENT-LEARNING($\boldsymbol{\theta}_1$)
2: $i \Leftarrow 1$
3: **repeat**
4: Environment randomly samples $\mathbf{s}_{i,1}$ from $p_e(\mathbf{s}_{i,1})$.
5: Learning Machine observes Agent select action $\mathbf{a}_{i,1}$ with probability $p(\mathbf{a}_{i,1}|\mathbf{s}_{i,1},\rho)$.
6: Environment randomly samples $\mathbf{s}_{i,2}$ from $p_e(\mathbf{s}_{i,2}|\mathbf{a}_{i,1},\mathbf{s}_{i,1})$.
7: $c([\mathbf{s}_{i,1},\mathbf{s}_{i,2}];\boldsymbol{\theta}_i) \Leftarrow \left(r_{i,1} - \left(\ddot{V}(\mathbf{s}_{i,1};\rho,\boldsymbol{\theta}_i) - \mu\ddot{V}(\mathbf{s}_{i,2};\rho,\boldsymbol{\theta}_i)\right)\right)^2$ for policy $\rho$
8: $\mathbf{g}([\mathbf{s}_{i,1},\mathbf{s}_{i,2}],\boldsymbol{\theta}_i) \Leftarrow dc([\mathbf{s}_{i,1},\mathbf{s}_{i,2}];\boldsymbol{\theta}_i)/d\boldsymbol{\theta}$
9: $\boldsymbol{\theta}_{i+1} \Leftarrow \boldsymbol{\theta}_i - \gamma_i\mathbf{g}([\mathbf{s}_{i,1},\mathbf{s}_{i,2}],\boldsymbol{\theta}_i)$
10: $i \Leftarrow i+1$.
11: **until** $\boldsymbol{\theta}_i \approx \boldsymbol{\theta}_{i-1}$
12: **return** $\{\boldsymbol{\theta}_i\}$
13: **end procedure**

**Example 1.7.2** (Linear Value Function Reinforcement Learning Algorithm)**.** Consider the Lunar Landing Reinforcement Problem discussed in Example 1.7.1. Let $a_i(t) = 1$ denote the action of applying thrust at time $t$ in trajectory $i$ and let $a_i(t) = 0$ denote the action of not applying thrust at time $t$ in trajectory $i$. It is assumed that thrust is applied at time t with some fixed probability P. In this example, it is assumed that the learning machine is observing either an expert pilot or a novice pilot land the space shuttle.

Assume the learning machine's value function for predicting the total future cumulative reinforcement the learning machine will receive when the machine is in environmental state $\mathbf{s}$ and follows policy $\rho$

$$\ddot{V}(\mathbf{s};\rho,\boldsymbol{\theta}) = \mathbf{s}^T\boldsymbol{\theta} \tag{1.49}$$

where $\mathbf{s}$ is the current state and $\boldsymbol{\theta}$ is a parameter vector used to specify a particular value function choice. Define the empirical risk function:

$$\hat{\ell}_n(\boldsymbol{\theta}) = (1/n)\sum_{i=1}^{n} c_i(\boldsymbol{\theta}) \tag{1.50}$$

where the loss function for the $i$th episode is obtained by substituting (1.49) into (1.46), (1.47), and (1.48) to obtain:

$$c_i(\boldsymbol{\theta}) = \left| r(\mathbf{s}_i(1), \mathbf{s}_i(2)) - \mathbf{s}_i(1)^T \boldsymbol{\theta} + \mu \mathbf{s}_i(2)^T \boldsymbol{\theta} \right|^2 \tag{1.51}$$

where $r(\mathbf{s}_i(1), \mathbf{s}_i(2))$ is the reinforcement signal observed at the beginning of episode $i$.

A gradient descent algorithm can then be used to minimize the empirical risk function in (1.50). This is a specific implementation of Algorithm 1.7.1.

Let $\hat{\boldsymbol{\theta}}_n$ denote the parameter vector learned by the learning machine that minimizes the empirical risk function in (1.50). After the learning process is completed, the estimated function $\ddot{V}(\mathbf{s}; \rho, \hat{\boldsymbol{\theta}}_n)$ may be used to guide the behavior of the learning machine by having the learning machine choose a particular action that will lead to a state $\mathbf{s}$ that minimizes $\ddot{V}(\mathbf{s}; \rho, \hat{\boldsymbol{\theta}}_n)$. △

### 1.7.3 Policy Gradient Reinforcement Reactive Learning

Policy gradient reinforcement methods assume a learning machine's actions influence the learning process in a reactive learning environment. Reinforcement signals from the environment are represented by a function $R$ that maps an episode $\mathbf{x}$ experienced by the learning machine into a reinforcement penalty denoted by $R(\mathbf{x}; \boldsymbol{\theta})$ where the parameter vector $\boldsymbol{\theta}$ is the learning machine's current state of knowledge. Because the reinforcement penalty is functionally dependent upon the learning machine's parameters, this implies that the learning machine has an implicit reinforcement model of the penalty incurred for experiencing episode $\mathbf{x}$ which evolves as the learning machine's parameter vector $\boldsymbol{\theta}$ evolves.

The loss function $c$ is defined such that the number $c(\mathbf{s}_{i,1}; \boldsymbol{\theta})$ specifies the loss the learning machine expects to receive for the entire episode $\mathbf{x}_i$ based only upon the machine's experience with the initial state $\mathbf{s}_{i,1}$ of episode $\mathbf{x}_i$. Let $\mathbf{h}_i$ denote all components of $\mathbf{x}_i$ except for $\mathbf{s}_{i,1}$. Then, the learning machine's expected loss based upon observing the initial state $\mathbf{s}_{i,1}$ of the $i$th episode $\mathbf{x}_i \equiv [\mathbf{s}_{i,1}, \mathbf{h}_i]$ is given by:

$$c(\mathbf{s}_{i,1}; \boldsymbol{\theta}) = \sum_{\mathbf{h}_i} R(\mathbf{x}_i; \boldsymbol{\theta}) p(\mathbf{h}_i | \mathbf{s}_{i,1}, \boldsymbol{\theta}).$$

The probability of episode $\mathbf{x}_i \equiv [\mathbf{s}_{i,1}, \mathbf{h}_i]$, $p(\mathbf{x}_i | \boldsymbol{\theta})$, is therefore functionally dependent upon three factors. First, the probability, $p_e(\mathbf{s}_{i,1})$, that the environment generates the initial state $\mathbf{s}_{i,1}$ of the episode. Second, the probability the learning machine generates action $\mathbf{a}_{i,t}$ given it experiences state $\mathbf{s}_{i,t}$ when the learning machine is in knowledge state $\boldsymbol{\theta}$. And third, the probability the environment generates state $\mathbf{s}_{i,t+1}$ when the learning machine generates action $\mathbf{a}_{i,t}$ in environmental state $\mathbf{s}_{i,t}$.

Intuitively, $c(\mathbf{s}_{i,1}; \boldsymbol{\theta})$ takes into account all possible sequences of actions and states which could originate from the initial state vector $\mathbf{s}_{i,1}$ of episode $\mathbf{x}_i$ and calculates the expected reinforcement penalty across all possibilities. The empirical risk function for learning $n$ episodes with respect to initial states $\mathbf{s}_{i,1}, \ldots, \mathbf{s}_{n,1}$ is then given by the formula:

$$\hat{\ell}_n(\boldsymbol{\theta}) = (1/n) \sum_{i=1}^{n} c(\mathbf{s}_{i,1}; \boldsymbol{\theta}). \tag{1.52}$$

Using the vector calculus methods in Chapter 5,

$$\frac{dc(\mathbf{s}_{i,1}; \boldsymbol{\theta})}{d\boldsymbol{\theta}} = \sum_{\mathbf{h}_i} \left( \frac{dR([\mathbf{s}_{i,1}, \mathbf{h}_i]; \boldsymbol{\theta})}{d\boldsymbol{\theta}} + R([\mathbf{s}_{i,1}, \mathbf{h}_i]; \boldsymbol{\theta}) \frac{d \log p(\mathbf{h}_i | \mathbf{s}_{i,1}; \boldsymbol{\theta})}{d\boldsymbol{\theta}} \right) p(\mathbf{h}_i | \mathbf{s}_{i,1}, \boldsymbol{\theta}). \tag{1.53}$$

Using the methods of stochastic approximation expectation maximization theory described in Chapter 12, an adaptive gradient descent algorithm which minimizes the expected value of the empirical risk function in (1.52) is specified by the gradient descent parameter update rule:

$$\boldsymbol{\theta}_{i+1} = \boldsymbol{\theta}_i - \gamma_i \frac{d\ddot{c}(\mathbf{x}_i; \boldsymbol{\theta})}{d\boldsymbol{\theta}} \tag{1.54}$$

where episode $\mathbf{x}_i \equiv [\mathbf{s}_{i,1}, \mathbf{h}_i]$,

$$\frac{d\ddot{c}(\mathbf{x}_i; \boldsymbol{\theta})}{d\boldsymbol{\theta}} = \frac{dR(\mathbf{x}_i; \boldsymbol{\theta})}{d\boldsymbol{\theta}} + R(\mathbf{x}_i; \boldsymbol{\theta}) \frac{d \log p(\mathbf{h}_i | \mathbf{s}_{i,1}; \boldsymbol{\theta})}{d\boldsymbol{\theta}}, \tag{1.55}$$

$\mathbf{s}_{i,1}$ is the initial state of the $i$th episode generated by the environment, and the sequence

$$\mathbf{h}_i \equiv [\mathbf{a}_{i,2}, \mathbf{s}_{i,2}, \mathbf{a}_{i,3}, \ldots, \mathbf{s}_{i,T_i}]$$

is the part of the $i$th episode which follows $\mathbf{s}_{i,1}$ with probability $p(\mathbf{h}_i | \mathbf{s}_{i,1}, \boldsymbol{\theta}_i)$. The stepsize $\gamma_i$ is a positive number which varies as a function of the current iteration $i$ of the learning algorithm.

**Algorithm 1.7.2 A Generic Policy Gradient Reinforcement Learning Algorithm.**

1: **procedure** POLICY-GRADIENT-REINFORCEMENT-LEARNING($\boldsymbol{\theta}_1$).
2: $i \Leftarrow 1$
3: **repeat**
4: $k \Leftarrow 1$
5: Observe $i$th initial state $\mathbf{s}_{i,1}$ of episode $i$ sampled from $p_e(\mathbf{s}_{i,1})$.
6: **repeat**
7: Machine Samples $\mathbf{a}_{i,k}$ from $p(\mathbf{a}_{i,k} | \mathbf{s}_{i,k}, \boldsymbol{\theta}_i)$
8: Environment Samples $\mathbf{s}_{i,k+1}$ from $p_e(\mathbf{s}_{i,k+1} | \mathbf{a}_{i,k}, \mathbf{s}_{i,k})$
9: Compute $T_i$ from partial episode $[\mathbf{s}_{i,1}, \mathbf{a}_{i,1}, \ldots, \mathbf{s}_{i,k+1}]$
10: $k \Leftarrow k + 1$;
11: **until** End of Episode $i$ (i.e., $k + 1 = T_i$)
12: Episode $\mathbf{x}_i \Leftarrow [\mathbf{s}_{i,1}, \mathbf{a}_{i,2}, \ldots, \mathbf{s}_{i,T_i}]$
13: $\mathbf{g}(\mathbf{x}_i, \boldsymbol{\theta}_i) \Leftarrow d\ddot{c}(\mathbf{x}_i; \boldsymbol{\theta}_i)/d\boldsymbol{\theta}$ where $d\ddot{c}/d\boldsymbol{\theta}$ is defined in (1.55)
14: $\boldsymbol{\theta}_{i+1} \Leftarrow \boldsymbol{\theta}_i - \gamma_i \mathbf{g}(\mathbf{x}_i, \boldsymbol{\theta}_i)$
15: $i \Leftarrow i + 1$.
16: **until** $\boldsymbol{\theta}_i \approx \boldsymbol{\theta}_{i-1}$
17: **return** $\{\boldsymbol{\theta}_i\}$
18: **end procedure**

An action $\mathbf{a}$ is generated by the learning machine by sampling from the *control law* or *policy* density $p(\mathbf{a}|\mathbf{s}, \boldsymbol{\theta})$ for a given environmental state $\mathbf{s}$ and $\boldsymbol{\theta}$. For example, suppose the learning machine is a multi-layer network architecture whose response is an action $\mathbf{a}$ given input pattern $\mathbf{s}$ and current parameter vector $\boldsymbol{\theta}$ so that $\mathbf{a}$ is a realization of a Gaussian random vector with mean $\mathbf{f}(\mathbf{s}, \boldsymbol{\theta})$ and some covariance matrix $\mathbf{C}$.

The learning machine's statistical environment is specified by two *environmental* probability density functions. The *initial environmental state* density generates the initial environmental state $\mathbf{s}_{i,1}$ for the $i$th episode by sampling from $p_e(\mathbf{s}_{i,1})$. And finally, the *environmental transition state density* generates the next state of the trajectory $\mathbf{s}_{i,k+1}$ by sampling from $p_e(\mathbf{s}_{i,k+1} | \mathbf{s}_{i,k}, \mathbf{a}_{i,k})$. Thus, the probability of the $i$th episode $\mathbf{x}_i$ is given by the formula:

$$p(\mathbf{x}_i | \boldsymbol{\theta}) = p_e(\mathbf{s}_{i,1}) \left[ \prod_{k=1}^{T_i - 1} p_e(\mathbf{s}_{i,k+1} | \mathbf{s}_{i,k}, \mathbf{a}_{i,k}) p(\mathbf{a}_{i,k} | \mathbf{s}_{i,k}, \boldsymbol{\theta}) \right].$$

Using these definitions with (1.53) and (1.54), gives rise to the Generic Policy Gradient Reinforcement Algorithm presented in Algorithm 1.7.2.

**Example 1.7.3** (Model-Based Policy Gradient Reinforcement Learning.). A *model-based policy gradient reinforcement learning machine* works by learning not only how to generate actions in its statistical environment but also how its statistical environment behaves. The ability to internally simulate the consequences of actions executed in the world before those actions are actually executed can greatly enhance a learning machine's ability to learn how to control its environment. For example, when learning to drive a new car, the responsiveness of the car to the turn of the steering wheel or to pressure on the brake pedal is quickly learned by the driver. This learned information may then be used by the driver to improve driving performance. This is an example of a reinforcement learning machine which not only learns to generate actions but also learns to anticipate the consequences of those actions. That is, it provides the reinforcement learning machine with the ability to mentally evaluate the consequences of its actions before those actions are actually executed.

More specifically, let $\ddot{\mathbf{s}}_{i,t+1}(\boldsymbol{\theta})$ be the learning machine's prediction of the environmental state $\mathbf{s}_{i,t+1}$ observed at time $t+1$ in episode $i$. The prediction $\ddot{\mathbf{s}}_{i,t+1}(\boldsymbol{\theta})$ is computed using the formula $\ddot{\mathbf{s}}_{i,t+1}(\boldsymbol{\theta}) = \mathbf{f}(\mathbf{a}_{i,t}, \mathbf{s}_{i,t}; \boldsymbol{\theta})$. The function $\mathbf{f}$ specifies the learning machine's mental model of how it expects its environment to change when action $\mathbf{a}_{i,t}$ is executed in state $\mathbf{s}_{i,t}$ when the learning machine's parameter vector is set equal to $\boldsymbol{\theta}$.

Thus, during the learning process, the parameter vector $\boldsymbol{\theta}$ not only identifies the learning machine's policy for choosing particular actions but also specifies the learning machine's theory about how its statistical environment will be altered when it executes particular actions in specific situations.

Now define a discrepancy function $D$ for computing the discrepancy between the observed value of the future environmental state $\mathbf{s}_{i,t+1}$ and the learning machine's prediction of the future environmental state, $\ddot{\mathbf{s}}_{i,t+1}(\boldsymbol{\theta})$ using the formula:

$$D(\mathbf{s}_{i,t+1}, \ddot{\mathbf{s}}_{i,t+1}(\boldsymbol{\theta})).$$

And finally, define the reinforcement per episode penalty, $R(\mathbf{x}_i; \boldsymbol{\theta})$, for the $i$th episode as:

$$R(\mathbf{x}_i; \boldsymbol{\theta}) = \sum_{t=1}^{T_i - 1} D(\mathbf{s}_{i,t+1}, \ddot{\mathbf{s}}_{i,t+1}(\boldsymbol{\theta})) + \sum_{t=1}^{T_i - 1} r_{i,t} \tag{1.56}$$

where $r_{i,t}$ is the received incremental reinforcement penalty from the environment delivered to the learning machine when the learning machine transitions from state $\mathbf{s}_{i,t}$ to $\mathbf{s}_{i,t+1}$. The total penalty incurred by the learning machine for experiencing a particular episode $\mathbf{x}_i$ depends not only upon the penalty received by the environment but the accuracy in which the learning machine is able to improve its predictions regarding the behavior of its environment.

By substituting the definition of $R(\mathbf{x}_i, \boldsymbol{\theta})$ in (1.56) into Policy Gradient Reinforcement Learning Algorithm 1.7.2, a model-based policy gradient reinforcement learning algorithm is obtained.

Show how to extend the above framework to the case where the learning machine's mental model is probabilistic so that the learning machine's mental model of its environment predicts the probability a particular environmental state $\mathbf{s}_{i,t+1}$ will occur when action $\mathbf{a}_{i,t}$ is executed in state $\mathbf{s}_{i,t}$ for a given parameter vector $\boldsymbol{\theta}$. △

**Example 1.7.4** (Lunar Lander Problem: Policy Gradient Reinforcement). In this example, a policy gradient control law is developed for the Lunar Lander Example (see Example 1.7.1). Unlike the predictive model example (see Example 1.7.3) where both the policy for generating actions and the predictive model are learned, in this lunar lander example the reinforcement learning machine only learns a control law policy.

*Lunar Lander Parametric Control Law.* Let the probability of applying the thrust $a_i(t) = 1$ given environmental state $\mathbf{s}_i(t)$ in trajectory $i$ be denoted by the probability $p_{a,i,t}(\boldsymbol{\theta})$ which is formally defined by

$$p_{a,i,t}(\boldsymbol{\theta}) = \mathcal{S}\left(f(\mathbf{s}_i(t), \boldsymbol{\theta}\right) \tag{1.57}$$

where larger values of the *action function* $f(\mathbf{s}_i(t), \boldsymbol{\theta})$ correspond to an increased probability of choosing action $a_i(t) = 1$. An example choice for the action function $f$ would be:

$$f(\mathbf{s}_i(t), \boldsymbol{\theta}) = \boldsymbol{\theta}^T \boldsymbol{\psi}(\mathbf{s}_i(t))$$

where $\boldsymbol{\psi}$ is a constant preprocessing transformation. The vector $\boldsymbol{\theta}$ is called the *control law parameter vector* with different choices of $\boldsymbol{\theta}$ corresponding to different specific "control laws". An alternative choice for the action function $f$ might be:

$$f(\mathbf{s}_i(t), \boldsymbol{\theta}) = \boldsymbol{\beta}^T \boldsymbol{\psi}(\mathbf{s}_i(t), \boldsymbol{\eta})$$

where $\boldsymbol{\psi}$ is a parameterized preprocessing transformation with parameters $\boldsymbol{\eta}$ and $\boldsymbol{\theta}^T \equiv [\boldsymbol{\beta}^T, \boldsymbol{\eta}^T]$.

*Lunar Lander Empirical Risk Function.* An empirical risk function can be formulated for the lunar lander temporal reinforcement problem. First, an initial state of the trajectory $\mathbf{s}(1)$ is chosen at random. Second, the learning machine uses its current parameter vector $\boldsymbol{\theta}$ to generate action $a(1)$. The environment then processes the action $a(1)$ in conjunction with $\mathbf{s}(1)$ to generate a new environmental state $\mathbf{s}(2)$. In addition, the environment calculates a reinforcement signal which is provided to the learning machine. The sequence $\mathbf{x}_i(t) \equiv [\mathbf{s}_i(1), a_i(1), \mathbf{s}_i(2), a_i(2), \mathbf{s}_i(3)]$ corresponds to a short segment of the longer lunar landing trajectory so it is called the $i$th "trajectory segment" or "episode segment". For a given $\boldsymbol{\theta}$, it is assumed that trajectory segments are independent and identically distributed. After each trajectory segment generated by the learning machine interacting with its environment, the parameter vector of the learning machine $\boldsymbol{\theta}$ is updated. As the learning machine's knowledge state $\boldsymbol{\theta}$ evolves, the learning machine will exhibit different behaviors. These different behaviors, in turn, generate different statistical environments for the learning machine.

Define the probability that action $a_i(t)$ is chosen at time $t$ in trajectory segment $i$, $p(a_i(t)|\mathbf{s}_i(t); \boldsymbol{\theta})$, by the formula:

$$p(a_i(t)|\mathbf{s}_i(t); \boldsymbol{\theta}) \equiv a_i(t) p_{a,i,t}(\boldsymbol{\theta}) + (1 - a_i(t))\,(1 - p_{a,i,t}(\boldsymbol{\theta}))$$

where $p_{a,i,t}(\boldsymbol{\theta})$ is defined as in (1.57).

The empirical risk function corresponding to (1.52) is given by the formula:

$$\ell_n(\boldsymbol{\theta}) = (1/n) \sum_{i=1}^{n} c(\mathbf{x}_i; \boldsymbol{\theta}). \tag{1.58}$$

where

$$c(\mathbf{x}_i; \boldsymbol{\theta}) = (1/2)\lambda_\theta |\boldsymbol{\theta}|^2 + r(\mathbf{x}_i) p\left(\mathbf{h}_i | \mathbf{s}_{i,1}, \boldsymbol{\theta}\right), \tag{1.59}$$

$\lambda_\theta$ is a positive number, and

$$p(\mathbf{h}_i|\mathbf{s}_{i,1}, \boldsymbol{\theta}) = p(\mathbf{a}_i(1)|\mathbf{s}_i(1); \boldsymbol{\theta}) p(\mathbf{s}_i(2)|\mathbf{s}_i(1), \mathbf{a}_i(1)) p(\mathbf{a}_i(2)|\mathbf{s}_i(2); \boldsymbol{\theta}).$$

Note that the reinforcement received $r(\mathbf{x}_i)$ for the $i$th episode $\mathbf{x}_i$ is not assumed to be functionally dependent upon $\boldsymbol{\theta}$. In addition, $\mathbf{s}_i(t+1)$ is a deterministic function of $\mathbf{s}_i(t)$ and $\mathbf{a}_i(t)$ so that $p(\mathbf{s}_i(2)|\mathbf{s}_i(1), \mathbf{a}_i(1)) = 1$.

*Lunar Lander Gradient Descent Algorithm.* A gradient descent algorithm can then be derived to minimize the empirical risk function in (1.58) using (1.54) and (1.55) by computing the gradient

$$\mathbf{g}(\mathbf{x}_i;\boldsymbol{\theta}) = \lambda_\theta \boldsymbol{\theta} + r(\mathbf{x}_i)\left[\frac{d\log p(\mathbf{a}_i(1)|\mathbf{s}_i(1);\boldsymbol{\theta})}{d\boldsymbol{\theta}} + \frac{d\log p(\mathbf{a}_i(2)|\mathbf{s}_i(2);\boldsymbol{\theta})}{d\boldsymbol{\theta}}\right]^T. \tag{1.60}$$

The gradient in (1.60) is then used to specify a gradient descent algorithm that minimizes the empirical risk function in (1.58). First, generate the next episode

$$\mathbf{x}_i = [\mathbf{s}_i(1), \mathbf{a}_i(1), \mathbf{s}_i(2), \mathbf{a}_i(2), \mathbf{s}_i(3)]$$

by: (1) sampling $\mathbf{s}_i(1)$ from $p_e(\mathbf{s}_i(1))$, (ii) sampling $\mathbf{a}_i(1)$ from $p(\mathbf{a}_i(1)|\mathbf{s}_i(1),\boldsymbol{\theta})$, (iii) sampling $\mathbf{s}_i(2)$ from $p(\mathbf{s}_i(2)|\mathbf{s}_i(1),\mathbf{a}_i(1))$, (iv) sampling $\mathbf{a}_i(2)$ from $p(\mathbf{a}_i(2)|\mathbf{s}_i(2);\boldsymbol{\theta})$, and sampling $\mathbf{s}_i(3)$ from $\mathbf{p}(\mathbf{s}_i(3)|\mathbf{s}_i(2),\mathbf{a}_i(2))$. Then, use the resulting episode $\mathbf{x}_i$ to compute the reinforcement penalty for that episode using the deterministic formula $r(\mathbf{x}_i)$ which provides feedback regarding the quality of the lunar landing. Second, use the update rule:

$$\boldsymbol{\theta}_{i+1} = \boldsymbol{\theta}_i - \gamma_i \mathbf{g}(\mathbf{x}_i;\boldsymbol{\theta}) \tag{1.61}$$

to update the parameters of the lunar lander control law after each episode is observed during the learning process. Third, repeat this process by getting the next subsequent non-overlapping episode. △

## Exercises

1.7-1. *Nonstationary Statistical Environment Experiments.* Use an adaptive learning machine to learn a statistical environment and then gradually change the characteristics of the statistical environment and evaluate how well the learning machine tracks the changing statistical environment.

1.7-2. *Value Function Reinforcement Learning Using Supervised Learning.* Explain how a value function reinforcement learning algorithm can be interpreted as a supervised learning algorithm.

1.7-3. *Policy Gradient Reinforcement Learning Lunar Lander Experiments.* Implement the lunar lander reinforcement learning algorithm and explore how the performance of the system varies as a function of initial conditions.

## 1.8 Further Readings

### Machine Learning Environments

#### Feature Engineering

Discussions of feature engineering methods are available in the books by Dong and Liu (2018) and Zheng and Casari (2018). Other useful references include the books by: Geron (2019), Muller and Guido (2017), and Raschka and Mirjalili (2019).

### Nonstationary Environments

The general analysis of nonstationary statistical environments requires more advanced mathematical tools than those encountered in this text, but after mastering this text the reader will be prepared to pursue more advanced study in this area. Nonstationary statistical environments involving complicated models of complex phenomena are frequently encountered in the econometrics literature which is a rich source of tools for machine learning algorithm analysis and design. White's (1994) seminal work *Estimation, Inference, and Specification Analysis* specifically deals with the issue of estimation and inference in situations where imperfections may possibly exist in the proposed probability model. Another useful advanced resource is Davidson's (2002) book titled *Stochastic Limit Theory*.

### Consequences of Misleading Prior Knowledge

In engineering applications, the introduction of prior knowledge constraints is essential yet the introduction of such prior knowledge constraints increases the chance that the learning machine's probabilistic model of its environment is wrong. A unique distinguishing feature of the empirical risk minimization framework presented in Chapter 10 of this textbook is that analyses of generalization performance explicitly allow for the possibility the probability model is not perfect. Chapters 13, 14, 15, and 16 provide important tools for detecting the presence of model misspecification and supporting robust estimation and inference in the presence of model misspecification using the methods of White (1982; also see Golden 1995, 2003; White 1994) and Golden, Henley, White, and Kashner (2013, 2016).

## Empirical Risk Minimization Framework

Early discussions of the empirical risk minimization framework within the context of machine learning problems may be found in Nilsson (1965, Chapter 3) and Amari (1967). Empirical risk minimization is also the basis of Vapnik's Statistical Learning Theory (Vapnik, 2000) which plays a central role in machine learning theory.

The unified framework presented in this text extends the empirical risk minimization framework discussed by Golden (1988a, 1988b, 1988c, 1996a, 1996b, 1996c) and White (1989a, 1989b) for supporting the analysis and design of machine learning algorithms. The mathematical foundations of this framework associated with the analysis of generalization performance can be attributed to the work of Huber (1967). Huber (1967) characterized the asymptotic behavior of empirical risk minimization parameter estimators which are called $M$-estimators in the mathematical statistics literature (e.g., Serfling 1980).

A critical assumption of the empirical risk minimization framework developed in Chapters 13, 14, 15, and 16 is that the parameter estimates are converging to a strict local minimizer of the expected empirical risk function. This type of condition is called a "local identifiability assumption". Regularization is an important approach for ensuring the local identifiability condition holds. Hoerl and Kennard (1970) provides an early discussion regarding the relevance of $L_2$ regularization for regression modeling. Tibshirani (1996) introduced $L_1$ regularization for identifying models with sparse structures. Zou and Hastie (2005) discussed a method for combining $L_1$ and $L_2$ regularization in order to benefit from the best features of both methods. Ramirez et al. (2014), Schmidt et al. (2007), and Bagul (2017) discuss various differentiable versions of $L_1$ regularization. Watanabe (2010) has developed a version of the empirical risk minimization framework that provide a more direct approach to addressing the "local identifiability" assumption.

### Theory-Based System Analysis and Design

The section on Theory-Based System Analysis and Design extends discussions by Golden (1988a, 1988b, 1988c, 1996a, 1996b, 1996c) as well as earlier discussions by Simon (1969) and Marr (1982).

### Artificial Neural Networks: From 1950 to 1980

Adaptive neural network learning machines gained popularity in the 1960s (e.g., Widrow and Hoff 1960; Rosenblatt 1962; Amari 1967). Rosenblatt (1962) discussed a variety of multilayer perceptron architectures supported by both theoretical analyses and empirical results. Werbos (1974; also see Werbos 1994) described how multi-layer and recurrent learning machines constructed from simple computing units could be trained using gradient descent methods. In addition, Werbos (1974; also see Werbos 1994) emphasized the important connection of multi-layer learning machine learning algorithms to control theory (e.g., Bellman 1961). Anderson and Rosenfeld (2000) interviewed many of the early machine learning pioneers in a series of fascinating interviews.

### Artificial Neural Networks and Machine Learning: From 1980 to 2000

In the mid-1980s, David B. Parker (1985), Yann LeCun (1985), and Rumelhart, Hinton, and Williams (1986) independently explored methods for training multi-layer "deep learning" neural networks using sigmoidal hidden unit basis functions. Many of the basic principles underlying modern convolutional neural networks (CNNs) also appeared at this time (Fukushima 1980; Fukushima and Miyake 1982). An early 1990s review of the basic theory of deep learning can be found in Rumelhart et al. (1996). An important collection of seminal papers in the field of artificial neural network modeling and machine learning can be found in Anderson and Rosenfeld (1998a) and Anderson and Rosenfeld (1998b) which covers time periods from the late 19th century through the 1980s.

### Deep Learning and Machine Learning: From 2000 to 2020

Reviews of unsupervised and supervised machine learning algorithms can be found in the books by Duda et al. (2001), Bishop (2006), Murphy (2012), and Hastie et al. (2001). Goodfellow et al. (2016) provides an introduction to the modern Deep Learning literature, while Schmidhuber (2015) provides a recent comprehensive review. Discussions of supervised learning are referred to as nonlinear regression models in the mathematical statistics literature (Bates and Watts 2007).

### Function Approximation Methods

The concept that an approximate arbitrary nonlinear function can be represented as a weighted sum of simpler basis functions is not new. For example, Fourier (1822) introduced Fourier Analysis for representing complicated time-varying signals as weighted sums of cosine and sine functions. Laying the groundwork for the development of the modern digital computer, McCulloch and Pitts (1943) showed that a network of simple neuron-like computing units (i.e., `AND` gates, `OR` gates, and `NOT` gates) could be connected together to realize any arbitrary logical function. In the multilayer perceptron literature in the 1990s, a series of important basis function approximation theorems were published (e.g., Cybenko 1989; Hornik et al. 1989; Hornik 1991; Leshno et al. 1993; Pinkus 1999) showed a multilayer perceptron with only one layer of hidden units could approximate an arbitrary smooth nonlinear functions provided the number of hidden units was sufficiently large. More recently, Guliyev and Ismailov (2018) showed that a multilayer perceptron network with two layers of hidden units and a finite number of parameters is capable of approximating an arbitrary smooth multivariate function. Furthermore, Guliyev and Ismailov (2018) discuss situations where such approximation capabilities are not achievable with a multilayer perceptron with only a single layer of hidden units. More recently, Imaizumi and Fukumizu (2019) have provided theorems regarding function approximation capabilities of non-smooth perceptrons with one, two, or more layers of hidden units.

### Unsupervised Learning Machines

Bishop (2006), Murphy (2012), Hastie et al. (2001), Duda et al. (2001), and Goodfellow et al. (2016) provide discussions of unsupervised learning machines. Duda et al. (2001, Ch. 10) and Hastie et al. (2001, Ch. 14) describe the within-clustering approach discussed in Section 1.6.4. An early discussion of clustering algorithms can be found in Hartigan (1975). Duda and Hart (1973) describe the general clustering algorithm discussed in Section 1.6.4. Useful reviews of Latent Semantic Indexing can be found in Landauer et al. (2014). Domingos and Lowd (2009) and Getoor and Taskar (2007) provide useful introductions to Markov Logic Nets.

### Reinforcement Learning Machines

Students interested in the topic of reinforcement learning will find the books by Bertsekas and Tsitsiklis (1996), Sugiyama (2015), Sutton and Barto (2018), and Wiering and van Otterlo (2012) helpful. The general reinforcement learning framework presented here is a minor variation of the general VAP (Value And Policy Search) framework developed by Baird and Moore (1999). Williams (1992) originally introduced the policy gradient method into the machine learning literature. Many reinforcement learning problems are naturally viewed within the frameworks of: deterministic control theory (e.g., Bellman 1961; Lewis et al. 2012), optimal control theory (Bertsekas and Shreve 2004; Sutton and Barto 2018) framework, and Markov decision processes (e.g., Bellman 1961; Sutton and Barto 2018).

# 2

# *Set Theory for Concept Modeling*

**Learning Objectives**

- Use set theory to represent concepts and logical formulas.
- Define directed and undirected graphs as sets.
- Define a function as a relation between sets.
- Use metric spaces to represent concept similarity.

An inference and learning machine must possess representations of the real world. Such representations are mathematical models and thus abstractions of an extremely complicated reality. For example, consider the problem of designing a learning machine capable of distinguishing photographs of dogs from photographs of cats. To achieve this goal, the concept DOG and the concept of CAT must somehow be represented in the learning machine yet the variations within and between these species is considerable (see Figure 2.1). A similar challenge is faced in the design of learning machines for processing auditory input signals (see Figure 2.2).

Define a $d$-dimensional feature space $\Omega$ such that each point in the feature space corresponds to a feature vector representing information in a particular photograph image. Let the concepts of DOG and CAT be modeled respectively as two distinct subsets $\Omega_{dog}$ and $\Omega_{cat}$ of $\Omega$. The similarity between two photograph images is modeled as the distance between two points in $\Omega$. Let $\mathbf{x}$ be a point in feature space. If $\mathbf{x} \in \Omega_{dog}$, then $\mathbf{x}$ is classified as an instance of the concept DOG. If $\mathbf{x} \in \Omega_{cat}$, then $\mathbf{x}$ is classified as an instance of the concept CAT.

Thus, set theory provides one possible approach for building an abstract model of a concept. Within the framework of set theory, a concept $C$ is defined as a set of instantiations of $C$. Methods for combining and transforming sets correspond to different information processing strategies. This concept-construction strategy often arises in machine learning applications during the construction of real-valued feature vectors. The elements of such feature vectors may be physically measurable (e.g., the intensity of a light stimulus at a particular pixel), abstract (e.g., the presence or absence of a line segment), or very abstract (e.g., happy or unhappy facial expression). Nevertheless, the feature vector represents a mapping from a set of events in space and time into a point in a finite-dimensional vector space. Thus, set theory plays an important role in the development of mathematical models of features in machine learning applications.

Furthermore, set theory is the foundation of all mathematics. Mathematical definitions are statements about equivalent sets. Mathematical theorems are assertions that specific conclusions follow from specific assumptions. Mathematical proofs correspond to a sequence

**FIGURE 2.1**
**Examples of different real-world images of cats and dogs.** The design of a learning machine that can distinguish between images of dogs and images of cats is a challenging task. This task might appear straightforward but, in fact, this is a challenging problem because it is virtually impossible to identify a set of observable features which are both necessary and sufficient to distinguish these two object classes.

of logical arguments. Definitions, theorems, and proofs are all based upon logical assertions. Logical assertions, in turn, can be reformulated as statements in set theory.

## 2.1 Set Theory and Logic

Standard set theory operators may be used to make assertions regarding relationships between concepts as well as define novel concepts. Let $A$ and $B$ be sets of objects which are subsets of a concept-universe $U$. The notation $A \subseteq U$ denotes that $A$ is a subset of $U$. The

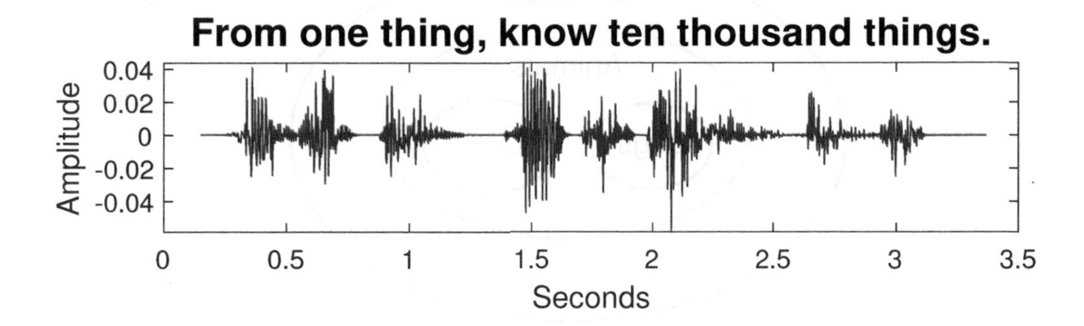


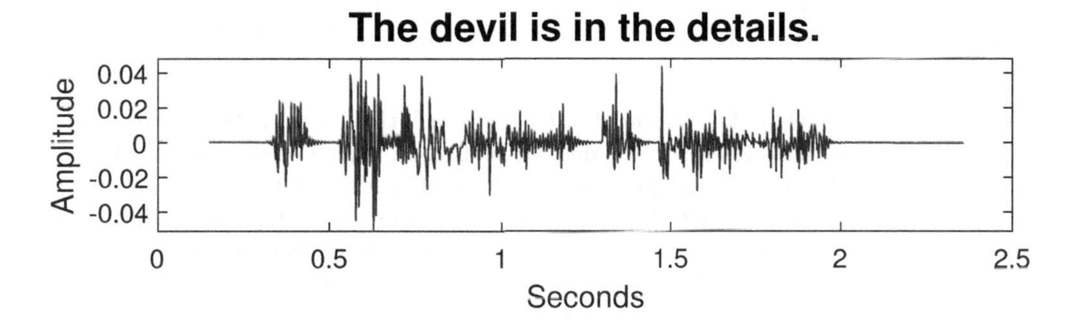


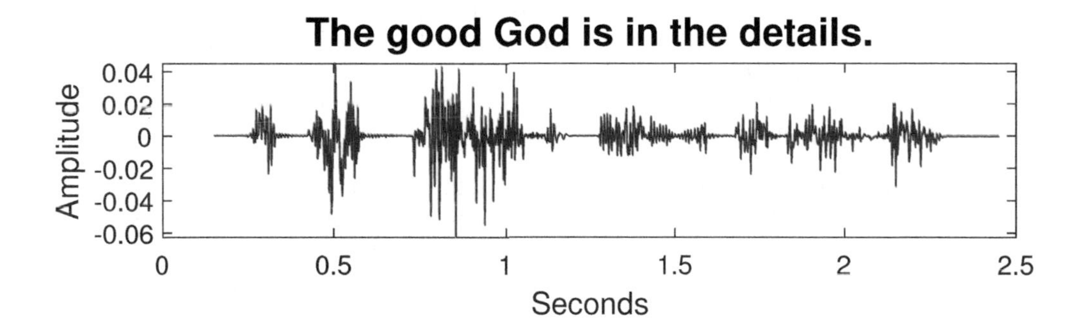


FIGURE 2.2
**Examples of time-varying sound-pressure induced voltage amplitude representations of three spoken sentences.** The auditory processing problem of distinguishing even clearly spoken distinct sentences is a challenging task. It is virtually impossible to identify a set of observable features that are necessary and sufficient for identifying individual spoken words or sentences. In this example, the sentences "*From one thing, know ten thousand things,*" "*The devil is the details,*" and "*The good God is in the details*" were spoken into a microphone. The microphone is a transducer that transforms sound pressure waves into changes in voltage amplitudes which vary over time. Even though the sentences were clearly spoken in a low-noise environment, it is not easy to identify consistent statistical regularities which detect specific words and word sequences.

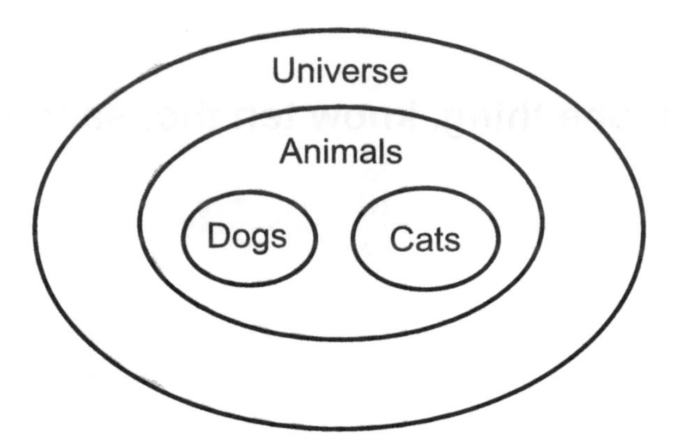


**FIGURE 2.3**
**Representation of logical assertions using set theory.** This figure depicts logical assertions such as: (i) `EVERY DOG IS AN ANIMAL`, (ii) `NO OBJECT IS BOTH A DOG AND A CAT`, and (iii) `IF AN OBJECT IS A DOG, THEN THE OBJECT IS AN ANIMAL`.

notation $A \subset U$ means that $A \subseteq U$ and $A \neq U$. If $A \subset U$, then $A$ is called a *proper subset* of $U$. The complementation operator $\backslash$ is defined such that: $A \backslash B$ is equivalent to $A \cap \neg B$.

For example, let $U$ be the concept-universe whose elements are the set of all objects. Let $D$ be the set of all dogs. Let $C$ be the set of all cats. Let $A$ be the set of all animals. Then, a new concept defining the set of all dogs and all cats is specified by $D \cup C$. The set of animals that are not dogs is specified by the expression $A \cap \neg D$ or equivalently $A \backslash D$.

In addition, set theory expressions are directly interpretable as logical formulas (see Figure 2.3). The assertion that $x \in D$ (i.e., x is a dog) may be assigned the value of `TRUE` or `FALSE`. The assertion that $x \in D$ `OR` $x \in C$ (i.e., x is either a dog or a cat) or equivalently $x \in D \cup C$ may also be assigned the value of `TRUE` or `FALSE`. The conditional assertion that `IF` $x \in D$ (i.e., x is a dog) `THEN` $x \in A$ (i.e., x is an animal) or equivalently $D \subseteq A$ may also be assigned the value of `TRUE` or `FALSE`. Equivalently, `IF` $x \in D$ `THEN` $x \in A$ can be expressed by the formula $\neg D \cup A$.

Note that throughout this text, the terminology "either A or B holds" means that: (1) A holds, (2) B holds, or (3) both A and B hold.

**Definition 2.1.1** (Natural Numbers)**.** Let $\mathbb{N} \equiv \{0, 1, 2, \ldots\}$ denote the *natural numbers* or the *non-negative integers*. Let $\mathbb{N}^+ \equiv \{1, 2, 3, \ldots\}$ denote the set of *positive integers*. □

**Definition 2.1.2** (Real Numbers)**.** Let $\mathcal{R} \equiv (-\infty, \infty)$ denote the set of *real numbers*. □

The notation $(a, b)$ denotes the subset of real numbers which are strictly greater than $a$ and strictly less than $b$. The notation $(a, b]$ denotes the subset of real numbers which are strictly greater than $a$ and less than or equal to $b$.

In many cases it will be useful to extend the real number line to include two additional symbols called $-\infty$ and $+\infty$ which denote the limits of a sequence of real numbers which approach $-\infty$ and $+\infty$ respectively.

**Definition 2.1.3** (Extended Real Numbers)**.** The *extended real numbers*, $[-\infty, +\infty]$, are defined as the set $\mathcal{R} \cup \{-\infty\} \cup \{+\infty\}$. □

Operations such as addition and multiplication involving symbols such as $-\infty$ and $+\infty$ are defined so that $\alpha + \infty = \infty$ and $\alpha\infty = \infty$ where $\alpha \in \mathcal{R}$. It is often convenient to additionally define $\infty$ multiplied by zero to equal zero.

**Definition 2.1.4** (One-to-One Correspondence). Let $X$ and $Y$ be sets. A *one-to-one correspondence* between $X$ and $Y$ is said to exist if every element of $X$ can be uniquely matched to an element of $Y$ such that there are no unpaired elements in the set $Y$. □

**Definition 2.1.5** (Finite Set). A set $E$ is called a *finite set* if either: (i) $E$ is the empty set, or (ii) there is a one-to-one correspondence between $E$ and $\{1, 2, \ldots, k\}$ for some finite positive integer $k$. □

**Definition 2.1.6** (Countably Infinite Set). A set $E$ is called a *countably infinite set* if there is a one-to-one correspondence between $E$ and $\mathbb{N}$. □

**Definition 2.1.7** (Uncountably Infinite Set). A set $E$ is called an *uncountably infinite set* if $E$ is not a countably infinite set. □

**Definition 2.1.8** (Countable Set). A set $E$ is called a *countable set* if $E$ is either a finite set or $E$ is a countably infinite set. □

**Definition 2.1.9** (Power Set). The *power set* of a set $E$ is the set of all subsets of $E$. □

**Definition 2.1.10** (Disjoint Sets). The sets $E$ and $F$ are said to be *mutually disjoint* if $E \cap F = \emptyset$. A collection of sets $E_1, E_2, \ldots$ is said to be *mutually disjoint* if every pair of elements in $\{E_1, E_2, \ldots, \}$ is mutually disjoint. □

**Definition 2.1.11** (Partition). Let $S$ be a set. A *partition* of $S$ is a set $G$ of nonempty mutually disjoint subsets of $S$ such that the union of all elements of $G$ is equal to $S$. If the number of elements in $G$ is a positive integer, then $G$ is called a *finite partition* of $S$. □

**Definition 2.1.12** (Cartesian Product). Let $E$ and $F$ be sets. The set

$$\{(x, y) : x \in E, y \in F\}$$

is called the *cartesian product* of $E$ and $F$. □

The notation $E \times F$ is also used to denote the cartesian product of $E$ and $F$.

Let $\mathcal{R}^d \equiv \times_{i=1}^{d} \mathcal{R}$ denote the set of $d$-dimensional real vectors where $d$ is a finite positive integer.

The notation $\{0, 1\}^d$ denotes the set of all possible $2^d$ $d$-dimensional binary vectors whose elements are either zero or one. Note that $\{0, 1\}^d \subset \mathcal{R}^d$.

## Exercises

2.1-1. Let $C$ be the set of cats. Let $D$ be the set of dogs. Let $A$ be the set of animals. Let $B$ be the set of objects larger than a golf ball. Use set theory notation to express the logical assertion that:

```
IF X is a cat or dog, THEN X is an animal larger than a golf ball.
```

2.1-2. Show that a phone number with an area code such as: `(214)909-1234` can be represented as an element of the set $S \times N^3 \times S \times N^3 \times S \times N^4$ where $S \equiv \{(, ), -\}$ and $N \equiv \{0, 1, 2, \ldots, 9\}$.

2.1-3. Let $A = \{1, 2, 3\}$. Let $B = \{4, 5\}$. Are $A$ and $B$ finite sets? Are $A$ and $B$ countably infinite sets? Are $A$ and $B$ countable sets? Are $A$ and $B$ disjoint sets? Is $\{A, B\}$ a finite partition of $A \cup B$?

2.1-4. Is the set $\{0, 1\}^d$ a countable set? Is the set $(0, 1)$ a countable set?

2.1-5. Is the set $\mathcal{R}$ a countable set?

## 2.2 Relations

A widely used method for knowledge representation is the "semantic network" (see Figure 2.4). A semantic network consists of a collection of nodes and a collection of arcs connecting the nodes. A connection of an arc from the $j$th node to the $i$th node typically specifies that node $j$ is semantically related to node $i$ according to the semantic relation $R$ associated with the semantic network. Different semantic relations correspond to different semantic networks. In this section, a mathematical formalism for representing these types of knowledge structures is provided.

### 2.2.1 Types of Relations

**Definition 2.2.1** (Relation). Let $X$ and $Y$ be sets. A *relation* $R$ is a subset of $X \times Y$. If $R$ is a subset of $X \times X$, then $R$ is called a *relation on* $X$. □

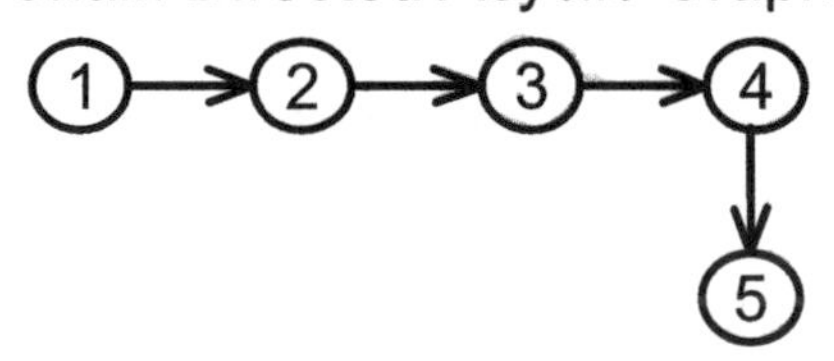


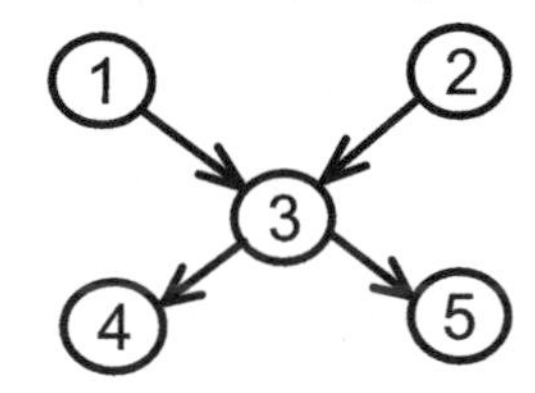


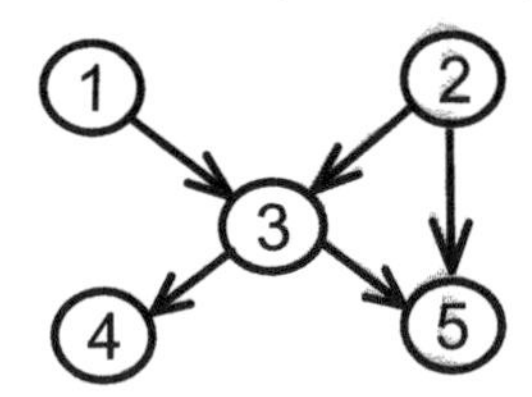


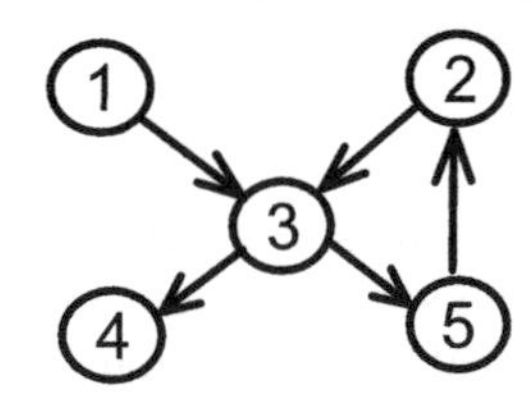


**FIGURE 2.4**
**Examples of different types of directed graphs.** The only graph in the figure which is not a directed acyclic graph is the directed graph with a directed cycle located in the lower right side of the figure.

**Definition 2.2.2** (Reflexive Relation). A *reflexive relation* $R$ on set $S$ has the property that $(x, x) \in R$ for all $x \in S$. □

Let $S \equiv \{1, 2, 3\}$. The relation $R = \{(1, 2), (2, 3), (1, 1)\}$ on $S$ is not reflexive, but the relation $R = \{(1, 2), (2, 3), (1, 1), (2, 2), (3, 3)\}$ on $S$ is reflexive.

**Definition 2.2.3** (Symmetric Relation). A *symmetric relation* $R$ on set $S$ has the property that for all $(x, y) \in \mathcal{R}$: $(x, y) \in R$ if and only if $(y, x) \in R$. □

Let $S \equiv \{1, 2, 3\}$. The relation $R = \{(1, 2), (2, 3), (1, 1)\}$ on $S$ is not symmetric, but the relation $R = \{(1, 2), (2, 3), (1, 1), (3, 2), (2, 1)\}$ on $S$ is symmetric.

**Definition 2.2.4** (Transitive Relation). A *transitive relation* $R$ on set $S$ has the property such that for all $(x, y) \in \mathcal{R}$: If $(x, y) \in R$ and $(y, z) \in R$, then $(x, z) \in R$. □

Let $S \equiv \{1, 2, 3\}$. The relation $R = \{(1, 2), (2, 3)\}$ on $S$ is not transitive, but the relation $R = \{(1, 2), (2, 3), (1, 3)\}$ on $S$ is transitive.

**Definition 2.2.5** (Equivalence Relation). A reflexive, symmetric, and transitive relation $R$ on set $S$ is called an *equivalence relation* on $S$. If $R$ is an equivalence relation, then the set $\{y : (x, y) \in R\}$ is called the *equivalence class* of $x$ with respect to the equivalence relation $R$. □

The notation $\equiv$ means that $\equiv$ is an equivalence relation. The notation $A \equiv B$ will often be used where the left-hand side of $A \equiv B$, $A$, is not defined and the right-hand side of $A \equiv B$, $B$, is defined. Thus, in this context, $A \equiv B$ may be interpreted as a statement which asserts that $A$ is defined as being equivalent to $B$. Alternative versions of such statements which are used in the formulation of formal definitions include both *$A$ is called $B$* or *$A$ is $B$*.

**Definition 2.2.6** (Complete Relation). A *complete relation* $R$ on set $S$ has the property that for all $x, y \in S$ either: (i) $(x, y) \in R$, (ii) $(y, x) \in R$, or (ii) both $(x, y) \in R$ and $(y, x) \in R$. □

Note that every complete relation is reflexive. To see this, simply note that if $R$ is a complete relation on $S$ then the definition of a complete relation when two elements $x, y \in S$ are the same (i.e., $x = y$) implies that $(x, x) \in R$.

Also note that if $S \equiv \{1, 2, 3\}$ and let $R \equiv \{(1, 2), (2, 3), (1, 1), (2, 2), (3, 3)\}$, then $R$ is not a complete relation on $S$ because neither $(1, 3)$ or $(3, 1)$ are elements of $R$. On the other hand, $Q \equiv \{(1, 2), (3, 1), (2, 3), (1, 1), (2, 2), (3, 3)\}$ is a complete relation on $\{1, 2, 3\}$.

As another example, the "less than or equal" operator denoted by $\leq$ is a complete relation on the real numbers since given any two real numbers $x, y \in \mathcal{R}$ either: (i) $x \leq y$, (ii) $y \leq x$, or (iii) $x = y$. On the other hand, the "less than" operator denoted by $<$ is not a complete relation on the real numbers. To see this, let $<$ be defined as the relation $R \equiv \{(x, y) \in \mathcal{R}^2 : x < y\}$. The set $R$ cannot be a complete relation because it does not contain the ordered pair $(x, y)$ when $x = y$.

**Definition 2.2.7** (Antisymmetric Relation). An *antisymmetric relation* $R$ on set $S$ has the property that for all $(x, y) \in R$: If $(x, y) \in R$ and $x \neq y$, then $(y, x) \notin R$. □

For example, if $S \equiv \{1, 2, 3\}$, then $R \equiv \{(1, 2), (2, 3), (1, 1)\}$ is an antisymmetric relation but the relation $R \equiv \{(1, 2), (2, 1), (2, 3), (1, 1)\}$ is not an antisymmetric relation on $\{1, 2, 3\}$.

**Definition 2.2.8** (Partial-Order Relation). If a relation $R$ on $S$ is reflexive, transitive, and antisymmetric then $R$ is called a *partial-order relation* on $S$. □

### 2.2.2 Directed Graphs

**Definition 2.2.9** (Directed Graph). Let $\mathcal{V}$ be a finite set of objects. Let $\mathcal{E}$ be a relation on $\mathcal{V}$. A *directed graph* $\mathcal{G}$ is the ordered pair $(\mathcal{V}, \mathcal{E})$ where an element of $\mathcal{V}$ is called a *vertex* and an element of $\mathcal{E}$ is called a *directed edge*. □

The terminology "node" is also often used to refer to a vertex in a directed graph. A directed edge $(A, B)$ in a relation $R$ may be graphically specified by drawing an arrow that originates from vertex $A$ and points to vertex $B$. Examples of different types of directed graphs are depicted in Figure 2.4.

Note that every relation $R$ on $S$ may be represented as a directed graph $\mathcal{G} \equiv (S, R)$. A complete relation corresponds to the concept of a directed graph which is defined such that there is a directed edge between any given pair of nodes on the graph $\mathcal{G}$.

Finally, let $\mathcal{G} \equiv (\mathcal{V}, \mathcal{E})$ be a directed graph. Let $v_1, v_2, \ldots$ be a sequence of elements of $\mathcal{V}$. If $(v_{k-1}, v_k) \in \mathcal{E}$ for $k = 2, 3, 4, \ldots$, then $v_1, v_2, \ldots$ is called a *path* through $\mathcal{G}$. In other words, traversals through the graph must follow the direction specified by the directed edges of the graph.

**Definition 2.2.10** (Parents). Let $\mathcal{G} \equiv (\mathcal{V}, \mathcal{E})$ be a directed graph. If the edge $(v_j, v_k) \in \mathcal{E}$, then vertex $v_j$ is called the *parent* of vertex $v_k$ and the vertex $v_k$ is called the *child* of vertex $v_j$. □

**Definition 2.2.11** (Self-loop). Let $\mathcal{G} \equiv (\mathcal{V}, \mathcal{E})$ be a directed graph. A *self-loop* is an edge $(v, v) \in \mathcal{E}$ of $\mathcal{G}$ where $v \in \mathcal{V}$. □

Thus, a self-loop is an edge on a graph which connects a particular node in the graph to itself.

**Definition 2.2.12** (Directed Acyclic Graph). Let $\mathcal{V} \equiv [v_1, v_2, \ldots, v_d]$ be an ordered set of objects. Let $\mathcal{G} \equiv (\mathcal{V}, \mathcal{E})$ be a directed graph. Suppose there exists an ordered sequence of integers $m_1, \ldots, m_d$ such that if $(v_{m_j}, v_{m_k}) \in \mathcal{E}$ then $m_j < m_k$ for all $j, k \in \{1, \ldots, d\}$. Then, the graph $\mathcal{G}$ is called a *directed acyclic graph* or *DAG*. The graph $\mathcal{G}$ is called a *topologically ordered DAG* provided that $(v_j, v_k) \in \mathcal{E}$ if and only if $j < k$ for all $j, k = 1, \ldots, d$. □

Assume there exists a path through a DAG $(\mathcal{V}, \mathcal{E})$ with the property that the path initiated at a vertex $v \in \mathcal{V}$ eventually returns to $v$ where only traversals in the direction of the arrows on the graph are permitted. Such a path is called a *directed cycle* of the DAG. A self-loop is a special case of a directed cycle. A directed graph is a directed acyclic graph provided the directed graph has no directed cycles. Some examples illustrating the concept of a directed graph are provided in Figure 2.4.

### 2.2.3 Undirected Graphs

**Definition 2.2.13** (Undirected Graph). Let $\mathcal{G} \equiv (\mathcal{V}, \mathcal{E})$ be a directed graph such that: (1) $\mathcal{G}$ does not contain self-loops, and (ii) $\mathcal{E}$ is a symmetric relation. Let $\bar{\mathcal{E}}$ be a set of unordered pairs such that $\{u, v\} \in \bar{\mathcal{E}}$ for each $(u, v)$ in $\mathcal{E}$. Then, $(\mathcal{V}, \bar{\mathcal{E}})$ is an *undirected graph* where $\mathcal{V}$ is the set of vertices and $\bar{\mathcal{E}}$ is the set of *undirected edges*. □

**Definition 2.2.14** (Neighbors). Let $\mathcal{V} \equiv \{v_1, \ldots, v_d\}$ be a set of objects. Let $\mathcal{G} \equiv (\mathcal{V}, \mathcal{E})$ be an undirected graph. For each edge $(v_j, v_k) \in \mathcal{E}$, the vertex $v_j$ is a *neighbor* of vertex $v_k$ and the vertex $v_k$ is a *neighbor* of vertex $v_j$. □

**Definition 2.2.15** (Clique). A subset, $\mathcal{C}$, of $\mathcal{V}$ is called a *clique* of an undirected graph $\mathcal{G} \equiv (\mathcal{V}, \mathcal{E})$ if (i) $\mathcal{C}$ contains exactly one element of $\mathcal{V}$, or (ii) for every $x \in \mathcal{C}$ all remaining elements in $\mathcal{C}$ are neighbors of $x$. A clique of $\mathcal{G}$ which is not a proper subset of some other clique of $\mathcal{G}$ is called a *maximal clique*. □

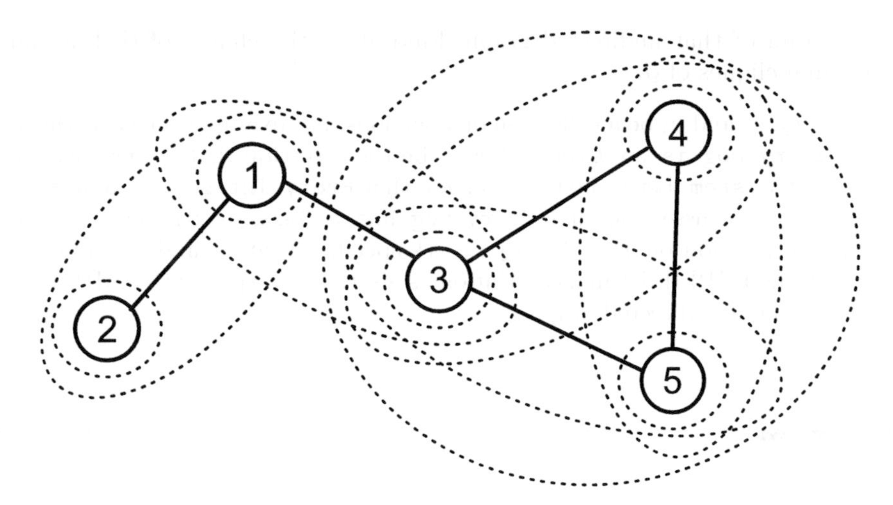


**FIGURE 2.5**
**An example of an undirected graph.** This figure shows an undirected graph $(\mathcal{V}, \mathcal{E})$ where $\mathcal{V} = \{1, 2, 3, 4, 5\}$ and $\mathcal{E} \equiv \{\{1, 2\}, \{1, 3\}, \{3, 4\}, \{3, 5\}, \{4, 5\}\}$. In addition, the eleven cliques of the graph are drawn as dashed circles which identify specific groups of nodes. This example has three maximal cliques $\{1, 2\}$, $\{1, 3\}$, and $\{3, 4, 5\}$.

Note that every nonempty subset of a clique is also a clique. Thus, all nonempty subsets of a maximal clique are cliques.

An undirected graph is graphically specified by drawing a set of nodes and then drawing a line without arrowheads connecting node $A$ to node $B$ if and only if $\{A, B\}$ is an undirected edge of the graph. Figure 2.5 illustrates how to draw an undirected graph and its cliques.

**Definition 2.2.16** (Fully Connected Graph). Let $\mathcal{G} \equiv (\mathcal{V}, \mathcal{E})$ be an undirected graph defined such that $\mathcal{E}$ contains all possible undirected edges. Then, $\mathcal{G}$ is called a *fully connected graph.* □

## Exercises

2.2-1. Let $S \equiv \{1, 2, 3, 4, 5\}$. Let $R \equiv \{(1, 2), (2, 4), (3, 5)\}$. Is $R$ reflexive on $S$? Is $R$ symmetric on $S$? Is $R$ transitive on $S$? Is $R$ an equivalence relation on $S$? Is $R$ an antisymmetric relation on $S$? Is $R$ a partial-order relation on $S$? Draw a directed graph of the relation $R$ on $S$.

2.2-2. Let $S \equiv \{A, B, C\}$. Let $R \equiv \{(A, B), (B, A), (C, A)\}$. Is $R$ symmetric on $S$? Is $R$ reflexive on $S$? Is $R$ complete on $S$? Is $R$ transitive on $S$? Is $R$ an equivalence relation on $S$? Is $R$ an antisymmetric relation on $S$? Is $R$ a partial-order relation on $S$? Draw a directed graph of the relation $R$ on $S$.

2.2-3. Let $\mathcal{V} \equiv \{1, 2, 3, 4\}$. Let $\mathcal{E} \equiv \{(1, 2), (2, 3), (3, 4), (3, 5)\}$. Let $\mathcal{G} \equiv (\mathcal{V}, \mathcal{E})$. Show that $\mathcal{G}$ is a directed acyclic graph. Draw the graph $\mathcal{G}$.

2.2-4. Let $\mathcal{V} \equiv \{1, 2, 3, 4\}$. Let $\mathcal{E} \equiv \{(1, 2), (2, 3), (2, 1), (3, 2)\}$. Let the directed graph $\mathcal{G} \equiv (\mathcal{V}, \mathcal{E})$. Show that $\mathcal{G}$ is a directed graph version of an undirected graph. Draw a diagram of that undirected graph. Find all of the cliques of $\mathcal{G}$. Find all of the maximal cliques of $\mathcal{G}$.

2.2-5. Let $v_1, v_2, v_3$ and $v_4$ be a collection of assertions. For example, let $v_1$ be the assertion that `Every Dog is an Animal`. Let $v_2$ be the assertion that `Every Animal has a Nervous System`. Let $v_3$ be the assertion that `Every Integer is a Real Number`. Let $v_4$ be the assertion that `Every Dog has a Nervous System`. The IMPLIES semantic relation on $\mathcal{V} \equiv \{v_1, v_2, v_3, v_4\}$ specifies that $v_1$ and $v_2$ imply $v_4$ holds. Specify the IMPLIES semantic relation as a set. Then, specify the IMPLIES semantic relation as a directed graph.

## 2.3 Functions

Computations by inference and learning machines are naturally expressed as "functions" which transform an input pattern into an output pattern. In this section, the fundamental concept of a "function" is formally developed and discussed.

**Definition 2.3.1** (Function). Let $X$ and $Y$ be sets. For each $x \in X$, a *function* $f : X \to Y$ specifies a unique element of $Y$ denoted as $f(x)$. The set $X$ is called the *domain* of $f$. The set $Y$ is called the *range* of $f$. □

In many cases, it will be useful to discuss collections of functions. The notation

$$\{(f_t : D \to R)|t = 1, 2, 3, \ldots, k\} \text{ or } \{(f_t : D \to R) : t = 1, 2, 3, \ldots, k\}$$

specifies a set of $k$ different functions $f_1, \ldots, f_k$ that have a common domain $D$ and range $R$.

Functions can also be used to specify constraints upon sets. For example, the notation $\{x \in G : f(x) = 0\}$ or equivalently the notation $\{x \in G|f(x) = 0\}$ means the set consisting of members of G such that the constraint f(x) = 0 holds for each member x in the set G.

**Definition 2.3.2** (Preimage). Let $f : X \to Y$. Let $S_Y$ be a subset of $Y$. The set

$$S_X \equiv \{x \in X : f(x) \in Y\}$$

is called the *preimage* of $S_Y$ under $f$. □

**Example 2.3.1** (Example Calculation of a Preimage). Let $\mathcal{S} : \mathcal{R} \to \mathcal{R}$ be defined such that for all $x \in \mathcal{R}$: $\mathcal{S}(x) = 1$ if $x > 1$, $\mathcal{S}(x) = x$ if $|x| \leq 1$, and $\mathcal{S}(x) = -1$ if $x < -1$. The preimage of $\{y \in \mathcal{R} : y = -1\}$ under $\mathcal{S}$ is the set $\{x : x \leq -1\}$. △

Figure 2.6 illustrates the important concept of the preimage of a set under a function.

**Definition 2.3.3** (Restriction and Extension of a Function). Let $f : D \to R$. A new function $f_d : d \to R$ can be defined such that $f_d(x) = f(x)$ for all $x \in d$. The function $f_d$ is called the *restriction* of $f$ to the domain $d$ where $d$ is a subset of the domain $D$ of $f$. The function $f$ is called the *extension* of $f_d$ from the domain $d$ to the superset of $d$ denoted as $D$. □

**Definition 2.3.4** (Crisp Membership Function). Let $\Omega$ be a set of objects. Let $S \subseteq \Omega$. A function $\phi_S : \Omega \to \{0, 1\}$ is called a *crisp set membership function* if $\phi_S(x) = 1$ when $x \in S$ and $\phi_S(x) = 0$ when $x \notin S$. □

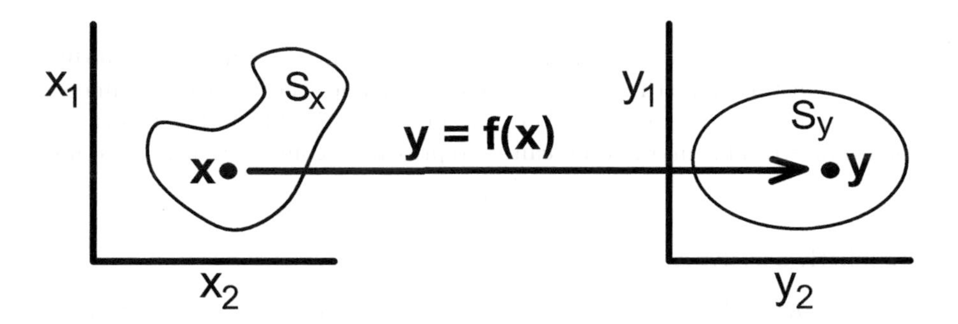


**FIGURE 2.6**
**The preimage of a set $S_y$ under a function f.** In this example, a vector-valued function $\mathbf{f}$ specifies a mapping from a point $\mathbf{x} = [x_1, x_2]$ in the domain of $\mathbf{f}$ into a point $\mathbf{y} = [y_1, y_2]$ in the range of $\mathbf{f}$. Let $S_x$ be some subset of points in the domain of $\mathbf{f}$. Let $S_y$ be the set of points in the range of $\mathbf{f}$ defined such that $\mathbf{f}(\mathbf{x}) \in S_y$ for all $\mathbf{x} \in S_x$. Then, $S_x$ is called the *preimage* of $S_y$ under $\mathbf{f}$.

Because the concept of a crisp membership function is widely used, the more common terminology for the *crisp membership function* defined in Definition (2.3.4) is simply *membership function.* The terminology *crisp set* simply refers to the standard definition of a set specified by a membership function.

**Definition 2.3.5** (Fuzzy Membership Function). Let $\Omega$ be a set of objects. A *fuzzy set* is an ordered pair $(S, \phi_S)$ where $S \subseteq \Omega$ and $\phi_s : \Omega \to [0, 1]$ is called the *fuzzy membership function for* $S$. If $\phi_S(\mathbf{x}) = 1$ for $\mathbf{x} \in \Omega$, then $\mathbf{x}$ is *fully included* in $S$. If $\phi_S(\mathbf{x}) = 0$ for $\mathbf{x} \in \Omega$, then $\mathbf{x}$ is *not included* in $S$. If $0 < \phi_S(\mathbf{x}) < 1$ for $\mathbf{x} \in \Omega$, then $\mathbf{x}$ is *partially included* in $S$ with *membership grade* $\phi_S(\mathbf{x})$. □

Note that the crisp membership function simply indicates whether or not an object is a member of a set. In fuzzy set theory, the fuzzy membership function is permitted to take on values between zero and one in order to explicitly represent varying degrees of set membership.

**Definition 2.3.6** (Bounded Function). Let $\Omega \subseteq \mathcal{R}^d$. A function $f : \Omega \to \mathcal{R}$ is bounded on $\Omega$ if there exists a finite real number $K$ such that for all $\mathbf{x} \in \Omega$: $|f(\mathbf{x})| \leq K$. □

A vector-valued function such as $\mathbf{f} : \mathcal{R}^d \to \mathcal{R}^c$ is a bounded function on $\Omega$ where $\Omega \subseteq \mathcal{R}^d$ if all elements of $\mathbf{f}$ are bounded functions.

**Definition 2.3.7** (Function Lower Bound). Let $\Omega \subseteq \mathcal{R}^d$ and $f : \Omega \to \mathcal{R}$. Assume there exists a real number $K$ such that: $f(\mathbf{x}) \geq K$ for all $\mathbf{x} \in \Omega$. The real number $K$ is a *lower bound for* $f$ *on* $\Omega$. □

**Definition 2.3.8** (Non-negative Function). A function $f : \Omega \to \mathcal{R}$ is a *non-negative function* if $f(\mathbf{x}) \geq 0$ for all $\mathbf{x} \in \Omega$. □

**Definition 2.3.9** (Strictly Positive Function). A function $f : \Omega \to \mathcal{R}$ is a *strictly positive function* if $f(\mathbf{x}) > 0$ for all $\mathbf{x} \in \Omega$. □

## Exercises

2.3-1. Let a function $f$ have two arguments where the first argument is a $d$-dimensional binary vector consisting of zeros and ones and the second argument is a $d$-dimensional vector of real numbers. Assume that the function $f$ returns the values of 1 and $-1$. This information can be represented compactly using the notation:

$$f : A \times B \rightarrow C.$$

Explicitly specify: A, B, and C.

2.3-2. Let $f : \mathcal{R} \times \mathcal{R} \rightarrow \mathcal{R}$. Let $g : \mathcal{R} \rightarrow \mathcal{R}$ be defined such that for all $x, y \in \mathcal{R}$: $g(y) = f(x, y)$. Now, if $f$ is the function $f(x, y) = 10x + \exp(y)$ for all $y \in \mathcal{R}$ and a particular $x \in \mathcal{R}$, then what is an explicit representation for $g$?

2.3-3. Let $V : \mathcal{R}^q \rightarrow \mathcal{R}$ be defined such that for all $\boldsymbol{\theta} \in \mathcal{R}^q$:

$$V(\boldsymbol{\theta}) = (y - \boldsymbol{\theta}^T \mathbf{s})^2$$

where the constant number $y \in \mathcal{R}$ and the constant vector $\mathbf{s} \in \mathcal{R}^q$. Is $V$ a bounded function on $\mathcal{R}^q$? Is $V$ a bounded function on $\{\boldsymbol{\theta} : |\boldsymbol{\theta}|^2 < 1000\}$? Does $V$ have a lower bound on $\mathcal{R}^q$? Does $V$ have a lower bound on $\{\boldsymbol{\theta} : |\boldsymbol{\theta}|^2 < 1000\}$? Is $V$ a non-negative function?

2.3-4. What is the preimage of $\{y : y < 1\}$ under the exponential function exp?

## 2.4 Metric Spaces

Consider a collection of patterns where each pattern corresponds to a $d$-dimensional vector of numbers. Such collections of patterns are called "vector spaces". For example, in an image processing application, an image might consist of 1,000 rows and 1,000 columns of pixels where the grey-level of a pixel is represented by a real number. A digital image could then be represented as a point in a $d$-dimensional real vector space where $d = 1,000^2 = 1,000,000$. A subset of nearby points in such a $d$-dimensional real vector space might correspond to possible representations that are similar but not exactly identical images.

As previously discussed, patterns of information and patterns of knowledge are naturally represented as lists of numbers or "vectors". Vectors, in turn, are typically defined with respect to a vector space. A vector space consists not only of a collection of vectors but additionally a distance measure for measuring the distance between two points in the vector space. And finally, a vector space is a special case of the more general concept of a metric space.

**Definition 2.4.1** (Metric Space). Let $E$ be a set. Let $\rho : E \times E \rightarrow [0, \infty)$. A *point* is an element of $E$. Assume: (i) $\rho(x, y) = 0$ if and only if $x = y$ for $x, y \in E$, (ii) $\rho(x, y) = \rho(y, x)$ for $x, y \in E$, and (iii) $\rho(x, z) \leq \rho(x, y) + \rho(y, z)$ for $x, y, z \in E$. Then, $\rho$ is called the *distance function* for the *metric space* $(E, \rho)$. □

When the identity of the distance function $\rho$ is implicitly clear from the context, the notation $E$ is used to refer to the metric space $(E, \rho)$. In addition, the terminology "subset of a metric space" refers to a subset of $E$ where $(E, \rho)$ is some metric space.

An important type of metric space is called the *Euclidean vector metric space* $(\mathcal{R}^d, \rho)$ where the *Euclidean vector distance function* $\rho : \mathcal{R}^d \times \mathcal{R}^d \to [0, \infty)$ is defined such that

$$\rho(\mathbf{x}, \mathbf{y}) = |\mathbf{x} - \mathbf{y}| = \left( \sum_{i=1}^{d} (x_i - y_i)^2 \right)^{1/2}$$

for all $\mathbf{x}, \mathbf{y} \in \mathcal{R}^d$. When the distance function $\rho$ for a metric space $(\mathcal{R}^d, \rho)$ has not been explicitly defined, then it will be assumed that $(\mathcal{R}^d, \rho)$ specifies a Euclidean vector metric space and $\rho$ is the Euclidean vector distance function.

Another example of a metric space is the *Hamming space* $(\{0,1\}^d, \rho)$ where the *Hamming distance function* $\rho : \{0,1\}^d \times \{0,1\}^d \to \{0, \ldots, d\}$ is defined such that for $\mathbf{x} = [x_1, \ldots, x_d] \in \{0,1\}^d$ and $\mathbf{y} = [y_1, \ldots, y_d] \in \{0,1\}^d$:

$$\rho(\mathbf{x}, \mathbf{y}) = \sum_{i=1}^{d} |x_i - y_i|.$$

That is, the Hamming distance function counts the number of unshared binary elements in the vectors $\mathbf{x}$ and $\mathbf{y}$.

Constructing an appropriate metric space for inference and learning is one of the most important topics in the field of machine learning. Assume that a learning machine observes an environmental event $\mathbf{x} \in \Omega$ where $\Omega$ is a $d$-dimensional Euclidean metric space. Let $\phi : \Omega \to \Gamma$ where $\Gamma \subseteq \mathcal{R}^m$. The function $\phi$ may be used to construct a new $m$-dimensional Euclidean metric space, $\Gamma$, such that each feature vector $\mathbf{y} = \phi(\mathbf{x})$ in $\Gamma$ is a recoded transformation of each feature vector $\mathbf{x} \in \Omega$.

The function $\phi$ is sometimes called a *preprocessing transformation*, *feature map*, or *recoding transformation*.

**Definition 2.4.2** (Ball). Let $(E, \rho)$ be a metric space. Let $x_0 \in E$. Let $r$ be a finite positive real number. The set $\{x \in E : \rho(x_0, x) < r\}$ is called an *open ball* centered at $\mathbf{x}_0$ with radius $r$. The set $\{x \in E : \rho(x_0, x) \leq r\}$ is called a *closed ball* centered at $\mathbf{x}_0$ with radius $r$. □

Note that an open ball centered at zero with radius $\delta$ in the Euclidean metric space $\mathcal{R}$ is specified by the set $\{x : -\delta < x < +\delta\}$.

However, an open ball centered at zero with radius $\delta$ in the Euclidean metric space $[0, 1]$ is specified by the set $\{x : 0 \leq x < +\delta\}$.

An open ball with radius $r$ centered at a point $\mathbf{x}$ in a metric space is called the *neighborhood* of the point $\mathbf{x}$. The *size* of the neighborhood is the radius $r$.

**Definition 2.4.3** (Open Set). A subset $\Omega$ of a metric space is *open* if for each $\mathbf{x} \in \Omega$, $\Omega$ contains some open ball centered at $\mathbf{x}$. □

For example, the set $(0, 1)$ is an example of an open set in $\mathcal{R}$, while $\mathcal{R}^d$ is an open set in $\mathcal{R}^d$. The set $[0, 1)$, however, is not an open set in $\mathcal{R}$ since an open ball centered at 0 cannot be constructed that contains only elements of $[0, 1)$. The empty set is also an open set in $\mathcal{R}$.

Similarly, the set $[0, 1]$ is also not an open set in $\mathcal{R}$ since an open ball centered at either 0 or 1 contains elements which are in $[0, 1]$ but are also in the intersection of $\mathcal{R}$ and the complement of $[0, 1]$. However, the set $[0, 1]$ is an open set in the metric space $[0, 1]$! To see this, note that for each $x_0$ in $[0, 1]$ the set

$$\{x \in [0, 1] : |x - x_0| < \delta\}$$

is a subset of $[0, 1]$ for a positive $\delta$.

**Definition 2.4.4** (Closed Set)**.** A subset $\Omega$ of a metric space $(E, \rho)$ is a *closed set* if the set of all points in $E$ which are not in $\Omega$ is an open set. □

The set $[0, 1]$ is an example of a closed set in $\mathcal{R}$. Also note that $\mathcal{R}^d$ is a closed set in $\mathcal{R}^d$. It is important to note that some sets are simultaneously neither open or closed. Thus, the set $[0, 1)$ is not an open set in $\mathcal{R}$, and the set $[0, 1)$ is not a closed set in $\mathcal{R}$. Similarly, some sets may be simultaneously open and closed such as the set $\mathcal{R}^d$ in $\mathcal{R}^d$. Finally, note that every finite set of points in $\mathcal{R}^d$ is a closed set. A closed ball is a closed set but an open ball is not a closed set.

**Definition 2.4.5** (Set Closure)**.** Let $\Omega$ be a subset of a metric space $(E, \rho)$. The *closure* of $\Omega$, $\bar{\Omega}$, in $E$ is the unique smallest closed set in $E$ containing $\Omega$. □

Also $\Omega$ is a closed set in $E$ if and only if $\Omega = \bar{\Omega}$. Thus, if $\Omega = [0, 1)$, then the closure of $\Omega$ in $\mathcal{R}$, $\bar{\Omega}$, would be given by the formula: $\bar{\Omega} = [0, 1]$.

**Definition 2.4.6** (Set Boundary)**.** Let $\Omega$ be a subset of a metric space. The *boundary* of $\Omega$ is the intersection of the closure of $\Omega$ and the closure of the complement of $\Omega$. □

Thus, the boundary for the set $[0, 3)$ would be $\{0, 3\}$. Geometrically, a closed set in $\mathcal{R}^d$ is a set which contains its boundary.

**Definition 2.4.7** (Set Interior)**.** Let $\Omega \subseteq \mathcal{R}^d$. Let $\mathbf{x} \in \Omega$. If there exists some open ball $B$ centered at $\mathbf{x}$ such that $B \subset \Omega$, then $\mathbf{x}$ is in the *interior* of $\Omega$. □

The notation *int* is defined such that *int* $\Omega$ is the set of points in the interior of $\Omega$.

**Definition 2.4.8** (Bounded Set)**.** If a set $\Omega$ in a metric space is contained in an open ball or a closed ball, then $\Omega$ is a *bounded set.* □

The set $\{x : |x| < 1\}$ is bounded. The set of real numbers is not bounded. An open ball is bounded. A closed ball is bounded.

**Definition 2.4.9** (Connected Set)**.** Let $\Omega$ be a set in a metric space. If $\Omega$ can be represented as the union of two nonempty disjoint sets $\Omega_1$ and $\Omega_2$ such that: $\bar{\Omega}_1 \cap \Omega_2 = \emptyset$ and $\Omega_1 \cap \bar{\Omega}_2 = \emptyset$, then the set $\Omega$ is called a *disconnected set.* If the set $\Omega$ is not a disconnected set, then the set $\Omega$ is called a *connected set.* □

Let $A \equiv [0, 1]$. Let $B \equiv (1, 2]$. Let $C = [100]$. Each of the sets $A$, $B$, and $C$ is connected. Each of the sets $A \cup B$, $A \cup C$, $B \cup C$ is disconnected. The set $\mathbb{N}$ is a disconnected set. The set $\mathcal{R}^d$ is a connected set.

**Definition 2.4.10** (Hyperrectangle)**.** The set

$$\{\mathbf{x} = [x_1, \dots, x_d] \in \mathcal{R}^d : -\infty < v_i < x_i < m_i < \infty, i = 1, \dots, d\}$$

is called an *open hyperrectangle.* The closure of an open hyperrectangle is called a *closed hyperrectangle.* The *volume* of either an open hyperrectangle or a closed hyperrectangle is: $\prod_{i=1}^{d}(m_i - v_i)$. □

**Definition 2.4.11** (Lower (Upper) Bound)**.** Let $\Omega \subseteq \mathcal{R}$. Assume there exists a finite real number $K$ such that: $x \geq K$ for all $x \in \Omega$. The real number $K$ is a *lower (upper) bound for the set* $\Omega$. If $\Omega$ has a lower (upper) bound, then $\Omega$ is said to be *bounded from below (above).* □

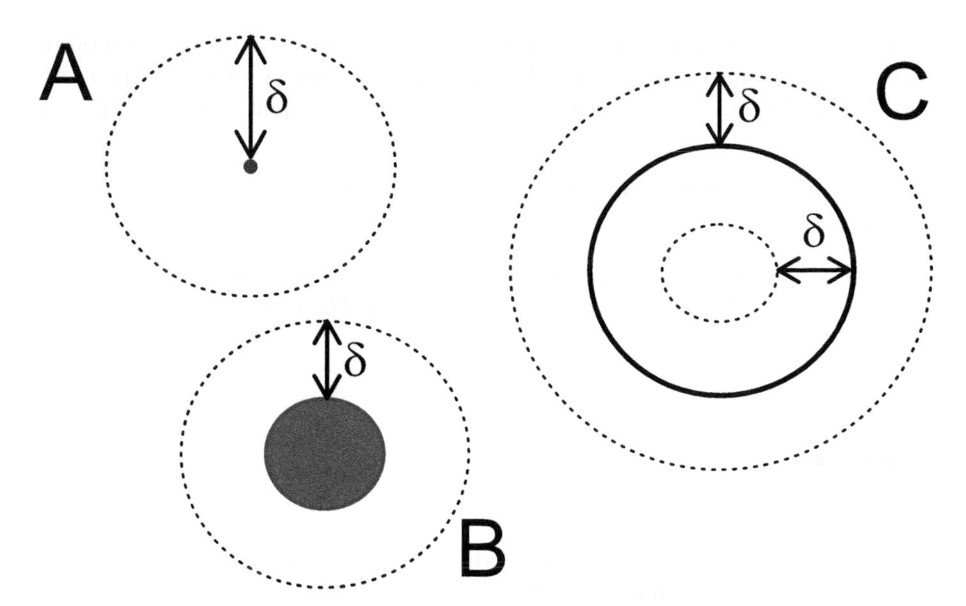


**FIGURE 2.7**
**Illustrating the concept of a $\delta$-neighborhood of a set in a Euclidean vector space.** Example (A) illustrates a $\delta$-neighborhood of a set containing a single point. Example (B) illustrates a $\delta$-neighborhood of a ball. Example (C) illustrates a $\delta$-neighborhood of the boundary of a ball.

The notation $a = \inf S$ means that $a$ is the greatest lower bound for the set $S$. For example, if $S$ has a smallest element, then $a = \inf S$ is a smallest element of $S$. If $S \equiv \{1, 1/2, 1/4, 1/8, 1/16, \ldots\}$, then $\inf S = 0$ but zero is not an element of $S$.

The notation $a = \sup S$ means that $a$ is the least upper bound of the set $S$. For example, if $S$ has a largest element, then $a = \sup S$ is a largest element of $S$.

**Definition 2.4.12** (Distance to a Set)**.** Let $(E, \rho)$ be a metric space. Let $\mathbf{x} \in E$. Let $Q$ be the power set of $E$. Let $\Gamma$ be a nonempty subset of $E$. The *distance* $d : E \times Q \to [0, \infty)$ from point $\mathbf{x} \in E$ to set $\Gamma$ is defined such that:

$$d(\mathbf{x}, \Gamma) = \inf\{\rho(\mathbf{x}, \mathbf{y}) \in [0, \infty) : \mathbf{y} \in \Gamma\}. \tag{2.1}$$

□

The short-hand notation

$$d(\mathbf{x}, \Gamma) \equiv \inf_{\mathbf{y} \in \Gamma} \rho(\mathbf{x}, \mathbf{y})$$

is used to denote $d(\mathbf{x}, \Gamma) \equiv \inf\{\rho(\mathbf{x}, \mathbf{y}) \in [0, \infty) : \mathbf{y} \in \Gamma\}$. Less formally, the distance from a point $\mathbf{x}$ to a set $\Gamma$ in a metric space is calculated by computing the distance of $\mathbf{x}$ to a point in $\Gamma$ which is the closest to $\mathbf{x}$ according to the metric space distance function.

**Definition 2.4.13** (Neighborhood of a Set)**.** Let $(E, \rho)$ be a metric space. Let $\Gamma$ be a non-empty subset of $E$. The *neighborhood*, $\mathcal{N}_\Gamma$, of the set $\Gamma$ is defined such that:

$$\mathcal{N}_\Gamma \equiv \{\mathbf{x} \in E : d(\mathbf{x}, \Gamma) < \delta\}$$

where $d(\mathbf{x}, \Gamma) \equiv \inf\{\rho(\mathbf{x}, \mathbf{y}) \in [0, \infty) : \mathbf{y} \in \Gamma\}$ and the strictly positive real number $\delta$ is the *size* of the neighborhood. If $\Gamma = \{\mathbf{x}^*\}$, then $\mathcal{N}_\Gamma$ is called the *neighborhood*, $\mathcal{N}_{\mathbf{x}^*}$, of $\mathbf{x}^*$. □

Figure 2.7 shows a geometric interpretation of the concept of the neighborhood of a set in a Euclidean vector space. Note that when $\Gamma$ contains a single point $\mathbf{y}$, then the open neighborhood of $\{\mathbf{y}\}$ in a Euclidean vector space is simply an open ball centered at $\mathbf{y}$ with radius $\delta$.

**Definition 2.4.14** (Convex Set). Let $\Omega \subseteq \mathcal{R}^d$. If for every $\mathbf{x}, \mathbf{y} \in \Omega$ and for every $\alpha \in [0, 1]$ : $\mathbf{x}\alpha + (1 - \alpha)\mathbf{y} \in \Omega$, then $\Omega$ is a *convex set.* □

Less formally, a convex region is a subset, $\Omega$, of $\mathcal{R}^d$ such that a line segment connecting any two points within $\Omega$ always remains entirely within $\Omega$. Thus, an open ball in $\mathcal{R}^d$ is always a convex set.

## Exercises

2.4-1. For each of the following sets in $\mathcal{R}^d$, indicate whether the set is: (i) an open set, (ii) a closed set, (iii) a bounded set, and (iv) a convex set.
(a) $[0, 1]^d$, (b) $(0, 1)^d$, (c), $\{0, 1\}^d$, (d) $\mathcal{R}^d$, (e) $\{\}$, (f) a closed ball of finite radius, and (g) an open hyperrectangle.

2.4-2. Draw a picture of a beer mug on a piece of paper. Then, draw an open $\delta$-neighborhood for the beer mug.

2.4-3. Provide an explicit expression for the set of points which are in the open $\delta$-neighborhood of the set

$$S \equiv \{\boldsymbol{\theta} \equiv [\theta_1, \ldots, \theta_q] \in \mathcal{R}^q : 0 \leq \theta_j \leq 1, j = 1, \ldots, q\}.$$

2.4-4. Formally show that $(\mathcal{R}^d, \rho)$ is a metric space where $\rho : \mathcal{R}^d \times \mathcal{R}^d \to [0, \infty)$ is defined such that $\rho(\mathbf{x}, \mathbf{y}) = |\mathbf{x} - \mathbf{y}|$.

2.4-5. Formally show that $(\mathcal{R}^d, \rho)$ is a metric space where $\rho : \mathcal{R}^d \times \mathcal{R}^d \to [0, \infty)$ is defined such that $\rho(\mathbf{x}, \mathbf{y}) = |\mathbf{x} - \mathbf{y}|^4$.

2.4-6. Formally show that $(\{0, 1\}^d, \rho)$ is a metric space where $\rho : (\{0, 1\}^d \times (\{0, 1\}^d \to [0, \infty)$ is defined such that $\rho(\mathbf{x}, \mathbf{y}) = \sum_{i=1}^{d} |x_i - y_i|$ for all $\mathbf{x} \equiv [x_1, \ldots, x_d] \in (\{0, 1\}^d$ and for all $\mathbf{y} \equiv [y_1, \ldots, y_d] \in (\{0, 1\}^d$.

2.4-7. A probability mass function on a finite sample space consisting of $d$ elements may be represented as a point in a set $S$ where

$$S \equiv \left\{ \mathbf{p} \equiv [p_1, \ldots, p_d] \in (0, 1)^d : \sum_{k=1}^{d} p_k = 1 \right\}.$$

Define the *Kullback-Leibler Divergence* $D(\cdot||\cdot) : S \times S \to \mathcal{R}$ such that

$$D(\mathbf{p}||\mathbf{q}) = -\sum_{i=1}^{d} p_i \log(q_i/p_i).$$

Show that $(S, D)$ is *not* a metric space.

2.4-8. Use a regression model to solve a supervised learning problem and evaluate performance of the model on both a training data set and a test data set. Now, recode some of the predictors of the model and examine how the performance of the model changes on both the training data and test data. Also discuss any differences associated with the amount of time required for learning in both cases.

2.4-9. Download a data set from the UCI Machine Learning Data Repository (*http://archive.ics.uci.edu/ml/*). Evaluate the performance of a linear regression model or some other machine learning algorithm on two versions of the data set which has been recoded in different ways.

2.4-10. Experiment with a clustering algorithm as defined in Example 1.6.4 where the similarity distance function $D$ is defined using the ellipsoidal cluster dissimilarity function $D(\mathbf{x}, \mathbf{y}) = (\mathbf{x}-\mathbf{y})^T\mathbf{M}(\mathbf{x}-\mathbf{y})$ where $\mathbf{M}$ is a positive definite matrix. Experiment with different data sets and different choices of the parameter vector $\boldsymbol{\theta}$ including the $K$-means choice where $\boldsymbol{\theta}$ is a vector of $K$ ones.

## 2.5 Further Readings

### Real Analysis

This chapter covered selected topics from elementary real analysis which will be used in later chapters. Goldberg (1964), Rosenlicht (1968), Kolmogorov and Fomin (1970), and Wade (1995) provide different systematic introductions to the real analysis topics discussed in this chapter. The presentation by Rosenlicht (1968) is particularly clear and concise. Discussions of the extended real number system can be found in Bartle (1966) and Wade (1995).

### Graph Theory

A discussion of graph theory relevant to machine learning can be found in Koller et al. (2009) and Lauritzen (1996).

### Fuzzy Logic and Fuzzy Relations

Readers interested in learning more about fuzzy logic, fuzzy set theory, and fuzzy relations should see Klir and Yuan (1995) and Klir and Folger (1988). McNeill and Freiburger (1994) provide an informal non-technical historical introduction.

# 3

# *Formal Machine Learning Algorithms*

**Learning Objectives**

- Specify machine environments.
- Specify machines as dynamical systems.
- Specify machines as intelligent rational decision makers.

The representation of a machine learning algorithm as a formal structure is a prerequisite to any mathematical analysis. Before a machine learning algorithm can be analyzed, it must be formally defined. Chapter 1 provided an informal discussion of examples of machine learning algorithms without providing a formal definition of such algorithms. The goal in this chapter is to attempt to provide a formal framework for specifying and defining machine learning algorithms.

Any computation that can be simulated on a digital computer can be simulated on a Turing machine. A Turing machine is a special type of computer that updates states in a discrete state space in discrete-time. The goal of this chapter is to provide a formal definition of a large class of machine learning algorithms that include Turing machine computers as an important special case. These computing machines operate in either continuous state spaces and possibly even continuous-time. Examples of computing machines that cannot be exactly represented as Turing machines are common in the machine learning literature. For example, gradient descent algorithms generate updates to a learning machine's parameter values in an uncountable infinite state space while a Turing machine (like a digital computer) assumes the state space is finite.

The chapter begins with a discussion regarding how to formally model both discrete-time and continuous-time learning environments for machine learning algorithms. Next, tools from dynamical systems theory are introduced to model the dynamical behavior of a large class of machine learning algorithms including algorithms which can only approximately be simulated on digital computers. The computational goal of a machine learning algorithm is formally defined in terms of the inference and learning machine's preferences. And finally, a linkage between the behavior of the machine learning algorithm and its preferences is provided that interprets the machine learning algorithm as a machine that makes rational decisions based upon its intrinsic preferences.

## 3.1 Environment Models

### 3.1.1 Time Models

**Definition 3.1.1** (Time Indices). A set of *time indices* is a subset of the real numbers. If $T \equiv \mathbb{N}$, then $T$ is a *discrete-time environment set of time indices.* If $T \equiv [0, \infty)$, then $T$ is a *continuous-time environment set of time indices.* □

When $T$ is the set of non-negative integers, this corresponds to the case where the environment is modeled as a *discrete-time environment.* An example of a discrete-time environment might be the environment of a machine that plays chess or tic-tac-toe. In this case, each time slice corresponds to a particular move in the game. Another example of a discrete-time environment might be a machine that is presented with a sequence of digital images and must learn to decide whether or not a particular digital image contains a particular object or face.

When $T$ is the set of non-negative real numbers this corresponds to the case where the environment is modeled as a *continuous-time environment.* An example of a continuous-time environment might be the environment of a machine that is designed to control a satellite. In this situation, real-time information must be continually processed and updated by the machine. Another example of a continuous-time environment might be the environment for a control system designed to increase or decrease the amount of medication provided to a patient based upon real-time physical measurements of the patient. Speech recognition machines that must process raw acoustic signals are also typically embedded within continuous-time environments.

Both discrete-time and continuous-time systems have their advantages and disadvantages for system modeling problems. Some systems are simply naturally modeled as either discrete-time or continuous-time systems. For example, an aircraft control system or speech processing system might be naturally specified as a continuous-time dynamical system, while a probabilistic reasoning system might be naturally specified as a discrete-time dynamical system.

Still, the appropriateness of a modeling methodology will dramatically vary depending upon what abstraction of reality is relevant to the modeling task. For example, discrete-time dynamical system models are more appropriate for modeling problems in quantum mechanics, and continuous-time dynamical system models may be more appropriate in the development of perceptual models of visual and speech information processing or motor control. This choice of representation is important because, in general, discrete-time and continuous-time representations of the same abstraction of reality will differ in their behavioral properties.

The ability to mathematically analyze a particular type of dynamical system may also influence whether a discrete-time or continuous-time representation is chosen. Consider a discrete-time dynamical system and continuous-time dynamical system that are both intended to model the same abstraction of reality. In some cases, the dynamical behavior of the discrete-time version of the model will be easier to mathematical analyze than the continuous-time version of the model. In other cases, the dynamical behavior of the continuous-time version of the model will be easier to analyze than the discrete-time version. Furthermore, as mentioned earlier, the asymptotic behaviors could be different which means that theoretical results regarding the behavior of a discrete-time dynamical system are not generally applicable to understanding the behavior of its continuous-time dynamical system analogue (and vice versa).

**Definition 3.1.2** (Time Interval). Let $T$ be a set of time indices. Let $t_0 \in T$ and $t_F \in T$. A *time interval* $[t_0, t_F)$ of $T$ is defined as the set $\{t \in T : t_0 \leq t < t_F\}$. □

For a discrete-time environment, a time interval specifies a sequence of time points where the first time index is $t_0$ and the last time index is $t_F - 1$. For a continuous-time environment, a time interval is an interval $[t_0, t_F)$ on the real number line that begins immediately at time index $t_0$ and ends just before time index $t_F$.

### 3.1.2 Event Environments

An event timeline function can be imagined as a type of world time-line which stretches from the beginning of time into the infinite future and specifies the order in which events in the learning machine's external environment are experienced by the learning machine.

**Definition 3.1.3** (Event Timeline Function). Let $T$ be a set of time indices. Let $\Omega_E \subseteq \mathcal{R}^e$ be a set of *events*. A function $\boldsymbol{\xi} : T \rightarrow \Omega_E$ is called an *event timeline function* generated from $T$ and $\Omega_E$. □

The $e$-dimensional real vector $\boldsymbol{\xi}(t)$ is an event in $\Omega_E$ that occurred at time index $t$. For a machine that is playing the game of tic-tac-toe, the notation $\boldsymbol{\xi}(k)$ might specify the board state for the $k$th move in the game. For example, in the game of tic-tac-toe the board state at move $k$ can be represented as a 9-dimensional vector $\boldsymbol{\xi}(k)$ where the jth element of $\boldsymbol{\xi}(k)$ is set equal to one if an X is present at board location $j$, set equal to zero if board location $j$ is empty, and set equal to $-1$ if an O is present at board location $j$, $j = 1, \ldots, 9$. The event timeline function $\boldsymbol{\xi}$ in this discrete-time case is a sequence of nine-dimensional vectors corresponding to a sequence of tic-tac-toe board states. The function $\boldsymbol{\xi}$ may be formally defined as:

$$\boldsymbol{\xi} : \mathbb{N} \rightarrow \{-1, 0, +1\}^9.$$

Or equivalently, $\boldsymbol{\xi}(0), \boldsymbol{\xi}(1), \ldots$ where $\boldsymbol{\xi}(t) \in \{-1, 0, +1\}^9$.

On the other hand, for a machine that is processing an auditory speech signal, the notation $\boldsymbol{\xi}(t)$ might be a real-valued electrical signal specifying the magnitude of electrical current flow at time $t$ in response to an acoustic event at time $t$. For example, $\boldsymbol{\xi} : [0, \infty) \rightarrow \mathcal{R}$ would be a possible representation for the event timeline function in this case.

The notation $\boldsymbol{\xi}_{t_a, t_b}$ is defined as the restriction of an event timeline function $\boldsymbol{\xi} : T \rightarrow \Omega_E$ to the domain $[t_a, t_b)$ when $T = [0, \infty)$. When $T = \mathbb{N}$, the notation $\boldsymbol{\xi}_{t_a, t_b}$ denotes the restriction of the event timeline function $\boldsymbol{\xi}$ to the domain $\{t_a, t_a + 1, t_a + 2, \ldots, t_b - 1\}$.

**Definition 3.1.4** (Event Environment). Let $T \equiv \mathbb{N}$ or $T \equiv [0, \infty)$. Let $E$ be a non-empty set of event timeline functions with common domain $T$ and common range $\Omega_E$. If $T \equiv \mathbb{N}$, then $E$ is called a *discrete-time event environment.* If $T \equiv [0, \infty)$, then $E$ is called a *continuous-time event environment.* □

## Exercises

3.1-1. Show that if $T$ is a discrete-time environment set of time indices, then a time interval $[1, 5)$ in $T$ may be represented as the ordered set $\{1, 2, 3, 4\}$.

3.1-2. A gambler flips a coin 1000 times and each time the coin comes up either "heads" or "tails". Let the time index $t$ correspond to coin flip number $t$ and define the event timeline function $\boldsymbol{\xi} : \{0, \ldots, 1000\} \rightarrow \{1, 0\}$ where $\boldsymbol{\xi}(t) = 1$ indicates the coin comes up heads on coin flip $t$ and $\boldsymbol{\xi}(t) = 0$ indicates the coin comes up tails on coin flip $t$. Construct an event environment that models these observations.

3.1-3. A race car driver observes the speed and RPM of his race car in continuous-time. Assume the velocity of the car ranges from zero to 180 miles per hour and the RPM of the car ranges from zero to 20,000 RPM. Construct an event environment that models observations in this situation.

## 3.2 Machine Models

### 3.2.1 Dynamical Systems

A key concept associated with any given system, is the concept of a "system state". Typically, the system state is a collection of real-valued variables called "state variables". Some collections of state variables are used to represent the feature vectors discussed in Section 1.2.1. Other collections of state variables might correspond to inferences, others might specify particular system outputs, and others might be used to represent system states that are never directly observable. A learning machine would also include state variables corresponding to the learning machine's parameters which are adjusted during the learning process.

**Definition 3.2.1** (Dynamical System). Let $T \equiv \mathbb{N}$ or $T \equiv [0, \infty)$. Let the *system state space* $\Omega \subseteq \mathcal{R}^d$. Let $E$ be an event environment generated from the time indices $T$ and events $\Omega_E$. Let

$$\boldsymbol{\Psi} : \Omega \times T \times T \times E \rightarrow \Omega.$$

- *Property 1: Boundary Conditions.* For all $\mathbf{x} \in \Omega$, for all $t_0 \in T$, and for all $\boldsymbol{\xi} \in E$:

$$\boldsymbol{\Psi}(\mathbf{x}, t_0, t_0, \boldsymbol{\xi}) = \mathbf{x}.$$

- *Property 2: Consistent Composition.* For all $t_a, t_b, t_c \in T$ such that $t_a < t_b < t_c$:

$$\boldsymbol{\Psi}(\mathbf{x}_a, t_a, t_c, \boldsymbol{\xi}) = \boldsymbol{\Psi}\left(\boldsymbol{\Psi}(\mathbf{x}_a, t_a, t_b, \boldsymbol{\xi}), t_b, t_c, \boldsymbol{\xi}\right)$$

  for all $\mathbf{x}_a \in \Omega$ and for all $\boldsymbol{\xi} \in E$.

- *Property 3: Causal Isolation.* If $\boldsymbol{\xi}, \boldsymbol{\zeta} \in E$ such that $\boldsymbol{\xi}_{t_a, t_b} = \boldsymbol{\zeta}_{t_a, t_b}$, then:

$$\boldsymbol{\Psi}(\mathbf{x}_a, t_a, t_b, \boldsymbol{\xi}) = \boldsymbol{\Psi}(\mathbf{x}_a, t_a, t_b, \boldsymbol{\zeta}).$$

The function $\boldsymbol{\Psi}$ is called a *dynamical system* whose *final state*

$$\mathbf{x}_f \equiv \boldsymbol{\Psi}\left(\mathbf{x}_0, t_0, t_f, \boldsymbol{\xi}\right)$$

observed at *final time* $t_f$ is determined by an *initial state* $\mathbf{x}_0$ at *initial time* $t_0$ and event timeline function $\boldsymbol{\xi} \in E$. □

The basic idea of a dynamical system is illustrated in Figure 3.1. A dynamical system is a function that maps an initial system state vector $\mathbf{x}_0$, an initial time index $t_0$, a final time index $t_f$, an event timeline function $\boldsymbol{\xi}$, and returns a final state $\mathbf{x}_f$. The properties

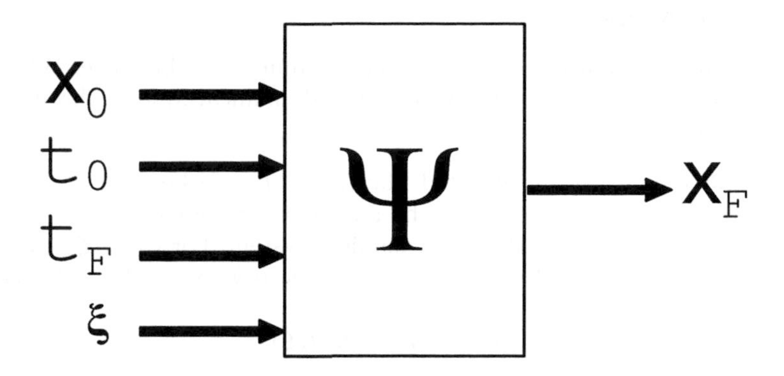


**FIGURE 3.1**
**The concept of a dynamical system.** A dynamical system generates a final state $\mathbf{x}_F$ given four inputs: (1) an initial state $\mathbf{x}_0$, (2) an initial time $t_0$, (3) a final time $t_F$, and (4) an event timeline function $\boldsymbol{\xi}$ which specifies the events experienced by the machine.

provided in the definition of a dynamical system are intended to specify certain features that the dynamical system must possess in order for the dynamical system to be physically realizable as a machine.

Property 1 in the definition of a dynamical system simply states that if the semantic interpretation of $t_f$ is to specify the system state at time $t_f$ and the semantic interpretation of $t_0$ is to specify the system state at time $t_0$, then logical consistency requires that the dynamical system function have the property that when $t_0 - t_f$ the initial and final system states are identical.

Property 2 in the definition of a dynamical system states that if the dynamical system machine is turned on for a time period from time $t_a$ to time $t_b$ and then the resulting final state is used as an initial condition for the machine running from time $t_b$ to time $t_c$, the resulting final state of the machine would be identical to the final state obtained from simply running the machine from time $t_a$ to time $t_c$. This property captures the idea that the current system state, rather than the past history of system states, governs the future behavior of the dynamical system machine.

And finally, Property 3 states that the behavior of a dynamical system over a time interval is only influenced by the values of the event timeline function on that time interval. Values of the event timeline function outside that time interval do not influence the dynamical system's behavior. Or, in other words, knowledge of past events which occurred before the initial state are irrelevant and knowledge of future environmental events which will occur after the final state of the system's behavior are irrelevant.

**Definition 3.2.2** (Discrete-Time and Continuous-Time Dynamical Systems)**.** Let $T$ be a set of time indices. Let $\Omega \subseteq \mathcal{R}^d$ be a set of *system states*. Let $E$ be an event environment generated from the time indices $T$ and events $\Omega_E$. The dynamical system

$$\boldsymbol{\Psi} : \Omega \times T \times T \times E \to \Omega$$

is called a *discrete-time dynamical system* when $T \equiv \mathbb{N}$. The dynamical system $\boldsymbol{\Psi}$ is called a *continuous-time dynamical system* when $T \equiv [0, \infty)$. □

### 3.2.2 Iterated Maps

In practice, the above definition of a dynamical system is cumbersome. When possible, it is usually more convenient to implicitly specify a dynamical system through a system of difference equations.

**Definition 3.2.3** (Time-Varying Iterated Map)**.** Let $T \equiv \mathbb{N}$ be a set of time indices. Let $\Omega \subseteq \mathcal{R}^d$ be a set of system states. Let $E$ be an event environment generated from $T$ and $\Omega_E$. Let $\mathbf{\Psi} : \Omega \times T \times T \times E \to \Omega$ be a dynamical system. For every $\boldsymbol{\xi} \in E$, every *initial state* $\mathbf{x}_0 \in \Omega$, and every *initial time* $t_0 \in T$, define the *orbit* $\mathbf{x}(\cdot) : T \to \Omega$ such that for all $t \in T$ such that $t \geq t_0$:

$$\mathbf{x}(t) \equiv \mathbf{\Psi}(\mathbf{x}_0, t_0, t, \boldsymbol{\xi}).$$

Assume a function $\mathbf{f} : \Omega \times T \times \Omega_e \to \Omega$ exists such that for all $\boldsymbol{\xi} \in E$ and for all $t \in T$:

$$\mathbf{x}(t+1) = \mathbf{f}(\mathbf{x}(t), t, \boldsymbol{\xi}(t)). \tag{3.1}$$

The function $\mathbf{f}$ is called a *time-varying iterated map* for $\mathbf{\Psi}$.

Note that (3.1) asserts that given the current state of the dynamical system at time $t$ denoted by $\mathbf{x}(t)$, the current time $t$, and the current environmental event (e.g., incoming information, a training pattern, or a disturbance) $\boldsymbol{\xi}(t)$ at time $t$ it is possible to uniquely compute the next system state $\mathbf{x}(t+1)$.

Also note that the time-varying property of the time-varying iterated map is due to two distinct sources. The second argument in $\mathbf{f}$ as defined in (3.1) is the time index $t$ that allows the internal mechanism of the machine to vary in time. On the other hand, the third argument in $\mathbf{f}$ specifies an external time-varying stimulus $\boldsymbol{\xi}(t)$ which has the potential to influence the system's dynamics. Thus, both internal and external time-varying sources are explicitly represented in the definition of a time-varying iterated map presented here.

**Example 3.2.1** (Batch Gradient Descent Iterated Map)**.** Let $\mathbf{x}_i \equiv (\mathbf{s}_i, y_i)$ for $i = 1, \ldots, n$. Let $\mathcal{D}_n \equiv \{\mathbf{x}_1 \ldots, \mathbf{x}_n\} \in \mathcal{R}^{dxn}$ be a data set. Let $\hat{\ell}_n : \mathcal{R}^q \to \mathcal{R}$ be an empirical risk function defined with respect to $\mathcal{D}_n$. Suppose a learning machine uses $\mathcal{D}_n$ to update its parameter vector $\boldsymbol{\theta}(t)$ at iteration $t$ using the formula:

$$\boldsymbol{\theta}(t+1) = \boldsymbol{\theta}(t) - \gamma(t)\frac{d\hat{\ell}_n(\boldsymbol{\theta}(t))}{d\boldsymbol{\theta}} \tag{3.2}$$

where $d\hat{\ell}_n(\boldsymbol{\theta})/d\boldsymbol{\theta}$ denotes the derivative of the empirical risk function for learning $\hat{\ell}_n(\boldsymbol{\theta})$ with respect to $\boldsymbol{\theta}$ and then evaluated at $\boldsymbol{\theta}(t)$. Assume that during the learning process, the data set $\mathcal{D}_n$ is always available to the learning machine so that the event timeline function

$$\xi(t) = \mathcal{D}_n$$

for all $t \in \mathbb{N}$. The learning dynamics of this learning algorithm can be represented as a discrete-time dynamical system specified by the time-varying iterated map

$$\mathbf{f} : \mathcal{R}^q \times \mathbb{N} \times E \to \mathcal{R}^q$$

defined such that for all $t \in \mathbb{N}$:

$$\mathbf{f}(\boldsymbol{\theta}, t, \boldsymbol{\xi}(t)) = \boldsymbol{\theta} - \gamma(t)\frac{d\hat{\ell}_n}{d\boldsymbol{\theta}}.$$

△

**Example 3.2.2** (Adaptive Gradient Descent Iterated Map). Let $p_e(\mathbf{s}, y)$ denote the probability that the environment generates an input pattern $\mathbf{s}$ and desired response $y$. Let $c((\mathbf{s}, y), \boldsymbol{\theta})$ denote the loss incurred by the learning machine in the situation where it observes $(\mathbf{s}, y)$ when its state of knowledge is $\boldsymbol{\theta}$. The expected loss (risk) incurred by the learning machine is given by the formula:

$$\ell(\boldsymbol{\theta}) = \sum_{(\mathbf{s},y)} c([\mathbf{s}, y], \boldsymbol{\theta}) \, p_e(\mathbf{s}, y).$$

Assume that during the learning process, the learning machine observes a sequence of environmental events specified by the event timeline function:

$$\xi(t) = [\mathbf{s}(t), y(t)]$$

for all $t \in \mathbb{N}$. Suppose an adaptive learning machine updates its $q$-dimensional parameter vector $\boldsymbol{\theta}(t)$ at iteration $t$ when it observes at time $t$ a $d$-dimensional input pattern vector $\mathbf{s}(t)$ and desired response $y(t)$ using the formula:

$$\boldsymbol{\theta}(t+1) = \boldsymbol{\theta}(t) - \gamma(t) \frac{d\left(c([\mathbf{s}(t), y(t)], \boldsymbol{\theta}(t)\right)}{d\boldsymbol{\theta}}$$

where the notation $dc([\mathbf{s}, y], \boldsymbol{\theta})/d\boldsymbol{\theta}$ denotes the derivative of the prediction error $c([\mathbf{s}, y], \boldsymbol{\theta})$ for training pattern $[\mathbf{s}, y]$ with respect to $\boldsymbol{\theta}$ and then evaluated at $\boldsymbol{\theta}(t)$.

The dynamics of this adaptive learning algorithm can be represented as a discrete-time iterated map

$$\mathbf{f} : \mathcal{R}^q \times \mathbb{N} \times E \to \mathcal{R}^q$$

defined such that for all $t \in \mathbb{N}$:

$$\mathbf{f}(\boldsymbol{\theta}, t, \xi(t)) = \boldsymbol{\theta} - \gamma(t) \frac{dc(\xi(t), \boldsymbol{\theta})}{d\boldsymbol{\theta}}.$$

△

When a discrete-time dynamical system is only state-dependent, it is convenient to define the following special type of iterated map.

**Definition 3.2.4** (Time-Invariant Iterated Map). Let $T \equiv \mathbb{N}$ be a set of time indices. Let $\Omega \subseteq \mathcal{R}^d$ be a set of system states. Let $\boldsymbol{\Psi} : \Omega \times T \to \Omega$ be a dynamical system. For every *initial state* $\mathbf{x}_0 \in \Omega$, define the *orbit* $\mathbf{x}(\cdot) : T \to \Omega$ such that for all $t \in T$:

$$\mathbf{x}(t) \equiv \boldsymbol{\Psi}(\mathbf{x}_0, t).$$

Assume a function $\mathbf{f} : \Omega \to \Omega$ exists such that:

$$\mathbf{x}(t+1) = \mathbf{f}(\mathbf{x}(t)) \tag{3.3}$$

for all $t \in T$. The function $\mathbf{f}$ is called a *time-invariant iterated map* for $\boldsymbol{\Psi}$.

**Example 3.2.3** (Turing Machine). A Turing machine is a simplified model of a real digital computer designed to explore issues of computation associated with real digital computers. In principle, every possible computation which can be accomplished using a real digital computer can also be accomplished using a Turing machine. A Turing machine is a device which has a finite set $Q$ of internal states and reads an infinitely long "tape" defined by a sequence of *symbols* where the symbol at position $k$ on the tape at time $t$ is denoted by $\xi(k, t)$ where $k$ is an integer and $t$ is a non-negative integer. The Turing machine's sensor is initially located with respect to a particular position $k$ on the tape. The Turing machine reads the symbol located at position $k$ on the tape, uses that symbol and its current internal state to both: (1) print a new symbol at position $k$ on the tape, and (2) relocate its sensor to either the left or right of its current position (corresponding to either tape position $k-1$ or $k+1$ respectively). The Turing machine's computation is completed when it reaches a *halt state*. Note that some computations can never be completed since a halt state may never be reached.

Formally, the Turing machine can be interpreted as an iterated map operating on a discrete-state space. In particular, let $Q$ be a finite set of *internal machine states*. Let $\Gamma$ be a finite set of *tape symbols*. Let $\mathcal{L}$ be the set of integers corresponding to the set of all possible sensor locations. Let $\Omega \equiv Q \times \mathcal{L} \times \Gamma$. Then, a Turing machine can be represented as a discrete-time dynamical system by the iterated map $\mathbf{f} : \Omega \rightarrow \Omega$ such that the new internal machine state $s(t+1)$, the new location of the machine's sensor $L(t+1)$, and the new symbol, $\xi(t+1)$ are computed from the previous internal machine state $s(t)$, the previous location of the machine's sensor $L(t)$, and the original symbol $\xi(t)$ using the formula:

$$\mathbf{x}(t+1) = \mathbf{f}(\mathbf{x}(t))$$

where $\mathbf{x}(t) \equiv [s(t), L(t), \xi(t)] \in \Omega$. △

### 3.2.3 Vector Fields

In the previous section, a notation for specifying a discrete-time dynamical system was introduced. In this section, an analogous notation for specifying a continuous-time dynamical system is developed as well.

Let $d\mathbf{x}/dt$ denote the derivative of the vector-valued function $\mathbf{x} : T \rightarrow \mathcal{R}^d$ whose $k$th element is $dx_k/dt$ where the $k$th element of $\mathbf{x}$ is the function $x_k : T \rightarrow \mathcal{R}$ for $k = 1, \ldots, d$.

**Definition 3.2.5** (Time-Varying Vector Field)**.** Let $T \equiv [0, \infty)$ be a set of time indices. Let $\Omega \subseteq \mathcal{R}^d$ be a set of system states. Let $E$ be an event environment generated from $T$ and $\Omega_E$. Let $\boldsymbol{\Psi} : \Omega \times T \times T \times E \rightarrow \Omega$ be a dynamical system. For every $\boldsymbol{\xi} \in E$, every *initial state* $\mathbf{x}_0 \in \Omega$, and every *initial time* $t_0 \in T$, define the *trajectory* $\mathbf{x} : T \rightarrow \Omega$ such that for all $t \in T$ such that $t \geq t_0$:

$$\mathbf{x}(t) \equiv \boldsymbol{\Psi}(\mathbf{x}_0, t_0, t, \boldsymbol{\xi}).$$

Assume a function $\mathbf{f} : \Omega \times T \times \Omega_E \rightarrow \Omega$ exists such that for all $\boldsymbol{\xi} \in E$ and for all $t \in T$:

$$\frac{d\mathbf{x}}{dt} = \mathbf{f}(\mathbf{x}(t), t, \boldsymbol{\xi}(t)) \tag{3.4}$$

on $T$. The function $\mathbf{f}$ is called a *time-varying vector field* for $\boldsymbol{\Psi}$. □

Note that (3.4) asserts that given the current state of the dynamical system at time $t$ denoted by $\mathbf{x}(t)$, the current time $t$, and the current environmental event (e.g., incoming information, a training pattern, or a disturbance) $\boldsymbol{\xi}(t)$ at time $t$ it is possible to uniquely compute the instantaneous change of the system state at time $t$.

**Example 3.2.4** (Batch Gradient Descent Time-Varying Vector Field)**.** Define a data set

$$\mathcal{D}_n \equiv \{\mathbf{x}_1 \ldots, \mathbf{x}_n\} \in \mathcal{R}^{d \times n}$$

where the $i$th training stimulus $\mathbf{x}_i \equiv (\mathbf{s}_i, y_i)$ for $i = 1, \ldots, n$. Let $\hat{\ell}_n(\boldsymbol{\theta})$ be an empirical risk function defined with respect to $\mathcal{D}_n$. Suppose a continuous-time supervised learning machine uses $\mathcal{D}_n$ to update its parameter vector $\boldsymbol{\theta}(t)$ at time $t$ using the formula:

$$d\boldsymbol{\theta}/dt = -\gamma(t)\frac{d\hat{\ell}_n}{d\boldsymbol{\theta}} \tag{3.5}$$

where $d\hat{\ell}_n/d\boldsymbol{\theta}$ denotes the derivative of the empirical risk function for learning $\hat{\ell}_n(\boldsymbol{\theta})$ with respect to $\boldsymbol{\theta}$ and then evaluated at $\boldsymbol{\theta}(t)$. Assume that during the learning process, the data set $\mathcal{D}_n$ is always available to the learning machine so that the event timeline function

$$\xi(t) = \mathcal{D}_n$$

for all $t \in [0, \infty)$. The learning dynamics of this supervised learning algorithm can be represented as a continuous-time dynamical system whose time-varying vector field $\mathbf{f}$ : $\Theta \times [0, \infty) \rightarrow \Theta$ is defined such that for all $\boldsymbol{\theta} \in \Theta$ and for all $t \in [0, \infty)$:

$$\mathbf{f}(\boldsymbol{\theta}, t) = -\gamma(t)\frac{d\hat{\ell}_n}{d\boldsymbol{\theta}}.$$

△

When a continuous-time dynamical system is only state-dependent, it is convenient to define the following special type of vector field.

**Definition 3.2.6** (Time-Invariant Vector Field)**.** Let $T \equiv [0, \infty)$ be a set of time indices. Let $\Omega \subseteq \mathcal{R}^d$ be a set of system states. Let $\boldsymbol{\Psi} : \Omega \times T \rightarrow \Omega$ be a dynamical system. For every *initial state* $\mathbf{x}_0 \in \Omega$, define the *trajectory* $\mathbf{x} : T \rightarrow \Omega$ such that for all $t \in T$:

$$\mathbf{x}(t) \equiv \boldsymbol{\Psi}(\mathbf{x}(0), t).$$

Assume a function $\mathbf{f} : \Omega \rightarrow \Omega$ exists such that:

$$\frac{d\mathbf{x}}{dt} = \mathbf{f}(\mathbf{x}) \tag{3.6}$$

on $T$. The function $\mathbf{f}$ is called a *time-invariant vector field* for $\boldsymbol{\Psi}$. □

**Example 3.2.5** (Batch Gradient Descent Time-Invariant Vector Field)**.** Note that if $\gamma(t) = \gamma_0$ where $\gamma_0$ is a positive real number for all $t \in [0, \infty)$ in (3.5), then the resulting dynamical system is formally specified by the batch gradient descent time-invariant vector field $\mathbf{f}$ : $\Theta \rightarrow \Theta$ defined such that for all $\boldsymbol{\theta} \in \Theta$:

$$\mathbf{f}(\boldsymbol{\theta}) = -\gamma_0\frac{d\hat{\ell}_n}{d\boldsymbol{\theta}}.$$

△

## Exercises

3.2-1.Let $\beta_0$ and $\beta_1$ be real numbers. What is the time-invariant iterated map implicitly specified by the difference equation:

$$x(t+1) = 1/[1 + \exp(-(\beta_1 x(t) + \beta_0))].$$

3.2-2.Let $\beta_0$ and $\beta_1$ be real numbers. What is the time-varying iterated map implicitly specified by the difference equation:

$$x(t+1) = 1/[1 + \exp(-(\beta_1 x(t) + \beta_0))] + \xi(t)$$

where $\xi(t)$ is a real-valued disturbance (possibly a random disturbance) from the environment. What is the event environment for the dynamical system?

3.2-3. Verify that the solution to the differential equation $dx/dt = -x$ with initial condition $x(0) \equiv x_0$ is given by $x(t) = x_0 \exp(-(t - t_0))$. Specify a dynamical system $\boldsymbol{\Psi}$ that generates a state $x(t)$ from an initial state $x_0$, an initial time $t_0$, and a final time $t$. What is the time-invariant vector field for $\boldsymbol{\Psi}$?

3.2-4. Specify a time-varying iterated map for the stock market prediction machine in Equations (1.5) and (1.6).

3.2-5. Specify a time-varying iterated map for the nonlinear regression gradient descent learning rule in (1.8).

3.2-6. Let $\gamma(t)$ be a positive number called the stepsize at iteration $t$ and $0 < \delta < 1$. Consider an adaptive learning algorithm which observes at iteration $t$ an input pattern vector $\mathbf{s}(t)$ and desired scalar response $y(t)$ and then updates the current parameter estimates denoted by the parameter vector $\boldsymbol{\theta}(t)$ according to the formula:

$$\boldsymbol{\theta}(t+1) = \boldsymbol{\theta}(t)(1-\delta) + \gamma(t)\left(y(t) - \boldsymbol{\theta}(t)^T \mathbf{s}(t)\right)\mathbf{s}(t)$$

Specify a time-varying iterated map for this learning rule.

## 3.3 Intelligent Machine Models

This section proposes necessary conditions for ensuring that a dynamical systems machine model implements a system of rational decision making for the purpose of selecting an appropriate output response. In addition, these same conditions are also relevant for ensuring that a dynamical systems machine model which updates the parameters of a learning machine is designed to select a parameter vector using rules of rational decision making as well.

The concept of a preference may be formalized using a particular type of relation called the "preference relation".

**Definition 3.3.1** (Preference Relation). Let $\mathcal{F}$ be a set. Let $\preceq \subseteq \mathcal{F} \times \mathcal{F}$. The relation $\preceq$ is called a *preference relation* $\preceq$ on $\mathcal{F}$ provided that for all $(x, y) \in \preceq$: $x \preceq y$ has the semantic interpretation that: "$y$ is at least as preferable as $x$".

The notation $x \sim y$ means that $x \preceq y$ and $y \preceq x$ (i.e., x and y are equally preferable with respect to the preference relation $\preceq$).

Many commonly used learning machines can be interpreted as minimizing a scalar-valued "risk" function. The following definition shows how minimizing a risk function is mathematically equivalent to searching for an element in the domain of the risk function which corresponds to the "most preferable" element in the function's domain where the concept of "preference" is formally defined by a preference relation. Thus, a formal linkage is established between the concept of optimization and the concept of decision making using a preference relation.

**Definition 3.3.2** (Utility Function Preference Relation). Let $\preceq$ be a preference relation on $\mathcal{F}$. Assume $\mathcal{U} : \mathcal{F} \to \mathcal{R}$ has the property that for all $x \in \mathcal{F}$ and for all $y \in \mathcal{F}$: $\mathcal{U}(x) \leq \mathcal{U}(y)$ if and only if $x \preceq y$. Then, $\mathcal{U}$ is called a *utility function* for $\preceq$. In addition, $-\mathcal{U}$ is called a *risk function* for $\preceq$. □

A commonly used rationality assumption is that the preference relation $\preceq$ on $\mathcal{F}$ is complete, which means in this context that all possible pairwise comparisons among the elements of $\mathcal{F}$ are well defined. If you ask a rational machine whether it prefers *beer* to *wine*, then it needs to have an answer which can be either: *Yes*, *No*, or *no preference*. This assumption implies that the preference relation is also reflexive because you could ask the rational machine whether it prefers *beer* to *beer*.

Another commonly used rationality assumption is that the preference relation $\preceq$ on $\mathcal{F}$ is transitive which means in this context that if the rational machine prefers *beer* to *bread* and additionally the rational machine prefers *bread* to *vegetables*, then the rational machine must prefer *beer* to *vegetables*.

**Definition 3.3.3** (Rational Preference Relation). A complete and transitive preference relation is called a *rational preference relation*. □

**Example 3.3.1** (Dinner with an Irrational Friend). You decide to invite your irrational friend out to dinner.

1. **You:** *The waiter just brought you a cup of coffee. Do you like it?*
   **Irrational Friend:** *No. I like the cup of coffee he brought you!*

2. **You:** *What do you want to order?*
   **Irrational Friend:** *Let's see. I like Lasagna more than Pizza and I like Pizza more than Shrimp but I definitely like Shrimp more than I like Lasagna.*

3. **You:** *You seem to be taking a long time to make your drink decision.*
   **Irrational Friend:** *I can't decide if I prefer coffee to tea and I definitely don't find coffee and tea to be equally preferable.*

For each of the above three irrational behaviors, indicate whether that irrational behavior is an example of a violation of the reflexive, complete, or transitivity property. △

The next theorem shows that every utility function preference relation is a rational preference relation.

**Theorem 3.3.1** (Rational Optimization). *If the function $\mathcal{U} : \mathcal{F} \to \mathcal{R}$ is a utility function for the preference relation $\preceq$, then the preference relation $\preceq$ is rational.*

*Proof.* For every pair of elements $x \in \mathcal{F}$ and $y \in \mathcal{F}$, by the definition of a utility function: $\mathcal{U}(x) < \mathcal{U}(y)$,$\mathcal{U}(y) < \mathcal{U}(x)$, or $U(x) = U(y)$. Thus, the relation $\preceq$ is complete. Suppose that $\preceq$ was not a transitive relation. Then, this would imply that there exists a $x \in \mathcal{F}$, $y \in \mathcal{F}$, and $z \in \mathcal{F}$ such that $x \preceq y$, $y \preceq z$, and $z \prec x$. By the definition of a utility function, this implies that $\mathcal{U}(x) \leq \mathcal{U}(y)$, $\mathcal{U}(y) \leq \mathcal{U}(z)$, and $\mathcal{U}(z) < \mathcal{U}(x)$ which is a contradiction. ■

A rational preference relation constructed from a risk function $-\mathcal{U}$ using the Rational Optimization Theorem (Theorem 3.3.1) is called a *rational preference relation for* $-\mathcal{U}$.

Let $V : \mathcal{F} \to \mathcal{R}$ where $\mathcal{F}$ is a collection of states. The Rational Optimization Theorem is important because it implies that if the goal of a learning (or inference) machine is to find a state $\mathbf{x}^*$ such that $V(\mathbf{x}^*) \leq V(\mathbf{x})$ for all $\mathbf{x}$ in the domain of $V$, then the goal of the learning (or inference) machine can be equivalently interpreted as searching for a state in the domain of $V$ which is maximally preferred with respect to a rational preference relation. This is important because our goal is to build artificially intelligent systems and the Rational Optimization Theorem shows that one way this can be achieved is through the development of numerical optimization algorithms.

**Example 3.3.2** (Learning as Selecting the Most Preferable Parameters). Example 1.3.3 describes the goal of learning a set of training data $\mathbf{x}_1, \ldots, \mathbf{x}_n$ as finding the minimum value $\hat{\boldsymbol{\theta}}_n$ of an empirical risk function $\hat{\ell}_n(\boldsymbol{\theta})$. Show how to construct a rational preference relation for $\hat{\ell}_n$ and interpret the learning machine's action as selecting the most preferable parameter values. △

**Example 3.3.3** (Classification as Selecting Most Preferable Category). Example 1.3.4 considers a situation where a learning machine must classify an input pattern $\mathbf{s}$ into one of two categories: $y = 1$ or $y = 0$. The learning machine believes that a particular input pattern $\mathbf{s}$ is a member of category $y = 1$ with probability $p(y|\mathbf{s})$. The goal of the learning machine is to find $y$ that maximizes $p(y|\mathbf{s})$. Show how to construct a rational preference relation $R$ for $p(y|\mathbf{s})$ for a given $\mathbf{s}$. Interpret the classification behavior of the learning machine as selecting the most preferable category for a given $\mathbf{s}$. Construct a risk function for the rational preference relation $R$. △

Many popular algorithms in the machine learning literature are interpretable as machines that are optimizing real-valued objective functions. The Rational Optimization Theorem (Theorem 3.3.1) provides a useful semantic interpretation of machines that are optimizing objective functions as rational decision makers. These observations motivate the following definition.

**Definition 3.3.4** ($\omega$-Intelligent Machine). Let $\Omega \subseteq \mathcal{R}^d$. Let $V : \Omega \rightarrow \mathcal{R}$ be a risk function for the rational preference relation $\omega \subseteq \Omega \times \Omega$. Let $E$ be an event environment generated from a set of time indices $T$ and events $\Omega_E$. Let

$$\boldsymbol{\Psi} : \Omega \times T \times T \times E \rightarrow \Omega$$

be a dynamical system. Assume, in addition, that there exists an *environmental niche* $S \subseteq \Omega$ such that for every initial state $\mathbf{x}_0 \in S$:

$$V\left(\boldsymbol{\Psi}(\mathbf{x}_0, t_0, t_f, \boldsymbol{\xi})\right) \leq V(\mathbf{x}_0)$$

for every initial time $t_0 \in T$, every final time $t_f \in T$ where $t_f \geq t_0$, and every event timeline function $\boldsymbol{\xi} \in E$. Then, $(\boldsymbol{\Psi}, \omega, S, E)$ is called an *$\omega$-Intelligent Machine* for the environmental niche $S$ and environment $E$.

**Definition 3.3.5** (Inference and Learning Machine). Let $E$ be an event environment defined with respect to a set of time indices $T$ and events $\Omega_E$. Let $(\boldsymbol{\Psi}, \omega, S, E)$ be an $\omega$-intelligent machine for an environmental niche $S$ and $E$. Let $\Omega_E \equiv \Omega_I \times \Omega_F$ where $\Omega_I$ is the set of *input pattern vectors* and $\Omega_F$ is the possibly empty set of *feedback pattern vectors.* Let $\Omega \equiv \Omega_o \times \Omega_H \times \Theta$ where $\Omega_o \subseteq \mathcal{R}^o$ is a set of *output pattern vectors,* $\Omega_H \subseteq \mathcal{R}^H$ is a possibly empty set of *hidden pattern vectors,* and $\Theta$ is the *parameter space.* If $\Theta = \emptyset$, then $(\boldsymbol{\Psi}, \omega, S, E)$ is an *inference machine.* If $\Theta \neq \emptyset$, then $(\boldsymbol{\Psi}, \omega, S, E)$ is an *inference and learning machine.* □

When the event environment for a learning machine uses the feedback pattern to always specify the desired output response for a given input pattern then the learning machine is called a *supervised learning machine.* When the event environment for a learning machine does not include feedback pattern vectors, then the learning machine is called an *unsupervised learning machine.* A learning machine which is not a supervised learning machine and which is not an unsupervised learning machine is called a *reinforcement learning machine.* An example of a reinforcement learning machine would be a supervised learning machine which does not always receive feedback regarding the appropriate desired response for a given input pattern vector. Another example of a reinforcement learning machine would be a learning machine which must generate a complicated response for a given input pattern but the environment does not provide a feedback regarding the identity of the appropriate desired response. Instead, in this latter case, the environment simply provides scalar feedback indicating the quality of the complicated response.

When the event environment for a learning machine is defined with respect to a singleton set of events, then that single element of the set of events is called the *training data set* and the learning machine is called a *batch learning machine.* A learning machine which is not a batch learning machine is called an *adaptive learning machine.*

When every event timeline function $\boldsymbol{\xi}$ in the environment is not functionally dependent upon the state of the inference and learning machine, then the environment is called a *passive learning environment.* A learning environment which is not a passive learning environment is called a *reactive learning environment.* In a reactive learning environment, the actions of the learning machine may modify the statistical characteristics of its learning environment (see Chapter 12 for further discussion).

## Exercises

3.3-1. Specify a time-varying iterated map for the dynamical system in Equations (1.5) and (1.6). Show how to construct a rational preference relation such that this learning machine may be interpreted as having the goal of computing the most preferable parameter values with respect to that rational preference relation.

3.3-2. Specify a time-varying iterated map for the gradient descent learning rule in (1.8). Show how to construct a rational preference relation such that this learning machine may be interpreted as having the goal of computing the most preferable parameter values with respect to that rational preference relation.

3.3-3. Specify an iterated map for a batch gradient descent algorithm with a constant learning rate that is designed to minimize an empirical risk function $\hat{\ell}_n : \mathcal{R}^q \rightarrow \mathcal{R}$. Show how to construct a rational preference relation such that this learning machine may be interpreted as having the goal of computing the most preferable parameter values with respect to that rational preference relation.

3.3-4. Specify a vector field for an adaptive gradient descent algorithm with a constant learning rate that is designed to minimize the risk function $\ell : \mathcal{R}^q \rightarrow \mathcal{R}$ in continuous-time. Show how to construct a rational preference relation such that this learning machine may be interpreted as having the goal of computing the most preferable parameter values with respect to that rational preference relation.

3.3-5. Specify the unsupervised clustering algorithm in Example 1.6.4 as a descent algorithm. Show how to construct a rational preference relation such that this clustering algorithm may be interpreted as having the goal of computing the most preferable clustering strategy with respect to that rational preference relation.

## 3.4 Further Readings

### Dynamical System Models of Machines

The general approach to modeling environments and machines as dynamical systems follows the approach of Kalman (1963) and Kalman, Falb, and Arbib (1969, Section 1.1). The concepts of iterated maps and vector fields are standard topics in dynamical systems theory. The texts by Luenberger (1979), Vidyasagar (1993), and Hirsch, Smale, and Devaney (2004) are recommended for supplemental introductions to dynamical systems modeling.

### Mathematical Models of Preferences

Von Neumann and Morgenstern ([1947] 1953) showed that an intelligent agent that selects actions in order to minimize an empirical risk function is implicitly behaving according to specific rational decision-making axioms. Grandmont (1972) extended the work of Von Neumann and Morgenstern ([1947] 1953) to more general statistical environments consisting of both discrete and continuous random variables. Cox (1946) provides an argument that shows the sense in which probability theory can be interpreted as the only inductive logic consistent with Boolean algebra deductive logic. Savage ([1954] 1972) provides a unified framework that derives expected loss decision rules from a set of rational decision-making axioms.

The Rational Optimization Theorem provides a simple argument for viewing the minimization of a scalar valued function as selecting the most preferred state in a state space. Various converse variations of the Rational Optimization Theorem have been discussed in the literature as well (Jaffray 1975; Mehta 1985; Bearden 1997).

### Complex Information Processing Systems

Marr (1982) has argued that complex information processing systems should be understood at multiple levels of description. Marr (1982; also see Simon 1969) proposed the computational level of description as a specification of the goal of an intelligent machine's computation, the algorithmic level of description as a specification of the algorithms that attempt to achieve that goal, and the implementational level of description as the details required to implement the algorithms. Golden (1988b, 1988c, 1996a, 1996b, 1996c), following Marr (1982), proposed that large classes of machine learning algorithms could be viewed as intelligent optimization algorithms.

### Turing Machines, Super-Turing Machines, and Hypercomputation

A review of the classical 1936 Turing machine model of computation may be found in Hopcraft et al. (2001) and Petzold (2008). Although the Turing machine is a discrete-time dynamical system operating in a finite state space, continuous-time and continuous-state dynamical systems can be simulated on a Turing machine with arbitrary precision. Thus, from this perspective, the limits of computation of continuous-time and continuous-state dynamical systems can be investigated within a classical Turing machine framework. Some researchers (Maclennan 2003; Cabessa and Siegelmann 2012), however, have proposed either alternative theories of computation or extensions of the original classical Turing machine model of computation for the purpose of obtaining complementary insights into the computational limits of analog computers (e.g., continuous-state numerical optimization algorithms and continuous-state continuous-time biological computers). Such extended theories of computation, however, are controversial (e.g., Davis 2006). A review of these issues can be found in the special issue on hypercomputation in the journal *Applied Mathematics and Computation* (Doria and Costa 2006).

# Part II

# Deterministic Learning Machines

# 4

# *Linear Algebra for Machine Learning*

**Learning Objectives**

- Use advanced matrix notation and operators.
- Use SVD to design linear hidden unit basis functions.
- Use SVD to design linear learning machines.

Many important and widely used machine learning algorithms are interpretable as linear machines. In addition, analyses of important and widely used nonlinear machine learning algorithms are based upon the analysis of linear machines. The effective analysis of linear machines often relies upon appropriate matrix notation in conjunction with key linear algebra concepts. Furthermore, the fundamental object in machine learning is the "pattern". Patterns of information are processed and manipulated by machine learning algorithms. A convenient and frequently used formalism for representing patterns of information is a list of numbers or "vector".

## 4.1 Matrix Notation and Operators

Let $\mathcal{R}^{m \times n}$ denote the set of rectangular $m$ by $n$ matrices of real numbers (i.e., matrices with $m$ rows and $n$ columns). An uppercase bold face letter (e.g., $\mathbf{A}$) will be used to denote an $m$ by $n$ dimensional *matrix*.

A lowercase boldface letter (e.g., $\mathbf{a}$) will denote a $d$-dimensional *column vector* which is also a matrix whose dimension is $d$ by 1. The set of $d$-dimensional column vectors is denoted by $\mathcal{R}^d$.

The notation $\mathbf{0}_d$ denotes a $d$-dimensional column vector of zeros. The notation $\mathbf{0}_{d \times d}$ denotes a $d$-dimensional matrix of zeros. The notation $\mathbf{1}_d$ denotes a $d$-dimensional column vector of ones. The matrix $\mathbf{I}_d$ is the *identity matrix* which is a $d$-dimensional matrix with ones on its main diagonal and zeros elsewhere.

Let $\mathbf{x} \in \mathcal{R}^d$ be a column vector. The notation $\mathbf{log}(\mathbf{x})$ is used to denote the inverse of $\mathbf{exp}(\mathbf{x})$ so that $\mathbf{log}(\mathbf{exp}(\mathbf{x})) = \mathbf{x}$. In addition, let $\mathbf{exp} : \mathcal{R}^d \rightarrow (0, \infty)$ be defined such that the $i$th element of the column vector $\mathbf{exp}(\mathbf{x})$ is $\exp(x_i)$, $i = 1, \ldots, d$. Let $\mathbf{log} : (0, \infty) \rightarrow \mathcal{R}^d$ be defined such that the column vector: $\mathbf{log}(\mathbf{exp}(\mathbf{x})) = \mathbf{x}$ for all $\mathbf{x} \in \mathcal{R}^d$.

**Definition 4.1.1** (Vector Transpose). The notation $\mathbf{a}^T \in \mathcal{R}^{1 \times d}$ denotes the *transpose* of the column vector $\mathbf{a} \in \mathcal{R}^d$. □

**Definition 4.1.2** (vec Function)**.** Let $\mathbf{W} \in \mathcal{R}^{m \times n}$ be a matrix defined such that $\mathbf{W} = [\mathbf{w}_1, \ldots, \mathbf{w}_n]$ where each column vector $\mathbf{w}_i \in \mathcal{R}^m$ for $i = 1, \ldots, n$. Then, $\mathbf{vec} : \mathcal{R}^{m \times n} \rightarrow \mathcal{R}^{mn}$ function is defined such that: $\mathbf{vec}(\mathbf{W}) = \left[\mathbf{w}_1^T, \ldots, \mathbf{w}_n^T\right]^T$. □

The **vec** function stacks the columns of a matrix $\mathbf{W}$ consisting of $m$ rows and $n$ columns into a column vector whose dimension is $mn$.

**Definition 4.1.3** ($\mathbf{vec}_m^{-1}$ Function)**.** The function $\mathbf{vec}_m^{-1} : \mathcal{R}^{mn} \rightarrow \mathcal{R}^{m \times n}$ is defined such that: $\mathbf{vec}_m^{-1}(\mathbf{vec}(\mathbf{W})) = \mathbf{W}$ for all $\mathbf{W} \in \mathcal{R}^{m \times n}$. □

**Definition 4.1.4** (Euclidean ($L_2$) Norm)**.** Let $\mathbf{v} \equiv [v_1, \ldots, v_d]^T \in \mathcal{R}^d$. The *Euclidean norm* or $L_2$ *norm* of $\mathbf{v}$ denoted by $|\mathbf{v}|$ is defined such that:

$$|\mathbf{v}| = \sqrt{\sum_{i=1}^{d} (v_i)^2}.$$

□

The Euclidean norm of a matrix $\mathbf{M} \in \mathcal{R}^{m \times n}$, $|\mathbf{M}|$, is defined by $|\mathbf{M}| = |\mathbf{vec}(\mathbf{M})|$. The Euclidean norm is also called the $L_2$ *norm*.

**Definition 4.1.5** ($L_1$ Norm)**.** Let $\mathbf{v} \equiv [v_1, \ldots, v_d]^T \in \mathcal{R}^d$. The $L_1$ *norm* of $\mathbf{v}$ denoted by $|\mathbf{v}|_1$ is defined such that:

$$|\mathbf{v}|_1 = \sum_{i=1}^{d} |v_i|.$$

□

The $L_1$ norm of a matrix $\mathbf{M} \in \mathcal{R}^{m \times n}$, $|\mathbf{M}|_1$, is defined by $|\mathbf{M}|_1 = |\mathbf{vec}(\mathbf{M})|_1$.

**Definition 4.1.6** (Infinity Norm)**.** Let $\mathbf{v} \equiv [v_1, \ldots, v_d]^T \in \mathcal{R}^d$. The *infinity norm* of $\mathbf{v}$ denoted by $|\mathbf{v}|_\infty$ is defined such that:

$$|\mathbf{v}|_\infty = \max\{|v_1|, \ldots, |v_d|\}.$$

□

The infinity norm of a matrix $\mathbf{M} \in \mathcal{R}^{m \times n}$, $|\mathbf{M}|_\infty$, is defined by $|\mathbf{M}|_\infty = |\mathbf{vec}(\mathbf{M})|_\infty$.

**Definition 4.1.7** (Dot Product)**.** Let $\mathbf{a} = [a_1, \ldots, a_d]^T \in \mathcal{R}^d$ and $\mathbf{b} = [b_1, \ldots, b_d]^T \in \mathcal{R}^d$. The *dot product* or *inner product* of two column vectors $\mathbf{a}, \mathbf{b} \in \mathcal{R}^d$ is defined by the formula:

$$\mathbf{a}^T\mathbf{b} = \sum_{i=1}^{d} a_i b_i.$$

□

Note that the linear basis function is defined by the dot product of a parameter vector $\boldsymbol{\theta}$ and an input pattern vector $\mathbf{s}$. In addition, the sigmoidal basis function and softplus basis function are both monotonically increasing functions of the dot product of a parameter vector and an input pattern vector. If $\mathbf{x}, \mathbf{y} \in \{0, 1\}^d$, then $\mathbf{x}^T\mathbf{y}$ counts the number of ones common to both $\mathbf{x}$ and $\mathbf{y}$.

**Definition 4.1.8** (Angular Separation between Two Vectors)**.** Let $\mathbf{x}$ and $\mathbf{y}$ be $d$-dimensional real column vectors with positive magnitudes. Let *normalized vector* $\ddot{\mathbf{x}} \equiv \mathbf{x}/|\mathbf{x}|$. Let $\ddot{\mathbf{y}} \equiv \mathbf{y}/|\mathbf{y}|$. Let $\cos^{-1} : \mathbf{R} \rightarrow [0, 360)$ be the inverse of the cosine function whose domain is in degrees. Then, the *angular separation* $\psi$ between the column vector $\mathbf{x}$ and column vector $\mathbf{y}$ is given by the formula

$$\psi = \cos^{-1}\left(\ddot{\mathbf{x}}^T \ddot{\mathbf{y}}\right).$$

□

The angular separation between two vectors can be interpreted as a type of similarity measure which indicates whether the *pattern of information* in the two vectors is similar. For example, the Euclidean distance between the vectors

$$\mathbf{x} \equiv [1, 0, 1, 0, 1, 0, 1, 0] \quad \text{and} \quad \mathbf{y} \equiv [101, 0, 101, 0, 101, 0, 101, 0]$$

is equal to 200 indicating $\mathbf{x}$ and $\mathbf{y}$ are different. However, the angular separation between $\mathbf{x}$ and $\mathbf{y}$ is zero indicating $\mathbf{x}$ and $\mathbf{y}$ are pointing in the same direction.

**Example 4.1.1** (Response of a Linear Basis Function)**.** Let $\boldsymbol{\theta} \in \mathcal{R}^d$ be a column parameter vector. Assume that $\boldsymbol{\theta}$ is chosen such that $|\boldsymbol{\theta}| = 1$. Let $\mathbf{s} \in \mathcal{R}^d$ be a column input pattern vector. The response $y = \boldsymbol{\theta}^T \mathbf{s}$ is defined as the output of a linear basis function unit for a given input pattern vector $\mathbf{s}$. For what input patterns will the linear basis function unit have the largest possible response? For what input patterns will the linear basis function have a zero response?

**Solution.** Let $\psi$ denote the angular separation between $\boldsymbol{\theta}$ and $\mathbf{s}$. Using the definition of angular separation,

$$y = \boldsymbol{\theta}^T \mathbf{s} - \cos(\psi)|\boldsymbol{\theta}||\mathbf{s}| \tag{4.1}$$

where $\cos(\psi)$ denotes the cosine of the angle $\psi$. When the input pattern vector $\mathbf{s}$ points in the same direction as the parameter vector $\boldsymbol{\theta}$, then the $\cos(\psi)$ is equal to one and the maximal response $y = |\boldsymbol{\theta}||\mathbf{s}|$ is obtained. When $\mathbf{s}$ is in the $d-1$-dimensional subspace whose elements are orthogonal to the $d$-dimensional parameter vector $\boldsymbol{\theta}$, the angular separation between parameter vector $\boldsymbol{\theta}$ and input pattern vector $\mathbf{s}$ is equal to 90 degrees so the response $y = 0$. △

**Definition 4.1.9** (Matrix Multiplication)**.** Let $\mathbf{A} \in \mathcal{R}^{m \times q}$ and $\mathbf{B} \in \mathcal{R}^{q \times r}$. The matrix $\mathbf{A}$ *multiplied* by the matrix $\mathbf{B}$ is called the *matrix product* $\mathbf{AB}$ which is a matrix with $m$ rows and $r$ columns whose $ij$th element is given by the dot product of the $i$th row of $\mathbf{A}$ with the $j$th column of $\mathbf{B}$, $i = 1, \ldots, m$, $j = 1, \ldots, r$. □

**Definition 4.1.10** (Conformable for Matrix Multiplication)**.** Let $\mathbf{A} \in \mathcal{R}^{m \times n}$ and $\mathbf{B} \in \mathcal{R}^{q \times r}$. The matrices $\mathbf{A}$ and $\mathbf{B}$ are *conformable* for $\mathbf{AB}$ if and only if $n = q$. □

Matrix multiplication is not commutative. Let $\mathbf{A}$ be a matrix consisting of two rows and three columns. Let $\mathbf{B}$ be a matrix with three rows and four columns. The matrices $\mathbf{A}$ and $\mathbf{B}$ are conformable for $\mathbf{AB}$ but they are not conformable for $\mathbf{BA}$.

Moreover, even when $\mathbf{A}$ and $\mathbf{B}$ are conformable for both $\mathbf{AB}$ and $\mathbf{BA}$, it is not necessarily the case that $\mathbf{AB} = \mathbf{BA}$. Let $\mathbf{x}$ be the column vector $[3, 4, 0]^T$. Then, $\mathbf{x}\mathbf{x}^T$ is a 3-dimensional matrix whose first row is $[9, 12, 0]$, whose second row is $[12, 16, 0]$, and whose third row is a row vector of zeros. On the other hand, $\mathbf{x}^T\mathbf{x}$ is the number 25.

**Example 4.1.2** (Linear Pattern Associator)**.** Consider an inference machine which generates an $m$-dimensional response vector $\mathbf{y}_i$ when presented with a $d$-dimensional input column pattern vector $\mathbf{s}_i$ for $i = 1, \ldots, n$. Assume, furthermore, the column vector $\mathbf{y}_i = \mathbf{W}\mathbf{s}_i$ for $i = 1, \ldots, n$ where the parameter vector of this inference machine is $\boldsymbol{\theta} \equiv \mathbf{vec}(\mathbf{W}^T)$. The relationship $\mathbf{y}_i = \mathbf{W}\mathbf{s}_i$ for $i = 1, \ldots, n$ may be more compactly written as $\mathbf{Y} = \mathbf{WS}$ where $\mathbf{Y} \equiv [\mathbf{y}_1, \ldots, \mathbf{y}_n]$ and $\mathbf{S} \equiv [\mathbf{s}_1, \ldots, \mathbf{s}_n]$. △

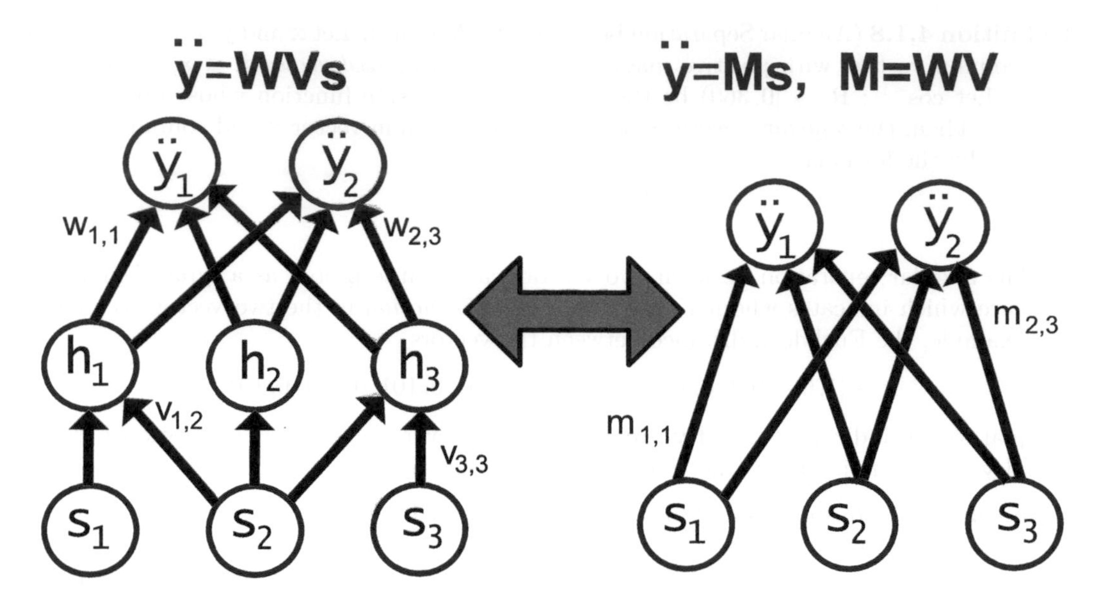


**FIGURE 4.1**
**A multilayer linear feedforward neural network is equivalent to a single-layer linear feedforward neural network.** The left-hand network depicts a linear response function which generates a response pattern $\ddot{\mathbf{y}}$ given input pattern $\mathbf{s}$ using the formulas $\ddot{\mathbf{y}} = \mathbf{Wh}$ and $\mathbf{h} = \mathbf{Vs}$. The right-hand network implements a linear response function which generates a response pattern $\ddot{\mathbf{y}}$ given input pattern $\mathbf{s}$ using the formula $\ddot{\mathbf{y}} = \mathbf{Ms}$. By choosing $\mathbf{M} = \mathbf{WV}$, both networks implement the same linear response function.

**Example 4.1.3** (Equivalence of Single-Layer and Multilayer Linear Machines)**.** Let $\mathbf{S} \equiv [\mathbf{s}_1, \ldots, \mathbf{s}_n]$ be a matrix whose columns are $n$ input pattern vectors. Let $\mathbf{H}$ be an $m$ by $n$ dimensional matrix whose columns are the $n$ respective responses to the $n$ input pattern vectors in $\mathbf{S}$ so that: $\mathbf{H} = \mathbf{VS}$. Let $\mathbf{Y}$ be a $q$ by $n$ dimensional matrix whose columns are the $n$ respective responses to the $n$ columns of $\mathbf{H}$ so that: $\mathbf{Y} = \mathbf{WH}$. This implements a multilayer network architecture as shown in Figure 4.1. Show that this multilayer linear network can be replaced with an equivalent single-layer linear network.

**Solution.** Substitute the relation $\mathbf{H} = \mathbf{VS}$ into $\mathbf{Y} = \mathbf{WH}$ to obtain: $\mathbf{Y} = \mathbf{WVS}$. Then, define $\mathbf{M} = \mathbf{WV}$ to obtain $\mathbf{Y} = \mathbf{MS}$. △

**Definition 4.1.11** (Commutation Matrix)**.** The *commutation matrix* $\boldsymbol{K}_{ur} \in \mathcal{R}^{ur \times ur}$ is defined such that for all $\mathbf{C} \in \mathcal{R}^{u \times r}$:

$$\boldsymbol{K}_{ur}\mathbf{vec}\left(\mathbf{C}\right) = \mathbf{vec}\left(\mathbf{C}^T\right).$$

□

Let $\mathbf{A}$ be a two by three matrix whose first row is $[1, 2, 3]$ and whose second row is $[4, 5, 6]$. Then, $\mathbf{vec}(\mathbf{A}^T) = [1, 2, 3, 4, 5, 6]^T$ and $\mathbf{vec}(\mathbf{A}) = [1, 4, 2, 5, 3, 6]^T$. In addition, there always exists a commutation matrix $\boldsymbol{K}_{2,3}$ such that $\boldsymbol{K}_{2,3}\mathbf{vec}(\mathbf{A}) = \mathbf{vec}(\mathbf{A}^T)$. In particular, if it is desired that the $j$th element of $\mathbf{vec}(\mathbf{A})$ is placed in the $k$th element of $\mathbf{vec}(\mathbf{A}^T)$, choose the $k$th row of the commutation matrix $\boldsymbol{K}_{2,3}$ to be the $j$th row of a 6-dimensional identity matrix.

**Definition 4.1.12** (DIAG Matrix Operator). The *diagonalization matrix operator* $\mathbf{DIAG} : \mathcal{R}^d \rightarrow \mathcal{R}^{d\times d}$ is defined such that for all $\mathbf{x} \in \mathcal{R}^d$: $\mathbf{DIAG}(\mathbf{x})$ is a $d$-dimensional square diagonal matrix whose $i$th on-diagonal element is the $i$th element of $\mathbf{x}$ for $i = 1, \ldots, d$. □

For example, $\mathbf{DIAG}(\mathbf{1}_q) = \mathbf{I}_q$.

**Definition 4.1.13** (Matrix Inverse (Square Matrix with Full Rank)). Let $\mathbf{M}$ be a $d$-dimensional square matrix. Assume, in addition, there exists a $d$-dimensional square matrix $\mathbf{M}^{-1}$ with full rank $d$ such that:

$$\mathbf{M}\mathbf{M}^{-1} = \mathbf{M}^{-1}\mathbf{M} = \mathbf{I}_d.$$

The matrix $\mathbf{M}^{-1}$ is called the *matrix inverse* of $\mathbf{M}$. □

When the matrix inverse of a square matrix $\mathbf{M}$ exists, then $\mathbf{M}$ is called an *invertible matrix*.

**Definition 4.1.14** (Trace Matrix Operator). Let $\mathbf{W} \in \mathcal{R}^{d\times d}$. The *trace* matrix operator $\mathrm{tr} : \mathcal{R}^{d\times d} \rightarrow \mathcal{R}$ is defined such that $\mathrm{tr}(\mathbf{W})$ is the sum of the on-diagonal elements of $\mathbf{W}$. □

**Theorem 4.1.1** (Cyclic Property of tr Operator). *Let* $\mathbf{A}, \mathbf{B}, \mathbf{C} \in \mathcal{R}^{d\times d}$. *Then,*

$$\mathrm{tr}(\mathbf{AB}) = \mathrm{tr}(\mathbf{BA}) \tag{4.2}$$

*and*

$$\mathrm{tr}(\mathbf{ABC}) = \mathrm{tr}(\mathbf{CAB}). \tag{4.3}$$

*Proof.* Let $a_{ij}$, $b_{ij}$, and $c_{ij}$ denote the $ij$th elements of $\mathbf{A}$, $\mathbf{B}$, and $\mathbf{C}$ respectively. Then,

$$\mathrm{tr}(\mathbf{AB}) = \sum_{i=1}^{d}\sum_{j=1}^{d} a_{ij}b_{ji} = \sum_{j=1}^{d}\sum_{i=1}^{d} b_{ji}a_{ij} = \mathrm{tr}(\mathbf{BA}) \tag{4.4}$$

which demonstrates (4.2). Let $\mathbf{R} = \mathbf{AB}$. Then, using the relation in (4.4), it follows that $\mathrm{tr}(\mathbf{RC}) = \mathrm{tr}(\mathbf{CR})$. Then, substitute the definition of $\mathbf{R}$ into this latter expression to obtain (4.3). ■

**Definition 4.1.15** (Hadamard Product). If $\mathbf{A} \in \mathcal{R}^{r\times s}$ and $\mathbf{B} \in \mathcal{R}^{r\times s}$, then $\mathbf{A} \odot \mathbf{B}$ is called the *Hadamard product* and is a matrix of dimension $r$ by $s$ whose $ij$th element is equal to the *ijth* element of $\mathbf{A}$ multiplied by the $ij$th element of $\mathbf{B}$, $i = 1, \ldots, r$, $j = 1, \ldots, s$. □

Let $\mathbf{a} = [1, 2, 3]$. Let $\mathbf{b} = [1, 0, -1]$. Then, $\mathbf{c} \equiv \mathbf{a} \odot \mathbf{b} = [1, 0, -3]$. Also note that $\mathbf{a}\mathbf{b}^T = \mathbf{1}_3^T[\mathbf{a} \odot \mathbf{b}]^T = -2$.

**Definition 4.1.16** (Symmetric Matrix). Let $\mathbf{W}$ be a square matrix. If $\mathbf{W} = \mathbf{W}^T$, then $\mathbf{W}$ is called a *symmetric matrix.* □

**Definition 4.1.17** (**vech** Function). Let $\mathbf{W} \in \mathcal{R}^{m\times m}$ be a symmetric matrix whose $ij$th element is denoted by $w_{ij}$, $i, j \in \{1, \ldots, m\}$. Let $k \equiv m(m+1)/2$. Then, $\mathbf{vech} : \mathcal{R}^{m\times m} \rightarrow \mathcal{R}^k$ function is defined such that:

$$\mathbf{vech}(\mathbf{W}) = \left[w_{1,1}, \ldots, w_{1,m}, w_{2,2}, \ldots, w_{2,m}, w_{3,3}, \ldots, \ldots w_{3,m}, \ldots, w_{m,m}\right]^T.$$

□

**Definition 4.1.18** ($\mathbf{vech}_m^{-1}$ **Function).** The function $\mathbf{vech}_m^{-1} : \mathcal{R}^{m(m+1)/2} \to \mathcal{R}^{m \times m}$ is defined such that: $\mathbf{vech}_m^{-1}(\mathbf{vech}(\mathbf{W})) = \mathbf{W}$ for every symmetric matrix $\mathbf{W} \in \mathcal{R}^{m \times m}$. □

**Definition 4.1.19** (Duplication Matrix)**.** The *duplication matrix* $\mathbf{D} \in \mathcal{R}^{m^2 \times m(m+1)/2}$ is defined such that for every symmetric matrix $\mathbf{W} \in \mathcal{R}^{m \times m}$:

$$\mathbf{vec}(\mathbf{W}) = \mathbf{D}\mathbf{vech}(\mathbf{W}).$$

□

Let $\mathbf{A}$ be a two by two matrix whose first row is $[1, 2]$ and whose second row is $[2, 3]$. Then, $\mathbf{vec}(\mathbf{A}) = [1, 2, 2, 3]^T$ and $\mathbf{vech}(\mathbf{A}) = [1, 2, 3]^T$. If

$$\mathbf{D} \equiv \begin{bmatrix} 1 & 0 & 0 \\ 0 & 1 & 0 \\ 0 & 1 & 0 \\ 0 & 0 & 1 \end{bmatrix}$$

then $\mathbf{D}$ is a duplication matrix since $\mathbf{vec}(\mathbf{A}) = \mathbf{D}[\mathbf{vech}(\mathbf{A})]$.

**Definition 4.1.20** (Square Root of a Real Symmetric Matrix)**.** The *square root* of a real symmetric matrix $\mathbf{M}$, $\mathbf{M}^{1/2}$, is defined such that: $\mathbf{M}^{1/2}\mathbf{M}^{1/2} = \mathbf{M}$. □

**Definition 4.1.21** (Kronecker Product)**.** If $\mathbf{A} \in \mathcal{R}^{r \times s}$ and $\mathbf{B} \in \mathcal{R}^{t \times u}$, then the *Kronecker product* $\mathbf{A} \otimes \mathbf{B} \in \mathcal{R}^{rt \times su}$ is a matrix consisting of $r$ rows and $t$ columns of submatrices such that the submatrix in the $i$th row and $j$th column is $a_{ij}\mathbf{B}$ where $a_{ij}$ is the $ij$th element of $\mathbf{A}$. □

Let $\mathbf{a}$ be a row vector defined such that $\mathbf{a} = [1, 2, 5]$. Let $\mathbf{B}$ be rectangular matrix with 5 rows and 4 columns. Then,

$$\mathbf{a} \otimes \mathbf{B} = [\mathbf{B}, 2\mathbf{B}, 5\mathbf{B}]$$

is a matrix with 5 rows and 12 columns. In general, $\mathbf{A} \otimes \mathbf{B} \neq \mathbf{B} \otimes \mathbf{A}$.

**Theorem 4.1.2** (Kronecker Matrix Product Identities)**.** *Let* $r, s, t, u, v, p, w, z \in \mathbb{N}^+$. *Let* $\mathbf{A} \in \mathcal{R}^{r \times s}$, $\mathbf{B} \in \mathcal{R}^{t \times u}$, $\mathbf{C} \in \mathcal{R}^{v \times p}$, $\mathbf{D} \in \mathcal{R}^{w \times z}$, $\mathbf{b} \in \mathcal{R}^t$, $\mathbf{d} \in \mathcal{R}^z$, *and* $\rho \in \mathcal{R}$. *Let* $\mathbf{K}_{ur} \in \mathcal{R}^{ur \times ur}$ *be the commutation matrix. Let* $\mathbf{I}_z$ *be the* $z$*-dimensional identity matrix.*

1. $(\mathbf{A} \otimes \mathbf{B})\rho = \rho\mathbf{A} \otimes \mathbf{B} = \mathbf{A} \otimes \rho\mathbf{B}$.
2. $\mathbf{A} \otimes (\mathbf{B} \otimes \mathbf{C}) = (\mathbf{A} \otimes \mathbf{B}) \otimes \mathbf{C}$.
3. $(\mathbf{A} \otimes \mathbf{B})^T = \mathbf{A}^T \otimes \mathbf{B}^T$.
4. $\mathbf{vec}(\mathbf{A} \otimes \mathbf{B}) = (\mathbf{I}_s \otimes \mathbf{K}_{ur} \otimes \mathbf{I}_t)(\mathbf{vec}(\mathbf{A}) \otimes \mathbf{vec}(\mathbf{B}))$.
5. *If* $\mathbf{A}$, $\mathbf{B}$ *and* $\mathbf{C}$ *are conformable for* $\mathbf{AB}$ *and* $\mathbf{BC}$, *then* $\mathbf{vec}(\mathbf{ABC}) = (\mathbf{C}^T \otimes \mathbf{A})\mathbf{vec}(\mathbf{B})$.
6. *If* $\mathbf{A}$, $\mathbf{B}$, $\mathbf{C}$ *and* $\mathbf{D}$ *are conformable for* $\mathbf{AC}$ *and* $\mathbf{BD}$, *then* $(\mathbf{A} \otimes \mathbf{B})(\mathbf{C} \otimes \mathbf{D}) = \mathbf{AC} \otimes \mathbf{BD}$.
7. *If* $\mathbf{A}$ *and* $\mathbf{C}$ *are conformable for* $\mathbf{AC}$, *then* $(\mathbf{A} \otimes \mathbf{b})\mathbf{C} = \mathbf{AC} \otimes \mathbf{b}$.
8. *If* $\mathbf{A}$ *and* $\mathbf{C}$ *are conformable for* $\mathbf{AC}$, *then* $\mathbf{A}(\mathbf{C} \otimes \mathbf{d}^T) = \mathbf{AC} \otimes \mathbf{d}^T$.
9. *If* $\mathbf{A}$ *and* $\mathbf{C}$ *are conformable for* $\mathbf{AC}$, *then* $(\mathbf{A} \otimes \mathbf{I}_z)(\mathbf{C} \otimes \mathbf{I}_z) = \mathbf{AC} \otimes \mathbf{I}_z$.
10. $\rho \otimes \mathbf{I}_z = \rho\mathbf{I}_z$.

*Proof.* See Exercise 4.1-11 for a proof of (1), (2), (3), and (10). See Magnus and Neudecker (2001, pp. 47-48, Theorem 10) for a proof of (4). See Magnus and Neudecker (2001, p. 30-31, Theorem 2) for a proof of (5). See Schott (2005) for a proof of (6). Properties (7), (8), and (9) follow directly from (6). ■

From a computational perspective, the Kronecker Product Operator needs to be carefully implemented. For example, suppose that $\mathbf{A}$ and $\mathbf{B}$ are 500-dimensional square matrices. Then, the Kronecker Product $\mathbf{C} = \mathbf{A} \otimes \mathbf{B}$ is a $250,000$-dimensional square matrix. If either $\mathbf{A}$ or $\mathbf{B}$ is a sparse matrix, then taking advantage of the sparse matrix structure by using only non-zero elements of these matrices for storage and computation is essential. The scientific programming language MATLAB®, for example, has a feature to support sparse matrix construction and manipulation.

Fortunately, however, the Kronecker Product operator can be efficiently calculated with a parallel processing algorithm involving $K$ parallel processors where $K$ is equal to the number of non-zero elements in the matrix $\mathbf{A}$. All $K$ parallel processors simultaneously compute $a_{ij}\mathbf{B}$ for each non-zero $a_{ij}$ where $a_{ij}$ is the $ij$th element of $\mathbf{A}$. Then, instead of storing one $25,000$-dimensional sparse square matrix, one stores $K$ sparse 500-dimensional square matrices $a_{1,1}\mathbf{B}$, $a_{1,2}\mathbf{B}$, and so on. The scientific programming language MATLAB provides support for parallel processing and multidimensional indexing to support such computations.

## Exercises

4.1-1. Let $\mathbf{x}_1 = [1 \ 3 \ 4]^T$, $\mathbf{x}_2 = [-4 \ 3 \ 5]^T$, and $\mathbf{x}_3 = [11 \ 3 \ 7]^T$. Let the row vector $\mathbf{x} = [\mathbf{x}_1^T \ \mathbf{x}_2^T \ \mathbf{x}_3^T]$. Let the matrix $\mathbf{W} = [\mathbf{x}_1 \ \mathbf{x}_2 \ \mathbf{x}_3]$. Let the matrix $\mathbf{Q} = \mathbf{x}_1(\mathbf{x}_1)^T + \mathbf{x}_3(\mathbf{x}_3)^T$. What is $\mathbf{vec}(\mathbf{W})$? What is $\mathbf{vech}(\mathbf{Q})$? What is $|\mathbf{vec}(\mathbf{W})|_\infty$? What is $\mathbf{x}_1 \odot \mathbf{x}_2$? What is $\mathbf{x}_1 \otimes \mathbf{W}$? What is $\mathbf{W} \otimes \mathbf{x}_1$? What is $\mathbf{QQ}$?

4.1-2. Let $\mathbf{x}_1 = [1 \ 2]$. Let $\mathbf{x}_2 = [0 \ 3]$. Let the matrix $\mathbf{A} = (\mathbf{x}_1)^T\mathbf{x}_1$. Let the row vector $\mathbf{b} = [\mathbf{x}_1, \ \mathbf{x}_2]$. What is $\mathbf{vec}(\mathbf{A} \otimes \mathbf{b})$? Let $\mathbf{I}_2$ be a 2-dimensional identity matrix and let $\mathbf{K}_{2,4}$ be the commutation matrix. Show that:

$$\mathbf{vec}(\mathbf{A} \otimes \mathbf{b}) = (\mathbf{I}_2 \otimes \mathbf{K}_{2,4})(\mathbf{vec}(\mathbf{A}) \otimes \mathbf{vec}(\mathbf{B})).$$

4.1-3. Let $\mathbf{x}_1 = [2, \ 2]^T$, $\mathbf{x}_2 = [1, \ 0]$, and $\mathbf{x}_3 = [1, \ 2, \ 3]$. Let $\mathbf{A} = \mathbf{x}_1\mathbf{x}_1^T$. Let $\mathbf{B} = \mathbf{x}_2\mathbf{x}_3^T$. Let $\mathbf{C} = \mathbf{x}_1\mathbf{x}_2^T$. Compute $\mathbf{vec}(\mathbf{ACB})$. Compute $(\mathbf{B}^T \otimes \mathbf{A})\mathbf{vec}(\mathbf{C})$. The two matrices you get should be equal to one another!

4.1-4. Let $\mathbf{E}$ be a matrix with $m$ rows and $d$ columns whose column vectors are $\mathbf{e}_1, \ldots, \mathbf{e}_d$. Let $\mathbf{F}$ be a matrix with $m$ rows and $d$ columns whose column vectors are $\mathbf{f}_1, \ldots, \mathbf{f}_d$. Let $\mathbf{D}$ be a $d$-dimensional diagonal matrix whose $i$th on-diagonal element is $\lambda_i$ for $i = 1, \ldots, d$. Let $\mathbf{W} = \mathbf{EDF}^T$. Show that

$$\mathbf{W} = \sum_{i=1}^{d} \lambda_i \mathbf{e}_i \mathbf{f}_i^T.$$

HINT: This might make more sense if you examine how this works for the case where $d = 1$ and then work out the case for $d = 2$.

4.1-5.Are the vectors $[0, 1, 2]$ and $[0, 1000, 2000]$ linearly independent?

4.1-6.Let $\mathbf{x}_1 = [1 \;\; 2]$. Let $\mathbf{x}_2 = [0 \;\; 3]$. Let $\mathbf{R} = \mathbf{x}_1^T \mathbf{x}_1$. What is the rank of $\mathbf{R}$?

4.1-7.Let $\mathbf{v} = [1, 2, 3]$. Let $\mathbf{w} = [1, -1, 2, 1]$. Let $\mathbf{V} = \mathbf{v}\mathbf{w}^T$. Construct a commutation matrix $\mathbf{K}$ such that $\mathbf{K}\mathbf{vec}(\mathbf{V}) = \mathbf{vec}(\mathbf{V}^T)$. Express your answer by defining $\mathbf{K}$ such that each row of $\mathbf{K}$ is a row vector of an identity matrix.

4.1-8.Let the empirical risk function $\hat{\ell}_n : \Theta \to \mathcal{R}$ be defined such that for all $\boldsymbol{\theta} \in \Theta \subseteq \mathcal{R}^q$:

$$\hat{\ell}_n(\boldsymbol{\theta}) = (1/n) \sum_{i=1}^{n} (y_i - \boldsymbol{\theta}^T \mathbf{s}_i)^2$$

where $y_i$ is the desired response of the learning machine given $q$-dimensional input pattern $\mathbf{s}_i$ for $i = 1, \ldots, n$. Show that $\hat{\ell}_n$ may be rewritten using matrix notation as:

$$\hat{\ell}_n(\boldsymbol{\theta}) = (1/n) \left[ (\mathbf{y}^T - \boldsymbol{\theta}^T \mathbf{S}) \odot (\mathbf{y}^T - \boldsymbol{\theta}^T \mathbf{S}) \right] \mathbf{1}_n$$

where $\mathbf{S} \equiv [\mathbf{s}_1, \ldots, \mathbf{s}_n]$ and $\mathbf{y} \equiv [y_1, \ldots, y_n]^T$.

4.1-9.Let $V : \mathcal{R}^d \to \mathcal{R}$. Let $\mathbf{a}^T = [a_1, \ldots, a_d]$. Let $\mathbf{B}$ be a $d$-dimensional square matrix whose $ij$th element is $b_{i,j}$. Let $\mathbf{x}^T = [x_1, \ldots, x_d]^T$. Show that if

$$V(\mathbf{x}) \equiv \sum_{i=1}^{d} a_i x_i + \sum_{i=1}^{d} \sum_{j=1}^{d} b_{i,j} x_i x_j$$

then

$$V(\mathbf{x}) = \mathbf{a}^T \mathbf{x} + \mathbf{x}^T \mathbf{B} \mathbf{x}.$$

4.1-10.Let $V : \mathcal{R}^d \to \mathcal{R}$. Let $\mathbf{a}^T = [a_1, \ldots, a_d]$. Let $\mathbf{B}$ be a $d$-dimensional square matrix whose $ij$th element is $b_{i,j}$. Let $\mathbf{C} \in \mathcal{R}^{d \times d^2}$ be a rectangular matrix whose elements are specified using the formula $c_{i,j,k} = \mathbf{u}_i^T \mathbf{C}(\mathbf{u}_j \otimes \mathbf{u}_k)$ where $\mathbf{u}_i$, $\mathbf{u}_j$, and $\mathbf{u}_k$ are the $i$th, $j$th, and $k$th columns of a $d$-dimensional identity matrix for $i, j, k = 1, \ldots, d$. Show that the expansion

$$V(\mathbf{x}) = \sum_{i=1}^{d} a_i x_i + \sum_{i=1}^{d} \sum_{j=1}^{d} b_{i,j} x_i x_j + \sum_{i=1}^{d} \sum_{j=1}^{d} \sum_{k=1}^{d} c_{i,j,k} x_i x_j x_k$$

can be rewritten as:

$$V(\mathbf{x}) = \mathbf{a}^T \mathbf{x} + \mathbf{x}^T \mathbf{B} \mathbf{x} + \mathbf{x}^T \mathbf{C}(\mathbf{x} \otimes \mathbf{x}).$$

4.1-11.Prove Properties (1), (2), (3), and (10) of the Kronecker Matrix Product Identities Theorem 4.1.2. Use Property (6) to prove Properties (7), (8), and (9).

## 4.2 Linear Subspace Projection Theorems

Let $\mathbf{s}$ denote a $d$-dimensional input pattern such that $\mathbf{s} \in \mathcal{R}^d$. Let the linear hidden unit basis function $h : \mathcal{R}^d \times \mathcal{R}^d \to \mathcal{R}$ be defined such that $h(\mathbf{s}, \mathbf{w}_k) \equiv \mathbf{w}_k^T \mathbf{s}$ for $k = 1, \ldots, m$.

Equivalently, one can define a linear transformation $\mathbf{h} = \mathbf{W}\mathbf{s}$ where the $k$th row of $\mathbf{W} \in \mathcal{R}^{m \times d}$ is defined as the row vector $\mathbf{w}_k^T$ for $k = 1, \ldots, m$. Thus, the $d$-dimensional input pattern $\mathbf{s}$ is transformed into an $m$-dimensional pattern $\mathbf{h}$. Methods for analyzing such linear transformations are discussed in this section.

**Definition 4.2.1** (Linearly Independent). A set of vectors $\mathbf{x}_1, \ldots, \mathbf{x}_k$ are called *linearly dependent* when there exists a set of numbers $c_1, \ldots, c_k$ such that: (1)

$$\sum_{i=1}^{k} c_i \mathbf{x}_i = \mathbf{0},$$

and (2) at least one of the numbers $c_1, \ldots, c_k$ is not equal to zero. A set of vectors which is not linearly dependent is called a *linearly independent* set of vectors. □

Less formally, a collection of vectors is linearly independent if no vector in that set can be rewritten as a (non-trivial) linear combination of the other vectors in the set. Note that $\{\mathbf{0}_d\}$ is a linearly dependent set, while $\{\mathbf{1}_d\}$ is a linearly independent set.

**Definition 4.2.2** (Rank of a Matrix). Let $\mathbf{A} \in \mathcal{R}^{m \times n}$. The number of linearly independent rows of a matrix $\mathbf{A}$ is called the *row rank* of $\mathbf{A}$. If the row rank of $\mathbf{A}$ is equal to $m$ then $\mathbf{A}$ is said to have *full row rank.* The number of linearly independent columns of a matrix $\mathbf{A}$ is called the *column rank* of $\mathbf{A}$. If the column rank of $\mathbf{A}$ is equal to $n$ then $\mathbf{A}$ is said to have *full column rank.* □

**Theorem 4.2.1** (Equivalence of Row Rank and Column Rank). *Let $\mathbf{A} \in \mathcal{R}^{m \times n}$. The row rank of $\mathbf{A}$ is equal to the column rank of $\mathbf{A}$.*

*Proof.* See Theorem 4.8 and Corollary 4.1 of Noble and Daniel (1977, pp. 126-127). ■

**Definition 4.2.3** (Subspace of a Matrix). Let the $n$-dimensional row vectors $\mathbf{a}_1, \ldots, \mathbf{a}_m$ denote the $m$ rows of a matrix $\mathbf{A} \in \mathcal{R}^{m \times n}$. The *row subspace* of the matrix $\mathbf{A}$ is the set:

$$S \equiv \left\{ \mathbf{x} \in \mathcal{R}^n : \mathbf{x} \equiv \sum_{j=1}^{m} c_j \mathbf{a}_j, c_j \in \mathcal{R}, j = 1, \ldots, m \right\}.$$

The *column subspace* of the matrix $\mathbf{A}$ is the row subspace of $\mathbf{A}^T$. The *dimensionality* of the row subspace or column subspace is defined as the rank of $\mathbf{A}$. The *null space* of the matrix $\mathbf{A} \in \mathcal{R}^{m \times n}$ is the set:

$$S = \{\mathbf{x} \in \mathcal{R}^n : \mathbf{A}\mathbf{x} = \mathbf{0}_m\}.$$

□

Less formally, the row subspace of a matrix $\mathbf{A}$ is the set of all vectors constructed from all possible weighted sums of the rows of $\mathbf{A}$.

**Definition 4.2.4** (Positive Definite and Positive Semidefinite Matrices). Let $\Omega \subseteq \mathcal{R}^d$. Let $\mathbf{W} \in \mathcal{R}^{d \times d}$. We say that $\mathbf{W}$ is *positive semidefinite* on $\Omega$ if $\mathbf{x}^T \mathbf{W} \mathbf{x} \geq 0$ for all $\mathbf{x} \in \Omega$. The matrix $\mathbf{W}$ is *positive definite* on $\Omega$ if $\mathbf{x}^T \mathbf{W} \mathbf{x} > 0$ for all $\mathbf{x} \in \Omega$ such that $\mathbf{x} \neq \mathbf{0}_d$. □

**Definition 4.2.5** (Eigenvectors and Eigenvalues). Let $\mathbf{W} \in \mathcal{R}^{d \times d}$. If there exists a complex number $\lambda$ and a $d$-dimensional (possibly complex) column vector $\mathbf{e}$ with non-zero magnitude such that

$$\mathbf{W}\mathbf{e} = \lambda \mathbf{e},$$

then $\mathbf{e}$ is an *eigenvector* of $\mathbf{W}$ with corresponding *eigenvalue* $\lambda$. If, in addition, $\mathbf{e}$ has the property that $|\mathbf{e}| = 1$, then $\mathbf{e}$ is called a *normalized eigenvector.* □

Note that if a collection of column vectors $\mathbf{e}_1, \ldots, \mathbf{e}_n$ are *orthonormal* this means that: (i) $|\mathbf{e}_k| = 1$ for $k = 1, \ldots, n$, and (ii) $\mathbf{e}_j^T \mathbf{e}_k = 0$ for all $j, k = 1, \ldots, n$ such that $j \neq k$.

**Theorem 4.2.2** (Eigendecomposition (Symmetric Matrix)). *If $\mathbf{W} \in \mathcal{R}^{d \times d}$ is a real symmetric matrix, then $\mathbf{W}$ can be rewritten as:*

$$\mathbf{W} = \sum_{i=1}^{d} \lambda_i \mathbf{e}_i \mathbf{e}_i^T$$

*where $\mathbf{e}_1, \ldots, \mathbf{e}_d$ are $d$ orthonormal real eigenvectors with respective real eigenvalues $\lambda_1, \ldots, \lambda_d$.*

*Proof.* See Magnus and Neudecker (2001, Theorem 13, pp. 16-17) for a proof. ■

The conclusion of Theorem 4.2.2 can be rewritten as follows. Let $\mathbf{D}$ be a $d$-dimensional diagonal matrix whose $i$th on-diagonal element is the $i$th eigenvalue of a real symmetric matrix $\mathbf{W}$. In addition, assume the $i$th column of the $d$-dimensional matrix $\mathbf{E} \equiv [\mathbf{e}_1, \ldots, \mathbf{e}_d]$ is the $i$th eigenvector, $\mathbf{e}_i$, corresponding to the $i$th on-diagonal element of $\mathbf{D}$. Then, Theorem 4.2.2 implies:

$$\mathbf{W} = \sum_{i=1}^{d} \lambda_i \mathbf{e}_i \mathbf{e}_i^T = \mathbf{E}\mathbf{D}\mathbf{E}^T.$$

**Example 4.2.1** (Alternative Eigenspectrum Definition of a Positive Definite Matrix). Let $\mathbf{W}$ be a real symmetric matrix. Then,

$$\mathbf{x}^T \mathbf{W} \mathbf{x} = \mathbf{x}^T \left[ \sum_{i=1}^{d} \lambda_i \mathbf{e}_i \mathbf{e}_i^T \right] \mathbf{x} = \sum_{i=1}^{d} \lambda_i (\mathbf{e}_i^T \mathbf{x})^2.$$

Thus, $\mathbf{W}$ is positive semidefinite when the eigenvalues of $\mathbf{W}$ are non-negative. Furthermore $\mathbf{W}$ is positive definite when all eigenvalues of $\mathbf{W}$ are strictly positive. △

**Example 4.2.2** (Linear Inference Machine Response (Symmetric Matrix)). The above Eigenspectrum Analysis theorem (Theorem 4.2.2) can be used to analyze a linear inference machine that generates a $d$-dimensional response vector $\mathbf{y}$ for a given $d$-dimensional input pattern vector $\mathbf{s}$ where $\mathbf{y} = \mathbf{W}\mathbf{s}$. Assume the $d$-dimensional $\mathbf{W}$ is a symmetric matrix of full rank $d$.

$$\mathbf{y} = \mathbf{W}\mathbf{s}$$

$$\mathbf{y} = \left[ \sum_{i=1}^{d} \lambda_i \mathbf{e}_i \mathbf{e}_i^T \right] \mathbf{s}$$

$$\mathbf{y} = \sum_{i=1}^{d} \lambda_i \left( \mathbf{e}_i^T \mathbf{s} \right) \mathbf{e}_i$$

which shows clearly how the linear response can be expressed in terms of the degree of similarity of $\mathbf{s}$ to the eigenvectors of $\mathbf{W}$ with the larger eigenvalues. Thus, the eigenvectors of $\mathbf{W}$ may be interpreted as abstract features of $\mathbf{W}$. This is an example of a linear subspace machine whose response $\mathbf{y}$ is interpretable as a projection of the original input pattern vector $\mathbf{s}$ into a linear subspace spanned by the eigenvectors of $\mathbf{W}$ with the largest eigenvalues. The dimensionality of this linear subspace is the rank, $r$, of $\mathbf{W}$. The rank $r$ is equal to the number of eigenvectors of $\mathbf{W}$ associated with nonzero eigenvalues. △

In applications, eigenvector analysis of a non-symmetric square full rank matrix is sometimes necessary.

**Theorem 4.2.3** (Eigenvalue Analysis (Square Non-symmetric Matrix)). *Let* $\mathbf{W}$ *be a* $d$*-dimensional square matrix. The matrix* $\mathbf{W}$ *has a linearly independent set of* $d$ *eigenvectors if and only if there exists an invertible* $d$*-dimensional square matrix* $\mathbf{E}$ *and a* $d$*-dimensional diagonal matrix* $\mathbf{D}$ *such that:* $\mathbf{W} = \mathbf{E}\mathbf{D}\mathbf{E}^{-1}$ *and where the* $i$*th column vector of* $\mathbf{E}$ *is an eigenvector associated with the* $i$*th eigenvalue,* $\lambda_i$*, located on the* $i$*th on-diagonal element of the matrix* $\mathbf{D}$.

*Proof.* For a proof, see Theorem 8.3 of Noble and Daniel (1977, p. 273). ■

Note that the eigenvalues of a $d$-dimensional square real matrix can be complex numbers.
The following non-standard definition of a determinant which specifies the determinant in terms of the eigenvalues of a square matrix is useful for the applications considered here.

**Definition 4.2.6** (Determinant of a Square Matrix). Let $\mathbf{M}$ be a square $d$-dimensional matrix with eigenvalues $\lambda_1, \ldots, \lambda_d$. The *determinant* of $\mathbf{M}$, $\det(\mathbf{M})$, is defined such that: $\det(\mathbf{M}) = \prod_{i=1}^{d} \lambda_i$. □

The determinant of a positive semidefinite matrix may be interpreted as a measure of the magnitude of that matrix since all of the eigenvalues are non-negative.

**Theorem 4.2.4** (Determinant Products). *Let* $\mathbf{A}$ *and* $\mathbf{B}$ *be square matrices. Then,*

$$\det(\mathbf{AB}) = \det(\mathbf{A})\det(\mathbf{B}).$$

*Proof.* See proof of Theorem 6.6 in Noble and Daniel (1977, pp. 203 204). ■

**Theorem 4.2.5** (Matrix Trace Is Sum of Eigenvalues). *The trace of a square matrix* $\mathbf{W}$ *is equal to the sum of the eigenvalues of* $\mathbf{W}$.

*Proof.* See Schott 2005 (Theorem 3.5) for proof. ■

**Definition 4.2.7** (Condition Number (Positive Definite Symmetric Matrix)). The *condition number* of a positive definite symmetric matrix $\mathbf{M}$ is the largest eigenvalue of $\mathbf{M}$ divided by the smallest eigenvalue of $\mathbf{M}$. □

In machine learning applications, a real symmetric matrix $\hat{\mathbf{M}}$ with positive eigenvalues is often used for the purposes of approximating another real symmetric matrix $\mathbf{M}^*$ and the rank of $\mathbf{M}^*$ is investigated by examining the eigenvalues of $\hat{\mathbf{M}}$. Evidence that $\mathbf{M}^*$ may not have full rank is obtained by: (i) checking that the largest eigenvalue of the approximation $\hat{\mathbf{M}}$ for $\mathbf{M}^*$ is greater than some small positive number, and (2) checking the condition number of the approximation $\hat{\mathbf{M}}$ for $\mathbf{M}^*$ is not extremely large.

For example, suppose that $\hat{\mathbf{M}}$ is an invertible matrix whose largest eigenvalue is 0.0001 and whose smallest eigenvalue is $10^{-15}$. All eigenvalues of $\hat{\mathbf{M}}$ are positive and $\hat{\mathbf{M}}_n$ is invertible but, from a practical perspective, $\hat{\mathbf{M}}_n$ is numerically non-invertible.

As another example, suppose $\hat{\mathbf{M}}$ is an invertible matrix such that the largest eigenvalue of $\hat{\mathbf{M}}$ is $10^{-15}$ and the smallest eigenvalue is $10^{-16}$. Although the condition number in this case is equal to 10 and all eigenvalues are strictly positive, again this corresponds to the case where $\hat{\mathbf{M}}$ is numerically a non-invertible matrix.

A more general matrix factorization can be done using SVD (singular-value decomposition) methods which generalize the notions of eigenvectors and eigenvalues to the case of rectangular matrices.

**Theorem 4.2.6** (Singular Value Decomposition (Rectangular Matrix)). *Let* $\mathbf{W} \in \mathcal{R}^{n \times m}$ *have positive rank* $k$. *Let the* $k$ *orthonormal eigenvectors of* $\mathbf{W}\mathbf{W}^T$, $\mathbf{u}_1, \ldots, \mathbf{u}_k$, *be arranged as* $n$*-dimensional column vectors in an* $n$ *by* $k$*-dimensional matrix* $\mathbf{U} = [\mathbf{u}_1, \ldots, \mathbf{u}_k]$. *Let the* $k$ *orthonormal eigenvectors of* $\mathbf{W}^T\mathbf{W}$, $\mathbf{v}_1, \ldots, \mathbf{v}_k$, *be arranged as* $m$*-dimensional column vectors in the matrix* $\mathbf{V} = [\mathbf{v}_1, \ldots, \mathbf{v}_k]$. *Let* $\mathbf{D} \in \mathcal{R}^{k \times k}$ *be a diagonal matrix defined such that* $d_{jj} = \mathbf{u}_j^T \mathbf{W} \mathbf{v}_j$ *for* $j = 1, \ldots, k$. *Then,* $\mathbf{W} = \mathbf{U}\mathbf{D}\mathbf{V}^T$ *and, in addition, the on-diagonal elements of* $\mathbf{D}$ *are strictly positive.*

*Proof.* For a proof see Noble and Daniel (1977, Theorem 9.7, p. 327). ■

Note that the columns of the matrix $\mathbf{U}$ in the Singular Value Decomposition Theorem are called the *left eigenvectors* of $\mathbf{W}$, the columns of the matrix $\mathbf{V}$ are called the *right eigenvectors* of $\mathbf{W}$, and the $i$th on-diagonal element of the diagonal matrix $\mathbf{D}$ is called the *singular value* associated with the $i$th left eigenvector $\mathbf{u}_i$ and the $i$th right eigenvector $\mathbf{v}_i$, $i = 1, \ldots, k$. Also note that the singular value decomposition can be equivalently expressed as:

$$\mathbf{W} = \mathbf{U}\mathbf{D}\mathbf{V}^T = \sum_{j=1}^{k} d_{j,j} \mathbf{u}_j \mathbf{v}_j^T.$$

In practice, the $\mathbf{U}$, $\mathbf{V}$ and $\mathbf{D}$ matrices may be calculated from $\mathbf{W}$ using existing computer software (e.g., using the function `SVD` in MATLAB).

**Example 4.2.3** (Image Compression Using SVD). Singular Value Decomposition is useful for solving image compression problems. Let an *image* $\mathbf{W} \in \mathcal{R}^{n \times m}$ be defined as a rectangular matrix of integers where each integer can take on $d$ possible values. Typical values of $m$, $n$, and $d$ might be $3 \times 10^4$, $5 \times 10^4$, and 256 respectively. The total number of bits required to store such an image is given by the formula:

$$-\log_2((1/d)^{mn}) = mn \log_2(d).$$

Now generate a singular value decomposition of $\mathbf{W}$ so that $\mathbf{W} = \mathbf{U}\mathbf{D}\mathbf{V}^T$ where the number of non-zero on-diagonal elements of $\mathbf{D}$ is equal to $k$. This means that the matrix $\mathbf{W}$ can be exactly represented using only the $k$ left eigenvectors and $k$ right eigenvectors associated with the non-zero singular values. Thus, only $k(m+n+1)$ numbers are required to represent $\mathbf{W}$ exactly or equivalently only

$$-\log_2((1/d)^{k(m+n+1)}) = k(m+n+1)\log_2(d)$$

information storage bits are required. If, in addition, $u$ of the non-zero on-diagonal elements of $\mathbf{D}$ are very small positive numbers relative to the other non-zero on-diagonal elements of $\mathbf{D}$, then these $\mathbf{D}$ can be approximated by setting $\ddot{\mathbf{D}}$ equal to $\mathbf{D}$ and then setting the $u$ small on-diagonal elements of $\ddot{\mathbf{D}}$ equal to zero. Show that only $(k-u)(m+n+1)$ numbers are required to approximately represent $\mathbf{W}$. △

**Example 4.2.4** (Reducing Computational Requirements Using LSI). In Latent Semantic Indexing (LSI), the term by document matrix is reconstructed using only a subset of the positive singular values corresponding to the most dominant and most reliable statistical regularities. This approximate reconstruction of the original term by document matrix is similar to the method used in the image compression example (see Example 4.2.3) and thus additionally avoids excessive storage and computation requirements when the term by document matrix is very large.

Suppose that a term by document matrix $\mathbf{W}$ constructed from $10^5$ terms and $10^7$ documents has an SVD decomposition $\mathbf{W} = \mathbf{U}\mathbf{D}\mathbf{V}^T$ where $\mathbf{D}$ is a diagonal matrix such that

10 of the on-diagonal elements of $\mathbf{D}$ are strictly greater than 0.01 and the remaining on-diagonal elements of $\mathbf{D}$ are less than $10^{-6}$. Show that an approximate reconstruction of $\mathbf{W}$ denoted by $\ddot{\mathbf{W}}$ is generated by the formula $\ddot{\mathbf{W}} = \mathbf{U}\ddot{\mathbf{D}}\mathbf{V}^T$ where the $j$th on-diagonal element of the diagonal matrix $\ddot{\mathbf{D}}$ is equal to the $j$th on-diagonal element of $\mathbf{D}$, $d_{ii}$, if $d_{ii} > 0.01$ and is set to zero otherwise.

Show that only 10,100,010 numbers are necessary to represent $\ddot{\mathbf{W}}$ which consists of $10^{12}$ numbers. Let $\mathbf{q}$ be a query vector as defined in Example 1.6.1. In order to compute the similarity of $\mathbf{q}$ to the documents stored in $\ddot{\mathbf{W}}$ one typically evaluates $\mathbf{q}^T\ddot{\mathbf{W}}$ since the $k$th element of $\mathbf{q}^T\ddot{\mathbf{W}}$ is equal to the dot product of $\mathbf{q}$ with the $k$th document in the term by document matrix $\ddot{\mathbf{W}}$. Show that $\mathbf{q}^T\ddot{\mathbf{W}}$ can be computed without explicitly representing the large matrix $\ddot{\mathbf{W}}$ and only using the 10 singular values, 10 left eigenvectors, and 10 right eigenvectors of $\ddot{\mathbf{W}}$. △

## Exercises

4.2-1. Let $\mathbf{M}$ be a real $d$-dimensional symmetric matrix. Let $\lambda_{min}$ and $\lambda_{max}$ be respectively the smallest and largest eigenvalues of $\mathbf{M}$. Show that for all $\mathbf{x} \in \mathcal{R}^d$:

$$\lambda_{min}|\mathbf{x}|^2 \leq \mathbf{x}^T\mathbf{M}\mathbf{x} \leq \lambda_{max}|\mathbf{x}|^2.$$

HINT: Let $\mathbf{M} = \mathbf{E}\mathbf{D}\mathbf{E}^T$ where $\mathbf{D}$ is a diagonal matrix with non-negative elements on the main diagonal and the columns of $\mathbf{E}$ are the eigenvectors of $\mathbf{M}$. Let $\mathbf{y}(\mathbf{x}) = \mathbf{E}^T\mathbf{x}$. Then, show that for all $\mathbf{x}$ that $\mathbf{y}(\mathbf{x})^T\mathbf{D}\mathbf{y}(\mathbf{x}) = \sum_{i=1}^{d} \lambda_i y_i(\mathbf{x})^2$ where $\lambda_{min} \leq \lambda_i \leq \lambda_{max}$.

4.2-2. Show how an eigenvector decomposition of a real symmetric matrix $\mathbf{W}$ obtained using Theorem 4.2.2 is related to a singular value decomposition of $\mathbf{W}$.

4.2-3. Define $\mathbf{V}(\mathbf{x}) = -\mathbf{x}^T\mathbf{W}\mathbf{x}$ where $\mathbf{W}$ is a positive semidefinite real symmetric matrix. Show that $\mathbf{V}(\mathbf{x})$ may be rewritten as:

$$V(\mathbf{x}) = -\sum_{i=1}^{d} \lambda_i(\mathbf{e}_i^T\mathbf{x})^2$$

where $\lambda_i$ is the $i$th eigenvalue of $\mathbf{W}$ and $\mathbf{e}_i$ is the $i$th eigenvector of $\mathbf{W}$. Interpret $V$ as a measure of how similar $\mathbf{x}$ is to the subspace spanned by the eigenvectors with the largest eigenvalues of $\mathbf{W}$.

4.2-4. *Linear Inference Machine Response (Rectangular Matrix).* Consider a linear inference machine that generates an $m$-dimensional response vector $\mathbf{y}$ for a given $d$-dimensional input pattern $\mathbf{s}$ where $\mathbf{y} = \mathbf{W}\mathbf{s}$ so that $\mathbf{W} \in \mathcal{R}^{m \times d}$ may be a rectangular matrix. An SVD of $\mathbf{W}$ is done using a computer program to obtain $\mathbf{W} = \mathbf{U}\mathbf{D}\mathbf{V}^T$ where the columns of $\mathbf{U} \equiv [\mathbf{u}_1, \ldots, \mathbf{u}_r]$ are orthonormal eigenvectors of $\mathbf{W}\mathbf{W}^T$, the columns of $\mathbf{V} \equiv [\mathbf{v}_1, \ldots, \mathbf{v}_r]$ are orthonormal eigenvectors of $\mathbf{W}^T\mathbf{W}$, and $\mathbf{D}$ is a diagonal matrix with $r$ non-zero on-diagonal elements $d_1, \ldots, d_r$. Use Singular Value Decomposition (SVD) to show that:

$$\mathbf{W} = \sum_{i=1}^{r} d_i \mathbf{u}_i \mathbf{v}_i^T$$

and

$$\mathbf{y} = \sum_{i=1}^{r} d_i \beta_i \mathbf{v}_i.$$

where $\beta_i = (\mathbf{v}_i^T \mathbf{s})$ is the dot-product similarity measure between the input pattern $\mathbf{s}$ and the $i$th right eigenvector $\mathbf{v}_i$. Thus, the response is a weighted sum of right eigenvectors where the contribution of each right eigenvector is directly proportional to both the presence of that eigenvector in the matrix $\mathbf{W}$ as measured by $d_i$ and the similarity of that right eigenvector to the input pattern vector. Assume $d_r$ is very small in magnitude relative to $d_1, \ldots, d_{r-1}$. Explain why setting $d_r = 0$ might result in: (1) reduced storage requirements, (ii) reduced computation, and (iii) improved generalization performance.

## 4.3 Linear System Solution Theorems

Let $\mathbf{A}$ be a known $n$-dimensional square invertible matrix and $\mathbf{b}$ be a known $n$-dimensional vector. Assume the goal is to solve for an unknown $n$-dimensional column vector $\mathbf{x}$ which satisfies the system of linear equations $\mathbf{Ax} = \mathbf{b}$. When an exact solution for $\mathbf{x}$ exists, then $\mathbf{x} = \mathbf{A}^{-1}\mathbf{b}$.

In general, $\mathbf{A}$ may not be invertible. For example, $\mathbf{A}$ may not have full rank or $\mathbf{A}$ may be a rectangular matrix. However, in such cases, it is still possible to introduce the concept of a "matrix pseudo-inverse" and then apply the matrix pseudo-inverse in a similar manner to obtain solutions to systems of linear equations in a more general setting.

Let $\mathbf{A} \in \mathcal{R}^{m \times n}$ be a rectangular matrix with full column rank $n$ and $\mathbf{b}$ be an $m$-dimensional vector. The equation $\mathbf{Ax} = \mathbf{b}$ specifies a system of $m$ linear equations where the $n$-dimensional vector $\mathbf{x}$ is the unknown variable.

Define

$$\mathbf{A}^\dagger \equiv (\mathbf{A}^T\mathbf{A})^{-1}\mathbf{A}^T.$$

and left multiply both sides of

$$\mathbf{Ax} = \mathbf{b} \tag{4.5}$$

by $\mathbf{A}^\dagger$ to obtain:

$$\mathbf{A}^\dagger\mathbf{Ax} = \mathbf{A}^\dagger\mathbf{b}.$$

Now substitute the formula for $\mathbf{A}^\dagger$ to obtain

$$(\mathbf{A}^T\mathbf{A})^{-1}\mathbf{A}^T\mathbf{Ax} = \mathbf{A}^\dagger\mathbf{b} \tag{4.6}$$

which gives:

$$\mathbf{x} = \mathbf{A}^\dagger\mathbf{b}$$

by recognizing the terms on the left-hand side in (4.6) cancel out since $\mathbf{A}^T\mathbf{A}$ is an invertible $n$-dimensional square matrix. Thus, we can think of the matrix $\mathbf{A}^\dagger$ as a more generalized concept of the inverse of a matrix called the "left pseudoinverse" since $\mathbf{A}^\dagger\mathbf{A} = \mathbf{I}_n$. Note that $\mathbf{AA}^\dagger$ is not necessarily equal to the identity matrix $\mathbf{I}_m$ when $\mathbf{A}$ has full column rank $n$.

Now assume a second case where $\mathbf{A} \in \mathcal{R}^{m \times n}$ is a rectangular matrix with full row rank $m$ and the goal is to solve the system of linear equations $\mathbf{x}^T\mathbf{A} = \mathbf{b}^T$ for an unknown $m$-dimensional vector $\mathbf{x}$ where $\mathbf{A}$ and $\mathbf{b} \in \mathcal{R}^n$ are known.

Define the "right pseudoinverse"

$$\mathbf{A}^\dagger = \mathbf{A}^T(\mathbf{A}\mathbf{A}^T)^{-1}$$

and multiply the right-hand sides of $\mathbf{x}^T\mathbf{A} = \mathbf{b}^T$ to obtain:

$$\mathbf{x}^T\mathbf{A}\mathbf{A}^\dagger = \mathbf{b}^T\mathbf{A}^\dagger$$

which implies $\mathbf{x}^T = \mathbf{b}^T\mathbf{A}^\dagger$ since $\mathbf{A}\mathbf{A}^\dagger = \mathbf{I}_m$. Note that $\mathbf{A}^\dagger\mathbf{A}$ is not necessarily equal to the identity matrix $\mathbf{I}_n$ when $\mathbf{A}$ has full row rank $m$.

Moreover, in the special case where the matrix $\mathbf{A}$ is a square invertible matrix with full row rank and full column rank, then the pseudoinverse $\mathbf{A}^\dagger = \mathbf{A}^{-1}$.

**Definition 4.3.1** (Left and Right Pseudoinverses)**.** If $\mathbf{A} \in \mathcal{R}^{m\times n}$ has full column rank, then

$$\mathbf{A}^\dagger \equiv (\mathbf{A}^T\mathbf{A})^{-1}\mathbf{A}^T$$

is called a *left pseudoinverse.* If $\mathbf{A} \in \mathcal{R}^{m\times n}$ has full row rank, then

$$\mathbf{A}^\dagger \equiv \mathbf{A}^T(\mathbf{A}\mathbf{A}^T)^{-1}$$

is called a *right pseudoinverse.* □

In the more general situation involving rectangular matrices that do not have full row or column rank, the following SVD Theorem may be used to define an even more general type of pseudoinverse matrix.

**Definition 4.3.2** (Moore-Penrose Pseudoinverse)**.** Let the matrix $\mathbf{A} \in \mathcal{R}^{n\times m}$ have positive rank $k$. Let the singular value decomposition of $\mathbf{A}$ be defined such that $\mathbf{A} = \mathbf{U}\mathbf{D}\mathbf{V}^T$ where $\mathbf{U} \in \mathcal{R}^{n\times k}$, $\mathbf{D} \in \mathcal{R}^{k\times k}$ is a diagonal matrix with positive on-diagonal elements, and $\mathbf{V} \in \mathcal{R}^{m\times k}$. The *Moore-Penrose Pseudoinverse* of $\mathbf{A}$ is defined by the formula:

$$\mathbf{A}^\dagger = \mathbf{V}\mathbf{D}^{-1}\mathbf{U}^T.$$

□

Note that $\mathbf{D}^{-1}$ in Definition 4.3.2 is a diagonal matrix whose $i$th on-diagonal element is the reciprocal of the $i$th on-diagonal element of $\mathbf{D}$.

**Example 4.3.1** (Verify a Pseudoinverse Property of the Moore-Penrose Pseudoinverse)**.** Let the matrix $\mathbf{A} \in \mathcal{R}^{n\times m}$ have rank $k$ where $0 < k \leq \min\{m, n\}$. Let $\mathbf{A}^\dagger$ denote the Moore-Penrose pseudoinverse of $\mathbf{A}$. Show that

$$\mathbf{A}\mathbf{A}^\dagger\mathbf{A} = \mathbf{A}. \tag{4.7}$$

**Solution.** Rewrite $\mathbf{A}$ using a singular value decomposition such that $\mathbf{A} = \mathbf{U}\mathbf{D}\mathbf{V}^T$ where the $k$ columns of $\mathbf{U}$ are the left eigenvectors, the $k$ columns of $\mathbf{V}$ are the right eigenvectors, and $\mathbf{D}$ is a $k$-dimensional diagonal matrix. Then, $\mathbf{A}^\dagger = \mathbf{V}\mathbf{D}^{-1}\mathbf{U}^T$. Substitute the formulas for $\mathbf{A}$ and $\mathbf{A}^\dagger$ into (4.7) to obtain:

$$\mathbf{A}\mathbf{A}^\dagger\mathbf{A} = \mathbf{U}\mathbf{D}\mathbf{V}^T\mathbf{V}\mathbf{D}^{-1}\mathbf{U}^T\mathbf{U}\mathbf{D}\mathbf{V}^T = \mathbf{A}$$

since $\mathbf{U}^T\mathbf{U} = \mathbf{I}_k$, $\mathbf{D}^{-1}\mathbf{D} = \mathbf{I}_k$, $\mathbf{V}^T\mathbf{V} = \mathbf{I}_k$. △

Also note that the pseudoinverse is computed in MATLAB using the function `PINV` and a singular value decomposition is computed in MATLAB using the function `SVD`.

**Theorem 4.3.1** (Least Squares Minimization Using Pseudoinverse). *Let* $\mathbf{A} \in \mathcal{R}^{m \times d}$ *and* $\mathbf{B} \in \mathcal{R}^{m \times n}$. *Let* $\ell : \mathcal{R}^{d \times n} \rightarrow \mathcal{R}$ *be defined such that for all* $\mathbf{X} \in \mathcal{R}^{d \times n}$:

$$\ell(\mathbf{X}) = |\mathbf{AX} - \mathbf{B}|^2.$$

*Then,* $\mathbf{X}^* = \mathbf{A}^\dagger \mathbf{B}$ *is in the set,* $G$, *of global minimizers of* $\ell$. *If there is more than one global minimizer in* $G$, *then* $\mathbf{X}^*$ *is a global minimizer in* $G$ *such that* $|\mathbf{X}^*| \leq |\mathbf{Y}|$ *for all* $\mathbf{Y} \in G$.

*Proof.* Let $\mathbf{x}_i$ and $\mathbf{b}_i$ be the $i$th columns of $\mathbf{X}$ and $\mathbf{B}$ respectively for $i = 1, \ldots, n$. Rewrite $\ell$ such that:

$$\ell(\mathbf{X}) = \sum_{i=1}^{n} |\mathbf{Ax}_i - \mathbf{b}_i|^2 \tag{4.8}$$

and then apply Noble and Daniel (1977; Corollary 9.1, p. 338) to minimize the $i$th term in (4.8) by choosing $\mathbf{x}_i = \mathbf{A}^\dagger \mathbf{b}_i$ for $i = 1, \ldots, n$. ■

**Example 4.3.2** (Linear Regression Parameter Estimation Using SVD). Let the training data $\mathcal{D}_n$ be a collection of ordered pairs defined such that:

$$\mathcal{D}_n \equiv \{(\mathbf{s}_1, \mathbf{y}_1), \ldots, (\mathbf{s}_n, \mathbf{y}_n)\}$$

where $\mathbf{y}_i \in \mathcal{R}^m$ and $\mathbf{s}_i \in \mathcal{R}^d$ for $i = 1, \ldots, n$. Let $\mathbf{Y}_n \equiv [\mathbf{y}_1, \ldots, \mathbf{y}_n]$. Let $\mathbf{S}_n \equiv [\mathbf{s}_1, \ldots, \mathbf{s}_n]$. The goal of the analysis is to minimize the objective function $\ell_n : \mathcal{R}^{m \times d} \rightarrow \mathcal{R}$ defined such that:

$$\ell_n(\mathbf{W}_n) = |\mathbf{Y}_n - \mathbf{W}_n \mathbf{S}_n|^2 \tag{4.9}$$

so that the predicted response $\ddot{\mathbf{y}}_i \equiv \mathbf{W}_n \mathbf{s}_i$ for input pattern $\mathbf{s}_i$ is as close as possible to the desired response $\mathbf{y}_i$ in a least squares sense.

Note that in the special case of "perfect learning" a matrix $\mathbf{W}_n^*$ is obtained such that $\ell(\mathbf{W}_n^*) = 0$, then it follows that for a given input vector $\mathbf{s}_k$ the desired response $\mathbf{y}_k$ is produced by the formula $\mathbf{y}_k = \mathbf{W}_n^* \mathbf{s}_k$. When this exact solution does not exist, then the minimum mean square error solution across the set of training data is obtained.

**Solution.** Use Theorem 4.3.1 to obtain

$$\mathbf{W}_n^* = \mathbf{Y}_n \left[\mathbf{S}_n\right]^\dagger. \tag{4.10}$$

△

**Theorem 4.3.2** (Sherman-Morrison Theorem). *Let* $\mathbf{W}$ *be an invertible square* $d$*-dimensional matrix. Let* $\mathbf{u}$ *and* $\mathbf{v}$ *be* $d$*-dimensional column vectors. Assume* $\mathbf{W} + \mathbf{uv}^T$ *is invertible. Then,*

$$\left(\mathbf{W} + \mathbf{uv}^T\right)^{-1} = \mathbf{W}^{-1} - \frac{\mathbf{W}^{-1}\mathbf{uv}^T\mathbf{W}^{-1}}{1 + \mathbf{v}^T\mathbf{W}^{-1}\mathbf{u}}.$$

*Proof.* See Bartlett (1951). ■

**Example 4.3.3** (Matrix Inversion Using Adaptive Learning). Consider a sequence of learning problems of the type encountered in Example 4.3.2. Let $\mathbf{S}_n \equiv [\mathbf{s}_1, \ldots, \mathbf{s}_n]$. Let $\mathbf{Y}_n \equiv [\mathbf{y}_1, \ldots, \mathbf{y}_n]$. Let $\mathcal{D}_n \equiv \{(\mathbf{s}_1, \mathbf{y}_1), \ldots, (\mathbf{s}_n, \mathbf{y}_n)\}$, where $\mathbf{y}_k$ is the desired response $\ddot{\mathbf{y}}(\mathbf{s}_k, \mathbf{W}) \equiv \mathbf{W}\mathbf{s}_k$ for input pattern $\mathbf{s}_k$. Given just data set $\mathcal{D}_n$, the goal of learning is to estimate the parameter matrix $\hat{\mathbf{W}}_n$ such that $\hat{\mathbf{W}}_n$ is a global minimizer of

$$\ell_n(\mathbf{W}_n) = |\mathbf{Y}_n - \mathbf{W}_n \mathbf{S}_n|^2. \tag{4.11}$$

Using the solution in (4.10) from Example 4.3.2, $\ell_n$ is minimized by choosing:

$$\hat{\mathbf{W}}_n \equiv \mathbf{Y}_n[\mathbf{S}_n]^{\dagger}.$$

Now use the formula for the right pseudoinverse in Definition 4.3.1 to obtain the formula:

$$[\mathbf{S}_n]^{\dagger} = \mathbf{S}_n^T \mathbf{R}_n^{-1}$$

where $\mathbf{R}_n \equiv \mathbf{S}_n \mathbf{S}_n^T$ is invertible.

It is computationally expensive to invert $\mathbf{R}_n$ each time $\mathbf{S}_n$ is updated so a method for computing the inverse of $\mathbf{R}_{n+1}$ given $\mathbf{s}_{n+1}$ and $\mathbf{R}_n^{-1}$ is desired.

That is, the goal is to compute a sequence of solutions $\hat{\mathbf{W}}_n, \hat{\mathbf{W}}_{n+1}, \ldots$ for a sequence of data sets $\mathcal{D}_n, \mathcal{D}_{n+1}, \ldots$ respectively where $\mathcal{D}_{n+1} \equiv [\mathcal{D}_n, (\mathbf{s}_{n+1}, \mathbf{y}_{n+1})]$ for $n = 1, 2, \ldots$.

**Solution.** Use the Sherman-Morrison Theorem to obtain the relation:

$$\mathbf{R}_{n+1}^{-1} = \mathbf{R}_n^{-1} - \frac{\mathbf{R}_n^{-1}\mathbf{s}_{n+1}\mathbf{s}_{n+1}^T \mathbf{R}_n^{-1}}{1 + \mathbf{s}_{n+1}^T \mathbf{R}_n^{-1} \mathbf{s}_{n+1}}.$$

The matrix $\mathbf{R}_{n+1}^{-1}$ can then be used to compute:

$$\mathbf{W}_{n+1} = \mathbf{Y}_{n+1}[\mathbf{S}_{n+1}]^{\dagger}$$

where

$$[\mathbf{S}_{n+1}]^{\dagger} = \mathbf{S}_{n+1}^T \mathbf{R}_{n+1}^{-1}.$$

Thus, (4.3.3) may be used to increase the computational efficiency of "updating" the input pattern correlation matrix $\mathbf{R}_n^{-1}$ to account for an additional data point $(\mathbf{s}_{n+1}, \mathbf{y}_{n+1})$ by avoiding a full matrix inversion calculation of $\mathbf{R}_n$, $\mathbf{R}_{n+1}$, and so on. △

## Exercises

4.3-1. *Finding a Linear Preprocessing Transformation Using SVD.* Let $\mathbf{s}_1, \ldots, \mathbf{s}_n$ be a collection of $d$-dimensional feature vectors. Assume $\mathbf{s}_i$ is recoded by a linear transformation $\mathbf{W} \in \mathcal{R}^{m \times d}$ such that $\mathbf{h}_i = \mathbf{W}\mathbf{s}_i$ where the positive integer $m$ is less than $d$. The pattern vector $\mathbf{h}_i$ is a linear recoding transformation of $\mathbf{s}_i$ for $i = 1, \ldots, n$. Use singular value decomposition to find a $\mathbf{W}$ such that the empirical risk function

$$\ell(\mathbf{W}) = (1/n)\sum_{i=1}^{n} |\mathbf{s}_i - \mathbf{W}^T\mathbf{h}_i|^2 = (1/n)\sum_{i=1}^{n} |\mathbf{s}_i - \mathbf{W}^T\mathbf{W}\mathbf{s}_i|^2$$

is minimized.

4.3-2. *Linear Autoencoder* Show how to interpret a gradient descent algorithm that minimizes (4.3-1) as a linear version of the autoencoder learning machine described in Example (1.6.2).

4.3-3. *Training a Linear Machine with Scalar Response.* Let the training data $\mathcal{D}_n$ be a collection of ordered pairs defined such that:

$$\mathcal{D}_n \equiv \{(y_1, \mathbf{s}_1)), \ldots, (y_n, \mathbf{s}_n)\}$$

where $y_i \in \mathcal{R}$ and $\mathbf{s}_i \in \mathcal{R}^d$. Assume $\mathbf{S} \equiv [\mathbf{s}_1, \ldots, \mathbf{s}_n]$ has full row rank. The goal of learning is to construct a column vector $\mathbf{w}^*$ which is a minimizer of

$$\ell(\mathbf{w}) \equiv \sum_{i=1}^{n} \left| y_i - \mathbf{w}^T \mathbf{s}_i \right|^2 .$$

Using the method of singular value decomposition, show how to derive a simple formula for computing $\mathbf{w}^*$.

4.3-4. *Multilayer Linear Network Interpretation of SVD.* Let $\mathbf{y}_i$ be the desired response column vector pattern for input column pattern vector $\mathbf{s}_i$ for $i = 1, \ldots, n$. Let $\ddot{\mathbf{y}}_i(\mathbf{W}, \mathbf{V}) \equiv \mathbf{W}\mathbf{h}_i(\mathbf{V})$ and $\mathbf{h}_i(\mathbf{V}) \equiv \mathbf{V}\mathbf{s}_i$. The column vector $\ddot{\mathbf{y}}_i(\mathbf{W}, \mathbf{V})$ is interpreted as the state of the output units of a multilayer linear feedforward network generated from the state of the hidden units, $\mathbf{h}_i(\mathbf{V})$, which in turn are generated from the input pattern $\mathbf{s}_i$. The goal of learning is to minimize the objective function $\ell$ defined such that:

$$\ell(\mathbf{W}, \mathbf{V}) = (1/n) \sum_{i=1}^{n} |\mathbf{y}_i - \ddot{\mathbf{y}}_i(\mathbf{W}, \mathbf{V})|^2.$$

Obtain a formula for a global minimizer of $\ell$ using singular value decomposition. In addition, discuss how this solution is essentially the same as the solution obtained from a gradient descent algorithm that minimizes $\ell$.

4.3-5. *Adaptive Natural Gradient Descent.* In some cases, the rate of convergence of the standard gradient descent algorithm can be improved by using the method of "natural gradient descent". An adaptive natural gradient descent algorithm may be defined as follows. Let $c(\mathbf{x}, \boldsymbol{\theta})$ denote the loss incurred by the learning machine when it observes the $d$-dimensional event $\mathbf{x}$ given that its parameter values are set to the value of the $q$-dimensional parameter vector $\boldsymbol{\theta}$. Let $\mathbf{g}(\mathbf{x}, \boldsymbol{\theta})$ be a column vector which is defined as the transpose of the derivative of $c(\mathbf{x}; \boldsymbol{\theta})$. Let $\mathbf{x}(0), \mathbf{x}(1), \ldots$ be a sequence of $d$-dimensional vectors observed by an adaptive gradient descent learning machine defined by the time-varying iterated map

$$\mathbf{f}(\boldsymbol{\theta}, t, \mathbf{x}(t)) = \boldsymbol{\theta} - \gamma_t \mathbf{G}_t^{-1} \mathbf{g}(\mathbf{x}(t), \boldsymbol{\theta})$$

where $\gamma_1, \gamma_2, \ldots$ is a sequence of strictly positive stepsizes, $\mathbf{G}_0 = \mathbf{I}_q$, and for all $t \in \mathcal{N}^+$ :

$$\mathbf{G}_t \equiv (1/T) \sum_{s=0}^{T-1} \mathbf{g}(\mathbf{x}(t-s), \boldsymbol{\theta})[\mathbf{g}(\mathbf{x}(t-s), \boldsymbol{\theta})]^T.$$

Show how to use the Sherman-Morrison Theorem to design a computationally efficient way of computing $\mathbf{G}_{t+1}^{-1}$ from $\mathbf{G}_t^{-1}$ and $\mathbf{x}(t+1)$ when $\mathbf{G}_t^{-1}$ exists. Express your solution in terms of a new time-varying iterated map.

## 4.4 Further Readings

### Linear Algebra

This chapter covered selected standard topics from linear algebra especially relevant to machine learning. Students interested in further discussion of these topics should check out the texts by Franklin (1968), Noble and Daniel (1977), Schott (2005), Banerjee et al. (2014), Strang (2016), and Harville (2018).

### Advanced Matrix Operations

Some of the more esoteric matrix operations such as the Hadamard and Kronecker Tensor Products are discussed in Magnus and Neudecker (2001), Schott (2005), and Marlow (2012).

# 5

# *Matrix Calculus for Machine Learning*

**Learning Objectives**

- Apply vector and matrix chain rules.
- Construct Taylor series function approximations.
- Design gradient descent algorithms.
- Design descent algorithm stopping criteria.
- Solve constrained optimization problems.

Vector calculus is an essential tool for machine learning algorithm analysis and design for several reasons. First, as previously discussed in Chapter 1, many commonly used high-dimensional machine learning algorithms can be interpreted as gradient descent type algorithms that minimize a smooth or almost smooth objective function. The design of such algorithms requires that one computes the objective function's derivative. Second, vector calculus methods can be used to investigate the shape of the objective function by locally approximating the objective function using a multidimensional Taylor series expansion. Such information regarding the objective function's shape can provide critical insights into the behavior of gradient descent type algorithms which use the objective function's shape to guide their search. In addition, (see Chapter 15 and Chapter 16), the approximation error associated with estimating an objective function using only a finite rather than an infinite amount of training data can also be investigated using multidimensional Tayler series expansions. This latter approximation error plays a crucial role in developing practical methods for characterizing the learning machine's generalization performance. And third, in many machine learning applications, one is interested in minimizing an objective function subject to a collection of nonlinear constraints. Lagrange multiplier methods for multidimensional vector spaces can be used to derive a new objective function that can be minimized using gradient descent type methods for problems that require the original objective function is minimized subject to specific nonlinear constraints.

## 5.1 Convergence and Continuity

### 5.1.1 Deterministic Convergence

**Definition 5.1.1** (Point Sequence). A *sequence* of points in a metric space $(E, \rho)$ is a function $\mathbf{x} : \mathbb{N} \to E$. □

A sequence of points in $\mathcal{R}^d$ may be denoted as either:
(i) $\mathbf{x}(0), \mathbf{x}(1), \mathbf{x}(2), \ldots$, (ii) $\{\mathbf{x}(t)\}_{t=0}^{\infty}$, or (iii) $\{\mathbf{x}(t)\}$.

**Definition 5.1.2** (Point Convergence). Let $\mathbf{x}(1), \mathbf{x}(2), \ldots$ be a sequence of points in a metric space $(E, \rho)$. Let $\mathbf{x}^* \in E$. Assume that for every positive real number $\epsilon$ there exists a positive integer $T(\epsilon)$ such that $\rho(\mathbf{x}(t), \mathbf{x}^*) < \epsilon$ whenever $t > T(\epsilon)$. Then, the sequence $\mathbf{x}(1), \mathbf{x}(2), \ldots$ is said to *converge* or *approach* the *limit point* $\mathbf{x}^*$. A set of limit points is called a *limit set.* The sequence $\mathbf{x}(1), \mathbf{x}(2), \ldots$ is called a *convergent sequence* to $\mathbf{x}^*$. □

The terminology $\mathbf{x}(t) \rightarrow \mathbf{x}^*$ as $t \rightarrow \infty$ means that $\mathbf{x}(1), \mathbf{x}(2), \ldots$ is a convergent sequence with *limit* $\mathbf{x}^*$. If the distance function $\rho$ for metric space is not explicitly specified, then it will always be assumed that $\rho$ is the Euclidean vector distance function and the metric space is the Euclidean vector metric space.

For example, the sequence $x(1), x(2), \ldots$ defined such that: $x(t) = (-1)^t$ is not a convergent sequence to some point $x^*$. The sequence defined by $x(t) = [1 - (0.5)^t]$, however, is a convergent sequence to 1.

The definition of a convergent sequence does not require $d_1 \geq d_2 \geq d_3 \geq \ldots$. For example, the sequence defined by $x(t) = (1/2)^t$ whose terms decrease in value converges to zero. However, the sequence defined by $x(t) = (-1/2)^t$ (i.e., $-1/2, 1/4, -1/8, 1/16, -1/32, \ldots$) also converges to zero.

**Definition 5.1.3** (lim inf and lim sup). Let $x_1, x_2, \ldots$ be a sequence of points in $\mathcal{R}$. Let $S_t \equiv \inf\{x_t, x_{t+1}, \ldots\}$. Let $R_t \equiv \sup\{x_t, x_{t+1}, \ldots\}$. The lim inf and lim sup are defined as the limit points of the sequences $S_1, S_2, \ldots$ and $R_1, R_2, \ldots$ respectively. □

For example, consider the sequence $S$ of numbers $x(1), x(2), \ldots$ where $x(t) = -1$ for $t = 1, 2, 3, 4$ and $x(t) = (1/2)^{t-5}$ for $t = 5, 6, 7, \ldots$. Note that $\inf S$ is equal to $-1$ but $\liminf S = 0$. Now define the sequence $R$ of numbers $x(1), x(2), \ldots$ such that $x(t) = (-1)^t$ for $t = 1, 2, \ldots$, then $\liminf R = -1$ and $\limsup R = 1$.

**Definition 5.1.4** (Convergence to Infinity). A sequence of numbers $x_1, x_2, \ldots$ *converges to* $+\infty$ if for every positive real number $K$: there exists a positive integer $N_K$ such that $x_n > K$ for all $n > N_K$. □

Note that if $-x_n \rightarrow +\infty$, then $x_1, x_2, \ldots$ *converges to* $-\infty$.

**Definition 5.1.5** (Bounded Sequence). A sequence of points $\mathbf{x}(1), \mathbf{x}(2), \ldots$ in $\mathcal{R}^d$ is *bounded* if there exists a finite number $K$ such that $|\mathbf{x}(t)| \leq K$ for $t = 1, 2, \ldots$. □

Note that a critical aspect of the definition of a bounded sequence is that the constant $K$ is not functionally dependent upon the time index $t$. In other words, the sequence of points $1, 2, 4, 8, 16, \ldots$ specified by the formula $x(t) = 2^{t-1}$ is not bounded but each point in the sequence is a finite number.

Note that a sequence which is not bounded does not necessarily converge to infinity (e.g., $0, 1, 0, 2, 0, 4, 0, 8, 0, 16, 0, 32, \ldots$). However, if a sequence converges to infinity, then the sequence is not a bounded sequence.

**Example 5.1.1** (A Bounded Sequence of Vectors Is Contained in a Ball). Assume a sequence of $d$-dimensional vectors $\mathbf{x}(1), \mathbf{x}(2), \ldots$ is bounded so that $|\mathbf{x}(t)| \leq K$ for $t = 1, 2, \ldots$. Then, for all $t \in \mathbb{N}^+$: $\mathbf{x}(t)$ is an element of a closed ball of radius $K$ centered at the origin. △

**Example 5.1.2** (A Convergent Sequence of Vectors Is a Bounded Sequence). Assume the sequence of $d$-dimensional real vectors $\mathbf{x}(1), \mathbf{x}(2), \ldots$ converges to $\mathbf{x}^*$. That is, assume for every positive number $\epsilon$ there exists a $T_\epsilon$ such that for all $t > T_\epsilon$: $|\mathbf{x}(t) - \mathbf{x}^*| < \epsilon$. Let

$$R_\epsilon \equiv \max\{|\mathbf{x}(1)|, \ldots, |\mathbf{x}(T_\epsilon)|\}.$$

Then, the sequence $\mathbf{x}(1), \mathbf{x}(2), \ldots$ is contained in a ball of radius $\max\{R_\epsilon, |\mathbf{x}^*| + \epsilon\}$. △

The concept of a subsequence is now introduced.

**Definition 5.1.6** (Subsequence). If $a_1, a_2, a_3, \ldots$, is a countably infinite sequence of objects and if $n_1, n_2, n_3, \ldots$ is a countably infinite sequence of positive integers such that $n_1 < n_2 < n_3 < \ldots$, then the countably infinite sequence $a_{n_1}, a_{n_2}, a_{n_3}, \ldots$, is called a *subsequence*. □

For example, given the sequence of vectors $\mathbf{x}_1, \mathbf{x}_2, \mathbf{x}_3, \ldots$, then the sequence $\mathbf{x}_2, \mathbf{x}_3, \mathbf{x}_{45}, \mathbf{x}_{99}, \mathbf{x}_{100}, \ldots$ is a subsequence of $\mathbf{x}_1, \mathbf{x}_2, \mathbf{x}_3, \ldots$.

**Theorem 5.1.1** (Properties of Convergent and Divergent Sequences).

*(i) If $S$ is a convergent sequence of points in a metric space with finite limit point $\mathbf{x}^*$, then every subsequence of $S$ is a convergent sequence with finite limit point $\mathbf{x}^*$.*

*(ii) Let $\Omega$ be a subset of a metric space. The set $\Omega$ is closed if and only if $\Omega$ contains the limit point of every convergent sequence of points in $\Omega$.*

*(iii) A convergent sequence of points in $\mathcal{R}^d$ is a bounded sequence of points.*

*(iv) If a sequence of points $\mathbf{x}_1, \mathbf{x}_2, \ldots$ does not converge to a point $\mathbf{x}^*$ in $\mathcal{R}^d$, then there exists at least one subsequence of $\mathbf{x}_1, \mathbf{x}_2, \ldots$ which does not converge to $\mathbf{x}^*$.*

*(v) Assume as $T \to \infty$ that $\sum_{t=1}^{T} a_t \to K$ where $K$ is a finite number. Then, $a_T \to 0$ as $T \to \infty$.*

*(vi) If $\mathbf{x}_1, \mathbf{x}_2, \ldots$ is not a bounded sequence, then there exists a subsequence $\mathbf{x}_{t_1}, \mathbf{x}_{t_2}, \ldots$ of $\mathbf{x}_1, \mathbf{x}_2, \ldots$ such that $|\mathbf{x}_{t_k}| \to \infty$ as $k \to \infty$.*

*(vii) Assume the sequence of numbers $x_1, x_2, \ldots$ is non-increasing so that $x_k \geq x_{k+1}$ for all integer $k$. If, in addition, $x_1, x_2, \ldots$ is bounded from below, then the sequence of numbers $x_1, x_2, \ldots$ converges to some number $x^*$.*

*Proof.* (i) Rosenlicht (1968, pp. 46-47, p. 56). (ii) Rosenlicht (1968, p. 47, 56). (iii) See Example 5.1.2. (iv) See Exercise 5.1-1. (v) Since $S_T = \sum_{t=1}^{T} a_t \to K$ as $T \to \infty$, then $a_T = S_T - S_{T-1} \to 0$ as $T \to \infty$. (vi) Construct a sequence such that $|\mathbf{x}_{t_k}|$ is not only increasing as $k \to \infty$ but additionally for a given finite constant $X_{max}$ there exists a positive integer $K$ which is possibly dependent upon $X_{max}$ such that for all $k \geq K$: $|\mathbf{x}_{t_k}| > X_{max}$. (vii) See Rosenlicht (1968, p. 50). ■

The following theorem is useful for determining if sums of real numbers converge to a finite number or converge to infinity. The terminology $\sum_{t=1}^{\infty} \eta_t = \infty$ means that as $n \to \infty$, $\sum_{t=1}^{n} \eta_t \to \infty$. The terminology $\sum_{t=1}^{\infty} \eta_t < \infty$ means that as $n \to \infty$, $\sum_{t=1}^{n} \eta_t \to K$ where $K$ is a finite number.

**Theorem 5.1.2** (Convergent and Divergent Series for Adaptive Learning).

$$\sum_{t=1}^{\infty} \frac{1}{t} = \infty. \tag{5.1}$$

$$\sum_{t=1}^{\infty} \frac{1}{t^2} < K < \infty. \tag{5.2}$$

*Proof.* See Wade (1995, p. 146, 154) for a proof. ■

The following limit comparison test theorem is useful for the analysis and design of step-size sequence choices for adaptive learning algorithms. Such adaptive learning algorithms are discussed in Chapter 12.

**Theorem 5.1.3** (Limit Comparison Test for Adaptive Learning). *Let $\gamma_1, \gamma_2, \ldots$ and $\eta_1, \eta_2, \ldots$ be two sequences of strictly positive numbers. Assume that as $t \to \infty$, that $\gamma_t/\eta_t \to K$ where $K$ is a finite positive number. (i) $\sum_{t=1}^{\infty} \eta_t = \infty$ if and only if $\sum_{t=1}^{\infty} \gamma_t = \infty$. (ii) $\sum_{t=1}^{\infty} \eta_t^2 < \infty$ if and only if $\sum_{t=1}^{\infty} \gamma_t^2 < \infty$.*

*Proof.* See Wade (1995, Theorem 4.7, p. 155). ■

**Example 5.1.3** (Adaptive Learning Rate Adjustment). In Chapter 12, a stochastic convergence theorem is provided which can be used to characterize the asymptotic behavior of discrete-time adaptive learning machines. In this theorem, the learning rate $\gamma_t$ of the learning machine is gradually decreased resulting in a sequence of learning rates $\gamma_1, \gamma_2, \ldots$. In particular, the sequence of learning rates must be chosen to decrease sufficiently fast so that:

$$\sum_{t=1}^{T} \gamma_t^2 \to K < \infty \tag{5.3}$$

as $T \to \infty$. In addition, the sequence of learning rates is also constrained to not decrease too rapidly by requiring that the condition:

$$\sum_{t=1}^{T} \gamma_t \to \infty \tag{5.4}$$

as $T \to \infty$ is also satisfied. Show that $\gamma_t = 1/(1+t)$ satisfies conditions (5.3) and(5.4).

**Solution.** Using Theorem 5.1.2, $\sum_{t=1}^{\infty} \eta_t = \infty$ and $\sum_{t=1}^{\infty} (\eta_t)^2 < \infty$ if $\eta_t \equiv (1/t)$. Since $\gamma_t/\eta_t = t/(t+1) \to 1$ as $t \to \infty$, the Limit Comparison Test Theorem 5.1.3 implies that $\sum \gamma_t = \infty$ since $\sum \eta_t = \infty$. Since $\gamma_t/\eta_t \to 1$ as $t \to \infty$, the Limit Comparison Test Theorem implies that $\sum \gamma_t^2 < \infty$ since $\sum \eta_t^2 < \infty$ by Theorem 5.1.2. △

**Definition 5.1.7** (Uniform Convergence of Functions). Let $f_n : \mathcal{R}^d \to \mathcal{R}$ for $n = 1, 2, \ldots$ be a sequence of functions. Let $f : \mathcal{R}^d \to \mathcal{R}$. Let $\Omega \subseteq \mathcal{R}^d$. If for each $\mathbf{x} \in \Omega$, $f_n(\mathbf{x}) \to f(\mathbf{x})$ then the sequence of functions $f_1, f_2, \ldots$ *converges pointwise to* $f$ on $\Omega$. The sequence $f_1, f_2, \ldots$ is said to *converge uniformly to* $f$ on $\Omega$ if for every positive $\epsilon$ there exists an $N_\epsilon$ such that for all $n > N_\epsilon$: $|f_n(\mathbf{x}) - f(\mathbf{x})| < \epsilon$ for all $\mathbf{x} \in \Omega$. □

For example, let $f_n(x) = 1/(1 + \exp(-x/n))$ and let $f(x) = 1$ if $x > 0$, $f(x) = 0$ if $x < 0$, and $f(0) = 1/2$. The sequence $f_1, f_2, \ldots$ converges pointwise on $\mathcal{R}$. The sequence $f_1, f_2, \ldots$ converges both pointwise and uniformly to $f$ on $\{a \leq x \leq b\}$ where $0 < a < b$. The sequence $f_1, f_2, \ldots$ does not converge uniformly on $\mathcal{R}$. To see this note that for every positive $\epsilon$ and for every $N_\epsilon$, there does not exist a sufficiently small positive number $\delta_\epsilon$ such that $|f_n(x) - f(0)| < \epsilon$ for all $|x| < \delta_\epsilon$.

**Definition 5.1.8** (Convergence to a Set). Let $\Gamma$ be a nonempty subset of $E$. Let $\mathbf{x}(1), \mathbf{x}(2), \ldots$ be a sequence of points in a metric space $(E, \rho)$ such that as $t \to \infty$,

$$\inf_{\mathbf{y} \in \Gamma} \rho(\mathbf{x}(t), \mathbf{y}) \to 0.$$

Then, the sequence $\mathbf{x}(1), \mathbf{x}(2), \ldots$ *converges to the set* $\Gamma$. □

The notation $\mathbf{x}(t) \to \Gamma$ is also used to denote the concept that the sequence $\mathbf{x}(1), \mathbf{x}(2), \ldots$ converges to the set $\Gamma$.

Less formally, $\mathbf{x}(t) \to \Gamma$ provided that the distance from $\mathbf{x}(t)$ to the "closest" point in $\mathcal{H}$ at iteration $t$ approaches zero as $t \to \infty$.

In the special case where $\mathbf{x}(t) \to \Gamma$ and $\Gamma$ consists of a single element (i.e., $\Gamma = \{\mathbf{x}^*\}$), then the definition of convergence to a subset, $\Gamma$, of $E$ coincides with the definition of convergence to a point.

The notion of convergence to a set is illustrated by the following example. Let $\Gamma_1 = \{-1, 1, 14\}$ and let $\Gamma_2 = \{1, 14\}$. If $x(t) \to 14$ as $t \to \infty$, then $x(t) \to \Gamma_1$ and $x(t) \to \Gamma_2$ as $t \to \infty$. If $x(t) = [1 - (0.5)^t](-1)^t$ for $t = 0, 1, 2, \ldots$, then $x(t) \to \Gamma_1$ as $t \to \infty$ but $\mathbf{x}(1), \mathbf{x}(2), \ldots$ does not converge to $\Gamma_2$.

**Definition 5.1.9** (Big O Notation). Let $f : \mathcal{R}^d \to \mathcal{R}$ and $g : \mathcal{R}^d \to \mathcal{R}$. The notation $f(\mathbf{x}) = O(g(\mathbf{x}))$ as $\mathbf{x} \to \mathbf{x}^*$ means that there exists positive number $K$ and positive number $M$ such that for all $\mathbf{x}$ such that $|\mathbf{x} - \mathbf{x}^*| < K$:

$$|f(\mathbf{x})| \leq M|\mathbf{g}(\mathbf{x})|.$$

□

Less formally, this means $f(\mathbf{x}) \to f(\mathbf{x}^*)$ at least as rapidly as $g(\mathbf{x}) \to g(\mathbf{x}^*)$ as $\mathbf{x} \to \mathbf{x}^*$.

**Definition 5.1.10** (Little O Notation). Let $f : \mathcal{R}^d \to \mathcal{R}$ and $g : \mathcal{R}^d \to \mathcal{R}$. The notation $f(\mathbf{x}) = o(g(\mathbf{x}))$ as $\mathbf{x} \to \mathbf{x}^*$ means that $|f(\mathbf{x})|/|g(\mathbf{x})| \to 0$ as $\mathbf{x} \to \mathbf{x}^*$. □

Less formally, this means $f(\mathbf{x}) \to f(\mathbf{x}^*)$ more rapidly than $g(\mathbf{x}) \to g(\mathbf{x}^*)$ as $\mathbf{x} \to \mathbf{x}^*$.

**Example 5.1.4** (Example Big O and Little O Calculations). Let $\mathbf{M}$ be a $d$-dimensional positive definite symmetric matrix and $f : \mathcal{R}^d \to \mathcal{R}$ be defined such that for all $\mathbf{x} \in \mathcal{R}^d$:

$$f(\mathbf{x}) = (\mathbf{x} - \mathbf{x}^*)^T \mathbf{M}(\mathbf{x} - \mathbf{x}^*). \tag{5.5}$$

Note that since $\mathbf{M}$ can be diagonalized that the inequality

$$(\mathbf{x} - \mathbf{x}^*)^T \mathbf{M}(\mathbf{x} - \mathbf{x}^*) \leq \lambda_{max}|\mathbf{x} - \mathbf{x}^*|^2$$

holds where $\lambda_{max}$ is the largest eigenvalue of $\mathbf{M}$.

It then follows that as $\mathbf{x} \to \mathbf{x}^*$ that using the definition of $f$ in (5.5):

$$f(\mathbf{x}) = O(1), f(\mathbf{x}) = O(|\mathbf{x} - \mathbf{x}^*|), f(\mathbf{x}) = O(|\mathbf{x} - \mathbf{x}^*|^2), f(\mathbf{x}) \neq O(|\mathbf{x} - \mathbf{x}^*|^3)$$

and

$$f(\mathbf{x}) = o(1), f(\mathbf{x}) = o(|\mathbf{x} - \mathbf{x}^*|), f(\mathbf{x}) \neq o(|\mathbf{x} - \mathbf{x}^*|^2), f(\mathbf{x}) \neq o(|\mathbf{x} - \mathbf{x}^*|^3).$$

Note that $f(\mathbf{x}) = o(|\mathbf{x} - \mathbf{x}^*|^2)$ implies $f(\mathbf{x}) = O(|\mathbf{x} - \mathbf{x}^*|^2)$ but the converse is not necessarily true. △

The following convergence rate definitions are useful for characterizing the rate of convergence of a descent algorithm in the final stages of convergence. Luenberger (1984) provides additional discussions of the features of these definitions.

**Definition 5.1.11** (Convergence Rate Order). Assume an algorithm $A$ generates a sequence of $d$-dimensional vectors $\mathbf{x}(0), \mathbf{x}(1), \mathbf{x}(2), \ldots$, that converge to a point $\mathbf{x}^*$ in $\mathcal{R}^d$. The *convergence rate order* of $\{\mathbf{x}(t)\}$ is defined as:

$$\rho^* = \sup \left\{ \rho \geq 0 : \limsup_{t \to \infty} \frac{|\mathbf{x}(t+1) - \mathbf{x}^*|}{|\mathbf{x}(t) - \mathbf{x}^*|^\rho} \leq \beta < \infty \right\}.$$

The terminology "order of convergence" is also commonly used to refer to the convergence rate order.

Less formally, the definition of the convergence rate order of an algorithm means that if an algorithm generates a sequence $\mathbf{x}(0), \mathbf{x}(1), \ldots$ with an order of convergence $\rho$, then for sufficiently large $t$ and some finite positive number $\beta$ the relation

$$|\mathbf{x}(t+1) - \mathbf{x}^*| \leq \beta |\mathbf{x}(t) - \mathbf{x}^*|^\rho$$

approximately holds.

Algorithms that have convergence rates of order 1 are prevalent in machine learning. The following definition identifies an important subclass of such algorithms.

**Definition 5.1.12** (Linear Convergence Rate). Assume an algorithm $A$ generates a sequence of $d$-dimensional vectors $\mathbf{x}(0), \mathbf{x}(1), \mathbf{x}(2), \ldots,$ that converge to a point $\mathbf{x}^*$ in $\mathcal{R}^d$ such that

$$\lim_{t \to \infty} \frac{|\mathbf{x}(t+1) - \mathbf{x}^*|}{|\mathbf{x}(t) - \mathbf{x}^*|} = \beta$$

where $\beta < 1$, then $\mathbf{x}(0), \mathbf{x}(1), \ldots$ *converges linearly* at an *asymptotic geometric convergence rate* to $\mathbf{x}^*$ with *convergence ratio* $\beta$. If $\beta = 0$ then $\mathbf{x}(0), \mathbf{x}(1), \ldots$ converges *superlinearly* to $\mathbf{x}^*$.

Note that an algorithm with a convergence rate of order one does not necessarily have a linear convergence rate because the convergence ratio may be greater than or equal to one. Furthermore, note that every algorithm with a convergence rate greater than order one converges at a superlinear convergence rate.

Definition 5.1.11 only characterizes the speed of convergence in the final stages of convergence when the system state $\mathbf{x}(t)$ is close to its asymptotic destination $\mathbf{x}^*$. These convergence rate definitions often do not provide useful insights regarding: (1) the time required to enter the vicinity of a particular minimizer, (2) the time required to complete the computation for an algorithm iteration, or (3) the time required to *reach* the destination $\mathbf{x}^*$.

### 5.1.2 Continuous Functions

**Definition 5.1.13** (Continuous Function). Let $\Omega$ and $\Gamma$ be metric spaces with respective distance functions $\rho_\Omega$ and $\rho_\Gamma$. Let $f : \Omega \to \Gamma$. The function $f$ is *continuous at* $\mathbf{x} \in \Omega$ if for every positive number $\epsilon$ there exists a positive number $\delta_\mathbf{x}$ such that $\rho_\Gamma(f(\mathbf{x}), f(\mathbf{y})) < \epsilon$ for all $\mathbf{y}$ satisfying $\rho_\Omega(\mathbf{x}, \mathbf{y}) < \delta_\mathbf{x}$. If $f$ is continuous at every point in $\Omega$ then $f$ is *continuous on* $\Omega$. □

**Example 5.1.5** (A Function Whose Domain Consists of Exactly One Point Is Continuous). Let $\mathbf{x}^*$ be a point in a Euclidean metric space. Let $f : \{\mathbf{x}^*\} \to \mathcal{R}$. The function $f$ is continuous at $\mathbf{x}^*$ using a Euclidean distance measure because for every positive number $\epsilon$, $|f(\mathbf{x}^*) - f(\mathbf{y})| < \epsilon$ for all $\mathbf{y} \in \{\mathbf{x}^*\}$ and, in addition, $|\mathbf{x}^* - \mathbf{y}| = 0 < \delta$ for all positive $\delta$ and for all $\mathbf{y} \in \{\mathbf{x}^*\}$. △

**Example 5.1.6** (Bounded Discontinuous Function). Let $h : \mathcal{R} \to \{0, 1\}$ be defined such that $h(x) = 1$ for all $x > 0$ and $h(x) = 0$ for all $x \leq 0$. The function $h$ is continuous on the intervals $(-\infty, 0)$ and $(0, +\infty)$ but is not continuous at $x = 0$. To see this, note that for every positive $\delta$ there exists a positive number $x$ in the interval $|x| < \delta$ such that:

$$|h(x) - h(0)| = |h(x)| = 1 > \epsilon.$$

△

**Example 5.1.7** (Unbounded Discontinuous Function). Let $f : \mathcal{R} \to \mathcal{R}$ be defined such that $f(x) = 1/x$ for all $x \neq 0$ and $f(x) = +\infty$ for $x = 0$. Note that the function $f(x) = 1/x$ is a continuous function on the intervals $(-\infty, 0)$ and $(0, +\infty)$ but is not a continuous function at 0 since $f(0) = +\infty$ and $f(x)$ is finite for all $x \neq 0$ which implies $|f(x) - f(0)| = \infty > \epsilon$. △

**Example 5.1.8** (Non-Differentiable Continuous Function). Let $\mathcal{S} : \mathcal{R} \to [-1, 1]$ be defined such that for all $|x| \leq 1$: $\mathcal{S}(x) = x$, $\mathcal{S}(x) = 1$ for $x > 1$, and $\mathcal{S}(x) = -1$ for $x < -1$. The function $\mathcal{S}$ is continuous on $\mathcal{R}$ but not differentiable at the points $x = -1$ and $x = +1$. △

There are two important alternative definitions of a continuous function which will be presented here as theorems.

The first theorem provides insights into how the concept of a continuous function is related to the concept of convergence.

**Theorem 5.1.4** (Continuous Function Convergence Property). *Let $\Omega$ and $\Gamma$ be metric spaces. Let $V : \Omega \to \Gamma$. Let $\mathbf{x}^* \in \Omega$. The function $V$ is continuous at $\mathbf{x}^*$ if and only if $V(\mathbf{x}_t) \to V(\mathbf{x}^*)$ for every sequence of points $\mathbf{x}_1, \mathbf{x}_2, \ldots$ in $\Omega$ which converges to $\mathbf{x}^*$.*

*Proof.* See Rosenlicht (1968 p. 74). ■

The Continuous Function Convergence Property is often used as a tool for analyzing the convergence of algorithms. If a learning machine is generating a sequence of $q$-dimensional parameter vectors $\boldsymbol{\theta}_1, \boldsymbol{\theta}_2, \ldots$ which are convergent to $\boldsymbol{\theta}^*$ and $V : \mathcal{R}^q \to \mathcal{R}$ is continuous, then it follows that $V(\boldsymbol{\theta}_t) \to V(\boldsymbol{\theta}^*)$.

**Theorem 5.1.5** (Level Set Convergence). *Let $\Omega \subseteq \mathcal{R}^d$ be a closed set. Assume $V : \Omega \to \mathcal{R}$ is a continuous function and that $V(\mathbf{x}_t) \to V^*$ as $t \to \infty$. Then,*

$$\mathbf{x}_t \to \Gamma \equiv \{\mathbf{x} \in \Omega : V(\mathbf{x}) = V^*\}$$

*as $t \to \infty$.*

*Proof.* Assume $\mathbf{x}_1, \mathbf{x}_2, \ldots$ is a sequence of points such that $V(\mathbf{x}_t) \to V^*$ as $t \to \infty$. That is, every subsequence of $\{V(\mathbf{x}_t)\}$ converges to $V^*$. Assume there exists a subsequence $\{\mathbf{x}_{t_k}\}$ in $\Omega$ which does not converge to $\Gamma$ as $k \to \infty$. Since $V$ is continuous on $\Omega$ then the subsequence $\{V(\mathbf{x}_{t_k})\}$ does not converge to $V^*$ as $k \to \infty$ but this contradicts the statement that every subsequence of $\{V(\mathbf{x}_t)\}$ converges to $V^*$. ■

An alternative definition of a continuous function can also be developed in a second way which is very relevant to machine learning applications.

**Theorem 5.1.6** (Continuous Function Set Mapping Property). *Let $\Omega$ and $\Gamma$ be metric spaces. Let $f : \Omega \to \Gamma$. (i) The function $f$ is a continuous function if and only if the preimage of every open subset of $\Gamma$ under $f$ is also an open subset of $\Omega$. (ii) The function $f$ is a continuous function if and only if the preimage of every closed subset of $\Gamma$ under $f$ is also a closed subset of $\Omega$.*

*Proof.* See Rosenlicht (1968 p. 70) for the proof of part (i). The proof of part (ii) follows from the observation that since (i) holds it follows that the complement of the preimage of an open set is the complement of an open set which yields the desired conclusion using Definition 2.4.4. ■

**Definition 5.1.14** (Uniformly Continuous Function). Let $\Omega$ and $\Gamma$ be metric spaces with respective distance functions $\rho_\Omega$ and $\rho_\Gamma$. Let $f : \Omega \to \Gamma$. The function $f$ is *uniformly continuous* on $\Omega$ if for every positive number $\epsilon$ there exists a positive number $\delta$ such that $\rho_\Gamma(f(\mathbf{x}), f(\mathbf{y})) < \epsilon$ for all $\mathbf{x}, \mathbf{y} \in \Omega$ such that $\rho_\Omega(\mathbf{x}, \mathbf{y}) < \delta$. □

**Example 5.1.9** (Continuous Function Which Is Not Uniformly Continuous). Let $h : \mathcal{R} \rightarrow (0, 1)$ be defined such that $h(x) = 1/x$ for all $x > 0$. The function $h$ is continuous for all positive $x$ but it is not uniformly continuous for all positive $x$. To see this, note that for a given positive $\epsilon$ and for a given $x$, one can choose $y$ sufficiently close to $x$ so that

$$|h(x) - h(y)| < \epsilon.$$

However, for every positive $\epsilon$ there does not exist a positive $\delta$ such that $|h(0) - h(\delta)| < \epsilon$ for $x$ sufficiently close to zero. △

**Definition 5.1.15** (Piecewise Continuous Function). Let $\{\Omega_1, \ldots, \Omega_M\}$ be a finite partition of $\Omega \subseteq \mathcal{R}^d$. Let $\bar{\Omega}_k$ be the closure of $\Omega_k$. Assume $f_k : \bar{\Omega}_k \rightarrow \bar{\Omega}_k$ is continuous for $k = 1, \ldots, M$. Let $\phi_k : \Omega \rightarrow \{0, 1\}$ be defined such that $\phi_k(\mathbf{x}) = 1$ if and only if $\mathbf{x} \in \Omega_k$ for $k = 1, \ldots, M$. Let $f : \Omega \rightarrow \mathcal{R}$ be defined such that for all $\mathbf{x} \in \Omega$:

$$f(\mathbf{x}) = \sum_{k=1}^{M} f_k(\mathbf{x})\phi_k(\mathbf{x}).$$

Then, $f$ is a *piecewise continuous function* on $\Omega$. □

Let $D \subseteq \mathcal{R}^d$. Note that if $f$ is a continuous function on $D$ then $f$ can be represented as a piecewise continuous function on the finite partition $\Omega \equiv \{D, \neg D\}$. Every continuous function is a piecewise continuous function.

**Example 5.1.10** (Piecewise Continuous Recoding Strategies). In many important situations in machine learning, it is necessary to recode an input pattern vector to improve the learning machine's generalization performance. For example, consider a one-input linear prediction model where the response variable $y = \theta S + \theta_0$ is functionally dependent upon the parameters of the model $\{\theta, \theta_0\}$ and the one-dimensional input variable $S$. If the correct relationship between $S$ and $y$ is, in fact, linear then a recoding of $S$ is not necessary. For example, as $S$ increases correct predictive performance requires that $y$ must also increase or decrease. However, suppose a nonlinear relationship exists between $y$ and $S$ so that $y$ takes on a particular value when $S \in \Omega$ and another value when $S \notin \Omega$ where $\Omega$ is a closed and bounded subset of $\mathcal{R}$. Let the function $\phi : \mathcal{R} \rightarrow \mathcal{R}$ be defined such that $\phi(x) = 1$ if $x \in \Omega$ and $\phi(x) = 0$ if $x \notin \Omega$. The linear prediction model can now be used in the nonlinear case if the original input pattern $S$ is recoded so that:

$$y = \theta\phi(S) + \theta_0.$$

This is called a piecewise continuous recoding strategy. Note that the function $\phi$ is not continuous but it is piecewise continuous since: (i) $\Omega$ and $\neg\Omega$ are disjoint sets, (ii) $\phi$ is a continuous function on $\Omega$ whose value is always equal to one, and (iii) $\phi$ is a continuous function on the closure of $\neg\Omega$ whose value is always equal to zero. △

A vector-valued function such as $\mathbf{f} : \mathcal{R}^d \rightarrow \mathcal{R}^c$ or a matrix-valued function such as $\mathbf{F} : \mathcal{R}^d \rightarrow \mathcal{R}^{d \times d}$ is piecewise continuous on some subset $\Omega$ where $\Omega \subseteq \mathcal{R}^d$ if the components of such functions (which are real-valued functions) are piecewise continuous on $\Omega$.

A *polynomial* function $f : \mathcal{R} \rightarrow \mathcal{R}$ is a function that can be expressed such that for all $x \in \mathcal{R}$:

$$f(x) = \sum_{j=1}^{k} \beta_j x^j$$

where $k$ is either $\infty$ or a non-negative integer and $\beta_1, \beta_2, \ldots$ are real numbers.

**Theorem 5.1.7** (Continuous Function Composition).
*(i) A polynomial function is a continuous function,*
*(ii) the exponential function is a continuous function,*
*(iii) the log function is a continuous function on the positive real numbers,*
*(iv) a weighted sum of continuous functions is a continuous function,*
*(v) a product of continuous functions is a continuous function,*
*(vi) a continuous function of a continuous function is a continuous function, and*
*(vii) if $a : \mathcal{R} \rightarrow \mathcal{R}$ and $b : \mathcal{R} \rightarrow \mathcal{R}$ are continuous functions, then $a/b$ is continuous at $(a, b)$ when $b \neq 0$.*

*Proof.* (i) See Rosenlicht (1968 p. 75). (ii) See Rosenlicht (1968 p. 129). (iii) See Rosenlicht (1968 p. 128). (iv) See Rosenlicht (1968 p. 75). (v) See Rosenlicht (1968 p. 75). (vi) See Rosenlicht (1968 p. 71). (vii) See Rosenlicht (1968 p. 75). ■

**Example 5.1.11** (Checking If an Empirical Risk Function Is Continuous). Let $(\mathbf{s}_1, \mathbf{y}_1), \ldots, \mathbf{s}_n, \mathbf{y}_n)$ be a set of known training vectors where $\mathbf{y}_i \in \mathcal{R}^m$ is the desired response of a multilayer feedforward perceptron for input pattern $\mathbf{s}_i \in \mathcal{R}^d$ for $i = 1, \ldots, n$. Let $\ddot{\mathbf{y}}(\mathbf{s}; \boldsymbol{\theta})$ denote the predicted response of the perceptron for input pattern $\mathbf{s}$ and parameter vector $\boldsymbol{\theta}$. The parameter vector $\boldsymbol{\theta}$ is defined by the formula $\boldsymbol{\theta}^T = [\mathbf{vec}(\mathbf{W}^T), \mathbf{vec}(\mathbf{V}^T)]^T$ where $\mathbf{w}_k^T$ is the $k$th row of $\mathbf{W}$ and $\mathbf{v}_r^T$ is the $r$th row of $\mathbf{V}$. Assume the $k$th element of $\ddot{\mathbf{y}}(\mathbf{s}; \boldsymbol{\theta})$, $\ddot{y}_k(\mathbf{s}; \boldsymbol{\theta})$, is defined such that

$$\ddot{y}_k(\mathbf{s}; \boldsymbol{\theta}) = \mathbf{w}_k^T \ddot{\mathbf{h}}(\mathbf{s}; \mathbf{V})$$

and the $r$th element of $\ddot{\mathbf{h}}(\mathbf{s}; \mathbf{V})$, $\ddot{h}_r(\mathbf{s}; \mathbf{V})$, is given by the formula:

$$\ddot{h}_r(\mathbf{s}; \mathbf{V}) = \phi(\mathbf{v}_r^T \mathbf{s})$$

where $\phi(\psi) = \psi$ if $\psi > 0$ and $\phi(\psi) = 0$ if $\psi \leq 0$. A researcher proposes an empirical risk function $\hat{\ell}_n : \mathcal{R}^q \rightarrow \mathcal{R}$ for representing a multilayer feedforward perceptron defined by:

$$\hat{\ell}_n(\boldsymbol{\theta}) = (1/n) \sum_{i=1}^{n} |\mathbf{y} - \ddot{\mathbf{y}}(\mathbf{s}; \boldsymbol{\theta})|^2.$$

Explain why this objective function is not continuous and then suggest some minor modifications to the objective function which will make it continuous.

**Solution.** The objective function $\hat{\ell}_n$ is not continuous because $\phi$ is not continuous. One strategy would be to replace $\phi$ with a smoothed version of $\phi$ such as the function $\ddot{\phi}$ defined by the formula:

$$\ddot{\phi}(\psi) = \tau \log\left(1 + \exp(\psi/\tau)\right)$$

where $\tau$ is a positive number. △

**Theorem 5.1.8** (Bounded Piecewise Continuous Functions). *Let $\Omega \subseteq \mathcal{R}^d$. If $f : \mathcal{R}^d \rightarrow \mathcal{R}$ is a piecewise continuous function on a non-empty closed and bounded set $\Omega$, then $f$ is bounded on $\Omega$.*

*Proof.* If $f$ is a piecewise continuous function on a finite partition $\Omega \equiv \{\Omega_1, \ldots, \Omega_M\}$ this implies that the restriction of $f$, $f_k$, to $\Omega_k$ is continuous on $\bar{\Omega}_k$ for $k = 1, \ldots, M$. Therefore $f_k$ is bounded on $\bar{\Omega}_k$ for $k = 1, \ldots, M$ by a standard real analysis theorem that a continuous function on a closed and bounded set is bounded (Rosenlicht 1968 p. 78). Since $f = \sum_{k=1}^{M} f_k \phi_k$ where $\phi_k : \Omega \rightarrow \{0, 1\}$ is a bounded function and $f_k$ is a bounded function for $k = 1, \ldots, M$, it follows that $f$ is a bounded function. ■

**Theorem 5.1.9** (Eigenvalues Are Continuous Functions of Matrix Elements). *The eigenvalues of a square matrix* $\mathbf{M}$ *are continuous functions of the elements of* $\mathbf{M}$.

*Proof.* See Franklin (1968 p. 191) for a proof. ■

## Exercises

5.1-1. Prove Theorem 5.1.1(iv) using Theorem 5.1.1(i).

5.1-2. Assume that the sequence of vectors $\mathbf{x}(1), \mathbf{x}(2), \ldots$ converges to some vector but the identity of that vector is unknown. Assume that, in addition, it is known that a countably infinite subsequence of the sequence $\mathbf{x}(1), \mathbf{x}(2), \ldots$ converges to $\mathbf{x}^*$. Use the results of Theorem 5.1.1 to prove $\mathbf{x}(t) \to \mathbf{x}^*$ as $t \to \infty$.

5.1-3. Show that the sequence $\eta_1, \eta_2, \ldots$ defined such that:

$$\eta_t = \frac{\eta_0 \left(1 + (t/\tau)\right)}{1 + (t/\tau) + (\eta_t/\tau)^2}$$

for $t > 100$ and $\eta_t = 1.0$ for $t \leq 100$ satisfies $\sum_{t=1}^{\infty} \eta_t = \infty$ and $\sum_{t=1}^{\infty} \eta_t^2 < \infty$. Plot a graph of $\eta_t$ versus $t$ for $\eta_0 = 1$ and $\tau = 40$.

5.1-4. Let $\Gamma \equiv \{\mathbf{q}, \mathbf{r}, \mathbf{s}\}$ in a $d$-dimensional vector space. Assume $\mathbf{x}(t) \to \Gamma$ as $t \to \infty$. Is $\mathbf{x}(t)$ guaranteed to converge to $\mathbf{q}$? Is $\mathbf{x}(t)$ guaranteed to visit each point in $\Gamma$ periodically? Is $\mathbf{x}(t)$ guaranteed to eventually equal $\mathbf{r}$ for $t > T$ where $T$ is some finite positive integer? Explain your reasoning.

5.1-5. Let

$$\eta_t = \frac{(t+1)^2}{1 + \beta t^2 + \gamma t^3}$$

where $\beta$ and $\gamma$ are positive numbers. Show that $\sum \eta_t = \infty$ and $\sum \eta_t^2 < \infty$. Is $\eta_t = O(1/t)$? Is $\eta_t = o(1/t)$?

5.1-6. Consider the following list of five functions: $V(x) = 1/x$, $V(x) = x^2$, $V(x) = \log(x)$, $V(x) = \exp(x)$, $V(x) = x$. Which of these functions is continuous on $\mathcal{R}$? Which of these functions is continuous on the interval $[1, 10]$? Which of these functions is a bounded continuous function on $\mathcal{R}$? Which of these functions is a bounded continuous function on $[1, 10]$?

5.1-7. Let $\mathbf{s} \in \mathcal{R}^d$ be a constant real column vector. Let the logistic sigmoidal function $\mathcal{S}$ be defined such that $\mathcal{S}(\phi) = (1 + \exp(-\phi))^{-1}$. Use Theorem 5.1.7 to prove that the function $\ddot{y}(\boldsymbol{\theta}) = \mathcal{S}(\boldsymbol{\theta}^T \mathbf{s})$ is a continuous function for each $\boldsymbol{\theta} \in \Theta \subseteq \mathcal{R}^d$.

5.1-8. Let $\ddot{y} : \mathcal{R}^d \to \mathcal{R}$. Let the logistic sigmoidal function $\mathcal{S}$ be defined such that $\mathcal{S}(\phi) = (1 + \exp(-\phi))^{-1}$. Is the function $\ddot{y}(\boldsymbol{\theta}) = \mathcal{S}(\boldsymbol{\theta}^T \mathbf{s})$ a bounded continuous function on $\mathcal{R}^d$?

5.1-9. Let $\ddot{y} : \mathcal{R}^d \to \mathcal{R}$. Is the function $\ddot{y}(\boldsymbol{\theta}) = \boldsymbol{\theta}^T \mathbf{s}$ a bounded continuous function on $\mathcal{R}^d$.

5.1-10. Let $\Omega$ be closed and bounded subset of $\mathcal{R}^d$. Let $\ddot{y} : \Omega \to \mathcal{R}$. Is the function $\ddot{y}(\boldsymbol{\theta}) = \boldsymbol{\theta}^T \mathbf{s}$ a bounded continuous function on $\Omega$?

## 5.2 Vector Derivatives

### 5.2.1 Vector Derivative Definitions

**Definition 5.2.1** (Vector Derivative). Let $\Omega \subseteq \mathcal{R}^n$ be an open set. Let $\mathbf{x} \in \Omega$. Let $\mathbf{f} : \Omega \to \mathcal{R}^m$. If there exists a unique matrix-valued function $\mathbf{D} : \Omega \to \mathcal{R}^{m \times n}$ such that

$$\frac{\mathbf{f}(\mathbf{x}+\mathbf{h}) - \mathbf{f}(\mathbf{x}) - \mathbf{D}(\mathbf{x})\mathbf{h}}{|\mathbf{h}|} \to \mathbf{0}_m$$

as $\mathbf{h} \to \mathbf{0}_n$, then $\mathbf{D}(\mathbf{x})$ is called the derivative of $\mathbf{f}$ evaluated at $\mathbf{x}$. The function $\mathbf{D}$ is called the *derivative of* $\mathbf{f}$ on $\Omega$. □

The derivative of a function $\mathbf{f} : \Omega \to \mathcal{R}^m$ where $\mathbf{x}$ is a variable taking on values in $\Omega$ is denoted by the notation $d\mathbf{f}/d\mathbf{x}$. If $\mathbf{g}$ is a function, then the notation $d\mathbf{f}/d\mathbf{g}$ denotes the derivative of $\mathbf{f}$ where the domain of the derivative is restricted to the range of the function $\mathbf{g}$.

If $\mathbf{f}$ is functionally dependent upon $\mathbf{x}$ as well as other variables and functions, then the notation $d\mathbf{f}/d\mathbf{x}$ is sometimes used as short-hand for the collection of partial derivatives in $\mathbf{f}$ with respect to the variables or functions in $\mathbf{x}$ to improve readability so that: $d\mathbf{f}/d\mathbf{x} \equiv \partial\mathbf{f}/\partial\mathbf{x}$.

The following theorem provides a practical methodology for determining when the derivative of $\mathbf{f}$ on $\Omega$ exists.

**Theorem 5.2.1** (Sufficient Condition for Differentiability). *Let $\Omega \subseteq \mathcal{R}^n$ be an open set. Let $\mathbf{f} : \Omega \to \mathcal{R}^m$ be continuous on $\Omega$. If all partial derivatives of each element of $\mathbf{f}$ exist and are continuous everywhere on $\Omega$, then the derivative of $\mathbf{f}$ exists on $\Omega$.*

*Proof.* See Rosenlicht (1968 p. 193). ■

The Sufficient Condition for Differentiability Theorem motivates the importance of the following definition.

**Definition 5.2.2** (Continuously Differentiable). Let $\Omega \subseteq \mathcal{R}^n$ be an open set. Let $\mathbf{f} : \Omega \to \mathcal{R}^m$ be continuous on $\Omega$. If all partial derivatives of each element of $\mathbf{f}$ exist and are continuous everywhere on $\Omega$, then $\mathbf{f}$ is said to be *continuously differentiable* on $\Omega$. □

The arrangement of the partial derivatives of a vector-valued function $\mathbf{f}$ with respect to a vector in the domain of $\mathbf{f}$ in a matrix format is called the "Jacobian Matrix".

**Definition 5.2.3** (Jacobian Matrix). Let $\Omega \subseteq \mathcal{R}^n$ be an open set whose elements are column vectors. The *Jacobian matrix* of a vector-valued function $\mathbf{f} : \Omega \to \mathcal{R}^m$ is a matrix-valued function $d\mathbf{f}/d\mathbf{x} : \Omega \to \mathcal{R}^{m \times n}$ whose element in the $i$th row and $j$th column is the partial derivative of the $i$th element of $\mathbf{f}$ with respect to the $j$th element of the $n$-dimensional vector variable $\mathbf{x}$. □

Assume all partial derivatives of $f_k : \mathcal{R}^n \to \mathcal{R}$ exist and are continuous for $k = 1, \ldots, m$. Equation (5.6) shows how the partial derivatives of the elements of a vector-valued function $\mathbf{f} \equiv [f_1, \ldots, f_m]^T$ are arranged in a Jacobian matrix.

$$\frac{d\mathbf{f}}{d\mathbf{x}} = \begin{bmatrix} \partial f_1/\partial x_1 & \ldots & \partial f_1/\partial x_j & \ldots & \partial f_1/\partial x_n \\ \vdots & \ddots & \vdots & \ddots & \vdots \\ \partial f_i/\partial x_1 & \ldots & \partial f_i/\partial x_j & \ldots & \partial f_i/\partial x_n \\ \vdots & \ddots & \vdots & \ddots & \vdots \\ \partial f_m/\partial x_1 & \ldots & \partial f_m/\partial x_j & \ldots & \partial f_m/\partial x_n \end{bmatrix} \tag{5.6}$$

If the Jacobian matrix is continuous on an open subset $\Omega$ of $\mathcal{R}^n$, then $f$ is continuously differentiable on $\Omega$ with derivative $d\mathbf{f}/d\mathbf{x}$.

Note that the derivative of a scalar-valued function with respect to a column vector, $df/d\mathbf{x}$, is a row vector. The transpose of $df/d\mathbf{x}$ will be called the *gradient* of $f$ with respect to $\mathbf{x}$. The notation $\nabla f$ or $\nabla_{\mathbf{x}} f$ may also be used to denote the gradient of $f$.

Also note that the derivative of the gradient of $f$ (which is the column vector $\nabla f$) is a symmetric square matrix denoted by the *second derivative* $d^2 f/d\mathbf{x}^2$. The second matrix derivative of $f$ with respect to $\mathbf{x}$ is called the *Hessian* of $f$ and may also be represented using the notation $\nabla^2 f$ or $\nabla^2_{\mathbf{x}} f$.

The notation $d\mathbf{f}(\mathbf{x}_0)/d\mathbf{x}$ means the derivative $d\mathbf{f}/d\mathbf{x}$ evaluated at the point $\mathbf{x}_0$.

The following definition shows how to compute the derivative of a matrix-valued function with respect to a matrix.

**Definition 5.2.4** (Matrix Derivative)**.** Let $\Omega \subseteq \mathcal{R}^{d \times q}$ be an open set whose elements are $d$ by $q$ dimensional rectangular matrices. Let $\Gamma \subseteq \mathcal{R}^{m \times n}$ be an open set whose elements are $m$ by $n$ dimensional rectangular matrices. Let $\mathbf{F} : \Omega \to \Gamma$. Then, (when it exists) the derivative of $\mathbf{F}$ on $\Omega$ is denoted as:

$$\frac{d\mathbf{F}}{d\mathbf{X}}$$

and defined such that:

$$\frac{d\mathbf{F}}{d\mathbf{X}} = \frac{d\mathbf{vec}(\mathbf{F}^T)}{d\mathbf{vec}(\mathbf{X}^T)}.$$

□

It is important to note that several definitions of Matrix Derivatives exist in the literature and identities derived using one definition may not be applicable when a different definition of a Matrix Derivative is used. The definition used here, unlike some Matrix Derivative definitions, conforms with good standards as advocated by Magnus and Neudecker (2001; also see Marlow, 2012) and is best-suited for many machine learning analyses.

### 5.2.2 Theorems for Computing Matrix Derivatives

**Theorem 5.2.2** (Vector Chain Rule Theorem)**.** *Let $U \subseteq \mathcal{R}^n$ and $V \subseteq \mathcal{R}^m$ be open sets of column vectors. Assume $\mathbf{g} : U \to V$ is continuously differentiable on $U$. Assume $\mathbf{f} : V \to \mathcal{R}^q$ is continuously differentiable on $V$. Then,*

$$\frac{d\mathbf{f}(\mathbf{g})}{d\mathbf{x}} = \left[\frac{d\mathbf{f}}{d\mathbf{g}}\right]\left[\frac{d\mathbf{g}}{d\mathbf{x}}\right]$$

*is a continuously differentiable matrix-valued function with $q$ rows and $n$ columns on $U$.*

*Proof.* See Marlow (2012 pp. 202-204). ■

The Vector Chain Rule Theorem provides explicit conditions for the composition of two continuously differentiable vector-valued functions to also be differentiable. Thus, if $\mathbf{f}$ and $\mathbf{g}$ are continuously differentiable vector-valued functions defined on open sets, then it follows from the Vector Chain Rule that $\mathbf{f}(\mathbf{g})$ is continuously differentiable. This conclusion does not require any explicit calculations of derivatives. As a bonus, however, the Vector Chain Rule Theorem also provides an explicit formula for computing the derivative of $\mathbf{f}(\mathbf{g})$ in terms of the derivatives of $\mathbf{f}$ and $\mathbf{g}$.

**Theorem 5.2.3** (Matrix Chain Rule). *Let $U$ be an open subset of $\mathcal{R}^{m \times p}$. Let $V$ be an open subset of $\mathcal{R}^{q \times r}$. Let $\mathbf{A} : U \rightarrow V$ and $\mathbf{B} : V \rightarrow \mathcal{R}^{u \times z}$ be continuously differentiable functions. Then,*

$$\frac{d\mathbf{B}(\mathbf{A})}{d\mathbf{X}} = \left(\frac{d\mathbf{B}}{d\mathbf{A}}\right) \frac{d\mathbf{A}}{d\mathbf{X}} \tag{5.7}$$

*is a continuously differentiable function on $U$.*

*Proof.* Let $\mathbf{Q} \equiv \mathbf{B}(\mathbf{A})$.

$$\frac{d\mathbf{Q}}{d\mathbf{X}} = \frac{d\mathbf{vec}(\mathbf{Q}^T)}{d\mathbf{vec}\,(\mathbf{X}^T)} = \left(\frac{d\mathbf{vec}(\mathbf{B}^T)}{d\mathbf{vec}(\mathbf{A}^T)}\right) \frac{d\mathbf{vec}(\mathbf{A}^T)}{d\mathbf{vec}(\mathbf{X}^T)} = \left(\frac{d\mathbf{B}}{d\mathbf{A}}\right) \frac{d\mathbf{A}}{d\mathbf{X}}.$$

■

Equation (5.7) is called the *matrix chain rule*.

**Recipe Box 5.1 Function Decomposition Using the Chain Rule (Theorem 5.2.3)** The following procedure is often used to compute derivatives of complicated functions using the chain rule.

- **Step 1: Rewrite a complicated function as a set of nested functions.** Let $\mathbf{f} : \mathcal{R}^d \rightarrow \mathcal{R}$. Assume $d\mathbf{f}/d\mathbf{x}$ is hard to compute. However, suppose one defines four new functions $\mathbf{f}_1$, $\mathbf{f}_2$, $\mathbf{f}_3$, and $\mathbf{f}_4$ such that:

$$\mathbf{f}(\mathbf{x}) = \mathbf{f}_1\left(\mathbf{f}_2\left(\mathbf{f}_3(\mathbf{f}_4(\mathbf{x}))\right)\right)$$

  where the Jacobians $d\mathbf{f}_1/d\mathbf{f}_2$, $d\mathbf{f}_2/d\mathbf{f}_3$, $d\mathbf{f}_3/d\mathbf{f}_4$, and $d\mathbf{f}_4/d\mathbf{x}$ are easy to compute.

- **Step 2: Apply the vector chain rule.** Then, it follows immediately from the Vector Chain Rule that:

$$\frac{d\mathbf{f}}{d\mathbf{x}} = \frac{d\mathbf{f}_1}{d\mathbf{f}_2} \frac{d\mathbf{f}_2}{d\mathbf{f}_3} \frac{d\mathbf{f}_3}{d\mathbf{f}_4} \frac{d\mathbf{f}_4}{d\mathbf{x}}.$$

**Theorem 5.2.4** (Matrix Product Derivative Rule). *Assume $\mathbf{A} : \mathcal{R}^d \rightarrow \mathcal{R}^{m \times n}$ and $\mathbf{B} : \mathcal{R}^d \rightarrow \mathcal{R}^{n \times q}$ are continuously differentiable functions on $\mathcal{R}^d$.*

$$\frac{d}{d\mathbf{x}}(\mathbf{AB}) = (\mathbf{A} \otimes \mathbf{I}_q) \frac{d\mathbf{B}}{d\mathbf{x}} + \left(\mathbf{I}_m \otimes \mathbf{B}^T\right) \frac{d\mathbf{A}}{d\mathbf{x}}. \tag{5.8}$$

*In addition, let $\mathbf{C} : \mathcal{R}^d \rightarrow \mathcal{R}^{q \times v}$ be a continuously differentiable function on $\mathcal{R}^d$. Then,*

$$\frac{d}{d\mathbf{x}}(\mathbf{ABC}) = (\mathbf{AB} \otimes \mathbf{I}_v) \frac{d\mathbf{C}}{d\mathbf{x}} + \left(\mathbf{A} \otimes \mathbf{C}^T\right) \frac{d\mathbf{B}}{d\mathbf{x}} + \left(\mathbf{I}_m \otimes \mathbf{C}^T\mathbf{B}^T\right) \frac{d\mathbf{A}}{d\mathbf{x}}. \tag{5.9}$$

*Proof.* See Marlow (2012 pp. 212-213) and Marlow (2012 p. 216) for a proof of (5.8). The Matrix Product Derivative Rule in (5.9) follows from application of (5.8) to the matrix product $\mathbf{AE}$ where $\mathbf{E} = \mathbf{BC}$ and then the application of (5.8) a second time to compute $d\mathbf{E}/d\mathbf{x} = d(\mathbf{BC})/d\mathbf{x}$. The final expression in (5.9) is then obtained using Property 6 of Theorem (4.1.2). ■

**Theorem 5.2.5** (Vector-Matrix Product Derivative Rule). *Assume $\mathbf{a} : \mathcal{R}^d \rightarrow \mathcal{R}^n$ and $\mathbf{B} : \mathcal{R}^d \rightarrow \mathcal{R}^{n \times q}$ are continuously differentiable functions on $\mathcal{R}^d$. Let the n-dimensional vector $\mathbf{b}_k$ be the kth column of $\mathbf{B}$, $k = 1, \ldots, q$.*

$$\frac{d}{d\mathbf{x}}\left(\mathbf{a}^T\mathbf{B}\right) = \left(\mathbf{a}^T \otimes \mathbf{I}_q\right)\frac{d\mathbf{B}}{d\mathbf{x}} + \mathbf{B}^T\frac{d\mathbf{a}}{d\mathbf{x}} \tag{5.10}$$

*or*

$$\frac{d}{d\mathbf{x}}\left(\mathbf{a}^T\mathbf{B}\right) = \left[\left(\frac{d\mathbf{b}_1}{d\mathbf{x}}^T\mathbf{a} + \frac{d\mathbf{a}}{d\mathbf{x}}^T\mathbf{b}_1\right), \ldots, \left(\frac{d\mathbf{b}_q}{d\mathbf{x}}^T\mathbf{a} + \frac{d\mathbf{a}}{d\mathbf{x}}^T\mathbf{b}_q\right)\right]^T. \tag{5.11}$$

*Proof.* Direct application of the Matrix Product Derivative Rule (see Theorem 5.2.4), gives (5.10). To obtain (5.11), rewrite $\mathbf{a}^T\mathbf{B}$ using the formula $\mathbf{a}^T[\mathbf{b}_1, \ldots, \mathbf{b}_q]$ to obtain:

$$\frac{d}{d\mathbf{x}}\left(\mathbf{a}^T\mathbf{B}\right) = \frac{d}{d\mathbf{x}}\left(\mathbf{a}^T[\mathbf{b}_1, \ldots, \mathbf{b}_q]\right) = \frac{d}{d\mathbf{x}}\left([\mathbf{a}^T\mathbf{b}_1, \ldots, \mathbf{a}^T\mathbf{b}_q]\right) \tag{5.12}$$

and then use the Dot Product Derivative Rule to obtain the formula for the $k$th element on the right-hand side of (5.12) given by the formula:

$$\frac{d}{d\mathbf{x}}\left(\mathbf{a}^T\mathbf{b}_k\right) = \mathbf{a}^T\frac{d\mathbf{b}_k}{d\mathbf{x}} + \mathbf{b}_k^T\frac{d\mathbf{a}}{d\mathbf{x}} \tag{5.13}$$

for $k = 1, \ldots, q$. ■

**Theorem 5.2.6** (Scalar-Vector Product Derivative Rule). *Assume the scalar-valued function $\psi : \mathcal{R}^d \rightarrow \mathcal{R}$ and the column vector-valued function $\mathbf{b} : \mathcal{R}^d \rightarrow \mathcal{R}^m$ are continuously differentiable on $\mathcal{R}^d$. then,*

$$\frac{d(\mathbf{b}\psi)}{d\mathbf{x}} = \frac{d\mathbf{b}}{d\mathbf{x}}\psi + \mathbf{b}\frac{d\psi}{d\mathbf{x}}$$

*Proof.* See Exercise 5.2-7. ■

**Theorem 5.2.7** (Dot Product Derivative Rule). *Assume $\mathbf{a} : \mathcal{R}^d \rightarrow \mathcal{R}^m$ and $\mathbf{b} : \mathcal{R}^d \rightarrow \mathcal{R}^m$ are continuously differentiable column vector-valued functions on $\mathcal{R}^d$. Then,*

$$\frac{d(\mathbf{a}^T\mathbf{b})}{d\mathbf{x}} = \mathbf{a}^T\frac{d\mathbf{b}}{d\mathbf{x}} + \mathbf{b}^T\frac{d\mathbf{a}}{d\mathbf{x}}$$

*Proof.* Let $\mathbf{a} = [a_1, \ldots, a_m]^T$. Let $\mathbf{b} = [b_1, \ldots, b_m]^T$. Let $\mathbf{x} = [x_1, \ldots, x_d]^T$. The $k$th element of the row vector-valued function

$$\frac{d(\mathbf{a}^T\mathbf{b})}{d\mathbf{x}}$$

is given by:

$$\frac{\partial(\mathbf{a}^T\mathbf{b})}{\partial x_k} = \frac{\partial\left[\sum_{i=1}^m (a_i b_i)\right]}{\partial x_k} = \sum_{i=1}^m \left[a_i\frac{\partial b_i}{\partial x_k} + \frac{\partial a_i}{\partial x_k}b_i\right] = \sum_{i=1}^m \left[a_i\frac{\partial b_i}{\partial x_k}\right] + \sum_{i=1}^m \left[b_i\frac{\partial a_i}{\partial x_k}\right].$$

■

**Theorem 5.2.8** (Outer Product Derivative Rule). *Assume $\mathbf{a} : \mathcal{R}^d \rightarrow \mathcal{R}^m$ and $\mathbf{b} : \mathcal{R}^d \rightarrow \mathcal{R}^q$ are continuously differentiable column vector-valued functions on $\mathcal{R}^d$. Then,*

$$\frac{d(\mathbf{a}\mathbf{b}^T)}{d\mathbf{x}} = (\mathbf{a} \otimes \mathbf{I}_q)\frac{d\mathbf{b}}{d\mathbf{x}} + (\mathbf{I}_m \otimes \mathbf{b})\frac{d\mathbf{a}}{d\mathbf{x}}.$$

*Proof.* See Exercise 5.2-8. ■

### 5.2.3 Useful Derivative Calculations for Deep Learning

**Example 5.2.1** (First and Second Derivatives of a Logistic Sigmoidal Function). Let $\mathcal{S}(\phi) = (1 + \exp(-\phi))^{-1}$. Then,

$$\frac{d\mathcal{S}}{d\phi} = \mathcal{S}(\phi)[1 - \mathcal{S}(\phi)].$$

In addition,

$$\frac{d^2\mathcal{S}}{d\phi^2} = [1 - 2\mathcal{S}(\phi)]\mathcal{S}(\phi)[1 - \mathcal{S}(\phi)].$$

△

**Example 5.2.2** (Derivative of a Hyperbolic Tangent Sigmoidal Function). Let $\mathcal{S}(\phi) = (1 + \exp(-\phi))^{-1}$. Let $\Psi(\phi) = 2\mathcal{S}(2\phi) - 1$. Then, $d\Psi/d\phi = 4\mathcal{S}(2\phi)[1 - \mathcal{S}(2\phi)]$. △

**Example 5.2.3** (Derivative of a Softplus Function). Let $\mathcal{J}(\psi) = \log(1 + \exp(\phi))$. Then,

$$d\mathcal{J}/d\phi = \exp(\phi)/(1 + \exp(\phi)) = \mathcal{S}(\phi)$$

where $\mathcal{S}(\phi) = (1 + \exp(-\phi))^{-1}$. △

**Example 5.2.4** (Derivative of a Radial Basis Function). Let $\Psi(\phi) = \exp\left(-\phi^2\right)$. Then,

$$d\Psi/d\phi = -2\phi\Psi(\phi).$$

△

**Example 5.2.5** (Dot Product Derivative of Weight Vector and Constant Vector). Let $\Psi(\boldsymbol{\theta}) = \boldsymbol{\theta}^T \boldsymbol{s}$ where $\boldsymbol{\theta}$ is a $q$-dimensional parameter column vector and s is a $q$-dimensional column vector of constants. Then,

$$d\Psi/d\boldsymbol{\theta} = \mathbf{s}^T.$$

△

**Example 5.2.6** (Derivative of a Cross-Entropy Error Function (Scalar Case)). Let $y \in \{0, 1\}$. Let $\mathcal{S}(\theta) \equiv (1+\exp(-\theta))^{-1}$. Let $p(\theta) \equiv \mathcal{S}(\theta)$. Now show that $dp/d\theta = p(\theta)(1-p(\theta))$. Let

$$\ell(\theta) = -y\log(p(\theta)) - (1 - y)\log(1 - p(\theta)).$$

$$\frac{d\ell}{d\theta} = \left(\frac{-y}{p(\theta)} + \frac{(1 - y)}{1 - p(\theta)}\right)\frac{dp}{d\theta}$$

and since $dp/d\theta = p(\theta)(1 - p(\theta))$ it follows that:

$$\frac{d\ell}{d\boldsymbol{\theta}} = \left(\frac{-y}{p(\theta)} + \frac{(1 - y)}{1 - p(\theta)}\right)p(\theta)(1 - p(\theta)) = -(y - p(\theta)).$$

△

**Example 5.2.7** (Derivative of a Vector-Valued Exponential Transformation).

$$\frac{d\mathbf{exp}(\mathbf{x})}{d\mathbf{x}} = \mathbf{DIAG}\left(\mathbf{exp}(\mathbf{x})\right).$$

△

**Example 5.2.8** (Derivative of a Vector-Valued Sigmoidal Transformation). Define the vector-valued function $\boldsymbol{\mathcal{S}} : \mathcal{R}^d \to (0,1)^d$ such that $\boldsymbol{\mathcal{S}}([x_1, \ldots, x_d]) = [\mathcal{S}(x_1), \ldots, \mathcal{S}(x_d)]$ where $\mathcal{S}(x_i) = 1/(1 + \exp(-x_i))$ for $i = 1, \ldots, d$.

$$\frac{d\boldsymbol{\mathcal{S}}(\mathbf{x})}{d\mathbf{x}} = \mathbf{DIAG}\left(\boldsymbol{\mathcal{S}}(\mathbf{x}) \odot (\mathbf{1}_d - \boldsymbol{\mathcal{S}}(\mathbf{x})\right).$$

△

**Example 5.2.9** (Derivative of a Vector-Valued Softplus Transformation). Let the function $\boldsymbol{\mathcal{J}} : \mathcal{R}^d \to (0,1)^d$ be defined such that the $i$th element of the $d$-dimensional column vector $\boldsymbol{\mathcal{J}}(\mathbf{x})$ is equal to $\log(1 + \exp(x_i))$ where $x_i$ is the $i$th element of $\mathbf{x}$. Then,

$$\frac{d\boldsymbol{\mathcal{J}}(\mathbf{x})}{dx} = \mathbf{DIAG}\left(\boldsymbol{\mathcal{S}}(\mathbf{x})\right)$$

where $\boldsymbol{\mathcal{S}}$ is the vector-valued sigmoidal function defined in Example 5.2.8. △

**Example 5.2.10** (Cross-Entropy Error Function Vector Derivative). Let $y \in \{0, 1\}$. Let $\mathcal{S}(\phi) \equiv (1 + \exp(-\phi))^{-1}$. Let $p(\boldsymbol{\theta}) \equiv \mathcal{S}(\boldsymbol{\theta}^T\mathbf{s})$ where $\boldsymbol{\theta} \in \mathcal{R}^d$. Let

$$\ell(\boldsymbol{\theta}) = -y \log(p(\boldsymbol{\theta}^T\mathbf{s})) - (1 - y)\log(1 - p(\boldsymbol{\theta}^T\mathbf{s})).$$

$$\frac{d\ell}{d\boldsymbol{\theta}} = \left(\frac{-y}{p(\boldsymbol{\theta}^T\mathbf{s})} + \frac{(1-y)}{1 - p(\boldsymbol{\theta}^T\mathbf{s})}\right)\frac{dp}{d\boldsymbol{\theta}}$$

$$\frac{d\ell}{d\boldsymbol{\theta}} = \left[\left(\frac{-y}{p(\boldsymbol{\theta}^T\mathbf{s})} + \frac{(1-y)}{1 - p(\boldsymbol{\theta}^T\mathbf{s})}\right)p(\boldsymbol{\theta}^T\mathbf{s})(1 - p(\boldsymbol{\theta}^T\mathbf{s}))\right]\frac{d(\boldsymbol{\theta}^T\mathbf{s})}{d\boldsymbol{\theta}} = -(y - p(\boldsymbol{\theta}^T\mathbf{s}))\mathbf{s}^T.$$

△

**Example 5.2.11** (Derivative of a Least Squares Regression Loss Function). Let the function $\ddot{y}$ be defined such that $\ddot{y}(\boldsymbol{\theta}) = \boldsymbol{\theta}^T\mathbf{s}$. Let the function $\ell$ be defined such that $\ell(\boldsymbol{\theta}) = |y - \ddot{y}(\boldsymbol{\theta})|^2$. Then,

$$\frac{d\ell}{d\boldsymbol{\theta}} = -2(y - \ddot{y}(\boldsymbol{\theta}))(d\ddot{y}/d\boldsymbol{\theta}) = -2(y - \ddot{y}(\boldsymbol{\theta}))\mathbf{s}^T.$$

△

**Example 5.2.12** (Derivative of a Vector-Valued Linear Transformation). Let $\mathbf{A} \in \mathcal{R}^{m \times n}$. Let $\mathbf{x} \in \mathcal{R}^n$. The goal of this example is to explain how to compute the derivative $d(\mathbf{Ax})/d\mathbf{x}$.

**Solution.** Two different ways of computing the derivative $d(\mathbf{Ax})/d\mathbf{x}$ are provided. The first way is the direct method:

$$\frac{d(\mathbf{Ax})}{d\mathbf{x}} = \mathbf{A}\frac{d(\mathbf{x})}{d\mathbf{x}} = \mathbf{AI} = \mathbf{A}. \tag{5.14}$$

The derivative $d(\mathbf{Ax})/d\mathbf{x}$ is now computed a second way using the Matrix Product Derivative Rule (Theorem 5.2.4) to illustrate how the Matrix Product Derivative Rule may be applied. Using the Matrix Product Derivative Rule:

$$\frac{d(\mathbf{Ax})}{d\mathbf{x}} = (\mathbf{A} \otimes \mathbf{I}_1)\frac{d\mathbf{x}}{d\mathbf{x}} + \left(\mathbf{I}_m \otimes \mathbf{x}^T\right)\frac{d\mathbf{A}}{d\mathbf{x}}.$$

Since $\mathbf{A}$ is a constant matrix, it follows that $d\mathbf{A}/d\mathbf{x}$ is a matrix of zeros. In addition, $d\mathbf{x}/d\mathbf{x}$ is an $n$-dimensional identity matrix and $\mathbf{I}_1 = 1$ so that the result in Equation (5.14) is obtained. △

**Example 5.2.13** (Computation of $d\mathbf{A}/d(\mathbf{A}^T)$). In applications where matrix-valued derivatives are involved, it is sometimes necessary to compute derivatives such as

$$\frac{d\mathbf{A}}{d(\mathbf{A}^T)}.$$

This can be accomplished by using the commutation matrix $\mathbf{K}$ described in Chapter 4 and then evaluating the formula:

$$\frac{d\mathbf{A}}{d\mathbf{A}^T} = \frac{d(\mathbf{vec}\,(\mathbf{A}^T))}{d\mathbf{vec}(\mathbf{A})} = \frac{d\,(\mathbf{Kvec}(\mathbf{A}))}{d\mathbf{vec}(\mathbf{A})} = \mathbf{K}\frac{d\mathbf{vec}(\mathbf{A})}{d\mathbf{vec}(\mathbf{A})} = \mathbf{KI} = \mathbf{K}.$$

△

**Example 5.2.14** (Exponential Family First Derivative). Let $p : \mathcal{R}^d \times \mathcal{R}^d \to \mathcal{R}$ be defined such that:

$$p(\mathbf{x};\boldsymbol{\theta}) = \frac{\exp(-\mathbf{x}^T\boldsymbol{\theta})}{\sum_{\mathbf{y}} \exp(-\mathbf{y}^T\boldsymbol{\theta})}.$$

$$\log p(\mathbf{x};\boldsymbol{\theta}) = -\mathbf{x}^T\boldsymbol{\theta} - \log\left(\sum_{\mathbf{y}} \exp(-\mathbf{y}^T\boldsymbol{\theta})\right).$$

Then,

$$\frac{d\log p(\mathbf{x};\boldsymbol{\theta})}{d\boldsymbol{\theta}} = -\boldsymbol{\theta}^T - \frac{\sum_{\mathbf{y}} \exp(-\mathbf{y}^T\boldsymbol{\theta})(-\mathbf{y}^T)}{\sum_{\mathbf{v}} \exp(-\mathbf{v}^T\boldsymbol{\theta})} = -\boldsymbol{\theta}^T + \sum_{\mathbf{y}} \mathbf{y}^T p(\mathbf{y};\boldsymbol{\theta}).$$

△

**Example 5.2.15** (Multinomial Logit (Softmax) Derivative). Let $c_k \equiv -\mathbf{y}_k^T \log(\mathbf{p})$ where $\mathbf{y}_k$ is the $k$th column of an $m$-dimensional identity matrix and the $k$th element, $p_k$, of the $m$-dimensional vector $\mathbf{p}$ is given by the formula

$$p_k = \frac{\exp(\phi_k)}{\sum_{j=1}^m \exp(\phi_j)}.$$

Then,

$$c_k = -\phi_k + \log\sum_{j=1}^m \exp(\phi_j).$$

If $u = k$, then

$$dc_k/d\phi_u = -1 + \frac{\exp(\phi_u)}{\sum_{j=1}^m \exp(\phi_j)} = -1 + p_u. \tag{5.15}$$

If $u \neq k$, then

$$dc_k/d\phi_u = \frac{\exp(\phi_u)}{\sum_{j=1}^m \exp(\phi_j)} = p_u. \tag{5.16}$$

Let $\boldsymbol{\phi} \equiv [\phi_1, \ldots, \phi_m]^T$. Combining (5.15) and (5.16) gives the following useful identity:

$$\frac{dc_k}{d\boldsymbol{\phi}} = -(\mathbf{y}_k - \mathbf{p})^T.$$

△

**Example 5.2.16** (Derivatives of Logistic Regression Objective Function). Assume the training data set $\mathcal{D}_n$ is defined such that $\mathcal{D}_n \equiv \{(\mathbf{s}_1, y_1), \ldots, (\mathbf{s}_n, y_n)\}$ where the logistic regression learning machine is supposed to generate response $y_i \in \{0, 1\}$ given input pattern $\mathbf{s}_i \in \mathcal{R}^d$ for $i = 1, \ldots, n$.

The scalar response of a logistic regression model is defined as:

$$\ddot{y}(\mathbf{s}, \boldsymbol{\theta}) = \mathcal{S}\left(\boldsymbol{\theta}^T [\mathbf{s}^T, 1]^T\right)$$

where $\ddot{y}(\mathbf{s}, \boldsymbol{\theta})$ is the predicted probability that $y = 1$ given input pattern vector $\mathbf{s}$ and parameter vector $\boldsymbol{\theta}$. The predicted probability that $y = 0$ given $\mathbf{s}$ and $\boldsymbol{\theta}$ is $1 - \ddot{y}(\mathbf{s}, \boldsymbol{\theta})$.

Assume the empirical risk function for learning $\hat{\ell}_n : \mathcal{R}^q \to \mathcal{R}$ is defined such that for all $\boldsymbol{\theta} \in \mathcal{R}^q$:

$$\hat{\ell}_n(\boldsymbol{\theta}) = -(1/n) \sum_{i=1}^{n} \left(y_i \log \ddot{y}(\mathbf{s}_i, \boldsymbol{\theta}) + (1 - y_i) \log(1 - \ddot{y}(\mathbf{s}_i, \boldsymbol{\theta})\right).$$

Compute the gradient and Hessian of $\hat{\ell}_n$.

**Solution.** Begin by defining a set of simpler functions whose composition yields the empirical risk function of interest. Let

$$\hat{\ell}_n(\boldsymbol{\theta}) = (1/n) \sum_{i=1}^{n} c_i(\ddot{y}).$$

Let

$$c_i(\ddot{y}) = -\left[y_i \log \ddot{y}_i + (1 - y_i) \log(1 - \ddot{y}_i)\right].$$

Let $\ddot{y}_i(\psi_i) = \mathcal{S}(\psi_i)$. Let $\psi_i(\boldsymbol{\theta}) \equiv \boldsymbol{\theta}^T [\mathbf{s}_i^T, 1]^T$. Then,

$$\frac{d\hat{\ell}_n(\boldsymbol{\theta})}{d\boldsymbol{\theta}} = (1/n) \sum_{i=1}^{n} \frac{dc_i}{d\boldsymbol{\theta}},$$

$$\frac{dc_i}{d\boldsymbol{\theta}} = \left(\frac{dc_i}{d\ddot{y}_i}\right)\left(\frac{d\ddot{y}_i}{d\psi_i}\right)\left(\frac{d\psi_i}{d\boldsymbol{\theta}}\right), \tag{5.17}$$

$$\frac{dc_i}{d\ddot{y}_i} = -(y_i - \ddot{y}_i)/(\ddot{y}_i(1 - \ddot{y}_i)), \tag{5.18}$$

$$\frac{d\ddot{y}_i}{d\psi_i} = \ddot{y}_i(1 - \ddot{y}_i), \tag{5.19}$$

and

$$\frac{d\psi_i}{d\boldsymbol{\theta}} = [\mathbf{s}_i^T, 1]. \tag{5.20}$$

Substitution of (5.18), (5.19), and (5.20) into (5.17) gives:

$$\frac{dc_i}{d\boldsymbol{\theta}} = -(y_i - \ddot{y}_i)[\mathbf{s}_i^T, 1].$$

Let $\mathbf{g}_i = -(y_i - \ddot{y}_i)[\mathbf{s}_i^T, 1]^T$. To compute the Hessian of $\hat{\ell}_n$, note that:

$$\nabla^2 \hat{\ell}_n(\boldsymbol{\theta}) = (1/n) \sum_{i=1}^{n} \frac{d\mathbf{g}_i}{d\boldsymbol{\theta}}.$$

Then, use the chain rule to show:

$$\frac{d\mathbf{g}_i}{d\boldsymbol{\theta}} = [\mathbf{s}_i^T, 1]^T \frac{d\ddot{y}_i}{d\boldsymbol{\theta}}$$

and since

$$\frac{d\ddot{y}_i}{d\boldsymbol{\theta}} = \left(\frac{d\ddot{y}_i}{d\psi_i}\right)\left(\frac{d\psi_i}{d\boldsymbol{\theta}}\right) = \ddot{y}_i(1-\ddot{y}_i)[\mathbf{s}_i^T, 1]$$

it follows that:

$$\frac{d\mathbf{g}_i}{d\boldsymbol{\theta}} = \ddot{y}_i(1-\ddot{y}_i)[\mathbf{s}_i^T, 1]^T[\mathbf{s}_i^T, 1].$$

△

**Example 5.2.17** (Gradient of a Perceptron Objective Function)**.** Let the column-vector valued function $\boldsymbol{\mathcal{J}} : \mathcal{R}^d \to (0,1)^d$ be defined such that the $i$th element of the $d$-dimensional column vector $\boldsymbol{\mathcal{J}}(\mathbf{x})$ is equal to $\log(1+\exp(x_i))$ where $x_i$ is the $i$th element of $\mathbf{x}$. Let $\mathcal{S} : \mathcal{R} \to (0,1)$ be defined such that $\mathcal{S}(\phi) = 1/(1+\exp(-\phi))$.

Assume the training data set $\mathcal{D}_n$ is defined such that $\mathcal{D}_n \equiv \{(\mathbf{s}_1, y_1), \ldots, (\mathbf{s}_n, y_n)\}$ where the perceptron is supposed to generate response $y_k \in \{0,1\}$ given input pattern $\mathbf{s}_k \in \mathcal{R}^d$ for $k = 1, \ldots, n$.

Let $\boldsymbol{\theta} \equiv \left(\mathbf{v}^T, \mathbf{vec}(\mathbf{W}^T)^T\right)^T$ where $\mathbf{v} \in \mathcal{R}^{h+1}$ and $\mathbf{W} \in \mathcal{R}^{h \times d}$. Then, the scalar response $\ddot{y}(\mathbf{s}, \boldsymbol{\theta})$ of a smooth multilayer perceptron with softplus hidden units and a sigmoidal output unit is defined as the probability

$$\ddot{y}(\mathbf{s}, \boldsymbol{\theta}) = \mathcal{S}\left(\mathbf{v}^T[(\boldsymbol{\mathcal{J}}(\mathbf{W}\mathbf{s}))^T, 1]^T\right)$$

that $y = 1$ given $\mathbf{s}$ and $\boldsymbol{\theta}$.

The number of free parameters in the perceptron is equal to $q \equiv (h+1) + hd$. Assume the empirical risk function for learning $\hat{\ell}_n : \mathcal{R}^q \to \mathcal{R}$ is defined such that for all $\boldsymbol{\theta} \in \mathcal{R}^q$:

$$\hat{\ell}_n(\boldsymbol{\theta}) = -(1/n)\sum_{i=1}^{n}\left[y_i \log \ddot{y}(\mathbf{s}_i, \boldsymbol{\theta}) + (1-y_i)\log(1-\ddot{y}(\mathbf{s}_i, \boldsymbol{\theta}))\right].$$

Compute the gradient of $\hat{\ell}_n$.

**Solution.** As previously mentioned, the main trick for taking derivatives in multilayer perceptrons is to strategically define simpler functions so that the more complicated function can be represented in terms of the composition of simpler functions.

Let $\ddot{y}_i(\boldsymbol{\theta}) \equiv \ddot{y}(\mathbf{s}_i, \boldsymbol{\theta})$. Let the vector-valued function $\boldsymbol{\phi}_i(\boldsymbol{\theta}) \equiv \mathbf{W}\mathbf{s}_i$. Define the vector-valued function

$$\mathbf{h}_i(\boldsymbol{\theta}) = [(\boldsymbol{\mathcal{J}}(\boldsymbol{\phi}_i(\boldsymbol{\theta}))^T, 1]^T.$$

In addition, define the scalar-valued function

$$c_i(\ddot{y}_i) \equiv -y_i \log \ddot{y} - (1-y_i)\log(1-\ddot{y}).$$

Let the scalar-valued function $\psi_i(\boldsymbol{\theta}) \equiv \mathbf{v}^T \mathbf{h}_i(\boldsymbol{\theta})$.

Then,

$$\frac{dc_i}{d\mathbf{v}} = \left[\frac{dc_i}{d\psi_i}\right]\left[\frac{d\psi_i}{d\mathbf{v}}\right].$$

Now note that:

$$\frac{dc_i}{d\psi_i} = -(y_i - \ddot{y}_i)$$

and

$$\frac{d\psi_i}{d\mathbf{v}} = \mathbf{h}_i^T$$

and this completes the formula for the derivative of the empirical risk function with respect to $\mathbf{v}$.

Let $\mathbf{w}_k$ denote the $k$th row of $\mathbf{W}$. Let $v_k$ denote the $k$th element of $\mathbf{v}$. Let $\phi_{i,k} = \mathbf{w}_k^T \mathbf{s}_i$. The $k$th element of $\mathbf{h}_i$ is defined as $h_{i,k} = \mathcal{J}(\phi_{i,k})$.

The derivative of the loss function with respect to $\mathbf{w}_k$ is now computed. Note that:

$$\frac{dc_i}{d\mathbf{w}_k} = \left[\frac{dc_i}{d\psi_i}\right]\left[\frac{d\psi_i}{dh_{i,k}}\right]\left[\frac{dh_{i,k}}{d\phi_{i,k}}\right]\left[\frac{d\phi_{i,k}}{d\mathbf{w}_k}\right].$$

Since $d\psi_i/dh_{i,k} = v_k$, $dh_{i,k}/d\phi_{i,k} = \mathcal{S}(\phi_{i,k})$, and $d\phi_{i,k}/d\mathbf{w}_k = \mathbf{s}_i^T$, it follows that:

$$\frac{dc_i}{d\mathbf{w}_k} = -(y_i - \ddot{y}_i)v_k \mathcal{S}(\phi_{i,k})\mathbf{s}_i^T.$$

△

### 5.2.4 Gradient Backpropagation for Deep Learning

In this section, an algorithm for computing the gradient for feedforward deep learning network architectures in a computationally efficient manner is provided. These methods can also be adapted for recurrent network calculations as well.

Let $\ddot{\mathbf{y}}(\mathbf{s}, \boldsymbol{\theta})$ denote the $r$-dimensional predicted response of a feedforward deep network architecture consisting of $m$ layers of hidden units for a given $d$-dimensional input pattern $\mathbf{s}$ with $q$-dimensional parameter vector $\boldsymbol{\theta} = [\boldsymbol{\theta}_1^T, \ldots, \boldsymbol{\theta}_m^T]^T$. It is assumed that the predicted response is computed by the function $\ddot{\mathbf{y}} : \mathcal{R}^d \times \mathcal{R}^q \rightarrow \mathcal{R}^r$. The prediction response function $\ddot{\mathbf{y}}$ is defined such that for a one hidden layer system (i.e., $m = 2$)

$$\ddot{\mathbf{y}}(\mathbf{s}, \boldsymbol{\theta}) = \ddot{\mathbf{h}}_2(\ddot{\mathbf{h}}_1(\mathbf{s}, \boldsymbol{\theta}_1), \boldsymbol{\theta}_2)$$

and for a two hidden layer system (i.e., $m = 3$)

$$\ddot{\mathbf{y}}(\mathbf{s}, \boldsymbol{\theta}) = \ddot{\mathbf{h}}_3(\ddot{\mathbf{h}}_2(\ddot{\mathbf{h}}_1(\mathbf{s}, \boldsymbol{\theta}_1), \boldsymbol{\theta}_2), \boldsymbol{\theta}_3).$$

The predicted response function $\ddot{y}$ is defined in a similar way for $m = 4, 5, \ldots$. Let the discrepancy $D : \mathcal{R}^r \times \mathcal{R}^r \rightarrow \mathcal{R}$ specify the prediction error $D(\mathbf{y}, \ddot{\mathbf{y}}(\mathbf{s}, \boldsymbol{\theta}))$ between the desired response $\mathbf{y}$ of the feedforward network and the predicted response $\ddot{\mathbf{y}}(\mathbf{s}, \boldsymbol{\theta})$ for a given input pattern $\mathbf{s}$ and parameter vector $\boldsymbol{\theta}$. The loss function for the empirical risk function minimized by the network is a loss function $c : \mathcal{R}^{d+r} \times \mathcal{R}^q \rightarrow \mathcal{R}$ defined such that the incurred loss for pattern $[\mathbf{s}, \mathbf{y}]$ is $c([\mathbf{s}, \mathbf{y}], \boldsymbol{\theta}) = D(\mathbf{y}, \ddot{\mathbf{y}}(\mathbf{s}, \boldsymbol{\theta}))$.

Algorithm 5.2.1 provides a general-purpose method for computing the gradient of this feedforward deep neural network which has $m - 1$ layers of hidden units.

**Example 5.2.18** (Smooth Multilayer Perceptron Gradient Descent Learning). Let $\mathcal{D}_n \equiv \{(\mathbf{s}_1, \mathbf{y}_1), \ldots, (\mathbf{s}_n, \mathbf{y}_n)\}$ be a collection of $n$ training vectors where $\mathbf{y}_i$ is the desired response of the perceptron to input pattern $\mathbf{s}_i$ for $i = 1, \ldots, n$.

Let $\ddot{\mathbf{y}}(\mathbf{s}, \boldsymbol{\theta})$ denote the predicted response of the perceptron to input pattern $\mathbf{s}$ for a given parameter vector $\boldsymbol{\theta}$ which is defined by the formula:

$$\ddot{\mathbf{y}}(\mathbf{s}, \boldsymbol{\theta}) = \ddot{\mathbf{h}}_4(\ddot{\mathbf{h}}_3(\ddot{\mathbf{h}}_2(\ddot{\mathbf{h}}_1(\mathbf{s}, \boldsymbol{\theta}_1), \boldsymbol{\theta}_2), \boldsymbol{\theta}_3), \boldsymbol{\theta}_4).$$

**Algorithm 5.2.1 Feedforward $m$-Layer Deep Net Gradient Backpropagation.**

1: **procedure** GRADBACKPROP( $D$, $dD/d\ddot{\mathbf{y}}$, $(\mathbf{s}, \mathbf{y})$, $\ddot{\mathbf{h}}_k$, $d\ddot{\mathbf{h}}_k/d\ddot{\mathbf{h}}_{k-1}$)
2: $\quad \mathbf{h}_0 \Leftarrow \mathbf{s}$ ▷ **Forward Pass Computation**
3: $\quad$ **for** $k = 1$ to $m$ **do**
4: $\quad\quad \mathbf{h}_k \Leftarrow \ddot{\mathbf{h}}_k(\mathbf{h}_{k-1}, \boldsymbol{\theta}_k)$ ▷ Compute Activity Levels Layer $k$
5: $\quad$ **end for**
6: $\quad \ddot{\mathbf{y}} \Leftarrow \mathbf{h}_m$ ▷ Output Prediction
7: ▷ **Backward Pass Computation**
8: $\quad \mathbf{e}_m^T = dD(\mathbf{y}, \ddot{\mathbf{y}})/d\ddot{\mathbf{y}}$ ▷ Compute Errors Layer $m$
9: $\quad$ **for** $k = m$ to 2 **do**
10: $\quad\quad \mathbf{e}_{k-1}^T \Leftarrow \mathbf{e}_k^T \left( d\ddot{\mathbf{h}}_k(\mathbf{h}_{k-1}, \boldsymbol{\theta}_k)/d\ddot{\mathbf{h}}_{k-1} \right)$ ▷ Compute Errors Layer $k-1$
11: $\quad$ **end for**
12: ▷ **Gradient Computation**
13: $\quad$ **for** $k = 1$ **to** $m$ **do**
14: $\quad\quad dc/\boldsymbol{\theta}_k^T \Leftarrow \mathbf{e}_k^T \left( d\ddot{\mathbf{h}}_k(\mathbf{h}_{k-1}, \boldsymbol{\theta}_k)/d\boldsymbol{\theta}_k \right)$
15: $\quad$ **end for**
16: $\quad$ **return** $\{\ddot{\mathbf{y}}, dc/d\boldsymbol{\theta}_1, \ldots, dc/d\boldsymbol{\theta}_m\}$ ▷ **Return Prediction $\ddot{\mathbf{y}}$ and Gradient $dc/d\boldsymbol{\theta}$**
17: **end procedure**

Let the states of the hidden units in layer $k$ be computed from the states of the hidden units in layer $k-1$ by the formula:

$$\ddot{\mathbf{h}}_k(\ddot{\mathbf{h}}_{k-1}, \boldsymbol{\theta}_k) = \boldsymbol{\mathcal{S}}(\mathbf{W}_k \ddot{\mathbf{h}}_{k-1} + \mathbf{b}_k)$$

where $\boldsymbol{\mathcal{S}}$ is a vector-valued sigmoidal function as defined in Example 5.2.8.

Derive a gradient descent learning algorithm that minimizes an empirical risk function with loss function $c$ defined by the formula:

$$c([\mathbf{s}, \mathbf{y}], \boldsymbol{\theta}) = D(\mathbf{y}, \ddot{\mathbf{y}}(\mathbf{s}, \boldsymbol{\theta})) = |\mathbf{y} - \ddot{\mathbf{y}}(\mathbf{s}, \boldsymbol{\theta})|^2.$$

**Solution.** The gradient descent algorithm is given by the iterative learning rule:

$$\boldsymbol{\theta}(t+1) = \boldsymbol{\theta}(t) - \gamma_t d\ell(\boldsymbol{\theta}(t))/d\boldsymbol{\theta}$$

where

$$d\ell/d\boldsymbol{\theta} = (1/n) \sum_{i=1}^{n} dc([\mathbf{s}_i, \mathbf{y}_i], \boldsymbol{\theta})/d\boldsymbol{\theta}.$$

Note that

$$\frac{dD(\mathbf{y}, \ddot{\mathbf{h}}_4(\mathbf{s}, \boldsymbol{\theta}))}{d\ddot{\mathbf{y}}} = -2(\mathbf{y} - \ddot{\mathbf{h}}_4) \tag{5.21}$$

and for $k = 2, 3, 4$:

$$d\ddot{\mathbf{h}}_k/d\ddot{\mathbf{h}}_{k-1} = \mathbf{DIAG}(\ddot{\mathbf{h}}_k \odot (\mathbf{1}_{d_k} - \ddot{\mathbf{h}}_k))\mathbf{W}_k^T \tag{5.22}$$

where $d_k$ is the number of hidden units in layer $k$. In addition,

$$d\ddot{\mathbf{h}}_k/d\mathbf{b}_k = \mathbf{DIAG}\left(\ddot{\mathbf{h}}_k \odot (\mathbf{1}_{d_k} - \ddot{\mathbf{h}}_k)\right). \tag{5.23}$$

Now let $\mathbf{w}_{k,j}^T$ denote the $j$th row of $\mathbf{W}_k$ and let $\ddot{h}_{k,j}$ denote the $j$th element of $\ddot{\mathbf{h}}_k$. Let $bfu_{d_j}$ be the $j$th column of the $d_j$-dimensional identity matrix $\mathbf{I}_{d_j}$. Then,

$$d\ddot{\mathbf{h}}_k/d\mathbf{w}_{k,j} = \left(\mathbf{DIAG}\left(\ddot{\mathbf{h}}_k \odot (\mathbf{1}_{d_k} - \ddot{\mathbf{h}}_k)\right)\right) \mathbf{u}_{d_j} \ddot{\mathbf{h}}_{k-1}^T. \tag{5.24}$$

Equations (5.21), (5.22), (5.23), and (5.24) are then substituted into Steps 8, 10, and 14 of Algorithm 5.2.1 to implement feedforward 4-layer deep learning gradient backpropagation algorithm. △

## Exercises

5.2-1. Show that $d(\mathbf{x}^T\mathbf{B}\mathbf{x})/d\mathbf{x} = [\mathbf{B} + \mathbf{B}^T]\mathbf{x}$ by differentiating $\mathbf{x}^T\mathbf{B}\mathbf{x} = \sum_i \sum_j b_{ij}x_i x_j$.

5.2-2. Show that $d(\mathbf{x}^T\mathbf{B}\mathbf{x})/d\mathbf{x} = 2\mathbf{B}\mathbf{x}$ when $\mathbf{B}$ is a symmetric matrix.

5.2-3. Show that $d(\mathbf{x}^T\mathbf{B}\mathbf{x})/d\mathbf{B} = \mathbf{vec}\left(\mathbf{x}\mathbf{x}^T\right)^T$.

5.2-4. Let $r_k(\mathbf{w}_k, \mathbf{u}) = \mathcal{S}(\mathbf{w}_k^T\mathbf{u})$ where $\mathcal{S}(\phi) = 1/(1 + \exp(-\phi))$. Let $\mathbf{r} = [r_1, \ldots, r_M]$ be a column vector of the $M$ functions $r_1, \ldots, r_M$. Show that:

$$d\mathbf{r}/d\mathbf{u} = \mathbf{DIAG}(\mathbf{r} \odot [\mathbf{1}_d - \mathbf{r}])\mathbf{W}$$

where the $k$th row of $\mathbf{W}$ is the row vector $\mathbf{w}_k$.

5.2-5. Let $r_k(\mathbf{w}_k, \mathbf{u}) = \mathcal{S}(\mathbf{w}_k^T\mathbf{u})$ where $\mathcal{S}(\phi) = 1/(1 + \exp(-\phi))$. Let $\mathbf{r} = [r_1, \ldots, r_M]$ be a column vector of the $M$ functions $r_1, \ldots, r_M$. Show that:

$$dr_k/d\mathbf{w}_k = r_k(1 - r_k)\mathbf{u}^T$$

and that $dr_k/d\mathbf{w}_j$ is a row vector of zeros when $k \neq j$. Show that

$$dr_k/d\mathbf{W} = \mathbf{e}_k^T \otimes dr_k/d\mathbf{w}_k$$

where $\mathbf{e}_k$ is the $k$th column of an $M$-dimensional identity matrix.

5.2-6. *Gradient and Hessian for Softmax Regression Model.* Compute the gradient and Hessian of the empirical risk function for the softmax regression model described in Example 1.5.3.

5.2-7. *Prove the Scalar Function Multiplied by Vector Function Derivative Theorem.* Prove Theorem 5.2.6 by computing the derivative of the $j$th element, $\psi(\boldsymbol{\theta})b_j(\boldsymbol{\theta})$, of the vector

$$[\psi(\boldsymbol{\theta})b_1(\boldsymbol{\theta}), \ldots, \psi(\boldsymbol{\theta})b_m(\boldsymbol{\theta})]$$

with respect to $\boldsymbol{\theta}$.

5.2-8. *Derive Outer Product Derivative Rule.* Use the Matrix Product Derivative Rule (Theorem 5.2.4) to prove the Outer Product Derivative Rule (Theorem 5.2.8).

5.2-9. *A Contrastive Divergence Identity.* The following identity will be useful in the discussion of contrastive divergence learning algorithm derivations in Chapter 12 (see Section 12.3.3). Let $V : \mathcal{R}^d \times \mathcal{R}^q \rightarrow \mathcal{R}$. Let

$$Z(\boldsymbol{\theta}) = \sum_{j=1}^{M} \exp\left(-V(\mathbf{x}_j, \boldsymbol{\theta})\right).$$

Show that the derivative of $\log Z$ with respect to $\boldsymbol{\theta}$ is given by the formula:

$$\frac{d \log Z}{d\boldsymbol{\theta}} = -\sum_{j=1}^{M} p_j(\boldsymbol{\theta}) \frac{dV(\mathbf{x}_j, \boldsymbol{\theta})}{d\boldsymbol{\theta}}$$

where

$$p_j(\boldsymbol{\theta}) = \frac{\exp\left(-V(\mathbf{x}_j, \boldsymbol{\theta})\right)}{\sum_{k=1}^{M} \exp\left(-V(\mathbf{x}_k, \boldsymbol{\theta})\right)}.$$

5.2-10. *Gradient for Multilayer Perceptron.* Compute the gradient of the empirical risk function for a multilayer perceptron with one layer of hidden units where the empirical risk function is defined as in Equation 1.15 in Example 1.5.4. Compute the gradient when the hidden units are radial basis function units, sigmoidal hidden units, and softplus hidden units.

5.2-11. *Gradient for Nonlinear Denoising Autoencoder.* Compute the gradient of the nonlinear denoising autoencoder loss function defined in Example 1.6.2.

5.2-12. *Gradient for Linear Value Function Reinforcement.* Derive the formulas used for a linear value function reinforcement learning algorithm by computing the derivative of the loss function of the empirical risk function defined in Example 1.7.2.

5.2-13. *Gradient for Simple Recurrent Network.* Compute the gradient of the empirical risk function for a Simple Recurrent Network as defined in Example 1.5.7.

5.2-14. *Value Function Reinforcement Using Perceptron.* Let $\ddot{V}$ be a function defined such that $\ddot{V}(\mathbf{s}^0; \boldsymbol{\theta})$ is the predicted total negative reinforcement received in the distant future when the learning machine encounters initial state $\mathbf{s}^0$ given the learning machine follows some particular constant policy $\rho$. Assume, in addition, that $\ddot{V}$ may be interpreted as the output of a feedforward network consisting of one layer of hidden units. Let $\boldsymbol{\mathcal{S}} : \mathcal{R}^m \to \mathcal{R}^m$ be defined such that the $k$th element of $\boldsymbol{\mathcal{S}}(\boldsymbol{\psi})$ is equal to $1/(1 + \exp(-\psi_k))$ where $\psi_k$ is the $k$th element of $\boldsymbol{\psi}$. In particular, assume that

$$\ddot{V}(\mathbf{s}; \boldsymbol{\theta}) = \mathbf{a}^T \boldsymbol{\mathcal{S}}(\mathbf{W}\mathbf{s} + \mathbf{b})$$

where $\boldsymbol{\theta} = [\mathbf{a}^T, \mathbf{vec}(\mathbf{W}^T), \mathbf{b}^T]^T$. Consider the empirical risk function defined by the formula:

$$\hat{\ell}_n(\boldsymbol{\theta}) = (1/n) \sum_{i=1}^{n} c([\mathbf{s}_i^0, \mathbf{s}_i^F, r_i]; \boldsymbol{\theta})$$

where $r_i$ is the incremental negative reinforcement penalty received by the learning machine when the environmental state changes from $\mathbf{s}_i^0$ to $\mathbf{s}_i^F$. Provide a semantic interpretation of the loss function

$$c([\mathbf{s}_i^0, \mathbf{s}_i^F, r_i]; \boldsymbol{\theta}) = \left(r_i - \ddot{V}(\mathbf{s}_i^0; \boldsymbol{\theta}) + \mu \ddot{V}(\mathbf{s}_i^F; \boldsymbol{\theta})\right)^2$$

as a Value Function Reinforcement Objective Function where $0 < \mu < 1$. Compute the gradient of $c$ with respect to $\boldsymbol{\theta}$ and show how to derive an iterative learning algorithm for implementing a gradient descent value function reinforcement learning algorithm.

5.2-15. *Stochastic Neighborhood Embedding Gradient Descent.* Consider a Stochastic Neighborhood Embedding (SNE) as discussed in Example 1.6.5 where the probability that $\mathbf{x}_j$ is a neighbor of $\mathbf{x}_i$ in the original feature space is defined by the formula:

$$p(\mathbf{x}_j|\mathbf{x}_i) = \frac{\exp(-D_x(\mathbf{x}_i, \mathbf{x}_j))}{\sum_{k \neq i} \exp(-D_x(\mathbf{x}_i, \mathbf{x}_k))}$$

where $D_x(\mathbf{x}_i, \mathbf{x}_k) = |\mathbf{x}_i - \mathbf{x}_k|^2$. In addition, assume that the probability that the reduced dimensionality feature vector $\mathbf{y}_j$ is a neighbor of $\mathbf{y}_i$ is given by the formula:

$$q(\mathbf{y}_j|\mathbf{y}_i) = \frac{\exp(-D_y(\mathbf{y}_i, \mathbf{y}_j))}{\sum_{k \neq i} \exp(-D_y(\mathbf{y}_i, \mathbf{x}_k))}$$

where $D_y(\mathbf{y}_i, \mathbf{y}_j) = |\mathbf{y}_i - \mathbf{y}_j|^2$. Let

$$V(\mathbf{y}_1, \ldots, \mathbf{y}_M) = -\sum_{i=1}^{M} \sum_{j=1}^{M} p(\mathbf{x}_j|\mathbf{x}_i) \log q(\mathbf{y}_j|\mathbf{y}_i)$$

and compute the derivative of $V$ with respect to $\mathbf{y}_k$ for the purpose of deriving a gradient descent algorithm that minimizes $V$.

5.2-16. Repeat Exercise 5.2-14 but replace the sigmoidal hidden units with softplus hidden units.

5.2-17. Repeat Exercise 5.2-14 but replace the sigmoidal hidden units with radial basis function hidden units.

5.2-18. Repeat Exercise 5.2-14 but add an $L_2$ regularization constant.

5.2-19. *Missing Information Principle Hessian.* Derive the Missing Information Principle Hessian for the missing data log-likelihood function in (13.52) and (13.53) by taking the derivative of (13.51).

## 5.3 Objective Function Analysis

### 5.3.1 Taylor Series Expansions

An important tool used throughout engineering mathematics as well as throughout this text is the Taylor series expansion because highly nonlinear sufficiently smooth objective functions can be locally approximated by linear or quadratic objective functions.

**Theorem 5.3.1** (Vector Taylor Series Expansion). *Let $U \subseteq \mathcal{R}^d$ be an open convex set. Let $\mathbf{x}, \mathbf{x}^* \in U$. Assume $V : U \rightarrow \mathcal{R}$ is a function whose gradient, $\mathbf{g} : U \rightarrow \mathcal{R}^d$, exists and is continuous on $U$. Then, there exists a real number $\theta \in [0, 1]$ such that for all $\mathbf{x} \in U$:*

$$V(\mathbf{x}) = V(\mathbf{x}^*) + R_1 \tag{5.25}$$

*where $R_1 = \mathbf{g}(\mathbf{c}_\theta)^T[\mathbf{x} - \mathbf{x}^*]$ with $\mathbf{c}_\theta = \mathbf{x}^* + \theta(\mathbf{x} - \mathbf{x}^*)$.*

*In addition, suppose that the Hessian of* $V$, $\mathbf{H} : U \to \mathcal{R}^{d \times d}$, *exists and is continuous on* $U$. *Then, there exists a real number* $\theta \in [0, 1]$ *such that for all* $\mathbf{x} \in U$*:*

$$V(\mathbf{x}) = V(\mathbf{x}^*) + \mathbf{g}(\mathbf{x}^*)^T [\mathbf{x} - \mathbf{x}^*] + R_2 \tag{5.26}$$

*where*

$$R_2 = (1/2)[\mathbf{x} - \mathbf{x}^*]^T \mathbf{H}(\mathbf{c}_\theta)[\mathbf{x} - \mathbf{x}^*]$$

*with* $\mathbf{c}_\theta = \mathbf{x}^* + \theta(\mathbf{x} - \mathbf{x}^*)$.

*In addition, suppose that the derivative of* $\mathbf{H}$, $d\mathbf{H}/d\mathbf{x}$, *exists and is continuous on* $U$. *Then, there exists a real number* $\theta \in [0, 1]$ *such that for all* $\mathbf{x} \in U$*:*

$$V(\mathbf{x}) = V(\mathbf{x}^*) + \mathbf{g}(\mathbf{x}^*)^T [\mathbf{x} - \mathbf{x}^*] + (1/2)[\mathbf{x} - \mathbf{x}^*]^T \mathbf{H}(\mathbf{x}^*)[\mathbf{x} - \mathbf{x}^*] + R_3 \tag{5.27}$$

*where*

$$R_3 = (1/6)\left[(\mathbf{x} - \mathbf{x}^*)^T \otimes (\mathbf{x} - \mathbf{x}^*)^T\right] \left[\frac{d\mathbf{H}}{d\mathbf{x}}(\mathbf{c}_\theta)\right] (\mathbf{x} - \mathbf{x}^*)$$

*with* $\mathbf{c}_\theta = \mathbf{x}^* + \theta(\mathbf{x} - \mathbf{x}^*)$.

*Proof.* See Marlow (2012 pp. 224-225). ■

Equation (5.25) is called the *zero-order* Taylor expansion of $V$ with remainder term $R_1$. Equation (5.25) is also called the *Mean-Value Theorem.* Equation (5.26) is called the *first-order* Taylor expansion of $V$ with remainder term $R_2$. Equation (5.27) is called the *second-order* Taylor expansion of $V$ with remainder term $R_3$.

The statement that $\mathbf{c}_\theta = \mathbf{x}^* + \theta(\mathbf{x} - \mathbf{x}^*)$ for some real number $\theta \in [0, 1]$ may be interpreted geometrically to mean that there exists a point $\mathbf{c}_\theta$ located on the line segment connecting the point $\mathbf{x}$ and the point $\mathbf{x}^*$. The assumption that $U$ is a convex set ensures that for every $\mathbf{x}$ and $\mathbf{x}^*$ which are elements of $U$, there exists a line segment connecting $\mathbf{x}$ and $\mathbf{x}^*$ which lies entirely in $U$.

Although higher-order Taylor expansions can be constructed in a similar manner, the first-order and second-order expansions will be adequate for covering most topics in this textbook.

### 5.3.2 Gradient Descent Type Algorithms

Many important algorithms used for inference and learning are naturally interpreted as descent algorithms that minimize an objective function $V$.

The following theorem discusses conditions for the system of difference equations:

$$\mathbf{x}(t + 1) = \mathbf{x}(t) + \gamma_t \mathbf{d}(\mathbf{x}(t))$$

to have the property that $V(\mathbf{x}(t+1)) < V(\mathbf{x}(t))$ if the gradient of $V$ evaluated at $\mathbf{x}(t)$ does not vanish.

**Theorem 5.3.2** (Descent Direction Theorem). *Let* $V : \mathcal{R}^d \to \mathcal{R}$ *be a twice continuously differentiable function. Let*

$$\Gamma \equiv \left\{\mathbf{x} \in \mathcal{R}^d : \frac{dV(\mathbf{x})}{d\mathbf{x}} = \mathbf{0}_d^T\right\}.$$

*Assume there exists a continuous column vector-valued function* $\mathbf{d} : \mathcal{R}^d \to \mathcal{R}^d$ *defined such that for all* $\mathbf{x} \notin \Gamma$*:*

$$[dV/d\mathbf{x}]\mathbf{d}(\mathbf{x}) < 0. \tag{5.28}$$

*Let* $\mathbf{x}(0) \in \mathcal{R}^d$ *and define the sequence* $\mathbf{x}(1), \mathbf{x}(2), \ldots$ *such that:*

$$\mathbf{x}(t+1) = \mathbf{x}(t) + \gamma_t \mathbf{d}(\mathbf{x}(t))$$

*where* $\gamma_1, \gamma_2, \ldots$ *is a sequence of positive numbers. Then, for each* $t \in \mathbb{N}$*, there exists a* $\gamma_t$ *such that* $V(\mathbf{x}(t+1)) < V(\mathbf{x}(t))$ *if* $\mathbf{x}(t) \notin \Gamma$.

*Proof.* Let $\mathbf{d}_t \equiv \mathbf{d}(\mathbf{x}(t))$. Without loss in generality, assume $\gamma : \mathbb{N} \rightarrow [0, \infty)$ is a bounded function. Expand $V$ in a first-order Taylor expansion about $\mathbf{x}(t)$ and evaluate at $\mathbf{x}(t) + \gamma_t \mathbf{d}_t$ to obtain for a given $\mathbf{x}(t)$:

$$V(\mathbf{x}(t) + \gamma_t \mathbf{d}_t) = V(\mathbf{x}(t)) + \gamma_t [dV(\mathbf{x}(t))/d\mathbf{x}] \mathbf{d}_t + R_t(\gamma_t^2) \tag{5.29}$$

where the remainder term

$$R_t(\gamma_t) \equiv [\gamma_t \mathbf{d}_t]^T \mathbf{H}(\mathbf{c}_t) [\gamma_t \mathbf{d}_t],$$

$\mathbf{H} \equiv \nabla^2 V$, and $\mathbf{c}_t$ is a point on the chord connecting $\mathbf{x}(t)$ and $\mathbf{x}(t) + \gamma_t \mathbf{d}_t$. Since $\mathbf{H}$ and $\mathbf{d}$ are continuous functions on the closure of the chord which is a bounded set, $\mathbf{H}$ and $\mathbf{d}$ are bounded functions on the chord.

Since $\mathbf{x}(t) \notin \Gamma$, the second term on the right-hand side of (5.29) is a strictly negative number because of the downhill condition in (5.28). In addition, the second term on the right-hand side of (5.29) is $O(\gamma_t)$ while the third term on the right-hand side of (5.29) is $O(\gamma_t^2)$ implying the second term dominates the third term for sufficiently small $\gamma_t$. Thus, it is always possible to choose $\gamma_t$ sufficiently small so that:

$$V(\mathbf{x}(t) + \gamma_t \mathbf{d}_t) < V(\mathbf{x}(t)). \tag{5.30}$$

■

The Descent Direction Theorem 5.3.2 states there exists a sequence of positive *stepsizes* $\gamma_0, \gamma_1, \gamma_2, \ldots$ such that the objective function $V$ is non-increasing on the trajectory $\mathbf{x}(0), \mathbf{x}(1), \mathbf{x}(2), \ldots$ (i.e., $V(\mathbf{x}(t+m)) < V(\mathbf{x}(t))$ for $m = 1, 2, \ldots$) provided that an acceptable sequence of search directions $\mathbf{d}_0, \mathbf{d}_1, \ldots$ can be found.

Note that if $\mathbf{M}$ is a positive definite $d$-dimensional matrix, a descent direction $\mathbf{d}$ chosen such that:

$$\mathbf{d} = -\mathbf{M}[dV/d\mathbf{x}]^T$$

satisfies the *downhill* condition in (5.28) since

$$[dV/d\mathbf{x}]\mathbf{d} = [dV/d\mathbf{x}]\mathbf{M}[dV/d\mathbf{x}]^T > 0.$$

In the special case where $\mathbf{M}$ is the identity matrix, the search direction

$$\mathbf{d} \equiv -\mathbf{M}[dV/d\mathbf{x}]^T = -[dV/d\mathbf{x}]^T$$

is called the *gradient descent search direction.*

The Descent Direction Theorem 5.3.2 is not really useful in practice and is presented mainly for pedagogical reasons. Let $\Gamma$ be a subset of $\mathcal{R}^d$ whose elements are the parameter vectors which are minimizers of $V$. The Descent Direction theorem provides conditions ensuring that $V$ decreases but does not establish conditions ensuring that $\mathbf{x}(t) \rightarrow \Gamma$ as $t \rightarrow \infty$. Conditions for $\mathbf{x}(t) \rightarrow \Gamma$ as $t \rightarrow \infty$ are discussed in Chapter 6 and Chapter 7.

**Example 5.3.1** (Unsupervised Batch Correlational (Hebbian) Learning). Let the training data be defined as a set of $n$ $d$-dimensional training vectors: $\{\mathbf{x}_1, \ldots, \mathbf{x}_n\}$ for an unsupervised learning machine that takes a $d$-dimensional input vector $\mathbf{x}$ and computes a $d$-dimensional response vector $\mathbf{r} = \mathbf{W}\mathbf{x}$. The magnitude of $\mathbf{r}$ measures the inference machine's familiarity with $\mathbf{x}$. So, for example, if $\mathbf{x}$ is in the null space of $\mathbf{W}$, then the inference machine does not recognize $\mathbf{x}$ since $\mathbf{r} = \mathbf{W}\mathbf{x} = \mathbf{0}_d$.

Let the function $\ell : \mathcal{R}^{d \times d} \rightarrow \mathcal{R}$ be defined such that for all $\mathbf{W} \in \mathcal{R}^{d \times d}$:

$$\ell(\mathbf{W}) = (\delta/2)|\mathbf{vec}(\mathbf{W})|^2 - (1/n)\sum_{i=1}^{n} \mathbf{x}_i^T \mathbf{W}\mathbf{x}_i$$

where $\delta$ is a small positive number. Derive a gradient descent learning algorithm that minimizes $\ell$.

**Solution.** First show that:

$$\frac{d\ell(\mathbf{W})}{d\mathbf{W}} = \left[\delta \mathbf{vec}(\mathbf{W}^T) - (1/n)\sum_{i=1}^{n} \mathbf{vec}\left(\mathbf{x}_i \mathbf{x}_i^T\right)\right]^T. \tag{5.31}$$

The gradient descent algorithm formula is:

$$\mathbf{vec}(\mathbf{W}_{t+1}^T) = \mathbf{vec}(\mathbf{W}_t^T) - \gamma_t \left[\frac{d\ell(\mathbf{W}_t)}{d\mathbf{W}}\right]^T \tag{5.32}$$

where the positive numbers $\gamma_1, \gamma_2, \ldots$ is a sequence of stepsizes chosen in an appropriate manner. Combining (5.31) and (5.32) gives:

$$\mathbf{W}_{t+1} = (1 - \delta\gamma_t)\mathbf{W}_t + (\gamma_t/n)\sum_{i=1}^{n} \mathbf{x}_i \mathbf{x}_i^T. \tag{5.33}$$

△

**Example 5.3.2** (Supervised Batch Linear Regression Learning as Gradient Descent). Let the training data be defined as a set of $n$ $d+1$-dimensional training vectors:

$$\{(y_1, \mathbf{s}_1), \ldots, (y_n, \mathbf{s}_n)\}$$

where $y \in \mathcal{R}$ is the desired response of a linear inference machine to $d$-dimensional input column pattern vector $\mathbf{s}$. Let $\ddot{y}(\mathbf{s}, \boldsymbol{\theta}) \equiv \boldsymbol{\theta}^T [\mathbf{s}^T \; 1]^T$ denote the learning machine's predicted response to input pattern $\mathbf{s}$ given its current parameter vector $\boldsymbol{\theta}$. Let the function $\ell : \mathcal{R}^{d+1} \rightarrow \mathcal{R}$ be defined such that for all $\boldsymbol{\theta} \in \mathcal{R}^{d+1}$:

$$\ell(\boldsymbol{\theta}) = (1/2)(1/n)\sum_{i=1}^{n} |y_i - \ddot{y}(\mathbf{s}_i, \boldsymbol{\theta})|^2.$$

Derive a gradient descent learning algorithm that will minimize $\ell$.

**Solution.** The derivative of $\ell$ is given by the formula:

$$\frac{d\ell}{d\boldsymbol{\theta}} = -(1/n)\sum_{i=1}^{n} (y_i - \ddot{y}(\mathbf{s}_i, \boldsymbol{\theta}))\, \mathbf{s}_i^T \tag{5.34}$$

and the gradient descent formula is given by:

$$\boldsymbol{\theta}_{t+1} = \boldsymbol{\theta}_t - \gamma_t \frac{d\ell(\boldsymbol{\theta}_t)}{d\boldsymbol{\theta}} \tag{5.35}$$

where the positive numbers $\gamma_1, \gamma_2, \ldots$ is a sequence of stepsizes chosen in an appropriate manner. Plugging the derivative of $\ell$ in (5.34) into the gradient descent formula in (5.35) gives:

$$\boldsymbol{\theta}_{t+1} = \boldsymbol{\theta}_t + (\gamma_t/n)\sum_{i=1}^{n} (y_i - \ddot{y}(\mathbf{s}_i, \boldsymbol{\theta}_t))\, \mathbf{s}_i^T. \tag{5.36}$$

△

### 5.3.3 Critical Point Classification

A critical point of an objective function corresponds to parameter estimates that cause the derivative of the objective function to vanish at that point. Methods for designing learning algorithms that generate sequences of parameter estimates which converge to the critical points of an objective function are discussed in Chapters 6, 7, and 12. For this reason, it is useful to develop procedures for classifying critical points as minima, maxima, and saddlepoints (see Figure 5.1).

#### 5.3.3.1 Identifying Critical Points

**Definition 5.3.1** (Critical Point). Let $\Omega \subseteq \mathcal{R}^d$ be an open set. Let $V : \Omega \rightarrow \mathcal{R}$ be continuously differentiable on $\Omega$. If $\mathbf{x}^* \in \Omega$ and $dV(\mathbf{x}^*)/d\mathbf{x} = \mathbf{0}_d^T$, then $\mathbf{x}^*$ is a *critical point* of $V$. □

Analyses of the asymptotic behavior of learning dynamical systems often establish that the system state $\mathbf{x}(t)$ will converge to a critical point $\mathbf{x}^*$ as $t \rightarrow \infty$. This implies that $|\mathbf{x}(t+1) - \mathbf{x}(t)|$ converges to zero as $t \rightarrow \infty$. However, such theorems do not prove that the trajectory $\mathbf{x}(1), \mathbf{x}(2), \mathbf{x}(3), \ldots$ will *reach* $\mathbf{x}^*$ (i.e., $\mathbf{x}(t) = \mathbf{x}^*$ when $t$ is sufficiently large). This means that $|\mathbf{x}(t+1) - \mathbf{x}(t)|$ may be strictly positive even when $t$ is sufficiently large.

These remarks identify a potential problem with assessing learning algorithm convergence by checking if $|\mathbf{x}(t+1) - \mathbf{x}(t)|$ is sufficiently small. In fact, since $\mathbf{x}(t+1)$ is generated by

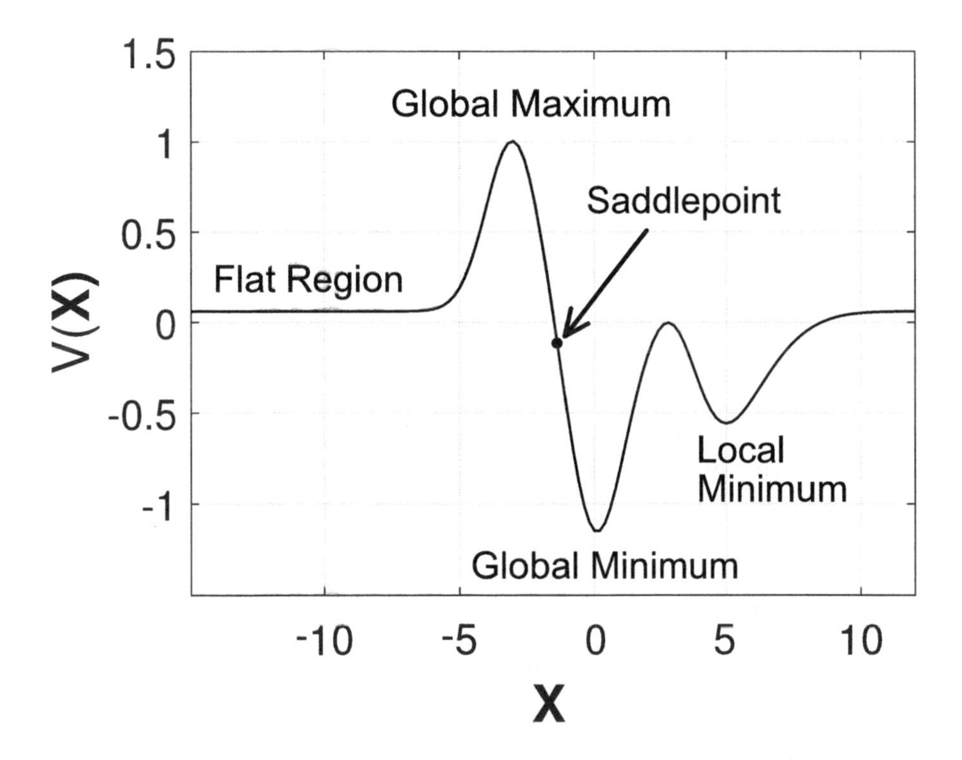


**FIGURE 5.1**
**Examples of critical points of an objective function on a one-dimensional parameter space.** The objective function $V : \mathcal{R} \rightarrow \mathcal{R}$ depicted in this figure provides a geometric illustration of saddlepoints, local minima, local maxima, and flat regions.

the learning algorithm from $\mathbf{x}(t)$, the quantity $|\mathbf{x}(t+1) - \mathbf{x}(t)|$ depends upon the properties of the learning algorithm.

A better method for deciding when $\mathbf{x}(t)$ is sufficiently close to $\mathbf{x}^*$ is to compute the gradient of $V$. If the infinity norm of the gradient evaluated at the point $\mathbf{x}(t)$ is sufficiently small, then one may conclude that $\mathbf{x}(t)$ is sufficiently close to $\mathbf{x}^*$. This provides a convenient algorithm-independent criteria for assessing if $\mathbf{x}(t)$ is sufficiently close to the critical point.

**Example 5.3.3** (Numerically Robust Check for a Critical Point). Assume that during the learning process $\hat{\ell}_n : \mathcal{R}^q \rightarrow \mathcal{R}$ is an empirical risk function which is an approximation for the risk function $\ell : \mathcal{R}^q \rightarrow \mathcal{R}$. Let $\hat{\boldsymbol{\theta}}_n$ be an approximate minimizer of $\hat{\ell}_n$. In addition, assume the quality of these approximations improves in some sense as $n \rightarrow \infty$. Provide a numerically robust method for checking if $\hat{\boldsymbol{\theta}}_n$ is a critical point of $\hat{\ell}_n$.

In practice, computing the gradient of the empirical risk function $\hat{\ell}_n$ at the parameter estimates $\hat{\boldsymbol{\theta}}_n$ should not be compared with a vector of zeros for the purpose of checking convergence to a critical point because $\hat{\boldsymbol{\theta}}_n$ might be converging to the critical point but might not actually be the critical point. Derive a numerically robust stopping criterion for checking if a learning algorithm is converging to a critical point of empirical risk function $\hat{\ell}_n$.

**Solution.** Let $\epsilon$ be a small positive number. If

$$\left| \frac{d\hat{\ell}_n(\hat{\boldsymbol{\theta}}_n)}{d\boldsymbol{\theta}} \right|_\infty < \epsilon,$$

then decide $\hat{\boldsymbol{\theta}}_n$ is a critical point. An improved version of this procedure is to repeat this procedure for $n/8, n/4, n/2$ in order to empirically investigate if the gradient of the empirical risk function is converging to zero as $n$ increases. △

#### 5.3.3.2 Identifying Flat Regions

**Definition 5.3.2** (Flat Region). Let $V : \mathcal{R}^d \rightarrow \mathcal{R}$. Let $\Omega \subseteq \mathcal{R}^d$ be a connected set which contains at least two points. If there exists a $K \in \mathcal{R}$ such that $V(\mathbf{x}) = K$ for all $\mathbf{x} \in \Omega$, then $\Omega$ is called a *flat region*. □

Note that $dV/d\mathbf{x} = \mathbf{0}_d^T$ on a flat region $\Omega$ so every point in the flat region $\Omega$ is a critical point. Flat regions arise frequently in many machine learning applications.

**Example 5.3.4** (Flat Regions in Linear Regression). Let $\mathcal{D}_n \equiv \{(\mathbf{s}_1, y_1), \ldots, (\mathbf{s}_n, y_n)\}$ be a training data set where the scalar number $y_i$ is the desired response of the supervised linear regression learning machine for a given $d$-dimensional input pattern vector $\mathbf{s}_i$ for $i = 1, \ldots, n$. Assume that a linear regression learning machine as defined in Example 1.5.1 is constructed to learn the data set $\mathcal{D}_n$ by minimizing the empirical risk function, $\hat{\ell}_n : \mathcal{R}^q \rightarrow \mathcal{R}$, in (1.12) where $q \equiv d + 1$. Does this risk function have a flat region? Under what conditions will a flat region exist?

**Solution.** The gradient of $\hat{\ell}_n$, $\mathbf{g}_n$, is given by the formula:

$$\mathbf{g}_n(\boldsymbol{\theta})^T = -(2/n) \sum_{i=1}^{n} \left( y_i - \boldsymbol{\theta}^T [\mathbf{s}_i^T, 1]^T \right) [\mathbf{s}_i^T, 1]. \tag{5.37}$$

If the system of equations $\mathbf{g}_n(\boldsymbol{\theta}) = \mathbf{0}_q$ has multiple solutions then a flat region will exist. This system of equations may be rewritten as:

$$\mathbf{g}_n(\boldsymbol{\theta})^T = -\mathbf{b}^T + \boldsymbol{\theta}^T \mathbf{C} \tag{5.38}$$

where $\mathbf{b}^T \equiv (2/n) \sum_{i=1}^{n} y_i [\mathbf{s}_i^T, 1]$ and $\mathbf{C} \equiv (2/n) \sum_{i=1}^{n} [\mathbf{s}_i^T, 1]^T [\mathbf{s}_i^T, 1]$.

If the system of equations $\mathbf{g}_n(\boldsymbol{\theta}) = \mathbf{0}_q$ is a connected set containing at least two points, then the set

$$\{\boldsymbol{\theta} : \mathbf{g}_n(\boldsymbol{\theta}) = \mathbf{0}_q\} \tag{5.39}$$

is a flat region. △

**Example 5.3.5** (Flat Regions in Multilayer Perceptron Empirical Risk Functions). Consider an empirical risk function $\hat{\ell}_n : \mathcal{R}^q \to \mathcal{R}$ for a multilayer perceptron with one layer of hidden units defined such that for all $\boldsymbol{\theta} = [w_1, \ldots, w_M, \mathbf{v}_1^T, \ldots, \mathbf{v}_M^T]^T$:

$$\hat{\ell}_n(\boldsymbol{\theta}) = (1/n) \sum_{i=1}^{n} c(\mathbf{x}_i, \boldsymbol{\theta})$$

where the loss function

$$c(\mathbf{x}_i, \boldsymbol{\theta}) = (y_i - \ddot{y}(\mathbf{s}_i; \boldsymbol{\theta}))^2$$

and the predicted response $\ddot{y}(\mathbf{s}_i; \boldsymbol{\theta})$ for $d$-dimensional input pattern $\mathbf{s}_i$ is defined such that:

$$\ddot{y}(\mathbf{s}_i; \boldsymbol{\theta}) = \sum_{k=1}^{M} w_k \mathcal{J}(\mathbf{v}_k^T \mathbf{s}_i)$$

where the softplus hidden unit response $\mathcal{J}(\mathbf{v}_k^T \mathbf{s}_i) \equiv \log(1 + \exp(\mathbf{v}_k^T \mathbf{s}_I))$. Show that if $\boldsymbol{\theta}^*$ is a critical point such that $w_k^* = 0$, then $\boldsymbol{\theta}^*$ is an element of a flat region of $\hat{\ell}_n$. Explicitly define that flat region.

**Solution.** If the output weight of the $k$th hidden unit, $w_k^* = 0$, then the choice of the input weights to the $k$th hidden unit $\mathbf{v}_k^*$ have no influence on the value of $\hat{\ell}_n$ because $w_k^* = 0$ essentially disconnects the $k$th hidden unit from the network. Thus, define the flat region for the critical point as the set of points $\Omega_k \subseteq \mathcal{R}^q$ such that:

$$\Omega_k = \left\{[w_1^*, \ldots, w_{k-1}^*, 0, w_{k+1}^*, \ldots, w_M, \mathbf{v}^{k-1}, (\mathbf{v}_k)^T, \mathbf{v}^{k+1}] : \mathbf{v}_k \in \mathcal{R}^d\right\}$$

where $\mathbf{v}^{k-1} \equiv [(\mathbf{v}_1^*)^T, \ldots, (\mathbf{v}_{k-1}^*)^T]$ and $\mathbf{v}^{k+1} \equiv [(\mathbf{v}_{k+1}^*)^T, \ldots, (\mathbf{v}_M^*)^T]$.

A similar analysis shows the presence of such flat regions is common to all feedforward networks with one or more layers of hidden units. Show that this particular argument regarding flat regions existence is not applicable for an empirical risk function with $L_2$ regularization defined such that:

$$\hat{\ell}_n(\boldsymbol{\theta}) = \lambda|\boldsymbol{\theta}|^2 + (1/n) \sum_{i=1}^{n} c(\mathbf{x}_i, \boldsymbol{\theta})$$

for some sufficiently large positive number $\lambda$. △

#### 5.3.3.3 Identifying Local Minimizers

**Definition 5.3.3** (Local Minimizer). Let $\Omega$ be a subset of $\mathcal{R}^d$ and let $\mathbf{x}^*$ be a point in the interior of $\Omega$. Let $V$ be a function such that $V : \Omega \to \mathcal{R}$. If there exists a $\delta$-neighborhood of $\mathbf{x}^*$, $\mathcal{N}_{\mathbf{x}^*}$, such that $V(\mathbf{y}) \geq V(\mathbf{x}^*)$ for all $\mathbf{y} \in \mathcal{N}_{\mathbf{x}^*}$, then $\mathbf{x}^*$ is a *local minimizer*. If, in addition, $V(\mathbf{y}) > V(\mathbf{x}^*)$ for all $\mathbf{y} \in \mathcal{N}_{\mathbf{x}^*}$ such that $\mathbf{y} \neq \mathbf{x}^*$, then $\mathbf{x}^*$ is a *strict local minimizer*. □

The following theorem provides sufficient conditions for identifying a point as a strict local minimizer.

**Theorem 5.3.3** (Strict Local Minimizer Test). *Let $\Omega$ be a subset of $\mathcal{R}^d$ and let $\mathbf{x}^*$ be a point in the interior of $\Omega$. Let $V : \Omega \rightarrow \mathcal{R}$ be twice-continuously differentiable on $\Omega$. The d-dimensional real vector $\mathbf{x}^*$ is a strict local minimizer if (i) $\mathbf{x}^* \in \mathcal{R}^d$ is a critical point of $V$, and (ii) the Hessian of $V$ evaluated at $\mathbf{x}^*$ is positive definite.*

*Proof.* Let $\mathbf{g}(\mathbf{x}^*)$ be the gradient of $V$ evaluated at $\mathbf{x}^*$. Assume that $\mathbf{g}(\mathbf{x}^*) = \mathbf{0}_d$ and let $\mathbf{H}(\mathbf{x}^*)$ be the Hessian of $V$ evaluated at $\mathbf{x}^*$. Expand $V$ in a Taylor expansion about $\mathbf{x}^*$ to obtain for $\mathbf{x}$ sufficiently close to $\mathbf{x}^*$:

$$V(\mathbf{x}) = V(\mathbf{x}^*) + \mathbf{g}(\mathbf{x}^*)^T(\mathbf{x} - \mathbf{x}^*) + (1/2)(\mathbf{x} - \mathbf{x}^*)^T\mathbf{H}(\mathbf{c})(\mathbf{x} - \mathbf{x}^*) \tag{5.40}$$

where $\mathbf{c}$ lies on the chord connecting $\mathbf{x}$ and $\mathbf{x}^*$.

Since $\mathbf{x}^*$ is a critical point in the interior of $\Omega$, $\mathbf{g}(\mathbf{x}^*) = \mathbf{0}_d$. Since the Hessian of $V$ is positive definite at $\mathbf{x}^*$ and the Hessian of $V$ is continuous on its domain, there exists a neighborhood of $\mathbf{x}^*$ such that the Hessian of $V$ is positive definite on that neighborhood. These two observations with (5.40) imply that $\mathbf{x}^*$ is a strict local minimizer. ■

Note that the converse of this theorem does not hold. To see this, consider the function $V(x) \equiv x^4$. The gradient of $V$ is $4x^3$ which vanishes at the point $x = 0$. The Hessian of $V$ is $12x^2$ which also vanishes at the point $x = 0$. However, $x = 0$ is a strict local minimizer of $V$.

Once a critical point $\mathbf{x}^*$ is reached, the Strict Local Minimizer Test Theorem may be used to then check if $\mathbf{x}^*$ is a strict local minimizer by computing the condition number of the Hessian of $V$ evaluated at the critical point $\mathbf{x}^*$.

**Example 5.3.6** (Numerical Stopping Criterion for a Strict Local Minimizer). In Example 5.3.3, a numerically robust procedure for checking if the current parameter estimates are converging to a critical point was proposed. Derive a numerically robust procedure for checking if the current parameter estimates $\hat{\boldsymbol{\theta}}_n$ are converging to a strict local minimizer of the objective function $\hat{\ell}_n$ as $n$ increases.

**Solution.** Let $\epsilon$ be a small positive number. Let $R_{max}$ be a large positive number. First, check if $\hat{\boldsymbol{\theta}}_n$ is a critical point by checking if:

$$\left| \frac{d\hat{\ell}_n(\hat{\boldsymbol{\theta}}_n)}{d\boldsymbol{\theta}} \right|_\infty < \epsilon.$$

Second, if it is decided that $\hat{\boldsymbol{\theta}}_n$ is a critical point, then compute

$$\mathbf{H}_n \equiv \frac{d\hat{\ell}_n^2(\hat{\boldsymbol{\theta}}_n)}{d\boldsymbol{\theta}^2}.$$

Let $\lambda_{max}$ and $\lambda_{min}$ be the largest and smallest eigenvalues of $\mathbf{H}_n$. If (i) $\lambda_{max} > \epsilon$, (ii) $\lambda_{min} > \epsilon$, and (iii) $(\lambda_{max}/\lambda_{min}) < R_{max}$, then decide $\hat{\boldsymbol{\theta}}_n$ is a strict local minimizer. An improved version of this procedure is to repeat this procedure for $n/8$, $n/4$, $n/2$ in order to empirically investigate if the Hessian of the empirical risk function evaluated in the vicinity of the critical point is converging to a positive definite matrix as $n$ increases. △

**Definition 5.3.4** (Local Maximizers). Let $\Omega \subseteq \mathcal{R}^d$. Let $V : \Omega \rightarrow \mathcal{R}$. If $\mathbf{x}^*$ is a (strict) local minimizer of $V$ on $\Omega$, then $\mathbf{x}^*$ is a (strict) *local maximizer* of $-V$ on $\Omega$. □

#### 5.3.3.4 Identifying Saddlepoints

**Definition 5.3.5** (Saddlepoint of a Function)**.** Let $\Omega$ be an open subset of $\mathcal{R}^d$. Let $V : \Omega \to \mathcal{R}$ be a continuously differentiable function. Let $\mathbf{x}^*$ be a critical point of $V$. If $\mathbf{x}^*$ is not a local minimizer and $\mathbf{x}^*$ is not a local maximizer, then $\mathbf{x}^*$ is called a *saddlepoint.* □

**Definition 5.3.6** (Saddlepoint Index of a Twice Differentiable Function)**.** Let $\Omega \subseteq \mathcal{R}^d$ be an open and convex set. Let $V : \Omega \to \mathcal{R}$ be a twice continuously differentiable function on $\Omega$. Assume $\mathbf{x}^*$ is a critical point of $V$ in the interior of $\Omega$. Let $\mathbf{H}^*$ denote the Hessian of $V$ evaluated at $\mathbf{x}^*$. The number of negative eigenvalues of $\mathbf{H}^*$ is called the *negative saddlepoint index* of $\mathbf{x}^*$. The number of positive eigenvalues of $\mathbf{H}^*$ is called the *positive saddlepoint index* of $\mathbf{x}^*$. In the special case where $\mathbf{H}^*$ is invertible, then the negative saddlepoint index of $\mathbf{H}^*$ is called the *saddlepoint index.* □

Saddlepoints have very interesting and important properties in high-dimensional state spaces (see Figure 5.2). The saddlepoint index specifies a particular qualitative characterization of the curvature of the objective function surface at a given critical point $\mathbf{x}^* \in \mathcal{R}^d$. If the positive saddlepoint index at $\mathbf{x}^*$ is $d$, then $\mathbf{x}^*$ is a strict local minimizer. If the positive saddlepoint index at $\mathbf{x}^*$ is zero, then $\mathbf{x}^*$ is a strict local maximizer. If the positive saddlepoint index at $\mathbf{x}^*$ is a positive integer $k$ which is less than $d$ and the negative saddlepoint index at $\mathbf{x}^*$ is a positive integer $d - k$, this corresponds to a saddlepoint where the local curvature points "upward" in $k$ dimensions and points "downward" in $d - k$ dimensions.

Thus, the number of potential types of saddlepoints grows exponentially in the dimensionality $d$ according to the formula: $2^d - 2$. To see this, note that a sufficiently small change in the value of each coordinate of a $d$-dimensional critical point may either decrease or increase the value of the objective function.

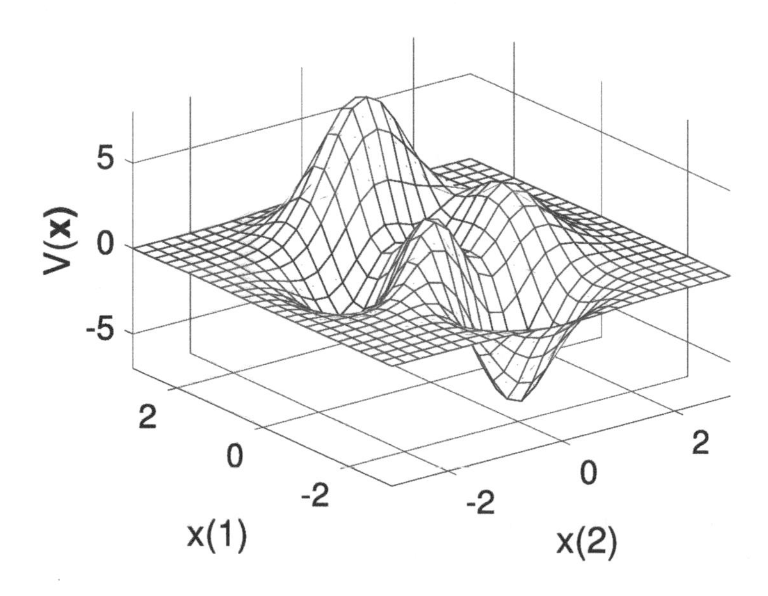


**FIGURE 5.2**
**Examples of saddlepoints of an objective function on a two-dimensional parameter space.** The objective function $V : \mathcal{R}^2 \to \mathcal{R}$ depicted in this figure provides a geometric illustration of saddlepoints as well as local minima, local maxima, and flat regions.

#### 5.3.3.5 Identifying Global Minimizers

**Definition 5.3.7** (Global Minimizer). Let $\Omega \subseteq \mathcal{R}^d$. Let $V : \Omega \to \mathcal{R}$. If for all $\mathbf{x} \in \Omega$: $V(\mathbf{x}^*) \leq V(\mathbf{x})$, then $\mathbf{x}^*$ is a *global minimizer* on $\Omega$. If for all $\mathbf{x} \in \Omega$ such that $\mathbf{x} \neq \mathbf{x}^*$: $V(\mathbf{x}^*) < V(\mathbf{x})$, then $\mathbf{x}^*$ is a *strict global minimizer* on $\Omega$. □

**Definition 5.3.8** (Convex Function). Let $\Omega$ be a convex region of $\mathcal{R}^d$. The function $V : \Omega \to \mathcal{R}$ is *convex* on $\Omega$ if for every $\alpha \in [0, 1]$:

$$V(\alpha\mathbf{x}_1 + (1 - \alpha)\mathbf{x}_2) \leq \alpha V(\mathbf{x}_1) + (1 - \alpha)V(\mathbf{x}_2)$$

for all $\mathbf{x}_1, \mathbf{x}_2 \in \Omega$. The function $V$ is *strictly convex* if for every $\alpha \in [0, 1]$:

$$V(\alpha\mathbf{x}_1 + (1 - \alpha)\mathbf{x}_2) < \alpha V(\mathbf{x}_1) + (1 - \alpha)V(\mathbf{x}_2)$$

for all $\mathbf{x}_1, \mathbf{x}_2 \in \Omega$. □

Figure 5.3 illustrates the concept of a convex function. A "bowl-shaped" function is therefore an example of a *convex function.*

**Theorem 5.3.4** (Convex Set Construction Using a Convex Function). *Let $\Omega \subseteq \mathcal{R}^d$. Assume $V : \mathcal{R}^d \to \mathcal{R}$ is a convex function on a convex subset, $\Gamma$, of $\mathcal{R}^d$. Then,*

$$\Omega \equiv \{\mathbf{x} \in \Gamma : V(\mathbf{x}) \leq K\}$$

*is a convex set.*

*Proof.* Since $V$ is convex on $\Gamma$:

$$V(\mathbf{z}_\alpha) \leq \alpha V(\mathbf{x}) + (1 - \alpha)V(\mathbf{y}) \tag{5.41}$$

where $\mathbf{z}_\alpha = \alpha\mathbf{x} + (1 - \alpha)\mathbf{y}$ for all $\mathbf{x}, \mathbf{y} \in \Gamma$ and for all $\alpha \in [0, 1]$.

If $V(\mathbf{x}) \leq K$ and $V(\mathbf{y}) \leq K$ this implies that $\mathbf{x}, \mathbf{y} \in \Omega$. If $\mathbf{x}, \mathbf{y} \in \Omega$, then $\mathbf{z}_\alpha \in \Omega$ (5.41) implies

$$V(\mathbf{z}_\alpha) \leq \alpha V(\mathbf{x}) + (1 - \alpha)V(\mathbf{y}) \leq \alpha K + (1 - \alpha)K \leq K.$$

Thus, $\Omega$ is a convex set. ■

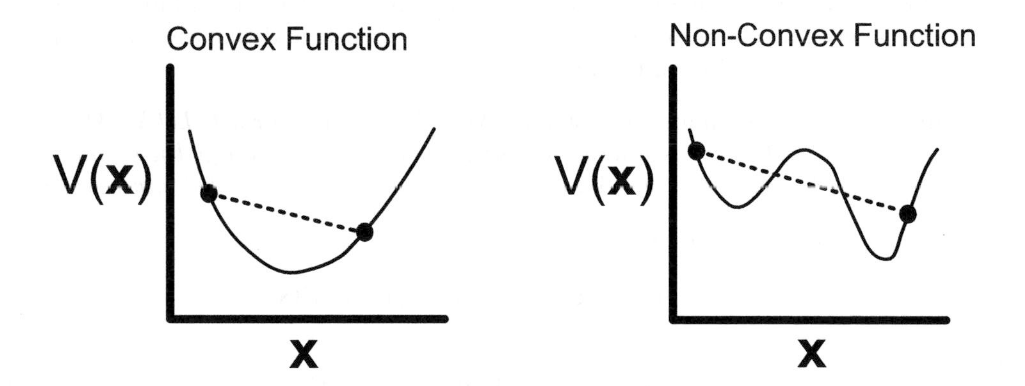


**FIGURE 5.3**
**Geometric interpretation of the definition of a convex function.** A convex function, $V$, is defined as a function such that the value of $V$ is less than or equal to a "linearized" version of $V$ evaluated along *any* line segment (e.g., dashed line in figure) connecting two points in the function's domain.

**Theorem 5.3.5** (Smooth Convex Function). *Let $\Omega$ be a convex subset of $\mathcal{R}^d$ which has a non-empty interior. Let $V : \Omega \to \mathcal{R}$ be a twice continuously differentiable function. The function $V$ is convex on $\Omega$ if and only if the Hessian of $V$ is positive semidefinite everywhere on $\Omega$.*

*Proof.* See Luenberger (1984, Proposition 5) for a proof. ■

It is possible for a function $V$ to be strictly convex yet the Hessian of $V$ is not positive definite everywhere on the domain of the function. For example, define the strictly convex function $V(x) = x^4$, then the first derivative of $V$, $\nabla V(x) = 4x^3$ vanishes and the second derivative of $V$, $\nabla^2 V(x) = 12x^2$ is positive semidefinite everywhere. This implies that $V$ is convex. However, $\nabla^2 V(x) = 12x^2 = 0$ for $x = 0$ which means that the second derivative of $V$ is not positive definite everywhere despite the fact that $V(x) = x^4$ is a strictly convex function with unique global minimizer at $x = 0$.

**Theorem 5.3.6** (Convex Function Composition).

*(i) If $f_1, \ldots, f_M$ are convex functions on a convex set $\Omega \subseteq \mathcal{R}^d$ and $w_1, \ldots, w_M$ are non-negative real numbers, then $\sum_{i=1}^{M} w_i f_i$ is a convex function on $\Omega$.*

*(ii) Let $\Gamma \subseteq \mathcal{R}$. Let $f : \Omega \to \Gamma$ be a convex function on $\Omega \subseteq \mathcal{R}^d$. Let $h : \Gamma \to \mathcal{R}$ be a convex function that is non-decreasing on $\Gamma$. Then, $h(f)$ is convex on $\Omega$.*

*Proof.* (i) See Exercise 5.3-10 for a proof. (ii) See Marlow (2012, p. 244) for a proof. ■

**Example 5.3.7** (Increasing Convex Function of a Convex Function). Use the Convex Function Composition Theorem to show that $V(x) = \exp\left(x^2\right) + 2x^4$ is a convex function.

**Solution.** Let $V(x) \equiv h(f(x)) + g(x)$ where $h(f) = \exp(f)$, $f(x) = x^2$, and $g(x) = 2x^4$. The function $h$ is an increasing function since $df/dh = 2h > 0$. The functions $f$ and $g$ are convex functions by the Smooth Convex Function Theorem (Theorem 5.3.5) since $d^2 f/x^2 = 2 \geq 0$ and $d^2 g/dx^2 = 24x^2 \geq 0$. Assertion (iii) of the Convex Function Composition Theorem 5.3.6 implies that $h(f(x))$ is a convex function on $(0, \infty)$. Assertion (ii) implies that $V(x) = h(f(x)) + g(x)$ is a convex function since $h(f(x))$ and $g(x)$ are convex functions. △

Identifying convex functions is important because any critical point of a convex function is a global minimizer of the function. Thus, if one can prove that an algorithm will converge to a critical point of a convex function, then one has demonstrated that the algorithm will converge to a global minimizer of the function.

**Theorem 5.3.7** (Global Minimizer Test). *Let $\Omega \subseteq \mathcal{R}^d$ be a convex set. Let $V : \Omega \to \mathcal{R}$ be a convex function on $\Omega$. If $\mathbf{x}^*$ is a local minimizer of $V$, then $\mathbf{x}^*$ is a global minimizer on $\Omega$. If $\mathbf{x}^*$ is a strict local minimizer of $V$, then $\mathbf{x}^*$ is the unique strict global minimizer on $\Omega$.*

*Proof.* Because $V$ is convex for every $\alpha \in (0, 1)$:

$$V\left(\alpha \mathbf{y} + (1 - \alpha)\mathbf{x}^*\right) \leq \alpha V\left(\mathbf{y}\right) + (1 - \alpha)V(\mathbf{x}^*)$$

which implies

$$V\left(\mathbf{x}^* + \alpha(\mathbf{y} - \mathbf{x}^*)\right) \leq \alpha V(\mathbf{y}) + (1 - \alpha)V(\mathbf{x}^*). \tag{5.42}$$

Assume $\mathbf{x}^*$ is a local minimizer of $V$ and $\mathbf{x}^*$ is not a global minimizer since there exists a $\mathbf{y} \in \Omega$ such that $V(\mathbf{y}) < V(\mathbf{x}^*)$. This latter assumption implies that the right-hand side of (5.42) is strictly less than $\alpha V(\mathbf{x}^*) + (1 - \alpha)V(\mathbf{x}^*) = V(\mathbf{x}^*)$ which contradicts the assumption that $\mathbf{x}^*$ is a local minimizer for all sufficiently small positive values of $\alpha$ since $V(\mathbf{x}^* + \alpha(\mathbf{y} - \mathbf{x}^*)) < V(\mathbf{x}^*)$.

Similarly, assume $\mathbf{x}^*$ is a strict local minimizer of $V$ and $\mathbf{x}^*$ is not the unique global minimizer of $V$ since there exists a $\mathbf{y} \in \Omega$ such that $V(\mathbf{y}) \leq V(\mathbf{x}^*)$. This latter assumption implies that the right-hand side of (5.42) satisfies for all $\alpha \in [0,1]$: $V(\mathbf{x}^*+\alpha(\mathbf{y}-\mathbf{x}^*)) \leq V(\mathbf{x}^*)$ which contradicts the assumption that $\mathbf{x}^*$ is a strict local minimizer for all sufficiently small values of $\alpha$ since $V(\mathbf{x}^* + \alpha(\mathbf{y} - \mathbf{x}^*)) \leq V(\mathbf{x}^*)$. ■

**Example 5.3.8** (Least Squares Solution for Linear Regression). Let $\mathbf{A} \in \mathcal{R}^{m \times n}$ have full column rank so that the matrix $\mathbf{A}$ has rank $n$ where $n \leq m$. Let $\mathbf{b} \in \mathcal{R}^m$. Let $V : \mathcal{R}^n \rightarrow [0, \infty)$ be defined such that for all $\mathbf{x} \in \mathcal{R}^n$ that

$$V(\mathbf{x}) = |\mathbf{A}\mathbf{x} - \mathbf{b}|^2.$$

Show that the formula:

$$\mathbf{x}^* = (\mathbf{A}^T\mathbf{A})^{-1}\mathbf{A}^T\mathbf{b}$$

is the unique strict global minimizer of $V$.

**Solution.** Let $\mathbf{r} = \mathbf{A}\mathbf{x} - \mathbf{b}$. Note that $d\mathbf{r}/d\mathbf{x} = \mathbf{A}$. The derivative of $V$ is given by the formula:

$$dV/d\mathbf{x} = (dV/d\mathbf{r})(d\mathbf{r}/d\mathbf{x}) = \left(2(\mathbf{A}\mathbf{x} - \mathbf{b})^T\right)\mathbf{A}.$$

Setting $dV(\mathbf{x}^*)/d\mathbf{x}$ equal to zero and solving for $\mathbf{x}^*$ gives the unique critical point:

$$\mathbf{x}^* = (\mathbf{A}^T\mathbf{A})^{-1}\mathbf{A}^T\mathbf{b}$$

where $(\mathbf{A}^T\mathbf{A})^{-1}$ is defined because $\mathbf{A}$ has full column rank $n$ where $n \leq m$.

Then, the second derivative of $V$ with respect to $\mathbf{x}$ is given by the formula:

$$d^2V/d\mathbf{x}^2 = 2\mathbf{A}^T\mathbf{A}.$$

Since $d^2V/d\mathbf{x}^2 = 2\mathbf{A}^T\mathbf{A}$ is positive definite because $\mathbf{A}$ has full column rank, by the Strict Local Minimizer test $\mathbf{x}^*$ is a strict local minimizer.

In addition, for every column vector $\mathbf{y}$ we have:

$$\mathbf{y}^T(d^2V/d\mathbf{x}^2)\mathbf{y} = \mathbf{y}^T 2\mathbf{A}^T\mathbf{A}\mathbf{y} = 2|\mathbf{A}\mathbf{y}|^2 \geq 0$$

which implies that $V$ is a convex function by Theorem 5.3.7. Thus, $\mathbf{x}^*$ is the unique strict global minimizer. △

**Example 5.3.9** (Linear Regression Global Minimizer Test). Let

$$\ddot{y}(\mathbf{s}, \boldsymbol{\theta}) \equiv \boldsymbol{\theta}^T[\mathbf{s}^T \;\; 1]^T$$

be the predicted response of a learning machine to input pattern vector $\mathbf{s}$ when the learning machine's internal state of knowledge is $\boldsymbol{\theta}$ as described in Example 5.3.2. A gradient descent algorithm minimizes the objective function

$$\ell(\boldsymbol{\theta}) = \frac{1}{2n}\sum_{i=1}^{n} |y_i - \ddot{y}(\mathbf{s}_i, \boldsymbol{\theta})|^2$$

which was described in Example 5.3.2. Assume a gradient descent algorithm converges to a point $\boldsymbol{\theta}^*$. Obtain explicit formulas for checking if the gradient evaluated at $\boldsymbol{\theta}^*$ is zero and for checking if the Hessian of $\ell$ evaluated at $\boldsymbol{\theta}^*$ is positive definite. Obtain an explicit formula for checking if $\boldsymbol{\theta}^*$ is the unique global minimizer.

**Solution.** To obtain an explicit formula for checking if $\boldsymbol{\theta}^*$ is a strict local minimizer, check that the gradient of $\ell$ evaluated at $\boldsymbol{\theta}^*$ is a vector of zeros and the Hessian of $\ell$

evaluated at $\boldsymbol{\theta}^*$ is positive definite, then $\boldsymbol{\theta}^*$ is a strict local minimizer of $\ell$ (Theorem 5.3.3). Let $\mathbf{u}_i \equiv [\mathbf{s}_i, 1]^T$. The gradient of $\ell$ is given by the formula:

$$d\ell/d\boldsymbol{\theta} = -(1/n)\sum_{i=1}^{n}(y_i - \ddot{y}(\mathbf{s}_i, \boldsymbol{\theta}))\mathbf{u}_i^T$$

which implies that $\boldsymbol{\theta}^*$ must satisfy the relation

$$\mathbf{0}^T = -(1/n)\sum_{i=1}^{n}(y_i - \ddot{y}(\mathbf{s}, \boldsymbol{\theta}^*)^T\mathbf{u}_i^T$$

to satisfy the condition that the gradient of $\ell$ evaluated at $\boldsymbol{\theta}^*$ is a vector of zeros.

The Hessian of $\ell$, $\mathbf{H}$, is given by the formula:

$$\mathbf{H} = (1/n)\sum_{i=1}^{n}\mathbf{u}_i\mathbf{u}_i^T.$$

Thus, the condition that the Hessian of $\ell$ evaluated at $\boldsymbol{\theta}^*$ is positive definite may be checked by evaluating the Hessian $\mathbf{H}$ at $\boldsymbol{\theta}^*$. Since $\mathbf{H}$ is real and symmetric, $\mathbf{H}$ is positive definite if all of its eigenvalues are strictly positive. So this is one way to check if $\mathbf{H}$ is positive definite. However, note that in this special case for all $\mathbf{q}$:

$$\mathbf{q}^T\mathbf{H}\mathbf{q} = \mathbf{q}^T\left[(1/n)\sum_{i=1}^{n}\mathbf{u}_i\mathbf{u}_i^T\right]\mathbf{q} = (1/n)\sum_{i=1}^{n}(\mathbf{q}^T\mathbf{u}_i)^2 \geq 0$$

which implies that not only is $\mathbf{H}$ positive semidefinite at $\boldsymbol{\theta}^*$, $\mathbf{H}$ is positive semidefinite for all possible values of $\boldsymbol{\theta}$. This implies that if $\mathbf{H}$ is positive definite at $\boldsymbol{\theta}^*$ and the gradient vanishes at $\boldsymbol{\theta}^*$, then $\boldsymbol{\theta}^*$ must be the unique global minimizer of $\ell$ (Theorem 5.3.7). △

### 5.3.4 Lagrange Multipliers

Let $\Omega \subseteq \mathcal{R}^d$. Let $V : \Omega \to \mathcal{R}$ be an objective function. Let $\boldsymbol{\phi} : \Omega \to \mathcal{R}^m$ be called the *constraint function* which specifies a hypersurface

$$C \equiv \{\mathbf{x} \in \Omega : \boldsymbol{\phi}(\mathbf{x}) = \mathbf{0}_m\}.$$

In this section, we describe how to find a local minimizer of $V$ on the constraint hypersurface $C$. That is, the goal is to minimize $V$ where the domain of $V$ is restricted to a hypersurface $C$ (see Figure 5.4).

The following theorem provides necessary but not sufficient conditions that a point $\mathbf{x}^*$ must satisfy if it is a local minimizer of the restriction of $V$ to the constraint hypersurface $C$.

**Theorem 5.3.8** (Lagrange Multiplier Theorem). *Let $\Omega$ be a convex open subset of $\mathcal{R}^d$. Let $\boldsymbol{\phi} : \Omega \to \mathcal{R}^m$ be continuously differentiable on $\Omega$ with $m < d$. Let $V : \Omega \to \mathcal{R}$ be a continuously differentiable function. Assume $\mathbf{x}^*$ is a local minimizer of the restriction of $V$ to*

$$C \equiv \{\mathbf{x} \in \Omega : \boldsymbol{\phi}(\mathbf{x}) = \mathbf{0}_m\}.$$

*Assume $d\boldsymbol{\phi}(\mathbf{x}^*)/d\mathbf{x}$ has full row rank $m$. Then, there exists a unique column vector $\boldsymbol{\lambda} \in \mathcal{R}^m$ such that:*

$$\frac{dV(\mathbf{x}^*)}{d\mathbf{x}} + \boldsymbol{\lambda}^T\frac{d\boldsymbol{\phi}(\mathbf{x}^*)}{d\mathbf{x}} = \mathbf{0}_d^T.$$

*Proof.* See Marlow (2012, p. 258). ■

## Recipe Box 5.2 Gradient Descent Algorithm Design

- **Step 1: Construct a smooth objective function.** Construct a twice continuously differentiable objective function $\ell : \Theta \rightarrow \mathcal{R}$ on the parameter space $\Theta \subseteq \mathcal{R}^q$ with the property that a parameter vector $\boldsymbol{\theta}$ is at least as preferable to a parameter vector $\boldsymbol{\psi}$ if and only if $\ell(\boldsymbol{\theta}) \leq \ell(\boldsymbol{\psi})$. Let $\mathbf{g}$ denote the gradient of $\ell$. Let $\mathbf{H}$ denote the Hessian of $\ell$.

- **Step 2: Derive gradient descent algorithm.** Let $\boldsymbol{\theta}(0)$ be the initial guess. Then,

$$\boldsymbol{\theta}(t+1) = \boldsymbol{\theta}(t) - \gamma_t \mathbf{g}(\boldsymbol{\theta}(t))$$

is the gradient descent algorithm which generates a sequence of parameter estimates $\boldsymbol{\theta}(0), \boldsymbol{\theta}(1), \boldsymbol{\theta}(2), \ldots$ during the learning process. The sequence of stepsizes $\gamma_1, \gamma_2, \ldots$ is chosen such that $\ell(\boldsymbol{\theta}(t+1)) < \ell(\boldsymbol{\theta}(t))$ for all $t \in \mathbb{N}$.

- **Step 3: Derive stopping criterion for descent algorithm.** Let $\epsilon$ be a small positive number (e.g., $\epsilon = 10^{-6}$). If $|\mathbf{g}(\boldsymbol{\theta}(t))|_\infty < \epsilon$, then stop iterating and classify $\boldsymbol{\theta}(t)$ as a critical point of $\ell$.

- **Step 4: Derive a formula for checking for strict local minimizer.** Let $\epsilon$ be a small positive number (e.g., $\epsilon = 10^{-6}$). Let $\boldsymbol{\theta}(t)$ satisfy the stopping criterion $|\mathbf{g}(\boldsymbol{\theta}(t))|_\infty < \epsilon$. Let $\lambda_{min}(t)$ and $\lambda_{max}(t)$ be the respective smallest and largest eigenvalues of $\mathbf{H}(\boldsymbol{\theta}(t))$. Let $r(t) \equiv \lambda_{max}(t)/\lambda_{min}(t)$. If both $\lambda_{min}(t)$ and $\lambda_{max}(t)$ are greater than $\epsilon$ and in addition $r(t) < (1/\epsilon)$, classify $\boldsymbol{\theta}(t)$ as a strict local minimizer.

- **Step 5: Check if solution is the unique global minimizer.** Let $\boldsymbol{\theta}(t)$ be a critical point classified as a strict local minimizer of $\ell$ on the parameter space $\Theta$. If, in addition, both: (i) $\Theta$ is convex, and (ii) the Hessian of $\ell$ evaluated at $\boldsymbol{\theta}$ is positive semidefinite for every $\boldsymbol{\theta} \in \Theta$, then $\boldsymbol{\theta}(t)$ is classified as the unique global minimizer of $\ell$ on $\Theta$.

The Lagrange Multiplier Theorem is often used to transform a constrained optimization problem into an unconstrained optimization problem that can be solved with a descent algorithm. For example, suppose that the goal is to minimize the restriction of a continuously differentiable function $V : \Omega \rightarrow \mathcal{R}^d$ to a smooth nonlinear hypersurface $\{\mathbf{x} \in \mathcal{R}^d : \boldsymbol{\phi}(\mathbf{x}) = \mathbf{0}_m\}$. This problem can be transformed into an unconstrained nonlinear optimization problem using the Lagrange Multiplier Theorem provided that $\boldsymbol{\phi}$ is continuously differentiable and the Jacobian of $\boldsymbol{\phi}$ evaluated at $\mathbf{x}^*$ has full row rank.

To solve this problem, define the *Lagrangian* function $\mathcal{L} : \Omega \times \mathcal{R}^m \rightarrow \mathcal{R}^d$ such that for all $\mathbf{x} \in \Omega$ and for all $\boldsymbol{\lambda} \in \mathcal{R}^m$:

$$\mathcal{L}(\mathbf{x}, \boldsymbol{\lambda}) = V(\mathbf{x}) + \boldsymbol{\lambda}^T \boldsymbol{\phi}(\mathbf{x}). \tag{5.43}$$

The vector $\boldsymbol{\lambda}$ is called the *Lagrange multiplier.* The Lagrange Multiplier Theorem states that a local minimizer $\mathbf{x}^*$ of the restriction of $V$ to the hypersurface $\{\mathbf{x} : \boldsymbol{\phi}(\mathbf{x}) = \mathbf{0}_m\}$ is a critical point (not necessarily a minimizer) of the Lagrangian in (5.43) provided that $d\boldsymbol{\phi}(\mathbf{x}^*)/d\mathbf{x}$ has full row rank $m$.

A convenient numerical check that $d\boldsymbol{\phi}(\mathbf{x}^*)/d\mathbf{x}$ has full row rank $m$ is to check that the largest eigenvalue of the $m$-dimensional real symmetric matrix $\mathbf{M}^* \equiv$

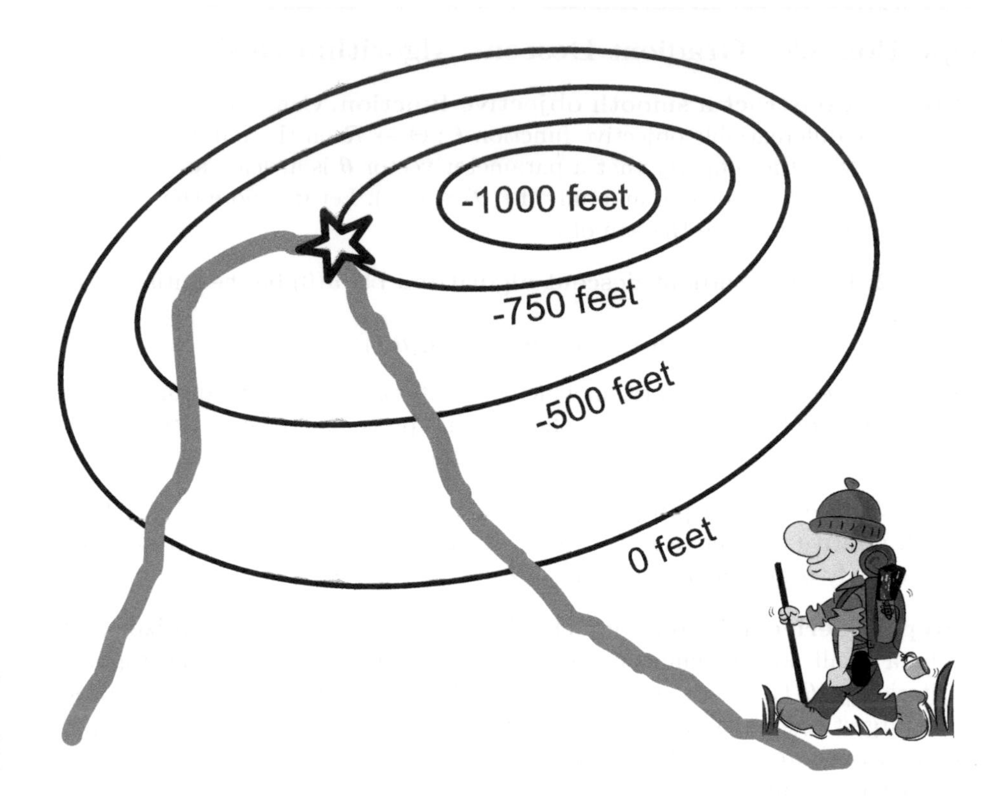


**FIGURE 5.4**
**Hiking on a canyon path as an example nonlinear constrained optimization problem.** The figure depicts a contour map of a canyon whose depth reaches 1000 feet below sea level (0 feet elevation). The hiker follows a hiking path which does not visit the bottom of the canyon but reaches a maximum depth of 750 feet before the hiking path takes the hiker out of the canyon. In a typical constrained optimization machine learning problem, the hiker's location on the hiking path specifies a point in the parameter space which satisfies a specified constraint. That is, if the hiker leaves the hiking path, then the constraint is violated. The depth of the canyon at the hiker's location corresponds to the objective function evaluated at the hiker's location.

$[d\boldsymbol{\phi}(\mathbf{x}^*)/d\mathbf{x}][d\boldsymbol{\phi}(\mathbf{x}^*)/d\mathbf{x}]^T$ is greater than a positive number $\epsilon$ and the condition number of $\mathbf{M}^*$ is less than some constant $K$ (e.g., $\epsilon = 10^{-7}$ and $K = 10^{15}$).

**Example 5.3.10** (Cross-Entropy Minimization). Let $S \equiv \{\mathbf{x}_1, \ldots, \mathbf{x}_m\}$ be a set of $m$ $d$-dimensional real vectors. Let $p : S \to (0, 1]$ be a probability mass function which assigns a positive probability to each element of $S$ such that $\sum_{j=1}^{m} p(\mathbf{x}_j) = 1$. Let $q : S \to (0, 1]$ be a probability mass function which assigns a positive probability to each element of $S$ such that $\sum_{j=1}^{m} q(\mathbf{x}_j) = 1$. Now define the probability vectors $\mathbf{p} = [p_1, \ldots, p_m]^T$ and $\mathbf{q} = [q_1, \ldots, q_m]^T$ such that $p_k = p(\mathbf{x}_k)$ and $q_k = q(\mathbf{x}_k)$.

Show that a critical point of the restriction of the cross-entropy function

$$\ell(\mathbf{q}) = -\sum_{j=1}^{m} p_j \log(q_j)$$

to the constraint surface

$$\Omega \equiv \left\{ \mathbf{q} \in (0,1)^m : \sum_{k=1}^{m} q_k = 1 \right\}$$

must satisfy the constraint $\mathbf{q} = \mathbf{p}$.

**Solution.** The conditions of the Lagrange Multiplier Theorem (Theorem 5.3.8) are satisfied since $\ell$ is continuously differentiable on the convex open set $(0, \infty)^m$ and because the Jacobian of $\left(\sum_{j=1}^{m} q_j - 1\right)$ is the row vector $\mathbf{1}_m^T$ which has full row rank. Define the Lagrangian $\mathcal{L}$ such that

$$\mathcal{L}(\mathbf{q}, \lambda) = \ell(\mathbf{q}) + \lambda \left( \sum_{j=1}^{m} q_j - 1 \right).$$

Set the partial derivative of $\mathcal{L}$ with respect to $q_k$ equal to zero to obtain:

$$d\mathcal{L}/dq_k = -(p_k/q_k) + \lambda = 0$$

which implies $p_k = \lambda q_k$. Since $\mathbf{q} \in \Omega$ and $\sum_{k=1}^{m} p_k = 1$, it follows that $\lambda = 1$ and thus $\mathbf{p} = \mathbf{q}$ is the unique critical point. △

**Example 5.3.11** (Principal Component Analysis (Unsupervised Learning)). Principal components analysis (PCA) is an important and widely used method of unsupervised learning. Let $\mathbf{X} \equiv [\mathbf{x}_1, \ldots, \mathbf{x}_n]$ be a data set consisting of $n$ $d$-dimensional feature vectors. Assume the column rank of $\mathbf{X}$ is $r$.

Let the mean $\mathbf{m} = (1/n)\mathbf{X}\mathbf{1}_n$. The *sample covariance matrix* associated with data set $\mathbf{X}$ is defined by the real symmetric matrix:

$$\mathbf{C}_x \equiv (1/(n-1)) \left(\mathbf{X} - \mathbf{m}\mathbf{1}_n^T\right) \left(\mathbf{X} - \mathbf{m}\mathbf{1}_n^T\right)^T. \tag{5.44}$$

Let $\mathbf{P} \in \mathcal{R}^{r \times d}$. Let $\mathbf{p}_k^T$ be the $k$th row of $\mathbf{P}$ such that $|\mathbf{p}_k|^2 = 1$ when $|\mathbf{p}_k| > 0$, $k = 1, \ldots, r$.

Create a new data set $\mathbf{Y} \equiv [\mathbf{y}_1, \ldots, \mathbf{y}_n]$ defined such that $\mathbf{y}_i = \mathbf{P}(\mathbf{x}_i - \mathbf{m})$. This new data $\mathbf{Y}$ is the projection of the original data set $[\mathbf{x}_1, \ldots, \mathbf{x}_n]$ into an $r$-dimensional linear subspace. In this context, $\mathbf{P}$ is called the *projection matrix*. Note that $\mathbf{Y} = [\mathbf{P}(\mathbf{x}_1 - \mathbf{m}), \ldots, \mathbf{P}(\mathbf{x}_n - \mathbf{m})]$ so that the sample covariance matrix associated with the projected data set $\mathbf{Y}$, $\mathbf{C}_y$, is defined by the formula:

$$\mathbf{C}_y = (1/(n-1))\mathbf{Y}\mathbf{Y}^T = \mathbf{P}\mathbf{C}_x\mathbf{P}^T.$$

The goal of principal components analysis (PCA) is to find a projection matrix such that the negative magnitude of the covariance matrix of the projection of the original data set, $\mathbf{C}_y$, is minimized. Intuitively, this means that the points in the projected data set, $\mathbf{y}_1, \ldots, \mathbf{y}_n$ are separated to the greatest extent possible subject to the constraint that they have been projected into a linear subspace of rank $r$. The magnitude of the covariance matrix $\mathbf{C}_y$ is defined as the sum of the on-diagonal variance elements of $\mathbf{C}_y$ or equivalently the trace of $\mathbf{C}_y$ which can be written as:

$$\operatorname{tr}(\mathbf{C}_y) = \operatorname{tr}\left(\mathbf{P}\mathbf{C}_x\mathbf{P}^T\right) = \sum_{k=1}^{r} \mathbf{p}_k^T \mathbf{C}_x \mathbf{p}_k.$$

Thus, more formally, the goal is to minimize $-\sum_{k=1}^{r} \mathbf{p}_k^T \mathbf{C}_x \mathbf{p}_k$ over all possible choices for the matrix $\mathbf{P}$ subject to the constraint that each row of $\mathbf{P} = [\mathbf{p}_1, \ldots, \mathbf{p}_r]^T$ is a normalized vector. Use the method of Lagrange Multipliers to solve this constrained optimization problem.

**Solution.** Let $\phi(\mathbf{P}) = [|\mathbf{p}_1|^2, \ldots, |\mathbf{p}_r|^2]^T - \mathbf{1}_r$. Let

$$\mathcal{L}(\mathbf{P}, \boldsymbol{\lambda}) = \boldsymbol{\lambda}^T \phi(\mathbf{P}) - \sum_{k=1}^{r} \mathbf{p}_k^T \mathbf{C}_x \mathbf{p}_k.$$

Now compute the derivative of $\mathcal{L}$ with respect to $\mathbf{p}_j$ to obtain for $j = 1, \ldots, r$:

$$\frac{d\mathcal{L}}{d\mathbf{p}_j} = 2\mathbf{p}_j^T \mathbf{C}_x - 2\lambda_j \mathbf{p}_j^T. \tag{5.45}$$

Setting the gradient of $\mathcal{L}$ in (5.45) equal to zero for $j = 1, \ldots, r$:

$$\mathbf{p}_j^T \mathbf{C}_x = \lambda_j \mathbf{p}_j^T.$$

Thus, a critical point $\mathbf{p}_j$ of $\mathcal{L}$ has the property that $\mathbf{p}_j$ is an eigenvector of $\mathbf{C}$. △

**Example 5.3.12** (Designing Linear Recoding Transformations). Let $\mathbf{x}_1, \ldots, \mathbf{x}_n$ be a set of $n$ $d$-dimensional feature vectors generated from $d$ feature detectors where $d$ is a large number (e.g., $d = 10^5$). Assume the $n$ feature vectors span an $r$-dimensional linear subspace which is relatively small (e.g., $r = 100$).

Suppose that some of the $d$ feature detectors are picking up redundant information or have responses that could be interpreted as linear combinations of the outputs of other feature detectors. Design a linear recoding transformation $\mathbf{P}$ that constructs a set of $n$ $r$-dimensional recoded feature vectors $\mathbf{y}_1, \ldots, \mathbf{y}_n$ such that $\mathbf{y}_i = \mathbf{P}\mathbf{x}_i$ for $i = 1, \ldots, n$. In addition, the $r$ rows of $\mathbf{P}$ should be orthonormal vectors.

**Solution.** The problem will be solved using the Principal Components Analysis method described in Example 5.3.11. Use (5.44) to construct the rank $r$ real symmetric covariance matrix $\mathbf{C}_x$. Let $\mathbf{e}_1, \ldots, \mathbf{e}_r$ be $r$ orthonormal eigenvectors of $\mathbf{C}_x$. Then, define the $j$th row of $\mathbf{P}$ to be equal to $\mathbf{e}_j^T$ for $j = 1, \ldots, r$.

By constructing $\mathbf{P}$ in this way, the set of feature vectors $\mathbf{y}_1, \ldots, \mathbf{y}_n$ span an $r$-dimensional linear subspace specified by weighted sums of the $r$ rows of the projection matrix $\mathbf{P}$.

If, in addition, only the $m$ eigenvectors associated with the $m$ largest eigenvalues of $\mathbf{C}_x$ are used where $m < r$, then the resulting projection matrix which has $m$ rows and $d$ columns projects the original $d$-dimensional feature vectors into an $m$-dimensional linear subspace that minimizes the trace of $\mathbf{C}_y$. △

**Example 5.3.13** (Slack Variable for Representing One Inequality Constraint). Let $\phi : \mathcal{R}^d \to \mathcal{R}$ be a continuously differentiable function. Assume that $\mathbf{x}^*$ is a strict local minimizer of a continuously differentiable function $V : \mathcal{R}^d \to \mathcal{R}$ which satisfies the constraint $\phi(\mathbf{x}) \leq 0$. Assume $d\phi/d\mathbf{x}$ evaluated at $\mathbf{x}^*$ has full row rank. Use the Lagrange Multiplier Theorem to rewrite this constrained nonlinear optimization problem with inequality constraints as an unconstrained nonlinear optimization problem.

**Solution.** Replace the inequality constraint $\phi(\mathbf{x}) \leq 0$ with the equality constraint $\phi(\mathbf{x}) + \eta^2 = 0$ where $\eta$ is an additional free parameter which is called a *slack variable*. Now define the Lagrangian as:

$$\mathcal{L}(\mathbf{x}, \eta, \lambda) = V(\mathbf{x}) + \lambda(\phi(\mathbf{x}) + \eta^2).$$

The set of critical points of $\mathcal{L}$ are then obtained by setting the derivatives

$$d\mathcal{L}/d\mathbf{x} = dV/d\mathbf{x} + \lambda d\phi/d\mathbf{x}, \tag{5.46}$$

$$d\mathcal{L}/d\eta = 2\lambda\eta, \tag{5.47}$$

and

$$d\mathcal{L}/d\lambda = \phi(\mathbf{x}) + \eta^2 \tag{5.48}$$

equal to zero.

Note that the condition that the derivative $d\mathcal{L}/d\eta = 2\lambda\eta = 0$ implies that either $\lambda = 0$ or $\eta = 0$. If $\eta = 0$, then the problem is mathematically equivalent to the constrained optimization problem with equality constraints described in the Lagrange Multiplier Theorem. In this case we say the constraint $\phi(\mathbf{x}) = 0$ is *active*. If $\eta \neq 0$, then $d\mathcal{L}/d\eta = 2\lambda\eta = 0$ only if $\lambda = 0$ which implies that we have an unconstrained optimization problem (see Equation 5.46) and the constraint $\phi(\mathbf{x}) = 0$ is *inactive* (because it is multiplied by $\lambda$).

Geometrically, when we have a constrained optimization problem involving inequality constraints, the inequality constraints define a region where the constraints are satisfied. If a critical point is in the interior of this region, then all of the constraints are inactive and the problem can be simply treated as an unconstrained optimization problem with no constraints! If the critical point falls on the boundary of this region, then the constrained optimization problem involving inequality constraints is equivalent to a constrained optimization problem involving equality constraints so that the Lagrange Multiplier Theorem (Theorem 5.3.8) is directly applicable. △

**Example 5.3.14** (Support Vector Machine Derivation (Soft Margin)). Let the training data set $\mathcal{D}_n \equiv \{(\mathbf{s}_1, y_1), \ldots, (\mathbf{s}_n, y_n)\}$ where $y_i \in \{-1, +1\}$ is the desired response for a given $d$-dimensional input pattern vector $\mathbf{s}_i$, $i = 1, \ldots, n$. Let the augmented input pattern vector $\mathbf{u}_i^T \equiv [\mathbf{s}_i^T, 1]$ for $i = 1, \ldots, n$. The *predicted response* of a *support vector machine classifier*, $\ddot{y}_i$, is defined such that for $i = 1, \ldots, n$:

$$\ddot{y}_i = 1 \text{ if } \mathbf{w}^T\mathbf{u}_i \geq \beta \quad \text{and} \quad \ddot{y}_i = -1 \text{ if } \mathbf{w}^T\mathbf{u}_i \leq -\beta \tag{5.49}$$

where the non-negative number $\beta$ specifies a *decision threshold* for classification.

Under the assumption that $\mathbf{w}$ should be chosen to ensure that each member of the training data set is correctly classified, it follows from (5.49) that $\mathbf{w}$ should be chosen so that for $i = 1, \ldots, n$:

$$y_i\mathbf{w}^T\mathbf{u}_i \geq \beta. \tag{5.50}$$

Equation (5.50) implicitly makes the assumption that a solution $\mathbf{w}$ exists which will satisfy (5.50) (i.e., the training data is "linearly separable"). Since this assumption may not always hold, a generalization of (5.50) is used to handle situations involving misclassifications. In particular, the revised goal is to select a $\mathbf{w}$ such that for $i = 1, \ldots, n$:

$$y_i\mathbf{w}^T\mathbf{u}_i \geq \beta - \eta_i \tag{5.51}$$

where the non-negative number $\eta_i$ is chosen to be as small as possible. When $\eta_i = 0$ for $i = 1, \ldots, n$, then Equation (5.51) reduces to the case of (5.50) where the training data set is, in fact, linearly separable. When $\eta_i$ is a strictly positive number, this corresponds to a violation of (5.50) and a possible classification error for training stimulus $(\mathbf{s}_i, y_i)$.

The boundaries of the decision regions in (5.50) correspond to two parallel hyperplanes which share the normal vector $\mathbf{w}$. Let $\mathbf{u}_0$ be a point on the hyperplane $\{\mathbf{u} : \mathbf{w}^T\mathbf{u} = -\beta\}$. Then, the closest point to $\mathbf{u}_0$, $\mathbf{u}_0^*$, located on the hyperplane $\{\mathbf{u} : \mathbf{w}^T\mathbf{u} = \beta\}$ must satisfy the formula

$$\mathbf{w}^T\mathbf{u}_0^* = \mathbf{w}^T(\mathbf{u}_0 + \alpha\mathbf{w}) = \beta \tag{5.52}$$

for some number $\alpha$. Equation (5.52) may be solved for $\alpha$ using the relation $\mathbf{w}^T\mathbf{u}_0 = -\beta$ to obtain $\alpha = 2\beta/|\mathbf{w}|^2$. Therefore, the distance between the two parallel hyperplanes is

$$|\mathbf{u}_0 - \mathbf{u}_0^*| = |\alpha\mathbf{w}| = 2\beta/|\mathbf{w}|.$$

The goal of learning in a support vector machine is typically to maximize the distance between the two decision hyperplanes while simultaneously trying to keep the individual data points in the correct decision regions by minimizing $\eta_1, \ldots, \eta_n$.

Consider the constrained optimization problem whose objective is to minimize for some positive constant $K$:

$$(1/2)|\mathbf{w}|^2 + (K/n)\sum_{i=1}^{n} \eta_i \tag{5.53}$$

subject to the constraint in (5.51) and the constraint $\eta_i \geq 0$ for $i = 1, \ldots, n$. Minimizing the first term in (5.53) has the effect of maximizing the distance between the separating hyperplanes.

Now define the Lagrangian using the slack variables $\gamma_1, \ldots, \gamma_n$ and $\delta_1, \ldots, \delta_n$ as:

$$\mathcal{L}\left(\mathbf{w}, \{\eta_i\}, \{\lambda_i\}, \{\gamma_i\}, \{\delta_i\}, \{\mu_i\}\right) = \frac{|\mathbf{w}|^2}{2} + \frac{K}{n}\sum_{i=1}^{n} \eta_i + \frac{1}{n}\sum_{i=1}^{n} \mu_i(\eta_i - \delta_i^2)$$

$$+\frac{1}{n}\sum_{i=1}^{n} \lambda_i \left(y_i \mathbf{w}^T \mathbf{u}_i - \beta + \eta_i - \gamma_i^2\right).$$

The critical points of the Lagrangian $\mathcal{L}$ can be obtained using gradient descent methods, gradient ascent methods, setting the gradient (or components of the gradient) equal to zero and solving the resulting equations, or various combinations of these strategies. For example, taking the derivative of $\mathcal{L}$ with respect to $\mathbf{w}$ and setting the result equal to zero gives the following key formula for $\mathbf{w}$:

$$\mathbf{w} = (1/n)\sum_{i=1}^{n} \lambda_i y_i \mathbf{u}_i. \tag{5.54}$$

which says that the normal vector to the two parallel decision hyperplanes is a weighted average of the augmented input pattern vectors $\mathbf{u}_1, \ldots, \mathbf{u}_n$.

The constraint that the derivative of the Lagrangian with respect to $\gamma_i$ is equal to zero implies that the constraint in (5.51) is inactive when $\lambda_i = 0$. A vector $\mathbf{u}_i$ which specifies an active constraint is called a *support vector*. Therefore, (5.54) implies that if a vector $\mathbf{u}_i$ is not a support vector (i.e., $\lambda_i = 0$), then it follows that training stimulus $(\mathbf{s}_i, y_i)$ plays no role in specifying the decision boundary surface for the support vector machine. Furthermore, inspection of (5.54) shows that a training stimulus $(\mathbf{s}_i, y_i)$ associated with a larger $\lambda_i$ plays a greater role in determining the support vector machine's classification behavior.

In summary, support vector machines are based upon the heuristic that the decision boundary of the classification machine should be chosen to maximize separation of the two training stimuli categories. Support vector machines also identify which training stimuli in the training data are most relevant for specifying the hyperplane decision boundaries and therefore are also helpful for providing explanations of the support vector machine's behavior in terms of training examples. △

**Example 5.3.15** (Multilayer Perceptron Gradient Descent Using Lagrangian). Given an input pattern $\mathbf{s}$, let the desired response of a feedforward network consisting of one layer of $M$ radial basis function hidden units be denoted as $r$. Let the connection weights from the layer of hidden units to the single output unit be specified by the parameter vector $\mathbf{v}$, and let the connection weights from the $d$ input units to the $k$th hidden unit be specified by the parameter vector $\mathbf{w}_k$ for $k = 1, \ldots, M$. The parameter vector for the network is denoted as $\boldsymbol{\theta} \equiv \left[\mathbf{v}^T, \mathbf{vec}(\mathbf{W}^T)^T\right]$. Let the predicted response, $r$, corresponding to the network's output for a given input pattern $\mathbf{s}$ be computed using the formula $r = \ddot{y}(\mathbf{s}, \boldsymbol{\theta})$ where

$$\ddot{y}(\mathbf{s}, \boldsymbol{\theta}) = \mathbf{v}^T \mathbf{h}. \tag{5.55}$$

The hidden unit response vector $\mathbf{h}$ in (5.55) is computed using the formula $\mathbf{h} = \ddot{\mathbf{h}}(\mathbf{s}, \mathbf{W})$ where $\ddot{\mathbf{h}}(\mathbf{s}, \mathbf{W}) = [h_1(\mathbf{s}, \mathbf{w}_1), \ldots, h_M(\mathbf{s}, \mathbf{w}_M)]^T$, $\mathbf{w}_j^T$ is the $j$th row of $\mathbf{W}$, and

$$h_j(\mathbf{s}, \mathbf{w}_j) = \exp\left(-|\mathbf{s} - \mathbf{w}_j|^2\right). \tag{5.56}$$

The goal of learning is to minimize $(y - \ddot{r}(\mathbf{s}, \boldsymbol{\theta}))^2$ given a particular $y$ and input pattern $\mathbf{s}$ so that the empirical risk function is defined by the formula:

$$\ell(\boldsymbol{\theta}) = \left(y - \mathbf{v}^T\mathbf{h}\right)^2.$$

A gradient descent learning algorithm can then be derived by computing the gradient of $\ell$.

However, rather than computing the gradient of $\ell$ directly, one can reformulate the goal of learning as a constrained nonlinear optimization problem.

To explore this approach, define the goal of learning as minimizing the function

$$D(r) = (y - r)^2$$

subject to the nonlinear architectural constraints imposed on $r$ by (5.55) and (5.56). Now define the Lagrangian $L$ such that:

$$\mathcal{L}([\mathbf{v}, \mathbf{W}, \lambda_r, \boldsymbol{\lambda}_h, r, \mathbf{h}]) = (y - r)^2 + \lambda_r[r - \mathbf{v}^T\mathbf{h}] + \boldsymbol{\lambda}_h^T[\mathbf{h} - \ddot{\mathbf{h}}(\mathbf{s}, \mathbf{W})].$$

Use this Lagrangian in conjunction with the Lagrange Multiplier Theorem to derive the backpropagation gradient descent algorithm for a feedforward network with only one layer of hidden units (see Algorithm 5.2.1).

**Solution.** To derive the backpropagation gradient descent algorithm which seeks out critical points of the objective function, the strategy will be to take the partial derivatives of $\lambda_r$, $\boldsymbol{\lambda}_n$, $r$, and $\mathbf{h}$ and set those partial derivatives equal to zero. The parameter vector $\boldsymbol{\theta}$, which is comprised of $\mathbf{v}$ and the rows of $\mathbf{W}$, is updated using the standard gradient descent algorithm. The details of this approach are now provided.

First, the constraint that the derivative of $\mathcal{L}$ with respect to $\lambda_r$ is equal to zero corresponds to the constraint in (5.55). Similarly, the constraint that the derivative of $\mathcal{L}$ with respect to $\boldsymbol{\lambda}_h$ is equal to zero corresponds to the constraint in (5.56).

Second, compute the derivative of $\mathcal{L}$ with respect to $r$ to obtain:

$$d\mathcal{L}/d\lambda_r = -2(y - r) + \lambda_r \tag{5.57}$$

and then set (5.57) equal to zero to obtain:

$$\lambda_r = 2(y - r). \tag{5.58}$$

Equation (5.58) is interpreted as computing an error signal, $\lambda_r$, for the output unit of the network.

Third, compute the derivative of $\mathcal{L}$ with respect to $\mathbf{h}$ to obtain:

$$d\mathcal{L}/d\boldsymbol{\lambda}_h = -\lambda_r\mathbf{v}^T + \boldsymbol{\lambda}_\mathbf{h} \tag{5.59}$$

and then set (5.59) equal to zero to obtain:

$$\boldsymbol{\lambda}_h = \lambda_r\mathbf{v}^T. \tag{5.60}$$

Equation (5.60) is interpreted as a formula for computing the error signal, $\boldsymbol{\lambda}_\mathbf{h}$, for the layer of hidden units by "backpropagating" the output error signal $\lambda_r$.

Fourth, the derivative of $\mathcal{L}$ with respect to $\mathbf{v}$ is given by:

$$d\mathcal{L}/d\mathbf{v} = -\lambda_r \mathbf{h}^T. \tag{5.61}$$

Fifth, the derivative of $\mathcal{L}$ with respect to $\mathbf{W}$ is given by:

$$d\mathcal{L}/d\mathbf{w}_k = -\boldsymbol{\lambda}_h^T \frac{d\mathbf{h}(\mathbf{s}, \mathbf{W})}{d\mathbf{w}_k} \tag{5.62}$$

where

$$\frac{d\mathbf{h}(\mathbf{s}, \mathbf{W})}{d\mathbf{w}_k} = \left(2h_k(\mathbf{s}, \mathbf{W})\mathbf{u}_k\right)\left(\mathbf{s} - \mathbf{w}_k\right)^T$$

where $\mathbf{u}_k$ is the $k$th column of an $M$-dimensional identity matrix.

Combining these results yields the following algorithm which is a special case of Algorithm 5.2.1.

- **Step 1.** Compute hidden unit responses using (5.56).
- **Step 2.** Compute output unit response using (5.55).
- **Step 3.** Compute error signal at output unit using (5.58).
- **Step 4.** Compute error signals for hidden units using (5.60).
- **Step 5.** Compute parameter update $\boldsymbol{\theta}(t+1) = \boldsymbol{\theta}(t) - \gamma_t d\mathcal{L}(\boldsymbol{\theta}(t))/d\boldsymbol{\theta}$ using (5.61) and (5.62) where $\gamma_t$ is a positive stepsize.

△

**Example 5.3.16** (Simple Recurrent Net Gradient Descent Using Lagrangian). Consider a temporal learning problem where the training data is represented as a collection of $n$ trajectories where the $k$th trajectory consists of the $T_k$ ordered pairs:

$$\mathbf{x}^k \equiv [(\mathbf{s}^k(1), \mathbf{y}^k(1)), \ldots, (\mathbf{s}^k(T_k), \mathbf{y}^k(T_k))]$$

where the binary scalar $\mathbf{y}^k(t) \in \{0, 1\}$ is the *desired response* given the past history $\mathbf{s}^k(1), \ldots, \mathbf{s}^k(t)$.

Define a simple recurrent network with discrepancy function $D$ for comparing the observed output $\mathbf{y}^k(t)$ with the predicted output $\ddot{\mathbf{y}}^k(t)$. For this example, choose

$$D(\mathbf{y}^k(t), \ddot{\mathbf{y}}^k(t)) = |\mathbf{y}^k(t) - \ddot{\mathbf{y}}^k(t)|^2.$$

In addition, the predicted $r$-dimensional response $\mathbf{y}^k(T_k)$ is computed from the current $m$-dimensional hidden state vector $\mathbf{h}^k(t)$ and $q$-dimensional parameter vector $\boldsymbol{\theta}$ using the function $\boldsymbol{\Phi} : \mathcal{R}^m \times \Theta \to \mathcal{R}^r$ defined such that $\mathbf{y}^k(T_k) = \boldsymbol{\Phi}(\mathbf{h}^k(T_k), \boldsymbol{\theta})$. Finally, define the function $\boldsymbol{\Psi} : \mathcal{R}^d \times \mathcal{R}^m \times \Theta \to \mathcal{R}^m$ which specifies how the current $d$-dimensional input vector $\mathbf{s}^k(t)$ and previous $m$-dimension hidden state vector $\mathbf{h}^k(t-1)$ are combined with the $q$-dimensional parameter vector $\boldsymbol{\theta}$ to generate the current $m$-dimensional hidden state vector $\mathbf{h}^k(t)$.

The objective function for learning $\ell_n : \mathcal{R}^q \to \mathcal{R}$ is defined such that for all $\boldsymbol{\theta} \in \mathcal{R}^q$:

$$\ell_n(\boldsymbol{\theta}) = (1/n) \sum_{k=1}^{n} c(\mathbf{x}^k, \boldsymbol{\theta}) \tag{5.63}$$

where the loss function $c$ is defined such that:

$$c(\mathbf{x}^k, \boldsymbol{\theta}) = \sum_{t=1}^{T_k} |\mathbf{y}^k(t) - \ddot{\mathbf{y}}^k(t)|^2.$$

Use the Lagrange multiplier approach (see Example 5.3.15) to derive a gradient descent backpropagation algorithm for this simple recurrent network.

**Solution.** The augmented objective function $\mathcal{L}^k$ is defined as:

$$\mathcal{L}^k = \sum_{t=1}^{T_k} |\mathbf{y}^k(t) - \ddot{\mathbf{y}}^k(t)|^2 + \sum_{t=1}^{T_k} \boldsymbol{\mu}^k(t)^T \left[\ddot{\mathbf{y}}^k(t) - \boldsymbol{\Phi}(\mathbf{h}^k(t), \boldsymbol{\theta})\right]$$

$$+ \sum_{t=1}^{T_k} \boldsymbol{\lambda}^k(t)^T [\ddot{\mathbf{h}}^k(t) - \boldsymbol{\Psi}(\mathbf{s}^k(t), \ddot{\mathbf{h}}^k(t-1), \boldsymbol{\theta})].$$

Take the derivative of $\mathcal{L}^k$ with respect to $\ddot{\mathbf{y}}^k(t)$ to obtain:

$$\frac{d\mathcal{L}^k}{d\ddot{\mathbf{y}}^k(t)} = -2(\mathbf{y}^k(t) - \ddot{\mathbf{y}}^k(t)) + \boldsymbol{\mu}^k(t). \tag{5.64}$$

Now set (5.64) equal to a vector of zeros and rearrange terms to derive a formula for the output unit error signals $\boldsymbol{\mu}^k(t)$ which is given by:

$$\boldsymbol{\mu}^k(t) = 2(\mathbf{y}^k(t) - \ddot{\mathbf{y}}^k(t)). \tag{5.65}$$

Taking the derivative of $\mathcal{L}^k$ with respect to $\ddot{\mathbf{h}}^k(t)$ to obtain:

$$\frac{d\mathcal{L}^k}{d\ddot{\mathbf{h}}^k(t)} = -\boldsymbol{\mu}^k(t)^T \left[\frac{d\boldsymbol{\Phi}(\mathbf{s}^k(t), \ddot{\mathbf{h}}^k(t), \boldsymbol{\theta})}{d\ddot{\mathbf{h}}^k(t)}\right] + \boldsymbol{\lambda}^k(t)^T$$

$$-\boldsymbol{\lambda}^k(t+1)^T \left[\frac{d\boldsymbol{\Psi}\left(\mathbf{s}^k(t+1), \ddot{\mathbf{h}}^k(t), \boldsymbol{\theta}\right)}{d\ddot{\mathbf{h}}^k(t)}\right] \tag{5.66}$$

Now set (5.66) equal to a vector of zeros and rearrange terms to derive the update formula:

$$\boldsymbol{\lambda}^k(t)^T = \boldsymbol{\mu}^k(t)^T \left[\frac{d\boldsymbol{\Phi}(\mathbf{s}^k(t), \ddot{\mathbf{h}}^k(t), \boldsymbol{\theta})}{d\ddot{\mathbf{h}}^k(t)}\right] + \boldsymbol{\lambda}^k(t+1)^T \left[\frac{d\boldsymbol{\Psi}\left(\mathbf{s}^k(t+1), \ddot{\mathbf{h}}^k(t), \boldsymbol{\theta}\right)}{d\ddot{\mathbf{h}}^k(t)}\right], \tag{5.67}$$

which specifies how to compute the hidden unit error signals for the hidden units at time $t$ (i.e., $\boldsymbol{\lambda}^k(t)$) from the hidden unit error signals at time $t+1$ (i.e., $\boldsymbol{\lambda}^k(t+1)$) and the output unit error signals $\boldsymbol{\mu}^k(t)$ at time $t$ (i.e., $\boldsymbol{\mu}^k(t)$).

The derivative of $\mathcal{L}^k$ with respect to $\boldsymbol{\theta}$ is given by the formula:

$$\frac{d\mathcal{L}^k}{d\boldsymbol{\theta}} = -\sum_{t=1}^{T_k} \boldsymbol{\mu}^k(t)^T \left[\frac{d\boldsymbol{\Phi}(\mathbf{s}^k(t), \ddot{\mathbf{h}}^k(t), \boldsymbol{\theta})}{d\boldsymbol{\theta}}\right] - \sum_{t=1}^{T_k} \boldsymbol{\lambda}^k(t)^T \left[\frac{d\boldsymbol{\Psi}(\mathbf{s}^k(t), \bar{\mathbf{h}}^k(t-1))}{d\boldsymbol{\theta}}\right] \tag{5.68}$$

and is then used to specify the gradient descent update rule:

$$\boldsymbol{\theta}(k+1) = \boldsymbol{\theta}(k) - \gamma_k d\mathcal{L}^k / d\boldsymbol{\theta}$$

where $d\mathcal{L}^k/d\boldsymbol{\theta}$ is evaluated at $(\mathbf{x}^k, \boldsymbol{\theta}(k))$.

The architectural constraints and the results in (5.65), (5.67), and (5.68) are then used to specify the Simple Recurrent Network Gradient Backpropagation Algorithm (see Algorithm 5.3.1). △

**Algorithm 5.3.1 Simple Recurrent Net Gradient Backpropagation.** This recurrent network gradient propagation algorithm is derived using the method of Lagrange Multipliers for learning the $k$th trajectory (see Example 5.3.16).

1: **procedure** RecurrentNetGrad( $\{\boldsymbol{\Phi}, \boldsymbol{\Psi}\}$)
2: $\quad \ddot{\mathbf{h}}^k(0) \Leftarrow \mathbf{0}$ ▷ **Forward Pass**
3: $\quad$ **for** $t = 1$ to $T_k$ **do**
4: $\quad\quad \ddot{\mathbf{h}}^k(t) \Leftarrow \boldsymbol{\Phi}(\mathbf{s}^k(t), \ddot{\mathbf{h}}^k(t-1), \boldsymbol{\theta})$
5: $\quad\quad \ddot{\mathbf{y}}^k(t) \Leftarrow \boldsymbol{\Psi}(\ddot{\mathbf{h}}^k(t), \boldsymbol{\theta})$
6: $\quad$ **end for**
7: ▷ **Backward Pass**
8: $\quad \boldsymbol{\lambda}^k(T_k + 1) \Leftarrow \mathbf{0}$
9: $\quad$ **for** $t = T_k$ to 1 **do**
10: $\quad\quad \boldsymbol{\mu}^k(t)^T \Leftarrow 2(\mathbf{y}^k(t) - \ddot{\mathbf{y}}^k(t))$.
11: $\quad\quad \boldsymbol{\lambda}^k(t)^T \Leftarrow \boldsymbol{\mu}^k(t)^T \left[\frac{d\boldsymbol{\Phi}(\ddot{\mathbf{h}}^k(t),\boldsymbol{\theta})}{d\ddot{\mathbf{h}}^k(t)}\right] + \boldsymbol{\lambda}^k(t+1)^T \left[\frac{d\boldsymbol{\Psi}(\mathbf{s}^k(t+1),\ddot{\mathbf{h}}^k(t),\boldsymbol{\theta})}{d\ddot{\mathbf{h}}^k(t)}\right]$
12: $\quad$ **end for**
13: ▷ **Gradient Computation**
14: $\quad \frac{d\mathcal{L}^k}{d\boldsymbol{\theta}} \Leftarrow \mathbf{0}^T$
15: $\quad$ **for** $t = 1$ **to** $T_k$ **do**
16: $\quad\quad \frac{d\mathcal{L}^k}{d\boldsymbol{\theta}} \Leftarrow \frac{d\mathcal{L}^k}{d\boldsymbol{\theta}} - \boldsymbol{\mu}^k(t)^T \left[\frac{\boldsymbol{\Phi}(\ddot{\mathbf{h}}^k(t),\boldsymbol{\theta})}{d\boldsymbol{\theta}}\right] - \boldsymbol{\lambda}^k(t)^T \left[\frac{\boldsymbol{\Psi}(\mathbf{s}^k(t),\ddot{\mathbf{h}}^k(t),\boldsymbol{\theta})}{d\boldsymbol{\theta}}\right]$
17: $\quad$ **end for**
18: $\quad$ **return** $\left\{\frac{d\mathcal{L}^k}{d\boldsymbol{\theta}}\right\}$ ▷ **Return Gradient of Loss Function**
19: **end procedure**

## Exercises

5.3-1. *Logistic Regression Gradient Descent.* Let the training data be defined as a set of $n$ $d+1$-dimensional training vectors:

$$\{(y_1, \mathbf{s}_1), \ldots, (y_n, \mathbf{s}_n)\}$$

where $y \in \{0, 1\}$ is the desired response of a linear inference machine to $d$-dimensional input column pattern vector $\mathbf{s}$. Let $\mathcal{S}(\phi) \equiv 1/(1 + \exp(-\phi))$. Let $\ddot{p}(\mathbf{s}, \boldsymbol{\theta}) \equiv \mathcal{S}\left(\boldsymbol{\theta}^T[\mathbf{s}^T \; 1]^T\right)$ denote the learning machine's prediction of the probability that $y = 1$ given input pattern $\mathbf{s}$ is presented and given its current parameter vector $\boldsymbol{\theta}$. Let the function $\ell : \mathcal{R}^{d+1} \to \mathcal{R}$ be defined such that for all $\boldsymbol{\theta} \in \mathcal{R}^{d+1}$:

$$\ell(\boldsymbol{\theta}) = -(1/n)\sum_{i=1}^{n}\left[y_i \log\left(\ddot{p}(\mathbf{s}_i, \boldsymbol{\theta})\right) + (1 - y_i)\log\left(1 - \ddot{p}(\mathbf{s}_i, \boldsymbol{\theta})\right)\right] + \lambda|\boldsymbol{\theta}|^2 \tag{5.69}$$

where $\lambda$ is a known positive number (e.g., $\lambda = 0.001$). Derive a gradient descent learning algorithm that will minimize $\ell$.

5.3-2. *Taylor Expansion of Logistic Regression Risk.* Construct a second-order Taylor series expansion of the empirical risk function $\ell$ as defined in Equation (5.69).

5.3-3. *Multinomial Logistic Regression Gradient Descent.* Let the training data be defined as a set of $n$ $d+m$-dimensional training vectors:

$$\{(\mathbf{s}_1, \mathbf{y}_1), \ldots, (\mathbf{s}_n, \mathbf{y}_n)\}$$

where $\mathbf{y}_i$ is the desired response of the learning machine to the $d$-dimensional input pattern vector $\mathbf{s}_i$ for $i = 1, \ldots, n$. It is assumed that the $m$-dimensional desired response vector $\mathbf{y}_i$ is a column of an $m$-dimensional identity matrix $\mathbf{I}_m$ whose semantic interpretation is that $\mathbf{y}_i$ is equal to the $k$th column of $\mathbf{I}_m$ when the desired response is category number $k$ out of the $m$ possible categories.

The learning machine's state of knowledge is specified by a parameter value matrix

$$\mathbf{W} \equiv [\mathbf{w}_1 \ldots \mathbf{w}_{m-1}]^T \in \mathcal{R}^{(m-1)\times d}$$

where $\mathbf{w}_k \in \mathcal{R}^d$, $k = 1, \ldots, m-1$.

The response vector of the learning machine to an input vector $\mathbf{s}$ for a given $\mathbf{W}$ is an $m$-dimensional column vector of probabilities

$$\mathbf{p}(\mathbf{s}, \mathbf{W}) \equiv [p_1(\mathbf{s}_k, \mathbf{W}), \ldots, p_m(\mathbf{s}_k, \mathbf{W})]^T$$

defined such that:

$$\mathbf{p}(\mathbf{s}, \mathbf{W}) = \frac{\mathbf{exp}\left([\mathbf{W}^T, \mathbf{0}_d]^T \mathbf{s}\right)}{\mathbf{1}_m^T \mathbf{exp}\left([\mathbf{W}^T, \mathbf{0}_d]^T \mathbf{s}\right)}.$$

Let the function $\ell : \mathcal{R}^{(m-1)\times d} \to \mathcal{R}$ be defined such that for all $\mathbf{W} \in \mathcal{R}^{(m-1)\times d}$:

$$\ell(\mathbf{W}) = -(1/n) \sum_{i=1}^{n} \mathbf{y}_i^T \mathbf{log}\left(\mathbf{p}(\mathbf{s}_i, \mathbf{W})\right).$$

Show that $d\ell/d\mathbf{w}_k$ is given by the formula

$$d\ell/d\mathbf{w}_k = -(1/n) \sum_{i=1}^{n} (\mathbf{y}_i - \mathbf{p}(\mathbf{s}_i, \mathbf{W}))^T \mathbf{u}_k \mathbf{s}_i^T$$

for $k = 1, \ldots, m-1$. Then, show how to use $d\ell/d\mathbf{w}_k$ to specify a gradient descent algorithm that minimizes $\ell$.

5.3-4. *Linear Regression with Sigmoidal Response.* Let $\mathcal{S}(\phi) = 1/(1 + \exp(-\phi))$. Note also that

$$d\mathcal{S}/d\phi = \mathcal{S}(\phi)(1 - \mathcal{S}(\phi)).$$

Now let $\ddot{y}(\boldsymbol{s}, \boldsymbol{\theta}) = \mathcal{S}(\boldsymbol{\theta}^T \boldsymbol{s})$ be defined as the probability that the response category is equal to 1 given input pattern $\mathbf{s}$ and parameter vector $\boldsymbol{\theta}$. Let the training data set $\mathcal{D}_n \equiv \{(\mathbf{s}_1, y_1), \ldots, (\mathbf{s}_n, y_n)\}$ where $\mathbf{s}_i$ is a $d$-dimensional vector and $y_i \in (0, 1)$. Assume the empirical risk function for learning $\hat{\ell}_n(\boldsymbol{\theta})$ is given by the formula:

$$\hat{\ell}_n(\boldsymbol{\theta}) = (1/n) \sum_{i=1}^{n} (y_i - \ddot{y}(\mathbf{s}_i, \boldsymbol{\theta}))^2.$$

Show the derivative of $\hat{\ell}_n$ is given by the formula:

$$d\hat{\ell}_n/d\boldsymbol{\theta} = -(2/n) \sum_{i=1}^{n} (y_i - \mathcal{S}(\boldsymbol{\theta}^T \mathbf{s}_i))\mathcal{S}(\boldsymbol{\theta}^T \mathbf{s}_i)[1 - \mathcal{S}(\boldsymbol{\theta}^T \mathbf{s}_i)]\mathbf{s}_i^T.$$

Design a supervised learning machine which "learns" by using a gradient descent method to update its parameter vector $\boldsymbol{\theta}$ to predict $y_i$ given $\mathbf{s}_i$ for $i = 1, \ldots, n$.

5.3-5. *Radial Basis Net Gradient Descent Chain Rule Derivation.* In a supervised learning paradigm, suppose the objective is to teach a smooth multilayer perceptron with one layer of $h$ hidden units to generate the response $y \in \mathcal{R}$ given the input pattern $\mathbf{s} \in \mathcal{R}^d$. Thus, the training data can be represented as the ordered pair $(\mathbf{s}, y)$.

Let $\ddot{y} : \mathcal{R}^h \times \mathcal{R}^h \to \mathcal{R}$ be defined such that: $\ddot{y}(\mathbf{h}, \mathbf{v}) = \mathbf{v}^T\mathbf{h}$ where $\mathbf{v} \in \mathcal{R}^h$ and $\mathbf{h} : \mathcal{R}^d \times \mathcal{R}^{d \times h} \to \mathcal{R}^h$ is defined such that the $j$th element of $\mathbf{h}$,

$$h_j(\mathbf{s}, \mathbf{w}_j) = \exp\left(-|\mathbf{s} - \mathbf{w}_j|^2\right)$$

where $\mathbf{w}_j^T$ is the $j$th row of $\mathbf{W}$ for $j = 1, \ldots, h$.

The objective function $\ell : \mathcal{R}^h \times \mathcal{R}^{d \times h} \to \mathcal{R}$ for a smooth multilayer perceptron is defined such that

$$\ell([\mathbf{v}, \mathbf{W}]) = (y - \ddot{y}(\mathbf{h}(\mathbf{s}, \mathbf{W}), \mathbf{v}))^2.$$

Let the discrepancy $D$ be defined such that $D(y, \ddot{y}(\mathbf{h}(\mathbf{s}, \mathbf{W}), \mathbf{v}) = \ell([\mathbf{v}, \mathbf{W}])$.

Use the vector chain rule to obtain a general formula for the gradient of $\ell$ by using the formulas:

$$\frac{d\ell}{d\mathbf{v}} = \left[\frac{dD}{d\ddot{y}}\right]\left[\frac{d\ddot{y}}{d\mathbf{v}}\right]$$

and

$$\frac{d\ell}{d\mathbf{w}_j} = \left[\frac{dD}{d\ddot{y}}\right]\left[\frac{d\ddot{y}}{d\mathbf{h}}\right]\left[\frac{d\mathbf{h}}{d\mathbf{w}_j}\right].$$

5.3-6. *Linear Value Function Reinforcement Learning.* Let $V : \mathcal{R}^d \times \mathcal{R}^q \to \mathcal{R}$. The goal of the learning process is to find a $q$-dimensional parameter vector $\boldsymbol{\theta}$ such that for each state $\mathbf{s} \in \mathcal{R}^d$, $V(\mathbf{s}; \boldsymbol{\theta})$ specifies the total cumulative reinforcement the learning machine expects to receive given the current state $\mathbf{s}$. In particular, assume that $V(\mathbf{s}; \boldsymbol{\theta}) = \boldsymbol{\theta}^T[\mathbf{s}^T, 1]^T$. In order to estimate $\boldsymbol{\theta}$, assume the training data $\mathcal{D}_n \equiv \{(\mathbf{s}_{1,1}, \mathbf{s}_{1,2}, r_1), \ldots, (\mathbf{s}_{n,1}, \mathbf{s}_{n,2}, r_n)\}$ where $r_i$ is the incremental reinforcement received by the learning machine when it observes that the state of its environment changes from the first $d$-dimensional state $\mathbf{s}_{i,1}$ to the second $d$-dimensional state $\mathbf{s}_{i,2}$ of the $i$th episode. Let

$$\hat{\ell}_n(\boldsymbol{\theta}) = (1/n)\sum_{i=1}^{n} c(\mathbf{x}_i, \boldsymbol{\theta})$$

be an empirical risk function with loss function

$$c(\mathbf{x}_i, \boldsymbol{\theta}) = (\ddot{r}(\mathbf{s}_o(i), \mathbf{s}_f(i)) - r(i))^2$$

where $\ddot{r}(\mathbf{s}_{i,1}, \mathbf{s}_{i,2}) = V(\mathbf{s}_{i,2}; \boldsymbol{\theta}) - \lambda V(\mathbf{s}_{i,1}; \boldsymbol{\theta})$ is an estimator of the expected incremental reinforcement received by the learning machine and $\lambda \in [0, 1]$ is the discount factor (see Section 1.7.2).

Derive a gradient descent algorithm for minimizing this empirical risk function.

5.3-7. Let $V : \mathcal{R}^d \to \mathcal{R}$ be defined such that $V(\mathbf{x}) = -\mathbf{x}^T\mathbf{W}\mathbf{x}$ where $\mathbf{W}$ is a real symmetric matrix with rank $k$. The gradient of $V$ is $-2\mathbf{W}\mathbf{x}$. Indicate what values of $k$ ensure a flat region of $V$ will exist?

5.3-8. *Logistic Regression Global Minimizer Convergence Test.* Derive formulas for determining if a given parameter vector for the logistic regression learning machine described in Exercise 5.3-1 is a strict global minimizer.

5.3-9. *Multinomial Logistic Regression Local Minimizer Convergence Test.* Derive formulas for determining if a given parameter vector for the logistic regression learning machine described in Exercise 5.3-3 is a strict global minimizer.

5.3-10. *Convexity of Linear Combination of Convex Functions.* Let $\Omega \subseteq \mathcal{R}^d$ be an open convex set. Let $f : \Omega \rightarrow \mathcal{R}$ and $g : \Omega \rightarrow \mathcal{R}$ be convex functions on $\Omega$. Let $h \equiv \alpha_1 f + \alpha_2 g$ where $\alpha_1 \geq 0$ and $\alpha_2 \geq 0$. Use the definition of a convex function to show that $h$ is also convex on $\Omega$.

5.3-11. *Smooth Convex Function Composition Theorem.* Let $\Omega \subseteq \mathcal{R}^d$ be an open convex set. Let $\Gamma \subseteq \mathcal{R}$ be an open convex set. Let $f : \Omega \rightarrow \Gamma$ be a twice continuously differentiable convex function on $\Omega$. Let $g : \Gamma \rightarrow \mathcal{R}$ be a twice continuously differentiable non-decreasing convex function on $\Gamma$. Use the Smooth Convex Function Theorem to show that $h = g(f)$ is also convex on $\Omega$.

5.3-12. *Stacked Denoising Autoencoder.* Denoising autoencoders can be stacked upon one another. They can be trained all at once. Or, they can be trained stage by stage. Define a stacked denoising autoencoder and explicitly specify an empirical risk function. Then, use the Feedforward M-Layer Deep Net Gradient Backpropagation Algorithm to derive a gradient descent learning algorithm for your stacked denoising autoencoder.

## 5.4 Further Readings

The topics covered in this chapter are a mixture of standard topics in elementary real analysis (e.g., Rosenlicht 1968), topics in advanced matrix calculus (Neudecker 1969; Marlow 2012; Magnus and Neudecker 2001; Magnus 2010), and numerical optimization theory (e.g., Luenberger 1984; Nocedal and Wright 1999).

### Matrix Calculus

It should be emphasized that the definition of the derivative of a vector-valued function on a vector space (i.e., the "Jacobian") is standard and uncontroversial among experts in the classical nonlinear optimization theory and matrix calculus literature.

On the other hand, a standard derivative of a matrix-valued function on a vector space or a space of matrices has not yet been established. Magnus and Neudecker (2001) as well as Magnus (2010) have discussed several different definitions of the derivative of a matrix on a vector space (e.g., Magnus and Neudecker 2001). All of these definitions are consistent with the "broad definition" of a Matrix Derivative because they are matrices corresponding to different physical arrangements of all the relevant partial derivatives (Magnus 2010).

However, there are serious problems with some matrix derivative definitions falling under the broad definition of a matrix derivative because such definitions are sometimes not mutually consistent with one another or may require different customized versions of the matrix chain rule. Therefore, Magnus and Neudecker (2001) and Magnus (2010) strongly advocate that for the purposes of matrix calculus derivations, one should use a "narrower definition" of the matrix derivative. The use of bad matrix derivative notation can lead to incorrectly derived matrix calculus expressions and hamper communications among investigators.

The definition of a matrix derivative on a space of matrices presented in this chapter is consistent with the narrow definition of a matrix derivative advocated by Magnus (2010; also see Magnus and Neudecker 2001). However, the specific narrow definition of $d\mathbf{A}/d\mathbf{X}$

presented here follows Marlow (2012) rather than Magnus (2010). Marlow's (2012) notation is more useful in deep learning and machine learning applications because it tends to arrange the partial derivatives physically in a matrix so that partial derivatives associated with a single computing unit are located physically together in the Jacobian rather than being scattered throughout the Jacobian.

Unfortunately, in the machine learning literature, many different definitions of matrix derivatives have been introduced that would be classified by Magnus (2010) and Magnus and Neudecker (2001) as "bad notation". Matrix derivative formulas and matrix chain rule calculations using this bad notation should be avoided. In addition, it should be emphasized matrix differentiation formulas may yield wrong results when combined with incompatible matrix derivative definitions.

### Backpropagation of Derivatives for Deep Learning Applications

Werbos (1974, 1994), Rumelhart, Hinton, and Williams (1986), Le Cun (1985), and Parker (1985) derived methods for backpropagating gradient information in feedforward and recurrent network architectures. Symbolic gradient propagation methods can also be derived using signal flow graph methods (Osowski 1994; Campolucci et al. 2000). Hoffmann (2016) and Baydin et al. (2017) provide an overview of automatic differentiation methods.

### Nonlinear Optimization Theory

In this chapter, only a brief overview of nonlinear optimization theory methods was presented. Additional nonlinear optimization theory analyses of gradient descent, BFGS (Broyden-Fletcher-Goldfarb-Shanno) methods, and Newton methods may be found in nonlinear optimization textbooks (e.g., Luenberger 1984; Nocedal and Wright 1999; Bertsekas 1996). Constrained gradient projection methods are important techniques for gradient descent on an objective function subject to possibly nonlinear constraints on the objective function's domain (e.g., Luenberger 1984; Bertsekas 1996; Nocedal and Wright 1999).

The specific convergence rate definitions presented here are based upon the discussion of Luenberger (1984) and are referred to as $Q$-convergence (i.e., "quotient" converge) rate definitions. Nocedal and Wright (1999) provide a different, but closely related, definition of $Q$-convergence.

The Lagrange Multiplier Theorem presented in this chapter is essentially an introduction to the Karush-Kuhn-Tucker (also known as the Kuhn-Tucker) theorem which handles inequality constraints more directly without the aid of slack variables. Discussions of the Karush-Kuhn-Tucker theorem may be found in Luenberger (1984), Bertsekas (1996), Magnus and Neudecker (2001), and Marlow (2012). Hiriart-Urruty and Lemarechal (1996) provide a discussion of subgradient (also known as subdifferential) methods for solving nonlinear optimization problems involving non-smooth convex functions.

### Support Vector Machines

Discussions of support vector machines can be found in Bishop (2006, Chapter 7) and Hastie et al. (2001, Chapter 12). The example provided in this chapter was intended simply to introduce a few key ideas. Other important principles such as in-depth discussions of kernel methods, which are often used in conjunction with support vector machines, can be found in Bishop (2006, Chapter 7) and Hastie et al. (2001, Chapter 12). Readers desiring a more in-depth examination of support vector machines should take a look at Chapter 4 of Mohri et al. (2012), Cristianini and Shawe-Taylor (2000), and the more advanced text by Scholkopf and Smola (2002).

# 6

# *Convergence of Time-Invariant Dynamical Systems*

**Learning Objectives**

- Identify system equilibrium points.
- Design convergent discrete-time and continuous-time machines.
- Design convergent clustering algorithms.
- Design convergent model search algorithms.

A large class of inference and learning algorithms can be represented as systems of nonlinear equations where parameters or state variables are updated in discrete-time. Let $\mathbf{x}(t)$ denote a collection of state variables associated with an inference or learning machine at time $t$. For example, the $i$th element of $\mathbf{x}(t)$ might correspond to a particular parameter of the learning machine which is adjusted during the learning process. Or, alternatively, the $i$th element of $\mathbf{x}(t)$ might correspond to the value of a state variable specifying whether or not a particular assertion is assumed to hold.

More specifically, the new system state $\mathbf{x}(t+1)$ might be some nonlinear function $\mathbf{f}$ of the previous system state $\mathbf{x}(t)$. This relationship is concisely expressed as:

$$\mathbf{x}(t+1) = \mathbf{f}(\mathbf{x}(t)). \tag{6.1}$$

As discussed in Section 3.1.1, some machine learning algorithm applications process information continuously received from sensors or continuously generate motor control signals. In these cases, it may be desirable to specify the behavior of the machine learning algorithm as a machine which processes information in continuous-time using the continuous-time version of (6.1) which is specified by the system of nonlinear differential equations:

$$\frac{d\mathbf{x}}{dt} = \mathbf{f}(\mathbf{x}(t)). \tag{6.2}$$

In addition, studying the continuous-time counterpart of a particular discrete-time dynamical system may simplify theoretical analyses leading to new insights into the original discrete-time dynamical system.

In this chapter, methods for studying the behavior of systems of nonlinear equations of the form of (6.1) and (6.2) are discussed. Specifically, some sufficient conditions are provided that ensure that a sequence of system states $\mathbf{x}(0), \mathbf{x}(1), \ldots$ generated by either a discrete-time or continuous-time dynamical system will eventually converge to a particular region of the state space.

## 6.1 Dynamical System Existence Theorems

Recall from Section 3.2, that a time-invariant dynamical system maps an initial system state $\mathbf{x}_0$ and scalar time-index $t$ into the system state $\mathbf{x}(t)$. It is not always possible to construct a time-invariant iterated map or time-invariant vector field for exactly representing a particular dynamical system $\boldsymbol{\Psi}$. The following theorems in this section provide sufficient conditions ensuring such constructions are possible.

**Theorem 6.1.1** (EUC (Discrete-Time System)). *Let $T \equiv \mathbb{N}$. Let $\Omega \subseteq \mathcal{R}^d$. Let $\mathbf{f} : \Omega \rightarrow \Omega$ be a function. (i) Then, there exists a function $\boldsymbol{\Psi} : \Omega \times T \rightarrow \Omega$ defined such that for all $t \in \mathbb{N}$:*

$$\mathbf{x}(t) = \boldsymbol{\Psi}(\mathbf{x}_0, t), \ \mathbf{x}_0 \in \Omega$$

*is the unique solution for the difference equation*

$$\mathbf{x}(t+1) = \mathbf{f}(\mathbf{x}(t))$$

*which satisfies $\mathbf{x}(0) = \mathbf{x}_0$. (ii) If, in addition, $\mathbf{f} : \Omega \rightarrow \Omega$ is continuous, then $\boldsymbol{\Psi}(\cdot, t)$ is continuous on $\Omega$ for all $t \in T$.*

*Proof.* The theorem is proved by constructing the solution $\boldsymbol{\Psi}$. (i) Define a function $\boldsymbol{\Psi} : \Omega \times T \rightarrow \Omega$ such that:

$$\boldsymbol{\Psi}(\mathbf{x}_0, 1) = \mathbf{f}(\mathbf{x}_0),$$

$$\boldsymbol{\Psi}(\mathbf{x}_0, 2) = \mathbf{f}(\mathbf{f}(\mathbf{x}_0)),$$

$$\boldsymbol{\Psi}(\mathbf{x}_0, 3) = \mathbf{f}(\mathbf{f}(\mathbf{f}(\mathbf{x}_0))),$$

and so on. The function $\boldsymbol{\Psi}$ is uniquely determined from $\mathbf{f}$ by construction. (ii) Since $\mathbf{f}$ is continuous and a continuous function of a continuous function is continuous (see Theorem 5.1.7), it follows that $\boldsymbol{\Psi}(\cdot, t)$ is continuous on $\Omega$ for all $t \in T$. ■

The EUC Discrete-Time System Theorem establishes conditions such that the iterated map $\mathbf{f} : \Omega \rightarrow \Omega$ specifies exactly one discrete-time dynamical system. In addition, the theorem establishes conditions on an iterated map $\mathbf{f}$ for a discrete-time dynamical system $\boldsymbol{\Psi} : \Omega \times \mathbb{N} \rightarrow \Omega$ that ensure $\boldsymbol{\Psi}(\cdot, t_f)$ is continuous on $\Omega$ for all $t = 0, 1, 2, \ldots$. The requirement that the current system state is a continuous function of the initial system state (i.e., $\boldsymbol{\Psi}$ is continuous in its first argument) greatly simplifies asymptotic behavioral analyses of a discrete-time dynamical system.

The next theorem provides sufficient but not necessary conditions for a given vector field $\mathbf{f} : \mathcal{R}^d \rightarrow \mathcal{R}^d$ to specify a unique continuous-time dynamical system with the property that the trajectories of the system are continuous functions of their respective initial conditions.

**Theorem 6.1.2** (EUC (Continuous-Time System)). *Let $T \equiv [0, \infty)$. Let $\mathbf{f} : \mathcal{R}^d \rightarrow \mathcal{R}^d$ be a continuously differentiable function. Then, there exists a function $\boldsymbol{\Psi} : \mathcal{R}^d \times [0, T_{max}) \rightarrow \mathcal{R}^d$ for some $T_{max} \in (0, \infty)$ which is continuous in its first argument and defined such that $\mathbf{x}(\cdot) = \boldsymbol{\Psi}(\mathbf{x}_0, \cdot)$ is the unique solution of the differential equation*

$$\frac{d\mathbf{x}}{dt} = \mathbf{f}(\mathbf{x}(t))$$

*which satisfies $\mathbf{x}(0) = \mathbf{x}_0$.*

*Proof.* See Hirsch et al. (2004, Chapter 17). ■

## Exercises

6.1-1. Let $f : \mathcal{R} \to \mathcal{R}$ be defined such that for all $x \in \mathcal{R}$: $f(x) = ax$ where $a$ is real number. Provide an explicit formula for the dynamical system $\Psi : \mathcal{R} \times \mathbb{N} \to \mathcal{R}$ specified by the time-invariant iterated map $f$. Show the solution for this system is a continuous function of its initial conditions. Repeat this problem for the case where $\Psi$ is specified by the time-invariant vector field $f$. Show that the solution $\boldsymbol{\Psi}(\cdot, t)$ is a continuous function of $\mathbf{x}(0)$ for all $t \in (0, \infty)$.

6.1-2. Let $\mathcal{S} : \mathcal{R} \to \mathcal{R}$ be defined such that $\mathcal{S}(\phi) = 1$ if $\phi > 1$, $\mathcal{S}(\phi) = \phi$ if $-1 \leq \phi \leq 1$, and $\mathcal{S}(\phi) = -1$ if $\phi < -1$. Let $f : \mathcal{R} \to \mathcal{R}$ be defined such that for all $x \in \mathcal{R}$: $f(x) = \mathcal{S}(ax)$ where $a$ is a real number. Do the assumptions of the EUC Theorem for Discrete-Time Systems hold when $f$ is an iterated map? Do the assumptions of the EUC Theorem for Continuous-Time Systems hold when $f$ is a vector field?

6.1-3. *EUC and Continuous-Time Gradient Descent.* Consider the continuous-time parameter updating scheme specified by a vector field $\mathbf{f} : \mathcal{R}^d \to \mathcal{R}^d$ defined such that

$$\mathbf{f}(\boldsymbol{\theta}) = -\gamma(\boldsymbol{\theta}) \left(d\ell/d\boldsymbol{\theta}\right)^T$$

where $\gamma : \mathcal{R}^d \to (0, \infty)$ is a positive-valued function. Explain why this vector field corresponds to a continuous-time gradient descent algorithm in a continuous state-space. Provide sufficient conditions on $\ell$ and $\gamma$ which ensure the EUC conditions are satisfied.

6.1-4. *EUC and Discrete-Time Gradient Descent.* Consider the discrete-time parameter updating scheme specified by the iterated map $\mathbf{f} : \mathcal{R}^d \to \mathcal{R}^d$ defined such that

$$\mathbf{f}(\boldsymbol{\theta}) = -\gamma(\boldsymbol{\theta}) \left(d\ell/d\boldsymbol{\theta}\right)^T$$

where $\gamma : \mathcal{R}^d \to (0, \infty)$ is a positive-valued function. Explain why this iterated map corresponds to a discrete-time gradient descent algorithm in a continuous state-space. Provide sufficient conditions on $\ell$ and $\gamma$ which ensure the EUC conditions are satisfied.

6.1-5. *Multilayer Perceptron Multiple Learning Rates Gradient Descent.* Let the predicted response of a feedforward network with one layer of hidden units for a given input pattern $\mathbf{s}$ be specified by the function $\ddot{y}$ defined such that:

$$\ddot{y}(\mathbf{s}, \boldsymbol{\theta}) = \sum_{k=1}^{m} v_k \mathcal{J}(\mathbf{w}_k^T \mathbf{s})$$

where $\boldsymbol{\theta} \equiv [v_1, \ldots, v_m, \mathbf{w}_1^T, \ldots, \mathbf{w}_m^T]^T$. Assume the scalar $y_i$ is the desired response for input pattern vector $\mathbf{s}_i$ for $i = 1, \ldots, n$. The risk function for learning, $\ell : \mathcal{R}^q \to \mathcal{R}$, is defined by the formula:

$$\ell(\boldsymbol{\theta}) = (1/n) \sum_{i=1}^{n} \left(y_i - \ddot{y}(\mathbf{s}_i, \boldsymbol{\theta})\right)^2.$$

Define a vector field $\mathbf{f} : \mathcal{R}^q \to \mathcal{R}^q$ such that for all $\boldsymbol{\theta} \in \mathcal{R}^q$:

$$\mathbf{f}(\boldsymbol{\theta}) = -\mathbf{M} \left(d\ell/d\boldsymbol{\theta}\right)^T$$

where $\mathbf{M}$ is a diagonal matrix whose on-diagonal elements are positive. Explain why the vector field specified by $\mathbf{f}$ can be interpreted as specifying a system of gradient descent update formulas. Show that the EUC conditions for a continuous time dynamical system specified by $\mathbf{f}$ are not satisfied for $\mathbf{f}$ when $\mathcal{J}$ is defined as a rectified linear transformation such that for all $x \in \mathcal{R}$: $\mathcal{J}(x) = x$ when $x \geq 0$ and $\mathcal{J}(x) = 0$ otherwise. Show that the EUC conditions for a continuous time dynamical system specified by $\mathbf{f}$ are satisfied for $\mathbf{f}$ when $\mathcal{J}$ is defined as a softplus transformation such that for all $x \in \mathcal{R}$: $\mathcal{J}(x) = \log(1 + \exp(x))$.

## 6.2 Invariant Sets

An important approach to the analysis of a nonlinear dynamical system's behavior involves identifying regions of the state space in which the dynamical system exhibits qualitatively different properties. For example, Figure 6.1 illustrates a state space for a dynamical system where trajectories of the dynamical system initiated at points $A$ and $B$ tend to converge to a particular region of the state space. The trajectory of the dynamical system initiated at point $C$ converges to one specific equilibrium point.

Toward this end, we now introduce the important concept of an *invariant set*. An invariant set is a region of the state space of a particular dynamical system with the property that any trajectory initiated in the invariant set will remain forever in that invariant set.

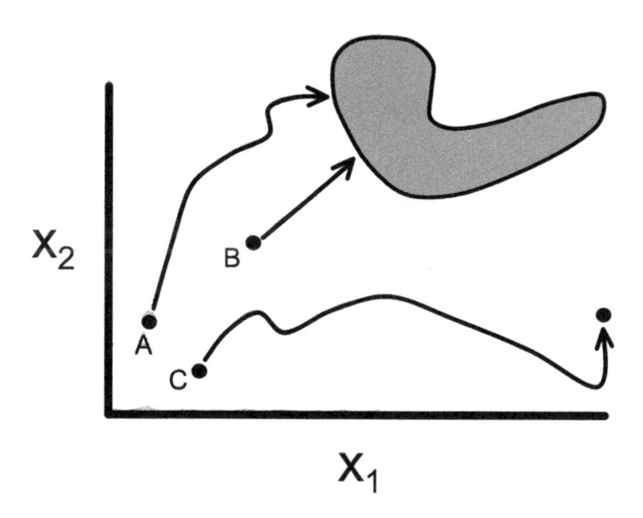


**FIGURE 6.1**
**Examples of convergence of trajectories to different types of invariant sets.** Three trajectories of a nonlinear dynamical system initiated at three distinct initial conditions labeled A, B, and C respectively in a two-dimensional state space. The trajectories with initial conditions A and B converge to a region of the state space consisting of multiple equilibrium points. The trajectory with initial condition C converges to one specific equilibrium point.

**Definition 6.2.1** (Invariant Set). Let $T \equiv \mathbb{N}$ or $T \equiv [0, \infty)$. Let $\Omega \subseteq \mathcal{R}^d$. Let $\mathbf{\Psi} : \Omega \times T \to \Omega$ be a time-invariant dynamical system. Let $\Gamma$ be a subset of $\Omega$. Then, $\Gamma$ is an *invariant set* with respect to $\mathbf{\Psi}$ if for every $\mathbf{x}_0 \in \Gamma$: $\mathbf{\Psi}(\mathbf{x}_0, t) \in \Gamma$ for all $t \in T$. □

Note that although the entire state space of any dynamical system is an invariant set, the concept of an invariant set is particularly useful when one can demonstrate that specific proper subsets of a dynamical system's state space are invariant sets.

Several important types of invariant sets for nonlinear systems analysis are now discussed.

**Definition 6.2.2** (Equilibrium Point). Let $\Gamma$ be an invariant set defined with respect to a time-invariant dynamical system $\mathbf{\Psi} : \Omega \times T \to \Omega$. Assume $\Gamma$ is a set containing exactly one point $\mathbf{x}^* \in \Omega$. If $\mathbf{\Psi}$ is a continuous-time dynamical system, then $\mathbf{x}^*$ is called an *equilibrium point.* If $\mathbf{\Psi}$ is a discrete-time dynamical system, then $\mathbf{x}^*$ is called a *fixed point.* □

If the trajectory of a dynamical system is initiated at an equilibrium point, then the system state does not change as a function of time. For a discrete-time dynamical system of the form $\mathbf{x}(t+1) = \mathbf{f}(\mathbf{x}(t))$ where $\mathbf{f} : \mathcal{R}^d \to \mathcal{R}^d$ is the iterated map, a fixed point $\mathbf{x}^*$ will sometimes be informally referred to as an equilibrium point as well. Also note that any collection of equilibrium points is also an invariant set.

Note that a fixed point, $\mathbf{x}^*$, of a time-invariant iterated map $\mathbf{f}$ has the property that:

$$\mathbf{x}^* = \mathbf{f}(\mathbf{x}^*). \tag{6.3}$$

The system of $d$ equations in (6.3) can thus be solved to find the equilibrium points associated with the iterated map $\mathbf{f}$.

Now consider a continuous-time dynamical system specified by the vector field $\mathbf{f} : \mathcal{R}^d \to \mathcal{R}^d$. An equilibrium point $\mathbf{x}^*$ for this dynamical system would have the property that

$$d\mathbf{x}^*/dt = \mathbf{0}_d$$

where the notation $d\mathbf{x}^*/dt$ indicates $d\mathbf{x}/dt$ evaluated at the point $\mathbf{x}^*$ and $\mathbf{0}_d$ is a $d$-dimensional vector of zeros. The system of $d$ equations:

$$d\mathbf{x}^*/dt = \mathbf{f}(\mathbf{x}^*) = \mathbf{0}_d$$

or equivalently $\mathbf{f}(\mathbf{x}^*) = \mathbf{0}_d$ can then be solved to find the equilibrium points of the time-invariant continuous-time dynamical system associated with the vector field $\mathbf{f}$.

**Definition 6.2.3** (Periodic Invariant Set). Let $T \equiv \mathbb{N}$ or $T \equiv [0, \infty)$. Let $\Omega \subseteq \mathcal{R}^d$. Let $\mathbf{\Psi} : \Omega \times T \to \Omega$ be a dynamical system. Let $\Gamma \subseteq \Omega$ be a nonempty invariant set with respect to $\mathbf{\Psi}$. The set $\Gamma$ is a *periodic invariant set* if there exists a finite positive number $\tau \in T$ such that for all $\mathbf{x}_0 \in \Gamma$:

$$\mathbf{\Psi}(\mathbf{x}_0, \tau) = \mathbf{x}_0. \tag{6.4}$$

The smallest number $\tau \in T$ which satisfies (6.4) is called the *period* of the periodic invariant set. □

When the system state is initiated within a periodic invariant set, the system state changes its values as a function of time in a periodic manner yet remains for all time within the periodic invariant set. A *limit cycle* is a special type of periodic invariant set. The union of a collection of equilibrium point invariant sets and a collection of periodic invariant sets is also a periodic invariant set according to Definition 6.2.3. However, in practice, the terminology "periodic invariant set" is not used to refer to an invariant set which only contains equilibrium points.

Now consider an invariant set which cannot be specified as a union of equilibrium point invariant sets and periodic invariant sets. In addition, suppose this invariant set has the property that every trajectory initiated within the set exhibits highly unpredictable asymptotic behavior then such an invariant set is sometimes called a *chaotic invariant set* (see Wiggins 2003 or Hirsch et al. 2004 for further discussion).

## Exercises

6.2-1. Let $\mathbf{x} : \mathcal{R} \rightarrow (0,1)^d$ be defined such that $\mathbf{x}(t) = [x_1(t), \ldots, x_d(t)]$ which satisfies the system of $d$ differential equations:

$$dx_i/dt = x_i(1 - x_i)$$

for $i = 1 \ldots d$. Let the state space $\Omega \equiv (0,1)^d$. Let the vector field $\mathbf{f} : \Omega \rightarrow \Omega$ for this dynamical system be defined such that

$$\mathbf{f}(\mathbf{x}) = [x_1(1 - x_1), \ldots, x_d(1 - x_d)]^T.$$

Find the $2^d$ equilibrium points of this dynamical system. Show that all $2^d$ equilibrium points are not in $\Omega$ but they are located in the closure of $\Omega$.

6.2-2. *Linear System Equilibrium Points.* Let $\mathbf{f} : \mathcal{R}^d \rightarrow \mathcal{R}^d$ be defined such that for all $\mathbf{x} \in \mathcal{R}^d$: $\mathbf{f}(\mathbf{x}) = \mathbf{W}\mathbf{x}$ where $\mathbf{W}$ is a real symmetric matrix. Find the set of equilibrium points of the time-invariant discrete-time system specified by the iterated map $\mathbf{f}$. Find the set of equilibrium points of the time-invariant continuous-time system specified by the vector field $\mathbf{f}$. Discuss the similarities and differences between these two sets of equilibrium points.

6.2-3. *Gradient Descent Equilibrium Points.* Let the continuously differentiable function $\mathbf{f} : \mathcal{R}^d \rightarrow \mathcal{R}^d$ be defined such that for all $\boldsymbol{\theta} \in \mathcal{R}^d$:

$$\mathbf{f}(\boldsymbol{\theta}) = -\gamma \left( d\ell/d\boldsymbol{\theta} \right)^T .$$

Show that $\boldsymbol{\theta}^*$ is a critical point of $\ell$ if and only if $\boldsymbol{\theta}^*$ is an equilibrium point of the continuous-time dynamical system specified by a vector field $\mathbf{f}$. Show that this is not the case for the discrete-time dynamical system specified by the iterated map $\mathbf{f}$. How are the equilibrium points of the discrete-time dynamical system specified by the iterated map $\mathbf{f}$ related to the critical points of $\ell$?

6.2-4. *Regularization and Gradient Descent Equilibrium Points.* Let the continuously differentiable function $\mathbf{f} : \mathcal{R}^d \rightarrow \mathcal{R}^d$ be defined such that for all $\boldsymbol{\theta} \in \mathcal{R}^d$:

$$\mathbf{f}(\boldsymbol{\theta}) = -\gamma\delta\boldsymbol{\theta} - \gamma \left( d\ell/d\boldsymbol{\theta} \right)^T$$

where $0 < \delta < 1$ and $\gamma > 0$. How are the equilibrium points of the continuous-time dynamical system specified by the vector field $\mathbf{f}$ related to the critical points of $\ell$? How are the equilibrium points of the discrete-time dynamical system specified by the iterated map $\mathbf{f}$ related to the critical points of $\ell$?

## 6.3 Lyapunov Convergence Theorems

### 6.3.1 Lyapunov Functions

If a dynamical system can be shown to be an algorithm which is minimizing an objective function, this information can sometimes provide important insights regarding the asymptotic behavior of the dynamical system. Such objective functions are called Lyapunov functions and typically have two important properties. First, the Lyapunov function $V : \mathcal{R}^d \to \mathcal{R}$ maps a state of the dynamical system into a number. And second, the Lyapunov function's value is non-increasing on the dynamical system's trajectories. With these properties, a dynamical system trajectory initiated at point $\mathbf{x}_0$ is constrained for all time to remain in the invariant set

$$\{\mathbf{x} \in \mathcal{R}^d : V(x) \leq V(\mathbf{x}_0)\}.$$

**Definition 6.3.1** (Discrete-Time Lyapunov Function). Let $T = \mathbb{N}$. Let $\Omega$ be a nonempty subset of $\mathcal{R}^d$. Let $\bar{\Omega}$ be the closure of $\Omega$. Let $\mathbf{f} : \Omega \to \Omega$ be a column-valued iterated map for the dynamical system $\mathbf{\Psi} : \Omega \times T \to \Omega$. Let $V : \bar{\Omega} \to \mathcal{R}$. Let the function $\dot{V} : \bar{\Omega} \to \mathcal{R}$ be defined such that for all $\mathbf{x} \in \bar{\Omega}$:

$$\dot{V}(\mathbf{x}) = V(\mathbf{f}(\mathbf{x})) - V(\mathbf{x}).$$

Assume $\dot{V}(\mathbf{x}) \leq 0$ for all $\mathbf{x} \in \bar{\Omega}$. And, if $\bar{\Omega}$ is not a finite set, additionally assume $\dot{V}$ is continuous on $\bar{\Omega}$. Then, $V$ is a *discrete-time Lyapunov function* on $\bar{\Omega}$ with respect to $\mathbf{\Psi}$. □

Note that the assumption that $\dot{V}(\mathbf{x}) \leq 0$ for all $\mathbf{x} \in \bar{\Omega}$ for the discrete time dynamical system specified by an iterated map $\mathbf{f}$ implies that the trajectory $\mathbf{x}(0), \mathbf{x}(1), \ldots$ generated by the iterated map has the property that: $V(\mathbf{x}(t+1)) \leq V(\mathbf{x}(t))$ for $t = 0, 1, 2, \ldots$.

**Example 6.3.1** (Discrete-Time Linear System Asymptotic Behavior). Consider a discrete-time dynamical system with iterated map $\mathbf{f}(\mathbf{x}) = \mathbf{W}\mathbf{x}$ where $\mathbf{W} \in \mathcal{R}^{d \times d}$ has full rank. Thus, the dynamical system is specified by the difference equation:

$$\mathbf{x}(t+1) = \mathbf{W}\mathbf{x}(t).$$

Show that $V(\mathbf{x}) = |\mathbf{x}|^2$ is a Lyapunov function for the iterated map.

**Solution.** Note that

$$\dot{V}(\mathbf{x}) = |\mathbf{W}\mathbf{x}|^2 - |\mathbf{x}|^2 = \mathbf{x}^T[\mathbf{W}^T\mathbf{W} - \mathbf{I}_d]\mathbf{x}.$$

First note that $V$ and $\dot{V}$ are continuous functions. Now it will be shown that $\dot{V} \leq 0$. Let $\lambda_1, \ldots, \lambda_d$ be the real eigenvalues of the symmetric matrix $\mathbf{W}^T\mathbf{W}$ with respective orthogonal eigenvectors $\mathbf{e}_1, \ldots, \mathbf{e}_d$. Then,

$$\dot{V}(\mathbf{x}) = \mathbf{x}^T[\mathbf{W}^T\mathbf{W} - \mathbf{I}_d]\mathbf{x} = \mathbf{x}^T\left[\sum_{i=1}^{d}(\lambda_i - 1)\mathbf{e}_i\mathbf{e}_i^T\right]\mathbf{x} = \sum_{i=1}^{d}(\lambda_i - 1)(\mathbf{e}_i^T\mathbf{x})^2. \tag{6.5}$$

Thus, referring to (6.5), $\dot{V} \leq 0$ if all eigenvalues of $\mathbf{W}^T\mathbf{W}$ are less than or equal to one. △

**Definition 6.3.2** (Continuous-Time Lyapunov Function). Let $T = [0, \infty)$. Let $\Omega$ be a nonempty subset of $\mathcal{R}^d$. Let $\bar{\Omega}$ be the closure of $\Omega$. Let $V : \bar{\Omega} \to \mathcal{R}$ be a continuously differentiable function on $\bar{\Omega}$. Let the continuous function $\mathbf{f} : \Omega \to \Omega$ be a column-valued vector field for the dynamical system $\boldsymbol{\Psi} : \Omega \times T \to \Omega$. Let the function $\dot{V} : \bar{\Omega} \to \mathcal{R}$ be defined such that for all $\mathbf{x} \in \bar{\Omega}$:

$$\dot{V}(\mathbf{x}) = [dV/d\mathbf{x}]\mathbf{f}(\mathbf{x}). \tag{6.6}$$

If $\dot{V}(\mathbf{x}) \leq 0$ for all $\mathbf{x} \in \bar{\Omega}$, then $V$ is a *continuous-time Lyapunov function* on $\bar{\Omega}$ with respect to $\boldsymbol{\Psi}$. □

A helpful semantic interpretation of $\dot{V}$ for the continuous-time case is obtained by defining a function $v : \mathcal{R} \to \mathcal{R}$ such that for all $t \in T$: $v(t) = V(\mathbf{x}(t))$. Then, substituting the relation $d\mathbf{x}/dt = \mathbf{f}$ into (6.6) to obtain:

$$\dot{V} = [dV/d\mathbf{x}]\mathbf{f}(\mathbf{x}) = [dV/d\mathbf{x}][d\mathbf{x}/dt] = dv/dt$$

which implies that the condition $\dot{V} \leq 0$ is equivalent to the condition that $dv/dt \leq 0$. That is, $V$ is non-increasing on the trajectories of the continuous-time dynamical system when $\dot{V} \leq 0$.

**Example 6.3.2** (Continuous-Time Gradient Descent). Let $\mathbf{f} : \mathcal{R}^d \to \mathcal{R}^d$ be a time-invariant vector field for a continuous-time gradient descent algorithm which minimizes a twice continuously differentiable function $\ell : \mathcal{R}^d \to \mathcal{R}$, defined such that for all $\boldsymbol{\theta} \in \mathcal{R}^d$:

$$\mathbf{f}(\boldsymbol{\theta}) = -\gamma d\ell/d\boldsymbol{\theta}.$$

Show that $\ell$ is a Lyapunov function for this continuous-time gradient descent algorithm.

**Solution.** Let $\dot{\ell} \equiv (d\ell/d\boldsymbol{\theta})\mathbf{f}$. The function $\ell$ is a Lyapunov function because both $\ell$ and $\dot{\ell}$ are continuous and, in addition,

$$\dot{\ell}(\boldsymbol{\theta}) = -\gamma \left|d\ell/d\boldsymbol{\theta}\right|^2 \leq 0.$$

△

## 6.3.2 Invariant Set Theorems

### 6.3.2.1 Finite State Space Convergence Analyses

When a Lyapunov function exists for a discrete-time dynamical system defined on a finite state space, convergence analyses are greatly simplified. The following theorem deals with this special situation.

**Theorem 6.3.1** (Finite State Space Invariant Set Theorem). *Let $T \equiv \mathbb{N}$. Let $\Omega$ be a nonempty finite subset of $\mathcal{R}^d$. Let $\mathbf{f} : \Omega \to \Omega$ be a time-invariant iterated map for a time-invariant dynamical system $\boldsymbol{\Psi} : \Omega \times T \to \Omega$. Let $V : \Omega \to \mathcal{R}$. Let $\dot{V}(\mathbf{x}) \equiv V(\mathbf{f}(\mathbf{x})) - V(\mathbf{x})$ for all $\mathbf{x} \in \Omega$. Let $\Gamma \equiv \{\mathbf{x} \in \Omega : \dot{V}(\mathbf{x}) = 0\}$. Assume for all $\mathbf{x} \in \Omega$ that:*

1. *$\dot{V}(\mathbf{x}) \leq 0$, and*
2. *$\dot{V}(\mathbf{x}) < 0$ if and only if $\mathbf{x} \notin \Gamma$.*

*Then, $\Gamma$ is nonempty and for every $\mathbf{x}_0 \in \Omega$ there exists some positive integer $T^*$ such that $\boldsymbol{\Psi}(\mathbf{x}_0, t) \in \Gamma$ for all $t > T^*$.*

*Proof.* The dynamical system generates a sequence of states $\mathbf{x}(1), \mathbf{x}(2), \ldots$. Each time a state $\mathbf{x}$ outside of $\Gamma$ is visited, the system cannot return to that state without violating the condition that $\dot{V}(\mathbf{x}) < 0$ for all $\mathbf{x} \notin \Gamma$. Since the total number of states in $\Omega$ is finite, this implies that a nonempty set $\Gamma$ must exist and all trajectories will reach $\Gamma$ in a finite number of iterations. ■

Given a Lyapunov function $V$ for a dynamical system $\mathbf{\Psi}$, the Finite State Space Invariant Set Theorem provides conditions ensuring that all trajectories of the dynamical system will eventually reach the largest invariant set $\Gamma$ where $\dot{V} = 0$ in a finite number of iterations. Although the conclusion of the theorem seems rather weak, it is usually possible in many important applications to characterize the nature of the invariant set $\Gamma$. If this can be done, then the Invariant Set Theorem is a powerful tool for deterministic dynamical system analysis and design.

For example, for many important applications in optimization theory, $V$ can be chosen such that the set of points where $\dot{V} = 0$ is exactly equal to the set of equilibrium points of $V$ or even a set containing only the unique global minimizer of $V$.

**Example 6.3.3** (Iterated Conditional Modes (ICM) Algorithm for Model Search). An important machine learning problem that often arises in practice involves finding the minimum of a function $V$ whose domain is a finite state space. Transforming a given pattern vector $\mathbf{m}(0)$ into a more preferable pattern vector $\mathbf{m}(t)$ where the system's preference for a state vector $\mathbf{m}$ is defined as $V(\mathbf{m})$.

For example, consider a learning problem where the $d$-dimensional input pattern vector is specified by a collection of $d$ input features. To avoid the curse of dimensionality discussed in Chapter 1, an investigator decides to compare models which either include or exclude different input features. Let $\mathbf{m} \in \{0,1\}^d$ specify which of the $d$ input features should be included in a particular model by setting the $k$th element of $\mathbf{m}$, $m_k$, equal to one if the $k$th input feature is included in the model and setting $m_k = 0$ if the $k$th input feature is excluded from the model. It is computationally intractable to estimate all $2^d$ possible models and compare their prediction errors when $d$ is large (e.g., $d = 1000$). Let $V(\mathbf{m})$ denote the prediction error on the training data using a model specified by the model specification vector $\mathbf{m}$. This type of problem is called the *model search problem*. The following algorithm called the ICM algorithm is now shown to converge to the set of local minimizers of $V$.

The first step involves specifying a probability mass function $p : \{0,1\}^d \to [0,1]$ that assigns a probability to each $\mathbf{m} \in \{0,1\}^d$ such that

$$p(\mathbf{m}) = \frac{\exp(-V(\mathbf{m}))}{\sum_{k-1}^{2^d} \exp(-V(\mathbf{m}_k))}.$$

Let the notation $\mathbf{m}^i(k) \equiv [m_1, \ldots, m_{i-1}, m_i = k, m_{i+1}, \ldots, m_d]$. The conditional probability that the discrete binary random variable $\tilde{m}_i = 1$ is then computed using the formula:

$$p(m_i = 1 | m_1, \ldots, m_{i-1}, m_{i+1}) = \frac{p\left(\mathbf{m}^i(1)\right)}{p\left(\mathbf{m}^i(0)\right) + p\left(\mathbf{m}^i(1)\right)} \tag{6.7}$$

which can be rewritten as:

$$p(m_i = 1 | m_1, \ldots, m_{i-1}, m_{i+1}, \ldots, m_d) = \mathcal{S}\left(V(\mathbf{m}^i(0) - V(\mathbf{m}^i(1))\right) \tag{6.8}$$

where $\mathcal{S}(\phi) = 1/(1 + \exp(-\phi))$. Note that (6.8) is typically easy to compute.

Now consider the following algorithm which is called the ICM (Iterated Conditional Modes) algorithm (see Besag 1986 or Winkler 2012). Let $\mathbf{m}(0)$ denote an initial guess for the local minimizer of $V$. Let the notation $\mathcal{N}_i(\mathbf{m}) \equiv [m_1, \ldots, m_{i-1}, m_{i+1}, \ldots, m_d]$. Next, for $i = 1, \ldots, d$: the $i$th element of $\mathbf{m}(t)$, $m_i(t)$, is selected and updated at iteration $t$. In particular, set

$$m_i(t+1) = \begin{cases} 1 & \text{if} \quad p\left(m_i(t) = 1 | \mathcal{N}_i(\mathbf{m}(t))\right) > 1/2 \\ 0 & \text{if} \quad p\left(m_i(t) = 1 | \mathcal{N}_i(\mathbf{m}(t))\right) < 1/2 \\ m_i(t) & \text{if} \quad p\left(m_i(t) = 1 | \mathcal{N}_i(\mathbf{m}(t))\right) = 1/2 \end{cases}$$

After the first $d$ iterations have been completed, then this is called a single *sweep*. Another sweep of updates is then applied to complete the second sweep. The sweeps are then continued until $\mathbf{m}(t)$ converges. Analyze the asymptotic behavior of this algorithm using the Finite State Space Invariant Set Theorem.

**Solution.** First note that if $\mathbf{m}(t+1) = \mathbf{m}(t)$ (i.e., the state is unchanged after a sweep), then the system state has reached an equilibrium point. Second note $p(\mathbf{m}(t+1)) > p(\mathbf{m}(t))$ if and only if $V(\mathbf{m}(t+1)) < V(\mathbf{m}(t))$. Therefore, if it can be shown that:

$$p(\mathbf{m}(t+1)) > p(\mathbf{m}(t)) \text{ when } \mathbf{m}(t+1) \neq \mathbf{m}(t)$$

then it immediately follows from the Finite State Space Invariant Set Theorem that all trajectories converge to the set of system equilibrium points which are local minimizers of $V$ since the domain of $V$ and $P$ is finite.

The update rule implies that when the system has not converged to an equilibrium point that within sweep $t$ the event:

$$p\left(m_i(t+1)|\mathcal{N}_i(\mathbf{m}(t)\right) > p\left(m_i(t)|\mathcal{N}_i(\mathbf{m}(t)\right) \tag{6.9}$$

will occur at least once. In addition, the ICM algorithm has the property that $\mathcal{N}_i(\mathbf{m}(t+1)) = \mathcal{N}_i(\mathbf{m}(t))$ when only the $i$th coordinate of $\mathbf{m}(t)$ is updated at sweep $t$. Equation (6.9) and the relation $\mathcal{N}_i(\mathbf{m}(t+1)) = \mathcal{N}_i(\mathbf{m}(t))$ imply that when $\mathbf{m}(t+1) \neq \mathbf{m}(t)$:

$$p\left(m_i(t+1)|\mathcal{N}_i(\mathbf{m}_i(t+1)\right) p\left(\mathcal{N}_i(\mathbf{m}(t+1)\right) > p\left(m_i(t)|\mathcal{N}_i(\mathbf{m}(t)\right) p\left(\mathcal{N}_i(\mathbf{m}(t)\right). \tag{6.10}$$

The desired conclusion that $p(\mathbf{m}(t+1)) > p(\mathbf{m}(t))$ is then obtained by realizing the left-hand side of (6.10) is $p(\mathbf{m}(t+1))$ and the right-hand side of (6.10) is $p(\mathbf{m}(t))$. △

#### 6.3.2.2 Continuous State Space Convergence Analyses

In many important applications in machine learning, the state space is not finite. For example, most machine learning algorithms are naturally represented as discrete-time dynamical systems that update a $q$-dimensional parameter state vector $\boldsymbol{\theta}(t)$ which is an element of $\mathcal{R}^q$, $q \in \mathbb{N}$. To address these more complex situations, the following theorem is introduced.

**Theorem 6.3.2** (Invariant Set Theorem). *Let $T \equiv \mathbb{N}$ or $T \equiv [0, \infty)$. Let $\bar{\Omega}$ be a nonempty, closed, bounded subset of $\mathcal{R}^d$. Let $\mathbf{\Psi} : \bar{\Omega} \times T \rightarrow \bar{\Omega}$ be a dynamical system. Let $\mathbf{\Psi}(\cdot, t)$ be continuous on $\bar{\Omega}$ for all $t \in T$. Let $V : \bar{\Omega} \rightarrow \mathcal{R}$. Let $V$ be a discrete-time Lyapunov function on $\bar{\Omega}$ with respect to $\mathbf{\Psi}$ when $T \equiv \mathbb{N}$. Let $V$ be a continuous-time Lyapunov function on $\bar{\Omega}$ with respect to $\mathbf{\Psi}$ when $T \equiv [0, \infty)$. Let $S$ denote the subset of $\bar{\Omega}$ defined such that:*

$$S = \{\mathbf{x} \in \bar{\Omega} : \dot{V}(\mathbf{x}) = 0\}.$$

*Let $\Gamma$ denote the largest invariant set contained in $S$ with respect to $\mathbf{\Psi}$. Then, $\Gamma$ is nonempty and for every $\mathbf{x}_0 \in \bar{\Omega}$: As $t \rightarrow \infty$, $\mathbf{\Psi}(\mathbf{x}_0, t) \rightarrow \Gamma$.*

*Proof.* The proof is divided into four parts. It follows the analyses by LaSalle (1960) and LaSalle (1976). The term "trajectory" should be read as "orbit" for the discrete-time case.

*Part 1: The set of limit points is non-empty and invariant.* Since the range of $\mathbf{\Psi}$ is a $\bar{\Omega}$ is a bounded set, every trajectory $\mathbf{x}(\cdot) \equiv \mathbf{\Psi}(\mathbf{x}_0, \cdot)$ for all $\mathbf{x}_0 \in \Omega$ is bounded. Thus, the limit set for the bounded trajectory $\mathbf{x}(\cdot)$ is nonempty by the Bolzano-Weierstrass Theorem (Knopp, 1956, p. 15) which states that every bounded infinite set contains at least one limit point.

Let $L$ be the nonempty set of limit points. If $\mathbf{x}^* \in L$, then this implies that there exists an $\mathbf{x}_0$ and an increasing sequence of positive integers $t_1 < t_2 < t_3 < \ldots$ such that:

$$\boldsymbol{\Psi}(\mathbf{x}_0, t_n) \to \mathbf{x}^* \tag{6.11}$$

as $n \to \infty$. Since $\boldsymbol{\Psi}$ is continuous in its first argument,

$$\boldsymbol{\Psi}(\boldsymbol{\Psi}(\mathbf{x}_0, t_n), s) \to \boldsymbol{\Psi}(\mathbf{x}^*, s) \tag{6.12}$$

for all $s \in T$ as $n \to \infty$. Using the Consistent Composition Property in the Definition of a Dynamical System provided in Definition 3.2.1 implies as $n \to \infty$ that for all $s \in T$:

$$\boldsymbol{\Psi}(\mathbf{x}_0, s + t_n) \to \boldsymbol{\Psi}(\mathbf{x}^*, s) \tag{6.13}$$

which, in turn, implies that $\boldsymbol{\Psi}(\mathbf{x}^*, s)$ is also an element of $L$. Since $\mathbf{x}^*$ is an element of $L$ and $\boldsymbol{\Psi}(\mathbf{x}^*, s)$ is also an element of $L$ for all $s \in T$, this implies that $L$ is an invariant set.

*Part 2: The Lyapunov function's value converges to a constant.* Since $V$ is a continuous function on a closed and bounded set $\bar{\Omega}$, $V$ has a finite lower bound on $\bar{\Omega}$. Since the Lyapunov function $V$ is non-increasing on the trajectories of the dynamical system $\boldsymbol{\Psi}$, it follows that $V(\mathbf{x}(t)) \to V^*$ as $t \to \infty$ where $V^*$ is a finite constant since $V$ has a finite lower bound. If $V(\mathbf{x}) = V^*$, this implies that $\dot{V}(\mathbf{x}) = 0$.

*Part 3: All trajectories converge to set where the Lyapunov function's value is constant.* Since $V(\mathbf{x}(t) \to V^*$ by Part 2 and $V$ is continuous, it follows from Theorem 5.1.1(viii) that $\mathbf{x}(t) \to S$ as $t \to \infty$.

*Part 4: All trajectories converge to the largest invariant set where the Lyapunov functions value is constant.* By part 1, the set of all limit points, $L$, is nonempty and invariant. By part 3, as $t \to \infty$, $\mathbf{x}(t) \to S$ where $S$ contains the set of limit points $L$. This implies that $\mathbf{x}(t) \to \Gamma$ where $\Gamma$ is the largest invariant set in $S$. ■

The assumptions that the system state is a continuous function of the initial state, that the trajectories are bounded, and that the Lyapunov function is continuous on the system state are key ingredients for proving the Invariant Set Theorem (Theorem 6.3.2) and thus establishing that the convergence of the Lyapunov function implies the convergence of the system state.

It is important to note that by choosing $\bar{\Omega}$ strategically, the Invariant Set Theorem can be used to investigate issues of global convergence (e.g., show that all trajectories converge to an invariant set) as well as issues of local convergence (e.g., provide sufficient conditions for all points in the vicinity of an invariant set $\Gamma$ to converge to $\Gamma$).

The Invariant Set Theorem is typically applied by constructing: (i) A closed, bounded, and invariant set $\bar{\Omega}$ and (ii) Lyapunov function $V$ on $\bar{\Omega}$. The assumption that $\bar{\Omega}$ is a closed, bounded, invariant set is usually satisfied in two distinct ways. First, the system state space may be closed, bounded, and invariant (e.g., each state variable has a maximum and minimum value). Thus, $\bar{\Omega}$ is by definition a closed, bounded, invariant set. A second common approach for constructing a set $\bar{\Omega}$ which is a closed, bounded, and invariant set is now described.

An orbit $\mathbf{x}(0), \mathbf{x}(1), \ldots$ is said to be a *bounded orbit* if there exists a finite number $K$ such that $|\mathbf{x}(t)| \leq K$ for all $t \in \mathbb{N}$. More generally, a trajectory $\mathbf{x} : T \to \mathcal{R}^d$ is called a *bounded trajectory* if for all $t \in \mathbb{N}$: $|\mathbf{x}(t)| \leq K$ for some finite number $K$.

**Theorem 6.3.3** (Bounded Trajectory Lemma). *Let $T = [0, \infty)$ or $T = \mathbb{N}$. Let $\mathbf{x} : T \to \mathcal{R}^d$.*

- *Let $V : \mathcal{R}^d \to \mathcal{R}$ be a continuous function.*
- *Let $\Omega_0 \equiv \{\mathbf{x} \in \mathcal{R}^d : V(\mathbf{x}) \leq V_0\}$ where $V_0 \equiv V(\mathbf{x}(0))$.*

- *For all $t_k, t_{k+1} \in T$ such that $t_{k+1} > t_k$, assume $V(\mathbf{x}(t_{k+1})) \leq V(\mathbf{x}(t_k))$.*
- *Assume $|\mathbf{x}| \rightarrow +\infty$ implies $V(\mathbf{x}) \rightarrow +\infty$.*

*Then, $\mathbf{x}(t) \in \Omega_0$ for all $t \in T$ where $\Omega_0$ is a closed and bounded set.*

*Proof.* (i) The construction of $\Omega_0$ and the assumption $V(\mathbf{x}(t_{k+1})) \leq V(\mathbf{x}(t_k))$ implies $\mathbf{x}(t) \in \Omega_0$ for all $t \in T$. (ii) By Theorem 5.1.6, $\Omega_0$ is closed since $V$ is continuous by assumption. (iii) To show that $\Omega_0$ is bounded, assume $\Omega_0$ is not a bounded set. If $\Omega_0$ is not a bounded set, then there exists a subsequence of points $\mathbf{x}(t_1), \mathbf{x}(t_2), \ldots$ in $\Omega_0$ such that $|\mathbf{x}(t_k)| \rightarrow \infty$ as $k \rightarrow \infty$ (see Theorem 5.1.1 (vi)). Since by assumption, $|\mathbf{x}(t_k)| \rightarrow \infty$ implies $V(\mathbf{x}(t_k)) \rightarrow +\infty$, it follows that $V(\mathbf{x}(t)) > V(\mathbf{x}(0))$ when $t$ is sufficiently large. That is, $\mathbf{x}(t) \notin \Omega_0$ when $t$ is sufficiently large. Thus, this proof by contradiction shows that $\Omega_0$ is a bounded set. ■

**Example 6.3.4** (Applying the Bounded Trajectory Lemma). Let $V : \mathcal{R} \rightarrow \mathcal{R}$ be defined such that for all $x \in \mathcal{R}$: $V(x) = x$. Assume $V$ is the Lyapunov function for some dynamical system $\boldsymbol{\Psi} : \mathcal{R} \times T \rightarrow \mathcal{R}$. Now construct a set $\Omega_s$ such that:

$$\Omega_s = \{x \in \mathcal{R} : V(x) \leq s\} = \{x \in \mathcal{R} : x \leq s\}.$$

The Bounded Trajectory Lemma is *not* applicable since as $|x| \rightarrow \infty$, $V(x) \rightarrow +\infty$ *or* $V(x) \rightarrow -\infty$. Direct inspection of $\Omega_s$ indicates that $\Omega_s$ is a closed set (since $V$ is continuous) and an invariant set (since $V$ is a Lyapunov function) but is not a bounded set. △

**Example 6.3.5** (Regularization Term for Ensuring Bounded Trajectories). In order to investigate learning dynamics, a common analysis strategy is to represent the learning dynamics as a dynamical system and attempt to formally show that the empirical risk function of the learning machine, $\hat{\ell}_n$, is a Lyapunov function for the learning dynamics. A regularization term can sometimes be effective for ensuring that the trajectories of the learning dynamical system are confined to a closed, bounded, and invariant set.

Let $\hat{\ell}_n : \mathcal{R}^q \rightarrow \mathcal{R}$ be a continuous function and assume the learning machine can be represented as a discrete-time dynamical system which has the property that

$$\hat{\ell}_n(\boldsymbol{\theta}(t+1)) \leq \hat{\ell}_n(\boldsymbol{\theta}(t))$$

for all $t \in \mathbb{N}$. In order to show that all trajectories in the closed set

$$\{\boldsymbol{\theta} : \hat{\ell}_n(\boldsymbol{\theta}) \leq \hat{\ell}(\boldsymbol{\theta}(0))\}$$

are bounded, it is sufficient to show that $\hat{\ell}_n(\boldsymbol{\theta}) \rightarrow +\infty$ as $|\boldsymbol{\theta}| \rightarrow +\infty$ (see Theorem 6.3.3). In many situations, a strategic choice of regularization term can help in ensuring this condition holds.

*Case 1: Loss function has a lower bound.* Consider the empirical risk function:

$$\hat{\ell}_n(\boldsymbol{\theta}) = \lambda|\boldsymbol{\theta}|^2 + (1/n)\sum_{i=1}^{n} c(\mathbf{x}_i, \boldsymbol{\theta}) \tag{6.14}$$

where $\lambda > 0$ and $c$ has a finite lower bound so that $c \geq K$ where $K$ is a finite number. Since $\hat{\ell}_n > \lambda|\boldsymbol{\theta}|^2 + K$, this implies that $\hat{\ell}_n \rightarrow +\infty$ as $|\boldsymbol{\theta}| \rightarrow +\infty$.

*Case 2: Loss function eventually dominated by regularization term.* Now suppose that the loss function for (6.14) is given by $c(\mathbf{x}_i, \boldsymbol{\theta}) = -\boldsymbol{\theta}^T\mathbf{x}_i$. In this case, $c$ does not have a lower bound.

## Recipe Box 6.1 Convergence Analysis Time-Invariant Systems (Theorem 6.3.2)

- **Step 1: Put dynamical system into canonical form.** Put the dynamical system into the form of a discrete-time dynamical system $\mathbf{x}(t+1) = \mathbf{f}(\mathbf{x}(t))$ where $T \equiv \mathbb{N}$ or continuous-time dynamical system $d\mathbf{x}/dt = \mathbf{f}(\mathbf{x})$ where $T \equiv [0, \infty)$.

- **Step 2 : Check conditions of relevant EUC Theorem.** Let $\Omega \subseteq \mathcal{R}^d$. If the dynamical system is a continuous-time dynamical system of the form $d\mathbf{x}/dt = \mathbf{f}(\mathbf{x})$, then check to see if $\mathbf{f} : \Omega \to \Omega$ is continuously differentiable on $\Omega$. If the dynamical system is a discrete-time dynamical system of the form $\mathbf{x}(t+1) = \mathbf{f}(\mathbf{x}(t))$, then check if $\mathbf{f} : \Omega \to \Omega$ is continuous on $\Omega$.

- **Step 3: Construct a Lyapunov function.** Let $\bar{\Omega}$ be the closure of $\Omega$. The candidate Lyapunov function $V : \bar{\Omega} \to \mathcal{R}$ must be continuous on $\bar{\Omega}$. In addition, the function $V$ must have the following properties. If the dynamical system is a discrete-time dynamical system, then

$$\dot{V}(\mathbf{x}) = V(f(\mathbf{x})) - V(\mathbf{x}) \leq 0$$

for all $\mathbf{x} \in \bar{\Omega}$. If the dynamical system is a continuous-time dynamical system, then:

$$\dot{V}(\mathbf{x}) = (dV/d\mathbf{x})\mathbf{f}(\mathbf{x}) \leq 0$$

for all $\mathbf{x} \in \bar{\Omega}$. In addition, function $\dot{V}$ must be continuous on $\bar{\Omega}$.

- **Step 4 : Construct a closed, bounded, invariant set.** There are two typical ways to do this. One approach involves constructing a closed and bounded set $\Omega$ such that $\mathbf{x}(t) \in \Omega$ for all $t \in T$. For example, if each state in the dynamical system has a minimum and maximum value, then this ensures the state space is closed and bounded. A second approach is applicable provided that the Lyapunov function $V$ has the property that $V(\mathbf{x}) \to +\infty$ as $|\mathbf{x}| \to +\infty$. If so, then all trajectories are constrained to the closed, bounded, and invariant set $\Omega = \{\mathbf{x} \in \mathcal{R}^d : V(\mathbf{x}) \leq V(\mathbf{x}(0))\}$.

- **Step 5: Conclude all trajectories converge to set S where $\dot{V} = 0$.** Let

$$S = \left\{\mathbf{x} \in \bar{\Omega} : \dot{V}(\mathbf{x}) = 0\right\}.$$

If the conditions of the Invariant Set Theorem are satisfied, conclude that all trajectories initiated in $\Omega$ converge to the largest invariant set, $\Gamma$, where $\Gamma \subseteq S$.

- **Step 6: Investigate the contents of $\Gamma$.** For example, given a vector field

$$\mathbf{f}(\mathbf{x}) = -[dV/d\mathbf{x}]^T$$

then the largest invariant set in $S$, $\Gamma$, is the set of system equilibrium points which is identical to the set of critical points of the function $V$ on $S$.

Assume $\{\mathbf{x}_1, \ldots, \mathbf{x}_n\}$ is a finite set of finite vectors so that there exists a number $K$ such that $|\mathbf{x}_i| \leq K$ for $i = 1, \ldots, n$. Since $|\boldsymbol{\theta}^T \mathbf{x}_i| \leq |\mathbf{x}_i||\boldsymbol{\theta}| \leq K|\boldsymbol{\theta}|$, it follows that:

$$\hat{\ell}_n(\boldsymbol{\theta}) \geq \lambda|\boldsymbol{\theta}|^2 - (1/n)\sum_{i=1}^{n} |\boldsymbol{\theta}||\mathbf{x}_i| \geq |\boldsymbol{\theta}|\left(\lambda|\boldsymbol{\theta}| - K\right).$$

Thus, $\hat{\ell}_n \to +\infty$ as $|\boldsymbol{\theta}| \to +\infty$ even though $c$ does not have a lower bound. △

**Example 6.3.6** (Discrete-Time Linear System Asymptotic Behavior)**.** Consider a discrete-time dynamical system with iterated map $\mathbf{f}(\mathbf{x}) = \mathbf{W}\mathbf{x}$ where $\mathbf{W} \in \mathcal{R}^{d \times d}$. Thus, the dynamical system is specified by the difference equation:

$$\mathbf{x}(t+1) = \mathbf{W}\mathbf{x}(t).$$

Let $\Omega \equiv \{\mathbf{x} : |\mathbf{x}|^2 \leq K\}$ where $K$ is some positive finite number. Show that if the magnitude of all eigenvalues of $\mathbf{W}^T\mathbf{W}$ are all strictly less than one, then all trajectories initiated in $\Omega$ either converge to the point $\mathbf{0}_d$ as $t \to \infty$ or converge to the null space of $\mathbf{W}^T\mathbf{W} = \mathbf{I}_d$ as $t \to \infty$.

**Solution.** First, identify the iterated map $\mathbf{f}$ defined such that $\mathbf{f}(\mathbf{x}) = \mathbf{W}\mathbf{x}$ and note that $\mathbf{f}$ is continuous and thus satisfies the EUC conditions. Second, let $V(\mathbf{x}) = |\mathbf{x}|^2$ be a candidate Lyapunov function since Example 6.3.1 shows that $\dot{V} \leq 0$. Also note that since $\dot{V} \leq 0$ and $|\mathbf{x}| \to \infty$ implies $V(\mathbf{x}) \to \infty$, then the Invariant Set Theorem holds and all trajectories converge to the largest invariant set $S$ where $\dot{V} = 0$. The set

$$S \equiv \{\mathbf{x} : \dot{V}(\mathbf{x}) = |\mathbf{W}\mathbf{x}|^2 - |\mathbf{x}|^2 = 0\}$$

is the union of $\mathbf{0}_d$ and the set of vectors in the null space of $\mathbf{W}^T\mathbf{W} - \mathbf{I}_d$. △

## Exercises

6.3-1. *Convergence to Hypercube Vertices.* Construct a vector field specifying a dynamical system whose behavior is characterized by the following $d$ differential equations:

$$dx_i/dt = x_i(1 - x_i) \tag{6.15}$$

for $i = 1 \ldots d$. Use the Invariant Set Theorem to prove that all trajectories converge to the vertices of a hypercube. In addition, use the Invariant Set Theorem to show that if a trajectory is initiated sufficiently close to a hypercube vertex, then the trajectory will converge to that vertex. HINT: Find the Lyapunov function by integrating the right-hand side of (6.15).

6.3-2. *Continuous-Time Gradient Descent with Multiple Learning Rates and Weight Decay.* In this problem, a continuous-time gradient descent algorithm with weight decay and multiple learning rates is analyzed. Let $\ell : \mathcal{R}^d \to \mathcal{R}$ be a twice continuously differentiable empirical risk function with a lower bound. To minimize $\ell$, define a continuous-time gradient descent algorithm such that:

$$\frac{d\boldsymbol{\theta}}{dt} = -\lambda \mathbf{DIAG}(\boldsymbol{\gamma})\,\boldsymbol{\theta}(t) - \mathbf{DIAG}(\boldsymbol{\gamma})\,\frac{d\ell(\boldsymbol{\theta}(t))}{d\boldsymbol{\theta}}$$

where the positive number $\lambda \in (0, \infty)$ is called the weight decay parameter, and $\boldsymbol{\gamma} \in (0, \infty)^d$ is a vector of positive numbers which indicate the respective rate of learning for each individual parameter in the learning machine.

Use the Invariant Set Theorem to prove all trajectories converge to a set $\Gamma$. Carefully characterize the contents of $\Gamma$.

6.3-3. *Convergence to Periodic Invariant Set.* Define a vector field specifying the continuous-time dynamical system associated with the differential equations:

$$dx/dt = y + x(1 - x^2 - y^2)$$

and

$$dy/dt = -x + y(1 - x^2 - y^2).$$

Show that all trajectories converge to an invariant set which includes the origin $(0, 0)$ and the set $Q \equiv \{(x, y) : x^2 + y^2 = 1\}$. Show that the point $(0, 0)$ is an equilibrium point and that $Q$ is a periodic invariant set. HINT: Note that if the initial condition of the dynamical system $dx/dt = y$ and $dy/dt = -x$ is the point $(x(0), y(0))$ which satisfies $x(0)^2 + y(0)^2 = 1$, then the resulting trajectory will be a function $(x(t), y(t))$ which satisfies $x(t)^2 + y(t)^2 = 1$ for all non-negative $t$.

6.3-4. *Solving Constraint Satisfaction Problems in Finite State Spaces.* Another important problem in the field of Artificial Intelligence is the *constraint satisfaction problem.* Let $V : \{0, 1\}^d \rightarrow \mathcal{R}$ be a function that specifies how effectively a set of $d$ constraints are satisfied. Assume, in addition, that each constraint can either be "satisfied" or "not satisfied". In particular, let $\mathbf{x} \in \{0, 1\}^d$ be defined such that the $k$th element of $\mathbf{x}$, $x_k$, is equal to one when the $k$th constraint is satisfied and $x_k = 0$ when the $k$th constraint is not satisfied. Show how to design an algorithm that will minimize $V$ using the Finite State Space Invariant Set Theorem.

6.3-5. *BSB Optimization Algorithm.* The Brain-State-in-a-Box (BSB) algorithm (Anderson et al. 1977; Golden 1986; Golden 1993) may be interpreted as a time-invariant discrete-time dynamical system that transforms an initial state vector $\mathbf{x}(0)$ into a more preferable state vector $\mathbf{x}(t)$. More specifically, the function $V(\mathbf{x}) = -\mathbf{x}^T\mathbf{W}\mathbf{x}$ takes on smaller values when $\mathbf{x}$ is more preferable. Assume the matrix $\mathbf{W}$ is a positive semidefinite symmetric matrix chosen such that vectors located in the subspace spanned by the eigenvectors with the larger eigenvalues are assumed to be maximally preferable. All state trajectories are confined to the origin-centered $d$-dimensional hypercube $[-1, 1]^d$.

In particular, let $\boldsymbol{\mathcal{S}} : \mathcal{R}^d \rightarrow [-1, +1]^d$ be defined such that

$$[q_1, \ldots, q_d]^T = \mathcal{S}([y_1, \ldots, y_d])$$

where $q_k = 1$ if $y_k > 1$, $q_k = -1$ if $y_k < -1$, and $q_k = y_k$ otherwise. The BSB algorithm is defined as the discrete-time dynamical system specified by the difference equation:

$$\mathbf{x}(t + 1) = \mathcal{S}(\boldsymbol{\Psi}(t))$$

where

$$\boldsymbol{\Psi}(t) = \mathbf{x}(t) + \gamma\mathbf{W}\mathbf{x}(t).$$

Use the Invariant Set Theorem to show that for every positive stepsize $\gamma$, every trajectory of the BSB dynamical system converges to the set of system equilibrium points. Next, show that if $\mathbf{x}^*$ is an eigenvector of $\mathbf{W}$ with a positive eigenvalue and, in addition, $\mathbf{x}^*$ is a hypercube vertex (i.e., $\mathbf{x}^* \in \{-1, +1\}^d$), then $\mathbf{x}^*$ is both an equilibrium point and local minimizer of $V$.

HINT: Show that the BSB algorithm can be rewritten as:

$$\mathbf{x}(t+1) = \mathbf{x}(t) + \gamma\left[\boldsymbol{\alpha}(\mathbf{x}(t), \boldsymbol{\Psi}(t))\right] \odot \mathbf{W}\mathbf{x}(t)$$

where $\boldsymbol{\alpha}$ is a continuous non-negative vector-valued function of $\mathbf{x}(t)$ and $\boldsymbol{\Psi}(t)$.

6.3-6. *AdaGrad Continuous-Time Dynamical System.* Let $\mathbf{g} : \mathcal{R}^q \to \mathcal{R}^q$ be a column vector-valued function corresponding to the first derivative of a twice continuously differentiable empirical risk function $\ell : \mathcal{R}^q \to \mathcal{R}$. Consider the continuous-time gradient descent learning algorithm defined by the differential equation:

$$\frac{d\boldsymbol{\theta}}{dt} = \frac{-\gamma \mathbf{g}(\boldsymbol{\theta})}{\sqrt{|\mathbf{g}(\boldsymbol{\theta})|^2 + \epsilon}}$$

where $\gamma$ is the learning rate and $\epsilon$ is a small positive number. Let

$$\Omega \equiv \{\boldsymbol{\theta} : \ell(\boldsymbol{\theta}) \leq \ell(\boldsymbol{\theta}(0))\}$$

where $\boldsymbol{\theta}(0)$ is the initial condition for the continuous-time dynamical system. Show that all trajectories converge to the set of critical points $\ell$. Notice that this method of normalizing the gradient has the effect of "speeding up" the learning algorithm when the current parameter values are located in a region of low curvature on the objective function surface. This algorithm can be viewed as a variation of the AdaGrad algorithm (Duchi et al. 2011).

6.3-7. *Convergence Analysis of a General Class of Clustering Algorithms.* Consider the general class of clustering algorithms discussed in Example 1.6.4 that maximize within-cluster similarity. Assume a clustering algorithm works by moving one feature vector to a different cluster at each iteration of the algorithm such that the objective function $V(\mathbf{S})$ decreases in value for each such move. Show, using the Finite State Invariant Set Theorem, that such a clustering algorithm will eventually converge to a set of minimizers of $V$. Note that the K-Means clustering algorithm is a special case of this class of clustering algorithms when the classification penalty $\rho(\mathbf{x}, C_k)$ is chosen to minimize the average distance of $\mathbf{x}$ to each element of $C_k$ where distance between $\mathbf{x}$ and a point $\mathbf{y} \in C_k$ is defined by the similarity distance function $D(\mathbf{x}, \mathbf{y}) = |\mathbf{x} - \mathbf{y}|^2$.

6.3-8. *Hopfield (1982) Neural Network Is an ICM Algorithm.* The Hopfield (1982) Neural Network consists of a collection of $d$ computing units whose respective states are $\mathbf{x}(t) = [x_1(t), \ldots, x_d(t)]^T$ at iteration $t$ of the algorithm. The first computing unit is selected and the dot product of the parameter vector $\mathbf{w}_1$ for the first unit with the current state $\mathbf{x}(t)$ is computed. If this dot product is greater than zero, then the unit's state is set to one. If the dot product is less than zero, then the unit's state is set to zero. If the dot product is equal to zero, then the unit's state is unchanged. The second computing unit is then selected and this process is repeated. Next, the third computing unit is selected, and the process is repeated until all $d$ units have been updated. Then, the states of all $d$ units are repeatedly updated in this manner. In this problem, this system is formally defined and its asymptotic behavior is analyzed.

The goal of the Hopfield (1982) Neural Network is to transform an initial binary state vector $\mathbf{x}(0)$ into a new binary state vector $\mathbf{x}(t)$ such that $\mathbf{x}(t)$ is more preferable than $\mathbf{x}(0)$ where the preference for $\mathbf{x}$ is given by the formula:

$$V(\mathbf{x}) = -\mathbf{x}^T \mathbf{W} \mathbf{x}. \tag{6.16}$$

The matrix $\mathbf{W}$ is a $d$-dimensional symmetric positive semidefinite matrix whose $k$th row is the $d$-dimensional row vector $\mathbf{w}_k^T$.

Let $\mathcal{S}(\phi) = 1$ if $\phi > 0$, $\mathcal{S}(\phi) = \phi$ if $\phi = 0$, and $\mathcal{S}(\phi) = 0$ if $\mathcal{S}(\phi) < 0$.

Let the time-invariant iterated map $\mathbf{f} : \{0,1\}^d \rightarrow \{0,1\}^d$ be defined such that the first element of $\mathbf{f}(\mathbf{x})$ is equal to $\psi_1 = \mathcal{S}(\mathbf{w}_1^T \mathbf{x})$, the second element of $\mathbf{f}(\mathbf{x})$ is equal to $\psi_2 = \mathcal{S}(\mathbf{w}_2^T [\psi_1, x_2, \ldots, x_d]^T)$, and the $j$th element of $\mathbf{f}(\mathbf{x})$ is equal to

$$\psi_j = \mathcal{S}(\mathbf{w}_j^T \mathbf{y}_j)$$

where $\mathbf{y}_j = [\psi_1, \ldots, \psi_{j-1}, x_j, \ldots, x_d]$ for $j = 1, \ldots, d$. Show that the dynamical system specified by the iterative equation

$$\mathbf{x}(t+1) = \mathbf{f}(\mathbf{x}(t))$$

converges to the set of equilibrium points of the dynamical system using the Finite State Space Invariant Set Theorem and the Lyapunov function $V$ in (6.16). Show that this algorithm is a special case of the ICM (Iterated Conditional Modes) algorithm described in Example 6.3.3.

6.3-9. *ICM Algorithm with Categorical Variables.* Example 6.3.3 discussed a special case of the ICM algorithm which assumes a probability mass function that assigns a probability to a $d$-dimensional vector $\mathbf{x} = [x_1, \ldots, x_d]$ such that the $k$th element of $\mathbf{x}$, $x_k$ is restricted to take on the values of either 0 or 1. Prove the more general version of the ICM algorithm following the method of Example 6.3.3 for the case where a probability mass function on a finite state space that assigns a probability to a $d$-dimensional vector $\mathbf{x} = [x_1, \ldots, x_d]$ such that the $k$th element of $\mathbf{x}$, $x_k$ can take on $M_k$ possible values for $k = 1, \ldots, d$. Thus, the ICM algorithm in this example may be used to search for the most probable state in an arbitrary high-dimensional finite state space.

## 6.4 Further Readings

### Nonlinear Dynamical Systems Theory

Hirsch et al. (2004) discuss both discrete-time and continuous-time dynamical systems theory at a level of technical detail comparable to this text. Hirsch et al. (2004) also discuss the existence and uniqueness of solutions to discrete-time and continuous-time systems as well as a discussion of chaotic dynamical systems. Further discussions of chaotic dynamical systems and their applications can be found in Wiggins (2003) and Bahi and Guyeux (2013).

### Invariant Set Theorem

A key contribution of this chapter is the Invariant Set Theorem (Theorem 6.3.2) which supports the analysis and design of a large class of commonly used machine learning algorithms. The statement and proof of Theorem 6.3.2 are based upon LaSalle's (1960) Invariant Set Theorem.

A detailed discussion of the discrete-time case as well as the continuous-time case can be found in LaSalle (1976). Luenberger (1979) provides an introductory discussion of the

Invariant Set Theorem for both the discrete-time and continuous-time case. Vidyasagar (1992) provides a detailed discussion of the Invariant Set Theorem for the continuous-time case.

## Iterated Conditional Modes

The ICM (Iterated Conditional Modes) algorithm is discussed in Besag (1986), Winkler (2003), and Varin et al. (2011).

# 7

# *Batch Learning Algorithm Convergence*

**Learning Objectives**

- Design automatic learning rate adjustment algorithms for batch learning.
- Design convergent batch learning algorithms.
- Design classical gradient descent type optimization algorithms.

As previously noted, many machine learning algorithms can be viewed as gradient descent type learning algorithms. Moreover, although the concept of gradient descent was introduced in Chapter 1 and discussed further in Chapters 5 and 6, sufficient conditions ensuring that discrete-time deterministic gradient descent type algorithms will converge to a desired solution set have not been formally provided. In this chapter, sufficient and practical conditions for ensuring convergence of discrete-time deterministic gradient descent type algorithms are provided.

In particular, this chapter focuses on the analysis of the asymptotic behavior of a class of discrete-time deterministic dynamical systems specified by the time-varying iterated map:

$$\mathbf{x}(t+1) = \mathbf{x}(t) + \gamma(t)\mathbf{d}(t) \tag{7.1}$$

where the strictly positive number $\gamma(t)$ is called the *stepsize* and the $d$-dimensional column vector $\mathbf{d}(t)$ is called the *search direction*. As noted in Chapter 1 and Chapter 5, a large class of important unsupervised, supervised, and reinforcement learning algorithms in the field of machine learning are typically represented using time-varying iterated maps of the form of (7.1).

In this chapter, the Zoutendijk-Wolfe Convergence Theorem is introduced that provides sufficient conditions for $\mathbf{x}(t)$ to converge to the set of critical points of a twice continuously differentiable function of $V : \mathcal{R}^d \rightarrow \mathcal{R}$. These sufficient conditions consist of flexible constraints on the choice of stepsize $\gamma(t)$ and search direction $\mathbf{d}(t)$ that ensure that $V$ is non-increasing on the trajectories of (7.1).

Using the Zoutendijk-Wolfe Convergence Theorem, it is possible to provide convergence analyses of a number of popular unconstrained nonlinear optimization algorithms. In particular, the latter part of this chapter will apply the Zoutendijk-Wolfe Convergence Theorem to establish convergence of the most important descent algorithms which include: gradient descent, gradient descent with momentum, Newton-Raphson, Levenberg-Marquardt, limited Memory BFGS, and Conjugate Gradient algorithms.

## 7.1 Search Direction and Stepsize Choices

### 7.1.1 Search Direction Selection

Consider an iterative algorithm which generates a sequence of $d$-dimensional state vectors $\mathbf{x}(0), \mathbf{x}(1), \ldots$ using the time-varying iterated map specified by:

$$\mathbf{x}(t+1) = \mathbf{x}(t) + \gamma(t)\mathbf{d}(t)$$

where $\gamma(0), \gamma(1), \ldots$ is a sequence of positive stepsizes and $\mathbf{d}(0), \mathbf{d}(1), \ldots$ is a sequence of *search directions.* In addition, let $\mathbf{g}(t)$ denote the gradient of a twice continuously differentiable objective function $V : \mathcal{R}^d \to \mathcal{R}$.

Using arguments similar to those discussed in the Discrete-Time Descent Direction Theorem (Theorem 5.3.2), the search direction $\mathbf{d}(t)$ is chosen such that the cosine of the angular separation, $\psi_t$, between $\mathbf{d}(t)$ and $-\mathbf{g}(t)$ is less than 90 degrees. That is, choose $\mathbf{d}(t)$ such that

$$\cos(\psi_t) \equiv \frac{-\mathbf{g}(t)^T \mathbf{d}(t)}{|\mathbf{g}(t)|\,|\mathbf{d}(t)|} \geq \delta > 0, \tag{7.2}$$

to ensure that $V(\mathbf{x}(t+1)) < V(\mathbf{x}(t))$. A search direction chosen in this manner is said to satisfy the *downhill condition* (see Figure 7.1).

In the important special case where $\mathbf{d}(t) = -\mathbf{g}(t)$, the search direction is called the *gradient descent direction* and the angular separation between $\mathbf{d}(t)$ and $-\mathbf{g}(t)$ is zero degrees. The constant $\delta$ in (7.2) is chosen to be a positive number in order to avoid situations where the sequence of search direction vectors converges to a search direction $\mathbf{d}(t)$ that is orthogonal to the gradient descent direction $-\mathbf{g}(t)$.

**Example 7.1.1** (Scaled Gradient Descent). In some cases, a linear scaling transformation of the gradient can dramatically improve the convergence rate of a gradient descent algorithm.

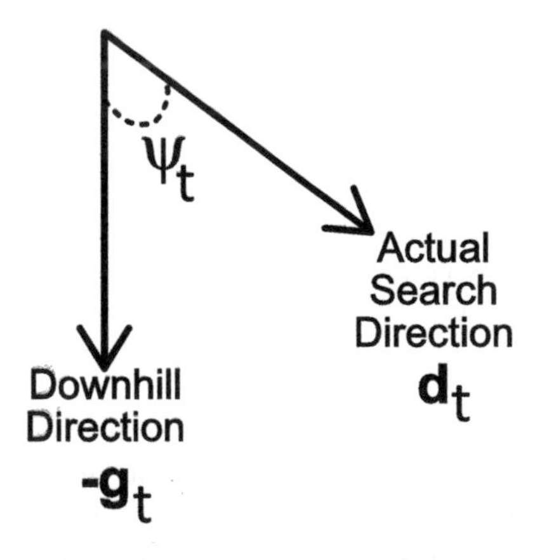


**FIGURE 7.1**
**Geometric interpretation of the downhill condition in deterministic nonlinear optimization theory.** The downhill search condition requires that the angular separation $\psi_t$ between the search direction $\mathbf{d}_t$ of the descent algorithm and the negative gradient of the objective function $-\mathbf{g}_t$ be less than 90 degrees.

Let $\mathbf{g} : \mathcal{R}^d \rightarrow \mathcal{R}^d$ denote the gradient of an objective function $\ell : \mathcal{R}^d \rightarrow \mathcal{R}$. Consider a descent algorithm of the form of (7.1) where the descent direction $\mathbf{d}(t) = -\mathbf{M}\mathbf{g}(\mathbf{x}(t))$, the stepsize $\gamma(t)$ is a positive number, and $\mathbf{M}$ is a positive definite matrix specifying the linear scaling transformation. Show that this algorithm generates a search direction $\mathbf{d}(t)$ that satisfies the downhill condition in (7.2). HINT: Use the identity

$$\lambda_{min}|\mathbf{g}|^2 \leq \mathbf{g}^T\mathbf{M}\mathbf{g} \leq \lambda_{max}|\mathbf{g}|^2.$$

**Solution.** Let $\lambda_{min}$ and $\lambda_{max}$ denote the smallest and largest eigenvalues of $\mathbf{M}$. Since $\mathbf{M}$ is positive definite, $\lambda_{max} \geq \lambda_{min} > 0$. Let $\mathbf{g}_t \equiv \mathbf{g}(\mathbf{x}(t))$. Let $\mathbf{d}_t \equiv \mathbf{d}(t)$. Assume $|\mathbf{g}_t| > 0$ and $|\mathbf{d}_t| > 0$. Then, the downhill condition is satisfied since

$$\frac{\mathbf{d}_t^T \mathbf{g}_t}{|\mathbf{d}_t|\;|\mathbf{g}_t|} = \frac{(-\mathbf{M}\mathbf{g}_t)^T \mathbf{g}_t}{|\mathbf{M}\mathbf{g}_t|\;|\mathbf{g}_t|} = \frac{-\mathbf{g}_t^T\mathbf{M}\mathbf{g}_t}{|\mathbf{g}_t|\sqrt{|\mathbf{M}\mathbf{g}_t|^2}} = \frac{-\mathbf{g}_t^T\mathbf{M}\mathbf{g}_t}{|\mathbf{g}_t|\sqrt{\mathbf{g}_t\mathbf{M}^2\mathbf{g}_t}} \leq -\lambda_{min}/\lambda_{max}.$$

△

**Example 7.1.2** (Gradient Descent Using Automatically Adjusted Momentum). In many deep learning applications, gradient descent with momentum remains a popular nonlinear optimization strategy (Polyak 1964; Rumelhart et al. 1986; Sutskever et al. 2013). The method is based upon the heuristic that the search direction computed for the current iteration should tend to point in the same direction as the search direction calculated for the previous iteration. The momentum method also has the advantage that it is not computationally demanding. Consider the following descent algorithm:

$$\mathbf{x}(t+1) = \mathbf{x}(t) + \gamma_t \mathbf{d}_t$$

where $\gamma_t$ is the stepsize and $\mathbf{d}_t$ is the search direction.

An automatic adaptive momentum method uses the following choice for the search direction $\mathbf{d}_t$ given by:

$$\mathbf{d}_t = -\mathbf{g}_t + \mu_t \mathbf{d}_{t-1} \tag{7.3}$$

where $\mathbf{d}_t$ is the search direction at iteration $t$ and $\mathbf{g}_t$ is the gradient of the objective function at iteration $t$. Let $D_{max}$ and $\mu$ be positive numbers. The non-negative number $\mu_t$ in (7.3) is chosen so that if $\mathbf{d}_{t-1}^T \mathbf{g}_t > 0$ and $|\mathbf{d}_t| \leq D_{max}|\mathbf{g}_t|$, then

$$\mu_t = -\mu|\mathbf{g}_t|^2/(\mathbf{d}_{t-1}^T\mathbf{g}_t); \tag{7.4}$$

otherwise choose $\mu_t = 0$.

Show that the search direction for this automatic adaptive momentum learning algorithm satisfies the downhill condition in (7.2).

**Solution.** Substitute (7.4) into (7.3) and then evaluate the downhill condition in (7.2) to obtain:

$$\frac{\mathbf{d}_t^T \mathbf{g}_t}{|\mathbf{d}_t|\;|\mathbf{g}_t|} = \frac{-|\mathbf{g}_t|^2 - \mu|\mathbf{g}_t|^2}{|\mathbf{d}_t|\;|\mathbf{g}_t|} = \frac{-(1+\mu)|\mathbf{g}_t|^2}{|\mathbf{d}_t|\;|\mathbf{g}_t|} \leq \frac{-(1+\mu)|\mathbf{g}_t|^2}{|\mathbf{g}_t|\;|\mathbf{g}_t|D_{max}} \leq -(1+\mu)/D_{max}.$$

△

### 7.1.2 Stepsize Selection

Given a current state $\mathbf{x}(t)$ and current search direction $\mathbf{d}(t)$, each possible choice of a stepsize $\gamma \in [0, \infty)$ results in a new state of the discrete-time descent dynamical system,

$\mathbf{x}(t+1) = \mathbf{x}(t) + \gamma\mathbf{d}(t)$. The objective function $V$ evaluated at this new state $\mathbf{x}(t+1)$ is equal to $V(\mathbf{x}(t+1)) = V(\mathbf{x}(t) + \gamma\mathbf{d}(t))$. The stepsize selection problem is to select a particular choice of $\gamma$ given a known system state $\mathbf{x}(t)$ and search direction $\mathbf{d}(t)$.

Since the goal of a descent algorithm is to minimize the objective function $V$, it makes sense to define the concept of a stepsize that decreases the objective function's value $V$.

**Definition 7.1.1** (Downhill Stepsize). Let $V : \mathcal{R}^d \rightarrow \mathcal{R}$ be an *objective function*. Let $\mathbf{x} \in \mathcal{R}^d$ be a *system state*. Let $\mathbf{d} \in \mathcal{R}^d$ be a *search direction*. A positive real number $\gamma$ is called a *downhill stepsize* if

$$V(\mathbf{x} + \gamma\mathbf{d}) \leq V(\mathbf{x}) \tag{7.5}$$

and is called a *strict downhill stepsize* if

$$V(\mathbf{x} + \gamma\mathbf{d}) < V(\mathbf{x}). \tag{7.6}$$

□

It will be convenient to define the *objective function projection* $V_{\mathbf{x},\mathbf{d}} : [0, \infty) \rightarrow \mathcal{R}$ which is the projection of $V$ onto a search direction $\mathbf{d}$. Specifically, the function $V_{\mathbf{x},\mathbf{d}}$ is defined such that for all $\gamma \in [0, \infty)$:

$$V_{\mathbf{x},\mathbf{d}}(\gamma) = V(\mathbf{x} + \gamma\mathbf{d}) \tag{7.7}$$

so that $V_{\mathbf{x},\mathbf{d}}(0) = V(\mathbf{x})$.

Using this notation, the definition of the downhill stepsize in (7.5) would be expressed as:

$$V_{\mathbf{x},\mathbf{d}}(\gamma) \leq V_{\mathbf{x},\mathbf{d}}(0).$$

**Definition 7.1.2** (Optimal Stepsize). Let $V_{\mathbf{x},\mathbf{d}} : [0, \infty) \rightarrow \mathcal{R}$ be an objective function projection with respect to some objective function $V : \mathcal{R}^d \rightarrow \mathcal{R}$ and some search direction $\mathbf{d} \in \mathcal{R}^d$. Let $\gamma_{max}$ be a positive number. A global minimizer of $V_{\mathbf{x},\mathbf{d}}$ on the interval $[0, \gamma_{max}]$ is an *optimal stepsize* on $[0, \gamma_{max}]$. □

Figure (7.2) illustrates the concept of an optimal stepsize. The optimal stepsize, $\gamma^*$, is a global minimizer of $V_{\mathbf{x},\mathbf{d}}$ defined with respect to some current state $\mathbf{x}$ and search direction $\mathbf{d}$. The system state, $\mathbf{x} + \gamma^*\mathbf{d}$, corresponding to choosing optimal stepsize $\gamma^*$ is not necessarily a global minimizer (or even a critical point) of $V$. Also, as illustrated in Figure (7.2) for a given $V_{\mathbf{x},\mathbf{d}}$, there might exist multiple local minimizers.

In many cases, selecting the optimal stepsize is a computationally desirable goal. However, in practice, it is often computationally expensive to compute the exact optimal stepsize at each iteration of a descent algorithm. For example, each objective function evaluation may require computations involving all of the training data. The benefits of an exact optimal stepsize are further diminished when the search takes place in a high-dimensional state space (e.g., $d > 10$ or $d > 100$) as opposed to a state space consisting of only a few variables (e.g., $d < 5$).

Furthermore, note that if the stepsize $\gamma(t)$ is chosen to be constant for some iteration $t$, then the system state might oscillate around a local minimizer of $V$ as illustrated in Figure 7.3. On the other hand, if the sequence $\gamma(1), \gamma(2), \ldots$ converges too rapidly, then the algorithm may take a long time to converge and may not even converge to a critical point of the function $V$ as shown in Figure 7.4. For example, suppose that $\gamma(t) = 1/t^{100}$. In this case, the decreasing sequence $\{\gamma(t)\}$ would most likely converge to zero before the sequence of states generated by a descent algorithm converges to a critical point! A critical issue, therefore, is the proper choice of stepsize for the descent algorithm at each iteration and realizing that a good choice of stepsize will vary as a function of the curvature of the function associated with the current state of the dynamical system descent algorithm.

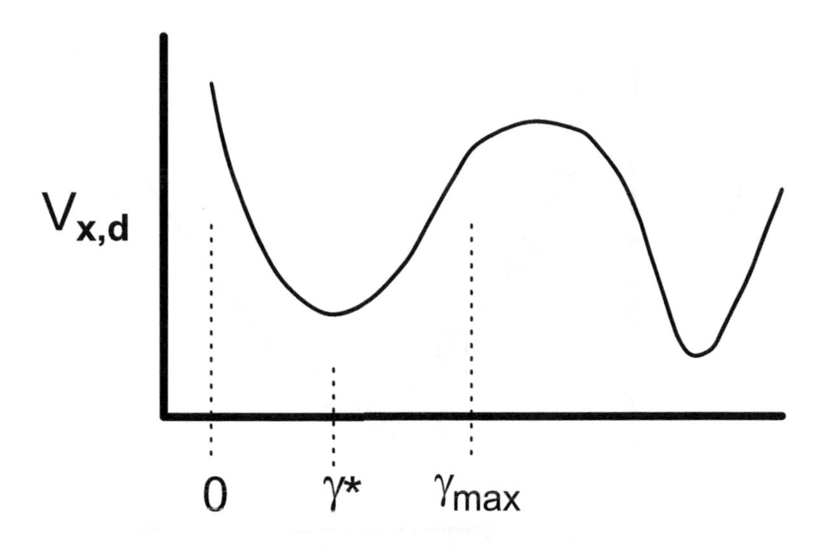


**FIGURE 7.2**
**Geometric interpretation of an optimal stepsize.** If $V : \mathcal{R}^d \rightarrow \mathcal{R}$ is an objective function, $\mathbf{x}$ is the current state of the search algorithm in state space, and let the search direction $\mathbf{d} \in \mathcal{R}^d$. Let $V_{\mathbf{x},\mathbf{d}} : [0, \infty) \rightarrow \mathcal{R}$ be defined such that $V_{\mathbf{x},\mathbf{d}}(\gamma) = V(\mathbf{x} + \gamma\mathbf{d})$. An optimal stepsize $\gamma^*$ decreases the objective function $V_{\mathbf{x},\mathbf{d}}$ by a maximum amount over a closed interval $[0, \gamma_{max}]$. The optimal stepsize $\gamma^*$ may change if $\gamma_{max}$ is changed.

The following concept of a sloppy, inexact stepsize (called a "Wolfe stepsize") is designed to address the problem of selecting a stepsize that decreases yet does not decrease too rapidly at every iteration in order to support convergence to critical points of the objective function. This flexibility for stepsize selection reduces the computational burden of computing an optimal stepsize at every iteration.

Let $\mathbf{g}$ be the gradient of $V$. Let $V_{\mathbf{x},\mathbf{d}}$ be an objective function projection as defined in (7.7). Using the chain rule, it follows that

$$dV_{\mathbf{x},\mathbf{d}}/d\gamma = \mathbf{g}(\mathbf{x} + \gamma\mathbf{d}(\mathbf{x}))^T\mathbf{d}(\mathbf{x}).$$

The notation $dV_{\mathbf{x},\mathbf{d}}(\gamma_0)/d\gamma$ is used to indicate $dV_{\mathbf{x},\mathbf{d}}/d\gamma$ evaluated at $\gamma_0$.

**Definition 7.1.3** (Wolfe Stepsize). Let $\mathbf{x}$ and $\mathbf{d}$ be $d$-dimensional real constant column vectors. Let $V : \mathcal{R}^d \rightarrow \mathcal{R}$ be a continuously differentiable function with a lower bound. Let $V_{\mathbf{x},\mathbf{d}} : [0, \infty) \rightarrow \mathcal{R}$ be defined as in (7.7) with respect to $V$, $\mathbf{x}$, and $\mathbf{d}$. Let $dV_{\mathbf{x},\mathbf{d}}(0)/d\gamma < 0$. A *Wolfe stepsize* is a positive number $\gamma$ that satisfies the *Wolfe conditions*

$$V_{\mathbf{x},\mathbf{d}}(\gamma) \leq V_{\mathbf{x},\mathbf{d}}(0) + \alpha\gamma dV_{\mathbf{x},\mathbf{d}}(0)/d\gamma \tag{7.8}$$

and

$$\frac{dV_{\mathbf{x},\mathbf{d}}(\gamma)}{d\gamma} \geq \beta\frac{dV_{\mathbf{x},\mathbf{d}}(0)}{d\gamma} \tag{7.9}$$

when $0 < \alpha < \beta < 1$. Now assume (7.8) holds and assume (7.9) is satisfied by:

$$\left|\frac{dV_{\mathbf{x},\mathbf{d}}(\gamma)}{d\gamma}\right| \leq \beta\left|\frac{dV_{\mathbf{x},\mathbf{d}}(0)}{d\gamma}\right| \tag{7.10}$$

then the stepsize $\gamma$ satisfies the *strong Wolfe conditions*. □

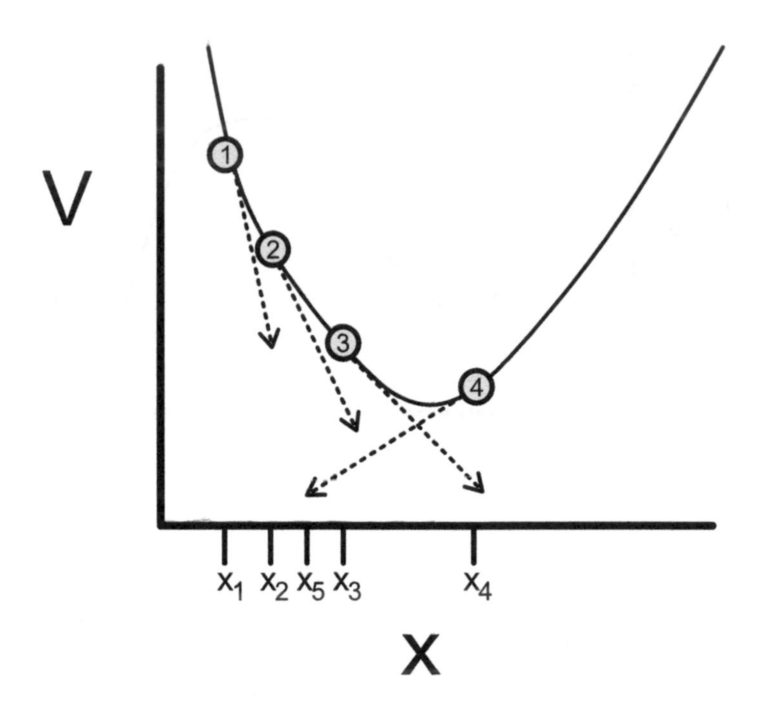


**FIGURE 7.3**
**Example of a descent algorithm which does not converge with constant stepsize.** If the stepsize for a descent algorithm is constant, then the descent algorithm generates a sequence of states which will tend to "jump over" the desired solution, resulting in a situation where the descent algorithm fails to converge.

The first condition in (7.8) requires that the stepsize $\gamma$ is chosen so that

$$V_{\mathbf{x},\mathbf{d}}(\gamma) - V_{\mathbf{x},\mathbf{d}}(0) \leq -\alpha\gamma|dV_{\mathbf{x},\mathbf{d}}(0)/d\gamma| \tag{7.11}$$

because $dV_{\mathbf{x},\mathbf{d}}(0)/d\gamma \leq 0$. Thus, the first condition in (7.8) is a requirement that $V_{\mathbf{x},\mathbf{d}}$ decrease in value by an amount at least equal to $\alpha\gamma|dV_{\mathbf{x},\mathbf{d}}(0)/d\gamma|$. The constant $\alpha$ in (7.10) and (7.11) is called the *sufficient decrease constant.*

Equation (7.9) is called the *curvature condition*, which ensures that the derivative of $V_{\mathbf{x},\mathbf{d}}(\gamma)$ is either less negative than the derivative of $V_{\mathbf{x},\mathbf{d}}(0)$ or possibly even positive. When (7.9) is satisfied by (7.10), this corresponds to the strong Wolfe condition case. In practice, the strong Wolfe condition in (7.10) is used instead of (7.9). The condition in (7.10) is a requirement that the stepsize $\gamma$ is chosen so the magnitude of the slope $dV_{\mathbf{x},\mathbf{d}}/d\gamma$ after taking a stepsize of length $\gamma$ is smaller than the percentage of the magnitude of the slope $dV_{\mathbf{x},\mathbf{d}}/d\gamma$ evaluated before the step is taken. An optimal stepsize would minimize $V_{\mathbf{x},\mathbf{d}}$ and thus require that the derivative of $V_{\mathbf{x},\mathbf{d}}$ after an optimal stepsize step must be zero. Rather than requiring that the stepsize $\gamma$ force the derivative of $V_{\mathbf{x},\mathbf{d}}$ to be equal to zero, the strong Wolfe condition (7.10) simply requires that the magnitude of $dV_{\mathbf{x},\mathbf{d}}/d\gamma$ decreases by a sufficient amount. The constant $\beta$ in (7.10) is called the *curvature constant.*

Figure 7.4 illustrates a situation where the objective function value is decreasing at each iteration of the automatic stepsize selection algorithm but the magnitude of the derivative of $V_{\mathbf{x},\mathbf{d}}$ is not decreasing by a sufficient amount at each step.

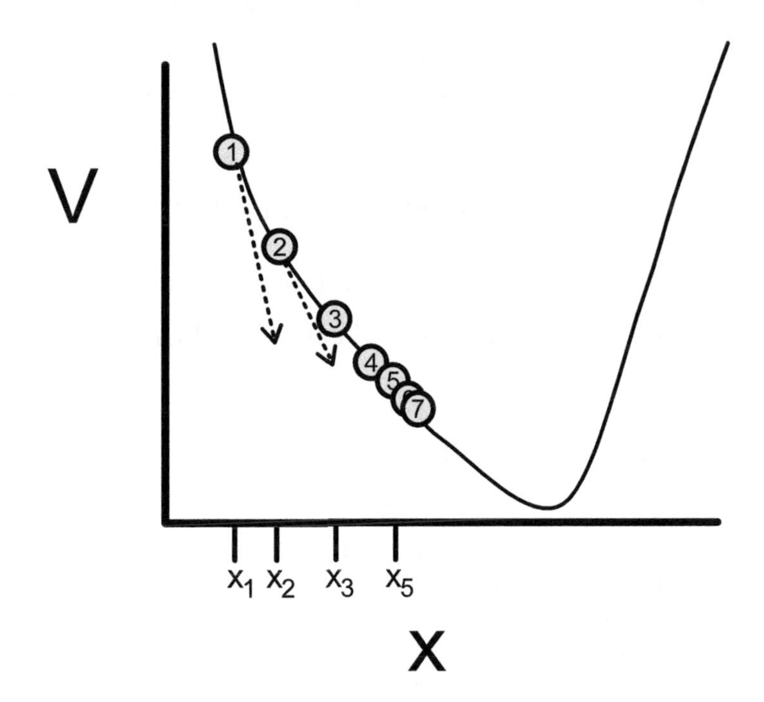


**FIGURE 7.4**
**Example of a descent algorithm which does not converge with decreasing stepsize.** If the stepsize for a descent algorithm is decreased too rapidly, then the descent algorithm might stop before it reaches a desirable solution.

In most machine learning applications, high-dimensional parameter vectors are updated and the goal is to seek a local minimizer of a complicated objective function characterized by multiple minima, maxima, and saddlepoints. In such situations, even small perturbations in the high-dimensional search direction may substantially impact how the value of the objective function and the curvature of the objective function changes at each algorithm iteration. Therefore, it is prudent to place as few constraints as possible on the choice of the stepsize. This can be achieved while still satisfying the conditions of the convergence theorem by choosing values of $\beta$ which are closer to one and values of $\alpha$ closer to zero. For many machine learning applications, a reasonable choice of $\beta$ might be $\beta = 0.9$ and setting $\alpha = 10^{-4}$. For more classical convex optimization problems, however, choosing a smaller value of $\beta$ such as $\beta = 0.1$ may be beneficial (Nocedal and Wright 1999, pp. 38-39).

The following theorem provides sufficient conditions for an open interval on the real number line to exist whose elements are Wolfe stepsizes.

**Theorem 7.1.1** (Wolfe Stepsize Existence). *Let* $\mathbf{x}$ *and* $\mathbf{d}$ *be d-dimensional real constant column vectors. Let* $V : \mathcal{R}^d \rightarrow \mathcal{R}$ *be a continuously differentiable function with a lower bound. Let* $V_{\mathbf{x},\mathbf{d}} : [0, \infty) \rightarrow \mathcal{R}$ *be defined as in (7.7) with respect to* $V$, $\mathbf{x}$, *and* $\mathbf{d}$. *There exists an open interval in* $\mathcal{R}$ *such that every stepsize in that open interval satisfies the strong Wolfe conditions defined by (7.8) and (7.10).*

*Proof.* The following proof follows the discussion in Nocedal and Wright (1999, p. 40, Lemma 3.1).

Since $\alpha dV_{\mathbf{x},\mathbf{d}}(0)/d\gamma < 0$ and $V$ has a lower bound, there exists a $\gamma'$ such that:

$$V_{\mathbf{x},\mathbf{d}}(\gamma') = V_{\mathbf{x},\mathbf{d}}(0) + \alpha\gamma' dV_{\mathbf{x},\mathbf{d}}(0)/d\gamma. \tag{7.12}$$

The sufficient decrease condition in (7.8) then holds for all $\gamma \in [0, \gamma']$ again using the fact that $\alpha dV_{\mathbf{x},\mathbf{d}}(0)/d\gamma < 0$.

Now apply the mean value theorem to $V_{\mathbf{x},\mathbf{d}}$ evaluated at $\gamma'$ to obtain:

$$V_{\mathbf{x},\mathbf{d}}(\gamma') = V_{\mathbf{x},\mathbf{d}}(0) + \gamma'[dV_{\mathbf{x},\mathbf{d}}(\gamma'')/d\gamma] \tag{7.13}$$

for some $\gamma'' \in [0, \gamma']$.

Setting the right-hand side of (7.12) equal to the right-hand side of (7.13) gives

$$V_{\mathbf{x},\mathbf{d}}(0) + \alpha\gamma'[dV_{\mathbf{x},\mathbf{d}}(0)/d\gamma] = V_{\mathbf{x},\mathbf{d}}(0) + \gamma'[dV_{\mathbf{x},\mathbf{d}}(\gamma'')/d\gamma]$$

which implies

$$\alpha dV_{\mathbf{x},\mathbf{d}}(0)/d\gamma = dV_{\mathbf{x},\mathbf{d}}(\gamma'')/d\gamma. \tag{7.14}$$

Since $\beta > \alpha$ and $dV_{\mathbf{x},\mathbf{d}}(0)/d\gamma < 0$, (7.14) implies

$$\beta dV_{\mathbf{x},\mathbf{d}}(0)/d\gamma < dV_{\mathbf{x},\mathbf{d}}(\gamma'')/d\gamma. \tag{7.15}$$

Equation (7.15) thus simultaneously satisfies both (7.8) and (7.9) for some $\gamma^* \in [0, \gamma'']$. In addition, the strong Wolfe conditions (7.8) and (7.10) are satisfied since (7.14) implies that the right-hand side of (7.15) is negative.

Since $V_{\mathbf{x},\mathbf{d}}$ and $dV_{\mathbf{x},\mathbf{d}}/d\gamma$ are continuous functions, it follows that there exists an open interval containing $\gamma^*$ that satisfies the strong Wolfe conditions (7.8) and (7.10) (see Theorem 5.2.3).

■

**Algorithm 7.1.1** Backtracking Wolfe Stepsize Selection

```
1: procedure BACKTRACKSTEPSIZE(x, d, g, V, α, β, ρ, γ_max, K_max)
2:     γ ⇐ γ_max
3:     γ_downhill ⇐ 0
4:     for k = 0 to K_max do
5:         if V_x,d(γ) ≤ V_x,d(0) + αγ dV_x,d(0)/dγ then
6:             if γ_downhill = 0 then γ_downhill ⇐ γ
7:             end if
8:             if |dV_x,d(γ)/dγ| ≤ β |dV_x,d(0)/dγ| then return {γ}
9:             end if
10:        end if
11:        γ ⇐ ργ;        ▷ Since 0 < ρ < 1, this step decreases current stepsize value.
12:    end for
13:    return {γ_downhill}        ▷ Search Failed. Return largest downhill stepsize.
14: end procedure
```

The Backtracking Wolfe Stepsize Selection Algorithm is a simple algorithm which illustrates a practical procedure for selecting a strong Wolfe stepsize. In practice, the number of attempts to find a stepsize must be kept very small because function evaluations are computationally expensive. The variable $K_{max}$ which governs the number of stepsize choices could be as small as $K_{max} = 2$ or as large as $K_{max} = 100$. The maximum stepsize $\gamma_{max}$ is usually set equal to one. A typical choice for $\rho$ would be $0.1 \leq \rho \leq 0.9$

Exercise 7.1-5, and Exercise 7.1-6 describe methods for computing "approximately optimal stepsizes". Such methods can be used to enhance the performance of the Backtracking Wolfe Stepsize Selection algorithm.

## Exercises

7.1-1. *Multiple Learning Rates.* Let $\mathbf{g} : \mathcal{R}^q \rightarrow \mathcal{R}^q$ denote the gradient of an objective function $\hat{\ell}_n(\boldsymbol{\theta})$. Let $\boldsymbol{\eta}$ be a $q$-dimensional vector of positive numbers. Define a descent algorithm with positive stepsize $\gamma(t)$ by the iterated map:

$$\boldsymbol{\theta}(t+1) = \boldsymbol{\theta}(t) - \gamma(t)\boldsymbol{\eta} \odot \mathbf{g}(\boldsymbol{\theta}(t)). \tag{7.16}$$

Show that this iterated map may be interpreted as a descent algorithm that minimizes $\hat{\ell}_n(\boldsymbol{\theta})$ such that each parameter in the learning machine has its own unique learning rate. Define the search direction $\mathbf{d}(t)$ of this descent algorithm and show the search direction of this descent algorithm satisfies the downhill direction for all $t \in \mathbb{N}$.

7.1-2. *Coordinate Gradient Descent.* Let $\mathbf{g} : \mathcal{R}^q \rightarrow \mathcal{R}^q$ denote the gradient of an objective function $\hat{\ell}_n(\boldsymbol{\theta})$. Let $\delta$ be a positive number. Let $\mathbf{u}(t)$ denote a randomly chosen column of a $q$-dimensional identity matrix for each $t \in \mathbb{N}$. Define a descent algorithm with positive stepsize $\gamma(t)$ by the iterated map:

$$\boldsymbol{\theta}(t+1) = \boldsymbol{\theta}(t) + \gamma(t)\mathbf{d}(t) \tag{7.17}$$

where the search direction $\mathbf{d}(t) = -\mathbf{u}(t) \odot \mathbf{g}(\boldsymbol{\theta}(t))$ if $|\mathbf{u}(t)^T\mathbf{g}(\boldsymbol{\theta}(t))| > \delta|\mathbf{g}(\boldsymbol{\theta}(t)|$ and $\mathbf{d}(t) = -\mathbf{g}(\boldsymbol{\theta}(t))$ otherwise. This descent algorithm is designed to update only one parameter at each iteration when the magnitude of the gradient is not small. Show this choice of search direction satisfies the downhill direction for $t \in \mathbb{N}$.

7.1-3. *An Optimal Stepsize is a Wolfe Stepsize.* Let $V_{\mathbf{x},\mathbf{d}} : [0, \infty) \rightarrow \mathcal{R}$ be an objective function projection which is continuously differentiable. If an optimal stepsize $\gamma^*$ exists and is chosen, then it follows that: (1) $V_{\mathbf{x},\mathbf{d}}(\gamma^*) < V_{\mathbf{x},\mathbf{d}}(0)$, and (2) the derivative of $V_{\mathbf{x},\mathbf{d}}$ evaluated at $\gamma^*$ is equal to zero. Show that these two properties of the optimal stepsize $\gamma^*$ imply that $\gamma^*$ is also a Wolfe stepsize.

7.1-4. *Exhaustive Backtracking Stepsize Search.* Modify the backtracking algorithm so that the algorithm takes an initial stepsize $\gamma_{max}$ and uses that to generate $K$ stepsizes $\gamma_1, \gamma_2, \ldots, \gamma_K$ where $\gamma_k = \gamma_{max}/K$. The algorithm then first checks $\gamma_K$, then checks $\gamma_{K-1}$, and so on until it finds the largest stepsize in the set of $K$ stepsizes satisfying both Wolfe conditions.

7.1-5. *Second-Order Accelerated Convergence Rate for Stepsize Search.* Modify the backtracking algorithm so that if the initial choice of the stepsize $\gamma_{max}$ does not satisfy both Wolfe conditions, then the algorithm uses the information from $V_{\mathbf{x},\mathbf{d}}(0)$, $dV_{\mathbf{x},\mathbf{d}}(\gamma_{max})$, $dV_{\mathbf{x},\mathbf{d}}(0)/d\gamma$, and $dV_{\mathbf{x},\mathbf{d}}(\gamma_{max})/d\gamma$ to explicitly compute the next choice of $\gamma$ under the assumption that $V_{\mathbf{x},\mathbf{d}}$ may be approximated using:

$$V_{\mathbf{x},\mathbf{d}}(\gamma) \approx V_{\mathbf{x},\mathbf{d}}(0) + \gamma dV_{\mathbf{x},\mathbf{d}}(\gamma)/d\gamma + (1/2)\gamma^2 Q$$

where $Q$ is a constant to be determined from $V_{\mathbf{x},\mathbf{d}}(0)$, $dV_{\mathbf{x},\mathbf{d}}(\gamma_{max})$, $dV_{\mathbf{x},\mathbf{d}}(0)/d\gamma$, and $dV_{\mathbf{x},\mathbf{d}}(\gamma_{max})/d\gamma$. Write out your new algorithm explicitly in a manner similar to the description of the Backtracking Algorithm.

7.1-6. *Third-Order Accelerated Stepsize Search.* In high-dimensional parameter search problems in complex search spaces, saddlepoints are frequently encountered and objective functions are not convex. Accordingly, extend the second-order Taylor series algorithm you developed in Exercise 7.1-5 so that $V_{\mathbf{x},\mathbf{d}}$ is locally approximated by a third-order Taylor series. This cubic approximation is usually sufficient to handle most problems commonly encountered in practice.

7.1-7. *Backtracking Algorithm Simulation Experiments.* Implement a backtracking algorithm and the gradient descent algorithm as a computer program. The gradient descent algorithm should implement gradient descent on the objective function $V([x_1, x_2]) = \mu_1(x-1)^2 + \mu_2(x-2)^2$. Examine the performance of the backtracking algorithm for different choices of $\rho$, $\alpha$, $\beta$ with $\gamma_{max} = 1$ and $K_{max} = 20$ when $\mu_1 = \mu_2 = 1$. Then, repeat your performance evaluation for the case where $\mu_1 = 10^3$ and $\mu_2 = 10^{-3}$.

## 7.2 Descent Algorithm Convergence Analysis

Given a Wolfe stepsize and a downhill search direction, the Zoutendijk-Wolfe Convergence Theorem can be used to investigate conditions which ensure convergence of the sequence of system states to the set of critical points of the objective function.

**Theorem 7.2.1** (Zoutendijk-Wolfe Convergence Theorem). *Let $T \equiv \mathbb{N}$. Let $V : \mathcal{R}^d \to \mathcal{R}$ be a twice continuously differentiable function with a lower bound on $\mathcal{R}^d$. Let $\mathbf{g} : \mathcal{R}^d \to \mathcal{R}^d$ be the gradient of $V$. Let $\mathbf{H} : \mathcal{R}^d \to \mathcal{R}^d$ be the Hessian of $V$. Assume $\mathbf{H}$ is continuous. Let $\mathbf{d} : T \to \mathcal{R}^d$. Let $\mathbf{d}_t \equiv \mathbf{d}(t)$. For all $t \in T$, let*

$$\mathbf{x}(t+1) = \mathbf{x}(t) + \gamma(t)\mathbf{d}_t \tag{7.18}$$

*where $\gamma(t)$ is a Wolfe stepsize as defined in Definition 7.1.3. Let $\mathbf{g}_t \equiv \mathbf{g}(\mathbf{x}(t))$. Let $\delta$ be a positive number. Assume for each $t \in \mathbb{N}$ either: (i)*

$$\frac{\mathbf{g}_t^T \mathbf{d}_t}{|\mathbf{g}_t|\,|\mathbf{d}_t|} \leq -\delta \tag{7.19}$$

*or (ii)*

$$|\mathbf{d}_t| = 0 \text{ if and only if } |\mathbf{g}_t| = 0. \tag{7.20}$$

*Assume that $\Omega$ is a closed and bounded subset of $\mathcal{R}^d$ such that there exists a positive integer $T$ such that for all $t > T$: $\mathbf{x}(t) \in \Omega$. Then, $\mathbf{x}(1), \mathbf{x}(2), \ldots$ converges to the set of critical points contained in $\Omega$.*

*Proof.* The following analysis is based upon the discussion of Zoutendijk's Lemma by Nocedal and Wright (1999, Theorem 3.2, pp. 43-44).

Using the mean value theorem and (7.18), expand $\mathbf{g}$ about $\mathbf{x}(t)$ and evaluate at $\mathbf{x}(t+1)$ to obtain:

$$\mathbf{g}_{t+1} - \mathbf{g}_t = \gamma(t)\mathbf{H}(\mathbf{u}_t)\mathbf{d}_t \tag{7.21}$$

where $\mathbf{H}$ is the Hessian of $V$ and the $k$th element of $\mathbf{u}_t$ is a point on a chord connecting $\mathbf{x}(t)$ and $\mathbf{x}(t+1)$. Multiply the left-hand side of (7.21) by $\mathbf{d}_t^T$ to obtain:

$$\mathbf{d}_t^T(\mathbf{g}_{t+1} - \mathbf{g}_t) = \gamma(t)\mathbf{d}_t^T\mathbf{H}(\mathbf{u}_t)\mathbf{d}_t. \tag{7.22}$$

Since $\mathbf{H}$ is a continuous function on the closed and bounded subset $\Omega$, there exists a positive number $H_{max}$ such that

$$\mathbf{d}_t^T\mathbf{H}(\mathbf{u}_t)\mathbf{d}_t \leq H_{max}|\mathbf{d}_t|^2. \tag{7.23}$$

Substituting (7.23) into the right-hand side of (7.22) gives:

$$\mathbf{d}_t^T(\mathbf{g}_{t+1} - \mathbf{g}_t) \le \gamma(t) H_{max} |\mathbf{d}_t|^2. \tag{7.24}$$

The second Wolfe condition in (7.9) may be restated as:

$$\beta \mathbf{g}_t^T \mathbf{d}_t \le \mathbf{g}_{t+1}^T \mathbf{d}_t. \tag{7.25}$$

Subtracting $\mathbf{g}_t^T \mathbf{d}_t$ from both sides of (7.25) gives:

$$(\beta - 1)\mathbf{g}_t^T \mathbf{d}_t \le [\mathbf{g}_{t+1} - \mathbf{g}_t]^T \mathbf{d}_t. \tag{7.26}$$

Substitute (7.24) into the right-hand side of (7.26) to obtain:

$$(\beta - 1)\mathbf{g}_t^T \mathbf{d}_t \le \gamma(t) H_{max} |\mathbf{d}_t|^2,$$

which may be rewritten as:

$$\gamma(t) \ge \frac{(\beta - 1)\mathbf{g}_t^T \mathbf{d}_t}{H_{max} |\mathbf{d}_t|^2} \tag{7.27}$$

where $|\mathbf{d}_t| > 0$.

If $|\mathbf{d}_t| = 0$, then (7.20) implies $|\mathbf{g}_t| = 0$ which implies $\mathbf{x}(t)$ is a critical point.

Now substitute (7.27) into the first Wolfe condition (7.8):

$$V(\mathbf{x}(t+1)) - V(\mathbf{x}(t)) \le \left[\alpha(\beta - 1)\mathbf{g}_t^T \mathbf{d}_t / \left(H_{max} |\mathbf{d}_t|^2\right)\right] \mathbf{g}_t^T \mathbf{d}_t. \tag{7.28}$$

Rearrange terms in (7.28) to obtain

$$V(\mathbf{x}(t+1)) - V(\mathbf{x}(t)) \le (\alpha / H_{max})(\beta - 1)|\mathbf{g}_t|^2 (\cos(\psi_t))^2. \tag{7.29}$$

where

$$\cos(\psi_t) = \frac{-\mathbf{g}_t^T \mathbf{d}_t}{|\mathbf{g}_t||\mathbf{d}_t|}.$$

Equation (7.29) and the relations $\cos(\psi_t) > \delta$ and $(\alpha / H_{max})(\beta - 1) \le 0$ imply:

$$V(\mathbf{x}(t+1)) - V(\mathbf{x}(t)) \le (\alpha / H_{max})(\beta - 1)\delta^2 |\mathbf{g}_t|^2. \tag{7.30}$$

Summing $V(\mathbf{x}(t+1)) - V(\mathbf{x}(t))$ over $t = 0, 1, 2, \ldots, M$ gives:

$$V(\mathbf{x}(M)) - V(\mathbf{x}(0)) = \sum_{t=0}^{M} \left[V(\mathbf{x}(t+1)) - V(\mathbf{x}(t))\right]. \tag{7.31}$$

Now substitute (7.30) into (7.31) to obtain:

$$V(\mathbf{x}(M)) - V(\mathbf{x}(0)) \le (\alpha / H_{max})(\beta - 1)\delta^2 \sum_{t=0}^{M} |\mathbf{g}_t|^2,$$

Thus, since $(\alpha / H_{max})(\beta - 1) \le 0$:

$$|V(\mathbf{x}(M)) - V(\mathbf{x}(0))| \ge |(\alpha / H_{max})(\beta - 1)\delta^2| \sum_{t=0}^{M} |\mathbf{g}_t|^2. \tag{7.32}$$

Since $V$ is bounded from below on $\Omega$, this implies that $|V(\mathbf{x}(M)) - V(\mathbf{x}(0))|$ is a finite number as $M \to \infty$. In addition, recall that $|(\alpha/H_{max})(\beta - 1)\delta^2|$ is bounded. Thus, (7.32) implies:

$$\sum_{t=0}^{\infty} |\mathbf{g}_t|^2 < \infty$$

which implies as $t \to \infty$:

$$|\mathbf{g}(\mathbf{x}(t))|^2 \to 0 \tag{7.33}$$

(see Theorem 5.1.5(v)).

Let $S$ be the set of critical points contained in $\Omega$. In order to show that $\mathbf{x}(t) \to S$ as $t \to \infty$, a proof by contradiction will be used. Assume that there exists a subsequence $\mathbf{x}(t_1), \mathbf{x}(t_2), \ldots$ of $\mathbf{x}(0), \mathbf{x}(1), \ldots$ such that $\mathbf{x}(t_k) \not\to S$ as $k \to \infty$. Since $|\mathbf{g}|^2$ is continuous on $\Omega$, it follows that the subsequence $|\mathbf{g}(\mathbf{x}(t_1))|^2, |\mathbf{g}(\mathbf{x}(t_2))|^2, \ldots$ does not converge to zero as $k \to \infty$. But this contradicts the result in (7.33) which states that every subsequence of $|\mathbf{g}(\mathbf{x}(0))|^2, |\mathbf{g}(\mathbf{x}(1))|^2, \ldots$ converges to zero. Therefore, $\mathbf{x}(t) \to S$ as $t \to \infty$. ■

**Example 7.2.1** (Convergence to Critical Points for Convex and Non-convex Cases). The conclusion of the Zoutendijk-Wolfe convergence theorem states that a descent algorithm that satisfies the conditions of the Zoutendijk-Wolfe convergence theorem will converge to the set of critical points in the state space $\Omega$ provided that the descent algorithm's trajectory eventually remains in the closed and bounded set $\Omega$ for all time. In many applications, this constraint can be achieved by constructing $\Omega$ in an appropriate way.

Let $\Omega \equiv \{\mathbf{x} : V(\mathbf{x}) \leq V(\mathbf{x}(0))\}$ where $\mathbf{x}(0)$ is the initial state of a descent algorithm. Then, by the Mapping Criteria for a Continuous Function Theorem (Theorem 5.1.6), the set $\Omega$ is a closed set when $V$ is continuous. In addition, any trajectory initiated in $\Omega$ remains in $\Omega$ for all time since $V$ is non-increasing on any trajectory $\mathbf{x}(0), \mathbf{x}(1), \ldots$ since it is assumed that the Wolfe stepsize conditions hold. Now, using arguments from the proof of the Bounded Trajectory Lemma (Theorem 6.3.3), if $V$ has the property that

$$V(\mathbf{x}) \to +\infty \text{ as } |\mathbf{x}| \to \infty, \tag{7.34}$$

then it follows that $\Omega$ is a bounded set. To see this, suppose that $\Omega$ was not a bounded set. This means that a subsequence $\mathbf{x}(t_1), \mathbf{x}(t_2), \ldots$ exists such that $|\mathbf{x}_{t_k}| \to +\infty$ as $k \to \infty$. But this would imply that $V(\mathbf{x}_{t_k}) \to +\infty$, which violates the assumption that $V(\mathbf{x}_{t_k}) \leq V(\mathbf{x}(0))$ for all $t_1, t_2, \ldots$. If $V$ does not have the property in (7.34), then introducing a regularization term modification to $V$ can be considered (see Example 6.3.5 and Example 7.2.3).

In situations where one is unable to show that all trajectories are asymptotically confined to a closed and bounded region of the state space, the conclusion of the Zoutendijk-Wolfe convergence theorem can simply be modified to state that either a particular trajectory is not asymptotically confined to a particular closed and bounded region of the state space or that trajectory converges to the set of critical points in that region.

The Zoutendijk-Wolfe convergence theorem can also be used to investigate convergence to a particular point in the state space. This is achieved when $\Omega$ is constructed such that it contains exactly one critical point.

Figure 7.5 shows how different choices of the set $\Omega$ can be used to partition the domain of a non-convex function in a variety of ways. The restriction of the function $V$ to $\Omega_1$ is a strictly convex function with exactly one critical point which is a unique global minimizer in $\Omega_1$. On the other hand, $\Omega_2$ contains one critical point corresponding to a global maximizer on $\Omega_2$ and a connected set of critical points forming a flat region. The region $\Omega_3$ contains a local maximizer and local minimizer. △

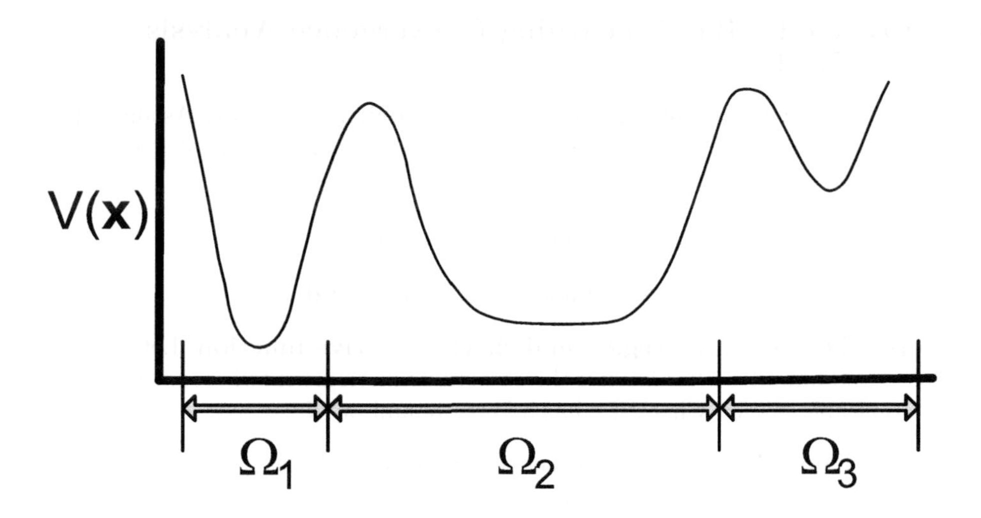


**FIGURE 7.5**
**Example partitions of the domain of a non-convex function.** This non-convex function has a unique global minimizer on $\Omega_1$, a global maximizer and flat on $\Omega_2$, and a global maximizer and global minimizer on $\Omega_3$.

**Example 7.2.2** (Convergence of Gradient Descent Batch Learning). Consider a machine learning algorithm which can be represented as a gradient descent algorithm that minimizes a twice continuously differentiable empirical risk function $\hat{\ell}_n : \mathcal{R}^q \to [0, \infty)$. The gradient descent batch learning algorithm is specified by the learning rule:

$$\boldsymbol{\theta}(t+1) = \boldsymbol{\theta}(t) - \gamma_t \mathbf{g}_t \tag{7.35}$$

where the gradient evaluated at $\boldsymbol{\theta}(t)$ is specified by the formula:

$$\mathbf{g}_t = \frac{d\hat{\ell}_n(\boldsymbol{\theta}(t))}{d\boldsymbol{\theta}}$$

and $\gamma_t$ is a Wolfe stepsize. Analyze the asymptotic behavior of this batch learning algorithm.

**Solution.** The asymptotic behavior is analyzed following the steps in Recipe Box 7.1.

Step 1: Equation (7.35) provides a specification of the learning rule in the standard descent algorithm format.

Step 2: The objective function $\hat{\ell}_n$ is twice continuously differentiable with a lower bound. The set $\Omega$ in this case is chosen to be an arbitrary convex, closed, and bounded set.

Step 3: To check the downhill condition with the search direction $\mathbf{d}_t = -\mathbf{g}_t$, compute

$$\frac{\mathbf{g}_t^T \mathbf{d}_t}{|\mathbf{g}_t| \, |\mathbf{d}_t|} = -1 < 0$$

so the downhill condition is satisfied.

Step 4: The Wolfe conditions are satisfied by assumption.

**Recipe Box 7.1 Batch Learning Convergence Analysis (Theorem 7.2.1)**

- **Step 1: Express algorithm in descent algorithm format.** Assume the initial state of the descent algorithm is $\boldsymbol{\theta}(0) \in \mathcal{R}^q$. The form of a descent algorithm is given for $t = 0, 1, 2, \ldots$:

$$\boldsymbol{\theta}(t+1) = \boldsymbol{\theta}(t) + \gamma(t)\mathbf{d}_t$$

where the stepsize $\gamma(t) \in [0, \infty)$ and descent direction $\mathbf{d} : \mathbb{N} \rightarrow \mathcal{R}^q$.

- **Step 2: Define search region and check objective function.** Let $\Theta$ be a closed and bounded set. Check if the objective function $\ell$ is twice continuously differentiable on $\mathcal{R}^q$. Check if $\ell$ has a lower bound on $\mathcal{R}^q$.

- **Step 3: Check if the descent direction is downhill.** Let $\mathbf{g}_t$ be the gradient of $\ell$ evaluated at $\boldsymbol{\theta}(t)$. Check that

$$\frac{\mathbf{g}_t^T \mathbf{d}_t}{|\mathbf{g}_t| \, |\mathbf{d}_t|} \leq -\delta$$

for some positive number $\delta$ when $|\mathbf{g}_t| > 0$ and $|\mathbf{d}_t| > 0$. In addition, check that $|\mathbf{d}_t| = 0$ if and only if $|\mathbf{g}_t| = 0$.

- **Step 4: Check if the stepsize satisfies the Wolfe conditions.** Check to make sure that all stepsizes satisfy the Wolfe conditions defined by Equation (7.8) and Equation (7.10).

- **Step 5: Conclude trajectories converge to critical points or trajectories are not bounded.** Let $S$ be the set of critical points of $\ell$ in the closed and bounded set $\Theta$. Conclude that if the trajectory $\boldsymbol{\theta}(t) \in \Theta$ for all $t \in \mathbb{N}$, then $\boldsymbol{\theta}(t) \rightarrow S$ as $t \rightarrow \infty$.

Step 5: Conclude that either: (i) the sequence of parameter vectors $\boldsymbol{\theta}(1), \boldsymbol{\theta}(2), \ldots$ generated by the learning algorithm converges to the set of critical points in $\Omega$, or (ii) the trajectory $\boldsymbol{\theta}(1), \boldsymbol{\theta}(2), \ldots$ is not asymptotically confined to $\Omega$ (i.e., it is not the case there exists a positive integer $T$ such that $\boldsymbol{\theta}(t) \in \Omega$ for all $t > T$). △

**Example 7.2.3** (Convergence of Gradient Descent Batch Learning with Regularization). Consider a machine learning algorithm which can be represented as a gradient descent algorithm that minimizes a twice continuously differentiable empirical risk function $\hat{\ell}_n : \mathcal{R}^q \rightarrow [0, \infty)$ which is defined such that:

$$\hat{\ell}_n(\boldsymbol{\theta}) = \ddot{\ell}_n(\boldsymbol{\theta}) + \lambda|\boldsymbol{\theta}|^2$$

where $\ddot{\ell}_n : \mathcal{R}^q \rightarrow [0, \infty)$ is a twice continuously differentiable function with a lower bound and $\lambda$ is a positive number. The gradient descent algorithm batch learning algorithm is specified by the learning rule:

$$\boldsymbol{\theta}(t+1) = \boldsymbol{\theta}(t) - \gamma_t \mathbf{g}_t \tag{7.36}$$

where the gradient evaluated at $\boldsymbol{\theta}(t)$ is specified by the formula:

$$\mathbf{g}_t \equiv \frac{d\hat{\ell}_n(\boldsymbol{\theta}(t))}{d\boldsymbol{\theta}}$$

and $\gamma_t$ is a Wolfe stepsize. Analyze the asymptotic behavior of this batch learning algorithm.

**Solution.** The asymptotic behavior is analyzed with exactly the same analysis as in Example 7.2.2. Thus, one concludes that either: (i) the sequence of parameter vectors $\boldsymbol{\theta}(1), \boldsymbol{\theta}(2), \ldots$ generated by the learning algorithm converges to the set of critical points in a region of the parameter space denoted by $\Omega_0$, or (ii) the trajectory $\boldsymbol{\theta}(1), \boldsymbol{\theta}(2), \ldots$ is not asymptotically confined to $\Omega_0$. However, in this example $\hat{\ell}_n$ has the additional known property that $|\boldsymbol{\theta}| \to \infty$ implies that $\hat{\ell}_n(\boldsymbol{\theta}) \to +\infty$. Now define $\Omega_0 \equiv \{\boldsymbol{\theta} : \hat{\ell}_n(\boldsymbol{\theta}) \leq \hat{\ell}_n(\boldsymbol{\theta}(0))\}$ where $\boldsymbol{\theta}(0) \in \mathcal{R}^q$. It then follows from the Bounded Trajectory Lemma 6.3.3, that $\boldsymbol{\theta}(1), \boldsymbol{\theta}(2), \ldots$ and its limit points must be asymptotically confined to $\Omega_0$ (see Bounded Trajectory Lemma 6.3.3). Thus, one obtains the stronger conclusion that $\boldsymbol{\theta}(t)$ converges to the set of critical points of $\hat{\ell}_n$ in $\Omega_0$ as $t \to \infty$. This result holds for complicated smooth non-convex objective functions with multiple minima, maxima, flats, and saddlepoints. △

**Example 7.2.4** (Convergence of Modified Gradient Descent Batch Learning). Let $\mathbf{g} : \mathcal{R}^q \to \mathcal{R}^q$ denote the gradient of an objective function $\hat{\ell}_n : \mathcal{R}^q \to \mathcal{R}$. Let $\mathbf{M}_1, \mathbf{M}_2, \ldots$ be a sequence of symmetric positive definite matrices such that there exists two positive numbers $\lambda_{min}$ and $\lambda_{max}$ such that for all $k \in \mathbb{N}$ and for all $\boldsymbol{\theta} \in \mathcal{R}^q$:

$$\lambda_{min}|\boldsymbol{\theta}|^2 \leq \boldsymbol{\theta}^T \mathbf{M}_k \boldsymbol{\theta} \leq \lambda_{max}|\boldsymbol{\theta}|^2.$$

Define a modified gradient descent batch learning rule which generates an updated parameter vector $\boldsymbol{\theta}(t+1)$ from the current parameter vector $\boldsymbol{\theta}(t)$, and Wolfe stepsize $\gamma(t)$ using the iterative formula

$$\boldsymbol{\theta}(t+1) = \boldsymbol{\theta}(t) - \gamma(t)\mathbf{M}_t \mathbf{g}(\boldsymbol{\theta}(t)).$$

Analyze the asymptotic behavior of this batch learning algorithm.

**Solution.** The asymptotic behavior is analyzed following the steps in Recipe Box 7.1. All Steps except for Step 3 are identical to the solution in Example 7.2.2. Let $\mathbf{g}_t \equiv \mathbf{g}(\boldsymbol{\theta}(t))$. Let $\mathbf{d}_t = -\mathbf{M}_t \mathbf{g}_t$. To verify the conditions associated with Step 3, note that the downhill condition:

$$\frac{\mathbf{g}_t^T \mathbf{d}_t}{|\mathbf{g}_t|\,|\mathbf{d}_t|} = \frac{-\mathbf{g}_t^T \mathbf{M}_t \mathbf{g}_t}{|\mathbf{g}_t)|\,|\mathbf{M}\mathbf{g}_t|} \leq \frac{-\lambda_{min}|\mathbf{g}_t|^2}{|\mathbf{g}_t|\,\sqrt{\mathbf{g}_t^T \mathbf{M}_t^2 \mathbf{g}_t}} \leq \frac{-\lambda_{min}|\mathbf{g}_t|^2}{|\mathbf{g}_t|\sqrt{\lambda_{max}^2|\mathbf{g}_t|^2}} = \frac{-\lambda_{min}}{\lambda_{max}} < 0.$$

△

**Example 7.2.5** (Convergence of Linear Regression Batch Gradient Descent Learning). Let a data set $\mathcal{D}_n$ be defined such that

$$\mathcal{D}_n = \{(\mathbf{s}_1, y_1), \ldots, (\mathbf{s}_n, y_n)\}$$

where $y_i$ is the desired response of the learning machine when the $d$-dimensional input pattern vector $\mathbf{s}_i$ is presented, $i = 1, \ldots, n$. The predicted response of the learning machine is $\ddot{y}(\mathbf{s}_i, \boldsymbol{\theta}) = \boldsymbol{\theta}^T \mathbf{s}_i$ for a given input pattern $\mathbf{s}_i$ and parameter vector $\boldsymbol{\theta}$. The goal of the learning process is to find a parameter vector $\hat{\boldsymbol{\theta}}_n$ which minimizes the empirical risk function

$$\hat{\ell}_n(\boldsymbol{\theta}) = (1/n)\sum_{i=1}^{n}(y_i - \ddot{y}(\mathbf{s}_i, \boldsymbol{\theta}))^2.$$

Design a gradient descent learning algorithm that converges to the set of global minimizers of this empirical risk function.

**Solution.** The analysis in Recipe Box 7.1 is followed.

Step 1: Let

$$\boldsymbol{\theta}(t+1) = \boldsymbol{\theta}(t) - \gamma_t \mathbf{g}_t$$

where $\gamma_t$ is a Wolfe stepsize and

$$\mathbf{g}_t = -(2/n)\sum_{i=1}^{n}(y_i - \ddot{y}(\mathbf{s}_i, \boldsymbol{\theta}(t))\mathbf{s}_i.$$

Step 2: Let $\Omega \equiv \{\boldsymbol{\theta} : \hat{\ell}_n(\boldsymbol{\theta}) \leq \hat{\ell}_n(\boldsymbol{\theta}(0))\}$. The set $\Omega$ is a closed set since $\hat{\ell}_n$ is continuous. Also since $\gamma_t$ is a Wolfe stepsize, $\hat{\ell}_n$ is non-increasing on the trajectories and so all trajectories are contained in $\Omega$. Note that $\hat{\ell}_n(\boldsymbol{\theta}) \to +\infty$ when $|\boldsymbol{\theta}| \to \infty$, which implies by Theorem 6.3.3 that the set $\Omega$ which contains all trajectories is also a bounded set. In addition, $\hat{\ell}_n$ has a lower bound of zero.

Step 3: The search direction is downhill since the search direction is a gradient descent direction so that $\mathbf{g}_t^T \mathbf{d}_t < \delta|\mathbf{g}_t||\mathbf{d}_t|$ where $\mathbf{d}_t = -\mathbf{g}_t$ for some positive number $\delta$.

Step 4: The Wolfe conditions are satisfied by assumption.

Step 5: The Hessian of the objective function evaluated at $\boldsymbol{\theta}(t)$ is given by the formula:

$$\mathbf{H}_t = (2/n)\sum_{i=1}^{n}\mathbf{s}_i\mathbf{s}_i^T.$$

The set $\Omega$ is also a convex set. Since $\mathbf{H}_t$ is a positive semidefinite matrix, this implies that the objective function is a convex function on the convex set $\Omega$, so every critical point of the objective function is a global minimizer. Thus, all trajectories initiated in $\Omega$ will converge to the set of global minimizers in $\Omega$. Furthermore, if $\mathbf{H}_t$ is positive definite on $\Omega$ and a critical point exists in the interior of $\Omega$, then all trajectories will converge to that critical point which will be the unique global minimizer of $V$ on $\Omega$. △

## Exercises

7.2-1. *Behavior of gradient descent algorithm with multiple parameter-specific learning rates.* Exercise 7.1-1 describes a gradient descent algorithm where each parameter has its own learning rate. Provide conditions using the Zoutendijk-Wolfe convergence theorem that ensure all trajectories converge to the set of critical points of the objective function.

7.2-2. *Behavior of coordinate gradient descent algorithm.* Provide conditions that ensure the coordinate gradient descent algorithm in Exercise 7.1-2 converges to a set $S$ which contains the critical points of the objective function.

## 7.3 Descent Strategies

### 7.3.1 Gradient and Steepest Descent

Algorithm 7.3.1 provides an example of a practical gradient descent algorithm which includes a stopping criterion based upon checking if the infinity norm of the gradient is less than some positive number $\epsilon$ and terminating after some maximum number of iterations $T_{max}$.

In practice, $\epsilon$ is a small number (e.g., $\epsilon = 10^{-15}$ or $\epsilon = 10^{-5}$). The maximum number of iterations $T_{max}$ should be chosen very large so that the algorithm is designed to terminate when the infinity norm of the gradient is less than $\epsilon$.

**Algorithm 7.3.1** Batch Gradient Descent Algorithm

1: **procedure** BATCHGRADIENTDESCENT($\mathbf{x}(0)$, $\epsilon$, $T_{max}$)
2: $t \Leftarrow 0$
3: **repeat**
4: $\mathbf{g}_t \Leftarrow \nabla V(\mathbf{x}(t))$
5: Compute Wolfe stepsize $\gamma_t$ from $\mathbf{x}(t)$ and $-\mathbf{g}_t$
6: $\mathbf{x}(t+1) \Leftarrow \mathbf{x}(t) - \gamma_t \mathbf{g}_t$
7: $t \Leftarrow t+1$
8: **until** $|\mathbf{g}_t|_\infty < \epsilon$ or $t = T_{max}$
9: **return** $\{\mathbf{x}(t), |\mathbf{g}_t|_\infty\}$
10: **end procedure**

A gradient descent algorithm will be called a *steepest descent* algorithm when the stepsize is always chosen to be optimal. It can be shown that the convergence rate of a steepest descent algorithm with an optimal stepsize in the vicinity of a minimizer of $V$, $\mathbf{x}^*$, has a linear convergence rate with an upper bound on the convergence ratio which increases as the condition number of the Hessian at $\mathbf{x}^*$ becomes larger (e.g., Luenberger 1984).

**Example 7.3.1** (Batch Gradient Descent for Nonlinear Regression). Let a training data set $\mathcal{D}_n \equiv \{(\mathbf{s}_1, y_1), \ldots, (\mathbf{s}_n, y_n)\}$ be defined such that $y_k$ is the desired response for a given input pattern $\mathbf{s}_k$. Let the notation $\boldsymbol{\mathcal{S}}$ be a vector-valued function defined such that the $k$th element of $\boldsymbol{\mathcal{S}}(\boldsymbol{\phi})$ is $(1+\exp(-\phi_k))^{-1}$ where $\phi_k$ is the $k$th element of $\boldsymbol{\phi}$. Consider a nonlinear regression modeling problem where the desired response $\ddot{y}$ is defined such that for each input pattern $\mathbf{s}$ and parameter vector $\boldsymbol{\theta}^T \equiv [\mathbf{w}^T, \mathbf{vec}(\mathbf{V}^T)^T]$:

$$\ddot{y}(\mathbf{s}, \boldsymbol{\theta}) = \mathcal{S}\left(\mathbf{w}^T \boldsymbol{\mathcal{S}}(\mathbf{V}\mathbf{s})\right)$$

where $\mathcal{S}(\phi) = (1+\exp(-\phi))^{-1}$. The goal of the learning process is to minimize the objective function:

$$\hat{\ell}_n(\boldsymbol{\theta}) = (1/n)\sum_{i=1}^{n}(y_i - \ddot{y}(\mathbf{s}_i; \boldsymbol{\theta}))^2 + \lambda|\boldsymbol{\theta}|^2$$

where $\lambda$ is a positive number.

A researcher proposes a batch gradient descent algorithm for minimizing $\hat{\ell}_n$ defined such that:

$$\boldsymbol{\theta}(t+1) = \boldsymbol{\theta}(t) - \gamma_t \mathbf{g}_n(\boldsymbol{\theta}(t))$$

where $\gamma_t$ is a Wolfe stepsize and $\mathbf{g}_n$ is the gradient of $\hat{\ell}_n$. Show that all trajectories converge to the set of critical points of $\hat{\ell}_n$.

**Solution.** Define the set

$$\Omega \equiv \left\{ \boldsymbol{\theta} \in \mathcal{R}^q : \hat{\ell}_n(\boldsymbol{\theta}) \leq \hat{\ell}_n(\boldsymbol{\theta}(0)) \right\}.$$

Since $\gamma_t$ is a Wolfe stepsize, $\hat{\ell}_n(\boldsymbol{\theta}(t+1)) \leq \hat{\ell}_n(\boldsymbol{\theta}(t))$, which implies that $\boldsymbol{\theta}(t) \in \Omega$ for all $t \geq 0$. In addition, $\Omega$ is a closed set since $\hat{\ell}_n$ is a continuous function by the Open Set Criteria for a Continuous Function Theorem 5.1.6. And, in addition, since $\hat{\ell}_n(\boldsymbol{\theta}) \to +\infty$ as $|\boldsymbol{\theta}| \to \infty$, it follows (see Bounded Trajectory Theorem 6.3.3) that $\Omega$ is a bounded set. Since $\hat{\ell}_n$ is twice continuously differentiable with a lower bound, the gradient descent direction $-\mathbf{g}_n$ is downhill since $-\mathbf{g}_n^T \mathbf{g}_n \leq \delta |\mathbf{g}_n|^2$ for $\delta = 1$ when $|\mathbf{g}_n| > 0$, and since the stepsize is a Wolfe stepsize, the conditions of the Zoutendijk-Wolfe Convergence Theorem hold. Thus, all trajectories converge to the set of critical points in $\Omega$. △

## 7.3.2 Newton-Type Descent

### 7.3.2.1 Newton-Raphson Algorithm

The Newton-Raphson algorithm (also called the Newton algorithm) works by choosing a search direction that forces the next system state to be a critical point of a local quadratic approximation to the objective function $V$. Let $\mathbf{g} : \mathcal{R}^d \to \mathcal{R}^d$ and $\mathbf{H} : \mathcal{R}^d \to \mathcal{R}^{d \times d}$ denote respectively the gradient and Hessian of the objective function $V$. Now expand $\mathbf{g}$ in a first-order Taylor expansion about the current state $\mathbf{x}(t)$ and evaluating at the next state $\mathbf{x}(t+1) = \mathbf{x}(t) + \gamma(t)\mathbf{d}(t)$ gives the relation:

$$\mathbf{g}(\mathbf{x}(t+1)) = \mathbf{g}(\mathbf{x}(t)) + \mathbf{H}(\mathbf{x}(t))(\mathbf{x}(t+1) - \mathbf{x}(t)) + O(\gamma(t)^2).$$

Using the relation $\gamma(t)\mathbf{d}(t) = \mathbf{x}(t+1) - \mathbf{x}(t)$:

$$\mathbf{g}(\mathbf{x}(t+1)) = \mathbf{g}(\mathbf{x}(t)) + \gamma(t)\mathbf{H}(\mathbf{x}(t))\mathbf{d}(t) + O(\gamma(t)^2). \tag{7.37}$$

It is desired to choose a search direction $\mathbf{d}(t)$ such that $\mathbf{g}(\mathbf{x}(t+1)) = \mathbf{0}_d$ since a necessary condition for a minimizer is that the gradient of $V$ vanishes.

Substituting $\mathbf{g}(\mathbf{x}(t+1)) = \mathbf{0}_d$ into (7.37) gives the relation:

$$\mathbf{0}_d = \mathbf{g}(\mathbf{x}(t)) + \gamma(t)\mathbf{H}(\mathbf{x}(t))\mathbf{d}(t) + O(\gamma(t)^2),$$

and solving for $\gamma(t)\mathbf{d}(t)$ gives the *Newton search direction:*

$$\gamma(t)\mathbf{d}(t) = -[\mathbf{H}(\mathbf{x}(t))]^{-1}\mathbf{g}(\mathbf{x}(t)) + O(\gamma(t)^2).$$

Note that the matrix inverse $[\mathbf{H}(\mathbf{x}(t))]^{-1}$ exists in a sufficiently small neighborhood of a strict local minimizer of $V$. Moreover, given that $\mathbf{x}(t)$ is sufficiently close to a strict local minimizer, $\mathbf{x}^*$, all eigenvalues of $[\mathbf{H}(\mathbf{x}(t))]^{-1}$, $H_{min}(t)$, will be finite positive numbers. Thus, the Zoutendijk-Wolfe convergence theorem can be applied (see Example 7.2.4) to show that $\mathbf{x}(t)$ converges to $\mathbf{x}^*$ when $\mathbf{x}(t)$ is sufficiently close to a strict local minimizer $\mathbf{x}^*$.

It can be shown that the Newton-Raphson algorithm has a quadratic convergence rate (e.g., Luenberger, 1984) which is a much faster convergence rate than the linear convergence rate of the steepest descent algorithm. This fast convergence rate is a very attractive feature of this algorithm. However, the Newton-Raphson algorithm requires more computations per iteration than the steepest descent algorithm since the Newton-Raphson algorithm requires both the storage and inversion of a $d$-dimensional matrix at each algorithm iteration. Moreover, if the objective function $V$ has local maximizers, the Newton-Raphson algorithm may use a search direction that points uphill instead of downhill! Furthermore, the matrix inverse $[\mathbf{H}(\mathbf{x}(t))]^{-1}$ is not guaranteed to exist when $\mathbf{x}(t)$ is not near a strict local minimizer. Finally, it should be emphasized that such convergence rate analyses are only applicable when the trajectory of the descent algorithm is in the vicinity of a strict local minimizer and can be misleading when the trajectory is not near a strict local minimizer.

#### 7.3.2.2 Levenberg-Marquardt Algorithm

A compromise between the classical Newton and steepest descent algorithm is the class of Levenberg-Marquardt descent algorithms. Let $\mathbf{g}_t$ and $\mathbf{H}_t$ be the gradient and Hessian of $V$ evaluated at $\mathbf{x}(t)$ respectively. The Levenberg-Marquardt descent algorithm has the form:

$$\mathbf{x}(t+1) = \mathbf{x}(t) + \gamma_t \mathbf{d}_t$$

where

$$\mathbf{d}_t = -(\mu_t \mathbf{I}_d + \mathbf{H}_t)^{-1} \mathbf{g}_t.$$

Notice that if $\mu_t = 0$, then $\mathbf{d}_t = -(\mathbf{H}_t)^{-1}\mathbf{g}_t$ corresponding to a Newton-Raphson search direction. On the other hand, if $\mu_t$ is a large positive number, then $\mathbf{d}_t \approx -(1/\mu_t)\mathbf{g}_t$ which is a gradient descent search direction.

The approach of Example 7.2.4 may be used to analyze the asymptotic behavior of the Levenberg-Marquardt algorithm.

**Algorithm 7.3.2** Levenberg-Marquardt Batch Learning Algorithm

1: **procedure** LEVENBERGMARQUARDTBATCHLEARNING($\mathbf{x}(0)$, $\delta$, $\epsilon$, $D_{max}$, $T_{max}$)
2: $\quad t \Leftarrow 0$
3: $\quad$ **repeat**
4: $\quad\quad \mathbf{g}_t \Leftarrow \nabla V(\mathbf{x}(t))$
5: $\quad\quad \mathbf{H}_t \Leftarrow \nabla^2 V(\mathbf{x}(t))$
6: $\quad\quad$ Let $\lambda_{min}(t)$ be the smallest eigenvalue of $\mathbf{H}_t$.
7: $\quad\quad$ **if** $\lambda_{min}(t) < \delta$ **then**
8: $\quad\quad\quad \mathbf{d}_t \Leftarrow -\left[(\delta - \lambda_{min}(t))\mathbf{I} + \mathbf{H}_t\right]^{-1} \mathbf{g}_t$
9: $\quad\quad$ **else**
10: $\quad\quad\quad \mathbf{d}_t \Leftarrow -\left[\mathbf{H}_t\right]^{-1} \mathbf{g}_t$
11: $\quad\quad$ **end if**
12: $\quad\quad$ **if** $|\mathbf{d}_t| > D_{max}$ **then**
13: $\quad\quad\quad \mathbf{d}_t \Leftarrow -\mathbf{g}_t$
14: $\quad\quad$ **end if**
15: $\quad\quad$ Compute Wolfe stepsize $\gamma_t$ from $\mathbf{x}(t)$ and $\mathbf{d}_t$
16: $\quad\quad \mathbf{x}(t+1) \Leftarrow \mathbf{x}(t) + \gamma_t \mathbf{d}_t$
17: $\quad\quad t \Leftarrow t+1$
18: $\quad$ **until** $|\mathbf{g}_t|_\infty < \epsilon$ or $t = T_{max}$
19: $\quad$ **return** $\{\mathbf{x}(t), |\mathbf{g}_t|_\infty\}$
20: **end procedure**

For complicated nonlinear objective functions, it makes sense to choose $\mu_t$ to be large when the algorithm is far away from a strict local minimizer and choose $\mu_t$ to be small in the vicinity of a strict local minimizer. Thus, the Levenberg-Marquardt algorithm, like the Newton-Raphson algorithm, has a quadratic convergence rate.

An important advantage of the Levenberg-Marquardt descent algorithm over the Newton-Raphson algorithm is that it does not generate "uphill" search directions for nonlinear objective functions with minima and maxima. Another important advantage of the

Levenberg-Marquardt search direction over the Newton-Raphson algorithm is that the matrix inverse is always defined.

The specific implementation of the Levenberg-Marquardt Algorithm (see Algorithm 7.3.2) should be compared with the specific implementation of the Gradient Descent Algorithm (see Algorithm 7.3.1). The constant $\mu$ is a small positive number (e.g., $\mu = 10^{-5}$).

For linear regression and logistic regression machine learning problems which are associated with convex objective functions, the Newton-Raphson algorithm is an attractive alternative. In situations, however, where the objective function is not convex and multiple local minimizers or multiple saddlepoints may exist the Newton-Raphson may have undesirable convergence properties if it is not applied judiciously.

**Example 7.3.2** (Logistic Regression Parameter Estimation)**.** Let $y_i \in \{0, 1\}$ be the desired response for input pattern vector $\mathbf{s}_i \in \mathcal{R}^d$ for $i = 1, \ldots, n$. Let $\mathbf{u}_i^T = [\mathbf{s}_i^T \; 1]$ for a given input pattern vector $\mathbf{s}_i$ for $i = 1, \ldots, n$. The predicted response

$$\ddot{y}(\mathbf{s}, \boldsymbol{\theta}) = \mathcal{S}\left(\boldsymbol{\theta}^T \mathbf{u}_i\right)$$

where $\mathcal{S}(\phi) = (1 + \exp(-\phi))^{-1}$.

The goal of the learning process is to minimize the empirical risk objective function

$$\hat{\ell}_n(\boldsymbol{\theta}) = -(1/n) \sum_{i=1}^{n} \left[ y_i \log \ddot{y}(\mathbf{s}_i; \boldsymbol{\theta}) + (1 - y_i) \log(1 - \ddot{y}(\mathbf{s}_i; \boldsymbol{\theta})) \right].$$

The gradient of $\hat{\ell}_n$, $\mathbf{g}_n$, is given by the formula:

$$\mathbf{g}_n(\boldsymbol{\theta}) = -(1/n) \sum_{i=1}^{n} \left[ y_i - \ddot{y}(\mathbf{s}_i; \boldsymbol{\theta}) \right] \mathbf{u}_i^T.$$

The Hessian of $\hat{\ell}_n$, $\mathbf{H}_n$, is given by the formula:

$$\mathbf{H}_n(\boldsymbol{\theta}) = (1/n) \sum_{i=1}^{n} \ddot{y}\,(\mathbf{s}_i; \boldsymbol{\theta})(1 - \ddot{y}(\mathbf{s}_i; \boldsymbol{\theta}))\, \mathbf{u}_i \mathbf{u}_i^T.$$

These formulas for the gradient and Hessian are substituted into Algorithm 7.3.2 in order to obtain a method for estimating parameters of a logistic regression model. Finally, note that since for every $\boldsymbol{\theta} \in \mathcal{R}^{d+1}$ that:

$$\boldsymbol{\theta}^T \mathbf{H}_n(\boldsymbol{\theta}) \boldsymbol{\theta} \geq 0$$

this implies that the objective function is convex on the parameter space by Theorem 5.3.5. △

### 7.3.3 L-BFGS and Conjugate Gradient Descent Methods

In this section, we introduce an advanced algorithm called the L-BFGS (Limited memory Broyden-Fletcher-Goldfarb-Shanno) algorithm that possesses a number of highly desirable features. First, unlike the Newton-Raphson and Levenberg-Marquardt algorithms, the L-BFGS algorithm does not have the computational requirements and memory requirements involved in storing or inverting a Hessian matrix. Second, the descent direction for the memoryless BFGS is always guaranteed to go downhill for nonlinear objective functions.

And third, the L-BFGS possesses a superlinear convergence rate in the vicinity of a strict local minimizer (e.g., Luenberger, 1984; Nocedal and Wright, 1999).

Intuitively, the L-BFGS works by computing a particular weighted sum of the gradient descent search direction, the previous search direction, and the previous gradient descent search direction. The underlying principle of L-BFGS is similar to a momentum type search direction as discussed in Example 7.1.2 yet the computational requirements are similar to a gradient descent algorithm. After $M$ iterations, the L-BFGS takes a gradient descent step and then over the next $M$ iterations, weighted sums of the current gradient descent direction with previous descent directions are computed again. It can be shown that after each inner cycle of $M$ iterations has been completed, the resulting search direction of the L-BFGS algorithm is a close approximation to the search direction for a Newton-Raphson algorithm. The number of inner iterations $M$ is always chosen to be less than or equal to the dimension of the system state $\mathbf{x}(t)$.

**Algorithm 7.3.3** L-BFGS Batch Learning Algorithm

1: **procedure** L-BFGS($\mathbf{x}(0)$, $\delta$, $\epsilon$, $T_{max}$, $M$)
2: $t \Leftarrow 0$
3: **repeat**
4: **if** $MOD(t, m) = 0$ **then** ▷ $MOD(t, m)$ is remainder of $t \div M$
5: $\mathbf{g}_t \Leftarrow \nabla V(\mathbf{x}(t))$
6: $\mathbf{d}_t \Leftarrow -\mathbf{g}_t$ ▷ Gradient Descent Direction
7: **else**
8: $\mathbf{g}_{t-1} \Leftarrow \mathbf{g}_t$
9: $\mathbf{g}_t \Leftarrow \nabla V(\mathbf{x}(t))$
10: $\mathbf{d}_{t-1} \Leftarrow \mathbf{d}_t$
11: $\gamma_{t-1} \Leftarrow \gamma_t$
12: $\mathbf{u}_t \Leftarrow \mathbf{g}_t - \mathbf{g}_{t-1}$
13: $a_t \Leftarrow \frac{\mathbf{d}_{t-1}^T \mathbf{g}_t}{\mathbf{d}_{t-1}^T \mathbf{u}_t}$
14: $b_t \Leftarrow \frac{\mathbf{u}_t^T \mathbf{g}_t}{\mathbf{d}_{t-1}^T \mathbf{u}_t}$
15: $c_t \Leftarrow \gamma_{t-1} + \frac{|\mathbf{u}_t|^2}{\mathbf{d}_{t-1}^T \mathbf{u}_t}$
16: $\mathbf{d}_t \Leftarrow -\mathbf{g}_t + a_t \mathbf{u}_t + (b_t - a_t c_t)\mathbf{d}_{t-1}$ ▷ L-BFGS Descent Direction
17: **end if**
18: **if** $\mathbf{g}_t^T \mathbf{d}_t > -\delta |\mathbf{g}_t| \, |\mathbf{d}_t|$ **then** ▷ Reset if Search Direction Not Downhill
19: $\mathbf{d}_t \Leftarrow -\mathbf{g}_t$
20: **end if**
21: Compute Wolfe stepsize $\gamma_t$ from $\mathbf{x}(t)$ and $\mathbf{d}_t$
22: $\mathbf{x}(t+1) \Leftarrow \mathbf{x}(t) + \gamma_t \mathbf{d}_t$
23: $t \Leftarrow t + 1$
24: **until** $|\mathbf{g}_t|_\infty < \epsilon$ or $t = T_{max}$
25: **return** $\{\mathbf{x}(t), |\mathbf{g}_t|_\infty\}$
26: **end procedure**

Exercise 7.3-6 shows how the L-BFGS Algorithm (Algorithm 7.3.3) can be analyzed using the Zoutendijk-Wolfe convergence theorem.

By definition, an optimal stepsize is a unique global minimizer and therefore is also a critical point, therefore the optimal stepsize must satisfy the condition that $dV_{\mathbf{x}_{t-1},\mathbf{d}_{t-1}}/d\gamma = 0$. Since $dV_{\mathbf{x}_{t-1},\mathbf{d}_{t-1}}/d\gamma = \mathbf{g}_t^T \mathbf{d}_{t-1} = 0$ this implies that $a_t = 0$. Substituting the formula $a_t = 0$ into the L-BFGS Algorithm (Algorithm 7.3.3) results in a *conjugate gradient descent algorithm* specified by the descent direction

$$\mathbf{d}_t = -\mathbf{g}_t + b_t \mathbf{d}_{t-1}.$$

Since this conjugate gradient descent derivation assumes an optimal rather than a Wolfe stepsize, this suggests that L-BFGS methods may be more robust than conjugate gradient descent methods in practice.

## Exercises

7.3-1. *Block Coordinate Gradient Descent.* In some situations small changes in one group of parameter values result in large changes in other parameter values. In other situations, the objective function is not convex on the entire parameter vector but is convex only on a subset of the parameters. In such cases, a very helpful strategy to facilitate learning is to update only a subset of the parameters of the learning machine at a time. For example, in a multilayer feedforward perceptron network, one strategy for learning is to update the connections for only one layer in the network at a time. Optimization methods of this type are called "block coordinate descent methods" in the nonlinear optimization literature. State and prove a theorem that characterizes the asymptotic behavior of descent algorithms that exploit block coordinate descent methods using the Zoutendijk-Wolfe convergence theorem.

7.3-2. *Batch Natural Gradient Descent Algorithm Behavior.* Let $c : \mathcal{R}^d \times \mathcal{R}^q \rightarrow \mathcal{R}$ be a loss function. Let the training data be a set of $n$ $d$-dimensional training vectors $\mathbf{x}_1, \ldots, \mathbf{x}_n$. Let $\ell_n(\boldsymbol{\theta}) = (1/n) \sum_{i=1}^{n} c(\mathbf{x}_i, \boldsymbol{\theta})$. Let $\mathbf{g}_i(\boldsymbol{\theta}) = (dc(\mathbf{x}_i, \boldsymbol{\theta})/d\boldsymbol{\theta})^T$. Let $\bar{\mathbf{g}}_n(\boldsymbol{\theta}) = [d\ell_n/d\boldsymbol{\theta}]^T$. Let

$$\mathbf{B}_n(\boldsymbol{\theta}) = (1/n) \sum_{i=1}^{n} \mathbf{g}_i(\boldsymbol{\theta}) \mathbf{g}_i(\boldsymbol{\theta})^T.$$

A version of the *batch natural gradient descent algorithm* (e.g., Amari 1998; Le Roux et al. 2008) intended to minimize the objective function $\hat{\ell}_n$ is then defined by the descent algorithm:

$$\boldsymbol{\theta}(k+1) = \boldsymbol{\theta}(k) - \gamma_k (\epsilon \mathbf{I}_q + \mathbf{B}_n(\boldsymbol{\theta}(k))^{-1} \bar{\mathbf{g}}_n(\boldsymbol{\theta}(k)),$$

where $\gamma_k$ is a positive stepsize, $\epsilon$ is a small positive number, and there exists a finite number $B_{max}$ such that $|\mathbf{B}_n(\boldsymbol{\theta}(k))| < B_{max}$ for all $k \in \mathbb{N}$. Analyze the asymptotic behavior of the batch natural gradient descent algorithm using the Zoutendijk-Wolfe Convergence Theorem by referring to Example 7.2.4.

7.3-3. *RMSPROP Algorithm Descent Algorithm Behavior.* In this exercise, the asymptotic behavior of a version of the RMSPROP algorithm (Goodfellow et al. 2016, Chapter 8) is investigated. Let $c : \mathcal{R}^d \times \mathcal{R}^q \rightarrow \mathcal{R}$ be a loss function. Let the training data be a set of $n$ $d$-dimensional training vectors $\mathbf{x}_1, \ldots, \mathbf{x}_n$. Let

$$\ell_n(\boldsymbol{\theta}) = (1/n) \sum_{i=1}^{n} c(\mathbf{x}_i, \boldsymbol{\theta}).$$

Let $\mathbf{g}_i(\boldsymbol{\theta}) = (dc(\mathbf{x}_i, \boldsymbol{\theta})/d\boldsymbol{\theta})^T$. Let $\bar{\mathbf{g}}_n(\boldsymbol{\theta}) = [d\ell_n/d\boldsymbol{\theta}]^T$. Let $\mathbf{r}(0) = \mathbf{0}_q$. Let $0 \leq \rho \leq 1$. Let $[x_1, \ldots, x_q]^{-1/2} \equiv [(x_1)^{-1/2}, \ldots, (x_q)^{-1/2}]$. Then, for $k = 1, 2, \ldots$:

$$\boldsymbol{\theta}(k+1) = \boldsymbol{\theta}(k) - \gamma_k \left(\epsilon \mathbf{1}_q + \mathbf{r}(k)\right)^{-1/2} \odot \bar{\mathbf{g}}_n(\boldsymbol{\theta}(k)).$$

where $\gamma_k$ is a positive stepsize and

$$\mathbf{r}(k) = \rho \mathbf{r}(k-1) + (1-\rho)\left[\bar{\mathbf{g}}_n(\boldsymbol{\theta}(k)) \odot \bar{\mathbf{g}}_n(\boldsymbol{\theta}(k))\right].$$

Analyze the asymptotic behavior of the RMSPROP algorithm using the Zoutendijk-Wolfe Convergence Theorem by referring to Example 7.2.4. In addition, explain how the RMSPROP algorithm is related to the special case of the batch natural gradient descent algorithm described in Exercise 7.3-2 for the special case where the off-diagonal elements of $\mathbf{B}_n$ are constrained to be zero.

7.3-4. *Levenberg-Marquadt Descent Algorithm Behavior.* State and prove a theorem that the Levenberg-Marquardt Algorithm (Algorithm 7.3.2) converges when $\delta = 10^{-5}$, $\epsilon = \infty$ and $T_{max} = \infty$. Use the Zoutendijk-Wolfe Convergence Theorem in this chapter to state and prove your theorem.

7.3-5. *Simulation Experiments with Search Direction Choices.* Implement the backtracking algorithm and either the L-BFGS algorithm or Levenberg-Marquardt algorithm as a computer program. Select one of the gradient descent algorithms used to implement either an unsupervised learning, supervised learning, or temporal reinforcement batch machine learning algorithm implemented as a batch learning algorithm in Chapter 5 and evaluate how the performance of learning improves when the gradient descent algorithm is replaced with either Conjugate Gradient, L-BFGS, or Levenberg-Marquardt and supplemented with the backtracking algorithm.

7.3-6. *Convergence of the L-BFGS Algorithm.* State and prove a theorem that the L-BFGS Algorithm converges when $\epsilon = 0$ and $T_{max} = \infty$. Use the Zoutendijk-Wolfe Convergence Theorem in this chapter to state and prove your theorem using the following procedure. First show that the search direction of the L-BFGS algorithm $\mathbf{d}_t$ in Algorithm 7.3.3 may be rewritten as $\mathbf{d}_t = -\mathbf{H}_t^{\dagger}\mathbf{g}_t$ when $\mathbf{d}_{t-1}^T\mathbf{u}_t > 0$ where:

$$\mathbf{H}_t^{\dagger} = \mathbf{W}_t^T\mathbf{W}_t + \eta_t \mathbf{d}_{t-1}\mathbf{d}_{t-1}^T, \tag{7.38}$$

$$\mathbf{W}_t = \mathbf{I} - \frac{\mathbf{u}_t\mathbf{d}_{t-1}^T}{\mathbf{d}_{t-1}^T\mathbf{u}_t}, \tag{7.39}$$

and

$$\eta_t = \frac{\gamma_{t-1}}{\mathbf{d}_{t-1}^T\mathbf{u}_t}. \tag{7.40}$$

Second, show that the matrix $\mathbf{H}_t^{\dagger}$ is a symmetric positive semidefinite matrix which has the property that when the curvature condition on the Wolfe stepsize $\gamma_t$ holds that $\mathbf{g}_t^T\mathbf{H}_t^{\dagger}\mathbf{g}_t > 0$.

Further discussion of the L-BFGS algorithm may be found in Luenberger (1984, p. 280) or Nocedal and Wright (1999, pp. 224, 228).

## 7.4 Further Readings

The Zoutendijk-Wolfe convergence theorem presented in this chapter is a straightforward extension of the well-known convergence theorem referred to as *Zoutendijk's Lemma* (Zoutendijk 1970) and the convergence theorems described by Wolfe (1969, 1971). Discussions of Zoutendijk's Lemma can be found in Dennis and Schanbel (1996), Nocedal and Wright (1999), and Bertsekas (1996). Both Wolfe conditions are discussed in Wolfe (1969, 1971) as well as Luenberger (1984), Bertsekas (1996), Nocedal and Wright (1999), and Dennis and Schnabel (1996). Luenberger (1984), Bertsekas (1996), Nocedal and Wright (1999), and Dennis and Schnabel (1996) also provide useful detailed discussions of convergence rate issues associated with steepest descent, Newton-Raphson, and BFGS methods that were not explored in this chapter.

# Part III

# Stochastic Learning Machines

# 8

# *Random Vectors and Random Functions*

**Learning Objectives**

- Define a mixed random vector as a measurable function.
- Use mixed random vector Radon-Nikodým probability densities.
- Compute expectations of random functions.
- Compute concentration inequality error bounds.

Learning in an environment characterized by uncertainty is a fundamental characteristic of all machine learning algorithms. Probability theory and statistics provide important tools for building mathematical models of statistical environments. Furthermore, such environments often are characterized as large vectors of random variables.

**Example 8.0.1** (Probability Mass Function). An example of a discrete random variable is the Bernoulli random variable $\tilde{y}$ which takes on the value of one with probability $p$ and takes on the value of zero with probability $1-p$ where $p \in [0,1]$. The probability mass function $P: \{0,1\} \rightarrow [0,1]$ for $\tilde{y}$ is defined such that $P(1) = p$ and $P(0) = 1-p$.

Now consider the case where the learning machine's statistical environment is modeled by assuming that at each learning trial, the statistical environment randomly samples with replacement from the set $S \equiv \{\mathbf{x}^1, \ldots, \mathbf{x}^M\}$. This situation can be interpreted as the learning machine observing a different value of a discrete random vector $\tilde{\mathbf{x}}$ at each learning trial where the probability mass function $P: S \rightarrow [0,1]$ is defined such that $P(\mathbf{x}^k) = (1/M)$ for all $k = 1, \ldots, M$. △

**Example 8.0.2** (Probability Density Function (Absolutely Continuous Variable)). An example of an absolutely continuous random variable is a Gaussian random variable $\tilde{x}$ with mean $\mu$ and variance $\sigma^2$. The probability that $\tilde{x}$ takes on a value in a subset, $\Omega$, of $\mathcal{R}$ is computed using a formula such as:

$$P(\tilde{x} \in \Omega) = \int_{x \in \Omega} p(x)dx$$

where the Gaussian probability density function $p: \mathcal{R} \rightarrow [0, \infty)$ for $\tilde{x}$ is defined such that for all $x \in \mathcal{R}$:

$$p(x) = (\sigma\sqrt{2\pi})^{-1} \exp\left(\frac{-(x-m)^2}{2\sigma^2}\right).$$

△

In many important machine learning applications, the probability distribution of a random vector which combines both discrete random variables and absolutely continuous random variables must be specified. Such random vectors are called "mixed random vectors".

For example, consider a speech processing problem where portions of a continuous-time signal are to be labeled. The continuous-time signal might be represented as a realization of a sequence of absolutely continuous correlated Gaussian random variables while the labeling of the continuous-time signal might be represented as a single discrete random variable which has a finite number of possible values. A vector containing both discrete random variables and absolutely continuous random variables is an example of a mixed random vector.

Another example of a mixed random vector arises frequently in linear and nonlinear regression modeling. Consider a linear prediction machine with two predictors $s_1$ and $s_2$. The predictor $s_1$ can take on the value of zero or one to indicate the patient's gender. The predictor $s_2$ is a real number indicating the patient's temperature. A linear regression model is used to compute a health score $y = \theta_1 s_1 + \theta_2 s_2$ for the patient where $\boldsymbol{\theta} \equiv [\theta_1, \theta_2]$ is the parameter vector for the learning machine. The data generating process which generates values for $s_1$ is modeled as a Bernoulli discrete random variable. The data generating process which generates values for $s_2$ is modeled as a Gaussian continuous random variable. The health score $y$ is a realization of a mixed random variable that is neither discrete or absolutely continuous.

Mixed random vectors, however, cannot be represented as either probability mass functions of the type discussed in Example 8.0.1 or probability density functions of the type discussed in Example 8.0.2.

A more general type of probability density function called the "Radon-Nikodým density" is required for the purpose of stating and proving theorems about discrete, absolutely continuous, and mixed random vectors. In this chapter, some essential mathematical machinery is introduced so that the concept of a Radon-Nikodým density can be discussed. The Radon-Nikodým density concept will then be used to define a large family of random vectors and random functions which are commonly encountered in engineering practice.

## 8.1 Probability Spaces

### 8.1.1 Sigma-Fields

**Definition 8.1.1** (Sigma-Field). Let $\Omega$ be a set of objects. Let $\mathcal{F}$ be a collection of subsets of $\Omega$. Then, $\mathcal{F}$ is called a *sigma-field* on $\Omega$ if:

1. $\Omega \in \mathcal{F}$,
2. $F \in \mathcal{F}$ implies $\neg F \in \mathcal{F}$ for all $F \in \mathcal{F}$,
3. $\cup_{i=1}^{\infty} F_i \in \mathcal{F}$ for every sequence $F_1, F_2, \ldots$ in $\mathcal{F}$.

Let $\mathcal{A}$ be a non-empty collection of subsets of $\Omega$. The smallest sigma-field containing $\mathcal{A}$ is called the *sigma-field generated by* $\mathcal{A}$. □

For example, let $\Omega = \{2, 19\}$. The set $\mathcal{F}_1 = \{\{\}, \Omega\}$ is a sigma-field. However, the set $\mathcal{F}_2 \equiv \{\{\}, \{2\}, \{19\}\}$ is not a sigma-field because it violates the condition that the complement of each element of a sigma-field must also be an element of that sigma-field. The sigma-field generated by $\mathcal{F}_1$ is simply the set $\mathcal{F}_1$. The sigma-field generated by $\mathcal{F}_2$ is the set $\mathcal{F}_2 \cup \{\Omega\}$.

Let $\Omega \equiv \{0,1\}^d$ be defined as the finite set consisting of all binary feature vectors. Let $\mathcal{F}$ be the power set of $\Omega$ so $\mathcal{F}$ includes both the empty set and $\Omega$. Since $\mathcal{F}$ is closed under the union operation, it follows that $\Omega$ is a sigma-field.

Note that a sigma-field is closed under the operations of intersection ($\cap$), union ($\cup$), and complementation ($\neg$). Thus, any set theoretic expression formed from the elements of a sigma-field is also a member of that sigma-field. If A and B are members of a sigma-field, it follows that $A \cup B$, $A \cap B$, $\neg A \cup B$, $A \cap \neg B$, $\neg A \cup \neg B$ and so on are also members of that same sigma-field. In addition, recall from Chapter 2 that each of these sets in the sigma-field may be interpreted as a logical assertion as well.

**Example 8.1.1** (Using a Sigma-Field as a Concept Model). Let $\Omega$ be the set of all possible dogs. Let $L$ be the set of Labrador Retriever dogs. The set $L$ is a proper subset of $\Omega$. In addition, all singleton sets which are proper subsets of $L$ correspond to different instances of the Labrador Retriever dog concept. The concept of this Labrador Retriever dog Fido instance is represented by the singleton set $\{d\}$ where $\{d\} \subset L \subset \Omega$.

Similarly, other pure breeds of dogs can be defined in this manner. For example, the set of Chow Chow dogs $C$, the set of Beagle dogs $B$, and the set of Terrier dogs $T$ are all subsets of $\Omega$. In addition, the set $C \cup B \cup T$ is a subset of the set of all pure-breed dogs. The set of mixed-breed dogs, $M$, is the complement of the set of pure-breed dogs. The set of puppy dogs less than 1 year old $P$ is also a subset of $\Omega$. The set of Labrador Retriever puppy dogs which are less than 1 year old is specified by the set $P \cap L$. Thus, by defining new concepts from unions, intersections, and complements, of other concepts, a concept heterarchy can be constructed in this way.

More generally, assume each element of $\Omega$ is a model of the state of the world at a particular physical location and at a particular instant in time. Then, an "event" in the world which takes place over some space-time region can be modeled as a subset of $\Omega$. In addition, unions, intersections, and complements of collections of events can be constructed to model a variety of events taking place in a space-time continuum. Thus, the elements of the sigma-field in this example may be interpreted as concepts. △

In many machine learning applications, different states of the worlds are represented as feature vectors which are elements of $\mathcal{R}^d$. Therefore, it is desirable to consider sigma-fields whose elements are subsets of $\mathcal{R}^d$.

**Definition 8.1.2** (Borel Sigma-Field). A sigma-field generated by the collection of open subsets of $\mathcal{R}^d$ is called the *Borel sigma-field.* A sigma-field generated by the collection of open subsets of $[-\infty, +\infty]^d$ is called the *extended Borel sigma-field.* □

The notation $\mathcal{B}^d$ is used to denote the Borel sigma-field generated from the collection of open subsets of $\mathcal{R}^d$. Thus, $\mathcal{B}^d$ includes the union and intersection of every possible open and closed set in $\mathcal{R}^d$. For engineering purposes, the Borel sigma-field includes every subset of $\mathcal{R}^d$ encountered in engineering practice. Probabilities cannot be assigned to some elements in the power set of $\mathcal{R}^d$ without causing paradoxical difficulties such as the Banach-Tarski Paradox (see Wapner 2005). Such difficulties are resolved by using the Borel sigma-field to specify subsets of $\mathcal{R}^d$ relevant in engineering practice.

**Definition 8.1.3** (Measurable Space). Let $\mathcal{F}$ be a sigma-field on $\Omega$. The ordered pair $(\Omega, \mathcal{F})$ is called a *measurable space.* □

The measurable space $(\mathcal{R}^d, \mathcal{B}^d)$ is called a *Borel-measurable space.*

### 8.1.2 Measures

**Definition 8.1.4** (Countably Additive Measure). Let $(\Omega, \mathcal{F})$ be a measurable space. A non-negative function $\nu : \mathcal{F} \to [0, \infty]$ is a *countably additive measure* on $(\Omega, \mathcal{F})$ if:

1. $\nu(\{\}) = 0$,
2. $\nu(F) \geq 0$ for all $F \in \mathcal{F}$,
3. $\nu\left(\cup_{i=1}^{\infty} F_i\right) = \sum_{i=1}^{\infty} \nu(F_i)$ for any disjoint sequence $F_1, F_2, \ldots$ in $\mathcal{F}$.

A *measure space* is the triplet $(\Omega, \mathcal{F}, \nu)$. □

Because countably additive measures are widely used in both integration and probability theory applications, the terminology "measure" is frequently used as a short-hand notation instead of the countably additive measure.

**Example 8.1.2** (Mathematical Model for Assigning Degrees of Belief to Concepts). Let a $d$-dimensional binary feature vector, $\mathbf{x}$, be defined such that the $k$th element of the feature vector $\mathbf{x}$, $x_k$, is equal to one if proposition $k$ is true and assume $x_k = 0$ when the truth-value of proposition $k$ is false. Assume the feature vector $\mathbf{x}$ has the semantic interpretation of a mathematical model of a possible state of the world. The set of all possible world states is $\Omega \equiv \{0,1\}^d$ and contains $2^d$ binary vectors. Let $\mathcal{F}$ denote the power set of $\Omega$ which consists of $2^{2^d}$ sets. The set $F \in \mathcal{F}$ is interpreted as a concept defined by the assertion that all elements of $F$ are instances of the concept $F$. The set $\mathcal{F}$ is the set of all possible concepts. A countably additive measure $\nu : \mathcal{F} \rightarrow [0,1]$ on the measurable space $(\Omega, \mathcal{F})$ is semantically interpreted as the *degree of belief* that an intelligent agent assigns to the concept $F \in \mathcal{F}$. △

**Definition 8.1.5** (Probability Measure). A *probability measure* $P$ on a measurable space $(\Omega, \mathcal{F})$ is a countably additive measure $P : \mathcal{F} \rightarrow [0,1]$ such that $P(\Omega) = 1$. □

**Definition 8.1.6** (Probability Space). A *probability space* is a measure space $(\Omega, \mathcal{F}, P)$ defined such that $P : \mathcal{F} \rightarrow [0,1]$ is a probability measure. An element of $\Omega$ is called a *sample point*. An element of $\mathcal{F}$ is called an *event*. □

An outcome of a coin tossing experiment is either `heads` or `tails`. Let $\Omega \equiv \{\texttt{heads}, \texttt{tails}\}$. The power set of $\Omega$ is $\mathcal{F} = \{\{\}, \{\texttt{heads}\}, \{\texttt{tails}\}, \{\texttt{heads}, \texttt{tails}\}\}$. A probability measure is proposed to model the probability of an outcome of the coin tossing experiment under the assumption that the coin toss is fair. To do this, define the probability measure $P : \mathcal{F} \rightarrow [0,1]$ such that $P(\{\}) = 0$, $P(\{\texttt{heads}\}) = 1/2$, $P(\{\texttt{tails}\}) = 1/2$, and $P(\{\texttt{heads}, \texttt{tails}\}) = 1$. The function $P$ in this example is a probability measure on the measurable space $(\Omega, \mathcal{F})$. The triplet $(\Omega, \mathcal{F}, P)$ is a probability space.

An important example of a probability space is $(\mathcal{R}^d, \mathcal{B}^d, P)$ which assigns probabilities to subsets of $\mathcal{R}^d$ in $\mathcal{B}^d$ using $P$.

**Example 8.1.3** (Constructing a Probability Measure). Let $\Omega \equiv \{\mathbf{x}_1, \ldots, \mathbf{x}_M\}$ where $\mathbf{x}_k \in \mathcal{R}^d$ for $k = 1, \ldots, M$. Let the probability mass function $P : \Omega \rightarrow (0,1]$ be defined such that $P(\mathbf{x}_k) = p_k$ for $k = 1, \ldots, M$. Let $\mathcal{F}$ be the power set of $\Omega$.

(i) Show how to construct a probability measure $\mu$ on the measurable space $(\Omega, \mathcal{F})$ such that $\mu$ represents the probability mass function $P$.

(ii) Show how to construct a probability measure $\nu$ on the Borel Measurable space $(\mathcal{R}^d, \mathcal{B}^d)$ such that $\nu$ represents the probability mass function $P$.

**Solution.**

(i) Let $\mathcal{F}$ be the power set of $\Omega$. Let $\mu : \mathcal{F} \rightarrow [0,1]$ be defined such that for every $F \in \mathcal{F}$:

$$\mu(F) = \sum_{\mathbf{x} \in F} P(\mathbf{x}).$$

(ii) Let $\nu : \mathcal{B}^d \to [0,1]$ be defined such that for every $B \in \mathcal{B}^d$:

$$\nu(B) = \sum_{\mathbf{x} \in B \cap \Omega} P(\mathbf{x}).$$

△

**Definition 8.1.7** (Lebesgue Measure). A *Lebesgue measure* is a countably additive measure, $\nu : \mathcal{B}^d \to [0,\infty]$, on the Borel measurable space $(\mathcal{R}, \mathcal{B})$ which is defined such that: $\nu([a,b]) = b - a$ for each $[a,b] \in \mathcal{B}$. □

If $S_1 = \{x \in \mathcal{R} : 1 < x < 4\}$, then the Lebesgue measure assigned to $S_1$ is equal to 3. If $S_2 = \{2\}$, then the Lebesgue measure assigned to $S_2$ is equal to 0. If $S_3 = \{2, 42\}$, then the Lebesgue measure assigned to $S_3$ is equal to 0. If $S_4 = \{x \in \mathcal{R} : x > 5\}$, then the Lebesgue measure assigned to $S_4$ is equal to $+\infty$.

**Definition 8.1.8** (Counting Measure). A *counting measure* is a countably additive measure, $\nu : \mathcal{F} \to [0,\infty]$, on the measurable space $(\Omega, \mathcal{F})$ defined such that for each $F \in \mathcal{F}$: (1) $\nu(F)$ is the number of elements in $F$ when $F$ is a finite set, and (2) $\nu(F) = +\infty$ when $F$ is not a finite set. □

If $S_1 = \{x \in \mathcal{R} : 1 < x < 4\}$, then the counting measure assigned to $S_1$ is equal to $+\infty$. If $S_2 = \{2\}$, then the counting measure assigned to $S_2$ is equal to 1. If $S_3 = \{2, 42\}$, then the counting measure assigned to $S_3$ is equal to 2.

In many situations, it will be necessary to assume that the measure space has the special property of being a complete measure space. Less formally, a measure space is complete if defined such that if a subset $S$ is assigned the measure zero in the original measure space then all subsets of $S$ are also assigned a measure of zero.

**Definition 8.1.9** (Complete Measure Space). Let $(\Omega, \mathcal{F}, \nu)$ be a measure space. Let $\mathcal{G}_F$ consist of all subsets of each $F \in \mathcal{F}$ when $F$ satisfies $\nu(F) = 0$. Let $\bar{\mathcal{F}}$ be the sigma-field generated by the union of $\mathcal{F}$ and $\mathcal{G}_F$. Let $\bar{\nu} : \bar{\mathcal{F}} \to [0,\infty]$ be defined such that for all $F \in \mathcal{F}$: $\bar{\nu}(G) = 0$ for all $G \in \mathcal{G}_F$ and $\bar{\nu}(F) = \nu(F)$. Then, the measure space $(\Omega, \bar{\mathcal{F}}, \bar{\nu})$ is called a *complete measure space*. In addition, $(\Omega, \bar{\mathcal{F}}, \bar{\nu})$ is called the *completion* of $(\Omega, \mathcal{F}, \nu)$. □

**Example 8.1.4** (Constructing a Complete Measure Space).

$$\text{Let } \Omega = \{1,2,3,4,5,6\}.$$

Let the sigma-field

$$\mathcal{F} \equiv \{\{\}, \Omega, \{1,2\}, \{3,4,5,6\}\}.$$

Let $\nu : \mathcal{F} \to [0,\infty]$ be defined such that $\nu(\{\}) = 0$, $\nu(\Omega) = 1$, $\nu(\{3,4,5,6\}) = 0.5$, and $\nu(\{1,2\}) = 0$. The measure space $(\Omega, \mathcal{F}, \nu)$ is not complete because $\{1\}$ and $\{2\}$ are not elements of $\mathcal{F}$. The completion of $(\Omega, \mathcal{F}, \nu)$, $(\Omega, \bar{\mathcal{F}}, \bar{\nu})$, is obtained by choosing $\bar{\mathcal{F}} \equiv \mathcal{F} \cup \{1\} \cup \{2\}$, $\bar{\nu}(F) = \nu(F)$ for all $F \in \mathcal{F}$, and $\bar{\nu}(G) = 0$ for all $G \in \{\{1\},\{2\}\}$. △

**Definition 8.1.10** (Sigma-Finite Measure). Let $(\Omega, \mathcal{F}, \nu)$ be a measure space. Assume there exists a sequence of sets $F_1, F_2, \ldots$ in $\mathcal{F}$ such that $\cup_{k=1}^{\infty} F_k = \Omega$ where $\nu(F_k) < \infty$ for all $k \in \mathbb{N}^+$, then $\nu : \mathcal{F} \to [0,\infty]$ is called a *sigma-finite measure*. □

For example, if a partition of $\Omega$ exists that consists of a countable set of elements in $\mathcal{F}$ and $\nu$ assigns a finite number to each element of the partition, then this means that $\nu$ is a sigma-finite measure. Probability measures and Lebesgue measures are examples of sigma-finite measures on $(\Omega, \mathcal{F})$ since for each of these measures there exists a way of partitioning $\Omega$ into a collection of sets such that the measure assigned to each set is a finite number.

The counting measure is also a sigma-finite measure $\nu$ when it is defined on a measurable space $(\Omega, \mathcal{F})$ where $\Omega$ is a countable set because one can choose a partition with a countable number of elements such that each element is a set containing exactly one element of $\Omega$. However, the counting measure $\nu$ is not a sigma-finite measure when it is defined on the Borel sigma field $(\mathcal{R}, \mathcal{B})$ because for any sequence $B_1, B_2, \ldots$ in $\mathcal{B}$ such that $\cup_{i=1}^{\infty} B_i = \mathcal{R}$ there will always exist an element of $B_1, B_2, \ldots$ that has an infinite number of elements.

However, suppose that $\mu$ is a counting measure on $(\Omega, \mathcal{F})$ where $\Omega$ is a countable set and $\mathcal{F}$ is the power set of $\mathcal{F}$. One can construct a sigma-finite measure $\nu$ on $(\mathcal{R}, \mathcal{B})$ such that $\nu(B) = \mu(B \cap \Omega)$ for each $B \in \mathcal{B}$. Thus, $\nu$ counts the number of points in a given subset, $B$, of $\mathcal{R}$ which are elements of $\Omega$ where $B \in \mathcal{B}$.

## Exercises

8.1-1. Which of the following sets is a measurable space?
(i) $(\{1,2\}, \{\{\}, \{1\}, \{1,2\}\})$,
(ii) $(\{1,2\}, \{\{\}, \{1\}, \{2\}, \{1,2\}\})$,
(iii) $(\{1,2\}, \{\{1\}, \{2\}\})$.

8.1-2. What is the sigma-field for a binary random variable, $\tilde{x}$, which takes on the values of 0 and 1?

8.1-3. Let $S \equiv \{x : 2 \leq x \leq 7\}$. If $\nu$ is a Lebesgue measure on the measurable space $(\mathcal{R}, \mathcal{B})$, what is $\nu(S)$?

8.1-4. Let $S \equiv \{x : 2 \leq x \leq 7\}$. If $\nu$ is a counting measure on the measurable space $(\mathcal{R}, \mathcal{B})$, what is $\nu(S)$?

8.1-5. Let $S \equiv \{1, 2, 3\}$ and $G \equiv \{\{\}, S, \{1\}, \{2, 3\}\}$. Which of the following is a measure space constructed from the measurable space $(S, G)$?
(i) $(S, G, \nu)$ where $\nu(\{x\}) = 1$ for all $\mathbf{x} \in S$.
(ii) $(S, G, \nu)$ where $\nu(\{\}) = 0$, $\nu(\{1\}) = 6$, $\nu(\{2,3\}) = 1$, and $\nu(S) = 7$.
(iii) $(S, G, \nu)$ where $\nu(\{\}) = 0$, $\nu(\{1\}) = 6$, $\nu(\{2,3\}) = 1$, and $\nu(S) = 8$.

8.1-6. Let $\Omega$ be a finite set. Explain why the counting measure on the measurable space $(\Omega, \mathcal{F})$ is a sigma-finite measure. Explain why the counting measure on the Borel measurable space $(\mathcal{R}, \mathcal{B})$ is not a sigma-finite measure. Give examples supporting your explanations.

8.1-7. Let $(S, G)$ be a measurable space where $S \equiv \{1, 2, 3\}$ and $G \equiv \{\{\}, S, \{1\}, \{2, 3\}\}$. Let $\nu : G \to [0, \infty]$ be a measure on $(S, G)$ defined such that $\nu(\{2, 3\}) = 0$. Construct the measure space which is the completion of $(S, G, \nu)$.

## 8.2 Random Vectors

### 8.2.1 Measurable Functions

**Definition 8.2.1** ($(\mathcal{F}_X, \mathcal{F}_Y)$-Measurable Function). Let $(\Omega_X, \mathcal{F}_X)$ and $(\Omega_Y, \mathcal{F}_Y)$ be two measurable spaces. The function $f : \Omega_X \to \Omega_Y$ is called a $(\mathcal{F}_X, \mathcal{F}_Y)$-*measurable function* if for every $E \in \mathcal{F}_Y$: $\{x \in \Omega_X : f(x) \in E\} \in \mathcal{F}_X$. □

If $f$ is an $(\mathcal{F}_X, \mathcal{F}_Y)$-measurable function then it is sometimes convenient to refer to $f$ using the short-hand terminology *measurable function*.

Here is an equivalent way of stating the definition of a measurable function. Let $(\Omega_X, \mathcal{F}_X)$ and $(\Omega_Y, \mathcal{F}_Y)$ be two measurable spaces. A function $f : \Omega_X \rightarrow \Omega_Y$ is called a $(\mathcal{F}_X, \mathcal{F}_Y)$-measurable function provided that the preimage (see Definition 2.3.2) under $f$ of every element of $\mathcal{F}_Y$ is also an element of $\mathcal{F}_X$ (see Figure 2.6).

**Definition 8.2.2** (Borel Measurable Function)**.** Let $(\Omega_X, \mathcal{F}_X)$ be a measurable space. Let $(\mathcal{R}^d, \mathcal{B}^d)$ be a Borel measurable space. A $(\mathcal{F}_X, \mathcal{B}^d)$-measurable function $\mathbf{f} : \Omega_X \rightarrow \mathcal{R}^d$ is called a *$\mathcal{F}_X$-measurable function.* If, in addition, $\Omega_X \equiv \mathcal{R}^k$ and $\mathcal{F}_X \equiv \mathcal{B}^k$, then $\mathbf{f}$ is called a *Borel-measurable* or *$\mathcal{B}^k$-measurable* function. □

From an engineering perspective, a Borel measurable function $\mathbf{f} : \mathcal{R}^k \rightarrow \mathcal{R}^d$ is a measurement device that maps a state of the physical world represented as a $k$-dimensional vector into a $d$-dimensional vector of measurements. This function, however, can be equivalently viewed as mapping events in the measurable space $(\mathcal{R}^k, \mathcal{B}^k)$ into events in the measurable space $(\mathcal{R}^d, \mathcal{B}^d)$. Note that it follows directly from the definition of a Borel measurable function that if both $\mathbf{f} : \mathcal{R}^k \rightarrow \mathcal{R}^d$ and $\mathbf{g} : \mathcal{R}^d \rightarrow \mathcal{R}^m$ are Borel measurable functions, then $\mathbf{g}(\mathbf{f})$ is a Borel measurable function. The following theorems are useful, in engineering practice, for determining if a function is Borel measurable.

**Theorem 8.2.1** (Borel Measurable Function Composition)**.** *(i) If $f$ and $q$ are scalar-valued Borel measurable functions, then $f + q$, $fq$, $|f|$ are Borel measurable functions. (ii) If the sequence of Borel measurable functions $h_1, h_2, \ldots$ converges to $h$, then $h$ is Borel measurable. (iii) If $h_1, h_2, \ldots$ is a sequence of Borel measurable functions, then* $\inf\{h_1, h_2, \ldots\}$ *and* $\liminf\{h_1, h_2, \ldots\}$ *are Borel measurable functions. (iv) If $\mathbf{f} : \mathcal{R}^d \rightarrow \mathcal{R}^k$ and $\mathbf{q} : \mathcal{R}^k \rightarrow \mathcal{R}^m$ are Borel measurable functions, then $\mathbf{q}(\mathbf{f})$ is a Borel measurable function.*

*Proof.* (i) See Lemma 2.6 of Bartle (1966, p. 12). (ii) See Corollary 2.10 of Bartle (1966, p. 12). (iii) See Lemma 2.9 of Bartle (1966, p. 12). (iv) See Exercise 8.2-2. ■

Piecewise continuous functions are often encountered in engineering practice. Such functions are Borel measurable.

**Theorem 8.2.2** (Piecewise Continuous Functions Are Borel Measurable)**.** *(i) A continuous function $f : \mathcal{R}^d \rightarrow \mathcal{R}$ is a Borel measurable function. (ii) A piecewise continuous function $f : \mathcal{R}^d \rightarrow \mathcal{R}$ is a Borel measurable function.*

*Proof.* Conclusion (i) is an immediate consequence of the Continuous Function Set Mapping Property (Theorem 5.1.6) and the Definition of a Borel Measurable Function (Definition 8.2.2).

Conclusion (ii) is an immediate consequence of Conclusion (i), the Definition of a Piecewise Continuous Function on a Finite Partition (Definition 5.1.15), and the Borel measurable Function Composition Theorem (Theorem 8.2.1). ■

For example, suppose that $S$ is a temperature measurement. A learning machine is designed such that it does not process $S$ directly but instead processes a nonlinear transformation of $S$ denoted by $\phi(S)$ where $\phi(S) = 1$ if $S \geq 100$ and $\phi(S) = 0$ if $S < 100$. The function $\phi$ is not continuous but it is a Borel measurable function.

**Definition 8.2.3** (Random Vector)**.** Let $(\Omega, \mathcal{F}, P)$ be a probability space. A function $\tilde{\mathbf{x}} : \Omega \rightarrow \mathcal{R}^d$ is called a *random vector* if it is an $\mathcal{F}$-measurable function. The vector $\mathbf{x} \equiv \tilde{\mathbf{x}}(\omega)$ for some $\omega \in \Omega$ is called a *realization* of $\tilde{\mathbf{x}}$. □

The notation $\tilde{\mathbf{x}}$ or the notation $\hat{\mathbf{x}}$ will be used to denote a random vector, while a realization of $\tilde{\mathbf{x}}$ or $\hat{\mathbf{x}}$ is simply represented by $\mathbf{x}$.

For example, let $(\Omega, \mathcal{F}, P)$ be a probability space. Let $\tilde{x} : \Omega \rightarrow \{0, 1\}$ be a function defined such that $\tilde{x}(\omega) = 1$ if $\omega \in F$ and $\tilde{x}(\omega) = 0$ if $\omega \notin F$ for some $F \in \mathcal{F}$. Then, the function $\tilde{x}$ is a random variable. If $\tilde{x}$ is evaluated at $\omega$ where $\omega \in \Omega$, then $\tilde{x}(\omega)$ is a realization of $\tilde{x}$ and may be denoted as $x$.

In practice, random vectors are usually defined on an underlying probability space $(\Omega, \mathcal{F}, \mathbb{F})$ where $\mathbb{F}$ is a probability measure defined such that $\mathbb{F}(F)$ is either the *observed frequency* or *expected frequency* that event $F$ occurs in the environment for each $F \in \mathcal{F}$.

The concept of a random vector is illustrated by Figure 1.2 which describes a function which maps a sample point in the environment into a sample point in the feature measurement space. Different events in the environment occur with different expected frequencies and a random vector simply maps a particular event in the environment into an event in feature space. Thus, a random vector is a deterministic measurement device where the randomness in the environment space essentially propagates through this deterministic device to induce randomness in the feature space.

**Definition 8.2.4** (Probability Measure for a Random Vector). Let $(\Omega, \mathcal{F}, P)$ be a probability space. Let $\tilde{\mathbf{x}} : \Omega \rightarrow \mathcal{R}^d$ be a random vector. Let $P_x : \mathcal{B}^d \rightarrow [0, 1]$ be defined such that for all $B \in \mathcal{B}^d$:

$$P_x(B) = P\left(\{\omega \in \mathcal{F} : \tilde{\mathbf{x}}(\omega) \in B\}\right).$$

The function $P_x$ is called the *probability measure for the random vector* $\tilde{\mathbf{x}}$. □

In practice, it is convenient to generate the Borel sigma-field using hyperrectangles in $\mathcal{R}^d$.

**Definition 8.2.5** (Cumulative Distribution Function for a Random Vector). Let $P_x : \mathcal{B}^d \rightarrow [0, 1]$ be the probability measure for the random vector $\tilde{\mathbf{x}}$. Let $F_{\mathbf{x}} : \mathcal{R}^d \rightarrow [0, 1]$ be defined such that for all $\mathbf{y} \equiv [y_1, \ldots, y_d] \in \mathcal{R}^d$:

$$F_{\mathbf{x}}(\mathbf{y}) = P_x(\{\tilde{x}_1 < y_1, \ldots, \tilde{x}_d < y_d\}).$$

The function $F_{\mathbf{x}}$ is called the *cumulative distribution function* or *probability distribution* for $\tilde{\mathbf{x}}$. □

**Definition 8.2.6** (Support for a Random Vector or Probability Measure). Let $P : \mathcal{B}^d \rightarrow [0, 1]$ be the probability measure for the random vector $\tilde{\mathbf{x}}$. The *support* for $\tilde{\mathbf{x}}$ is the smallest closed set $S \in \mathcal{B}^d$ such that $P(\neg S) = 0$. The support for $\tilde{\mathbf{x}}$ is the same as the support for the probability measure $P$. □

Equivalently, the support, $S$, for a random vector $\tilde{\mathbf{x}}$ with probability measure $P$ is the smallest closed set such that $P(S) = 1$. Less formally, the support of a random vector specifies the set of possible realizations of that random vector.

If $\tilde{\mathbf{x}}$ is discrete random vector with probability mass function $P$, then the support of $\tilde{\mathbf{x}}$ is the set $\{\mathbf{x} : P(\mathbf{x} > 0\}$. That is, the support for a discrete random variable $\tilde{\mathbf{x}}$ is all possible values for $\tilde{\mathbf{x}}$ such that the probability of every realization of $\tilde{\mathbf{x}}$ is strictly positive. For a binary random variable that takes on the value of zero or one with strictly positive probability, the support would be: $\{0, 1\}$. A binary random variable that always takes on the value of one with probability one and the value of zero with probability zero would have support: $\{1\}$.

If $\tilde{\mathbf{x}}$ is a continuous random vector with probability density function $p$, then the support of $\tilde{\mathbf{x}}$ is the set $\{\mathbf{x} : p(\mathbf{x}) > 0\}$. A probability density function with support $\mathcal{R}$ assigns a strictly positive probability to every open interval in $\mathcal{R}$.

A *proposition* $\phi : \Omega \rightarrow \{0, 1\}$ is TRUE on $\Omega$ if and only if $\phi(x) = 1$ for all $x \in \Omega$. The next useful definition provides a weaker version of such a statement.

**Definition 8.2.7** ($\nu$-Almost Everywhere). Let $(\Omega, \mathcal{F}, \nu)$ be a complete measure space. Let $\phi : \Omega \rightarrow \{0, 1\}$ be a proposition. Let $F \in \mathcal{F}$ with the property that $\nu(F) = 0$. Assume that for all $x \in \neg F \cap \Omega$ that $\phi(x) = 1$. Then, the proposition $\phi$ is true *$\nu$-almost everywhere.* □

**Example 8.2.1** (Two Almost-Everywhere Deterministic Functions). When a proposition $\phi$ is true almost everywhere but the measure space $(\Omega, \mathcal{F}, \nu)$ is not explicitly defined, it is typically assumed $(\Omega, \mathcal{F}, \nu)$ is a complete Lebesgue measure space. Let $f : \mathcal{R} \rightarrow \mathcal{R}$ be defined such that for all $x \in \mathcal{R}$: $f(x) = x^2$. Let $g : \mathcal{R} \rightarrow \mathcal{R}$ be defined such that for all $x \in \mathcal{R}$ such that $x \notin \{1, 2, 3\}$: $g(x) = x^2$ and additionally $g(1) = g(2) = g(3) = 42$. Then, $f = g$ $\nu$-almost everywhere. △

**Example 8.2.2** (Extremely Improbable Events). Consider a sample point in a set $\Omega$ which corresponds to the outcomes of an infinite sequence of tosses of a fair coin. Let $S$ denote the elements in $\Omega$ that correspond to fair coin toss outcome sequences where approximately half of the outcomes are "heads" and approximately half of the outcomes are "tails", let $\neg S$ denote the strange outlier extremely improbable sequences where the fair coin toss outcomes are not approximately 50%. For example, $\neg S$ contains the infinite sequence of coin toss outcomes where the coin comes up heads every time in the infinite sequence. The set $\neg S$ also contains the infinite sequence of coin toss outcomes where the coin comes up heads for the first coin toss, heads for the second toss, tails for the third coin toss, heads for the fourth coin toss, heads for fifth coin toss, tails for the sixth coin toss, and so on, so that the percentage of coin tosses which are heads is equal to 66% rather than 50%. The coin toss sequence outcomes in $\neg S$ could occur in the real world but it would be virtually impossible that they would actually occur in the real world.

Let $\mathcal{F}$ be a sigma-field generated from subsets of $S$. Let the probability measure $P : \mathcal{F} \rightarrow [0, 1]$ on the measurable space $(\Omega, \mathcal{F})$ be defined such that $P(S) = 1$ and $P(\neg S) = 0$. Then, with respect to the measure space $(\Omega, \mathcal{F}, P)$, the percentage of coin toss outcomes which are heads in an infinite sequence of coin tosses is approximately 50% $P$-almost everywhere. △

The notation $\nu$-a.e. means *$\nu$-almost everywhere.* When $\nu$ is a probability measure, the terminology *$\nu$-with probability one* or *with probability one* is commonly used instead of the terminology $\nu$-almost everywhere.

**Definition 8.2.8** (Bounded Random Vector). Let $P$ be a probability measure for a random vector $\tilde{\mathbf{x}} : \Omega \rightarrow \mathcal{R}^d$ with the property that there exists a finite positive number $K$ such that:

$$P(\{\omega : |\tilde{\mathbf{x}}(\omega)| \leq K\}) = 1.$$

Then, $\tilde{\mathbf{x}}$ is called a *bounded random vector.* □

If $\tilde{\mathbf{x}}$ is a bounded random vector, then $|\tilde{\mathbf{x}}| \leq K$ with probability one. For example, every random vector whose magnitude never exceeds some finite number $K$ is a bounded random vector. Every discrete random vector whose support is a finite set of vectors is a bounded random vector but a Gaussian random variable is not a bounded random variable. On the other hand, in practice, Gaussian random variables are sometimes approximated by averaging together a finite number of bounded random variables. Such an approximate Gaussian random variable is bounded. In general, the physically plausible assumption a random vector is bounded dramatically simplifies many important asymptotic statistical theory analyses.

### 8.2.2 Discrete, Continuous, and Mixed Random Vectors

#### Examples of Discrete, Absolutely Continuous, and Mixed Random Vectors

**Definition 8.2.9** (Discrete Random Vector). Let $S \in \mathcal{B}^d$ be a countable set. Let $P : \mathcal{B}^d \rightarrow [0, 1]$ be the probability measure for a random vector $\tilde{\mathbf{x}}$ with support $S$. When it exists, the *probability mass function (pmf)* $p : \mathcal{R}^d \rightarrow (0, 1]$ for the *discrete random vector* $\tilde{\mathbf{x}}$ is defined such that for all $B \in \mathcal{B}^d$:

$$P(B) = \sum_{\mathbf{x} \in B \cap S} p(\mathbf{x}). \tag{8.1}$$

□

**Example 8.2.3** (Support for a Discrete Random Vector on Finite Set). Let $p_k$ be defined such that $0 < p_k < 1$ for $k = 1, \ldots, M$. In addition, assume that $\sum_{k=1}^{M} p_k = 1$. Assume a random vector $\tilde{\mathbf{x}}$ takes on the value of the $d$-dimensional column vector $\mathbf{x}^k$ with probability $p_k$. The pmf for $\tilde{\mathbf{x}}$ is the function $p : \Omega \rightarrow [0, 1]$ where $\Omega \equiv \{\mathbf{x}^1, \ldots, \mathbf{x}^M\}$. The support of $\tilde{\mathbf{x}}$ is the set $\Omega$. △

**Definition 8.2.10** (Absolutely Continuous Random Vector). Let $P : \mathcal{B}^d \rightarrow [0, 1]$ be the probability measure for a random vector $\tilde{\mathbf{x}}$ with support $\Omega \subseteq \mathcal{R}^d$. When it exists, the *probability density function* $p : \Omega \rightarrow [0, \infty)$ for the *absolutely continuous random vector* $\tilde{\mathbf{x}}$ is defined such that for all $B \in \mathcal{B}^d$:

$$P(B) = \int_{\mathbf{x} \in B} p(\mathbf{x}) d\mathbf{x}. \tag{8.2}$$

□

**Example 8.2.4** (Support for a Uniformly Distributed Random Vector). Let $B$ be a closed ball of known radius $R$ in $\mathcal{R}^d$. Assume $\tilde{\mathbf{x}}$ has the probability density function $p : \mathcal{R}^d \rightarrow [0, \infty)$ defined such that $p(\mathbf{x}) = K$ for all $\mathbf{x}$ in $B$ and $\int_B p(\mathbf{x}) d\mathbf{x} = 1$. The support of $\tilde{\mathbf{x}}$ is $B$. Show how to calculate $K$ in terms of the radius $R$. △

**Example 8.2.5** (Support for an Absolutely Continuous Gaussian Random Vector). Let $\tilde{\mathbf{x}} \equiv [\tilde{x}_1, \ldots, \tilde{x}_d]$ with probability density function $p : \mathcal{R}^d \rightarrow [0, \infty)$ be defined such that

$$p(\mathbf{x}) = \prod_{i=1}^{d} p_i(x_i)$$

where $p_i : \mathcal{R} \rightarrow [0, \infty)$ is defined such that:

$$p_i(x_i) = (\sqrt{2\pi})^{-1} \exp\left(-(1/2)(x_i)^2\right)$$

for $i = 1, \ldots, d$. The support of $\tilde{\mathbf{x}}$ is $\mathcal{R}^d$. △

As previously noted, probabilistic models in machine learning often involve random vectors whose elements include both discrete and absolutely continuous random variables. Such random vectors are called "mixed-component random vectors".

**Definition 8.2.11** (Mixed-Component Random Vector). Let $\tilde{\mathbf{x}}$ be a $d$-dimensional random vector $\tilde{\mathbf{x}}$ consisting of $d$ random variables $\tilde{x}_1, \ldots, \tilde{x}_d$. If at least one element of $\tilde{\mathbf{x}}$ is a discrete random variable and at least one element of $\tilde{\mathbf{x}}$ is an absolutely continuous random variable, then $\tilde{\mathbf{x}}$ is called a *mixed-component random vector*. □

**Example 8.2.6** (Support for a Mixed-Component Random Vector). Let $\tilde{\mathbf{y}}$ be an $m$-dimensional discrete random vector with support $\Omega_y$ consisting of the $m$ discrete random variables in $\tilde{\mathbf{x}}$ and let $\tilde{\mathbf{z}}$ be a $d-m$-dimensional absolutely continuous random vector with support $\Omega_z$ consisting of the $d-m$ absolutely continuous random variables in $\tilde{\mathbf{x}}$. Assume $\tilde{\mathbf{x}} \equiv [\tilde{\mathbf{y}}, \tilde{\mathbf{z}}]$ is a $d$-dimensional mixed-component random vector with probability measure $P : \mathcal{B}^d \rightarrow [0,1]$ with support $S \equiv \Omega_y \times \Omega_z$. Let $p_{y,z} : \mathcal{R}^m \times \mathcal{R}^{d-m} \rightarrow [0, \infty)$ be a function defined such that for all $B \in \mathcal{B}^d$:

$$P(B) = \sum_{\mathbf{y} \in \Omega_y} \int_{\mathbf{z} \in \Omega_z} p_{y,z}(\mathbf{y}, \mathbf{z}) \phi_B(\mathbf{y}, \mathbf{z}) d\mathbf{z} \tag{8.3}$$

where $\phi_B(\mathbf{y}, \mathbf{z}) = 1$ when $(\mathbf{y}, \mathbf{z}) \in B$ and $\phi_B(\mathbf{y}, \mathbf{z}) = 0$ when $(\mathbf{y}, \mathbf{z}) \notin B$. By choosing the function $p_{y,z}$ in this way, one can specify the probability measure $P$ for the mixed random vector $\tilde{\mathbf{x}} \equiv [\tilde{\mathbf{y}}, \tilde{\mathbf{z}}]$ which is neither discrete or absolutely continuous! △

**Definition 8.2.12** (Mixed Random Vector). A *mixed random variable* $\tilde{x}$ is a random variable such that $\tilde{x}$ is not a discrete random variable, and $\tilde{x}$ is not an absolutely continuous random variable. A *mixed random vector* is defined as either: (1) a mixed-component random vector, or (2) a random vector containing at least one mixed random variable. □

**Example 8.2.7** (Mixed Random Variable Examples). An example of a mixed random variable $\tilde{x}$ relevant in a machine learning context would be a random variable that takes on values corresponding to the settings on an analog volume control of a smart phone. Assume the volume control adjusts sound volume which varies continuously from a sound level of zero to a maximum sound level of $M$. It is desired to represent information such that a learning machine can be built that is capable of learning a probability distribution that assigns probabilities to particular sound level intervals.

A discrete random variable representation is not adequate because there are infinitely many possible volume levels and the set of possible volume levels is not countable. That is, the probability of a particular volume level $x$ is equal to zero. An absolutely continuous random variable representation is not adequate because the probability that the sound volume control is adjusted to the maximum sound level $M$ will tend to be some positive probability which is not allowable for absolutely continuous random variables.

Another example of a mixed random variable is a random variable that is constructed from sums and products of discrete random variables and absolutely continuous random variables. Again, such situations are commonly present in statistical machine learning where information from both discrete random variables and absolutely continuous random variables is often combined.

For example, let a discrete random variable $\tilde{x}_1$ be a binary random variable with support $\{0, 1\}$. Let a continuous random variable $\tilde{x}_2$ be an absolutely continuous random variable with support $\mathcal{R}$. Let $\tilde{x}_3 = \tilde{x}_1 + \tilde{x}_2$. The random variable $\tilde{x}_3$ is not a discrete random variable since its support is not a countable set. On the other hand, the random variable $\tilde{x}_3$ is not a continuous random variable since $\tilde{x}_3$ has a strictly positive probability of taking on values in the support of the discrete random variable $\tilde{x}_1$. The random variable $\tilde{x}_3$ is thus an example of a mixed random variable. The probability distribution of mixed random variables may be specified using the Radon-Nikodým density function. △

### Radon-Nikodým Density Notation for Specifying Probability Distributions

The notation used for calculating a probability in (8.3) is very awkward. A more compact notation for specifying the probability distribution of a random vector would greatly facilitate the statement and proof of theorems that include discrete, continuous, and mixed random vectors as special cases. The key ingredient for such a notation involves realizing that some method is required for specifying the support for a random vector. For example, if the support is a finite set of points, then the random vector is a discrete random vector. On the other hand, an absolutely continuous random vector has support which is not a countable set.

**Definition 8.2.13** (Absolutely Continuous). Let $\tilde{\mathbf{x}}$ be a $d$-dimensional random vector with probability measure $P$ on $(\mathcal{R}^d, \mathcal{B}^d)$. Let $\nu$ be a sigma-finite measure on the measurable space $(\mathcal{R}^d, \mathcal{B}^d)$. Assume that for every $B \in \mathcal{B}^d$, $\nu(B) = 0 \implies P(B) = 0$. The probability measure $P$ is said to be *absolutely continuous* with respect to the sigma-finite measure $\nu$. The sigma-finite measure $\nu$ is called a *support specification measure* for $P$ or $\tilde{\mathbf{x}}$. □

For example, define a sigma-finite measure, $\nu$, for $\tilde{\mathbf{x}}$ that assigns the value of zero to the set $\{x \in \mathcal{R} : x \notin \{0, 1\}\}$. Now define the random variable $\tilde{x}$ which takes on the value of one with probability $P$ and the value of zero with probability $1 - P$ where $0 \leq p \leq 1$. Then, $P$ is absolutely continuous with respect to $\nu$.

Let $\nu$ and $P$ respectively specify a sigma-finite measure and probability measure on the measurable space $(\mathcal{R}^d, \mathcal{B}^d)$ such that $P$ is absolutely continuous with respect to $\nu$. The sigma-finite measure $\nu$ specifies the support of the probability measure $P$ in the sense that an event $B \in \mathcal{B}^d$ must have probability zero (i.e., $P(B) = 0$) when $\nu(B) = 0$. This is the motivation for referring to the sigma-finite measure $\nu$ in this context as a "support specification measure".

The Radon-Nikodým notation

$$P(B) = \int_B p(\mathbf{x}) d\nu(\mathbf{x}) \tag{8.4}$$

will be used to specify the operations of computing $P(B)$ which are explicitly presented in (8.1), (8.2), (8.3). The function $p$ in (8.4) will be used to denote a more general type of probability density function for specifying the probability distribution of random vectors which may include both discrete and absolutely continuous random variables. The more general density $p$ in (8.4) is called a "Radon-Nikodým density", while the integral in (8.4) is called a "Lebesgue integral".

If $\tilde{\mathbf{x}}$ is a discrete random vector with respect to a support specification measure $\nu$, the support specification measure $\nu$ specifies the support of $P$ as a countable set so that the integral in (8.4) may be expressed as:

$$P(B) = \sum_{\mathbf{x} \in B} p(\mathbf{x}).$$

If $\tilde{\mathbf{x}}$ is an absolutely continuous random vector, the support specification measure $\nu$ specifies the support of $P$ such that the integral in (8.4) is interpretable as a Riemann integral:

$$P(B) = \int_{\mathbf{x} \in B} p(\mathbf{x}) d\mathbf{x}.$$

If $\tilde{\mathbf{x}}$ contains $m$ discrete random variables and $d - m$ absolutely continuous random variables as specified above, the integral in (8.4) is interpreted as a summation over the support of the $m$ discrete random variables $\Omega_y$ and a Riemann integration over the support of the $d - m$ absolutely continuous random variables $\Omega_z$ as in (8.3). In this case, $\nu$ specifies the support of $P$ as $S \equiv \Omega_y \times \Omega_z$.

For example, the notation in (8.4) is also used to specify particular regions of integration and summation. Let

$$\tilde{\mathbf{x}} = [\tilde{x}_1, \tilde{x}_2, \tilde{x}_3, \tilde{x}_4]$$

be a four-dimensional mixed-component random vector consisting of two discrete random variables $\tilde{x}_1$ and $\tilde{x}_2$ and two absolutely continuous random variables $\tilde{x}_3$ and $\tilde{x}_4$. Let $p : \mathcal{R}^d \rightarrow [0, \infty)$ be the Radon-Nikodým density for $\tilde{\mathbf{x}}$. The density $p(x_2, x_3)$ would then be computed using the formula:

$$p(x_2, x_3) = \sum_{x_1} \int_{x_4} p(x_1, x_2, x_3, x_4) dx_4$$

and represented using the Radon-Nikodým notation in (8.4) as:

$$p(x_2, x_3) = \int p(x_1, x_2, x_3, x_4) d\nu(x_1, x_4)$$

where the support specification measure $\nu$ is used to specify that the summation operation should be used for $x_1$ and the integration operation should be used for $x_4$.

## Exercises

8.2-1. Which of the following functions is a Borel measurable function on $\mathcal{R}^d$?
(i) Define $f$ such that $f(\mathbf{x}) = |\mathbf{x}|^2$ for all $\mathbf{x} \in \mathcal{R}^d$.
(ii) Define $f$ such that $f(\mathbf{x}) = |\mathbf{x}|$ for all $\mathbf{x} \in \mathcal{R}^d$.
(iii) Define $f$ such that $f(\mathbf{x}) = |\mathbf{x}|^2$ if $|\mathbf{x}| > 5$ and $f(\mathbf{x}) = \exp(|\mathbf{x}|) + 7$ if $|\mathbf{x}| \leq 5$.

8.2-2. Use the definition of a Borel measurable function to prove Property (iv) of Theorem 8.2.1.

8.2-3. Explain how any Borel measurable function may be interpreted as a random vector.

8.2-4. Define the domain and range of a random vector that is used to model the outcome of an experiment where a die is thrown and the observed value of the die is observed to be equal to an element in the set $\{1, 2, 3, 4, 5, 6\}$.

8.2-5. Describe the domain and range of a random vector that is used to model the outcome of an experiment where a die is thrown and the observed value of the die is observed to be either "odd" or "even".

8.2-6. Consider a binary random variable $\tilde{x}$ used to model the outcome of tossing a weighted coin where $\tilde{x} = 1$ with probability $p$ and $\tilde{x} = 0$ with probability $1 - p$ for $0 \leq p \leq 1$. (a) Define the probability space for $\tilde{x}$. (b) What is the support of $\tilde{x}$? (c) What is the probability measure for $\tilde{x}$? (d) What is the cumulative distribution function for $\tilde{x}$? (e) Is $\tilde{x}$ an absolutely continuous random variable?

8.2-7. Give an example of a mixed random vector which was not mentioned in the text which might be encountered in a practical machine learning application.

8.2-8. What is the support for a random variable used to model the outcome of a die rolling experiment?

8.2-9. What is the support for a random variable used to model the outcome of a physical measurement of someone's height under the assumption that heights are strictly positive?

8.2-10. Let $\tilde{\mathbf{x}} = [\tilde{x}_1, \tilde{x}_2, \tilde{x}_3]$ where $\tilde{x}_1$ is a discrete random variable and $\tilde{x}_2$ and $\tilde{x}_3$ are absolutely continuous random variables. Let $p : \Omega \rightarrow [0, \infty)$ be a Radon-Nikodým density function for $\tilde{\mathbf{x}}$. Using a combination of both summation notation and Riemann integral notation explicitly calculate the marginal densities: $p(x_1)$, $p(x_2)$, and $p(x_3)$. Now show how to represent the marginal densities using the Radon-Nikodým density.

8.2-11. Let $\tilde{\mathbf{x}} = [\tilde{x}_1, \tilde{x}_2, \tilde{x}_3]$ be a three-dimensional random vector whose first element, $\tilde{x}_1$, is a discrete random variable with support $\{0, 1\}$ and whose remaining elements $\tilde{x}_2$ and $\tilde{x}_3$ are absolutely continuous random variables with support $[0, \infty)$. Let $p : \Omega \rightarrow [0, \infty)$ be a Radon-Nikodým density function for $\tilde{\mathbf{x}}$ with support specification measure $\nu$. Let $V : \mathcal{R}^3 \rightarrow \mathcal{R}$ be defined such that $V(\mathbf{x}) = |\mathbf{x}|^2$ for all $\mathbf{x} \in \mathcal{R}^3$. Let $B$ be a closed and bounded subset of $\Omega$. Show how to express

$$\int_B V(\mathbf{x}) p(\mathbf{x}) d\nu(\mathbf{x})$$

using only conventional summations and Riemann integrals.

## 8.3 Existence of the Radon-Nikodým Density (Optional Reading)

In the previous section, the Radon-Nikodým Density was introduced as a convenient notation which we will extensively use throughout the text for the purposes of stating theorems applicable to discrete, absolutely continuous, and mixed random vectors. However, some critical issues were ignored.

Given an arbitrary probability space $(\Omega, \mathcal{F}, P)$, is it possible to construct a Radon-Nikodým probability density $p : \Omega \rightarrow [0, \infty)$ and support specification measure $\nu : \mathcal{F} \rightarrow [0, \infty]$ such that for each $F \in \mathcal{F}$:

$$P(F) = \int_{\mathbf{x} \in F} p(\mathbf{x}) d\nu(\mathbf{x})?$$

And, if such a construction is possible, then under what conditions is such a construction possible?

The answer to these questions is very important in practical engineering applications because engineers typically analyze and design systems using probability density functions rather than probability spaces! Fortunately, under very general conditions this construction is possible provided one uses a generalization of the Riemann integral (e.g., $\int p(x)dx)$) which is called the "Lebesgue Integral".

### 8.3.1 Lebesgue Integral

**Definition 8.3.1** (Lebesgue Integral of a Non-negative Simple Function). Let $(\mathcal{R}^d, \mathcal{B}^d)$ be a measurable space. Let $\omega_1, \ldots, \omega_m$ be a finite partition of $\mathcal{R}^d$ such that $\omega_1, \ldots, \omega_m$ are elements of $\mathcal{B}^d$. Let $f : \mathcal{R}^d \rightarrow [0, \infty]$ be defined such that $f(\mathbf{x}) = f_j$ for all $\mathbf{x} \in \omega_j$, $j = 1, \ldots, m$. The function $f$ is called a *non-negative simple function.* Let $\nu$ be a countably additive measure on $(\mathcal{R}^d, \mathcal{B}^d)$. The *Lebesgue Integral of* $f$, $\int f(\mathbf{x}) d\nu(\mathbf{x})$, is defined such that:

$$\int f(\mathbf{x}) d\nu(\mathbf{x}) = \sum_{j=1}^{m} f_j \nu(\omega_j)$$

where $0\nu(\omega_j)$ is defined as 0 when $\nu(\omega_j) = +\infty$. □

**Definition 8.3.2** (Lebesgue Integral of a Non-negative Function). Let $(\mathcal{R}^d, \mathcal{B}^d, \nu)$ be a measure space. Let $([-\infty, +\infty], \mathcal{B}_\infty)$ denote the extended Borel sigma-field. For a particular $(\mathcal{B}^d, \mathcal{B}_\infty)$-measurable function $f : \mathcal{R}^d \to [0, \infty]$, define

$$M_f \equiv \{\phi : 0 \leq \phi(\mathbf{x}) \leq f(\mathbf{x}),\ \mathbf{x} \in \mathcal{R}^d\}$$

as a subset of all simple functions with domain $\mathcal{R}^d$ and range $[0, \infty]$. The *Lebesgue integral of the non-negative function* $f$ with respect to $\nu$ is the extended real number

$$\int f(\mathbf{x}) d\nu(\mathbf{x}) \equiv \sup \left\{ \int \phi(\mathbf{x}) d\nu(\mathbf{x}) : \phi \in M_f \right\}.$$

□

**Definition 8.3.3** (Integrable Function). Let $(\mathcal{R}^d, \mathcal{B}^d, \nu)$ be a measure space. Let $f : \mathcal{R}^d \to \mathcal{R}$ be a Borel measurable function. Let $f^+$ be defined such that $f^+ = f$ for all $\mathbf{x}$ such that $f(\mathbf{x}) > 0$ and let $f^-$ be defined such that $f^- = |f|$ for all $\mathbf{x}$ such that $f(\mathbf{x}) < 0$. The *Lebesgue integral* of $f$, $\int f(\mathbf{x}) d\nu(\mathbf{x})$, is then defined by the formula:

$$\int f(\mathbf{x}) d\nu(\mathbf{x}) \equiv \int f^+(\mathbf{x}) d\nu(\mathbf{x}) - \int f^-(\mathbf{x}) d\nu(\mathbf{x}).$$

If, in addition, $\int f(\mathbf{x}) d\nu(\mathbf{x}) < \infty$, then $f$ is called *integrable.* □

Let $\psi(\mathbf{x}) = 1$ for all $\mathbf{x} \in B$ and let $\psi(\mathbf{x}) = 0$ for all $\mathbf{x} \notin B$. Then, the notation

$$\int_B f(\mathbf{x}) d\nu(\mathbf{x}) = \int f(\mathbf{x}) \psi(\mathbf{x}) d\nu(\mathbf{x}).$$

Also note the assertion $f = g$ $\nu$-a.e. implies

$$\int_B f(\mathbf{x}) d\nu(\mathbf{x}) = \int_B g(\mathbf{x}) d\nu(\mathbf{x}).$$

Many of the standard operations which hold for the Riemann integral also hold for the Lebesgue integral.

**Theorem 8.3.1** (Properties of the Lebesgue Integral). *Let $\nu$ be a measure on the Borel measurable space $(\mathcal{R}^d, \mathcal{B}^d)$. Let $f : \mathcal{R}^d \to \mathcal{R}$ and $g : \mathcal{R}^d \to \mathcal{R}$ be Borel measurable functions. Then, the following assertions hold.*

- *(i) If $|f| \leq |g|$ and $g$ is integrable, then $f$ is integrable.*
- *(ii) If $f$ is integrable and $C \in \mathcal{R}$, then $\int Cf(\mathbf{x}) d\nu(\mathbf{x}) = C \int f(\mathbf{x}) d\nu(\mathbf{x})$.*
- *(iii) If $f$ and $g$ are integrable, then $\int [f(\mathbf{x}) + g(\mathbf{x})] d\nu(\mathbf{x}) = \int f(\mathbf{x}) d\nu(\mathbf{x}) + \int g(\mathbf{x}) d\nu(\mathbf{x})$.*
- *(iv) Assume $f_1, f_2, \ldots$ is a sequence of integrable functions that converge $\nu$-almost everywhere to $f$. If, in addition, $g$ is integrable such that $|f_k| \leq g$ for all $k = 1, 2, \ldots$, then $f$ is integrable and $\int f_k(\mathbf{x}) d\nu(\mathbf{x}) \to \int f(\mathbf{x}) d\nu(\mathbf{x})$ as $k \to \infty$.*
- *(v) Assume $f : \mathcal{R} \to \mathcal{R}$ is a non-negative continuous function on $[a, b]$. Then, $\int_{[a,b]} f(x) d\nu(x) = \int_a^b f(x) dx$ where $\nu$ is the Lebesgue measure on the Borel measurable space $(\mathcal{R}, \mathcal{B})$.*

*Proof.* (i) See Corollary 5.4 of Bartle (1966, p. 43). (ii) See Theorem 5.5 of Bartle (1966, p. 43). (iii) See Theorem 5.5 of Bartle (1966, p. 43). (iv) See Lebesgue Dominated Convergence Theorem (Theorem 5.6 of Bartle, 1966, p. 44). (v) See Problem 4.L of Bartle (1966, p. 38). ■

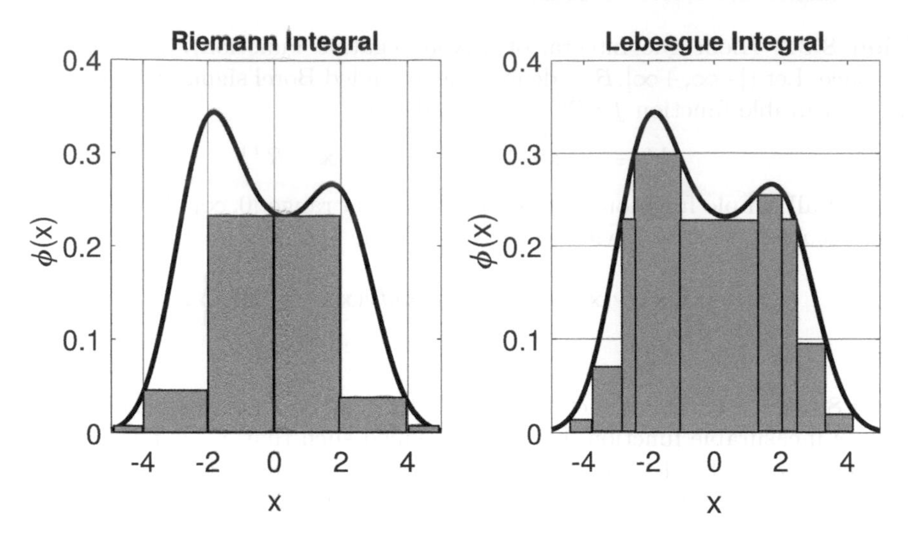


**FIGURE 8.1**
**Comparison of the Riemann integral with the Lebesgue integral.** Riemann integral covers the area under the curve using rectangles with some fixed width, while the Lebesgue integral covers the area under the curve using rectangles of varying widths.

The Lebesgue integral with respect to $\nu$ is a natural generalization of the classical Riemann integral (see Property (v) of Theorem 8.3.1). Essentially the Lebesgue integral does not try to shrink the width of all rectangles so that they are small enough to cover the space under the curve like the Riemann integral. Rather, the Lebesgue integral tries to select the best combination of simple shapes (which can be varying widths) that best cover the space under the curve. A comparison of the Lebesgue integral with the Riemann integral is provided in Figure 8.1

### 8.3.2 The Radon-Nikodým Probability Density Function

The Radon-Nikodým Density is now defined using the Lebesgue integral concept introduced in the previous section.

**Theorem 8.3.2** (Radon-Nikodým Density Existence). *Let $(\Omega, \mathcal{F})$ be a measurable space. Let $P : \mathcal{F} \to [0, 1]$ be a probability measure and let $\nu : \mathcal{F} \to [0, \infty]$ be a sigma-finite measure defined on $(\Omega, \mathcal{F})$. Assume that for all $F \in \mathcal{F}$: $\nu(F) = 0$ implies $P(F) = 0$. Then, there exists a measurable function $p : \Omega \to [0, \infty]$ such that for all $F \in \mathcal{F}$:*

$$P(F) = \int_{\mathbf{x} \in F} p(\mathbf{x}) d\nu(\mathbf{x}).$$

*In addition, $p$ is uniquely determined $\nu$-almost everywhere.*

*Proof.* See Bartle, 1966, p. 85, Theorem 8.9. ■

The function $p$ in the Radon-Nikodým Density Theorem is called a "*Radon-Nikodým density function* $p$ defined with respect to $\nu$" with the understanding that $\nu$ is a support specification measure on a measurable space $(\Omega, \mathcal{F})$ which specifies a particular set of impossible events that must be assigned the probability of zero.

The Radon-Nikodým Density Theorem is important because it provides weak conditions that ensure a Radon-Nikodým density function $p$ can be constructed which can assign probabilities to the elements of $\mathcal{B}^d$ according to a given desired probability space specification $(\mathcal{R}^d, \mathcal{B}^d, P)$.

### 8.3.3 Vector Support Specification Measures

For continuous scalar random variables, the Lebesgue measure on $(\mathcal{R}, \mathcal{B})$ assigns a number to an open interval equal to the length of that interval in $\mathcal{B}$. With this choice of $\nu$ as the Lebesgue measure, the Radon-Nikodým density $p$ with respect to $\nu$ specifies the probability of event $B$, $P(B)$, by the formula:

$$P(B) = \int_B p(\mathbf{x}) d\nu(\mathbf{x}) = \int_B p(\mathbf{x}) d\mathbf{x} \text{ for each } B \in \mathcal{B}.$$

For the case of discrete scalar random variables, suppose that a discrete random variable can take on a countable number of possible values corresponding to a countable subset $\Omega$ of $\mathcal{R}$. Let $\mathcal{P}(\Omega)$ denote the power set of $\Omega$. Define a sigma-finite measure $\nu$ on $(\mathcal{R}, \mathcal{B})$ with respect to $\Omega$ such that $\nu(B) = \mu(B \cap \Omega)$ where $\mu$ is the counting measure on $(\Omega, \mathcal{P}(\Omega))$. With this choice of $\nu$ as sigma-finite measure on $(\mathcal{R}, \mathcal{B})$, the Radon-Nikodým density $p$ with respect to $\nu$ specifies the probability of event $B$, $P(B)$, by the formula:

$$P(B) = \int_B p(\mathbf{x}) d\nu(\mathbf{x}) = \sum_{B \cap \Omega} p(\mathbf{x}) \text{ for each } B \in \mathcal{B}.$$

**Recipe Box 8.1 Applying the Radon-Nikodým Theorem (Theorem 8.3.2)**

- **Step 1: Identify support of each vector component.** Identify the support, $S_i$, for the $i$th random variable, $\tilde{x}_i$, of a random vector $\tilde{\mathbf{x}} = [\tilde{x}_1, \ldots, \tilde{x}_d]$ defined with respect to the probability space $(\mathcal{R}^d, \mathcal{B}^d, P)$.

- **Step 2: Define a sigma-finite measure for each vector component.** If $\tilde{x}_i$ is a discrete random variable, define the sigma-finite measure $\nu_i$ such that $\nu_i(B)$ is the number of elements in $B \cap S_i$ for all $B \in \mathcal{B}$. If $\tilde{x}_i$ is an absolutely continuous random variable whose support $S_i$ is a convex subset of $\mathcal{R}$, define the sigma-finite Lebesgue measure $\nu_i$ such that $\nu_i(B)$ is the length of the line segment $B \cap S_i$ for all $B \in \mathcal{B}$.

- **Step 3: Construct a sigma-finite measure for the random vector.** Let $\nu : \mathcal{B}^d \to [0, \infty)$ be defined such that for each $B \equiv [B_1, \ldots, B_d] \in \mathcal{B}^d$:

$$\nu(B) = \prod_{i=1}^{d} \nu_i(B_i).$$

- **Step 4: Complete the measure space.** Let $\bar{\nu} : \mathcal{B}^d \to [0, \infty]$ be defined such that for each $B \in \mathcal{B}^d$: (i) $\bar{\nu}(B) = \nu(B)$ when $\nu(B) > 0$, and (ii) $\bar{\nu}(S) = 0$ for all $S \subset B$ when $\nu(B) = 0$.

- **Step 5: Construct the Radon-Nikodým density.** Use the Radon-Nikodým theorem to conclude a Radon-Nikodým density $p : \mathcal{R}^d \to [0, \infty)$ with respect to $\bar{\nu}$ exists for specifying the probability measure $P$.

For the case of mixed random variables, a support specification measure for a scalar random variable can be constructed in a straightforward manner. Let $\nu$ be a support specification measure for an absolutely continuous scalar random variable on $(\mathcal{R}, \mathcal{B})$. Let $\mu$ be a support specification measure for a discrete scalar random variable on $(\mathcal{R}, \mathcal{B})$. Then, $\nu + \mu$ is a support specification measure for a mixed random variable on $(\mathcal{R}, \mathcal{B})$.

The more general case of a $d$-dimensional random vector is now discussed. The following theorem is used to construct a support specification measure for a $d$-dimensional random vector comprised of $d$ scalar random variables.

**Theorem 8.3.3** (Product Measure). *Let $\nu_1$ and $\nu_2$ be sigma-finite measures defined respectively on the measurable spaces $(\Omega_1, \mathcal{F}_1)$ and $(\Omega_2, \mathcal{F}_2)$. Then, there exists a unique measure $\nu$ on the measurable space $(\Omega_1 \times \Omega_2, \mathcal{F}_1 \times \mathcal{F}_2)$ defined such that for all $F_1 \in \mathcal{F}_1$ and for all $F_2 \in \mathcal{F}_2$:*

$$\nu(F) \equiv \nu_1(F_1)\nu_2(F_2).$$

*Proof.* See Bartle (1966, pp. 114-115). ■

The Product Measure Theorem ensures that if $\nu_1, \ldots, \nu_d$ are sigma-finite measures on $(\mathcal{R}, \mathcal{B})$ specifying the respective support of scalar random variables $\tilde{x}_1, \ldots, \tilde{x}_d$, then there exists a unique measure $\nu \equiv \prod_{i=1}^{d} \nu_i$ on $(\mathcal{R}^d, \mathcal{B}^d)$. Thus, the Product Measure Theorem (Bartle, p. 114) provides a straightforward construction of measures for random vectors on $(\mathcal{R}^d, \mathcal{B}^d)$.

After the Product Measure Theorem has been used to construct the measure space $(\mathcal{R}^d, \mathcal{B}^d, \nu)$, a complete measure space $(\mathcal{R}^d, \mathcal{B}^d, \bar{\nu})$ is formed as discussed in the Definition of a Complete Measure Space (see Definition 8.1.9). It is important to form a complete measure space because this ensures that if $\bar{\nu}$ is a support specification measure for a probability measure $P$ then every subset, $Q$, of an impossible event $E$ (i.e., $P(E) = 0$) must also be impossible (i.e., $P(Q) = 0$). This assumption prevents misinterpretations of statements that particular assertions hold $\nu$-almost everywhere.

## Exercises

8.3-1. Explain how to construct a support specification measure for a Radon-Nikodým density $p$ for a $d$-dimensional discrete random vector $\tilde{\mathbf{x}}$ which takes on the value of $\mathbf{x}^k$ with probability $p_k$, $k = 1, \ldots, M$.

8.3-2. Explain how to construct a support specification measure for a Radon-Nikodým density $p$ for an absolutely continuous $d$-dimensional random vector consisting of $d$ Gaussian random variables with common mean zero and common variance $\sigma^2$.

8.3-3. Explain how to construct a support specification measure for a Radon-Nikodým density $p$ for a two-dimensional random vector $[\tilde{x}_1, \tilde{x}_2]$ where the discrete random variable $\tilde{x}_1$ has support $\{0, 1\}$ and the absolutely continuous random variable $\tilde{x}_2$ is uniformly distributed on the interval $[-1, +1]$.

## 8.4 Expectation Operations

The Lebesgue integral notation illustrated in (8.4) provides a unified framework for discussing expectation operations involving many different types of random vectors and functions which may be discrete, absolutely continuous, or mixed.

**Definition 8.4.1** (Expectation). Let $p : \mathcal{R}^d \rightarrow [0, \infty)$ be a Radon-Nikodým density with respect to $\nu$ for the $d$-dimensional random vector $\tilde{\mathbf{x}}$. Let $h : \mathcal{R}^d \rightarrow \mathcal{R}$ be a Borel measurable function. When it exists, the *expectation* of $h(\tilde{\mathbf{x}})$ with respect to $\tilde{\mathbf{x}}$, $E\{h(\tilde{\mathbf{x}})\}$, is defined such that:

$$E\{h(\tilde{\mathbf{x}})\} = \int h(\mathbf{x})p(\mathbf{x})d\nu(\mathbf{x}) \tag{8.5}$$

□

For example, the expected value, $E\{\tilde{\mathbf{y}}\}$, of a $d$-dimensional discrete random vector $\tilde{\mathbf{y}}$ is explicitly calculated using the formula:

$$E\{\tilde{\mathbf{y}}\} = \int \mathbf{y}p_y(\mathbf{y})d\nu(\mathbf{y}) = \sum_{\mathbf{y} \in \Omega_y} \mathbf{y}p_y(\mathbf{y})$$

where $\Omega_y$ is the support of $\tilde{\mathbf{y}}$ and $p_y : \mathcal{R}^d \rightarrow [0, \infty)$ is a Radon-Nikodým density.

The expected value, $E\{\tilde{\mathbf{z}}\}$, of an absolutely continuous $d$-dimensional random vector $\tilde{\mathbf{z}}$ is explicitly calculated using the formula:

$$E\{\tilde{\mathbf{z}}\} = \int \mathbf{z}p_z(\mathbf{z})d\nu(\mathbf{z}) = \int_{\mathbf{z} \in \Omega_z} \mathbf{z}p_z(\mathbf{z})d\mathbf{z}$$

where $\Omega_z$ is the support of $\tilde{\mathbf{z}}$ and $p_z : \mathcal{R}^d \rightarrow [0, \infty)$ is a Radon-Nikodým density.

For the case of a mixed-component 3-dimensional random vector $\tilde{\mathbf{x}}$ whose first element is a discrete random variable and whose second and third elements are absolutely continuous random variables $\tilde{x}_2$ and $\tilde{x}_3$, the expected value of $\tilde{\mathbf{x}}$ is given by the formula:

$$E\{\tilde{\mathbf{x}}\} = \int \mathbf{x}p(\mathbf{x})d\nu(\mathbf{x}) = \sum_{x_1 \in \Omega_1} \int_{x_2 \in \Omega_2} \int_{x_3 \in \Omega_3} [x_1, x_2, x_3]p(x_1, x_2, x_3)dx_2dx_3$$

where $\Omega_k$ is the support of the $k$th random variable $\tilde{x}_k$ for $k = 1, 2, 3$ and $p : \mathcal{R}^3 \rightarrow [0, \infty)$ is a Radon-Nikodým density defined with respect to the measure $\nu$.

Instead of specifying $E\{h(\tilde{\mathbf{x}})\}$ using (8.5), the alternative formula

$$E\{h(\tilde{\mathbf{x}})\} = \int h(\mathbf{x})dP(\mathbf{x}) \tag{8.6}$$

is sometimes used which specifies the expectation operator directly in terms of the probability measure $P$ rather than representing $P$ indirectly using a Radon-Nikodým density $p$ defined with respect to some support specification measure $\nu$ as in (8.5). Equation (8.5) is a preferred representation in engineering applications since probability distributions are typically specified as density functions rather than probability measures. However, (8.6) may be preferable for supporting certain theoretical mathematical investigations.

A sufficient condition for ensuring that $E\{h(\tilde{\mathbf{x}})\}$ is finite is obtained by first showing that $|h(\tilde{\mathbf{x}})| \leq K$ with probability one which implies that:

$$|E\{h(\tilde{\mathbf{x}})\}| = \left|\int h(\mathbf{x})p(\mathbf{x})d\nu(\mathbf{x})\right| \leq \int |h(\mathbf{x})|p(\mathbf{x})d\nu(\mathbf{x}) \leq K \int p(\mathbf{x})d\nu(\mathbf{x}) = K.$$

Note the condition $|h(\tilde{\mathbf{x}})| \leq K$ with probability one can be established if $h$ is bounded but also can be established if $h$ is piecewise continuous and $\tilde{\mathbf{x}}$ is a bounded random vector.

**Example 8.4.1** (Expectation of the Empirical Risk). Assume the observed training data $\mathbf{x}_1, \ldots, \mathbf{x}_n$ is a realization of the stochastic sequence of an independent and identically distributed sequence of $d$-dimensional random vectors $\tilde{\mathbf{x}}_1, \ldots, \tilde{\mathbf{x}}_n$ with common Radon-Nikodým density $p_e$ defined with respect to a support specification measure $\nu$. Consider a learning machine which observes a collection of $n$ $d$-dimensional vectors which correspond to the training data. The goal of the learning machine is to minimize the risk function $\ell : \mathcal{R}^q \rightarrow \mathcal{R}$ which is defined such that for all $\boldsymbol{\theta} \in \mathcal{R}^q$:

$$\ell(\boldsymbol{\theta}) = E\{c(\tilde{\mathbf{x}}_i, \boldsymbol{\theta})\} \tag{8.7}$$

where the continuous function $c : \mathcal{R}^d \times \mathcal{R}^q \rightarrow \mathcal{R}$ is the loss function specifying the empirical risk function

$$\hat{\ell}_n(\boldsymbol{\theta}) = (1/n) \sum_{i=1}^{n} c(\tilde{\mathbf{x}}_i, \boldsymbol{\theta}),$$

where the notation $\hat{\ell}_n(\boldsymbol{\theta})$ denotes a continuous function $\ell_n : \mathcal{R}^{dn} \times \mathcal{R}^q \rightarrow \mathcal{R}$ evaluated at $([\tilde{\mathbf{x}}_1, \ldots, \tilde{\mathbf{x}}_n], \boldsymbol{\theta})$.

The expected value of $\hat{\ell}_n(\boldsymbol{\theta})$, $E\{\hat{\ell}_n(\boldsymbol{\theta})\}$, may be explicitly computed using the formula:

$$E\{\hat{\ell}_n(\boldsymbol{\theta})\} = (1/n) \sum_{i=1}^{n} E\{c(\tilde{\mathbf{x}}_i, \boldsymbol{\theta})\} = E\{c(\tilde{\mathbf{x}}_i, \boldsymbol{\theta})\} = \int c(\mathbf{x}, \boldsymbol{\theta}) p_e(\mathbf{x}) d\nu(\mathbf{x}). \tag{8.8}$$

In the special case where $p_e$ is a probability mass function and $\tilde{\mathbf{x}}_i$ is a discrete random vector with support $\Omega$, then (8.8) becomes:

$$E\{\hat{\ell}_n(\boldsymbol{\theta})\} = \int c(\mathbf{x}, \boldsymbol{\theta}) p_e(\mathbf{x}) d\nu(\mathbf{x}) = \sum_{\mathbf{x} \in \Omega} c(\mathbf{x}, \boldsymbol{\theta}) p_e(\mathbf{x}).$$

In the special case where $p_e$ is a probability density function and $\tilde{\mathbf{x}}_i$ is an absolutely continuous random vector with support $\Omega$, then (8.8) becomes:

$$E\{\hat{\ell}_n(\boldsymbol{\theta})\} = \int c(\mathbf{x}, \boldsymbol{\theta}) p_e(\mathbf{x}) d\nu(\mathbf{x}) = \int_{\mathbf{x} \in \Omega} c(\mathbf{x}, \boldsymbol{\theta}) p_e(\mathbf{x}) d\mathbf{x}.$$

In the special case where $\tilde{\mathbf{x}}_i$ is a mixed random vector that may be partitioned such that $\tilde{\mathbf{x}}_i = [\tilde{\mathbf{y}}_i, \tilde{\mathbf{z}}_i]$ where $\tilde{\mathbf{y}}_i$ is a $k$-dimensional discrete random vector with support $\Omega_y$ and $\tilde{\mathbf{z}}_i$ is a $d-k$-dimensional absolutely continuous random vector with support $\Omega_z$ for $i = 1, 2, \ldots$, then (8.8) becomes:

$$E\{\hat{\ell}_n(\boldsymbol{\theta})\} = \int c(\mathbf{x}, \boldsymbol{\theta}) p_e(\mathbf{x}) d\nu(\mathbf{x}) = \sum_{\mathbf{y} \in \Omega_y} \int_{\mathbf{z} \in \Omega_z} c([\mathbf{y}, \mathbf{z}], \boldsymbol{\theta}) p_e([\mathbf{y}, \mathbf{z}]) d\mathbf{z}.$$

△

**Example 8.4.2** (Covariance Matrix Estimation). Let $\tilde{\mathbf{x}}_1, \ldots, \tilde{\mathbf{x}}_n$ be a sequence of independent and identically distributed $d$-dimensional random vectors with common $d$-dimensional mean vector $\boldsymbol{\mu} \equiv E\{\tilde{\mathbf{x}}_i\}$ and $d$-dimensional covariance matrix

$$\mathbf{C} \equiv E\left\{(\tilde{\mathbf{x}}_i - \boldsymbol{\mu})(\tilde{\mathbf{x}}_i - \boldsymbol{\mu})^T\right\}.$$

The *sample mean* $\tilde{\mathbf{m}}_n \equiv (1/n)\sum_{i=1}^{n} \tilde{\mathbf{x}}_i$. The expected value of the sample mean is computed using the formula:

$$E\{\tilde{\mathbf{m}}_n\} = (1/n)\sum_{i=1}^{n} E\{\tilde{\mathbf{x}}_i\} = \boldsymbol{\mu}.$$

If $\boldsymbol{\mu}$ is known, then the covariance matrix of $\tilde{\mathbf{x}}_i$ is estimated using the formula:

$$\tilde{\mathbf{C}}_n = \frac{1}{n}\sum_{i=1}^{n} (\tilde{\mathbf{x}}_i - \boldsymbol{\mu})(\tilde{\mathbf{x}}_i - \boldsymbol{\mu})^T$$

since $E\{\tilde{\mathbf{C}}_n\} = \mathbf{C}$.

In practice, however, $\boldsymbol{\mu}$ is not known and is often estimated using the sample mean $\tilde{\mathbf{m}}_n$. Show that the estimator

$$\hat{\mathbf{C}}_n \equiv \frac{1}{n-1}\sum_{i=1}^{n} (\tilde{\mathbf{x}}_i - \tilde{\mathbf{m}}_n)(\tilde{\mathbf{x}}_i - \tilde{\mathbf{m}}_n)^T$$

is an unbiased estimator of $\mathbf{C}$ (i.e., $E\{\hat{\mathbf{C}}_n\} = \mathbf{C}$).

**Solution.** The covariance matrix of the sample mean $\tilde{\mathbf{m}}_n$ is defined by the formula:

$$\mathbf{cov}(\tilde{\mathbf{m}}_n) \equiv E\left\{(\tilde{\mathbf{m}}_n - \boldsymbol{\mu})(\tilde{\mathbf{m}}_n - \boldsymbol{\mu})^T\right\}. \tag{8.9}$$

In addition, using the relation that the covariance of the sum of independent random vectors is the sum of the covariances:

$$\mathbf{cov}(\tilde{\mathbf{m}}_n) = \frac{1}{n^2}\sum_{i=1}^{n} \left[E\{\tilde{\mathbf{x}}_i\tilde{\mathbf{x}}_i^T\} - \boldsymbol{\mu}\boldsymbol{\mu}^T\right] = \mathbf{C}/n. \tag{8.10}$$

Now rewrite $\hat{\mathbf{C}}_n$ as:

$$\hat{\mathbf{C}}_n = \frac{1}{n-1}\sum_{i=1}^{n} ((\tilde{\mathbf{x}}_i - \boldsymbol{\mu}) - (\tilde{\mathbf{m}}_n - \boldsymbol{\mu}))((\tilde{\mathbf{x}}_i - \boldsymbol{\mu}) - (\tilde{\mathbf{m}}_n - \boldsymbol{\mu}))^T. \tag{8.11}$$

Now take expectations of (8.11) and arrange terms to obtain:

$$E\{\hat{\mathbf{C}}_n\} = \frac{1}{n-1}\left[n\mathbf{C} - n\mathbf{cov}(\tilde{\mathbf{m}}_n)\right] \tag{8.12}$$

which directly implies that:

$$E\{\hat{\mathbf{C}}_n\} = \frac{n}{n-1}\mathbf{C} - \frac{n}{n-1}\frac{\mathbf{C}}{n} = \frac{n-1}{n-1}\mathbf{C} = \mathbf{C}. \tag{8.13}$$

△

### 8.4.1 Random Functions

The concept of a random function frequently arises in mathematical statistics and statistical machine learning. For example, the prediction error of a learning machine is a function of the machine's parameters $\boldsymbol{\theta}$ as well as the training data which is a random vector. The prediction error in this context is an example of a random function in the sense that the prediction error is a function of the parameter values but the randomness is due to variation generated from the random training data.

**Definition 8.4.2** (Random Function). Let $\tilde{\mathbf{x}}$ be a $d$-dimensional random vector with support $\Omega \subseteq \mathcal{R}^d$. Let $c : \Omega \times \Theta \to \mathcal{R}$. Assume $c(\cdot, \boldsymbol{\theta})$ is Borel measurable for each $\boldsymbol{\theta} \in \Theta$. Then, the function $c : \Omega \times \Theta \to \mathcal{R}$ is called a *random function* with respect to $\tilde{\mathbf{x}}$. □

Recall that a sufficient condition for $c(\cdot, \boldsymbol{\theta})$ to be Borel measurable for each $\boldsymbol{\theta} \in \Theta$ is that $c(\cdot, \boldsymbol{\theta})$ is a piecewise continuous function on a finite partition of $\Omega$ for each $\boldsymbol{\theta} \in \Theta$ (see Theorem 8.2.2).

The notation $\tilde{c}(\cdot) \equiv c(\tilde{\mathbf{x}}, \cdot)$ or $\hat{c}(\cdot) \equiv c(\tilde{\mathbf{x}}, \cdot)$ is used as a short-hand notation to refer to a random function.

**Definition 8.4.3** (Piecewise-Continuous Random Function). Let $c : \Omega \times \Theta \to \mathcal{R}$ be a random function defined such that $c(\mathbf{x}, \cdot)$ is piecewise continuous on a finite partition of $\Theta$ for each $\mathbf{x} \in \Omega$. Then, the function $c : \Omega \times \Theta \to \mathcal{R}$ is called a *piecewise-continuous random function.* □

**Definition 8.4.4** (Continuous Random Function). Let $c : \Omega \times \Theta \to \mathcal{R}$ be a random function defined such that $c(\mathbf{x}, \cdot)$ is continuous on $\Theta$ for each $\mathbf{x} \in \Omega$. Then, the function $c : \Omega \times \Theta \to \mathcal{R}$ is called a *continuous random function.* □

**Definition 8.4.5** (Continuously Differentiable Random Function). Let $c : \Omega \times \Theta \to \mathcal{R}$ be a function defined such that the $k$th derivative of $c$ with respect to the domain $\Theta$ is a continuous random function for $k = 1, 2, \ldots, m$. Then, the function $c$ is called an *$m$-times continuously differentiable random function.* □

For example, if $c(\mathbf{x}; \boldsymbol{\theta})$ denotes a continuous random function $c$ and $dc/d\boldsymbol{\theta}$ is a continuous random function, then $c$ is called a continuously differentiable random function. If, in addition, $d^2c/d\boldsymbol{\theta}^2$ is continuous, then $d^2c/d\boldsymbol{\theta}^2$ is called a twice continuously differentiable function.

## 8.4.2 Expectations of Random Functions

**Definition 8.4.6** (Dominated by an Integrable Function). Let $\tilde{\mathbf{x}}$ be a $d$-dimensional random vector with Radon-Nikodým density $p : \mathcal{R}^d \to [0, \infty)$. Let $\Theta \subseteq \mathcal{R}^q$. Let $\mathbf{F} : \mathcal{R}^d \times \Theta \to \mathcal{R}^{m \times n}$ be a random function with respect to $\tilde{\mathbf{x}}$. Assume there exists a function $C : \mathcal{R}^d \to [0, \infty)$ such that the absolute value of each element of $\mathbf{F}(\mathbf{x}, \boldsymbol{\theta})$ is less than or equal to $C(\mathbf{x})$ for all $\boldsymbol{\theta} \in \Theta$ and for all $\mathbf{x}$ in the support of $\tilde{\mathbf{x}}$. In addition, assume $\int C(\mathbf{x})p(\mathbf{x})\nu(\mathbf{x})$ is finite. Then, the random function $\mathbf{F}$ is *dominated by an integrable function* on $\Theta$ with respect to $p$. □

**Example 8.4.3** (Ensuring an Expectation Exists). Explain how the statement that a random function $f : \mathcal{R}^d \times \mathcal{R}^q \to \mathcal{R}$ is dominated by an integrable function on $\mathcal{R}^q$ with respect to $p$ implies the expectation $E\{\mathbf{f}(\tilde{\mathbf{x}}, \boldsymbol{\theta})\}$ is finite for each $\boldsymbol{\theta} \in \mathcal{R}^q$.

**Solution.**

$$|E\{\mathbf{f}(\tilde{\mathbf{x}}, \boldsymbol{\theta})\}| = \left| \int \mathbf{f}(\mathbf{x}, \boldsymbol{\theta})p(\mathbf{x})d\nu(\mathbf{x}) \right| \leq \int |\mathbf{f}(\mathbf{x}, \boldsymbol{\theta})|p(\mathbf{x})d\nu(\mathbf{x}) \leq \int C(\mathbf{x})p(\mathbf{x})d\nu(\mathbf{x}) < \infty$$

provided that there exists a function $C : \mathcal{R}^d \to \mathcal{R}$ with the property that $|\mathbf{f}(\mathbf{x}, \boldsymbol{\theta})| \leq C(\mathbf{x})$ for all $\boldsymbol{\theta} \in \mathcal{R}^q$ and $\int C(\mathbf{x})p(\mathbf{x})d\nu(\mathbf{x}) < \infty$. △

The following theorem shows that the expectation of a continuous random function is dominated by an integrable function is continuous.

**Theorem 8.4.1** (Expectation of a Continuous Random Function Is Continuous). *Let $\tilde{\mathbf{x}}$ be a $d$-dimensional random vector with Radon-Nikodým density $p : \mathcal{R}^d \to [0, \infty)$ defined with* *respect to support specification measure $\nu$. Let $\Theta$ be a closed and bounded subset of $\mathcal{R}^q$. If a continuous random function $c : \mathcal{R}^d \times \Theta \rightarrow \mathcal{R}$ is dominated by an integrable function on $\Theta$ with respect to p, then*

$$E\{c(\tilde{\mathbf{x}}, \cdot)\} \equiv \int c(\mathbf{x}, \cdot) p(\mathbf{x}) d\nu(\mathbf{x})$$

*is continuous on $\Theta$.*

*Proof.* See Bartle (1966, Corollary 5.8). ■

Many machine learning applications involve training data generated from bounded random vectors. In such cases, the following theorem provides useful sufficient conditions for establishing such random functions are dominated by integrable functions.

**Theorem 8.4.2** (Dominated by a Bounded Function). *Let $\tilde{\mathbf{x}}$ be a d-dimensional random vector with Radon-Nikodým density $p : \mathcal{R}^d \rightarrow [0, \infty)$. Let $\Theta \subseteq \mathcal{R}^q$ be a closed and bounded set. Let $\mathbf{F} : \mathcal{R}^d \times \Theta \rightarrow \mathcal{R}^{m \times n}$ be a continuous random function with respect to $\tilde{\mathbf{x}}$. Assume either: (i) each element of $\mathbf{F}$ is a bounded function, or (ii) each element of $\mathbf{F}(\cdot, \cdot)$ is a piecewise continuous function on $\mathcal{R}^d \times \Theta$ and $\tilde{\mathbf{x}}$ is a bounded random vector. Then, $\mathbf{F}$ is dominated by an integrable function on $\Theta$ with respect to p.*

*Proof.* If either condition (i) or condition (ii) holds, then there exists a finite constant $K$ such that the absolute value of the $ij$th element of $\mathbf{F}(\mathbf{x}, \boldsymbol{\theta})$, $f_{ij}(\mathbf{x}, \boldsymbol{\theta})$, is less than or equal to $K$ with probability one. And since $p$ is a probability density,

$$\left| \int f_{ij}(\mathbf{x}, \boldsymbol{\theta}) p(\mathbf{x}) d\nu(\mathbf{x}) \right| \leq \int |f_{ij}(\mathbf{x}, \boldsymbol{\theta})| p(\mathbf{x}) d\nu(\mathbf{x}) \leq \int K p(\mathbf{x}) d\nu(\mathbf{x}) = K < \infty.$$

■

**Example 8.4.4** (Simple Conditions to Ensure Domination by an Integrable Function). If $\mathbf{F} : \mathcal{R}^d \times \Theta \rightarrow \mathcal{R}^{m \times n}$ is a piecewise continuous random function on $\mathcal{R}^d \times \Theta$ and $p : \mathcal{R}^d \rightarrow [0, 1]$ is a probability mass function on a finite sample space, then $\mathbf{F}$ is dominated by an integrable function on $\Theta$ with respect to $p$. △

The next theorem provides conditions ensuring the global minimizer of a continuous random function is a random vector.

**Theorem 8.4.3** (Continuous Random Function Minimizers Are Random Vectors). *Let $\Theta$ be a closed and bounded subset of $\mathcal{R}^q$. Let $c : \mathcal{R}^d \times \Theta \rightarrow \mathcal{R}$ be a continuous random function. There exists a Borel measurable function $\tilde{\boldsymbol{\theta}} : \mathcal{R}^d \rightarrow \mathcal{R}^q$ such that $\tilde{\boldsymbol{\theta}}(\mathbf{x})$ is a global minimizer of $c(\mathbf{x}, \cdot)$ on $\Theta$ for all $\mathbf{x} \in \mathcal{R}^d$.*

*Proof.* See Jennrich (1969, Lemma 2, p. 637). ■

Note that Theorem 8.4.3 also shows a local minimizer of a continuous random function on $\Theta$ is a random vector when $\Theta$ is chosen to be a sufficiently small neighborhood of that local minimizer.

**Example 8.4.5** (Global Minimizer of Empirical Risk Function Is a Random Vector). Let $\Theta$ be a closed and bounded subset of $\mathcal{R}^q$. Let $\tilde{\mathbf{x}}_i, \ldots, \tilde{\mathbf{x}}_n$ be a stochastic sequence of independent and identically distributed $d$-dimensional random vectors. Let $\hat{\ell}_n$ be an empirical risk function defined such that for all $\boldsymbol{\theta} \in \Theta$:

$$\hat{\ell}_n(\boldsymbol{\theta}) = (1/n) \sum_{i=1}^{n} c(\tilde{\mathbf{x}}_i, \boldsymbol{\theta})$$

where $c$ is a continuous random function. Show that a global minimizer of the random function $\hat{\ell}_n$ on $\Theta$ is a random vector.

**Solution.** Since $\hat{\ell}_n$ is a weighted sum of continuous random functions, it follows that $\hat{\ell}_n$ is a continuous random function and the desired conclusion follows from Theorem 8.4.3. △

The next theorem provides conditions for interchanging integral and differentiation operators.

**Theorem 8.4.4** (Interchange of Derivative and Integral Operators). *Let $\tilde{\mathbf{x}}$ be a $d$-dimensional random vector with Radon-Nikodým density $p : \mathcal{R}^d \to [0, \infty)$ defined with respect to support specification measure $\nu$. Let $\Theta$ be a closed and bounded subset of $\mathcal{R}^q$. Let the gradient of a continuously differentiable random function $c : \mathcal{R}^d \times \Theta \to \mathcal{R}$ be dominated by an integrable function on $\Theta$ with respect to $p$. Then,*

$$\nabla \int c(\mathbf{x}, \cdot)p(\mathbf{x})d\nu(\mathbf{x}) = \int \nabla c(\mathbf{x}, \cdot)p(\mathbf{x})d\nu(\mathbf{x}).$$

*Proof.* See Bartle, 1966, Corollary 5.9, p. 46. ■

### 8.4.3 Conditional Expectation and Independence

**Definition 8.4.7** (Conditional Expectation). Let $\tilde{\mathbf{x}}$ and $\tilde{\mathbf{y}}$ be $d-k$ and $k$-dimensional random vectors with joint Radon-Nikodým density $p : \mathcal{R}^d \to [0, \infty)$ and support specification measure $\nu(\mathbf{x})$. The *marginal density* of $\tilde{\mathbf{y}}$, $p_y : \mathcal{R}^k \to [0, \infty)$, is defined such that for each $\mathbf{y}$ in the support of $\tilde{\mathbf{y}}$:

$$p_y(\mathbf{y}) \equiv \int p(\mathbf{x}, \mathbf{y})d\nu(\mathbf{x}).$$

The *conditional probability density*, $p_{x|y}(\cdot|\mathbf{y}) : \mathcal{R}^{d-k} \to [0, \infty)$ is defined such that

$$p_{x|y}(\mathbf{x}|\mathbf{y}) = \frac{p(\mathbf{x}, \mathbf{y})}{p_y(\mathbf{y})} \tag{8.14}$$

for each $\mathbf{x}$ in the support of $\tilde{\mathbf{x}}$ and for each $\mathbf{y}$ in the support of $\tilde{\mathbf{y}}$. □

The terminology *marginal distribution* and *conditional probability distribution* refer to the cumulative distribution functions specified respectively by the marginal and conditional probability density functions. Note that this definition of the concept of conditional probability permits the definition of quantities such as the probability measure of the random vector $\tilde{\mathbf{x}}$ for a given value of $\tilde{\mathbf{y}}$ which is calculated using the formula for each $B \in \mathcal{B}^d$:

$$P(\tilde{\mathbf{x}} \in B|\mathbf{y}) = \int_B p(\mathbf{x}|\mathbf{y})d\nu(\mathbf{x}).$$

Semantically, the *conditional probability* $P(\tilde{\mathbf{x}} \in B|\tilde{\mathbf{y}} = \mathbf{y})$ is interpreted as the probability that $\tilde{\mathbf{x}} \in B$ given prior knowledge that $\tilde{\mathbf{y}} = \mathbf{y}$. For a given $\mathbf{y}$, $P(\cdot|\mathbf{y})$ specifies a new probability space on $(\mathcal{R}^d, \mathcal{B}^d)$ which has been adjusted to take into account prior knowledge that $\tilde{\mathbf{y}} = \mathbf{y}$.

Note that when $\tilde{\mathbf{y}}$ is an absolutely continuous random vector, the semantic interpretation of the density $p(\mathbf{x}|\mathbf{y})$ might seem puzzling since the probability of the event $\tilde{\mathbf{y}} = \mathbf{y}$ is zero. There are two possible semantic interpretations in this case. First, one can simply view $\mathbf{y}$ as a realization of $\tilde{\mathbf{y}}$ so that $p(\cdot|\mathbf{y})$ implicitly specifies a family of probability density functions indexed by the choice of $\mathbf{y}$. Second, one can interpret $p(\mathbf{x}|\tilde{\mathbf{y}})$ as a special type of random function with respect to the random vector $\tilde{\mathbf{y}}$ called a *random conditional density*.

Let $h : \mathcal{R}^d \rightarrow \mathcal{R}$ be a Borel measurable function. For each $\mathbf{y}$ in the support of $\tilde{\mathbf{y}}$, $E\{h(\tilde{\mathbf{x}}, \mathbf{y})|\mathbf{y})\}$ is defined (when it exists) such that

$$E\{h(\tilde{\mathbf{x}}, \mathbf{y})|\mathbf{y})\} = \int h(\mathbf{x}, \mathbf{y}) p_{x|y}(\mathbf{x}|\mathbf{y}) d\nu(\mathbf{x})$$

where $p_{x|y}(\cdot|\mathbf{y})$ is the Radon-Nikodým density for $P(\cdot|\mathbf{y})$. The quantity $E\{h(\tilde{\mathbf{x}}, \mathbf{y})|\mathbf{y})\}$ is called the *conditional expectation* of $h(\tilde{\mathbf{x}}, \mathbf{y})$ given $\mathbf{y}$.

In addition,

$$E\{h(\mathbf{x}, \tilde{\mathbf{y}})|\tilde{\mathbf{y}})\} \equiv \int h(\mathbf{x}, \tilde{\mathbf{y}}) p_{x|y}(\mathbf{x}|\tilde{\mathbf{y}}) d\nu(\mathbf{x})$$

may be interpreted as a Borel measurable function of $\tilde{\mathbf{y}}$ (i.e., a new random variable).

Suppose that knowledge of a realization of $\tilde{\mathbf{y}}$ has no influence on the probability distribution of $\tilde{\mathbf{x}}$. That is, for all $[\mathbf{x}, \mathbf{y}]$ in the support of $[\tilde{\mathbf{x}}, \tilde{\mathbf{y}}]$

$$p_{x|y}(\mathbf{x}|\mathbf{y}) = p_x(\mathbf{x}) \tag{8.15}$$

which means that knowledge that $\tilde{\mathbf{y}}$ takes on the value of $\mathbf{y}$ does not influence the probability distribution of $\tilde{\mathbf{x}}$. It follows from the conditional independence assumption in (8.15) that:

$$p_{x|y}(\mathbf{x}|\mathbf{y}) = \frac{p_{x,y}(\mathbf{x}, \mathbf{y})}{p_y(\mathbf{y})} = p_x(\mathbf{x})$$

and hence

$$p_{x,y}(\mathbf{x}, \mathbf{y}) = p_x(\mathbf{x}) p_y(\mathbf{y})$$

for all $[\mathbf{x}, \mathbf{y}]$ in the support of $[\tilde{\mathbf{x}}, \tilde{\mathbf{y}}]$. The random vectors $\tilde{\mathbf{x}}$ and $\tilde{\mathbf{y}}$ in this situation are called "independent random vectors".

**Definition 8.4.8** (Independent Random Vectors). Let $\tilde{\mathbf{x}} \equiv [\tilde{\mathbf{x}}_1, \ldots, \tilde{\mathbf{x}}_n]$ be a finite set of $n$ $d$-dimensional random vectors with probability distribution specified by Radon-Nikodým density $p : \mathcal{R}^{dn} \rightarrow [0, \infty)$ with support specification measure $\nu$. Let the marginal density for $\tilde{\mathbf{x}}_i$ be denoted as the Radon-Nikodým density $p_i : \mathcal{R}^d \rightarrow [0, \infty)$ defined such that:

$$p_i(\mathbf{x}_i) = \int p(\mathbf{x}_1, \ldots, \mathbf{x}_n) d\nu(\mathbf{x}_1, \ldots, \mathbf{x}_{i-1}, \mathbf{x}_{i+1}, \ldots, \mathbf{x}_n),$$

$i = 1, \ldots, n$. The random vectors $\tilde{\mathbf{x}}_1, \ldots, \tilde{\mathbf{x}}_d$ are *independent* if

$$p(\mathbf{x}_1, \ldots, \mathbf{x}_d) = \prod_{i=1}^{d} p_i(\mathbf{x}_i).$$

A countably infinite set $S$ of random vectors is *independent* if every finite subset of $S$ consists of independent random vectors. □

**Example 8.4.6** (Adaptive Learning Algorithm Analysis). Let $\mathbf{x}_1, \mathbf{x}_2, \ldots$ be a sequence of $d$-dimensional feature vectors which is a realization of a sequence $\tilde{\mathbf{x}}_1, \tilde{\mathbf{x}}_2, \ldots$ of independent and identically distributed $d$-dimensional random vectors with common Radon-Nikodým density $p_e : \mathcal{R}^d \rightarrow [0, \infty)$ defined with respect to a support specification measure $\nu$.

Let $c : \mathcal{R}^d \times \mathcal{R}^q \rightarrow \mathcal{R}$ be a continuously differentiable random function defined such that $c(\mathbf{x}, \boldsymbol{\theta})$ is the loss incurred by the learning machine when it encounters a training vector $\mathbf{x}$ for a given $q$-dimensional parameter vector $\boldsymbol{\theta}$. Assume, in addition, $c$ is piecewise continuous in its first argument. Let $\ell(\boldsymbol{\theta}) = E\{c(\tilde{\mathbf{x}}_t, \boldsymbol{\theta})\}$.

The adaptive learning algorithm works by first choosing an initial guess for the parameter vector $\tilde{\theta}(0)$ which is a realization of the random vector $\tilde{\theta}(0)$. Next, the adaptive learning algorithm updates its current random parameter vector $\tilde{\theta}(t)$ to obtain a new random parameter vector using the formula:

$$\tilde{\theta}(t+1) = \tilde{\theta}(t) - \gamma_t \left( \frac{dc(\tilde{\mathbf{x}}_t, \tilde{\theta}(t))}{d\theta} \right)^T \tag{8.16}$$

where the $\gamma_1, \gamma_2, \ldots$ is a sequence of positive learning rates. Note that the stochastic sequence $\tilde{\theta}(0), \tilde{\theta}(1), \ldots$ is not a sequence of independent and identically distributed random vectors.

Show that (8.16) may be rewritten as a "noisy" batch gradient descent learning algorithm that is minimizing $\ell(\theta)$ by rewriting (8.16) such that:

$$\tilde{\theta}(t+1) = \tilde{\theta}(t) - \gamma_t \left( \frac{d\ell(\theta(t))}{d\theta} \right)^T - \gamma_t \tilde{\mathbf{n}}(t) \tag{8.17}$$

where $\tilde{\mathbf{n}}(t)$ is a random vector whose time-varying mean vector $\mu_t \equiv E\{\tilde{\mathbf{n}}(t)|\theta\}$ is a vector of zeros and whose covariance matrix $\mathbf{C}_t \equiv E\{(\tilde{\mathbf{n}}(t) - \mu_t)(\tilde{\mathbf{n}}(t) - \mu)^T|\theta\}$ is proportional to $\gamma_t^2$.

**Solution.** Substitute

$$\tilde{\mathbf{n}}(t) = \left( \frac{dc(\tilde{\mathbf{x}}_t, \tilde{\theta})}{d\theta} - \frac{d\ell(\tilde{\theta}(t))}{d\theta} \right)^T$$

into (8.17) to obtain (8.16).

Then,

$$\mu(t) = E\{\tilde{\mathbf{n}}(t)|\theta\} = E\left\{ \frac{dc(\tilde{\mathbf{x}}_t, \theta)}{d\theta} \middle| \theta \right\} - \frac{d\ell(\theta)}{d\theta} = \mathbf{0}_q$$

for all $\theta$. Then,

$$\mu_t = \int E\{\tilde{\mathbf{n}}(t)|\theta\} p_{\theta(t)}(\theta) d\theta = \mathbf{0}_q,$$

where $p_{\theta(t)}$ is the density specifying the distribution of $\tilde{\theta}(t)$ at iteration $t$. Using the result that $\mu_t = \mathbf{0}_q$, the covariance matrix

$$\mathbf{C}_t = E\{(\gamma_t \tilde{\mathbf{n}}(t))(\gamma_t \tilde{\mathbf{n}}(t))^T|\theta\} = \gamma_t^2 E\{\tilde{\mathbf{n}}(t)\tilde{\mathbf{n}}(t)^T|\theta\}.$$

## Exercises

8.4-1. Let $\tilde{x}$ be a random variable with support $\Omega \subseteq \mathcal{R}$. Assume $\Theta$ is a closed and bounded subset of $\mathcal{R}$. Let $g : \Omega \times \Theta \to \mathcal{R}$ be defined such that $g(x, \theta) = x\theta$ for all $(x, \theta) \in \mathcal{R} \times \mathcal{R}$. Show that the function $g$ is a random function. Is $g$ a twice continuously differentiable random function?

8.4-2. What additional assumptions would be sufficient to ensure that the random function $g$ defined in Exercise 8.4-1 is dominated by an integrable function? Are additional assumptions required to ensure that $E\{g(\tilde{x}, \cdot)\}$ is continuous on $\Theta$? If so, list those assumptions. Are additional assumptions required to ensure that $dE\{g(\tilde{x}, \cdot)\}/d\theta = E\{dg(\tilde{x}, \cdot)/d\theta\}$? If so, list those assumptions.

## 8.5 Concentration Inequalities

Concentration inequalities have an important role in the field of statistical machine learning. The following example of a "concentration inequality" is the Markov inequality.

**Theorem 8.5.1** (Markov Inequality). *If $\tilde{x}$ is a random variable with Radon-Nikodým density $p : \mathcal{R} \to [0, \infty)$ such that*

$$E[|\tilde{x}|^r] \equiv \int |x|^r p(x) d\nu(x)$$

*is finite, then for every positive number $\epsilon$ and every positive number $r$:*

$$p(|\tilde{x}| \geq \epsilon) \leq \frac{E[|\tilde{x}|^r]}{\epsilon^r}. \tag{8.18}$$

*Proof.* Let $\phi_\epsilon : \mathcal{R} \to \mathcal{R}$ be defined such that $\phi_\epsilon(a) = 1$ if $a \geq \epsilon$ and $\phi_\epsilon(a) = 0$ if $a < \epsilon$. Let $\tilde{y} = \phi_\epsilon(\tilde{x})$ where $\tilde{x}$ is a random variable such that $E\{|\tilde{x}|^r\} < C$. Also note that

$$|\tilde{x}|^r \geq |\tilde{x}|^r \tilde{y} \geq \epsilon^r \tilde{y}.$$

Thus, $E\{|\tilde{x}|^r\} \geq \epsilon^r E\{\tilde{y}\}$. And since:

$$E\{\tilde{y}\} = (1)p(|\tilde{x}| \geq \epsilon) + (0)p(|\tilde{x}| < \epsilon),$$

$$p(|\tilde{x}| \geq \epsilon) \leq \frac{E\{|\tilde{x}|^r\}}{\epsilon^r}.$$

■

Note that when $r = 2$ in (8.18), then (8.18) is referred to as the Chebyshev Inequality.

**Example 8.5.1** (Chebyshev Inequality Generalization Error Bound). Assume the prediction errors $\tilde{c}_1, \ldots, \tilde{c}_n$ for $n$ training stimuli are bounded, independent, and identically distributed. The average prediction error computed with respect to the training data is given by the formula:

$$\tilde{\ell}_n = (1/n) \sum_{i=1}^{n} \tilde{c}_i$$

and the expected value of $\tilde{\ell}_n$ is denoted by $\ell^*$. Use the Markov Inequality with $r = 2$ to estimate how large of a sample size $n$ will be sufficient to ensure that the probability that $|\tilde{\ell}_n - \ell^*| < \epsilon$ is greater than $(1 - \alpha)100\%$ where $0 < \alpha < 1$.

**Solution.** The variance of $\tilde{\ell}_n$, $\sigma^2$, may be estimated using the formula $(1/n)\hat{\sigma}^2$ where

$$\hat{\sigma}^2 = \frac{1}{n-1} \sum_{i=1}^{n} (\tilde{c}_i - \tilde{\ell}_n)^2. \tag{8.19}$$

Using the Chebyshev Inequality, note that:

$$p\left(|\tilde{\ell}_n - \ell^*| < \epsilon\right) \geq 1 - \frac{\hat{\sigma}^2}{n\epsilon^2}. \tag{8.20}$$

A sufficient condition for ensuring the left-hand side of (8.20) is greater than $1 - \alpha$ is to require that the right-hand side of (8.20) is greater than $1 - \alpha$. This implies that:

$$1 - \frac{\hat{\sigma}^2}{n\epsilon^2} > 1 - \alpha \text{ or equivalently that } n > \frac{\hat{\sigma}^2}{\alpha\epsilon^2}.$$

△

**Theorem 8.5.2** (Hoeffding's Inequality). *Let $\tilde{x}_1, \tilde{x}_2, \ldots, \tilde{x}_n$ be a sequence of $n$ independent scalar random variables with common mean $\mu \equiv E\{\tilde{m}_n\}$. Assume, in addition, that $X_{min} \leq \tilde{x}_i \leq X_{max}$ for finite numbers $X_{min}$ and $X_{max}$ for $i = 1, \ldots, n$. Let $\tilde{m}_n \equiv (1/n) \sum_{i=1}^{n} \tilde{x}_i$. Then, for every positive number $\epsilon$:*

$$P(|\tilde{m}_n - \mu| \geq \epsilon) \leq 2 \exp\left( \frac{-2n\epsilon^2}{(X_{max} - X_{min})^2} \right).$$

*Proof.* Follows from Theorem D.1 of Mohri et al. (2018). ■

**Example 8.5.2** (Hoeffding Inequality Generalization Error Bound). Rework Example 8.5.1 but use the Hoeffding Inequality to estimate how large of a sample size $n$ is sufficient for the probability that $|\tilde{\ell}_n - \ell^*| < \epsilon$ is greater than $(1-\alpha)100\%$ where $0 < \alpha < 1$. Assume the difference, $D$, between the largest possible value of $\tilde{c}_i$ and the smallest possible value of $\tilde{c}_i$ is equal to $2\hat{\sigma}$ where $\hat{\sigma}$ by (8.19).

**Solution.** Using the Hoeffding Inequality, note that:

$$p\left(|\tilde{\ell}_n - \ell^*| < \epsilon\right) \geq 1 - 2\exp\left(\frac{-2n\epsilon^2}{(2\hat{\sigma})^2}\right). \tag{8.21}$$

A sufficient condition for ensuring that the left-hand side of (8.21) is greater than $1-\alpha$ is to require that the right-hand side of (8.21) is greater than $1-\alpha$. Using this constraint,

$$1 - 2\exp\left(-\frac{2n\epsilon^2}{4\sigma^2}\right) > 1 - \alpha \text{ or equivalently that } n > \frac{2\log(2/\alpha)\sigma^2}{\epsilon^2}.$$

Notice that the lower bound for the sample size $n$ increases at a rate of $1/\alpha$ for the Chebyshev Inequality Generalization Error Bound in Example 8.5.1, but increases only at a rate of $\log(1/\alpha)$ for the Hoeffding Inequality. For example, with $\alpha = 10^{-6}$, $1/\alpha = 10^6$ and $\log(1/\alpha) \approx 6$. That is, when applicable, the Hoeffding Inequality tends to give much tighter bounds than the Chebyshev Inequality. △

## Exercises

8.5-1. In Example 8.5.1 and Example 8.5.2, it was assumed that $\epsilon$ was given and the goal was to calculate a lower bound for the sample size $n$ using both the Markov and Hoeffding Inequalities. Rework the calculations in Example 8.5.1 and Example 8.5.2 but assume that the sample size $n$ and $\alpha$ are known and that the goal is to find a choice of $\epsilon$ using the Markov and Hoeffding Inequalities.

8.5-2. In Example 8.5.1 and Example 8.5.2, it was assumed that $\epsilon$ and $\alpha$ were given and the goal was to calculate a lower bound for the sample size $n$ using both the Markov and Hoeffding Inequalities. Rework the calculations in Example 8.5.1 and Example 8.5.2 but assume that the sample size $n$ and $\epsilon$ are known and that the goal is to find $\alpha$ using the Markov and Hoeffding Inequalities.

## 8.6 Further Readings

### Measure Theory

The topics covered in this chapter are typically taught in a graduate-level course in measure theory (e.g., Bartle 1966), probability theory (e.g., Karr 1993; Davidson 2002; Lehmann and Casella 1998; Lehmann and Roman 2005; Billingsley 2012; Rosenthal 2016), or advanced econometrics (e.g., White, 1994, 2001). The presentation in this chapter is relatively unique because it assumes no prior background in measure theory or Lebesgue integration yet attempts to provide a concise introduction to random vectors, random functions, the Radon-Nikodým density, and expectation operations using the Radon-Nikodým density. The goal of this presentation is to provide enough relevant background to support usual applications of the density for probabilistic and statistical inference. These topics are relevant because subsequent theorems in the text make specific claims about mixed random vectors as well as discrete and absolutely continuous random vectors and these claims cannot be easily communicated without this more general formalism.

The goal of this chapter was to provide some relevant technical details to support correct usage and understanding of the Radon-Nikodým concept as well as encourage further reading in this area. The books by Bartle (1966) and Rosenthal (2016) are recommended for readers desiring a more comprehensive introduction.

The non-standard terminology *support specification measure* is introduced here for pedagogical reasons and to emphasize the functional role of the sigma-finite measure used to specify the Radon-Nikodým density.

### Banach-Tarski Paradox Illustrates Importance of Borel Sigma-Fields

Pathological subsets of $\mathcal{R}^d$ which are not members of $\mathcal{B}^d$ are rarely encountered in engineering practice. Such pathological subsets have undesirable properties that greatly complicate the development of theories that assign probabilities to subsets of $\mathcal{R}^d$. Note that assigning a probability to a subset in a $d$-dimensional space is essentially equivalent to integrating a function over that subset. Without the assumption that integration is only permitted over a member of the Borel sigma-field, the resulting mathematical theory of integration is not a proper model of volume invariance.

For example, the Banach-Tarski Theorem, states that a solid sphere can be divided into a finite number of pieces and then those pieces can be rearranged to form two solid spheres each of the same volume and shape (e.g., see Wapner 2005 for an introductory review). The conclusion of the Banach-Tarski Theorem does not hold if one models a physical sphere as a collection of a <u>finite</u> number of electrons, neutrons, and protons. The conclusion of the Banach-Tarski Theorem also does not hold if one models a physical sphere as an infinite number of points provided that the "pieces of the sphere" are elements of the Borel sigma-field generated from the open subsets of $\mathcal{R}^3$.

### Concentration Inequalities

Concentration inequalities play an important role in discussions of machine learning behavior in computer science (e.g., see Mohri et al. 2012 and Shalev-Shwartz and Ben-David 2014). Boucheron, Lugosi, and Massart (2016) provides a comprehensive discussion of concentration inequalities.

# 9

# *Stochastic Sequences*

**Learning Objectives**

- Describe a data generating process as a stochastic sequence.
- Describe the major types of stochastic convergence.
- Apply law of large numbers for mixed random vectors.
- Apply central limit theorem for mixed random vectors.

In Chapter 5, the concept of a sequence of vectors converging deterministically to a particular vector was introduced and then the dynamics of deterministic systems were analyzed in Chapters 6 and 7. In many real-world applications, however, it is necessary to discuss situations involving the convergence of a *sequence of random vectors* which might converge to either a particular vector or a specific random vector. For example, consider a deterministic learning machine which updates its parameters when an environmental event is observed. Although the learning machine is deterministic, the parameter updating process is driven by events which occur with different probabilities in the learning machine's environment. Thus, an investigation into the asymptotic behavior of the sequence of parameter updates for a learning machine corresponds to an investigation of the convergence of a sequence of random vectors (e.g., see Chapter 12). The analysis of methods for implementing stochastic search or approximating expectations (e.g., see Chapter 11) also require a theory of convergence of random vectors.

Concepts of convergence of random vectors are also fundamentally important in studies of generalization performance (see Chapters 13, 14, 15, 16). Assume the training data for a learning machine is a set of $n$ training vectors which may be interpreted as a realization of $n$ random vectors with some common probability distribution. The $n$ random vectors are called the "data generating process". If the learning machine is trained with multiple realizations of the data generating process, this results in multiple parameter estimates. The generalization performance of the learning machine can be characterized by the variation in these estimated parameter values across these data sets. In other words, the generalization performance of a learning machine can be investigated by characterizing the probability distribution of its estimated parameter values in terms of the learning machine's statistical environment (see Figure 1.1; also see Section 14.1).

## 9.1 Types of Stochastic Sequences

**Definition 9.1.1** (Stochastic Sequence). Let $(\Omega, \mathcal{F}, P)$ be a probability space. Let $\tilde{\mathbf{x}}_1 : \Omega \rightarrow \mathcal{R}^d, \tilde{\mathbf{x}}_2 : \Omega \rightarrow \mathcal{R}^d, \ldots$ be a sequence of $d$-dimensional random vectors. The sequence of random vectors $\tilde{\mathbf{x}}_1, \tilde{\mathbf{x}}_2, \ldots$ is called a *stochastic sequence* on $(\Omega, \mathcal{F}, P)$. If $\omega \in \Omega$, then $\tilde{\mathbf{x}}_1(\omega), \tilde{\mathbf{x}}_2(\omega), \ldots$ is called a *sample path* of $\tilde{\mathbf{x}}_1, \tilde{\mathbf{x}}_2, \ldots$. □

A sample path of a stochastic sequence may also be called a *realization* of the stochastic sequence. A realization of a stochastic sequence is sometimes interpretable as an orbit of a discrete-time deterministic dynamical system. Sample paths may also be referred to as trajectories in some contexts.

**Definition 9.1.2** (Stationary Stochastic Sequence). Let $\tilde{\mathbf{x}}_1, \tilde{\mathbf{x}}_2, \ldots$ be a stochastic sequence of $d$-dimensional random vectors. If for every positive integer $n$: the joint distribution of $\tilde{\mathbf{x}}_1, \ldots, \tilde{\mathbf{x}}_n$ is identical to the joint distribution of $\tilde{\mathbf{x}}_{k+1}, \ldots, \tilde{\mathbf{x}}_{k+n}$ for $k = 1, 2, 3, \ldots$, then $\tilde{\mathbf{x}}_1, \tilde{\mathbf{x}}_2, \ldots$ is called a *stationary stochastic sequence.* □

Every stationary stochastic sequence of random vectors consists of identically distributed random vectors but a sequence of identically distributed random vectors is not necessarily stationary. The terminology *strict-sense stationary stochastic sequence* is also used to denote a stationary stochastic sequence.

Some nonstationary stochastic sequences share some key characteristics with stationary stochastic sequences. An important example of such stochastic sequences is now discussed.

**Definition 9.1.3** (Cross-Covariance Function). Let $\tilde{\mathbf{x}}_1, \tilde{\mathbf{x}}_2, \ldots$ be a stochastic sequence of $d$-dimensional random vectors. Let $t, \tau \in \mathbb{N}$. Assume the *mean function* $\mathbf{m}_t \equiv E\{\tilde{\mathbf{x}}_t\}$ is finite. Assume a function $\mathbf{R}_t : \mathbb{N}^+ \rightarrow \mathcal{R}^{d \times d}$ exists such that the elements of

$$\mathbf{R}_t(\tau) = E\{(\tilde{\mathbf{x}}_{t-\tau} - \mathbf{m}_{t-\tau})(\tilde{\mathbf{x}}_t - \mathbf{m}_t)^T\}$$

are finite numbers for all $\tau \in \mathbb{N}$. Then, the function $\mathbf{R}_t$ is called the *cross-covariance function* at time $t$. The function $\mathbf{R}_t(0)$ is called the *covariance matrix* at time $t$. □

**Definition 9.1.4** (Wide-Sense Stationary Stochastic Sequence). Let $\tilde{\mathbf{x}}_1, \tilde{\mathbf{x}}_2, \ldots$ be a stochastic sequence of $d$-dimensional random vectors. Assume that $E\{\tilde{\mathbf{x}}_t\}$ is equal to a finite constant $\mathbf{m}$ for all $t \in \mathbb{N}^+$. Assume that a function $\mathbf{R} : \mathbb{N}^+ \rightarrow \mathcal{R}^d$ exists such that for all $t, \tau \in \mathbb{N}^+$:

$$E\{(\tilde{\mathbf{x}}_{t-\tau} - \mathbf{m}_{t-\tau})(\tilde{\mathbf{x}}_t - \mathbf{m}_t)^T\} = \mathbf{R}(\tau).$$

Then, the stochastic sequence $\tilde{\mathbf{x}}_1, \tilde{\mathbf{x}}_2, \ldots$ is called a *wide-sense stationary stochastic sequence* with *autocovariance function* $\mathbf{R}$. In, addition, $\mathbf{R}(0)$ is called the *covariance matrix* for the stochastic sequence. □

The terminology *weak-sense stationary stochastic sequence* is also used to refer to a wide-sense stationary stochastic sequence. If a stochastic sequence is stationary, then the stochastic sequence is wide-sense stationary but the converse does not necessarily hold. The covariance matrix $\tilde{\mathbf{x}}(t)$ for a wide-sense stationary stochastic sequence with autocovariance function $\mathbf{R}$ is given by the formula $\mathbf{R}(0)$.

**Definition 9.1.5** ($\tau$-Dependent Stochastic Sequence). Let $(\Omega, \mathcal{F}, P)$ be a probability space. Let $\tilde{\mathbf{x}}_1 : \Omega \rightarrow \mathcal{R}^d, \tilde{\mathbf{x}}_2 : \Omega \rightarrow \mathcal{R}^d, \ldots$ be a sequence of $d$-dimensional random vectors. Assume that for all non-negative integer $t$ and for all non-negative integer $s$: $\tilde{\mathbf{x}}_t$ and $\tilde{\mathbf{x}}_s$ are independent for all $|t - s| > \tau$ where $\tau$ is a non-negative integer. □

Note that a $\tau$-dependent stochastic sequence of random vectors with $\tau = 0$ is a stochastic sequence of independent random vectors.

**Example 9.1.1** (Stochastic Sequence that Is Not $\tau$-Dependent). Let $\tilde{x}_1, \tilde{x}_2, \ldots$ be a stochastic sequence of random variables defined such that $\tilde{x}_0$ is a random variable with a uniform probability distribution on the interval $[0, 1]$. In addition, assume that $\tilde{x}_t = (0.5)\tilde{x}_{t-1}$ for $t = 1, 2, 3, \ldots$. This stochastic sequence is not $\tau$-dependent. To see this, note $\tilde{x}_t = (0.5)^t \tilde{x}_0$ implies that there does not exist some finite integer $\tau$ such that if $t > \tau$, then $\tilde{x}_t$ and $\tilde{x}_0$ are independent. On the other hand, from a practical perspective, if $t = 100$, for example, then we have $\tilde{x}_{100} = (0.5)^{100}\tilde{x}_0$ showing that the random variables $\tilde{x}_{100}$ and $\tilde{x}_0$ are "approximately $\tau$-dependent" for $\tau = 100$ but not technically $\tau$-dependent for any positive integer $\tau$. △

**Definition 9.1.6** (I.I.D. Stochastic Sequence). A stochastic sequence of *independent and identically distributed*, or i.i.d., random vectors is a stochastic sequence of independent random vectors such that each random vector in the sequence has the same probability distribution. □

**Example 9.1.2** (Sampling with Replacement Generates I.I.D. Data). Let

$$S \equiv \{\mathbf{x}^1, \ldots, \mathbf{x}^M\}$$

be a set of $M$ feature vectors. Assume the statistical environment samples with replacement from the set $S$ $n$ times to generate a training data set $\mathcal{D}_n \equiv \{\mathbf{x}_1, \ldots, \mathbf{x}_n\}$ where $\mathbf{x}_i$ is the realization of the $i$th sampling event for $i = 1, \ldots, n$. The outcome of the $i$th sampling event is an element of the set $S$. Define a stochastic sequence such that the observed data set $\mathcal{D}_n$ is a realization of some random data set $\tilde{\mathcal{D}}_n$.

**Solution.** Let $\tilde{\mathcal{D}}_n \equiv \{\tilde{\mathbf{x}}(1), \ldots, \tilde{\mathbf{x}}(n)\}$ be a stochastic sequence of $n$ i.i.d. random vectors with common probability mass function $P : S \to \{1/M\}$ defined such that $P(\mathbf{x}^k) = 1/M$ for $k = 1, \ldots, M$. △

**Definition 9.1.7** (Bounded Stochastic Sequence). A stochastic sequence $\tilde{\mathbf{x}}_1, \tilde{\mathbf{x}}_2, \ldots$ of random vectors is *bounded* if there exists a finite positive number $C$ such that $|\tilde{\mathbf{x}}_k| \leq C$ with probability one for $k = 1, 2, \ldots$. □

It is important to emphasize that the number $C$ in the definition of a Bounded Stochastic Sequence is a constant. For example, let $\tilde{x}_k$ be a binary random variable taking on the values of 0 and 1 for $k = 1, 2, \ldots$. The stochastic sequence $\{\tilde{x}_k\}$ is bounded since there exists a finite number (e.g., 2) such that $|\tilde{x}_k| \leq 2$ for $k = 1, 2, \ldots$. Suppose, however, $\tilde{x}_k$ is an absolutely continuous random variable with support $\mathcal{R}$ such as a Gaussian random variable for $k = 1, 2, \ldots$. Although it is true that $|\tilde{x}_k| < \infty$ for $k = 1, 2, \ldots$, there does not exist a finite number $C$ such that $|\tilde{x}_k| \leq C$ in this case. For a third example, let $\tilde{y}_k = k\tilde{x}_k$ where $\tilde{x}_k$ is a binary random variable taking on the values of 0 or 1. Although it is true that $\tilde{y}_k$ is a bounded random variable for a particular value of $k$, the stochastic sequence $\{\tilde{y}_k\}$ is not bounded. In machine learning, stochastic sequences of discrete random vectors on a finite sample space are common and such sequences are bounded stochastic sequences.

The following weaker version of a bounded stochastic sequence is also useful in applications.

**Definition 9.1.8** (Bounded in Probability Stochastic Sequence). A stochastic sequence $\tilde{\mathbf{x}}_1, \tilde{\mathbf{x}}_2, \ldots$ of random vectors defined on a probability space $(\Omega, \mathcal{F}, P)$ is *bounded in probability* if for every positive number $\epsilon$ less than one there exists a finite positive number $C_\epsilon$ and there exists a finite positive integer $T_\epsilon$ such that for all positive integer $t$ greater than $T_\epsilon$:

$$P\left(|\tilde{\mathbf{x}}_t| \leq C_\epsilon\right) \geq 1 - \epsilon.$$

□

## Exercises

9.1-1.A deterministic sequence of $n$ feature vectors are generated by sampling with replacement from a finite set $S$ where $S$ consists of $M$ distinct training stimuli where each feature vector is an element of $\{0,1\}^d$. The sequence of $n$ training stimuli is interpreted as the realization of a stochastic sequence of $n$ random feature vectors. Is this stochastic sequence a stationary stochastic sequence? Is this stochastic sequence a $\tau$-dependent stochastic sequence? Is this stochastic sequence *i.i.d.*? Is this stochastic sequence a bounded stochastic sequence?

9.1-2.Let $\tilde{\mathbf{s}}(1), \tilde{\mathbf{s}}(2), \ldots$ be a bounded stochastic sequence of independent and identically distributed random vectors. Let $\gamma_1, \gamma_2, \ldots$ be a sequence of positive numbers. Consider the adaptive learning algorithm defined such that

$$\tilde{\boldsymbol{\theta}}(t+1) = \tilde{\boldsymbol{\theta}}(t) - \gamma_t \mathbf{g}(\tilde{\boldsymbol{\theta}}(t), \tilde{\mathbf{s}}(t))$$

for $t = 1, 2, \ldots$. Assume $\tilde{\boldsymbol{\theta}}(1)$ is a vector of zeros. Show that the stochastic sequence $\tilde{\boldsymbol{\theta}}(1), \tilde{\boldsymbol{\theta}}(2), \ldots$ is a stochastic sequence which is not stationary and not $\tau$-dependent.

## 9.2 Partially Observable Stochastic Sequences

In many applications of machine learning, partially observable data or latent (i.e., completely unobservable) variables are encountered. Suppose that $\mathbf{x}_t$ is a 4-dimensional vector whose various components are not always fully observable. Thus, the first training stimulus might be denoted as $\mathbf{x}_1 = [12, 15, 16]$, the second training stimulus might be denoted as $\mathbf{x}_2 = [?, -16, 27]$, and a third training stimulus might be denoted as $\mathbf{x}_3 = [5, ?, ?]$. Thus, only for the first training stimulus are all of the features observable. A convenient mathematical model for representing this important class of data generating processes which generate partially observable data is to assume that the statistical environment first generates a "complete data" training stimulus realized as $\mathbf{x}_t = [x_1, x_2, x_3]$ according to some probability distribution and then generates a "mask" realized as $\mathbf{m}_t = [m_1, m_2, m_3]$ where the probability distribution of observing a particular mask $\mathbf{m}_t$ is only functionally dependent upon $\mathbf{x}_t$. The mask $\mathbf{m}_t$ is a vector of zeros and ones where the presence of a zero for the $j$th element of $\mathbf{m}_t$ indicates the $j$th element of $\mathbf{x}_t$ is not observable and the presence of a one for the $j$th element of $\mathbf{m}_t$ indicates that the $j$th element of $\mathbf{x}_t$ is observable.

The following definition of a partially observable data generating process follows Golden, Henley, White, and Kashner (2019).

**Definition 9.2.1** (I.I.D. Partially Observable Data Generating Process). Let

$$(\tilde{\mathbf{x}}_1, \tilde{\mathbf{m}}_1), (\tilde{\mathbf{x}}_2, \tilde{\mathbf{m}}_2), \ldots \tag{9.1}$$

be a stochastic sequence consisting of independent and identically distributed $2d$-dimensional random vectors with common joint density

$$p_e : \mathcal{R}^d \times \{0,1\}^d \to [0, \infty)$$

defined with respect to some support specification measure $\nu([\mathbf{x}, \mathbf{m}])$. The stochastic sequence $(\tilde{\mathbf{x}}_1, \tilde{\mathbf{m}}_1), (\tilde{\mathbf{x}}_2, \tilde{\mathbf{m}}_2), \ldots$ is called an *i.i.d. partially observable data process.* A realization of $(\tilde{\mathbf{x}}_t, \tilde{\mathbf{m}}_t)$ is called a *partially observable data record.* A realization of $\tilde{\mathbf{x}}_t$ is called a *complete data vector* and a realization of the binary random vector $\tilde{\mathbf{m}}_t$ is called a *mask vector*, $t = 1, 2, \ldots$. □

Let $d_{\mathbf{m}} \equiv \mathbf{1}_d^T \mathbf{m}$. The *observable-data selection function* $\mathbf{S} : \{0,1\}^d \rightarrow \{0,1\}^{d_{\mathbf{m}} \times d}$ is defined such that the element in row $j$ and column $k$ of $\mathbf{S}(\mathbf{m})$ is equal to one if the $j$th non-zero element in $\mathbf{m}$ is the $k$th element of $\mathbf{m}$. Let the *unobservable-data selection function* $\bar{\mathbf{S}} : \{0,1\}^d \rightarrow \{0,1\}^{(d-d_{\mathbf{m}}) \times d}$ be defined such that the element in row $j$ and column $k$ of $\bar{\mathbf{S}}(\mathbf{m})$ is equal to one if the $j$th element in $\mathbf{m}$ with value zero is the $k$th element of $\mathbf{m}$.

The components

$$\tilde{\mathbf{v}}_t \equiv [\mathbf{S}(\tilde{\mathbf{m}}_t)]\tilde{\mathbf{x}}_t \text{ and } \tilde{\mathbf{h}}_t \equiv [\bar{\mathbf{S}}(\tilde{\mathbf{m}}_t)]\tilde{\mathbf{x}}_t$$

respectively correspond to the *observable* and *unobservable* components of $(\tilde{\mathbf{x}}_t, \tilde{\mathbf{m}}_t)$. This construction of the unobservable and observable components of the partially observable data generating process implies that the *observable data stochastic sequence* $(\tilde{\mathbf{v}}_1, \tilde{\mathbf{m}}_1), (\tilde{\mathbf{v}}_2, \tilde{\mathbf{m}}_2), \ldots$ is i.i.d.

Some important i.i.d. partially observable data processes in the field of machine learning have the property that a subset of the elements of the complete data vector $\tilde{\mathbf{x}}_t$ are *never* observable. In such cases, those elements of $\tilde{\mathbf{x}}_t$ are called *hidden random variables.* An element of $\tilde{\mathbf{x}}_t$ that is *always* observable is called a *visible random variable* or a *fully observable random variable.*

The conditional density $p_e(\mathbf{x}, \mathbf{m})$ of the full specification of a partially observable data record may be factored such that $p_e(\mathbf{x}, \mathbf{m}) = p_x(\mathbf{x}) p_{m|x}(\mathbf{m}|\mathbf{x})$. The density $p_x$ is called the *complete-data density.* The density $p_{m|x}$ is called the *missing-data mechanism.*

This factorization may be semantically interpreted as the following specific algorithm for generating partially observable data. First, a complete-data record $\mathbf{x}$ is sampled from the complete data density $p_x$. Second, this complete-data record $\mathbf{x}$ and the missing-data mechanism $p_{m|x}$ are used to generate a missing-data mask $\mathbf{m}$ by sampling from $p_{m|x}(\cdot|\mathbf{x})$. Third, if the $k$th element of the generated mask $\mathbf{m}$ is equal to zero then the $k$th element of $\mathbf{x}$ is deleted.

If the missing-data mechanism $p_{m|x}$ has the property that $p_{m|x}(\mathbf{m}|\mathbf{x}) = p_{m|x}(\mathbf{m}|\mathbf{v}_{\mathbf{m}})$ where $\mathbf{v}_{\mathbf{m}}$ contains only the elements of $\mathbf{x}$ which are observable as specified by $\mathbf{m}$ for all $(\mathbf{x}, \mathbf{m})$ in the support of $(\tilde{\mathbf{x}}_t, \tilde{\mathbf{m}}_t)$, then the missing-data mechanism $p_{m|x}$ is called *MAR (Missing at Random).*

If, in addition, a MAR missing-data mechanism $p_{m|x}$ has the property that $p_{m|x}(\mathbf{m}|\mathbf{x})$ is only functionally dependent upon $\mathbf{m}$, then the missing-data mechanism $p_{m|x}$ is called *MCAR (Missing Completely at Random).*

A missing-data mechanism $p_{m|x}$ which is not MAR is called *MNAR (Missing Not at Random).*

**Example 9.2.1** (Examples of MCAR, MAR, and MNAR Mechanisms). Consider a regression model that predicts the likelihood of a heart attack given the values of the predictor variables *blood pressure* and *age*. Suppose that the blood pressure data is stored on a computer and a hard drive failure destroys 5% of the blood pressure readings. This mechanism for generating missing data is an MCAR missing-data mechanism because the probability a particular blood pressure measurement is missing is not functionally dependent upon the content of the values of the predictors *blood pressure* and *age*.

Assume older adults are more likely to miss their blood pressure appointments than younger adults and that the probability a blood pressure measurement is missing can be determined by the patient's age. If the ages of all of the adults in the study are always fully observable, this corresponds to a missing data situation with a MAR but not MCAR missing-data mechanism. If the ages of all of the adults in the study are also sometimes missing, this corresponds to a missing data situation with an MNAR missing-data mechanism.

Another example of a situation involving MNAR missing-data mechanisms corresponds to the case where the instrument used to measure blood pressure has a defect. In particular, assume that if someone's systolic blood pressure measurement is above 140 then the instrument fails and does not generate a blood pressure measurement. The resulting blood pressure data which is collected has been generated by an MNAR missing-data mechanism because the probability a blood pressure measurement is missing is functionally dependent upon the original unobservable value of that blood pressure reading. △

The above definition of an i.i.d. partially observable data stochastic sequence may be used to represent statistical environments for unsupervised learning algorithms and reinforcement learning algorithms where the element in the training stimulus corresponding to the "correct response" is not always fully observable. It may also be used to represent active learning environments where the network needs to generate a request for information but is not provided explicit instructions regarding when to generate such requests (see Exercise 9.2-3).

**Example 9.2.2** (Representing Different Length Trajectories as i.i.d. Random Vectors). The concept of an i.i.d. partially observable data generating process is also useful for representing reinforcement learning problems within an i.i.d. framework (see discussion of Reinforcement Learning in Chapter 1). A reinforcement learning machine experiences sequences of observations consisting of varying lengths during the learning process. This type of data generating process is naturally represented within this theoretical framework by assuming that the i.i.d. partially observable data stochastic sequence $\{(\tilde{\mathbf{x}}_i, \tilde{\mathbf{m}}_i)\}$ is defined such that the completely observable $i$th trajectory is defined as

$$\tilde{\mathbf{x}}_i \equiv [\tilde{\mathbf{x}}_i(1), \ldots, \tilde{\mathbf{x}}_i(T_{max})]$$

where the finite positive integer $T_{max}$ is the maximum length of a temporal sequence, and the mask vector for the $i$th trajectory is defined as

$$\tilde{\mathbf{m}}_i \equiv [\tilde{\mathbf{m}}_i(1), \ldots, \tilde{\mathbf{m}}_i(T_{max})]$$

where $\tilde{\mathbf{m}}_i(t)$ not only specifies which state variables are observable at different points in the trajectory but also can be used to specify the length of the trajectory. For example, if the observed trajectory

$$\tilde{\mathbf{y}}_i \equiv [\tilde{\mathbf{x}}_i(1), \ldots, \tilde{\mathbf{x}}_i(T_i)]$$

where $T_i \leq T_{max}$, then this is modeled by setting $\tilde{\mathbf{m}}_i(t) = \mathbf{0}_d$ for all $t = T_{i+1}, \ldots, T_{max}$. △

## Exercises

9.2-1. *Passive Semi-supervised Learning DGP.* In a semi-supervised learning problem, the learning machine sometimes receives feedback from the environment regarding the desired response $\mathbf{y}(t)$ for a given input pattern $\mathbf{s}(t)$ while other times simply observes the input pattern $\mathbf{s}(t)$. Show how to construct a statistical environment which represents the semi-supervised learning problem using an i.i.d. partially observable data process.

9.2-2. *Passive Reinforcement Learning DGP.* In a reinforcement learning problem, a learning machine observes a $d$-dimensional state vector $\mathbf{s}(t)$ and then generates a $q$-dimensional response vector $\mathbf{y}(t)$. Sometimes the environment indicates that response is correct by providing a binary feedback signal of $r(t) \in \{0, 1, ?\}$ where $r(t) = 1$ indicates the response was correct, $r(t) = 0$ indicates the response was wrong, and $r(t) =?$ indicates that the environment did not provide feedback to the learning machine at learning trial $t$. It is assumed that the probability law used by the statistical environment is the same for each learning trial. In addition, it is assumed that the random events generated by the environment for each learning trial are independent. Using the Definition of an I.I.D. Partially Observable Data Process provided in the text, formally define the data generating process. Assume the responses of the learning machine do not alter the characteristics of the learning machine's statistical environment. That is, assume a passive statistical environment.

9.2-3. *Active Semi-supervised Learning DGP.* Assume the stochastic sequence of $d$-dimensional input patterns $\tilde{\mathbf{s}}(1), \tilde{\mathbf{s}}(2), \ldots$ generated by the statistical environment has a common environmental density $p_e(\mathbf{s})$. A learning machine with a parameter vector $\boldsymbol{\theta}$ has a deterministic decision function $\psi : \mathcal{R}^d \times \mathcal{R}^q \rightarrow \{0, 1\}$ defined such that $\psi(\mathbf{s}(t), \boldsymbol{\theta}) = 1$ is a request to the statistical environment to generate the desired response $\mathbf{y}(t)$ for a given $\mathbf{s}(t)$ and $\psi(\mathbf{s}(t), \boldsymbol{\theta}) = 0$ means that the learning machine is informing the statistical environment that the desired response $\mathbf{y}(t)$ for a given $\mathbf{s}(t)$ is not required. Explain how to model an i.i.d. partially observable data process for this active statistical environment by constructing the i.i.d. partially observable DGP assuming a particular fixed value of $\boldsymbol{\theta}$. That is, the i.i.d. partially observable DGP is conditionally dependent upon the learning machine's parameter vector $\boldsymbol{\theta}$.

## 9.3 Stochastic Convergence

Consider a stochastic sequence of independent binary random variables which have a common probability distribution $\tilde{x}_1, \tilde{x}_2, \ldots$ representing an experiment consisting of a successive sequence of coin tosses. The binary random variable $\tilde{x}_t$ models the outcome of an experiment associated with the $t$th coin toss which can take on the value of either "heads" (i.e., $\tilde{x}_t = 1$) or "tails" (i.e., $\tilde{x}_t = 0$) with equal probability, $t = 1, 2, \ldots$. Since $|\tilde{x}_t| \leq 1$ for all $t$, the stochastic sequence is bounded. The expected value of each coin toss experiment in the sequence is $1/2$. The percentage of times the coin toss outcome is "heads" after $n$ tosses is given by the stochastic sequence $\tilde{f}_1, \tilde{f}_2, \ldots$ defined such that:

$$\tilde{f}_n = (1/n) \sum_{t=1}^{n} \tilde{x}_t \tag{9.2}$$

for all non-negative integer $n$.

Despite the fact that a fair coin is used, every sample path $f_1, f_2, \ldots$ of the stochastic sequence $\tilde{f}_1, \tilde{f}_2, \ldots$ does not converge to $1/2$. For example, a sample path such as: $f_k = 1$ for $k = 0, 1, 2, \ldots$ is an example of a sample path that converges to the value of 1. Another possible sample path of $\tilde{f}_1, \tilde{f}_2, \ldots$ is the sample path where $f_n = 1$ for $n = 0, 3, 6, 9, \ldots$ and $f_n = 0$ for $k = 1, 2, 4, 5, 7, 8, 10, 11, \ldots$ which converges to the value of $1/3$. In fact, there are an infinite number of sample paths of the stochastic sequence $\tilde{f}_1 \tilde{f}_2, \ldots$ that *do not* converge to $1/2$. Thus, it is not the case that the stochastic sequence $\tilde{f}_1, \tilde{f}_2, \ldots$ in (9.2) converges deterministically as $k \rightarrow \infty$.

If, on the other hand, *all* of the sample paths of the stochastic sequence $\tilde{f}_1, \tilde{f}_2, \ldots$ did converge to $\tilde{f}^*$, then this situation would correspond to the deterministic convergence case where $\tilde{f}_t \rightarrow \tilde{f}^*$ as $t \rightarrow \infty$.

In general, however, very few stochastic sequences occur in practice where *all* sample paths of such stochastic sequences converge to the same value. Thus, the standard deterministic concept of convergence is essentially useless for the direct analysis of stochastic sequences. Therefore, to support such analyses it is helpful to consider alternative notions of convergence which are called stochastic convergence definitions. Essentially all definitions of stochastic convergence can be organized in terms of four major categories: (i) convergence with probability one, (ii) convergence in mean square, (iii) convergence in probability, and (iv) convergence in distribution.

### 9.3.1 Convergence with Probability One

**Definition 9.3.1** (Convergence with Probability One). Let $\tilde{\mathbf{x}}_1, \tilde{\mathbf{x}}_2, \ldots$ be a stochastic sequence of $d$-dimensional random vectors on the probability space $(\Omega, \mathcal{F}, P)$. Let $\tilde{\mathbf{x}}^*$ be a $d$-dimensional random vector on $(\Omega, \mathcal{F}, P)$. The stochastic sequence $\tilde{\mathbf{x}}_1, \tilde{\mathbf{x}}_2, \ldots$ *converges with probability 1 (w.p.1)*, *almost surely (a.s.)*, or *almost everywhere (a.e.)* to $\tilde{\mathbf{x}}^*$ if

$$P\left(\{\omega : \tilde{\mathbf{x}}_t(\omega) \rightarrow \tilde{\mathbf{x}}^*(\omega) \quad as \quad t \rightarrow \infty\}\right) = 1.$$

□

Figure 9.1 illustrates the concept of convergence with probability one. The probability of the infinite set of realizations which converge is equal to one and the probability of the infinite set of realizations which do not converge is equal to zero.

Recall that $\tilde{\mathbf{x}}_1, \tilde{\mathbf{x}}_2, \ldots$ is actually a sequence of functions. If this sequence of functions deterministically converges to a function $\tilde{\mathbf{x}}^*$, then it immediately follows that $\tilde{\mathbf{x}}_t \rightarrow \tilde{\mathbf{x}}^*$ with probability one as $t \rightarrow \infty$. That is, deterministic convergence implies convergence with probability one.

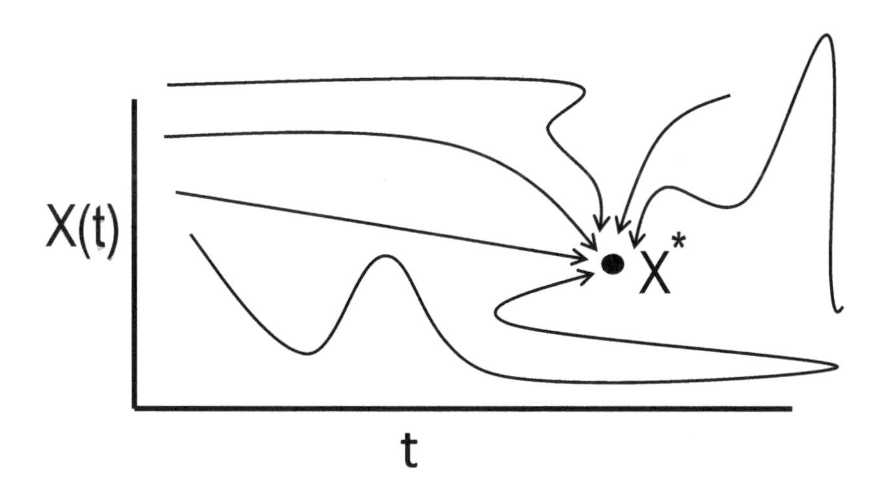


**FIGURE 9.1**
**Graphical illustration of convergence with probability one.** Let $S$ be the set of realizations of a stochastic sequence which converge and the complement of $S$, $\neg S$, be the set of realizations which do not converge. Convergence with probability one means that the probability that a realization is in $S$ is exactly equal to one and the probability that a realization is in $\neg S$ is exactly equal to zero.

Also note that the definition of convergence with probability one allows for the stochastic sequence to converge to a random vector rather than simply a constant. For example, let $\tilde{x}_1, \tilde{x}_2, \ldots$ be a sequence of random variables defined such that

$$\tilde{x}_t = (1 - (1/t))\tilde{z}$$

where $\tilde{z}$ is a random variable. Then, $\tilde{x}_t \to \tilde{z}$ as $t \to \infty$ with probability one.

An important example application of the convergence with probability one concept is the Strong Law of Large Numbers.

**Theorem 9.3.1** (Strong Law of Large Numbers). *Let $\tilde{\mathbf{x}}_1, \tilde{\mathbf{x}}_2, \ldots$ be a sequence of i.i.d. d-dimensional random vectors with common finite d-dimensional mean vector $\mathbf{m} \equiv E\{\tilde{\mathbf{x}}_t\}$. Then, as $n \to \infty$, $\tilde{\mathbf{f}}_n \to \mathbf{m}$ with probability one where*

$$\tilde{\mathbf{f}}_n = (1/n) \sum_{t=1}^{n} \tilde{\mathbf{x}}_t.$$

*Proof.* See Lukacs (1975; pp. 101-103. Theorem 4.3.3). ■

In the coin toss example, the stochastic sequence of binary-valued i.i.d. random variables $\tilde{x}_1, \tilde{x}_2, \ldots$ satisfies the conditions of the Strong Law of Large Numbers since a binary-valued random variable is a random variable whose absolute value is bounded by one. Thus, $E\{|\tilde{x}_t|\} < 1$ for every non-negative integer $t$. Thus, the Strong Law of Large Numbers may be used to assert that as $n \to \infty$:

$$\tilde{f}_n = (1/n) \sum_{t=1}^{n} \tilde{x}_t \to 1/2$$

with probability one.

**Example 9.3.1** (Importance Sampling Algorithm). In some applications, one wishes to evaluate a multidimensional integral such as:

$$\int f(\mathbf{x})p(\mathbf{x})d\nu(\mathbf{x})$$

where $p$ is a known Radon-Nikodým density defined with respect to $\nu$ but it is computationally expensive to evaluate the integral directly. Suppose that it is computationally inexpensive to sample from another Radon-Nikodým density $q$ whose support is a superset of the support of $p$. Then, one can use the Strong Law of Large Numbers to evaluate

$$\tilde{S}_n \equiv (1/n) \sum_{t=1}^{n} \frac{f(\tilde{\mathbf{x}}_t)p(\tilde{\mathbf{x}}_t)}{q(\tilde{\mathbf{x}}_t)} \tag{9.3}$$

where $\tilde{\mathbf{x}}_1, \tilde{\mathbf{x}}_2, \ldots$ is a sequence of independent and identically distributed random vectors with common probability density $q$ defined with respect to support specification measure $\nu$. By the strong law of large numbers,

$$\tilde{S}_n \to \int \left[\frac{f(\mathbf{x})p(\mathbf{x})}{q(\mathbf{x})}\right] q(\mathbf{x})d\nu(\mathbf{x}) = \int f(\mathbf{x})p(\mathbf{x})d\nu(\mathbf{x})$$

with probability one as $n \to \infty$. Thus, $\tilde{S}_n$ may be used as an approximation for $\int f(\mathbf{x})p(\mathbf{x})d\nu(\mathbf{x})$. Describe how to use this result to evaluate the multidimensional integral $\int f(\mathbf{x})p(\mathbf{x})d\nu(\mathbf{x})$.

**Solution.** Sample $n$ times from the probability density function $q$ to obtain $\mathbf{x}_1, \ldots, \mathbf{x}_n$ and then approximate $\int f(\mathbf{x})p(\mathbf{x})d\nu(\mathbf{x})$ by using the estimator:

$$(1/n)\sum_{j=1}^{n}\frac{f(\mathbf{x}_j)p(\mathbf{x}_j)}{q(\mathbf{x}_j)}.$$

△

The next key theorem is a generalization of the Kolmogorov Strong Law of Large Numbers to situations involving averages of random functions rather than random vectors. The Kolmogorov Strong Law of Large Numbers can be used to establish pointwise convergence but not uniform convergence for a sequence of random functions.

**Theorem 9.3.2** (Uniform Law of Large Numbers (ULLN)). *Let $\tilde{\mathbf{x}}_1, \ldots, \tilde{\mathbf{x}}_n$ be a sequence of independent and identically distributed d-dimensional random vectors with common Radon-Nikodým density $p : \mathcal{R}^d \rightarrow [0, \infty)$ with support specification measure $\nu$. Let $\Theta \subseteq \mathcal{R}^q$. Let the continuous random function $c : \mathcal{R}^d \times \Theta \rightarrow \mathcal{R}$ be dominated by an integrable function on $\Theta$ with respect to $p$. Let the continuous random function $\ell_n : \mathcal{R}^{d \times n} \times \Theta \rightarrow \mathcal{R}$ be defined such that:*

$$\tilde{\ell}_n(\boldsymbol{\theta}) = (1/n)\sum_{t=1}^{n} c(\tilde{\mathbf{x}}_t, \boldsymbol{\theta}).$$

*Then, as $n \rightarrow \infty$, $\tilde{\ell}_n$ uniformly converges on $\Theta$ with probability one to the continuous function $\ell : \Theta \rightarrow \mathcal{R}$ where*

$$\ell(\boldsymbol{\theta}) = \int c(\mathbf{x}, \boldsymbol{\theta})p(\mathbf{x})d\nu(\mathbf{x}).$$

*Proof.* See Jennrich (1969, Theorem 2; also see White 1994, Appendix 2, Theorem A.2.2). ■

Another important Uniform Law of Large Numbers which will be useful for approximating arbitrary probability distributions using simulation methods (see Chapter 14) is the Glivenko-Cantelli Theorem.

Let the notation $\mathbf{x} < \mathbf{y}$ mean that $x_k < y_k$ for $k = 1, \ldots, d$.

**Definition 9.3.2** (Empirical Distribution). Let $[\mathbf{x}_1, \ldots, \mathbf{x}_n]$ be a realization of a sequence of *i.i.d.* $d$-dimensional random vectors $[\tilde{\mathbf{x}}_1, \ldots, \tilde{\mathbf{x}}_n]$ with common cumulative probability distribution $P_e$. Let the *empirical distribution function* $F_n : \mathcal{R}^d \times \mathcal{R}^{d \times n} \rightarrow [0, 1]$ be defined such that for each $\mathbf{x}$ in the support of $\tilde{\mathcal{D}}_n \equiv [\tilde{\mathbf{x}}_1, \ldots, \tilde{\mathbf{x}}_n]$: $F_n(\mathbf{x}, \tilde{\mathcal{D}}_n)$ is the percentage of elements in $\tilde{\mathcal{D}}_n$ such that $\tilde{\mathbf{x}} < \mathbf{x}$. The probability mass function that specifies the empirical distribution function is called the *empirical mass function.* □

Note that the empirical mass function for a data set $\mathcal{D}_n \equiv [\mathbf{x}_1, \ldots, \mathbf{x}_n]$ may be specified by a probability mass function $p_e : \{\mathbf{x}_1, \ldots, \mathbf{x}_n\} \rightarrow [0, 1]$ defined such that $p_e(\mathbf{x}_i) = 1/n$ for $i = 1, \ldots, n$.

It immediately follows from the Strong Law of Large Numbers that the empirical distribution function $\tilde{F}_n$ converges pointwise with probability one to a cumulative probability distribution function $P_e$. However, this convergence is not guaranteed to be uniform.

**Theorem 9.3.3** (Glivenko-Cantelli Uniform Law of Large Numbers). *Let $\tilde{\mathcal{D}}_n \equiv [\tilde{\mathbf{x}}_1, \ldots, \tilde{\mathbf{x}}_n]$ be a stochastic sequence of n i.i.d. d-dimensional random vectors with common cumulative distribution $P_e : \mathcal{R}^d \rightarrow [0, 1]$. Let $\tilde{F}_n$ be the empirical distribution function defined with respect to $\tilde{\mathcal{D}}_n$. Then, the empirical distribution function $\tilde{F}_n(\cdot) \rightarrow P_e(\cdot)$ uniformly with probability one as $n \rightarrow \infty$.*

*Proof.* See Lukacs (1975, Theorem 4.3.4, p. 105). ■

The following theorem is useful in the analysis of stochastic adaptive learning algorithms and will play a critical role in the proof of the main convergence theorem for analyzing adaptive learning algorithms in Chapter 12.

**Theorem 9.3.4** (Almost Supermartingale Lemma). *Let $\tilde{\mathbf{x}}_1, \tilde{\mathbf{x}}_2, \ldots$ be a sequence of d-dimensional random vectors. Let $V : \mathcal{R}^d \rightarrow \mathcal{R}$, $Q : \mathcal{R}^d \rightarrow [0, \infty)$, and $R : \mathcal{R}^d \rightarrow [0, \infty)$ be piecewise continuous functions. Assume $V$ has a finite lower bound. Let $\tilde{r}_t \equiv R(\tilde{\mathbf{x}}_t)$, $\tilde{v}_t \equiv V(\tilde{\mathbf{x}}_t)$, and $\tilde{q}_t \equiv Q(\tilde{\mathbf{x}}_t)$. Assume there exists a finite number $K_r$ such that as $T \rightarrow \infty$: $\sum_{t=1}^{T} \tilde{r}_t \rightarrow K_r$ with probability one. Assume, in addition, that for $t = 0, 1, 2, \ldots$:*

$$E\{\tilde{v}_{t+1} | \tilde{\mathbf{x}}_t\} \leq \tilde{v}_t - \tilde{q}_t + \tilde{r}_t. \quad (9.4)$$

*Then,*

*(i) $\tilde{v}_t$ converges with probability one to a random variable as $t \rightarrow \infty$, and*

*(ii) there exists a finite number $K_q$ such that as $T \rightarrow \infty$: $\sum_{t=1}^{T} \tilde{q}_t < K_q$ with probability one.*

*Proof.* This theorem is a special case of the original Robbins-Siegmund Lemma (Robbins and Siegmund 1971). See Beneviste, Metivier, and Priouret (1990 Appendix to Part II) for the proof. ■

### 9.3.2 Convergence in Mean Square

**Definition 9.3.3** (Convergence in Mean Square). A stochastic sequence of $d$-dimensional random vectors $\tilde{\mathbf{x}}_1, \tilde{\mathbf{x}}_2, \ldots$ *converges in mean-square (m.s.)* to a random vector $\tilde{\mathbf{x}}^*$ if both (1) for some finite positive number $C$:

$$E\{|\tilde{\mathbf{x}}_t - \tilde{\mathbf{x}}^*|^2\} < C$$

for $t = 1, 2, \ldots$, and (2) as $t \rightarrow \infty$,

$$E\{|\tilde{\mathbf{x}}_t - \tilde{\mathbf{x}}^*|^2\} \rightarrow 0.$$

□

Figure 9.2 illustrates the concept of convergence in mean square.

**Theorem 9.3.5** (Weak (Mean Square) Law of Large Numbers). *Let $\tilde{\mathbf{x}}_1, \tilde{\mathbf{x}}_2, \ldots$ be a sequence of i.i.d. d-dimensional random vectors with common finite mean $\boldsymbol{\mu}$. Assume $E\{|\tilde{\mathbf{x}}_t - \boldsymbol{\mu}|^2\}$ is less than some finite number for $t = 1, 2, \ldots$. Let*

$$\tilde{\mathbf{f}}_n = (1/n) \sum_{t=1}^{n} \tilde{\mathbf{x}}_t.$$

*Then, as $n \rightarrow \infty$, $\tilde{\mathbf{f}}_n \rightarrow \boldsymbol{\mu}$ in mean square as $n \rightarrow \infty$.*

*Proof.*

$$E\{|\tilde{\mathbf{f}}_n - \boldsymbol{\mu}|^2\} = E\left\{\left|(1/n)\sum_{t=1}^{n} \tilde{\mathbf{x}}_t - \boldsymbol{\mu}\right|^2\right\} \leq E\left\{\frac{\sum_{t=1}^{n} |\tilde{\mathbf{x}}_t - \boldsymbol{\mu}|^2}{n^2}\right\} = \frac{E\{|\tilde{\mathbf{x}}_t - \boldsymbol{\mu}|^2\}}{n}. \quad (9.5)$$

Since $E\{|\tilde{\mathbf{x}}_t - \boldsymbol{\mu}|^2\}$ is less than some finite number, the term on the right-hand side of (9.5) converges to zero as $n \rightarrow \infty$. ■

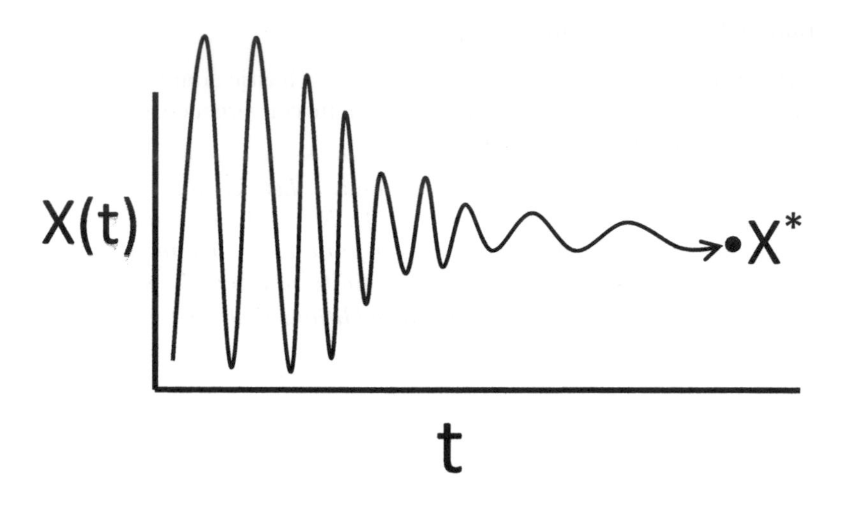


**FIGURE 9.2**
**Graphical illustration of convergence in mean square.** Convergence in mean square means that the expected variation of a trajectory from its destination tends to zero as time increases.

### 9.3.3 Convergence in Probability

**Definition 9.3.4** (Convergence in Probability). Let $\tilde{\mathbf{x}}_1, \tilde{\mathbf{x}}_2, \ldots$ be a stochastic sequence of $d$-dimensional random vectors on the probability space $(\Omega, \mathcal{F}, P)$. Let $\tilde{\mathbf{x}}^*$ be a $d$-dimensional random vector on $(\Omega, \mathcal{F}, P)$. The stochastic sequence $\tilde{\mathbf{x}}_1, \tilde{\mathbf{x}}_2, \ldots$ *converges in probability* to the random vector $\tilde{\mathbf{x}}^*$ if for every strictly positive real number $\epsilon$

$$P\left(\{\omega : |\tilde{\mathbf{x}}_t(\omega) - \tilde{\mathbf{x}}^*(\omega)| > \epsilon\}\right) \rightarrow 0 \text{ as } t \rightarrow \infty.$$

□

An equivalent definition of convergence in probability is that for every positive $\epsilon$,

$$P\left(\{\omega : |\tilde{\mathbf{x}}_t(\omega) - \tilde{\mathbf{x}}^*(\omega)| \leq \epsilon\}\right) \rightarrow 1 \text{ as } t \rightarrow \infty.$$

Figure 9.3 illustrates the concept of convergence in probability.

**Theorem 9.3.6** (MS Convergence Implies Convergence in Probability). *If $\tilde{\mathbf{x}}_1, \tilde{\mathbf{x}}_2, \ldots$ converges in mean square to a random vector $\tilde{\mathbf{x}}^*$, then the stochastic sequence converges in probability to $\tilde{\mathbf{x}}^*$.*

*Proof.* Let $\epsilon$ be a positive number. Application of the Markov Inequality

$$P\left(|\tilde{\mathbf{x}}_t - \tilde{\mathbf{x}}^*| \geq \epsilon\right) \leq E[|\tilde{\mathbf{x}}_t - \tilde{\mathbf{x}}^*|^2]/\epsilon^2.$$

Thus, if $E[|\tilde{\mathbf{x}}_t - \tilde{\mathbf{x}}^*|^2] \rightarrow 0$ as $t \rightarrow \infty$, then

$$P\left(|\tilde{\mathbf{x}}_t - \tilde{\mathbf{x}}^*| \geq \epsilon\right) \rightarrow 0$$

as $t \rightarrow \infty$. Thus, convergence in mean square implies convergence in probability. ■

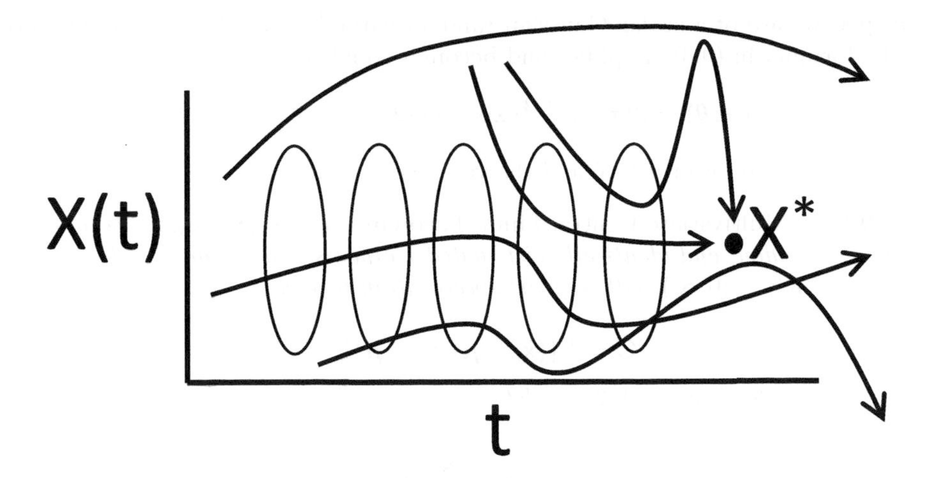


**FIGURE 9.3**
**Graphical illustration of the concept of convergence in probability.** Convergence in probability means that for a ball of given radius centered at the trajectory's destination, the probability that a trajectory is in that ball tends to one as time increases.

### 9.3.4 Convergence in Distribution

**Definition 9.3.5** (Convergence in Distribution)**.** Let $\tilde{\mathbf{x}}_1, \tilde{\mathbf{x}}_2, \ldots$ be a stochastic sequence where the $t$th $d$-dimensional random vector in the sequence, $\tilde{\mathbf{x}}_t$, has the cumulative distribution function $F_t : \mathcal{R}^d \to [0, 1]$. Let $F^*$ be the cumulative distribution function for a random vector $\tilde{\mathbf{x}}^*$. If $F_t(\mathbf{a}) \to F^*(\mathbf{a})$ as $t \to \infty$ for all $\mathbf{a}$ such that $F^*$ is continuous at $\mathbf{a}$, then the stochastic sequence $\tilde{\mathbf{x}}_1, \tilde{\mathbf{x}}_2, \ldots$ is said to *converge in distribution* or *converge in law* to $\tilde{\mathbf{x}}^*$ as $t \to \infty$. □

Note that the definition of convergence in distribution does not require that $F_t(\mathbf{a}) \to F^*(\mathbf{a})$ as $t \to \infty$ for all $\mathbf{a}$ in the domain of $F^*$ but only requires convergence at points in the domain of $F^*$ where $F^*$ is continuous. By only requiring convergence at points of continuity, this allows for the representation of cumulative distribution functions which arise in many practical applications. For example, every cumulative distribution function for a discrete random variable is discontinuous at a countable number of points.

The following multivariate Central Limit Theorem provides an example application of the concept of convergence in distribution.

**Definition 9.3.6** (Multivariate Gaussian Density)**.** Let $\tilde{\mathbf{x}}$ be a $d$-dimensional absolutely continuous random vector with support $\mathcal{R}^d$ and probability density function $p(\cdot; \boldsymbol{\theta})$ defined such that:

$$p(\mathbf{x}; \boldsymbol{\theta}) = (2\pi)^{-d/2} \left(\det(\mathbf{C})\right)^{-1/2} \exp\left[-(\mathbf{x} - \boldsymbol{\mu})^T \mathbf{C}^{-1} (\mathbf{x} - \boldsymbol{\mu})\right], \tag{9.6}$$

where the first $d$ elements of the $d(d + 1/2)$-dimensional parameter vector $\boldsymbol{\theta}$ are the $d$ elements of the *mean vector* $\boldsymbol{\mu}$ and the remaining $d(d-1)/2$ elements of $\boldsymbol{\theta}$ are equal to covariance parameter vector $\mathbf{vech}(\mathbf{C})$, which specifies the unique elements of the positive definite symmetric *covariance matrix* $\mathbf{C}$. □

In the special case of a scalar Gaussian random variable, $\tilde{x}$ with mean $\mu$ and variance $\sigma^2$, then the formula in (9.6) simplifies and becomes equal to:

$$p(x;\boldsymbol{\theta}) = (2\sigma^2\pi)^{-1/2} \exp\left[-(1/2)(x-\mu)^2/(2\sigma^2)\right]$$

where $\boldsymbol{\theta} = [\mu \;\; \sigma^2]$ with mean $\mu \in \mathcal{R}$ and the variance $\sigma^2 \in (0,\infty)$.

**Theorem 9.3.7** (Multivariate Central Limit Theorem). *Let $\tilde{\mathbf{x}}_1, \tilde{\mathbf{x}}_2, \tilde{\mathbf{x}}_3, \ldots$ be a stochastic sequence of independent and identically distributed d-dimensional random vectors with common density $p_e : \mathcal{R}^d \rightarrow [0,\infty)$ with support specification measure $\nu$. Assume $\boldsymbol{\mu} = E\{\tilde{\mathbf{x}}_t\}$ is finite. Assume*

$$\mathbf{C} = E\{(\tilde{\mathbf{x}}_t - \boldsymbol{\mu})(\tilde{\mathbf{x}}_t - \boldsymbol{\mu})^T\}$$

*is finite and invertible. For $n = 1, 2, \ldots$, let*

$$\tilde{\mathbf{f}}_n = (1/n)\sum_{t-1}^{n} \tilde{\mathbf{x}}_t.$$

*Then, as $n \rightarrow \infty$: $\sqrt{n}\left(\tilde{\mathbf{f}}_n - \boldsymbol{\mu}\right)$ converges in distribution to a Gaussian random vector with mean vector $\mathbf{0}_d$ and covariance matrix $\mathbf{C}$.*

*Proof.* For a proof see discussion by White (2001, p.114-115) which describes the Cramér-Wold device (White 2001, Proposition 5.1) and Lindeberg-Lévy Theorem (White 2001, Theorem 5.2). Also see Billingsley (2012, Section 27) for additional details regarding the proof of the Lindeberg-Lévy Theorem. ■

An equivalent statement of the Multivariate Central Limit Theorem is that the average of $n$ independent and identically distributed random vectors with common finite mean $\boldsymbol{\mu}$ and common invertible covariance matrix $\mathbf{C}$ converges in distribution to a Gaussian random vector with mean $\boldsymbol{\mu}$ and covariance matrix $\mathbf{C}/n$ as $n \rightarrow \infty$.

**Example 9.3.2** (Truncated Approximate Gaussian Random Variable). A random variable $\tilde{y}$ is called a *Truncated Approximate Gaussian random variable* if its probability distribution is very similar to the distribution of a Gaussian random variable but the random variable $\tilde{y}$ is bounded. Use the Multivariate Central Limit Theorem to show that the average of $K$ independent and identically distributed random variables with a common uniform distribution on the finite interval $[0,1]$, then $\tilde{y}$ converges in distribution to a Gaussian random variable with mean $1/2$ and variance $(12K)^{-1}$ as $K \rightarrow \infty$. Use this result to show how to write a computer program to generate a Truncated Approximate Gaussian random variable with mean $\mu$ and variance $\sigma^2$.

**Solution.** Note that a random variable with a uniform distribution on the interval $[0,1]$ has variance:

$$\int_0^1 (x-0.5)^2 dx = \frac{(1-0.5)^3}{3} - \frac{(0-0.5)^3}{3} = 1/12.$$

Write a computer program which generates a uniformly distributed random number on the interval $[0,1]$. Call this computer program $K$ times to generate the $K$ uniformly distributed random variables on the interval $[0,1]$ which will be denoted as $\tilde{u}_1, \ldots, \tilde{u}_K$. Let $\tilde{y} = (1/K)\sum_{k=1}^{K} \tilde{u}_k$. Let

$$\tilde{z} = \mu + \sigma\sqrt{12K}(\tilde{y} - (1/2)).$$

Then, $\tilde{z}$ will be a Truncated Approximate Gaussian Random Variable with mean $\mu$ and variance $\sigma^2$. △

### 9.3.5 Stochastic Convergence Relationships

The following theorem discusses the relationships among the four types of stochastic convergence.

**Theorem 9.3.8** (Stochastic Convergence Relationships). *Let $\tilde{\mathbf{x}}_1, \tilde{\mathbf{x}}_2, \ldots$ be a stochastic sequence of d-dimensional random vectors. Let $\tilde{\mathbf{x}}^*$ be a d-dimensional random vector.*

- *(i) If $\tilde{\mathbf{x}}_t \to \tilde{\mathbf{x}}^*$, then $\tilde{\mathbf{x}}_t \to \tilde{\mathbf{x}}^*$ with probability one.*
- *(ii) If $\tilde{\mathbf{x}}_t \to \tilde{\mathbf{x}}^*$ with probability one, then $\tilde{\mathbf{x}}_t \to \tilde{\mathbf{x}}^*$ in probability.*
- *(iii) If $\tilde{\mathbf{x}}_t \to \tilde{\mathbf{x}}^*$ in mean-square, then $\tilde{\mathbf{x}}_t \to \tilde{\mathbf{x}}^*$ in probability.*
- *(iv) If $\tilde{\mathbf{x}}_t \to \tilde{\mathbf{x}}^*$ in probability, then $\tilde{\mathbf{x}}_t \to \tilde{\mathbf{x}}^*$ in distribution.*
- *(v) If $\tilde{\mathbf{x}}_1, \tilde{\mathbf{x}}_2, \ldots$ is a bounded stochastic sequence and $\tilde{\mathbf{x}}_t \to \tilde{\mathbf{x}}^*$ in probability, then $\tilde{\mathbf{x}}_t \to \tilde{\mathbf{x}}^*$ in mean-square.*
- *(vi) If $\tilde{\mathbf{x}}_1, \tilde{\mathbf{x}}_2, \ldots$ converges in distribution to a constant $K$, then $\tilde{\mathbf{x}}_t \to K$ in probability.*

*Proof.* (i) Follows directly from the definition of Convergence with Probability One. (ii) See Lukacs (1975, pp. 33-34, Theorem 2.2.1). (iii) See Lukacs (1975, pp. 33-34, Theorems 2.2.2 and 2.2.4). (iv) See Lukacs (1975, pp. 33-24, Theorem 2.2.3). (v) Direct consequence of Serfling (1980, Theorem 1.3.6, p. 11). (vi) See Karr (1993, Proposition 5.14). ■

Figure 9.4 summarizes the Stochastic Convergence Relationships Theorem.

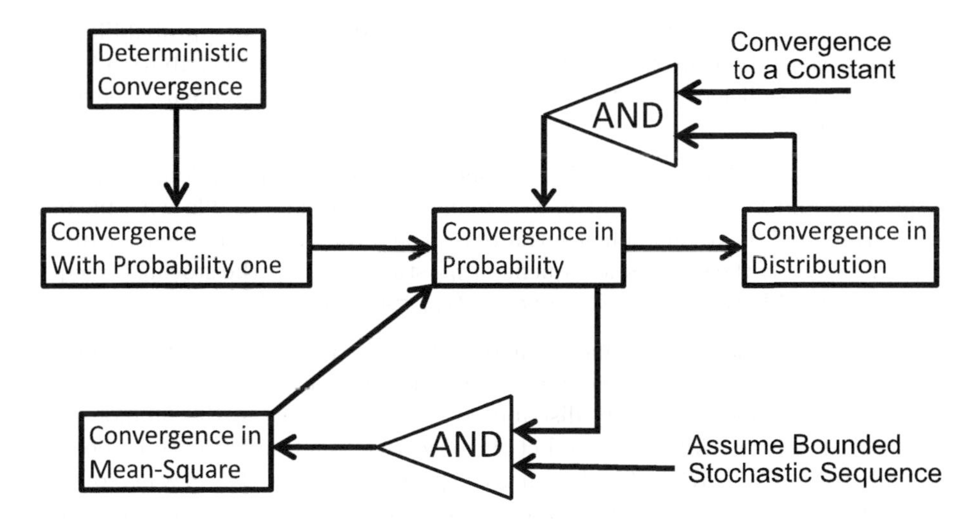


**FIGURE 9.4**
**Relationships among types of stochastic convergence.** Deterministic convergence implies convergence with probability one. Either convergence in mean-square or convergence with probability one implies convergence in probability. Convergence in probability implies convergence in distribution. Convergence in probability for a bounded stochastic sequence implies convergence in mean-square.

## Exercises

9.3-1. The definition of convergence in distribution to a target cumulative distribution function does not require convergence at points of discontinuity of the target cumulative distribution. Show that the cumulative distribution of the probability mass function for a random variable modeling the roll of a die has five points of discontinuity. Note that the roll of the die is modeled as a discrete random variable which takes on the values of $\{1, 2, 3, 4, 5, 6\}$ with equal probability.

9.3-2. Let $\tilde{\mathbf{x}}(1), \tilde{\mathbf{x}}(2), \ldots$ be a stochastic sequence of *i.i.d.* random $d$-dimensional vectors with common probability mass function $p$ that assigns probability mass $p_k$ to $\mathbf{x}^k$ for $k = 1 \ldots M$ $(M < \infty)$. Define for $t = 1, 2, 3, \ldots$:

$$\tilde{V}_t = (1/t) \sum_{i=1}^{t} V(\tilde{\mathbf{x}}(i))$$

where $V : \mathcal{R}^d \to \mathcal{R}$ is a continuous function. Prove the stochastic sequence of scalars $\tilde{V}_1, \tilde{V}_2, \tilde{V}_3, \ldots$ converges with probability one to some constant K. Give an explicit formula for the constant $K$. HINT: Use the strong law of large numbers.

9.3-3. Show that the stochastic sequence of scalars $\tilde{V}_1, \tilde{V}_2, \tilde{V}_3, \ldots$ defined in Exercise 9.3-2 converges in mean-square to some constant K. Give an explicit formula for the constant $K$.

9.3-4. Use the result obtained in Exercise 9.3-2 and the Markov Inequality to show that the stochastic sequence of scalars $\tilde{V}_1, \tilde{V}_2, \tilde{V}_3, \ldots$ converges in probability to some constant K. Give an explicit formula for the constant $K$.

9.3-5. Let $\tilde{\mathbf{x}}(1), \tilde{\mathbf{x}}(2), \ldots$ be a stochastic sequence which converges with probability one to a random variable. Does the stochastic sequence converge in distribution? Provide an additional sufficient condition that ensures this stochastic sequence converges in mean-square.

9.3-6. Let $\tilde{y}_{t+1} = \tilde{y}_t + \tilde{n}$ for $t = 1 \ldots M - 1$ where $\tilde{n}$ is a Gaussian random variable with mean zero and variable $\sigma^2$ and $\tilde{y}_1$ is a Gaussian random variable with mean zero and variable $\sigma^2$. Give an explicit formula for the joint probability density function: $p(y_1, \ldots, y_M)$. Is the stochastic sequence $\{\tilde{y}_t\}$ stationary? Is the stochastic sequence $\{\tilde{y}_t\}$ bounded? Is the stochastic sequence $\{\tilde{y}_t\}$ $\tau$-dependent?

9.3-7. Consider a statistical environment where a training data sequence $\mathbf{x}(0), \mathbf{x}(1), \ldots$ is modeled as a realization of a bounded stochastic sequence $\tilde{\mathbf{x}}(0), \tilde{\mathbf{x}}(1), \ldots$ consisting of independent and identically distributed random vectors with common Radon-Nikodým density $p : \mathcal{R}^d \to [0, \infty)$. Let the objective function for learning be defined as a function $c : \mathcal{R}^d \times \mathcal{R}^q \to \mathcal{R}$ which has the properties that: (1) $c(\cdot, \mathbf{w})$ is continuous on $\mathcal{R}^d$ for all $\mathbf{w} \in \mathcal{R}^q$, and (2) $c(\mathbf{x}, \cdot)$ is twice continuously differentiable on $\mathcal{R}^q$ for all $\mathbf{x} \in \mathcal{R}^d$. Let $g(\mathbf{x}, \mathbf{w}) \equiv (dc(\mathbf{x}, \mathbf{w})/d\mathbf{w})^T$. Show that:

$$(1/n) \sum_{i=1}^{n} g(\mathbf{x}(i), \mathbf{w}^*) \to E\{g(\tilde{\mathbf{x}}(i), \mathbf{w}^*)\}$$

with probability one as $n \to \infty$. Redo this problem and additionally establish convergence in mean square, convergence in probability, and convergence in distribution. Discuss how to modify your solution when $c$ is measurable in its first argument. Discuss how to modify your solutions when $c$ is piecewise continuous in its first argument.

## 9.4 Combining and Transforming Stochastic Sequences

The following theorems are helpful when a particular stochastic sequence (whose convergence properties are unknown) may be expressed as a function of one or more stochastic processes whose convergence properties are known.

**Theorem 9.4.1** (Functions of Stochastic Sequences). *Let $g : \mathcal{R}^d \to \mathcal{R}$ be a continuous function. Let $\tilde{\mathbf{x}}_1, \tilde{\mathbf{x}}_2, \ldots$ be a stochastic sequence of d-dimensional random vectors.*

- *(i) If $\tilde{\mathbf{x}}_t \to \tilde{\mathbf{x}}^*$ with probability one as $t \to \infty$, then $g(\tilde{\mathbf{x}}_t) \to g(\tilde{\mathbf{x}}^*)$ with probability one as $t \to \infty$.*
- *(ii) If $\tilde{\mathbf{x}}_t \to \tilde{\mathbf{x}}^*$ in probability as $t \to \infty$, then $g(\tilde{\mathbf{x}}_t) \to g(\tilde{\mathbf{x}}^*)$ in probability as $t \to \infty$.*
- *(iii) If $\tilde{\mathbf{x}}_t \to \tilde{\mathbf{x}}^*$ in distribution as $t \to \infty$, then $g(\tilde{\mathbf{x}}_t) \to g(\tilde{\mathbf{x}}^*)$ in distribution as $t \to \infty$.*

*Proof.* Follows directly from Serfling (1980 pp. 24-25). ■

**Example 9.4.1** (Convergence of Estimated Error Measures). Assume that the observed data $\mathbf{x}_1, \mathbf{x}_2, \ldots$ is a realization of a bounded stochastic sequence of independent and identically distributed random vectors $\tilde{\mathbf{x}}_1, \tilde{\mathbf{x}}_2, \ldots$ with common Radon-Nikodým density $p_e : \mathcal{R}^d \to \mathcal{R}$ with respect to support specification measure $\nu$. Let $\Theta$ be a closed and bounded subset of $\mathcal{R}^q$. Let the loss function $c : \mathcal{R}^d \times \Theta \to \mathcal{R}$ be defined such that: (i) $c(\cdot, \boldsymbol{\theta})$ is piecewise continuous on a finite partition of $\mathcal{R}^d$ for all $\boldsymbol{\theta} \in \Theta$, and (ii) $c(\mathbf{x}, \cdot)$ is a continuous function on $\Theta$ for all $\mathbf{x} \in \mathcal{R}^d$. Consider a learning machine where the expected loss function $\ell(\boldsymbol{\theta}) = \int c(\mathbf{x}, \boldsymbol{\theta}) p_e(\mathbf{x}) d\nu(\mathbf{x})$ is approximated by the empirical risk function

$$\hat{\ell}_n(\boldsymbol{\theta}) = (1/n) \sum_{i=1}^{n} c(\tilde{\mathbf{x}}, \boldsymbol{\theta}).$$

In addition, assume that $\hat{\boldsymbol{\theta}}_n$ is an estimator with the property that $\hat{\boldsymbol{\theta}}_n \to \boldsymbol{\theta}^*$ with probability one as $n \to \infty$. Show that $\hat{\ell}_n(\hat{\boldsymbol{\theta}}_n) \to \ell(\boldsymbol{\theta}^*)$ with probability one as $n \to \infty$.

**Solution.** Since the data generating sequence is bounded, $c$ is a continuous random function which is piecewise continuous on a finite partition, and the parameter space $\Theta$ is closed and bounded, it follows that $c$ is dominated by an integrable function on $\Theta$ with respect to $p_e$. Therefore, the Uniform Law of Large Numbers (Theorem 9.3.2) may be used to show that: (i) $\hat{\ell}_n(\boldsymbol{\theta}) \to \ell(\boldsymbol{\theta})$ with probability one for each $\boldsymbol{\theta} \in \Theta$, and (ii) $\ell$ is a continuous function on $\Theta$. Now note that

$$\hat{\ell}_n(\hat{\boldsymbol{\theta}}_n) - \ell(\boldsymbol{\theta}^*) = \left( \hat{\ell}_n(\hat{\boldsymbol{\theta}}_n) - \ell(\hat{\boldsymbol{\theta}}_n) \right) + \left( \ell(\hat{\boldsymbol{\theta}}_n) - \ell(\boldsymbol{\theta}^*) \right). \tag{9.7}$$

The first term on the right-hand side of (9.7) converges to zero with probability one since $\hat{\ell}_n \to \ell$ on $\Theta$ with probability one. The second term on the right-hand side of (9.7) converges to zero with probability one since $\ell$ is a continuous function and $\hat{\boldsymbol{\theta}}_n \to \boldsymbol{\theta}^*$ with probability one. It is important to note that the condition that $c(\mathbf{x}, \cdot)$ is a continuous function on $\Theta$ plays an important role in this argument for ensuring convergence. △

**Theorem 9.4.2** (Linear Transformation of Gaussian Random Variables). *Let $\tilde{\mathbf{x}}$ be a d-dimensional Gaussian random vector with d-dimensional mean vector $\boldsymbol{\mu} \equiv E\{\tilde{\mathbf{x}}\}$ and d-dimensional positive definite symmetric covariance matrix $\mathbf{C} \equiv E\left\{(\tilde{\mathbf{x}} - \boldsymbol{\mu})(\tilde{\mathbf{x}} - \boldsymbol{\mu})^T\right\}$. Let $\mathbf{A} \in \mathcal{R}^{r \times d}$ with full row rank r. Let $\tilde{\mathbf{y}} \equiv \mathbf{A}\tilde{\mathbf{x}}$. Then, $\tilde{\mathbf{y}}$ is an r-dimensional Gaussian random vector with mean $\mathbf{A}\boldsymbol{\mu}$ and r-dimensional positive definite symmetric covariance matrix $\mathbf{A}\mathbf{C}\mathbf{A}^T$.*

*Proof.* See Larson and Shubert (1979, Theorem 6.23, pp. 388-389) for a proof. ■

**Definition 9.4.1** (Gamma Function). The *gamma function* $\Gamma : [0, \infty) \to [0, \infty)$ is defined such that for each $k \in [0, \infty)$:

$$\Gamma(k) = \int_0^\infty x^{k-1} \exp(-x) dx.$$

□

Note that $\Gamma(k) = (k-1)!$ when $k$ is a positive integer.

**Definition 9.4.2** (Chi-Squared Density). Let $\Gamma : [0, \infty) \to [0, \infty)$ be the gamma function as defined in Definition 9.4.1. A *chi-squared density* with $d$ degrees of freedom is an absolutely continuous density $p : [0, \infty) \to [0, \infty)$ defined such that for all $x \geq 0$,

$$p(x) = \frac{x^{(d/2)-1} \exp(-x/2)}{2^{d/2}\Gamma(d/2)}.$$

The support of $p$ is $[0, \infty)$ for $d > 1$. The support of $p$ is $(0, \infty)$ for $d = 1$. □

**Theorem 9.4.3** (Sum of Squared Gaussian Random Variables Is Chi-Squared). *Let $\tilde{z}_1, \ldots, \tilde{z}_d$ be $d$ independent and identically distributed Gaussian random variables with common mean zero and common variance one. The random variable $\tilde{y} = \sum_{k=1}^{d} (\tilde{z}_k)^2$ has a* chi-squared distribution *with $d$ degrees of freedom.*

*Proof.* See Davidson (2002, pp. 123-124). ■

**Theorem 9.4.4** (Addition and Multiplication of Stochastic Sequences). *Let $\tilde{\mathbf{x}}_1, \tilde{\mathbf{x}}_2, \ldots$ and $\tilde{\mathbf{y}}_1, \tilde{\mathbf{y}}_2, \ldots$ be stochastic sequences consisting of $d$-dimensional random vectors.*

1. *If $\tilde{\mathbf{x}}_t \to \tilde{\mathbf{x}}^*$ in probability and $\tilde{\mathbf{y}}_t \to \tilde{\mathbf{y}}^*$ in probability as $t \to \infty$, then $\tilde{\mathbf{x}}_t + \tilde{\mathbf{y}}_t \to \tilde{\mathbf{x}}^* + \tilde{\mathbf{y}}^*$ in probability and $\tilde{\mathbf{x}}_t \odot \tilde{\mathbf{y}}_t \to \tilde{\mathbf{x}}^* \odot \tilde{\mathbf{y}}^*$ in probability as $t \to \infty$.*
2. *If $\tilde{\mathbf{x}}_t \to \tilde{\mathbf{x}}^*$ with probability one and $\tilde{\mathbf{y}}_t \to \tilde{\mathbf{y}}^*$ with probability one as $t \to \infty$, then $\tilde{\mathbf{x}}_t + \tilde{\mathbf{y}}_t \to \tilde{\mathbf{x}}^* + \tilde{\mathbf{y}}^*$ with probability one and $\tilde{\mathbf{x}}_t \odot \tilde{\mathbf{y}}_t \to \tilde{\mathbf{x}}^* \odot \tilde{\mathbf{y}}^*$ with probability one as $t \to \infty$.*

*Proof.* See Serfling (1980, p. 26). ■

In general, however, if $\tilde{x}_t \to \tilde{x}^*$ in distribution and $\tilde{y}_t \to \tilde{y}^*$ in distribution as $t \to \infty$, then it is not necessarily true that either $\tilde{x}_t\tilde{y}_t \to \tilde{x}^*\tilde{y}^*$ in distribution or $\tilde{x}_t + \tilde{y}_t \to \tilde{x}^* + \tilde{y}^*$ in distribution as $t \to \infty$.

The following theorem, however, is a useful tool for dealing with sums and products of stochastic sequences where one stochastic sequence converges in distribution to a random variable and the other stochastic sequence converges in probability (or in distribution) to a constant real number.

**Theorem 9.4.5** (Slutsky's Theorem). *Let $\tilde{x}_1, \tilde{x}_2, \ldots$ be a stochastic sequence of real-valued random variables which converges in distribution to the random variable $\tilde{x}^*$. Let $\tilde{y}_1, \tilde{y}_2, \ldots$ be a stochastic sequence of real-valued random variables which converges in probability to the finite real number $K$. Then, $\tilde{x}_t + \tilde{y}_t \to \tilde{x}^* + K$ in distribution as $t \to \infty$ and $\tilde{x}_t\tilde{y}_t \to K\tilde{x}^*$ in distribution as $t \to \infty$.*

*Proof.* See Serfling (1980, p. 19). ■

The next theorem is less commonly discussed but is a very useful theorem which is similar in spirit to Slutsky's Theorem.

**Theorem 9.4.6** (Bounded Probability Convergence Theorem). *Let $\tilde{x}_1, \tilde{x}_2, \ldots$ be a stochastic sequence of real-valued random variables which is bounded in probability. Let $\tilde{y}_1, \tilde{y}_2, \ldots$ be a stochastic sequence of real-valued random variables which converges with probability one to zero. Then, $\tilde{x}_t \tilde{y}_t \to 0$ with probability one as $t \to \infty$.*

*Proof.* See Davidson (2002, Theorem 18.12, p. 287). ■

The following notation, which is analogous to the deterministic Big O notation and Little O notation described in Chapter 4, is useful for the analysis of stochastic sequences.

**Definition 9.4.3** (Stochastic O Notation (almost sure)). Let $\rho$ be a non-negative number. The stochastic sequence $\tilde{\mathbf{x}}_1, \tilde{\mathbf{x}}_2, \ldots$ is $O_{a.s.}(n^{-\rho})$ with *order of convergence $n^{-\rho}$ with probability one* if there exists a finite number $K$ and a finite number $N$ such that for all $n \geq N$: $n^{\rho}|\tilde{\mathbf{x}}_n| \leq K$ with probability one. If, in addition, $n^{\rho}|\tilde{\mathbf{x}}_n| \to 0$ with probability one as $n \to \infty$, then the stochastic sequence $\tilde{\mathbf{x}}_1, \tilde{\mathbf{x}}_2, \ldots$ is $o_{a.s.}(n^{-\rho})$. □

**Definition 9.4.4** (Stochastic O Notation (in probability)). Let $\rho$ be a non-negative number. Let $\tilde{\mathbf{x}}_1, \tilde{\mathbf{x}}_2, \ldots$ be a sequence of random vectors with respective probability distributions $P_1, P_2, \ldots$. The stochastic sequence $\tilde{\mathbf{x}}_1, \tilde{\mathbf{x}}_2, \ldots$ is $O_p(n^{-\rho})$ with *order of convergence $n^{-\rho}$ in probability* if for every positive $\epsilon$ there exists a finite positive number $K_\epsilon$ and a finite number $N$ such that for all $n \geq N$: $P_n(n^{\rho}|\tilde{\mathbf{x}}_n| \leq K_\epsilon) > 1 - \epsilon$. If, in addition, $n^{\rho}|\tilde{\mathbf{x}}_n| \to 0$ in probability as $n \to \infty$, then the stochastic sequence $\tilde{\mathbf{x}}_1, \tilde{\mathbf{x}}_2, \ldots$ is $o_p(n^{-\rho})$. □

## Exercises

9.4-1. Let $\tilde{x}_1, \tilde{x}_2, \ldots$ and $\tilde{y}_1, \tilde{y}_2, \ldots$ be two stochastic sequences which converge in distribution and in probability respectively. Using the theorems provided in this section, what additional conditions are required to ensure that both the stochastic sequence $\tilde{x}_1 + \tilde{y}_1, \tilde{x}_2 + \tilde{y}_2, \ldots$ and the stochastic sequence $\tilde{x}_1\tilde{y}_1, \tilde{x}_2\tilde{y}_2, \ldots$ also converge in distribution?

9.4-2. Let $\tilde{x}_1, \tilde{x}_2, \ldots$ and $\tilde{y}_1, \tilde{y}_2, \ldots$ be two stochastic sequences which are bounded in probability and converges with probability one respectively. Using the theorems provided in this section, what additional conditions are required to ensure that the stochastic sequence $\tilde{x}_1\tilde{y}_1, \tilde{x}_2\tilde{y}_2, \ldots$ converges with probability one?

9.4-3. Assume $\mathbf{M} \in \mathcal{R}^{m \times n}$ has rank $m$ (where $m \leq n$) and $\tilde{\mathbf{x}}$ is a Gaussian random $n$-dimensional vector with mean $\boldsymbol{\mu}$ and $n$-dimensional positive definite real symmetric covariance matrix $\mathbf{Q}$. What is the probability density function for the random vector $\tilde{\mathbf{y}} \equiv \mathbf{M}\tilde{\mathbf{x}}$?

9.4-4. Show that $o_p(n^{-\rho})$ implies $O_p(n^{-\rho})$ where $\rho$ is a non-negative number.

9.4-5. Show $O_p(n^{-\rho})$ implies $o_p(1)$ when $\rho$ is a positive number.

9.4-6. Show that a stochastic sequence which is bounded in probability is $O_p(1)$.

9.4-7. Show that a stochastic sequence which converges in distribution is $O_p(1)$. That is, a stochastic sequence which converges in distribution is bounded in probability.

9.4-8. Show that if a stochastic sequence is $O_p(1)o_p(1)$ then it is also $o_p(1)$.

9.4-9. Show that if a stochastic sequence is $O_p(n^{-1/2})o_p(1)O_p(n^{-1/2})$ then it is also $o_p(1/n)$.

9.4-10. Show that if a stochastic sequence is $O_{a.s.}(1)o_{a.s.}(1)$ then it is also $o_p(1)$.

9.4-11. Show that if a stochastic sequence is $O_{a.s.}(n^{-1/2})O_{a.s.}(n^{-1/2})$ then it is also $O_p(1/n)$.

## 9.5 Further Readings

The reader should understand that only a small fraction of relevant work in stochastic sequences was covered in this chapter. The particular choice of topics discussed here was carefully chosen to provide a helpful foundation for topics covered in Chapters 11, 12, 13, 14, 15, and 16.

### Stochastic Processes

Readers wishing to pursue topics covered in this chapter can refer to graduate texts in econometrics (Davidson 1994; White 1994, 2001), graduate texts written from an electrical engineering perspective (Larson and Shubert 1979), graduate texts in probability theory (Karr 1993; Rosenthal 2016; Billingsley 2012) and mathematical statistics (Lukacs 1975; Serfling 1980; van der Vaart 1998; Lehmann and Casella 1998; Lehmann and Roman 2005).

The collection of papers in White (2004) provides an important collection of advanced theoretical tools from the field of econometrics for supporting the analysis of stochastic sequences commonly arising in statistical machine learning applications. For example, conditions that ensure the Strong Law of Large Numbers and the Multivariate Central Limit Theorem hold for nonstationary and $\tau$-dependent stochastic sequences have been developed (e.g., Serfling 1980; White and Domowitz 1984; White 1994; White 2001; Davidson 2002 ).

### Stochastic Convergence

Additional discussions of stochastic convergence may be found in Lukacs (1975), Larson and Shubert (1979), Serfling (1980), Karr (1993), and White (2001).

### Partially Observable Data Generating Processes

Golden, Henley, White, and Kashner (2019) introduce the definition of an i.i.d. partially observable data generating process that is discussed here for the purpose of investigating the consequences of maximum likelihood estimation in the simultaneous presence of model misspecification and missing data.

# 10

# *Probability Models of Data Generation*

**Learning Objectives**

- Provide learnability conditions for probability models.
- Describe examples of probability models.
- Analyze and design Bayes net probability models.
- Analyze and design Markov random field probability models.

A learning machine's probabilistic model of the likelihood of events in its environment is represented by a "probability model" or equivalently a collection of possible probabilistic laws. The objective of the learning process is to identify a probabilistic law that provides a good approximation to the probability distribution which actually generated the environmental events in the learning machine's environment (see Figure 1.1). Thus, the goal of learning may be formulated as estimating a set of parameter values that index a particular probabilistic law in the learning machine's probability model. In this chapter, tools and techniques for representing, analyzing, and designing probability models are provided.

## 10.1 Learnability of Probability Models

In this section, we discuss necessary conditions for a probability model to be "learnable". If these learnability conditions are not satisfied, then a learning machine whose goal is to estimate the parameters of a specific probability model of the data generating process will never achieve its absolute objective of perfectly representing the data generating process.

### 10.1.1 Correctly Specified and Misspecified Models

**Definition 10.1.1** (I.I.D. Data Generating Process (DGP)). A stochastic sequence of independent and identically distributed $d$-dimensional random vectors with common *Data Generating Process (DGP) cumulative probability distribution* $P_e$ is called an *I.I.D. DGP.* A subset of $n$ elements of the I.I.D. DGP, $\tilde{\mathcal{D}}_n \equiv [\tilde{\mathbf{x}}_1, \ldots, \tilde{\mathbf{x}}_n]$ is called a *random sample* of $n$ *observations* from $P_e$. A realization of a random sample, $\mathcal{D}_n \equiv [\mathbf{x}_1, \ldots, \mathbf{x}_n]$, is called a *data set* of $n$ *data records.* □

The DGP cumulative distribution function $P_e$, which generates the observed data, is also called the *environmental distribution.* A Radon-Nikodým density $p_e$, used to represent $P_e$, is called the *DGP density* or *environmental density.*

The terminology *sample n times from* either the probability distribution $P_e$ or its corresponding density specification to obtain $\mathbf{x}_1, \ldots, \mathbf{x}_n$ means that $\mathbf{x}_1, \ldots, \mathbf{x}_n$ is a realization of a sequence of independent and identically distributed random vectors $\tilde{\mathbf{x}}_1, \ldots, \tilde{\mathbf{x}}_n$ with common probability distribution $P_e$.

**Definition 10.1.2** (Probability Model). A *probability model,* $\mathcal{M}$, is a collection of cumulative probability distributions. □

In real-world engineering applications, most proposed probability models are incapable of perfectly representing their statistical environments. Furthermore, prior knowledge in the form of feature vector representations (see Section 1.2.1) and structural constraints (see Section 1.2.4) are often incorporated into machine learning algorithms to improve learning speed and generalization performance. Such prior knowledge constraints can be extremely effective in some statistical environments yet harmful in others. To address these issues, it is important that a general mathematical theory of machine learning allows for imperfections in the learning machine's probabilistic representation of its statistical environment.

**Definition 10.1.3** (Misspecified Model). Let $\tilde{\mathbf{x}}_1, \tilde{\mathbf{x}}_2, \ldots$ be an I.I.D. DGP with common cumulative probability distribution $P_e$. Let $\mathcal{M}$ be a probability model. If $P_e \notin \mathcal{M}$, then $\mathcal{M}$ is *misspecified* with respect to $P_e$. If $\mathcal{M}$ is not misspecified with respect to $P_e$, then $\mathcal{M}$ is *correctly specified* with respect to $P_e$. □

If a probability model $\mathcal{M}$ is misspecified with respect to the distribution $P_e$ which generated the data, then $\mathcal{M}$ can never perfectly represent $P_e$ because $P_e \notin \mathcal{M}$. Although situations involving model misspecification often correspond to situations where a model exhibits poor predictive performance, it is important to emphasize that the concept of model misspecification is not equivalent to the concept of poor predictive performance. In many important cases in machine learning, a misspecified model with excellent predictive performance and excellent generalization performance might be quite acceptable when a correctly specified model is not available.

In practice, it is often more convenient to work with the following alternative specification of a probability model which is defined in terms of a collection of Radon-Nikodým densities. This does not present any major conceptual problems except one must be careful because a Radon-Nikodým density defined with respect to a support specification measure $\nu$ is only unique $\nu$-almost everywhere.

**Definition 10.1.4** (Model Specification). A *model specification* is a collection $\mathcal{M}$ of Radon-Nikodým density functions defined with respect to a common support specification measure $\nu$. □

**Definition 10.1.5** (Classical Parametric Model Specification). Let $\Theta \subseteq \mathcal{R}^q$. Let $p : \Omega \times \Theta \to [0, \infty)$ be defined such that for each *parameter vector* $\boldsymbol{\theta}$ in the *parameter space* $\Theta$, $p(\cdot; \boldsymbol{\theta})$ is a Radon-Nikodým density with respect to a support specification measure $\nu$. The model specification $\mathcal{M} \equiv \{p(\cdot; \boldsymbol{\theta}) : \boldsymbol{\theta} \in \Theta\}$ is called a *parametric model specification* with respect to $\nu$. □

The less formal terminology *probability model* or *parametric probability model* is often used to refer to a classical parametric model specification. Also note that the assumption that all densities in a parametric model specification have the same support was chosen for notational convenience but is not a crucial characteristic.

**Definition 10.1.6** (Bayesian Parametric Model Specification). Let $\Theta \subseteq \mathcal{R}^q$. Let $p : \Omega \times \Theta \rightarrow [0, \infty)$ be defined such that for each *parameter vector* $\boldsymbol{\theta}$ in the *parameter space* $\Theta$, $p(\cdot|\boldsymbol{\theta})$ is a Radon-Nikodým density with respect to a support specification measure $\nu$. The model specification $\mathcal{M} \equiv \{p(\cdot|\boldsymbol{\theta}) : \boldsymbol{\theta} \in \Theta\}$ is called a *Bayesian parametric model specification* with respect to $\nu$. □

Each element of a Bayesian probability model is a conditional probability density $p(\mathbf{x}|\boldsymbol{\theta})$ rather than an (unconditional) probability density $p(\mathbf{x}; \boldsymbol{\theta})$. One may define a Bayesian probability model $\ddot{\mathcal{M}}$ that is equivalent to a classical probability model $\mathcal{M} \equiv \{p(\mathbf{x}; \boldsymbol{\theta}) : \boldsymbol{\theta} \in \Theta\}$ by choosing $\ddot{p}(\mathbf{x}|\boldsymbol{\theta}) \in \ddot{\mathcal{M}}$ if and only if $p(\mathbf{x}|\boldsymbol{\theta}) = p(\mathbf{x}; \boldsymbol{\theta})$ for all $\boldsymbol{\theta} \in \Theta$. Similarly, any Bayesian probability model may be redefined as a classical probability model. Therefore, in the following discussions, formal statements expressed in terms of classical probability models are applicable to their equivalent Bayesian probability model counterparts and vice versa.

The definition of model misspecification is slightly more complicated when a probability model specification is specified as a collection of Radon-Nikodým density functions. For example, suppose the data is generated from the density $p_e$ and the researcher's model of $p_e$ is the model specification $\mathcal{M} \equiv \{p_m\}$ and both $p_e$ and $p_m$ are defined with respect to the same support specification measure $\nu$. Assume that $p_m = p_e$ $\nu$ almost everywhere. Then, $p_e$ is not necessarily in $\mathcal{M}$ but both $p_e$ and $p_m$ specify exactly the same cumulative probability distribution function of the data generating process! The following definition addresses this issue by explicitly defining the concept of a misspecified model specification.

**Definition 10.1.7** (Misspecified Model Specification). Let the sequence of $d$-dimensional random vectors $\tilde{\mathbf{x}}_1, \tilde{\mathbf{x}}_2, \ldots$ be an I.I.D. DGP with common Radon-Nikodým density $p_e : \mathcal{R}^d \rightarrow [0, \infty)$ defined with respect to support specification measure $\nu$. A model specification $\mathcal{M}$ is *correctly specified* with respect to $p_e$ if there exists a $p \in \mathcal{M}$ such that $p = p_e$ $\nu$-almost everywhere. If $\mathcal{M}$ is not correctly specified with respect to $p_e$, then $\mathcal{M}$ is *misspecified* with respect to $p_e$. □

Note the condition that $p = p_e$ $\nu$-almost everywhere in Definition 10.1.7 automatically holds when $p$ and $p_e$ are identical (i.e., $p = p_e$). For example, let $S$ be a finite subset of $\mathcal{R}$ and let $p_e : \mathcal{R} \rightarrow (0, \infty)$ be a Gaussian probability density function defined with respect to a Lebesgue support specification measure $\nu$. And, in addition, define $p$ such that: (i) $p_e(x) = p(x)$ for all $x \in \mathcal{R} \cap \neg S$, and (ii) $p_e(x) \neq p(x)$ for all $x \in S$. Even though $p \neq p_e$, the respective cumulative distribution functions for $p$ and $p_e$ equivalent.

Example 10.1.1 provides an example of a correctly specified model that is not predictive. Example 10.1.2 provides an example of a predictive model that is misspecified.

**Example 10.1.1** (Correctly Specified Models are not Necessarily Predictive). To illustrate the distinction between a predictive model and a misspecified model, consider a nonlinear regression model

$$\mathcal{M} \equiv \{p(y|\mathbf{s}; \boldsymbol{\theta}, \sigma^2) : \boldsymbol{\theta} \in \mathcal{R}^q, \sigma^2 > 0\}$$

defined such that $p(y|\mathbf{s}; \boldsymbol{\theta}, \sigma^2)$ is a conditional Gaussian density function with mean $f(\mathbf{s}, \boldsymbol{\theta})$ and positive variance $\sigma^2$. A particular pair $(\boldsymbol{\theta}, \sigma^2)$ specifies a particular conditional density $p(y|\mathbf{s}; \boldsymbol{\theta}, \sigma^2) \in \mathcal{M}$.

Suppose the observed data $\{(y_1, \mathbf{s}_1), \ldots, (y_n, \mathbf{s}_n)\}$ was generated from a data generating process specified by:

$$\tilde{y} = f(\mathbf{s}, \boldsymbol{\theta}_e) + \tilde{n}_e \tag{10.1}$$

for some $\boldsymbol{\theta}_e \in \mathcal{R}^q$ and where $\tilde{n}_e$ is a zero-mean random variable with variance $\sigma_e^2$. If, in addition, $\tilde{n}_e$ is a Gaussian random variable then $\mathcal{M}$ is correctly specified. Suppose, however, that the variance $\sigma_e^2$ is very large; then although $\mathcal{M}$ is correctly specified, the predictive fit of $\mathcal{M}$ is poor. △

**Example 10.1.2** (Predictive Models Are Not Necessarily Correctly Specified). Now consider the situation presented in Example 10.1.1 but assume that $\tilde{n}_e$ is a zero-mean random variable with a very small variance but $\tilde{n}_e$ is not a Gaussian random variable. The model $\mathcal{M}$ in Example 10.1.1 is misspecified but would have very good predictive performance. △

The critical characteristic of a parametric probability model is that the model has a finite number of free parameters which does not change during the learning process. Probability models which allow for the number of free parameters to grow with the number of data points are considered *nonparametric probability models.* For example, consider a learning machine which initially has no basis functions (see Chapter 1) and each time it observes a feature vector in its environment, it creates a new basis function and that basis function has some additional free parameters. As the learning process continues, the number of free parameters of the learning machine increases. This is an example of a nonparametric probability model.

Such nonparametric probability models play an important role in current research in machine learning because they provide a mechanism for representing arbitrary probability distributions. On the other hand, a very flexible parametric probability model can have many of the desirable properties of a nonparametric probability model (see Example 10.1.4). More generally, if $\mathcal{M}$ is a probability model which contains an infinite number of probability distributions, then $\mathcal{M}$ can always be represented by a probability model with one free real-valued parameter which assigns a different parameter value to each distinct probability distribution in $\mathcal{M}$ (see Example 10.1.5).

**Example 10.1.3** (Reparameterization of Probability Models). Let the probability density function $p$ for a $d$-dimensional random vector $\tilde{\mathbf{x}}$ be specified by the formula:

$$p(\mathbf{x};\sigma) = \frac{\exp\left[-|\mathbf{x}|^2/(2\sigma^2)\right]}{\left(\sigma\sqrt{2\pi}\right)^d}.$$

The density $p(\mathbf{x};\sigma)$ is an element of model specification $\mathcal{M} \equiv \{p(\mathbf{x};\sigma) : \sigma \in (0,\infty)\}$ which has parameter space $(0,\infty)$. Construct another probability model specification $\ddot{\mathcal{M}}$ which is equivalent to $\mathcal{M}$ so that $\ddot{\mathcal{M}} = \mathcal{M}$ but: (1) the formula for an element of $\ddot{\mathcal{M}}$ is different than the formula for an element of $\mathcal{M}$, and (2) the parameter space for $\ddot{M}$ is different from $\mathcal{M}$.

**Solution.** Define the density

$$\ddot{p}(\mathbf{x};\theta) = \frac{\exp\left[-|\mathbf{x}|^2/(2(\exp(\theta))^2)\right]}{\left(\exp(\theta)\sqrt{2\pi}\right)^d}$$

as an element of the probability model specification $\ddot{\mathcal{M}} \equiv \{\ddot{p}(\mathbf{x};\theta) : \theta \in \mathcal{R}\}$ whose parameter space is $\mathcal{R}$. △

**Example 10.1.4** (Example of a Flexible (Saturated) Parametric Probability Model). Let $\tilde{\mathbf{x}}$ be a discrete random vector with support $S \equiv \{\mathbf{x}^1, \ldots, \mathbf{x}^M\} \subset \mathcal{R}^d$. Thus, $\tilde{\mathbf{x}}$ is a discrete random vector which can take on one of $M$ possible values with one of $M$ positive probabilities. Construct a parametric model specification which is flexible enough to represent any arbitrary probability mass function with support $S$.

**Solution.** Let $\boldsymbol{\theta} \equiv [\theta_1, \ldots, \theta_M]$ be an $M$-dimensional vector whose elements are non-negative and which has the property that the sum of the elements of $\boldsymbol{\theta}$ is equal to one. The parameter vector $\boldsymbol{\theta}$ specifies a probability mass function which assigns a probability mass of $\theta_k$ to $\mathbf{x}^k$. Any arbitrary probability mass function on $\mathbf{S}$ can be represented by choosing $\boldsymbol{\theta}$ appropriately. The parametric probability model $\mathcal{M}$ can then be defined so that $\mathcal{M} \equiv \{p(\mathbf{x};\boldsymbol{\theta}) : \boldsymbol{\theta} \in \Theta\}$ where $p(\mathbf{x};\boldsymbol{\theta}) = \theta_k$ when $\mathbf{x} = \mathbf{x}^k$ for $k = 1, \ldots, M$. △

**Example 10.1.5** (Example of a Flexible One Parameter Parametric Probability Model). Let a $q$ parameter probability model $\mathcal{M}_\theta$ be defined such that

$$\mathcal{M}_\theta \equiv \{P_\theta(\cdot;\boldsymbol{\theta}) : \boldsymbol{\theta} \in \mathcal{R}^q\}$$

where for a given $\boldsymbol{\theta}$: $P_\theta(\cdot;\boldsymbol{\theta})$ is the probability distribution for a random vector $\tilde{\mathbf{x}}$.

Construct a one-parameter probability model such that

$$\mathcal{M}_\eta \equiv \{P_\eta(\cdot;\eta) : \eta \in \mathcal{R}\}$$

where $\mathcal{M}_\theta = \mathcal{M}_\eta$. That is, the model $\mathcal{M}_\theta$ with a $q$-dimensional parameter space is exactly the same as the probability model $\mathcal{M}_\eta$ with a one-dimensional parameter space.

**Solution.** Note there is a one-to-one correspondence between a point $\boldsymbol{\theta} \in \mathcal{R}^q$ and a point $\boldsymbol{\psi} \in (0,1)^q$ by defining the $k$th element of $\boldsymbol{\psi}$, $\psi_k$, as $\psi_k = 1/(1+\exp(-\theta_k))$ where $\theta_k$ is the $k$th element of $\boldsymbol{\theta}$.

Now define a number $\eta$ such that the first digit of $\eta$ is the first digit of the number $\psi_1$ after the decimal point, the second digit of $\eta$ is the first digit of the number $\psi_2$ after the decimal point, and so until the $q$th digit of $\eta$ is assigned the first digit of $\psi_q$ after the decimal point. Now let the $(q+1)$st digit of $\eta$ be the second digit of the number $\theta_1$ after the decimal point, the $(q+2)$nd digit of $\eta$ is the second digit of the number $\theta_2$ after the decimal point, and so on. In this way we can show that a value of $\eta$ has a one-to-one correspondence with each value of $\boldsymbol{\theta}$. By assigning the density indexed by $\boldsymbol{\theta}$ in $\mathcal{M}_\theta$ to an element of $\mathcal{M}_\eta$ indexed by the value $\eta$, one can construct a new model $\mathcal{M}_\eta$ which has only one free parameter yet contains all of the densities in the original $q$-dimensional probability model $\mathcal{M}_\theta$. △

### 10.1.2 Smooth Parametric Probability Models

Although the many-parameter probability model $\mathcal{M}_\theta$ and the one-parameter probability model $\mathcal{M}_\eta$ in Example 10.1.5 are equivalent (i.e., $\mathcal{M}_\theta = \mathcal{M}_\eta$), nevertheless there is an important difference between $\mathcal{M}_\theta$ and $\mathcal{M}_\eta$. Assume $\boldsymbol{\theta}^* \in \mathcal{R}^q$ specifies a particular probability distribution $P_\theta(\cdot;\boldsymbol{\theta}^*) \in \mathcal{M}_\theta$. In addition, define $\eta^* \in \mathcal{R}$ such that $P_\eta(\cdot;\eta^*) = P_\theta(\cdot;\boldsymbol{\theta}^*)$. That is, $\boldsymbol{\theta}^*$ and $\eta^*$ index the same probability distribution which belongs to both the $q$-dimensional and 1-dimensional probability models. However, the set of probability distributions in a $\delta$-neighborhood of $\boldsymbol{\theta}^*$ for $q$-dimensional model $\mathcal{M}_\theta$ will, in general, be different from the set of probability distributions in the $\delta$-neighborhood of $\boldsymbol{\eta}^*$ for 1-dimensional model $\mathcal{M}_\eta$. In other words, an important intrinsic characteristic of a parametric probability model is that it has a "smoothness" property. That is, similar probability distributions in the model will have similar parameter values. Such similarity relationships are not, in general, preserved when a $q$-dimensional model is recoded as a 1-dimensional model.

Indeed, a machine learning algorithm adjusts its parameter values as a reflection of its experiences and a particular parameter vector corresponds to a particular probability density specifying the machine learning algorithm's beliefs about its probabilistic environment. Small changes in the machine learning algorithm's parameter values should not radically change the machine learning algorithm's beliefs about the likelihood of events in its environment.

**Definition 10.1.8** (Smooth Parametric Model Specification). Let the parameter space $\Theta \subseteq \mathcal{R}^q$. Let $\mathcal{M} \equiv \{p(\cdot;\boldsymbol{\theta}) : \boldsymbol{\theta} \in \Theta\}$ be a parametric model specification with respect to $\nu$. If, in addition, $p(\mathbf{x},\cdot)$ is continuous on $\Theta$ for all $\mathbf{x}$, then $\mathcal{M}$ is called a *smooth parametric model specification.* □

### 10.1.3 Local Probability Models

A standard assumption in classical statistics is that exactly one optimal solution exists (see Assumption **A8** in Definition 15.1.1 in this textbook). However, in the real world, multiple optimal solutions often exist. For example, consider the problem of setting the parameter values of a robot arm so that it can pick a cup off a table. There is really no optimal way to pick a cup off the table. Therefore, it would not be surprising if acceptable solutions corresponded to different local minimizers of some objective function.

To develop a statistical theory of generalization for these more complicated situations it is convenient to introduce the concept of a "local probability model" which is the restriction of the original probability model defined on the entire parameter space to a small region $\Gamma$ of the entire parameter space $\Theta$. This small region $\Gamma$ is often chosen so that it contains exactly one optimal solution. This construction permits the application of classical statistical theory to the complicated multi-modal probability models often encountered in machine learning applications.

**Definition 10.1.9** (Local Probability Models). Let $\mathcal{M} \equiv \{p(\cdot;\boldsymbol{\theta}) : \boldsymbol{\theta} \in \Theta\}$ be a smooth parametric model specification. Let $\mathcal{M}_\Gamma \equiv \{p(\cdot;\boldsymbol{\theta}) : \boldsymbol{\theta} \in \Gamma \subset \Theta\}$ where $\Gamma$ is a closed, bounded, and convex set. The probability model $\mathcal{M}_\Gamma$ is called a *local probability model.* □

### 10.1.4 Missing-Data Probability Models

In many practical machine learning applications, some of the state variables generated by the data generating process are only partially observable. Statistical environments of this type are specified as i.i.d. partially observable data generating processes.

**Definition 10.1.10** (Missing-Data Probability Model). Let $\tilde{\mathcal{D}}_n \equiv \{(\tilde{\mathbf{x}}_1, \tilde{\mathbf{m}}_1), (\tilde{\mathbf{x}}_2, \tilde{\mathbf{m}}_2), \ldots\}$ be an i.i.d. partially observable stochastic sequence of random vectors with common Radon-Nikodým density $p_e : \mathcal{R}^d \times \{0,1\}^d \rightarrow [0, \infty)$ with respect to a support specification measure $\nu$. As discussed in Section 9.2, the notation $(\mathbf{x}, \mathbf{m})$ means that the $i$th element of $\mathbf{x}$ is observable if and only if the $i$th element of $\mathbf{m}$ is equal to one.

The *missing-data probability model* for $\tilde{\mathcal{D}}_n$ is the set

$$\mathcal{M} \equiv \{p(\mathbf{x}, \mathbf{m}; \boldsymbol{\theta}) : \boldsymbol{\theta} \in \Theta\}$$

where $p(\cdot, \cdot; \boldsymbol{\theta}) : \mathcal{R}^d \times \{0,1\}^d \rightarrow [0, \infty)$ is a Radon-Nikodým density with respect to support specification measure $\nu$ for each $\boldsymbol{\theta}$ in the parameter space $\Theta \subseteq \mathcal{R}^q$. □

By factoring $p(\mathbf{x}, \mathbf{m}; \boldsymbol{\theta})$ such that $p(\mathbf{x}, \mathbf{m}; \boldsymbol{\theta}) = p(\mathbf{m}|\mathbf{x}; \boldsymbol{\theta})p(\mathbf{x}; \boldsymbol{\theta})$, a missing-data probability model $\mathcal{M} \equiv \{p(\mathbf{x}, \mathbf{m}; \boldsymbol{\theta}) : \boldsymbol{\theta} \in \Theta\}$ can be specified by a set of two probability models $(\mathcal{M}_c, \mathcal{M}_m)$. The first probability model $\mathcal{M}_c \equiv \{p(\mathbf{x}; \boldsymbol{\theta}) : \boldsymbol{\theta} \in \Theta\}$ is called the *complete-data probability model.* The density $p(\mathbf{x}; \boldsymbol{\theta})$ is called the *complete-data model density.* The second probability model $\mathcal{M}_m \equiv \{p(\mathbf{m}|\mathbf{x}; \boldsymbol{\theta}) : \boldsymbol{\theta} \in \Theta\}$ is called the *missing-data mechanism model.* The density $p(\mathbf{m}|\mathbf{x}; \boldsymbol{\theta})$ is called the *missing-data mechanism.*

This factorization specifies the model of the partially observable data generating process used by the learning machine. First, the learning machine assumes a data record without missing data $\mathbf{x}$ is generated by sampling from a complete-data model density $p(\mathbf{x}; \boldsymbol{\theta})$. Second, the learning machine assumes that a binary mask $\mathbf{m}$ is generated by sampling from the missing-data mechanism $p(\mathbf{m}|\mathbf{x}; \boldsymbol{\theta})$. Third, the learning machine then assumes the generated binary mask $\mathbf{m}$ was used to selectively decimate the original complete-data record $\mathbf{x}$ to create the observable data.

For a given mask vector $\mathbf{m}$, the complete-data model density $p(\mathbf{x}; \boldsymbol{\theta})$ can be alternatively represented using the notation

$$p(\mathbf{x}; \boldsymbol{\theta}) = p_{\mathbf{m}}(\mathbf{v}_{\mathbf{m}}, \mathbf{h}_{\mathbf{m}}; \boldsymbol{\theta}) \tag{10.2}$$

where $\mathbf{v}_{\mathbf{m}}$ consists of the $\mathbf{m}^T \mathbf{1}_d$ observable components of $\mathbf{x}$ and $\mathbf{h}_{\mathbf{m}}$ consists of the $d - \mathbf{m}^T \mathbf{1}_d$ unobservable components of $\mathbf{x}$ for a given $\mathbf{m} \in \{0,1\}^d$. With this representation of the complete-data probability model, a different choice of mask vector $\mathbf{m}$ corresponds to a different probability density $p_{\mathbf{m}}$ since the dimensions of $\mathbf{v}_{\mathbf{m}}$ and $\mathbf{h}_{\mathbf{m}}$ change as the pattern of missingness $\mathbf{m}$ changes.

**Definition 10.1.11** (Ignorable Missing-Data Mechanism). Let $\mathcal{M} \equiv \{p(\mathbf{x}, \mathbf{m}; \boldsymbol{\theta}) : \boldsymbol{\theta} \in \Theta\}$ be a missing-data probability model with missing-data mechanism $p(\mathbf{m}|\mathbf{x}; \boldsymbol{\theta})$ and complete-data density $p(\mathbf{x}; \boldsymbol{\theta})$. Let $\mathbf{v}_{\mathbf{m}}$ denote the observable components of $\mathbf{x}$ for a given mask vector $\mathbf{m}$. Assume for each $\boldsymbol{\theta} \equiv [\boldsymbol{\beta}, \boldsymbol{\psi}] \in \Theta$ and for all $(\mathbf{x}, \mathbf{m})$ in the support of $p(\mathbf{x}, \mathbf{m}; \boldsymbol{\theta})$:

$$p(\mathbf{m}|\mathbf{x}; \boldsymbol{\theta}) = p(\mathbf{m}|\mathbf{v}_{\mathbf{m}}; \boldsymbol{\beta}) \text{ and } p(\mathbf{x}; \boldsymbol{\theta}) = p(\mathbf{x}; \boldsymbol{\psi}).$$

The density $p(\mathbf{m}|\mathbf{v}_{\mathbf{m}}; \boldsymbol{\beta})$ is called the *ignorable missing-data mechanism density.* □

Less formally, an ignorable missing-data mechanism density $p_{\mathbf{m}}(\mathbf{m}|\mathbf{v}_{\mathbf{m}}; \boldsymbol{\beta})$ specifies that the probability density of the pattern of missingness is only functionally dependent upon the observable complete-data component $\mathbf{v}_{\mathbf{m}}$ and its parameters are distinct from the complete-data probability model.

## Exercises

10.1-1. Assume that the data generating process for a learning machine consists of sampling with replacement an infinite number of times from a box which contains $M$ distinct training stimuli. The data generating process generates a stochastic sequence of training stimuli $(\tilde{\mathbf{s}}_1, \tilde{y}_1), (\tilde{\mathbf{s}}_2, \tilde{y}_2), \ldots$ where $(\tilde{\mathbf{s}}_t, \tilde{y}_t)$ is a discrete random vector which can take on $M$ possible values at time $t$ for $t = 1, 2, \ldots$. Explicitly define the probability distribution used to specify the data generating process for the learning machine.

10.1-2. Consider a learning machine which generates a scalar response $\ddot{y}(\mathbf{s}, \boldsymbol{\theta})$ which is a deterministic function of a $d$-dimensional input pattern vector $\mathbf{s}$ and parameter vector $\boldsymbol{\theta}$. Let

$$\ddot{y}(\mathbf{s}, \boldsymbol{\theta}) = \mathbf{w}^T \mathcal{S}(\mathbf{V}\mathbf{s})$$

where $\mathcal{S}$ is a vector valued sigmoidal function and $\mathbf{w}$ and $\mathbf{V}$ are defined such that: $\boldsymbol{\theta}^T = [\mathbf{w}^T, \ \mathbf{vec}(\mathbf{V}^T)^T]$. In addition, assume that the learning machine assumes that the density of an observed response $\tilde{y}$ for a given $\mathbf{s}$ and a given $\boldsymbol{\theta}$ is a conditional univariate Gaussian density centered at $\ddot{y}(\mathbf{s}, \boldsymbol{\theta})$ with variance equal to one. Let $p_s : \mathcal{R}^d \to [0,1]$ be the probability that $\mathbf{s}$ is observed by the learning machine. Write an explicit formula for the learning machine's probability model $\mathcal{M}$ using set theory notation using both the conditional univariate Gaussian density and $p_s$. Is the learning machine's probability model $\mathcal{M}$ misspecified with respect to the data generating process in Exercise 10.1-1?

10.1-3. Let probability model specification $\mathcal{M}_1$ be defined with respect to support specification measure $\nu$ such that

$$\mathcal{M}_1 \equiv \left\{ p(\cdot|m,\sigma) : m \in \mathcal{R}, \sigma \in (0,\infty),\ p(x|m,\sigma) = \frac{\exp(-(x-m)^2/(2\sigma^2))}{\sigma\sqrt{2\pi}} \right\}.$$

Let probability model specification $\mathcal{M}_2$ be defined with respect to $\nu$ such that

$$\mathcal{M}_2 \equiv \left\{ p(\cdot|a,b) : a \in \mathcal{R}, b \in \mathcal{R},\ p(x|a,b) = c\exp(-ax^2 + 2b) \right\}.$$

Assume $c$ is chosen such that $\int p(x|a,b)dx = 1$. Are the probability model specifications $\mathcal{M}_1$ and $\mathcal{M}_2$ equivalent in the sense they are specifying exactly the same probability model?

10.1-4. A researcher has a parametric probability model $\mathcal{M}_q$ whose elements are indexed by a $q$-dimensional parameter vector and a parametric probability model $\mathcal{M}_r$ whose elements are indexed by an $r$-dimensional parameter vector where $q \neq r$. Show how to construct a parametric probability model $\mathcal{M}_k$ whose elements are indexed by a $k$-dimensional parameter vector such that $\mathcal{M}_k \equiv \mathcal{M}_q \cup \mathcal{M}_r$.

## 10.2 Gibbs Probability Models

Most commonly used probability models may be interpreted as collections of a special class of Radon-Nikodým probability density functions called Gibbs density functions.

**Definition 10.2.1** (Gibbs Radon-Nikodým Density). Let $\tilde{\mathbf{x}} = [\tilde{x}_1, \ldots, \tilde{x}_d]$ be a $d$-dimensional random vector with support $S \subseteq \mathcal{R}^d$. Let $\mathcal{V} \equiv \{1, \ldots, d\}$. Let the *energy function* $V : \mathcal{R}^d \rightarrow \mathcal{R}$. Define the *normalization constant*

$$Z \equiv \int_{\mathbf{x} \in S} \exp\left(\frac{-V(\mathbf{x})}{\mathcal{T}}\right) d\nu(\mathbf{x})$$

such that $Z$ is finite. The positive number $\mathcal{T}$ is called the *temperature constant.* A *Gibbs Radon-Nikodým Density* $p : \mathcal{R}^d \rightarrow (0,\infty)$ for $\tilde{\mathbf{x}}$ is defined such that for all $\mathbf{x} \in S$:

$$p(\mathbf{x}) = \frac{1}{Z} \exp\left(\frac{-V(\mathbf{x})}{\mathcal{T}}\right)$$

with respect to support specification $\nu$. □

**Definition 10.2.2** (Gibbs Probability Model). Let $\Theta \subseteq \mathcal{R}^q$. Let $p : \mathcal{R}^d \times \Theta \rightarrow [0,\infty)$ be defined such that for each parameter vector $\boldsymbol{\theta}$ in the parameter space $\Theta$, $p(\cdot;\boldsymbol{\theta}) : \mathcal{R}^d \rightarrow [0,\infty)$ is a Gibbs Radon-Nikodým density with respect to support specification measure $\nu$. Then, $\mathcal{M} \equiv \{p(\cdot;\boldsymbol{\theta}) : \boldsymbol{\theta} \in \Theta\}$ is called a *Gibbs model specification* defined with respect to $\nu$. □

An important and widely used type of Gibbs probability model is the exponential family probability model.

**Definition 10.2.3** (Exponential Family Probability Model). Let $\boldsymbol{\Psi} : \mathcal{R}^q \rightarrow \mathcal{R}^k$. Let $\boldsymbol{\Upsilon} : \mathcal{R}^d \rightarrow \mathcal{R}^k$ be a Borel-measurable function. Let $a : \mathcal{R}^q \rightarrow \mathcal{R}$ and let $b : \mathcal{R}^d \rightarrow \mathcal{R}$. An *exponential family model specification* is a Gibbs model specification

$$\mathcal{M} \equiv \{p(\cdot;\boldsymbol{\theta}) : \boldsymbol{\theta} \in \Theta\}$$

such that each Gibbs density $p(\cdot;\boldsymbol{\theta}) \in \mathcal{M}$ is specified by an energy function $V : \mathcal{R}^d \rightarrow \mathcal{R}$ where

$$V(\mathbf{x};\boldsymbol{\theta}) = -\boldsymbol{\Psi}(\boldsymbol{\theta})^T \boldsymbol{\Upsilon}(\mathbf{x}) + a(\boldsymbol{\theta}) + b(\mathbf{x}) \quad (10.3)$$

for all $\mathbf{x}$ in the support of $p(\cdot;\boldsymbol{\theta})$. For each $\mathbf{x} \in S$, $\boldsymbol{\Upsilon}(\mathbf{x})$ is called the *natural sufficient statistic.* In the special case where $\boldsymbol{\Psi}$ is defined such that $\boldsymbol{\Psi}(\boldsymbol{\theta}) = \boldsymbol{\theta}$ for an exponential family probability model $\mathcal{M}$, then $\mathcal{M}$ is called a *canonical exponential family model specification* and $\boldsymbol{\theta}$ is called the *natural parameter vector.* □

An element of a canonical exponential family model specification is called a *canonical exponential family density.* The set of all such densities is called the *canonical exponential family.* The shorthand terminology exponential family model may be used to refer to an exponential family model specification.

The form of an exponential family model specification greatly facilitates the design of learning algorithms for two reasons. First, each element of an exponential family model is a Gibbs density whose energy function $V$ is essentially a dot product of a recoding of the parameter vector and a recoding of the data. As shown in Chapter 13 (see Theorem 13.2.2), probability models with this structure often yield convex objective functions when the goal is to find the parameters that maximize the likelihood of the observed data.

Second, an exponential family model has the property that the observed data $\mathbf{x}$ can be projected into a low-dimensional subspace without any loss of information. For example, if $\tilde{\mathbf{x}}$ is a $d$-dimensional random vector, then the statistic $\Upsilon(\mathbf{x})$ in Equation 10.3 provides a sufficient statistic for computing the probability of $\mathbf{x}$ using only the projection $\Upsilon(\mathbf{x})$.

Every exponential family model consists of a set of Gibbs densities such that each Gibbs density has energy function $V(\cdot;\boldsymbol{\theta})$ for each $\boldsymbol{\theta} \in \mathcal{R}^q$ where

$$V(\mathbf{x};\boldsymbol{\theta}) \equiv \boldsymbol{\Psi}(\boldsymbol{\theta})^T \boldsymbol{\Upsilon}(\mathbf{x})$$

may be rewritten as a canonical exponential family model. The canonical exponential family model is a set of Gibbs densities such that each density has energy function $V(\cdot;\boldsymbol{\Psi})$ for each $\boldsymbol{\Psi} \in \mathcal{R}^k$ where

$$V(\mathbf{x};\boldsymbol{\Psi}) = \boldsymbol{\Psi}^T \boldsymbol{\Upsilon}(\mathbf{x}).$$

It is often advantageous to use the canonical exponential family model representation if possible since the energy function for the Gibbs density is now a simpler linear function of the model's parameters.

Multinomial logistic probability models also known as "softmax" models in the machine learning literature are often used in both statistics and machine learning to specify the probability distributions for a categorical random variable. In a multinomial logistic probability model, the one-hot encoding scheme (see Section 1.2.1) is a convenient choice for representing a categorical random variable.

**Definition 10.2.4** (Multinomial Logistic Probability Model). A *multinomial logistic probability model specification* is a Gibbs model specification $\mathcal{M} \equiv \{p(\cdot;\boldsymbol{\theta}) : \boldsymbol{\theta} \in \Theta \subseteq \mathcal{R}^{d-1}\}$ such that each $p(\cdot|\boldsymbol{\theta}) \in \mathcal{M}$ has support equal to the set of column vectors of $\mathbf{I}_d$ with energy function $V(\cdot;\boldsymbol{\theta})$ defined such that $V(\mathbf{x};\boldsymbol{\theta}) = -\mathbf{x}^T[\boldsymbol{\theta}^T,\ 0]^T$, and normalization constant $Z(\boldsymbol{\theta}) = \mathbf{1}_q^T \mathbf{exp}\left([\boldsymbol{\theta}^T,\ 0]^T\right)$. In addition, if $d = 2$, then $\mathcal{M}$ is called a *binary logistic probability model specification.* □

**Definition 10.2.5** (Poisson Probability Model). Let $\Theta \subseteq (0, \infty)$. A *Poisson probability model specification* is a Gibbs model specification $\mathcal{M} \equiv \{p(\cdot;\theta) : \theta \in \Theta\}$ such that for each $\theta \in \Theta$: $p(\cdot;\theta)$ has support $\mathbb{N}$ with energy function $V(\cdot;\theta)$ defined such that for all $k \in \mathbb{N}$:

$$V(k;\theta) = -k\log(\theta) + \log(k!),$$

and normalization constant $Z(\theta) = \exp(\theta)$. The parameter $\theta$ is called the *rate parameter.* □

The mean and variance of a Poisson random variable is equal to the rate parameter for the Poisson random variable. The Poisson random variable is sometimes a useful probabilistic model of the length of a trajectory.

**Definition 10.2.6** (Multivariate Gaussian Probability Model). A *multivariate Gaussian probability model specification* is a Gibbs probability model $\mathcal{M} \equiv \{p(\cdot;\boldsymbol{\theta}) : \boldsymbol{\theta} \in \Theta\}$ such that each density $p(\cdot;\boldsymbol{\theta}) \in \mathcal{M}$ has support $\mathcal{R}^d$ with energy function $V$ defined such that $V(\mathbf{x};\boldsymbol{\theta}) \equiv (1/2)(\mathbf{x}-\mathbf{m})^T\mathbf{C}^{-1}(\mathbf{x}-\mathbf{m})$, and normalization constant $Z(\boldsymbol{\theta}) \equiv (2\pi)^{d/2}(\det(\mathbf{C}))^{1/2}$ for all $\boldsymbol{\theta} \equiv [\mathbf{m}^T, \mathbf{vech}(\mathbf{C})^T] \in \Theta$. □

The multivariate Gaussian density (see Definition 9.3.6) is an element of a multivariate Gaussian probability model.

A probability mass function on a finite sample space of $d$ possible values can be represented as a $d$-dimensional vector $\mathbf{x}$ where the $i$th element of $\mathbf{x}$ specifies the probability of observing the $i$th outcome of the discrete random variable. Thus, the probability distribution $p$ of a $d$-dimensional random vector $\tilde{\mathbf{x}} \in (0,1)^d$ defined such that the sum of its elements is always exactly equal to one can be interpreted as a probability distribution of probability mass functions. In many applications, it is helpful to have a method for specifying the probability distribution of a collection of probability mass functions. The Dirichlet Probability Model, which is also a member of the exponential family, is a popular model for achieving this objective. The Dirichlet Probability Model is now defined using the definition of the $\Gamma$ function (see Definition 9.4.1).

**Definition 10.2.7** (Dirichlet Probability Model). Let $\Theta \subset (0,\infty)^d$. A *Dirichlet probability model specification* is a Gibbs probability model specification

$$\mathcal{M} \equiv \{p(\cdot;\boldsymbol{\theta}) : \boldsymbol{\theta} \equiv [\theta_1,\ldots,\theta_d] \in \Theta\},$$

such that each density $p(\cdot;\boldsymbol{\theta}) \in \mathcal{M}$ has support $S \equiv \left\{\mathbf{x} \equiv [x_1,\ldots,x_d] \in (0,1)^d : \sum_{i=1}^d x_i = 1\right\}$, energy function $V(\mathbf{x};\boldsymbol{\theta}) = \sum_{i=1}^d (1-\theta_i)\log x_i$, and normalization constant

$$Z(\boldsymbol{\theta}) = \frac{\prod_{i=1}^d \Gamma(\theta_i)}{\Gamma(\boldsymbol{\theta}^T\mathbf{1}_d)}$$

where $\Gamma$ is the Gamma function as defined in Definition 9.4.1. □

**Definition 10.2.8** (Gamma Probability Model). A *Gamma probability model specification* is a Gibbs probability model specification $\mathcal{M} \equiv \{p(\cdot;\boldsymbol{\theta}) : \boldsymbol{\theta} \equiv (\alpha,\beta) \in [0,\infty)^2\}$, such that each density $p(\cdot;\boldsymbol{\theta}) \in \mathcal{M}$ has support $[0,\infty)$ with energy function

$$V(x;\boldsymbol{\theta}) = \beta x - (\alpha-1)\log(x)$$

and normalization constant $Z(\boldsymbol{\theta}) = \beta^{-\alpha}\Gamma(\alpha)$ where $\Gamma$ is the Gamma function as defined in Definition 9.4.1. The parameter $\alpha$ is called the *shape parameter* and the parameter $\beta$ is called the *rate parameter*. □

**Definition 10.2.9** (Chi-Squared Probability Model). A *Chi-Squared probability model specification* with *degrees of freedom* parameter $k$ is a Gamma probability model specification $\mathcal{M} \equiv \{p(\cdot;\boldsymbol{\theta}) : \boldsymbol{\theta} \equiv (\alpha,\beta) \in [0,\infty)^2\}$ where $\alpha = k/2$ and $\beta = 2$. □

An element of a chi-squared probability model is a chi-squared density as defined in Definition 9.4.2. Figure 10.1 illustrates the shape of the chi-squared density for different choices of degrees of freedom.

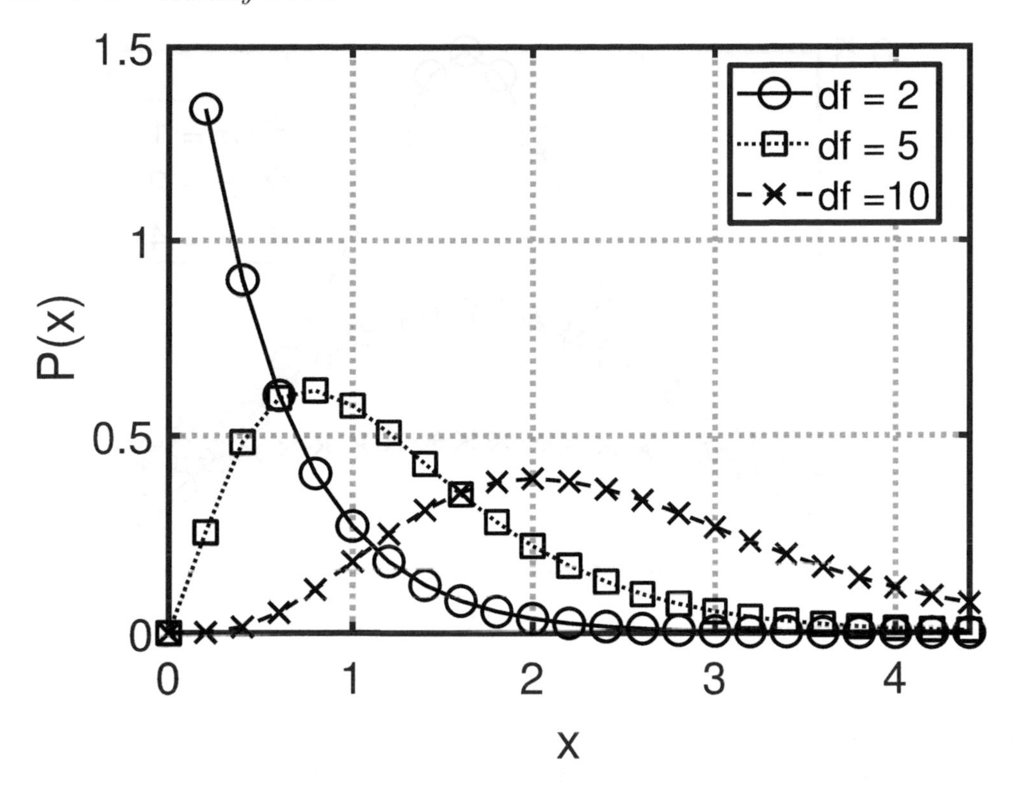


**FIGURE 10.1**
**Chi-squared density for different choices of degrees of freedom.** When the degrees of freedom are large, then the chi-squared probability density is similar in form to a Gaussian probability density. The chi-squared density is a special case of the gamma density function when the shape parameter is chosen to be one-half of the degrees of freedom of the chi-squared density and the rate parameter is chosen to be equal to 2. Definition 9.4.2 provides the formula used to plot the chi-squared density in this figure.

**Definition 10.2.10** (Laplace Probability Model). A *Laplace probability model specification* is a Gibbs model specification

$$\mathcal{M} \equiv \{p(\cdot;\boldsymbol{\theta}) : \boldsymbol{\theta} \equiv (\mu, \sigma) \in \mathcal{R} \times (0, \infty)\}$$

such that each element of $p(\cdot;\boldsymbol{\theta}) \in \mathcal{M}$ has support $\mathcal{R}$ with energy function $V(x;\boldsymbol{\theta}) = |x - \mu|/\sigma$ and normalization constant $Z = 2\sigma$. The parameter $\mu$ is called the *location parameter* and the positive parameter $\sigma$ is called the *scale parameter.* □

Each element of a Laplace probability model specification is a Laplace density of the form:

$$p(x;\boldsymbol{\theta}) = (2\sigma)^{-1} \exp\left(-|x - \mu|/\sigma\right)$$

which is not differentiable at the point $x = \mu$. In some applications, it is helpful to have a differentiable version of a Laplace density. In such cases, it is helpful to consider the following smooth approximation to the Laplace density when $\sigma$ is a small positive number.

**Definition 10.2.11** (Hyperbolic Secant Probability Model). A *hyperbolic secant probability model* is a Gibbs model specification $\mathcal{M} \equiv \{p(\cdot;\boldsymbol{\theta}) : \boldsymbol{\theta} \equiv (\mu, \sigma) \in \mathcal{R} \times (0, \infty)\}$ for each $p(\cdot|\boldsymbol{\theta}) \in \mathcal{M}$ with support $\mathcal{R}$, energy function

$$V(x;\boldsymbol{\theta}) = -\log\left(\left[\exp\left((\pi/2)(x - \mu)/\sigma\right) + \exp\left(-(\pi/2)(x - \mu)/\sigma\right)\right]\right),$$

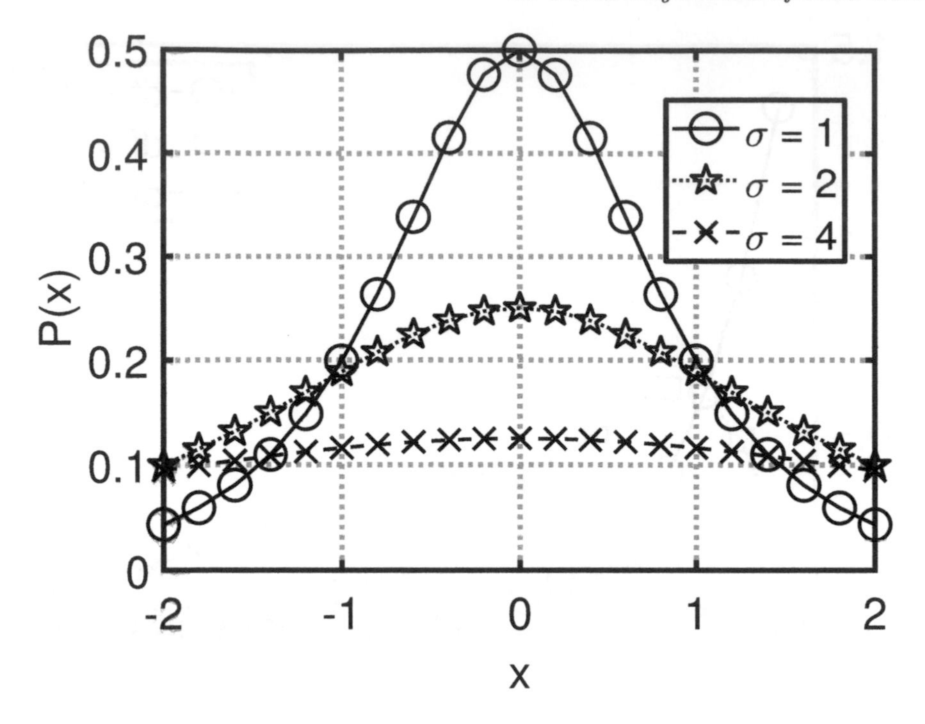


**FIGURE 10.2**
**Hyperbolic secant density as a function of its standard deviation.** This is a useful differentiable probability density since it closely approximates the non-differentiable Laplace probability density when the standard deviation is chosen to be sufficiently small.

and normalization constant $Z = \sigma$. The parameter $\mu$ is called the *location parameter* and the parameter $\sigma$ is called the *scale parameter.* □

Figure 10.2 shows how the shape of the hyperbolic secant density changes as a function of its standard deviation.

## Exercises

10.2-1. Let $\Theta \subseteq \mathcal{R}^q$. Let $\Upsilon : \mathcal{R}^d \to \mathcal{R}^q$. Let $Z : \Theta \to (0, \infty)$ be defined such that

$$Z(\boldsymbol{\theta}) \equiv \int_{\Theta} \exp\left(-\boldsymbol{\theta}^T \Upsilon(\mathbf{y})\right) d\nu(\mathbf{y}).$$

Let $p(\cdot;\cdot) : \mathcal{R}^d \times \Theta \to [0, \infty)$ be defined such that

$$p(\mathbf{x}; \boldsymbol{\theta}) = \frac{\exp\left(-\boldsymbol{\theta}^T \Upsilon(\mathbf{y})\right)}{Z(\boldsymbol{\theta})}$$

is a Radon-Nikodým density for a random vector $\tilde{\mathbf{x}}$ defined with respect to support specification measure $\nu$. Show that the gradient and Hessian of $\log(Z(\boldsymbol{\theta}))$ are, respectively, the mean and covariance matrix of $\Upsilon(\tilde{\mathbf{x}})$.

10.2-2. Derive explicit formulas for the mean and variance of: (1) nonparametric probability mass function, (2) multinomial logistic probability mass function, (3) binomial probability mass function, (4) uniform probability density, (5) univariate Gaussian density, (6) multivariate Gaussian density, (7) Gamma density, (8) exponential density, (9) chi-squared density, (10) Laplace density, and (11) hyperbolic secant density.

10.2-3. *Evaluating the Incomplete Gamma Function.* There is a function in MATLAB®called `gammainc` defined such that:

$$\texttt{gammainc}(x, a) = (1/\Gamma(a)) \int_0^x t^{a-1} \exp(-t) dt$$

where $a$ and $x$ are positive numbers and $\Gamma$ is defined as in Definition 9.4.1. Show that the MATLAB command `y = gammainc(y0/2,df/2)` computes the probability that a chi-squared random variable with *df* degrees of freedom is less than the critical value of $y0$.

## 10.3 Bayesian Networks

In statistical machine learning problems, the probability density functions typically used by the learning machine to approximate its statistical environment may specify the joint distribution of many random variables. Very powerful tools for the analysis and design of such probability density functions are based upon the principle of *factorization* where a complicated global joint probability density function is represented as a collection of local conditional probability density functions. For example, it may be challenging to construct a probability distribution which specifies the global (joint) probability of a particular long sequence of sunny days and rainy days. It is much easier to specify the local conditional probability that a sunny day will follow a rainy day and the local conditional probability that a rainy day will follow a sunny day. These local conditional probabilities are easier to semantically understand and easier to estimate in many cases yet provide an equivalent specification of the global joint distribution.

As a first example of factorizing a joint density, consider the sequence of $d$-dimensional random vectors $\tilde{\mathbf{x}}_1, \ldots, \tilde{\mathbf{x}}_n$ where a realization of $\tilde{\mathbf{x}}_j$ is a $d$-dimensional feature vector specifying the weather on the $j$th day of the year, $j = 1, \ldots, n$. Assume the goal of an engineer is to develop a probabilistic model of this sequence of $n$ random vectors. Direct specification of a candidate probability density $p(\mathbf{x}_1, \ldots, \mathbf{x}_n)$ is not straightforward especially in the case where $n$ is large. However, this global joint density function can be represented by local conditional density functions that specify the likelihood of the state of the weather on the $j$th day of the year given knowledge of the weather on days $1, \ldots, j-1$.

For example when $n = 3$,

$$p(\mathbf{x}_1, \mathbf{x}_2, \mathbf{x}_3) = p(\mathbf{x}_1) \left( \frac{p(\mathbf{x}_1, \mathbf{x}_2)}{p(\mathbf{x}_1)} \right) \left( \frac{p(\mathbf{x}_1, \mathbf{x}_2, \mathbf{x}_3)}{p(\mathbf{x}_1, \mathbf{x}_2)} \right) \tag{10.4}$$

which can be rewritten as:

$$p(\mathbf{x}_1, \mathbf{x}_2, \mathbf{x}_3) = p(\mathbf{x}_1) p(\mathbf{x}_2 | \mathbf{x}_1) p(\mathbf{x}_3 | \mathbf{x}_2, \mathbf{x}_1).$$

It is also important to emphasize that many possible factorizations of a particular joint density are possible. This is an important observation which can be effectively exploited in applications. So, for example, the joint density $p(\mathbf{x}_1, \mathbf{x}_2, \mathbf{x}_3)$ in (10.4) can be alternatively represented using the following entirely different factorization:

$$p(\mathbf{x}_1, \mathbf{x}_2, \mathbf{x}_3) = p(\mathbf{x}_3)\left(\frac{p(\mathbf{x}_2, \mathbf{x}_3)}{p(\mathbf{x}_3)}\right)\left(\frac{p(\mathbf{x}_1, \mathbf{x}_2, \mathbf{x}_3)}{p(\mathbf{x}_2, \mathbf{x}_3)}\right) \tag{10.5}$$

which can be rewritten as:

$$p(\mathbf{x}_1, \mathbf{x}_2, \mathbf{x}_3) = p(\mathbf{x}_3)p(\mathbf{x}_2|\mathbf{x}_3)p(\mathbf{x}_1|\mathbf{x}_2, \mathbf{x}_3).$$

In rule-based inference systems, knowledge is represented in terms of logical constraints such as:

$$\texttt{IF } \tilde{x}_1 = 1 \texttt{ AND } \tilde{x}_2 = 0, \texttt{ THEN } \tilde{x}_3 = 1. \tag{10.6}$$

In solving a particular logic problem using such constraints, there may not be a sufficient number of logical constraints to ensure a unique (or even a small number of conclusions) for a given set of initial conditions. Moreover, the collection of logical constraints may contain contradictory information.

In a probabilistic reasoning system, one may have probabilistic versions of logical constraints such as (10.6) so that if the premise of (10.6) holds, then the conclusion of (10.6) holds with some probability $p$. This is an example of a "probabilistic logical constraint". Like a typical rule-based inference system, a system based upon probabilistic logical constraints faces similar challenges. Specifically, are there a sufficient number of probabilistic logical constraints to ensure unique inferences? And, additionally, are the assumed set of probabilistic logical constraints internally consistent? Fortunately, both of these issues are easily resolved if one begins by specifying a joint distribution of all random variables $\tilde{x}_1, \tilde{x}_2, \ldots, \tilde{x}_d$ and then factors this joint distribution into a collection of probabilistic local constraints. Or, alternatively, correctly constructs a particular joint distribution from a given collection of probabilistic local constraints.

### 10.3.1 Factoring a Chain

**Theorem 10.3.1** (Chain Factorization Theorem). *Let $\tilde{\mathbf{x}} = [\tilde{x}_1, \ldots, \tilde{x}_d]$ be a d-dimensional random vector with Radon-Nikodým density $p : \mathcal{R}^d \rightarrow [0, \infty)$. Then, for all $\mathbf{x}$ such that $p(\mathbf{x}) > 0$ and $d > 1$:*

$$p(\mathbf{x}) = p(x_1)\prod_{i=2}^{d} p(x_i|x_{i-1}, \ldots, x_1). \tag{10.7}$$

*Proof.* The proof is done by induction. For the case of $d = 2$,

$$p(x_1, x_2) = p(x_1)p(x_2|x_1).$$

For the case of $d = 3$,

$$p(x_1, x_2, x_3) = \left[\frac{p(x_1, x_2, x_3)}{p(x_1, x_2)}\right]\left[\frac{p(x_1, x_2)}{p(x_1)}\right]p(x_1).$$

Given the joint density $p(x_1, \ldots, x_{d-1})$ has been factored, then the factorization of $p(x_1, \ldots, x_d)$ is computed using the formula:

$$p(x_1, \ldots, x_d) = p(x_d|x_{d-1}, \ldots, x_1)p(x_1, \ldots, x_{d-1}).$$

■

As previously noted, the "ordering" of the random variables is arbitrary. For example, one can rewrite (10.7) as:

$$p(\mathbf{x}) = p(x_d) \prod_{i=1}^{d-1} p(x_i | x_{i+1}, \ldots, x_d).$$

In fact, there are $d!$ different ways of expressing the joint density of $d$ random variables as a product of conditional probabilities using the chain factorization theorem.

### 10.3.2 Bayesian Network Factorization

**Definition 10.3.1** (Parent Function). Let $\mathcal{V} \equiv [1, \ldots, d]$. Let $\mathcal{G} \equiv (\mathcal{V}, \mathcal{E})$ be a directed acyclic graph. Let $S_i$ be the set of $m_i$ parents of the $i$th vertex in $\mathcal{G}$. Let $\mathcal{P}_i : \mathcal{R}^d \to \mathcal{R}^{m_i}$ be defined such that for all $\mathbf{x} \in \mathcal{R}^d$:

$$\mathcal{P}_i(\mathbf{x}) = \{x_j : (j, i) \in \mathcal{E}\}, \ i = 1, \ldots, d.$$

The function $\mathcal{P}_i$ is called the *parent function* for vertex $i$ for $i = 1, \ldots, d$. The set $\{\mathcal{P}_1, \ldots, \mathcal{P}_d\}$ is called the *parent function set* on $\mathcal{R}^d$ with respect to $\mathcal{G}$. □

Notice that a parent function $\mathcal{P}_i$ takes a vector as input and returns a subset of the elements of that vector which are parents of $v_i$ as specified by the directed acyclic graph $\mathcal{G}$, $i = 1, \ldots, d$.

**Definition 10.3.2** (Bayesian Network). Let $\tilde{\mathbf{x}} = [\tilde{x}_1, \ldots, \tilde{x}_d]$ be a $d$-dimensional random vector with Radon-Nikodým density $p : \mathcal{R}^d \to [0, \infty)$. Let $\mathcal{V} \equiv [1, \ldots, d]$. Let $\mathcal{G} \equiv (\mathcal{V}, \mathcal{E})$ be a directed acyclic graph. Let $\{\mathcal{P}_1, \ldots, \mathcal{P}_d\}$ be a parent function set on $\mathcal{R}^d$ defined with respect to $\mathcal{G}$. Define a sequence of distinct integers $m_1, \ldots, m_d$ in $\{1, \ldots, d\}$ such that if $(m_j, m_k) \in \mathcal{E}$, then $m_j < m_k$ for all $j, k \in \mathcal{V}$ and

$$p(x_{m_i} | x_{m_1}, \ldots, x_{m_{i-1}}) = p(x_{m_i} | \mathcal{P}_{m_i}(\mathbf{x}))$$

for $i = 1, \ldots, d$. The random vector $\tilde{\mathbf{x}}$ is called a *Bayesian network* specified by $(p, \mathcal{G})$ with *conditional dependency graph* $\mathcal{G}$. □

The assertion in Definition 10.3.2 that

$$p(x_{m_i} | x_{m_1}, \ldots, x_{m_{i-1}}) = p(x_{m_i} | \mathcal{P}_{m_i}(\mathbf{x}))$$

when $\mathcal{P}_{m_i}(\mathbf{x}) = \{\}$ simply means that $p(x_{m_i} | x_{m_1}, \ldots, x_{m_{i-1}}) = p(x_{m_i})$.

A key feature of the definition of a Bayesian network is that the probability distribution of a random variable $\tilde{x}_i$ in the Bayesian network is only functionally dependent upon the parents of that random variable.

Bayesian networks are widely used in many areas of artificial intelligence concerned with the representation and use of causal knowledge. An example of a Bayesian network for medical diagnosis is provided in Figure 10.3.

**Example 10.3.1** (First-Order Markov Chain (Finite Length)). An important special case of a Bayesian Network is the first-order Markov Chain whose conditional dependency graph is specified by a linear chain of nodes (see Figure 10.4). Let the $d$-dimensional random vector $\tilde{\mathbf{x}} = [\tilde{x}_1, \ldots, \tilde{x}_d]$ be a Bayesian network associated with the conditional dependency graph $\mathcal{G} = (\{1, \ldots, d\}, \mathcal{E})$ where

$$\mathcal{E} \equiv \{(1, 2), (2, 3), (3, 4), \ldots, (d - 1, d)\}.$$

Then, $\mathcal{G}$ is called a *first-order Markov chain*. △

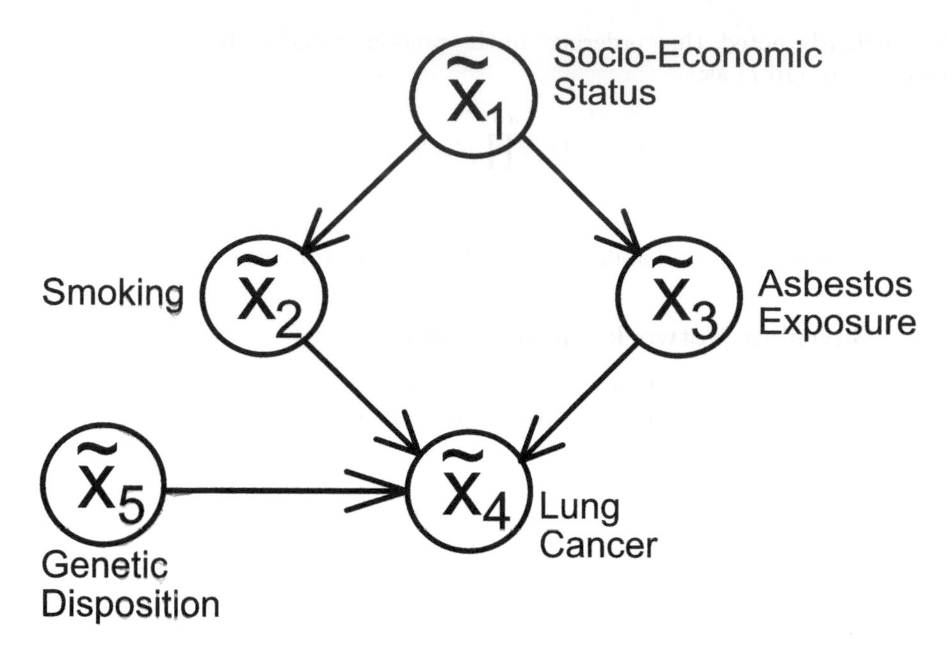


**FIGURE 10.3**
**A Bayesian network for representing knowledge for a medical diagnosis problem.** The Bayesian network consists of five binary-valued random variables which indicate the presence or absence of `Smoking`, `Lung Cancer`, `Genetic Disposition`, `Asbestos Exposure`, and `Socio-Economic Status`. The Bayesian network conditional independence graph indicates that the probability distributions of `Smoking` and `Asbestos Exposure` are functionally dependent only upon `Socio-Economic Status`. The probability distribution of `Lung Cancer` is only functionally dependent upon `Smoking`, `Asbestos Exposure`, and `Genetic Disposition`.

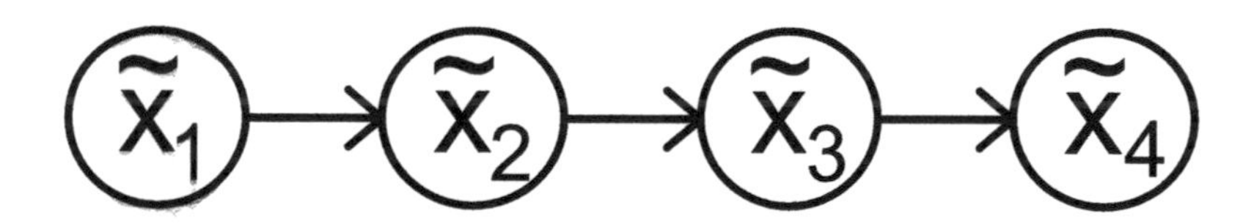


**FIGURE 10.4**
**Bayes network representation of a first-order Markov chain.** Conditional independence relationships between the random variables in a first-order Markov chain may be represented as a conditional dependency graph in which every node has both a parent node and child node with the possible exception of the end nodes.

**Definition 10.3.3** (Markov Blanket). Let $(p, \mathcal{G})$ be a Bayesian network with conditional dependency graph $\mathcal{G} \equiv \{\mathcal{V}, \mathcal{E}\}$. Let $v \in \mathcal{V}$. Let $P$ be the set of parents of $v$ with respect to $\mathcal{G}$. Let $C$ be the set of children of $v$ with respect to $\mathcal{G}$. Let $R$ be the set consisting of all parents of the elements of $C$ except for $v$. Then, the *Markov Blanket* for $v$ is defined as $P \cup C \cup R$. □

The Markov blanket of a random variable $\tilde{x}$ in a Bayesian network is the minimal set of all random variables which must be observable in order to completely specify the behavior of $\tilde{x}$. A learning machine, for example, can learn the probabilistic rules governing the behavior of a random variable $\tilde{x}$ in a Bayesian network by *only* observing the joint frequency distribution of $\tilde{x}$ and the random variables in the Markov blanket of $\tilde{x}$.

The Markov blanket for vertex $j$ in the first-order Markov chain would consist of the vertices $j-1$ and $j+1$.

**Theorem 10.3.2** (Bayesian Network Factorization)**.** *Let $\mathcal{V} \equiv [1,\ldots,d]$ where $d > 1$. Let $\mathcal{G} \equiv (\mathcal{V},\mathcal{E})$ be a directed acyclic graph. Let $\tilde{\mathbf{x}}$ be a Bayesian network with respect to Radon-Nikodým density $p : \mathcal{R}^d \to [0,\infty)$ and $\mathcal{G}$ is the conditional dependency graph for the Bayesian network. Define a sequence of distinct integers $m_1,\ldots,m_d$ in $\{1,\ldots,d\}$ such that if $(m_j,m_k) \in \mathcal{E}$ then $m_j < m_k$ for all $j,k \in \{1,\ldots,d\}$. Then, for all $\mathbf{x}$ in the support of $\tilde{\mathbf{x}}$:*

$$p(\mathbf{x}) = \prod_{i=1}^{d} p\left(x_{m_i}|\mathcal{P}_{m_i}(\mathbf{x})\right) \tag{10.8}$$

*where $\{\mathcal{P}_1,\ldots,\mathcal{P}_d\}$ is the parent function set on $\mathcal{R}^d$ defined with respect to $\mathcal{G}$.*

*Proof.* The chain factorization theorem provides the relation:

$$p(\mathbf{x}) = p(x_{m_1}) \prod_{i=2}^{d} p(x_{m_i}|x_{m_{i-1}},\ldots,x_{m_1}). \tag{10.9}$$

The conditional independence assumptions of the Bayesian Network specify that:

$$p(x_{m_i}|x_{m_{i-1}},\ldots,x_{m_1}) = p(x_{m_i}|\mathcal{P}_{m_i}(\mathbf{x})),\ i = 1,\ldots,d. \tag{10.10}$$

Substituting (10.10) into (10.9) gives (10.8). ■

An important special case of a Bayesian network is a regression model which corresponds to a Bayesian network defined by a directed conditional dependency graph which has $d$ nodes corresponding to the random variables comprising the $d$-dimensional input pattern vector random vector and $m$ nodes corresponding to the random variables comprising an $m$-dimensional pattern vector (see Figure (10.5)).

**Definition 10.3.4** (Regression Model)**.** Let $S \subseteq \mathcal{R}^d$. Let $\boldsymbol{\theta} \in \Theta \subseteq \mathcal{R}^q$. Let $p : S \times \Theta \to [0,\infty)$ be defined such that $p(\cdot;\mathbf{s},\boldsymbol{\theta})$ is a Radon-Nikodým density with respect to support specification $\nu$ for each *input pattern vector* $\mathbf{s}$ and for each parameter vector $\boldsymbol{\theta}$ in the *parameter space* $\Theta$. Then,

$$\mathcal{M} \equiv \{p(\cdot;\mathbf{s},\boldsymbol{\theta}) : (\mathbf{s},\boldsymbol{\theta}) \in S \times \Theta\}$$

is called a *regression probability model specification.* □

**Example 10.3.2** (Linear and Logistic Regression Probability Models)**.** For example, when an element, $p(\cdot|m(\mathbf{s}),\sigma^2)$, in a regression probability model specification is a univariate Gaussian density with mean $m(\mathbf{s}) \equiv \boldsymbol{\theta}^T\mathbf{s}$ and variance $\sigma^2$, then the regression probability model specification is called a *linear regression model.* As a second example, assume the

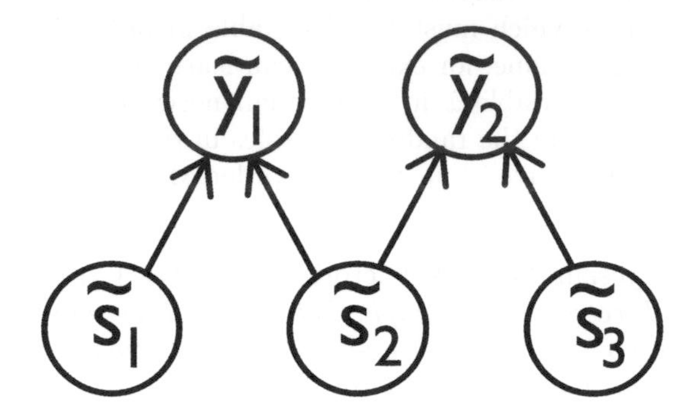


**FIGURE 10.5**
**A Bayesian network representation of a linear or nonlinear regression model.** A linear or nonlinear regression model is a Bayesian network where the predictor random variables in the $d$-dimensional input pattern $\tilde{s}_1, \ldots, \tilde{s}_d$ are the causal antecedents of the response random variables $\tilde{\mathbf{y}}_1, \ldots, \tilde{\mathbf{y}}_m$.

support of an element, $p(\cdot|\mathbf{s}, \boldsymbol{\theta})$, in a regression probability model specification $\mathcal{M}$ is $\{0, 1\}$ where $p(\tilde{y} = 1|\mathbf{s}, \boldsymbol{\theta})$ is defined such that

$$p(\tilde{y} = 1|\mathbf{s}, \boldsymbol{\theta}) = \mathcal{S}(\boldsymbol{\theta}^T \mathbf{s}), \; \mathcal{S}(\psi) = 1/(1 + \exp(-\psi))$$

then $\mathcal{M}$ is called a *logistic regression model.* Both linear and logistic regression probability models may be interpreted as Bayesian Networks (see Figure 10.5). △

**Example 10.3.3** (Hidden Markov Model)**.** Hidden Markov Models (HMMs) on finite state spaces have been applied in a number of areas including speech recognition, speech synthesis, part-of-speech tagging, text mining, DNA analysis, and plan recognition for a sequence of actions. Figure (10.6) depicts a conditional dependency graph for a Bayesian network corresponding to a Hidden Markov model designed to solve a part-of-speech tagging language processing problem. The 5-dimensional random vector $\tilde{\mathbf{h}}_k$ models the syntactic category label of the $k$th word in a sequence of three words, $k = 1, 2, 3$. In particular, the support of the 5-dimensional random vector $\tilde{\mathbf{h}}_k$ is the set of columns of identity matrix $\mathbf{I}_5$ which respectively correspond to the five category labels: `start`, `noun`, `verb`, `adjective`, `determiner`) for $k = 1, 2, 3$. The 4-dimensional random vector $\tilde{\mathbf{o}}_k$ models the specific words encountered in the three-word sentence where the support of $\tilde{\mathbf{o}}_k$ is the set of four column vectors of identity matrix $\mathbf{I}_4$ that respectively represent the possible words: *flies, like, superman, honey.*

Let the local emission probability, $p(\mathbf{o}_k|\mathbf{h}_k, \boldsymbol{\eta})$, specify the conditional probability of a word $\mathbf{o}_k$ given its category label $\mathbf{h}_k$ for a given parameter vector $\boldsymbol{\eta}$. Let the local transition probability, $p(\mathbf{h}_{k+1}|\mathbf{h}_k, \boldsymbol{\psi})$, specify the conditional probability that category label $\mathbf{h}_{k+1}$ immediately follows category label $\mathbf{h}_k$ for a given parameter vector $\boldsymbol{\psi}$. The goal of the learning process is to find parameter vectors $\boldsymbol{\eta}$ and $\boldsymbol{\psi}$ that maximize the probability of the word sequence $\mathbf{o}_1$, $\mathbf{o}_2$, $\mathbf{o}_3$ given parameters $\boldsymbol{\eta}$ and $\boldsymbol{\psi}$.

To achieve this objective, provide an explicit formula for $p(\mathbf{o}_1, \mathbf{o}_2, \mathbf{o}_3|\boldsymbol{\psi}, \boldsymbol{\eta})$.

**Solution.** Let $\mathbf{h}_0$ be chosen to have value of the `start` label. Then,

$$p(\mathbf{o}_1, \mathbf{o}_2, \mathbf{o}_3|\boldsymbol{\psi}, \boldsymbol{\eta}) = \sum_{\mathbf{h}_1, \mathbf{h}_2, \mathbf{h}_3} \left[ \prod_{j=1}^{3} p(\mathbf{o}_j|\mathbf{h}_j; \boldsymbol{\eta}) p(\mathbf{h}_j|\mathbf{h}_{j-1}; \boldsymbol{\psi}) \right]. \tag{10.11}$$

△

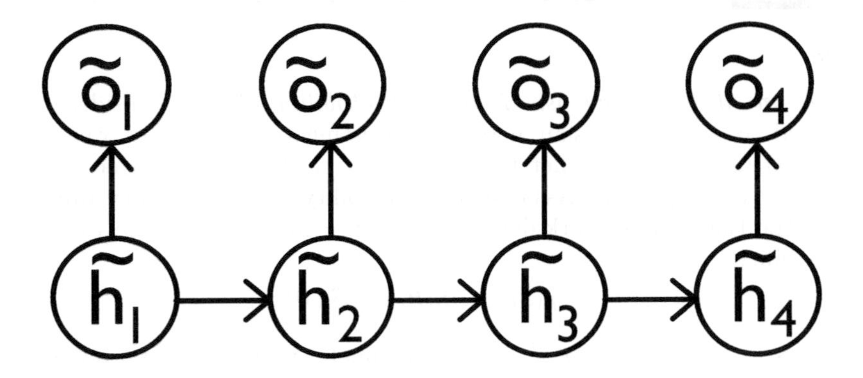


**FIGURE 10.6**
**A Bayesian network representation of a hidden Markov model.** The random variables $\tilde{\mathbf{h}}_1$, $\tilde{\mathbf{h}}_2$, and $\tilde{\mathbf{h}}_3$ are not directly observable and correspond to hidden system states. The probability distribution of $\tilde{\mathbf{h}}_{t+1}$ is functionally dependent only upon $\tilde{\mathbf{h}}_t$. The probability distribution of an observable random variable $\tilde{\mathbf{o}}_t$ is functionally dependent only upon its corresponding hidden counterpart $\tilde{\mathbf{h}}_t$.

**Example 10.3.4** (Continuous-State Hidden Markov Model)**.** Consider a navigation problem where the unobservable state of a satellite at time $t$ is denoted by the state vector $\mathbf{x}_t$ which includes state variables such as: position, velocity, atmospheric density, gravitational influences, and so on. This is analogous to the Hidden Markov Model problem in Example 10.3.3.

A researcher's model of how the state of the satellite evolves in time is specified by the state dynamics: $\tilde{\mathbf{x}}_t = \mathbf{A}\mathbf{x}_{t-1} + \tilde{\mathbf{w}}_{t-1}$ where $\mathbf{A}$ is a known $d$-dimensional state transition matrix, and the Gaussian random vector $\tilde{\mathbf{w}}_{t-1}$ is the *system noise* consisting of a combination of internal and external disturbances which influence the system dynamics. It is assumed that the Gaussian random vector $\tilde{\mathbf{w}}_{t-1}$ has mean zero and positive definite symmetric covariance matrix $\mathbf{Q} \in \mathcal{R}^{d \times d}$. The parameter vector for the hidden state dynamics of the satellite is defined as $\boldsymbol{\psi} \equiv [\mathbf{vec}(\mathbf{A}^T), \ \mathbf{vech}(\mathbf{Q})]$.

The researcher's model of how the unobservable state of the satellite influences the observable measurements is specified by the measurement dynamics: $\tilde{\mathbf{z}}_t = \mathbf{H}\mathbf{x}_t + \tilde{\mathbf{v}}_t$ where $\mathbf{H} \in \mathcal{R}^{m \times d}$ specifies how the system state $\mathbf{x}_t$ influences the measurements and the Gaussian random vector $\tilde{\mathbf{v}}_t$ is called the *measurement noise*. It is assumed that the Gaussian random vector $\tilde{\mathbf{v}}_{t-1}$ has mean zero and positive definite symmetric covariance matrix $\mathbf{R} \in \mathcal{R}^{m \times m}$. The parameter vector for the measurement dynamics of the satellite is defined as $\boldsymbol{\eta} \equiv [\mathbf{vec}(\mathbf{H}^T), \ \mathbf{vech}(\mathbf{R})]$.

Explicitly specify the local emission multivariate Gaussian density model $\{p(\mathbf{z}_t|\mathbf{x}_t; \boldsymbol{\eta})\}$ and local transition multivariate Gaussian density model $\{p(\mathbf{x}_{t+1}|\mathbf{x}_t; \boldsymbol{\psi})\}$. Then, show how to explicitly calculate the likelihood of an observed sequence of states $\tilde{\mathbf{z}}_1, \tilde{\mathbf{z}}_2, \ldots, \tilde{\mathbf{z}}_T$ given the parameter vectors $\boldsymbol{\eta}$ and $\boldsymbol{\psi}$. △

## Exercises

10.3-1. *Bayesian Medical Diagnosis Network Probability Model.* Consider the example medical diagnosis Bayesian network consisting of five binary random variables depicted in Figure 10.3. Let the function $\mathcal{S}$ be defined such that for all $\psi \in \mathcal{R}$: $\mathcal{S}(\psi) = (1 + \exp(-\psi))$. Assume the probability that the binary random variable $\tilde{x}_k$ in the Bayesian network depicted in Figure 10.3 takes on the value of one with probability

$$p(\tilde{x}_k = 1|\mathbf{x}; \boldsymbol{\theta}_k) = \mathcal{S}\left(\boldsymbol{\theta}_k^T \mathbf{x}\right)$$

where $\mathbf{x} = [x_1, x_2, x_3, x_4, x_5]^T$ and the parameter vector for the probability model is $\boldsymbol{\theta} = [\boldsymbol{\theta}_1^T, \ldots, \boldsymbol{\theta}_5^T]^T$. Place restrictions on $\boldsymbol{\theta}$ by constraining specific elements of $\boldsymbol{\theta}$ to be zero so that the resulting probability assignments specify the Bayesian network in Figure 10.3. Show the joint probability density of that Bayesian network is a canonical exponential family probability density.

10.3-2. *Bayesian Networks Whose Nodes Are Random Vectors.* Show that a Bayesian network can be defined such that each node in the Bayesian network is a random vector of different dimensionality rather than a random variable by rewriting the random vector Bayesian network as an equivalent but different Bayesian network whose nodes correspond to scalar random variables.

10.3-3. *Markov Model Interpretation of Recurrent Networks.* Interpret the recurrent network with gated units in Section 1.5.3 as specifying a conditional probability density. Next, show how to write the probability density of a sequence processed by the recurrent network as a product of the conditional densities. Explain how the conditional independence assumptions for a recurrent network are similar and different from the conditional independence assumptions for a first-order Markov model. Use this explanation to specify a class of statistical environments where the Gated Recurrent Network will be more effective than the first-order Markov model. Then, specify a class of statistical environments where the first-order Markov model and Gated Recurrent Network are expected to have similar performance.

## 10.4 Markov Random Fields

The key idea in Bayesian Networks is that a joint density $p(x_1, \ldots, x_d)$ of a collection of random variables $x_1, x_2, \ldots, x_d$ can be represented as a collection of simpler local conditional density functions such as $p(x_i|x_{i-1}, x_{i-2})$. In order to achieve this objective within a Bayesian Network framework, it is necessary that an indexing scheme for the collection of random variables exists such that the conditional probability distribution of the $i$th random variable, $\tilde{x}_i$, in the collection of $d$ random variables is only conditionally dependent upon a proper subset of $\{\tilde{x}_1, \ldots, \tilde{x}_{i-1}\}$.

The Markov random field (MRF) framework allows for more general indexing schemes and more general methods for factoring the joint density $p(x_1, \ldots, x_d)$. The conditional probability distribution of the $i$th random variable $\tilde{x}_i$ may be conditionally dependent upon an arbitrary subset of all of the remaining random variables $\{\tilde{x}_1, \tilde{x}_2, \tilde{x}_{i-1}, \tilde{x}_{i+1}, \ldots, \tilde{x}_d\}$. However, the classic MRF framework requires that the density $p(x_1, \ldots, x_d)$ satisfies a special condition called the "positivity condition". This condition, however, is not especially restrictive since every Gibbs density satisfies the positivity condition. Thus, if a Gibbs density can be factored as a Bayesian network, then that same density can also be factored as a Markov random field.

Markov random fields are widely used in image processing probabilistic models. Consider an image processing problem where each pixel in an image is represented as a random variable whose value is an intensity measurement. Assume the local conditional probability density of the $i$th pixel random variable is functionally dependent upon the values of the $i$th pixel random variable values surrounding the $i$th pixel random variable. For example, a typical local constraint might be that the probability the $i$th pixel random variable value takes on the intensity level of black will be larger when the values of the pixel random variables surrounding the $i$th pixel also have values corresponding to darker intensity values. Notice this is an example of a *homogeneous Markov random field* since the parametric form of the local conditional density for the $i$th pixel random variable in the field is exactly the same for all other pixel random variables in the image.

Markov random fields are also widely used to support probabilistic inference in rule-based systems (e.g., see Example 1.6.6). In a probabilistic inference problem, the random field might consist of a collection of binary random variables where the $i$th random variable in the field, $x_i$, takes on the value of one if and only if the $i$th proposition holds and takes on the value of zero otherwise. A probabilistic rule would then have the form that if propositions $j$, $k$, and $m$ hold, then the probability that proposition $i$ holds is specified by the probability mass function $p(\tilde{x}_i = 1|x_j, x_k, x_m)$. This is another example of a case where the Bayesian network assumption that the random variables in the random field can be ordered such that the probability distribution of $\tilde{x}_i$ is only functionally dependent upon a subset of the random variables $\tilde{x}_{i-1}, \ldots, \tilde{x}_1$ is too restrictive. Furthermore, in this situation the Markov random field is not homogeneous and is referred to as a *heterogeneous Markov random field.*

The choice of using a Bayesian network or a Markov random field should be based upon which methodology results in a collection of local conditional densities that are either more semantically transparent or more computationally convenient. In general, the joint density for every Bayesian network can be represented as the joint density for a Markov random field and vice versa. Bayesian networks and Markov random fields simply correspond to different ways of factoring the same joint density.

## 10.4.1 The Markov Random Field Concept

**Definition 10.4.1** (Positivity Condition). Let the Radon-Nikodým density $p : \mathcal{R}^d \rightarrow [0, \infty)$ specify the probability distribution for a $d$-dimensional random vector $\tilde{\mathbf{x}} \equiv [\tilde{x}_1, \ldots, \tilde{x}_d]$. Let $S_i$ denote the support of $\tilde{x}_i$, $i = 1, \ldots, d$. Let $\Omega \equiv \times_{i=1}^{d} S_i$. The *positivity condition* holds for the density $p$ if the support of $\tilde{\mathbf{x}}$ is equal to $\Omega$. □

Less formally, the positivity condition can be restated as the assertion that if it is possible for an isolated random variable $\tilde{x}_i$, to take on a particular value $x_i$, then regardless of the values of the other random variables in the field it is always possible for $\tilde{x}_i$ to take on the value of $x_i$.

Let $\Omega \subseteq \mathcal{R}^d$. Let $p : \Omega \rightarrow [0, \infty)$ be a Radon-Nikodým density function where $\Omega$ is the support for $p$. A sufficient condition for the positivity condition to hold is simply that $p(\mathbf{x}) > 0$ for all $\mathbf{x} \in \Omega$. This sufficient condition is satisfied by every Gibbs density function.

**Example 10.4.1** (Positivity Condition Violation by Global Constraints). Consider a three-dimensional binary random vector $\tilde{\mathbf{x}}$ consisting of three discrete random variables each with support $\{0,1\}$ with probability mass function $p : \{0,1\}^3 \rightarrow [0,1]$ defined such that:

$$p(\tilde{\mathbf{x}} = [0,1,1]) = 0.5, \quad p(\tilde{\mathbf{x}} = [0,1,0]) = 0.2, \tag{10.12}$$

$$p(\tilde{\mathbf{x}} = [1,0,0]) = 0.1, \quad \text{and} \quad p(\tilde{\mathbf{x}} = [0,0,1]) = 0.2. \tag{10.13}$$

This is an example where the pmf $p$ does not satisfy the positivity condition since

$$p(\tilde{\mathbf{x}} = [0,0,0]) = p(\tilde{\mathbf{x}} = [1,0,1]) = p(\tilde{\mathbf{x}} = [1,1,0]) = p(\tilde{\mathbf{x}} = [1,1,1]) = 0$$

yet all four realizations $[0,0,0]$, $[1,0,1]$, $[1,1,0]$, and $[1,1,1]$ of $\tilde{\mathbf{x}}$ are elements of the cartesian product of the support of the three discrete random variables (i.e., $\{0,1\} \times \{0,1\} \times \{0,1\}$). △

**Example 10.4.2** (Preventing Positivity Violation Using Smoothing). Although the probability mass function in Example 10.4.1 does not satisfy the positivity condition, it is not difficult to construct an alternative "smoothed pmf" which closely approximates the original pmf in (10.12) and (10.13).

In particular, the constraints in (10.12) and (10.13) are replaced respectively with the new constraints:

$$p(\tilde{\mathbf{x}} = [0,1,1]) = 0.49, \quad p(\tilde{\mathbf{x}} = [0,1,0]) = 0.19, \tag{10.14}$$

$$p(\tilde{\mathbf{x}} = [1,0,0]) = 0.09, \quad \text{and} \quad p(\tilde{\mathbf{x}} = [0,0,1]) = 0.19 \tag{10.15}$$

and

$$p(\tilde{\mathbf{x}} = [0,0,0]) = p(\tilde{\mathbf{x}} = [1,0,1]) = p(\tilde{\mathbf{x}} = [1,1,0]) = p(\tilde{\mathbf{x}} = [1,1,1]) = 0.01$$

so that all states are assigned strictly positive probabilities. The pmf specified by (10.14) and (10.15) closely approximates the pmf in (10.12) and (10.13) but nevertheless satisfies the positivity condition. △

**Example 10.4.3** (Positivity Condition Violation by Local Constraints). Example 10.4.1 specified constraints on the joint distribution of $\tilde{\mathbf{x}}$ that violated the positivity condition. In this example, a violation of the positivity condition is provided that is derived from local constraints on the relationships among individual elements of $\tilde{\mathbf{x}}$.

Assume $\tilde{x}_1$ and $\tilde{x}_2$ are binary random variables with support $S \equiv \{0,1\}$. Assume, in addition, that $\tilde{x}_2 = 1 - \tilde{x}_1$. Show that the positivity condition does not hold in this situation.

**Solution.** The local constraint $\tilde{x}_2 = 1 - \tilde{x}_1$ implies that situations where $\tilde{x}_1$ and $\tilde{x}_2$ both take on the same value cannot occur. For example, the situation $\tilde{x}_1 = 1$ and $\tilde{x}_2 = 1$ is impossible. Therefore the support of $\tilde{\mathbf{x}}$ is $\{[0,1],[1,0]\}$. The positivity condition is violated since $\Omega \equiv S \times S$ is not equal to the support of $\tilde{\mathbf{x}}$. △

**Definition 10.4.2** (Neighborhood Function). Let $\mathcal{V} \equiv \{1,\ldots,d\}$ be a collection of *sites*. Let $\mathcal{G} = (\mathcal{V},\mathcal{E})$ be an undirected graph. Let $S_i$ be the neighbors of vertex $i$ in the graph $\mathcal{G}$.

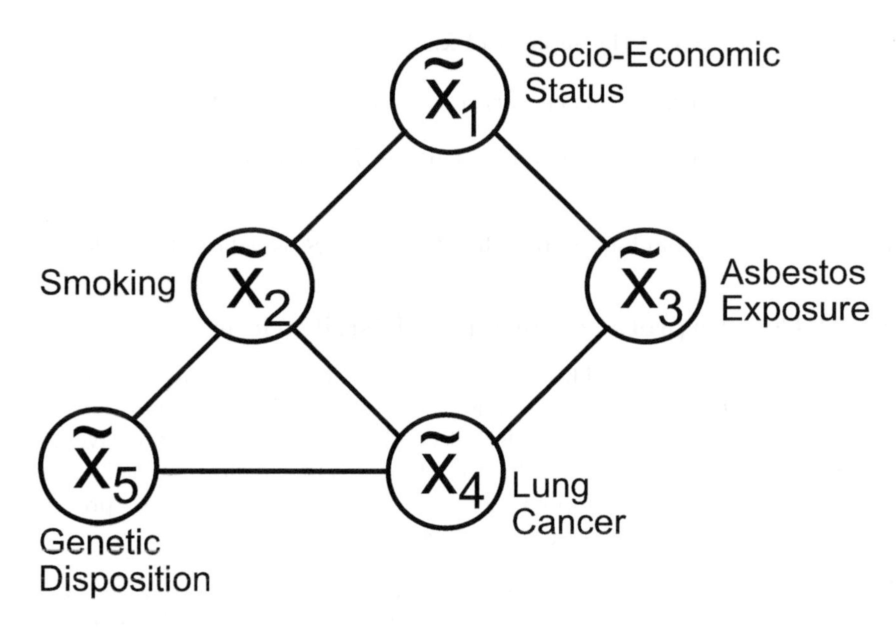


**FIGURE 10.7**
**Markov random field representation of medical knowledge.** This figure depicts the neighborhood graph for a Markov random field designed to represent the probabilistic knowledge representation depicted by the Bayes net in Figure 10.3 *plus* additional knowledge that genetic disposition and lung cancer can influence smoking behaviors. This additional knowledge is modeled by requiring the probability of $\tilde{x}_2$ (smoking) to be functionally dependent upon both the random variable $\tilde{x}_5$ (genetic disposition) and the random variable $\tilde{x}_4$ (lung cancer). Given these two additional probabilistic constraints, the collection of probabilistic constraints cannot be specified as a Bayesian network with a directed acyclic conditional dependency graph. However, such probabilistic constraints are naturally represented as a neighborhood graph for a Markov random field.

Let $m_i$ denote the number of elements of $S_i$. Let $\mathcal{N}_i : \mathcal{R}^d \to \mathcal{R}^{m_i}$ be called the *neighborhood function* for site $i$ which is defined such that for all $\mathbf{x} = [x_1, \ldots, x_d] \in \mathcal{R}^d$:

$$\mathcal{N}_i(\mathbf{x}) = \{x_j : (j, i) \in \mathcal{E}\},$$

$i = 1, \ldots, d$. The set $\{\mathcal{N}_1, \ldots, \mathcal{N}_d\}$ is called the *neighborhood system.* □

Notice that a neighborhood function $\mathcal{N}_i$ takes as input a vector and returns a subset of the elements of that vector which are neighbors of site $i$ with respect to the undirected graph $\mathcal{G}$ which defines the neighborhood system.

The essential idea of a Markov random field is that the conditional probability density of a random variable $\tilde{x}_i$ in the Markov random field $\tilde{\mathbf{x}} = [\tilde{x}_1, \ldots, \tilde{x}_d]$ is only functionally dependent upon the neighbors of the random variable $\tilde{x}_i$ for $i = 1, \ldots, d$. The neighbors of the random variables in the field are specified by an undirected graph as illustrated in Figure 10.7. Or equivalently, the conditional independence assumptions of the Markov random field are specified by the undirected graph.

**Definition 10.4.3** (Markov Random Field). Let $\tilde{\mathbf{x}} = [\tilde{x}_1, \ldots, \tilde{x}_d]$ be a $d$-dimensional random vector with Radon-Nikodým density $p : \mathcal{R}^d \to [0, \infty)$. Let $\mathcal{V} \equiv \{1, \ldots, d\}$. Let $\{\mathcal{N}_1, \ldots, \mathcal{N}_d\}$ be a neighborhood system specified by an undirected graph $\mathcal{G} = (\mathcal{V}, \mathcal{E})$.
(i) Assume $p$ satisfies the positivity condition.

(ii) Assume for each $\mathbf{x}$ in the support of $\tilde{\mathbf{x}}$:

$$p(x_i|x_1, \ldots, x_{i-1}, x_{i+1}, \ldots, x_d) = p_i(x_i|\mathcal{N}_i(\mathbf{x})).$$

Then, $\tilde{\mathbf{x}}$ is a *Markov random field* specified by $(p, \mathcal{G})$ where $\mathcal{G}$ is called the *neighborhood graph.* □

When $\tilde{\mathbf{x}}$ is a discrete random vector, then $\Omega \equiv \times_{i=1}^{d} S_i$ is called the *configuration space.*

### 10.4.2 MRF Interpretation of Gibbs Distributions

The Hammersley-Clifford (HC) Theorem is important for several reasons. First, it provides a methodology for deriving the local conditional independence assumptions of an MRF given the global joint density for the MRF. Second, it provides a constructive methodology for deriving the global joint density for an MRF given a collection of local conditional independence assumptions. The former methodology is called the analysis process while the latter methodology is called the synthesis process.

**Definition 10.4.4** (Clique Potential Function). Let $\Omega \subseteq \mathcal{R}^d$. Let $\mathcal{V} \equiv \{1, \ldots, d\}$. Let $\mathcal{G} = (\mathcal{V}, \mathcal{E})$ be an undirected graph. Let $c_1, \ldots, c_C$ be the cliques of $\mathcal{G}$. For each $j = 1, \ldots, C$, let $V_j : \Omega \rightarrow \mathcal{R}$ be defined such that for all $\mathbf{x} \in \Omega$ and for all $\mathbf{y} \in S_j$: $V_j(\mathbf{x}) = V_j(\mathbf{y})$ where

$$S_j \equiv \{[y_1, \ldots, y_d] \in \Omega : y_i = x_i \text{ for all } i \in c_j\}.$$

The functions $V_1, \ldots, V_C$ are called the *clique potential functions* on *clique specification graph* $\mathcal{G}$ with respect to $\Omega$. □

The definition of a clique potential function $V_j$ for an undirected clique specification graph $\mathcal{G}$ can be informally rephrased as follows. The clique potential function, $V_j$, for the $j$th clique is only functionally dependent upon the elements of $\mathbf{x}$ that are indexed by the nodes in the $j$th clique.

**Example 10.4.4** (Clique Potential Function Example). Let $\Omega \subseteq \mathcal{R}^5$. Let $\mathcal{G}$ be an undirected graph with nodes $\{1, \ldots, 5\}$ and edges $\{1, 2\}$, $\{2, 4\}$, $\{1, 3\}$, $\{4, 5\}$, and $\{2, 5\}$. Let the first clique of the graph be denoted as $c_1 \equiv \{2, 4, 5\}$. Then, $V_1(\mathbf{x}) = x_2 x_3 x_4 x_5$ is not a clique potential function for clique $c_1$ with respect to $\Omega$ because $V_1$ is functionally dependent upon $x_3$ and $c_1$ does not contain 3. However, $V_1(\mathbf{x}) = (x_2 x_4 x_5)^2$ is a clique potential function for clique $c_1$ with respect to $\Omega$ since $V_1$ is only functionally dependent on $x_2$, $x_4$, and $x_5$ and $2, 4, 5 \in c_1$. △

Note that an undirected graph can be constructed from its singleton and doubleton cliques which respectively specify the nodes and edges of the undirected graph. Therefore, if the maximal cliques of an undirected graph are identified, it is possible to identify the singleton and doubleton cliques for the purposes of finding the edges of the undirected graph.

**Example 10.4.5** (Construct an Undirected Graph from Its Cliques). Let $\mathcal{G}$ be an undirected graph with nodes $\{1, \ldots, 5\}$. Assume that it is known that $\mathcal{G}$ has only one maximal clique $\{2, 4, 5\}$. Find the edges of $\mathcal{G}$.

**Solution.** Since $\{2, 4, 5\}$ is a clique then $\{2, 4\}$, $\{4, 5\}$, and $\{2, 5\}$ are cliques. Therefore, $\{2, 4\}$, $\{4, 5\}$, and $\{2, 5\}$ are edges of $\mathcal{G}$. △

**Recipe Box 10.1 Constructing an MRF from a Gibbs Density (Theorem 10.4.1)**

- **Step 1: Identify clique potential functions.** Let the Gibbs density

$$p(\mathbf{x}) = \frac{\exp(-V(\mathbf{x}))}{\int \exp(-V(\mathbf{y}))d\nu(\mathbf{y})}, \quad V(\mathbf{x}) = \sum_{j=1}^{C} V_j(\mathbf{x})$$

where the local potential function $V_j : \mathcal{R}^d \rightarrow \mathcal{R}$ is functionally dependent upon some subset of the elements of $\mathbf{x}$.

- **Step 2: Construct neighborhood graph using clique specification graph.** Use the local potential functions $V_1, \ldots, V_C$ to construct the clique specification graph. Use Theorem 10.4.1 to conclude the clique specification graph is the neighborhood graph for the Markov random field.

- **Step 3: Derive energy difference for each random variable.** Let $\mathbf{y}$ be another realization of the random field. Let $V^{(i)}$ be the sum of all local potential functions that are functionally dependent upon $x_i$ for $i = 1, \ldots, d$.

$$\Delta_i(\mathbf{x}) = V^{(i)}(\mathbf{x}) - V^{(i)}(x_1, x_2, \ldots, x_{i-1}, y_i, x_{i+1}, \ldots, x_d), i = 1, \ldots, d.$$

- **Step 4: Construct local conditional densities.** Compute

$$p(x_i|\mathcal{N}_i(\mathbf{x})) = \frac{\exp(-\Delta_i(\mathbf{x}))}{\int \exp(-\Delta_i(\mathbf{y}))d\nu(y_i)}$$

where the set $\mathcal{N}_i(\mathbf{x})$ is the set of random variables in $\mathbf{x}$ which are neighbors of $x_i$ as specified by the neighborhood graph. For example, if the support of $\tilde{x}_i$ is $\{0, 1\}$, then

$$p(x_i|\mathcal{N}_i(\mathbf{x})) = \frac{\exp(-\Delta_i(\mathbf{x}))}{\exp(-\Delta_i(\mathbf{x}^1)) + \exp(-\Delta_i(\mathbf{x}^0))}$$

where $\mathbf{x}^1 = [x_1, \ldots, x_{i-1}, 1, x_{i+1}, \ldots, x_d]$, $\mathbf{x}^0 = [x_1, \ldots, x_{i-1}, 0, x_{i+1}, \ldots, x_d]$.

A version of the Hammersley-Clifford Theorem applicable to random fields comprised of discrete, continuous, and mixed random variables is now presented as two separate theorems.

The first theorem shows that every Gibbs Radon-Nikodým density can be interpreted as a Markov random field. It is called an "analysis" theorem because it explains how a particular global description specifying the likelihood of patterns in the random field can be equivalently expressed as a collection of local probabilistic interactions among the random variables in the random field.

**Theorem 10.4.1** (Hammersley-Clifford Theorem (Analysis)). *Let $p : \mathcal{R}^d \rightarrow (0, \infty)$ be a Radon-Nikodým density with respect to some support specification measure $\nu$ for a $d$-dimensional random vector $\tilde{\mathbf{x}} = [\tilde{x}_1, \ldots, \tilde{x}_d]$. In addition, assume $p$ is a Gibbs density with an energy function $V : \mathcal{R}^d \rightarrow \mathcal{R}$ defined such that for all $\mathbf{x} \in \mathcal{R}^d$:*

$$V(\mathbf{x}) = \sum_{j=1}^{C} V_j(\mathbf{x})$$

*where* $V_j : \mathcal{R}^d \rightarrow \mathcal{R}$ *is a clique potential function for the* $j$*th clique,* $c_j$*, of some clique specification graph* $\mathcal{G}$ *for* $j = 1, \ldots, C$*. Then,* $\tilde{\mathbf{x}}$ *is an MRF with a neighborhood system* $\{\mathcal{N}_1, \ldots, \mathcal{N}_d\}$ *specified by* $\mathcal{G}$*. And, in addition, for all* $\mathbf{x}$ *in the support of* $\tilde{\mathbf{x}}$*:*

$$p(x_k|\mathcal{N}_k(\mathbf{x})) = \frac{\exp(-V^{(k)}(\mathbf{x}))}{\int_{y_k} \exp(-V^{(k)}([x_1, \ldots, x_{k-1}, y_k, x_{k+1}, \ldots, x_d])d\nu(y_k)} \tag{10.16}$$

*where*

$$V^{(k)}(\mathbf{x}) = \sum_{c \in \mathcal{I}_k} V_c(\mathbf{x}), \quad \mathcal{I}_k \equiv \{j : k \in c_j\}.$$

*Proof.* By definition, the Gibbs Radon-Nikodým density $p$ is a strictly positive function which implies the positivity condition is satisfied. Let $\mathbf{x} = [x_1, \ldots, x_d]$ be a realization of $\tilde{\mathbf{x}}$. Let $\mathbf{y}^k \equiv [x_1, \ldots, x_{k-1}, y_k, x_{k+1}, \ldots, x_d]$ be defined such that the $k$th element of $\mathbf{y}^k$ is equal to a variable $y_k$ which is not equal to $x_k$ and the $j$th element of $\mathbf{y}^k$ is equal to $x_j$ for $j \neq k$. Note that:

$$\frac{p(x_k|x_1, \ldots, x_{k-1}, x_{k+1}, \ldots, x_d)}{p(y_k|x_1, \ldots, x_{k-1}, x_{k+1}, \ldots, x_d)} = \frac{p(\mathbf{x})}{p(\mathbf{y}^k)} = \exp\left[V(\mathbf{y}^k) - V(\mathbf{x})\right]$$

or equivalently

$$p(x_k|x_1, \ldots, x_{k-1}, x_{k+1}, \ldots, x_d) = Z_k^{-1} \exp[V(\mathbf{y}^k) - V(\mathbf{x})], \tag{10.17}$$

where $Z_k^{-1} = p(y_k|x_1, \ldots, x_{k-1}, x_{k+1}, \ldots, x_d)$.

Note that $V_c(\mathbf{y}^k) - V_c(\mathbf{x}) = 0$ when $c \notin \mathcal{I}_k$ which implies the right-hand side of (10.17) has the form of (10.16). Finally, since the right-hand side of (10.16) is only functionally dependent upon the neighbors of site $k$ in $\mathcal{G}$, this establishes that

$$p(x_k|x_1, \ldots, x_{k-1}, x_{k+1}, \ldots, x_d) = p(x_k|\mathcal{N}_k(\mathbf{x})).$$

■

**Example 10.4.6** (Derivation of Markov Logic Net Local Conditional PMF.)**.** Show that the Markov logic net probability mass function specified by (1.36) and (1.37) in Example 1.6.6 is a Gibbs probability mass function and therefore a Markov random field. Then, derive an explicit formula for computing the probability proposition $i$ holds given the other propositions in the random field hold. That is, obtain an explicit formula for the local conditional probability mass function $p(\tilde{x}_k = 1|\mathcal{N}_k(\mathbf{x}))$.

**Solution.** Note that:

$$\frac{p(\tilde{x}_k = 1|\mathcal{N}_k(\mathbf{x}))}{p(\tilde{x}_k = 0|\mathcal{N}_k(\mathbf{x}))} = \frac{p(x_1, \ldots, x_{k-1}, x_k = 1, x_{k+1}, \ldots, x_d)}{p(x_1, \ldots, x_{k-1}, x_k = 0, x_{k+1}, \ldots, x_d)}$$

$$\frac{p(\tilde{x}_k = 1|\mathcal{N}_k(\mathbf{x}))}{p(\tilde{x}_k = 0|\mathcal{N}_k(\mathbf{x}))} = \frac{(1/Z)\exp(-V([x_1, \ldots, x_{k-1}, x_k = 1, x_{k+1}, \ldots, x_d]))}{(1/Z)\exp(-V([x_1, \ldots, x_{k-1}, x_k = 0, x_{k+1}, \ldots, x_d]))} = \exp(\psi_k(\mathbf{x}))$$

where

$$\psi_k(\mathbf{x}) = -V([x_1, \ldots, x_{k-1}, x_k = 1, x_{k+1}, \ldots, x_d]) + V([x_1, \ldots, x_{k-1}, x_k = 0, x_{k+1}, \ldots, x_d]).$$

Then, show that:

$$p(\tilde{x}_k = 1|\mathcal{N}_k(\mathbf{x})) = (1 - p(\tilde{x}_k = 1|\mathcal{N}_k(\mathbf{x}))) \exp(\psi_k(\mathbf{x}))$$

which implies that:

$$p(\tilde{x}_k = 1|\mathcal{N}_k(\mathbf{x})) = \mathcal{S}(\psi_k(\mathbf{x})) \text{ where } \mathcal{S}(\psi_k) = (1 + \exp(-\psi_k))^{-1}.$$

Finally, show that:

$$\psi_k(\mathbf{x}) = \sum_{k=1}^{q} \theta_k \phi_k([x_1, \ldots, x_{k-1}, x_k = 1, x_{k+1}, \ldots, x_d]) - \sum_{k=1}^{q} \theta_k \phi_k([x_1, \ldots, x_{k-1}, x_k = 0, x_{k+1}, \ldots, x_d])$$

$$= \sum_{k \in S_k} \theta_k (\phi_k([x_1, \ldots, x_{k-1}, x_k = 1, x_{k+1}, \ldots, x_d]) - \phi_k([x_1, \ldots, x_{k-1}, x_k = 0, x_{k+1}, \ldots, x_d]))$$

where $S_k$ is defined such that $j \in S_k$ if and only if $\phi_j$ is functionally dependent upon $x_k$. △

The next theorem shows that the probability distribution for a Markov random field $\tilde{\mathbf{x}}$ may be represented as a Gibbs density. It is called a "synthesis" theorem because it explains how local descriptions of relationships among random variables in the field imply a particular global description.

**Theorem 10.4.2** (Hammersley-Clifford Theorem (Synthesis)). *Let $\tilde{\mathbf{x}} = [\tilde{x}_1, \ldots, \tilde{x}_d]$ be a Markov random field specified by a neighborhood graph $\mathcal{G} \equiv (\mathcal{V}, \mathcal{E})$ whose joint distribution is specified by a Radon-Nikodým density p defined with respect to support specification $\nu$. Then, p may be represented as a Gibbs density with energy function $V(\mathbf{x}) \equiv \sum_{c=1}^{C} V_c(\mathbf{x})$ where $V_1, \ldots, V_C$ are the clique potential functions defined with respect to the clique specification $\mathcal{G}$.*

*Proof.* The proof is based upon the discussion of the Hammersley-Clifford Theorem in Besag (1974).

*Step 1: Show that a Gibbs density with fully connected graph can represent the MRF density.* Let $Q : \mathcal{R}^d \rightarrow \mathcal{R}$ be defined such that for all $\mathbf{x}, \mathbf{y}$ in the support of $p$ such that $\mathbf{x} \neq \mathbf{y}$:

$$Q(\mathbf{x}) = -\log[p(\mathbf{x})/p(\mathbf{y})]$$

since $p(\mathbf{x}) > 0$ and $p(\mathbf{y}) > 0$.

Let $p$ be a Gibbs density defined with respect to the undirected graph $\mathcal{G}_G \equiv (\mathcal{V}, \mathcal{E})$ where $\mathcal{E}$ contains all possible undirected edges. Let $\mathcal{C}_i$ be the $i$th clique of the graph $\mathcal{G}_G$, $i = 1, \ldots, (2^d - 1)$.

Let $G_i : \mathcal{C}_i \rightarrow \mathcal{R}$, $G_{i,j} : \mathcal{C}_i \times \mathcal{C}_j \rightarrow \mathcal{R}$, and $G_{i,j,k} : \mathcal{C}_i \times \mathcal{C}_j \times \mathcal{C}_k \rightarrow \mathcal{R}$, and so on.

It will now be shown that the function $Q$ can be represented without loss in generality for all $\mathbf{x}$ in the support of $p$ as:

$$Q(\mathbf{x}) = \sum_{i=1}^{d} (x_i - y_i) G_i(x_i) + \sum_{i=1}^{d} \sum_{j>i} (x_i - y_i)(x_j - y_j) G_{i,j}(x_i, x_j) + \ldots$$

$$+ \sum_{i=1}^{d} \sum_{j>i} \sum_{k>j} (x_i - y_i)(x_j - y_j)(x_k - y_k) G_{i,j,k}(x_i, x_j, x_k)$$

$$+ (x_1 - y_1)(x_2 - y_2) \ldots (x_d - y_d) G_{1,2,\ldots,d}(x_1, \ldots, x_d). \quad (10.18)$$

Let $\mathbf{x}^i = [y_1, \ldots, y_{i-1}, x_i, y_{i+1}, \ldots, y_d]$ (which exists by the positivity assumption). Substituting $\mathbf{x}^i$ into (10.18) and solving for $G_i$ gives

$$G_i(x_i) = Q(\mathbf{x}^i)/(x_i - y_i)$$

by noting all third-order and higher-order terms vanish.

After $G_1, \ldots, G_d$ are defined in terms of $Q$, then substitute vectors of the form (which exist by the positivity assumption):

$$\mathbf{x}^{i,j} = [y_1, y_2, \ldots, y_{i-1}, x_i, y_{i+1}, \ldots, y_{j-1}, x_j, y_{j+1}, \ldots, y_d]$$

into (10.18) and solved for $G_{i,j}$ to obtain:

$$G_{i,j}(x_i, x_j) = \frac{Q(\mathbf{x}^{i,j}) - (x_i - y_i)G_i(x_i) - (x_j - y_j)G_j(x_j)}{(x_i - y_i)(x_j - y_j)}.$$

After $\{G_{i,j}\}$ have been defined in terms of $Q$ and $\{G_i\}$, then substitute vectors of the form (which exist by the positivity assumption):

$$\mathbf{x}^{i,j,k} = [y_1, y_2, \ldots, y_{i-1}, x_i, y_{i+1}, \ldots, y_{j-1}, x_j, y_{j+1}, \ldots, y_{k-1}, x_k, \ldots y_d]$$

into (10.18) and solve for $G_{i,j,k}$ in a similar manner. Thus, any choice of $Q$ function can be represented by (10.18). This observation and the definition of the $Q$ function implies any choice of MRF density $p$ can be represented by (10.18) which is a representation of a Gibbs density with a fully connected neighborhood graph.

*Step 2: Compute local conditional Gibbs density on a fully connected graph.* Let $\mathbf{x}$ and $\mathbf{y}^k \equiv [x_1, \ldots, x_{k-1}, y_k, x_{k+1}, \ldots, x_d]$ be distinct vectors in the support of $\tilde{\mathbf{x}}$. Note that:

$$\frac{p(x_k|x_1, \ldots, x_{k-1}, x_{k+1}, \ldots, x_d)}{p(y_k|x_1, \ldots, x_{k-1}, x_{k+1}, \ldots, x_d)} = \frac{p(\mathbf{x})}{p(\mathbf{y}^k)} = \exp[Q(\mathbf{y}^k) - Q(\mathbf{x})]$$

or equivalently

$$p(x_k|x_1, \ldots, x_{k-1}, x_{k+1}, \ldots, x_d) = (1/Z_k)\exp[Q(\mathbf{y}^k) - Q(\mathbf{x})] \quad (10.19)$$

where $(1/Z_k) = p(y_k|x_1, \ldots, x_{k-1}, x_{k+1}, \ldots, x_d)$.

Let $V_c$ be the local clique potential function for the $c$th clique defined whose functional form is chosen to be either:

$$V_c(\mathbf{x}) \equiv (x_i - y_i)G_i,$$
$$V_c(\mathbf{x}) \equiv (x_i - y_i)(x_j - y_j)G_{i,j},$$
$$V_c(\mathbf{x}) \equiv (x_i - y_i)(x_j - y_j)(x_k - y_k)G_{i,j,k},$$

and so on. Then, $Q$ may be rewritten using the notation:

$$Q(\mathbf{x}) = \sum_{c=1}^{2^d-1} V_c(\mathbf{x})$$

and interpreted as an energy function for a Gibbs density that represents $p$ using the fully connected graph $\mathcal{G}_G$.

Now note:

$$Q(\mathbf{y}^k) - Q(\mathbf{x}) = \sum_{c=1}^{2^d-1} [V_c(\mathbf{y}^k) - V_c(\mathbf{x})] \quad (10.20)$$

and substituting (10.20) into (10.19) and using the relation

$$p(x_k|x_1, \ldots, x_{k-1}, x_{k+1}, \ldots, x_d) = p\left(x_k|\mathcal{N}_k(\mathbf{x})\right)$$

it follows that:

$$p(x_k|\mathcal{N}_k(\mathbf{x})) = p(y_k|\mathcal{N}_k(\mathbf{x}))\exp\left(-\sum_{c=1}^{2^d-1}[V_c(\mathbf{y}^k) - V_c(\mathbf{x})]\right). \quad (10.21)$$

Equation (10.21) can only be satisfied if the right-hand side of (10.21) is not functionally dependent upon random variables in the field which are not neighbors of $\tilde{x}_k$. This can only be accomplished by setting $V_c = 0$ when $k$ is not a member of clique $c$ for $c = 1, \ldots, 2^d - 1$. ■

**Example 10.4.7.** *Markov Random Field Comprised of Binary Random Variables.* Let the $d$-dimensional random vector $\tilde{\mathbf{x}} = [\tilde{x}_1, \ldots, \tilde{x}_d]$ be a Markov random field consisting of $d$ binary random variables so that $\tilde{x}_k$ has support $\{0, 1\}$ for $k = 1, \ldots, d$. Following the proof of Theorem 10.4.2, verify that the Gibbs pmf $p(\mathbf{x}; \boldsymbol{\theta}) = Z^{-1} \exp(-V(\mathbf{x}; \boldsymbol{\theta}))$ specified by the energy function

$$V(\mathbf{x}; \boldsymbol{\theta}) = \sum_{i=1}^{d} \theta_i x_i + \sum_i \sum_j \theta_{ij} x_i x_j + \sum_i \sum_{j>i} \sum_{k>j} \theta_{ijk} x_i x_j x_k + \ldots \theta_{1,\ldots,d} \prod_{u=1}^{d} x_u \quad (10.22)$$

is a member of the canonical exponential family.

Now given an arbitrary probability mass function $p_e : \{0, 1\} \rightarrow (0, 1]$ whose domain is the state of $d$-dimensional binary vectors and whose range is the set of positive numbers, show that it is always possible to strategically choose the parameters $\theta_i$, $\theta_{i,j}$, $\theta_{i,j,k}$ so that $p(\mathbf{x}; \boldsymbol{\theta}) = p_e(\mathbf{x})$ for all $\mathbf{x}$ in the support of $p_e$.

**Solution.** Since the energy function $V$ in (10.22) may be written as the dot product of a parameter vector $\boldsymbol{\theta}$ and a nonlinear transformation of $\mathbf{x}$, then it follows that the MRF specified by $V$ is a member of the canonical exponential family.

Now the Hammersley-Clifford Theorem (Synthesis) (Theorem 10.4.2) is applied to show that the form of the energy function in (10.22) is sufficiently flexible that it can represent any arbitrary pmf over a sample space of $d$-dimensional binary random vectors which satisfies the positivity condition. The analysis follows the proof of Theorem 10.4.2.

Note that if the energy function $V$ on the set of $d$-dimensional binary vectors can be arbitrarily specified, then this is sufficient to ensure that the Gibbs pmf with energy function $V$ can be arbitrarily specified. To see this, note that $p(\mathbf{x}; \boldsymbol{\theta}) = p(\mathbf{0}; \boldsymbol{\theta}) \exp(-V(\mathbf{x}))$ since $V(\mathbf{0}) = 0$ which means that the probability $p(\mathbf{x}; \boldsymbol{\theta})$ is a invertible function of $V(\mathbf{x})$.

From Theorem 10.4.2, any arbitrary pmf that satisfies the positivity condition can be represented by an energy function with clique potential functions $V_i(\mathbf{x}), V_{ij}(\mathbf{x}), V_{ijk}(\mathbf{x}), \ldots$ for $i, j, k = 1, \ldots, d$. Assume the local potential functions are chosen as in (10.22). If $\mathbf{x}$ is a $d$-dimensional binary vector with exactly one nonzero element in the $k$th position, then it follows that $V(\mathbf{x}) = \theta_k$ (e.g., $V([0, 1, 0, 0]) = \theta_2$). Thus, the energy function $V$ assigned to the set of $d$ $d$-dimensional binary vectors which have exactly one nonzero element can be arbitrarily chosen by choosing $\theta_1, \ldots, \theta_d$ appropriately.

If $\mathbf{x}$ is a $d$-dimensional binary vector with exactly two nonzero elements in the $j$th and $k$th positions, then it follows that $V(\mathbf{x}) = \theta_j + \theta_k + \theta_{j,k}$ where the free parameter $\theta_{j,k}$ may be chosen to ensure that $V$ evaluated at the binary vector whose $j$th and $k$th elements are nonzero takes on a particular desired value.

If $\mathbf{x}$ is a $d$-dimensional binary vector with exactly three nonzero elements in the $j$th, $k$th, and $r$th positions, then it follows that the energy function specified in (10.22) is given by:

$$V(\mathbf{x}) = \theta_j + \theta_k + \theta_r + \theta_{jk} + \theta_{jr} + \theta_{jkr}$$

where the only free parameter $\theta_{jkr}$ may be chosen arbitrarily to ensure that $V$ takes on any arbitrary value when $\mathbf{x}$ has exactly 3 non-zero elements.

Similar arguments are used to complete the proof. △

**Recipe Box 10.2 Constructing a Gibbs Density from an MRF (Theorem 10.4.2)**

- **Step 1: Specify conditional independence assumptions.** The first step is to specify conditional independence assumptions regarding the collection of random variables using a neighborhood graph as in Figure 10.7.

- **Step 2: Identify cliques of the neighborhood graph.** In Figure 10.7, the cliques are: $\{\tilde{x}_1\}, \{\tilde{x}_2\}, \{\tilde{x}_3\}, \{\tilde{x}_4\}, \{\tilde{x}_5\}, \{\tilde{x}_1, \tilde{x}_2\}, \{\tilde{x}_1, \tilde{x}_3\}$, $\{\tilde{x}_3, \tilde{x}_4\}, \{\tilde{x}_2, \tilde{x}_4\}$, $\{\tilde{x}_2, \tilde{x}_5\}$, $\{\tilde{x}_4, \tilde{x}_5\}$, $\{\tilde{x}_2, \tilde{x}_4, \tilde{x}_5\}$.

- **Step 3: Assign a local potential function to each clique.** Choose a local potential function for each clique which is only functionally dependent upon the clique's members to ensure the joint density has the conditional dependence structure specified by the neighborhood graph. For example, the twelve cliques in Figure 10.7 would thus be assigned to local potential functions: $V_1(x_1)$, $V_2(x_2)$, $V_3(x_3)$, $V_4(x_4)$, $V_5(x_5)$, $V_6(x_1, x_2)$, $V_7(x_1, x_3)$, $V_8(x_3, x_4)$, $V_9(x_2, x_4)$, $V_{10}(x_2, x_5), V_{11}(x_4, x_5), V_{12}(x_2, x_4, x_5)$. All local potential functions are then added together to obtain the energy function V. So, in this example, we have:
$$V = \sum_{k=1}^{12} V_k.$$

- **Step 4: Construct joint density and check normalization constant.** By Theorem 10.4.2, $p$ in (10.23) specifies the probability density of the Markov random field $\tilde{\mathbf{x}}$ using the neighborhood graph of the random field as the clique specification graph. Construct joint density using the formula:
$$p(\mathbf{x}) = Z^{-1} \exp(-V(\mathbf{x})), \quad Z = \int \exp(-V(\mathbf{x})) d\nu(\mathbf{x}). \tag{10.23}$$
The integral used to compute the normalization constant $Z$ may not exist and so it is necessary to check that $Z$ is finite. The normalization constant is finite provided either: (i) $V$ is a piecewise continuous function and $\tilde{\mathbf{x}}$ is a bounded random vector, or (ii) $\tilde{\mathbf{x}}$ is a discrete random vector taking on a finite number of values.

**Example 10.4.8.** *Markov Random Field Comprised of Categorical Random Variables.* Let the $d$-dimensional random vector $\tilde{\mathbf{x}} = [\tilde{\mathbf{x}}_1, \ldots, \tilde{\mathbf{x}}_d]$ be a Markov random field consisting of $d$ categorical random variables such that the $j$th categorical random variable is represented as the $d_j$-dimensional random subvector $\tilde{\mathbf{x}}_j$ with $d = \sum_{j=1}^{d} d_j$. The support of the $j$th categorical random variable in the MRF consists of the $d_j$ columns of a $d_j$-dimensional identity matrix for $j = 1, \ldots, d$. That is, the $j$th categorical random variable in the MRF can take on $d_j$ possible values and has support associated with a categorical coding scheme.

Define the energy function $V(\cdot;\boldsymbol{\theta}) : \mathcal{R}^d \rightarrow \mathcal{R}$ for each column vector $\boldsymbol{\theta} \in \mathcal{R}^q$ such that for all $\mathbf{x}$ in the support of $\tilde{\mathbf{x}}$:

$$V(\mathbf{x};\boldsymbol{\theta}) = \sum_{i=1}^{d} \boldsymbol{\theta}_i^T \mathbf{x}_i + \sum_i \sum_{j>i} \boldsymbol{\theta}_{i,j}^T(\mathbf{x}_i \otimes \mathbf{x}_j) + \sum_i \sum_{j>i} \sum_{k>j} \boldsymbol{\theta}_{i,j,k}^T(\mathbf{x}_i \otimes \mathbf{x}_j \otimes \mathbf{x}_k) \tag{10.24}$$

$$+ \sum_i \sum_{j>i} \sum_{k>j} \sum_{m>k} \boldsymbol{\theta}_{i,j,k,m}^T(\mathbf{x}_i \otimes \mathbf{x}_j \otimes \mathbf{x}_k \otimes \mathbf{x}_m) + \ldots + \boldsymbol{\theta}_{1,\ldots,d}^T(\mathbf{x}_1 \otimes \mathbf{x}_2 \ldots \otimes \mathbf{x}_d)$$

where

$$\boldsymbol{\theta}^T = [\boldsymbol{\theta}_1^T, \ldots, \boldsymbol{\theta}_d^T, \boldsymbol{\theta}_{i,j}^T \ldots \boldsymbol{\theta}_{i,j,k}^T \ldots \boldsymbol{\theta}_{i,j,k,m}^T \ldots, \boldsymbol{\theta}_{1,\ldots,d}^T].$$

Following the approach of Example 10.4.7, show that an MRF for $\tilde{\mathbf{x}}$ with the energy function in (10.24) is a member of the canonical exponential family. In addition, show that by appropriate choice of the parameter vectors $\boldsymbol{\theta}_i, \boldsymbol{\theta}_{ij}, \boldsymbol{\theta}_{i,j,k}, \ldots, \boldsymbol{\theta}_{1,\ldots,d}$ that an MRF with an energy function of the form of (10.24) may be used to represent every possible pmf for $\tilde{\mathbf{x}}$. △

**Example 10.4.9** (Markov Random Field Analysis of the Markov Chain)**.** Let $d$ be a collection of random variables $\tilde{x}_1, \ldots, \tilde{x}_d$ such that $\tilde{x}_i$ has support $\Omega_i$. Assume that the positivity condition holds. In addition, assume that the first-order Markov chain conditional independence relationship $p(x_i|x_1, \ldots, x_{i-1}) = p(x_i|x_{i-1})$ holds. Thus, the joint density $p(x_1, \ldots, x_d)$ may be expressed as:

$$p(x_1, \ldots, x_d) = \prod_{i=1}^{d} p(x_i|x_{i-1})$$

where for convenience assume $x_0 = 0$. Show how to analyze a first-order Markov chain as a Markov random field by providing formulas for the local conditional densities and the joint density.

**Solution.** The first-order Markov chain conditional independence assumption, implies that $2d-1$ cliques of the Markov random field consisting of the $d$ cliques $\{x_1\}, \ldots, \{x_d\}$ and the $d-1$ cliques $\{x_1, x_2\}, \ldots, \{x_{d-1}, x_d\}$ so there are $2d-1$ potential functions: $V_1(x_1), \ldots, V_d(x_d)$ and $V_{d+1}(x_1, x_2), \ldots, V_{2d-1}(x_{d-1}, x_d)$. The joint density is a Gibbs density of the form:

$$p(x_1, \ldots, x_d) = (1/Z) \exp\left(-\sum_{i=1}^{d} V_i(x_i) - \sum_{i=2}^{d} V_{d+i-1}(x_{i-1}, x_i)\right)$$

where $Z$ is the normalization constant. The local conditional densities are

$$p(x_1|x_2) = \frac{\exp(-V_1(x_1) - V_{d+1}(x_1, x_2))}{Z_1},$$

$$p(x_i|x_{i-1}, x_{i+1}) = \frac{\exp(-V_i(x_i) - V_{d+i-1}(x_{i-1}, x_i) - V_{d+i}(x_i, x_{i+1}))}{Z_i},$$

for $i = 2, \ldots, d-1$, and

$$p(x_d|x_{d-1}) = \frac{\exp(-V_d(x_d) - V_{2d-1}(x_{d-1}, x_d))}{Z_d}.$$

△

**Example 10.4.10** (Conditional Random Fields). Consider a regression modeling problem where one desires to represent the conditional joint probability density of a response vector $\tilde{\mathbf{y}}$ consisting of discrete random variables given an input pattern vector $\boldsymbol{s}$ and parameter vector $\boldsymbol{\theta}$. In this case, one may use a Markov random field to model conditional independence relationships among the elements of the random response vector $\tilde{\mathbf{y}}$ for a given choice of $\mathbf{s}$ and $\boldsymbol{\theta}$. In particular, let

$$p(\mathbf{y}|\mathbf{s}, \boldsymbol{\theta}) = \frac{\exp\left(-V(\mathbf{y}; \mathbf{s}, \boldsymbol{\theta})\right)}{Z(\mathbf{s}; \boldsymbol{\theta})}$$

where

$$Z(\mathbf{s}; \boldsymbol{\theta}) = \sum_{\mathbf{y}} \exp\left(-V(\mathbf{y}; \mathbf{s}, \boldsymbol{\theta})\right).$$

A Markov random field that is conditioned in this way upon the external pattern vector $\mathbf{s}$ is called a *conditional random field.* Exercise 10.4-6 discusses the analysis and design of a conditional medical diagnosis MRF. △

**Example 10.4.11** (Mixture of Random Fields). Let $\tilde{\mathbf{x}}$ be a $d$-dimensional random vector with support $\Omega$. Assume that in some situations the behavior of $\tilde{\mathbf{x}}$ is best modeled as an MRF specified by a Gibbs density, $p_1(\cdot; \boldsymbol{\theta}_1)$, with energy function $V_1(\cdot; \boldsymbol{\theta}_1)$ while in other situations the behavior of $\tilde{\mathbf{x}}$ is best modeled as an MRF specified by Gibbs density, $p_2(\cdot : \boldsymbol{\theta}_2)$, with energy function $V_2(\cdot; \boldsymbol{\theta}_2)$ where $\boldsymbol{\theta}_1$ and $\boldsymbol{\theta}_2$ are the respective parameters of $p_1$ and $p_2$. In order to develop a probability model which incorporates the features of $p_1$ and $p_2$, define $p = \mathcal{S}(\beta)p_1 + (1 - \mathcal{S}(\beta))p_2$ where $\mathcal{S}(\beta) = 1/(1 + \exp(-\beta))$ and $\beta$ is an additional free parameter. That is,

$$p(\mathbf{x}; \boldsymbol{\theta}_1, \boldsymbol{\theta}_2, \beta) = \frac{\mathcal{S}(\beta)\exp(-V_1(\mathbf{x}; \boldsymbol{\theta}_1))}{Z_1(\boldsymbol{\theta}_1)} + \frac{(1 - \mathcal{S}(\beta))\exp(-V_2(\mathbf{x}; \boldsymbol{\theta}_2))}{Z_2(\boldsymbol{\theta}_2)}$$

where $Z_1(\boldsymbol{\theta}_1)$ and $Z_2(\boldsymbol{\theta}_2)$ are the respective normalization constants for $p_1$ and $p_2$. △

## Exercises

10.4-1. *Example where Positivity Condition Does Not Hold.* Assume a pmf $p : \{0,1\}^2 \to [0,1]$ for a two-dimensional random vector $[\tilde{x}_1, \tilde{x}_2]$ defined such that:
$p(\tilde{x}_1 = 0, \tilde{x}_2 = 0) = 0.25$,
$p(\tilde{x}_1 = 0, \tilde{x}_2 = 1) = 0.25$,
$p(\tilde{x}_1 = 1, \tilde{x}_2 = 0) = 0$,
$p(\tilde{x}_1 = 1, \tilde{x}_2 = 1) = 0.50$.

The support of the marginal pmf for $\tilde{x}_1$ is $\Omega_1 = \{0,1\}$ and the support for the marginal pmf for $\tilde{x}_2$ is $\Omega_2 = \{0,1\}$. The support for the joint pmf for $[\tilde{x}_1, \tilde{x}_2]$ is $\Omega = \{(0,0), (0,1), (1,1)\}$. Show the positivity condition is not satisfied for the joint pmf.

10.4-2. *Example where Positivity Condition Holds.* Consider a smoothed version of the joint pmf in Exercise 10.4-1. Let $p : \{0,1\}^2 \to [0,1]$ for a two-dimensional random vector $[\tilde{x}_1, \tilde{x}_2]$ defined such that:
$p(\tilde{x}_1 = 0, \tilde{x}_2 = 0) = 0.24999$,
$p(\tilde{x}_1 = 0, \tilde{x}_2 = 1) = 0.25$,
$p(\tilde{x}_1 = 1, \tilde{x}_2 = 0) = 0.00001$,
$p(\tilde{x}_1 = 1, \tilde{x}_2 = 1) = 0.50$.

The support of the marginal pmf for $\tilde{x}_1$ is $\Omega_1 = \{0,1\}$ and the support for the marginal pmf for $\tilde{x}_2$ is $\Omega_2 = \{0,1\}$. The support for the joint pmf for $[\tilde{x}_1, \tilde{x}_2]$ is $\Omega = \{(0,0),(0,1),(1,0),(1,1)\}$. Show the positivity condition is satisfied for this pmf.

10.4-3. *Markov Random Fields for Modeling Image Texture.* Markov random fields are widely used in image processing applications where the goal is to "sharpen" up the image by filtering out high spatial frequency noise. Let $\tilde{\mathbf{X}}$ be a random $d$-dimensional matrix whose element in the $i$th row and $j$th column is denoted by the random variable $\tilde{x}_{i,j}$. Assume that each random variable in the matrix can take on a finite number of values. In addition, assume that there is a local potential function, $V_{i,j}$, associated with each random variable $\tilde{x}_{ij}$ defined such that for all $i = 2, \ldots, d-1$ and for all $j = 2, \ldots, d-1$:

$$V_{i,j}(x_{i,j}) = (x_{i,j} - x_{i,j-1})^2 + (x_{i,j} - x_{i,j+1})^2 + (x_{i,j} - x_{i-1,j})^2 + (x_{i,j} - x_{i+1,j})^2.$$

The local potential function $V_{i,j}$ is essentially a local constraint designed to provide a formal specification of the idea that the value of image pixel random variable $\tilde{x}_{i,j}$ should be similar to its neighbors $\tilde{x}_{i-1,j}$, $\tilde{x}_{i+1,j}$, $\tilde{x}_{i,j-1}$, and $\tilde{x}_{i,j+1}$. Use the local potential function specification $V_{i,j}$ to derive the neighborhood graph for the MRF. In addition, show how to explicitly compute $p(x_{i,j}|\mathcal{N}_{i,j}(\mathbf{x}))$ for each $i, j = 2, \ldots, d-1$.

10.4-4. *Equivalence of Hidden Markov Field and Hidden Markov Model.* In Example 10.3.3, a Hidden Markov Model was proposed to specify a probability model for a part-of-speech tagging problem. Construct a Hidden Markov Random Field (HMRF) which has the same joint distribution and the same conditional independence assumptions as the Hidden Markov Model(HMM) in Example 10.3.3 subject to the constraint that the positivity condition holds. In some applications, the Hidden Markov Random Field probabilistic representation may have advantages because it specifies the probability of the current part-of-speech tag, $\mathbf{h}_k$, based not only upon the preceding part-of-speech tag, $\mathbf{h}_{k-1}$, but also the subsequent part-of-speech tag $\mathbf{h}_{k+1}$ as well as the observed word $\mathbf{w}_k$. However, the HMRF representation may require more computationally intensive sampling methods (see Chapter 11 and Chapter 12) than the HMM representation.

10.4-5. *Using Binary Markov Fields to Represent Non-Causal Knowledge.* Show how to explicitly specify the most general form of the joint probability mass function of an MRF comprised of binary random variables that satisfies the conditional dependency relationships in Figure 10.7. In addition, use the results of Example 10.4.7 to show that the joint probability mass function can be represented as a member of the canonical exponential family.

10.4-6. *Conditional Random Field for Medical Diagnosis.* The Markov random field in Exercise 10.4-5 provides a mechanism for representing generic knowledge of the likelihood of acquiring `lung cancer` given the presence or absence of other factors such as `asbestos exposure`. Let $p(\mathbf{x})$ denote the joint probability mass function for the Markov random field (MRF) in Exercise 10.4-5. In order to account for individual differences, one might imagine that the probabilistic relationships among the random variables in the Markov random field specified by $p(\mathbf{x})$ might be altered when provided a $d$-dimensional vector of patient characteristics which will be denoted as $\mathbf{s}$. An element of $\mathbf{s}$ might be binary (e.g., an element of $\mathbf{s}$ might equal one if the individual is a female and zero otherwise) or an element of $\mathbf{s}$ might be numerical (e.g., an element of $\mathbf{s}$ might specify the patient's age). Discuss how to extend the MRF specified by $p(\mathbf{x})$ so that it is a conditional MRF whose joint probability mass function is specified by $p(\mathbf{x}|\mathbf{s})$.

10.4-7. *Deriving a Local Conditional Density of a Gaussian Markov Random Field.* Let $\mathbf{Q}$ be a positive definite symmetric $d$-dimensional matrix. Let $\mathbf{M} \in \{0,1\}^{d \times d}$ be a symmetric matrix defined such that the $ij$th element of $\mathbf{M}$, $m_{ij}$, is equal to one if and only if the $ij$th element of $\mathbf{Q}$, $q_{ij}$, is non-zero. Let the random vector $\tilde{\mathbf{x}}$ be specified by the multivariate Gaussian probability density function:

$$p(\mathbf{x}) = \left[(2\pi)^d \det(\mathbf{Q})\right]^{-1/2} \exp\left(-(1/2)\mathbf{x}^T \mathbf{Q}\mathbf{x}\right)$$

where $\mathbf{x} = [x_1, \ldots, x_d]$ is in the support of $\tilde{\mathbf{x}}$. Provide an explicit formula for $p(x_i|\mathcal{N}_i(\mathbf{x})$ where $\mathcal{N}_i$ is the neighborhood function associated with the $i$th random variable in the MRF.

10.4-8. *MRF for Representing Clustering Dissimilarity Measures.* Construct a Markov random field for the clustering algorithm described in Example 1.6.4 which consists of $n$ categorical random variables corresponding to the $n$ feature vectors to be mapped into the clusters $C_1, \ldots, C_K$. Assume each random variable in the field can take on one of $K$ possible values. Interpret the clustering algorithm as finding the most probable assignment of feature vectors to clusters where the probability of a clustering assignment is specified by the Gibbs pmf for the random field.

10.4-9. *MRF for Stochastic Neighborhood Embedding.* Consider a stochastic neighborhood embedding clustering algorithm which maps a set of $n$ state vectors $\mathbf{x}_1, \ldots, \mathbf{x}_n$ into a set of $n$ binary feature vectors $\mathbf{y}_1, \ldots, \mathbf{y}_n$ as described in Example 1.6.5. Construct an energy function $V(\mathbf{y}_1, \ldots, \mathbf{y}_n)$ for a Markov random field $\tilde{\mathbf{y}}_1, \ldots \tilde{\mathbf{y}}_n$ such that a minimizer of $V$ corresponds to a solution to the stochastic neighborhood embedding clustering algorithm. Interpret the stochastic neighborhood embedding clustering algorithm as finding the most probable recoded vectors $\mathbf{y}_1, \ldots, \mathbf{y}_n$ with respect to the Gibbs probability mass function specifying the Markov random field.

## 10.5 Further Readings

### Model Misspecification Framework

Machine learning researchers are often faced with the problem of developing complex mathematical models of complex real-world phenomena. In practice, the complexity of the probability model often implies that the resulting model will tend to be misspecified. However, methods for estimation and inference in the possible presence of model misspecification are unfortunately not emphasized in popular graduate texts in mathematical statistics and machine learning.

In this textbook, all of the mathematical theory developed is intentionally designed to account for the possible presence of model misspecification. Reviews of methods for estimation and inference in the presence of possible model misspecification may be found in White (1982, 1994, 2001), Golden (1995, 2003), and Golden, Henley, White, and Kashner (2013, 2016, 2019).

## Common Probability Distributions

The commonly used probability distributions discussed in Section 10.2 are well known. Discussions of commonly encountered probability distributions may be found in Manoukian (1986) and Lehmann and Casella (1998).

## Bayesian Inference Algorithms

In some situations, deterministic computationally efficient methods may be used to compute conditional probabilities for joint distributions specified by Bayesian networks (e.g., Cowell 1996a, 1999b). Bishop (2006, Chapter 8) provides a discussion of the sum-product and max-sum algorithms for, respectively, integrating over nuisance random variables and finding the most probable values of random variables of interest. Further discussions can be found in Koller et al. (2009) and Cowell et al. (2006).

## Hammersley-Clifford Theorem

Unfortunately, Hammersley and Clifford never published the Hammersley-Clifford theorem because they hoped to relax the positivity condition. The paper by Besag (1974) is typically cited as the first published statement and proof of the Hammersley-Clifford Theorem. Recent progress has been made in relaxing the positivity assumption of the Hammersley-Clifford Theorem (e.g., Kaiser and Cressie 2000; Kopciuszewski 2004).

A unique feature of this chapter is a new statement and proof of the Hammersley-Clifford Theorem for random fields consisting of absolutely continuous, discrete, and mixed random variables such that the statement and proof are broadly accessible. Related developments using a Radon-Nikodým density framework have been discussed in Lauritzen (1996), Kaiser and Cressie (2000), Kopciusewski (2004), and Cernuschi-Frias (2007). The investigation of Markov random fields consisting of both discrete and absolutely continuous random variables is an active research area in machine learning and statistics (e.g., Yang et al. 2014; Lee and Hastie 2015).

# 11

# *Monte Carlo Markov Chain Algorithm Convergence*

**Learning Objectives**

- Design Metropolis-Hastings MCMC sampling algorithms.
- Design MCMC algorithms for minimizing functions.
- Design MCMC algorithms for satisfying constraints.
- Design MCMC algorithms for computing expectations.

Monte Carlo Markov Chain sampling methods provide a novel approach to finding minimizers of a function $V$ whose domain is a high-dimensional finite state space. Even though the domain of $V : \Omega \rightarrow \mathcal{R}$ is finite, an exhaustive search for a global minimizer is not computationally tractable. For example, if $\Omega \equiv \{0,1\}^d$ and $d = 100$ then an exhaustive search of the $2^d$ potential solutions is unlikely to be computationally possible.

The problem of function minimization on a finite state space arises frequently in the field of machine learning. Assume each element of a finite set $\Omega$ specifies a different machine learning architecture for a machine learning algorithm design problem. Assume, in addition, there exists a function $V : \Omega \rightarrow \mathcal{R}$ that assigns a number to each network architecture in $\Omega$ such that $V(\mathbf{x}) < V(\mathbf{y})$ implies architecture $\mathbf{x}$ is superior to architecture $\mathbf{y}$. Finding a minimizer of $V$ then corresponds to selecting a good machine learning architecture choice. In practice, there could be billions of different architectures in $\Omega$ (see Example 11.2.5).

Constraint satisfaction problems are also frequently encountered in the field of artificial intelligence. Constraint satisfaction problems are often represented as a vector of $d$ variables that takes on values in some set $\Omega$ which is a finite subset of $\mathcal{R}^d$, where each variable takes on a finite number of values. In this case, the function $V : \Omega \rightarrow \mathcal{R}$ measures the degree to which the constraints are satisfied. Finding a minimizer of the function $V$ corresponds to finding a good solution to the constraint satisfaction problem specified by $V$ (e.g., Markov Logic Net Example 1.6.6 and Example 11.2.4).

The basic strategy for solving such optimization problems is as follows. Define a Gibbs pmf $p : \Omega \rightarrow [0,1]$ with energy function $V : \Omega \rightarrow \mathcal{R}$ so that $p(\mathbf{x}) = (1/Z)\exp(-V(\mathbf{x}))$ where $Z$ is a normalization constant. Then, choose an initial random guess $\mathbf{x}(0)$ which corresponds to a realization of a random vector $\tilde{\mathbf{x}}(0)$. Next, random perturbations are applied to $\mathbf{x}(0)$ in order to transform $\tilde{\mathbf{x}}(0)$ into $\tilde{\mathbf{x}}(1)$. This process is then repeated multiple times so that a sequence of random vectors $\tilde{\mathbf{x}}(0), \tilde{\mathbf{x}}(1), \tilde{\mathbf{x}}(2), \ldots$ is generated. If the random perturbations are chosen in an appropriate way, then it can be shown that the random vectors which are eventually generated will be approximately independent and identically distributed with common density $p$. Consequently, global maximizers of $p$ will be more frequently generated.

Because $V$ is a monotonically decreasing function of $p$, it then follows that global minimizers of $V$ will be more frequently generated.

Monte Carlo Markov Chain sampling methods are also useful for computing approximations of computationally intractable expectation operations. Let $\phi : \mathcal{R}^M \to \mathcal{R}$. Suppose the goal is to compute the expected value of $\phi(\tilde{\mathbf{x}})$ where $\tilde{\mathbf{x}}$ is a discrete random vector whose probability mass function is a Gibbs pmf $p : \Omega \to (0, 1]$ on a high-dimensional finite state space $\Omega$. The exact computation of the expectation:

$$E\{\phi(\tilde{\mathbf{x}})\} = \sum_{\mathbf{x} \in \Omega} \phi(\mathbf{x}) p(\mathbf{x})$$

is computationally intractable when $\Omega$ is very large (e.g., suppose $\Omega \equiv \{0,1\}^d$ and $d = 100$).

However, the expectation $E\{\phi(\tilde{\mathbf{x}})\}$ may be approximated using a Law of Large Numbers by:

$$E\{\phi(\tilde{\mathbf{x}})\} = \sum_{\mathbf{x} \in \Omega} \phi(\mathbf{x}) p(\mathbf{x}) \approx (1/T) \sum_{t=1}^{T} \phi(\tilde{\mathbf{x}}(t))$$

provided that the $T$ random vectors $\tilde{\mathbf{x}}(1), \ldots \tilde{\mathbf{x}}(T)$ are independent and identically distributed with common pmf $p$ and $T$ is sufficiently large. Thus, knowledge of how to sample from the probability mass function $p : \Omega \to (0, 1]$ may also be used to design algorithms for approximately computing expectations with respect to $p$.

Such approximate methods for computing expectations frequently arise in the field of machine learning. Example machine learning applications include computing local conditional probabilities in probabilistic logic nets (e.g., Example 1.6.6), model averaging (e.g., Example 11.2.3), and image processing (e.g., Example 10.4-3).

## 11.1 Monte Carlo Markov Chain (MCMC) Algorithms

### 11.1.1 Countably Infinite First-Order Chains on Finite State Spaces

**Definition 11.1.1** (First-Order Markov Chain (Finite State Space)). Let $\Omega \equiv \{\mathbf{e}^1, \ldots, \mathbf{e}^M\}$ be a finite subset of $\mathcal{R}^d$. A *first-order Markov chain* is a stochastic sequence of $d$-dimensional random vectors $\tilde{\mathbf{x}}(0), \tilde{\mathbf{x}}(1), \tilde{\mathbf{x}}(2), \ldots,$ such that the support of $\tilde{\mathbf{x}}(t)$ is $\Omega$ for all $t \in \mathbb{N}$. The *transition matrix* for the first-order Markov chain is an $M$-dimensional matrix whose element, $p_{j,k}$, in the $j$th row and $k$th column is defined such that:

$$p_{j,k} = p\left(\tilde{\mathbf{x}}(t) = \mathbf{e}^k | \tilde{\mathbf{x}}(t-1) = \mathbf{e}^j\right)$$

for all $j, k \in \{1, \ldots, M\}$. □

In this chapter, the terminology "Markov chain" will be used to refer to a first-order Markov chain on a finite state space as defined in Definition 11.1.1. Note that this definition differs from the discussion of finite length Markov chains in Chapter 10 because it allows for Markov chains of countably infinite length. On the other hand, the Markov chain in Example 10.3.1 can consist of absolutely continuous, mixed, as well as discrete random variables while the Markov chain in Definition 11.1.1 assumes all random variables in the chain are discrete random variables which can take on only a finite number of values.

**Definition 11.1.2** (Probability Vector). Let $\Omega \equiv \{\mathbf{e}^1, \ldots, \mathbf{e}^M\}$ be a finite subset of $\mathcal{R}^d$. Let $p : \Omega \to [0, 1]$ be a probability mass function. Let $\mathbf{p} \in [0, 1]^M$ be defined such that the

$k$th element of the column vector $\mathbf{p}$ is equal to $p(\mathbf{e}^k)$ for $k = 1, \ldots, M$. Then, $\mathbf{p}$ is called a *probability vector.* □

**Definition 11.1.3** (State Transition Graph for a First-Order Markov Chain). Let $\mathbf{P}$ be the $M$-dimensional transition matrix for a first-order Markov chain on a finite state space consisting of $M$ states. Let $\mathcal{G} \equiv (\mathcal{V}, \mathcal{E})$ be a directed graph defined such that the set of vertices $\mathcal{V} = \{1, \ldots, M\}$ and the set of edges $\mathcal{E}$ contains the edge $(j, k)$ if and only if the element in row $j$ and column $k$ of $\mathbf{P}$ is strictly positive. The directed graph $\mathcal{G}$ is called the *state transition graph* for the first-order Markov chain specified by $\mathbf{P}$. □

A Markov chain is a stochastic sequence of random variables. The number of nodes in a Bayesian network directed graph representation (see Figure 10.4) or Markov random field undirected graph representation (see Example 10.4.9) correspond to the number of random variables in this stochastic sequence while the edges specify conditional independence assumptions. In contrast, the number of nodes in a state transition graph is equal to the dimensionality of the state transition matrix for the first-order Markov chain and the edges of the state transition graph (see Figure 11.2) specify which state transitions in the first-order Markov chain are possible. Let the transition matrix for a 3-state first-order Markov chain model of the likelihood of tomorrow's weather given today's weather state be specified as in Figure 11.1 with the state transition graph specified in Figure 11.2.

Let the stochastic sequence of $d$-dimensional random vectors $\tilde{\mathbf{x}}(0), \tilde{\mathbf{x}}(1), \ldots$ be defined such that the support of $\tilde{\mathbf{x}}(t)$ is a finite set $\Omega \equiv \{\mathbf{e}^1, \ldots, \mathbf{e}^M\}$ of $M$ $d$-dimensional vectors. Assume that $\tilde{\mathbf{x}}(0), \tilde{\mathbf{x}}(1), \ldots$ is a first-order Markov chain with $M$-dimensional transition matrix $\mathbf{P}$.

Let the probability distribution of $\tilde{\mathbf{x}}(t)$ be specified by an $M$-dimensional probability vector $\mathbf{p}(t) = [p_1(t), \ldots, p_M(t)]^T$ so that the elements of $\mathbf{p}(t)$ are non negative and sum to one. The $jk$th element of $\mathbf{P}$, $p_{j,k}$, is the probability of transitioning from state $\mathbf{e}^j$ at time $t-1$ to state $\mathbf{e}^k$ at time $t$. Therefore,

$$p_k(t) = \sum_{j=1}^{M} p_j(t-1)p_{j,k}. \tag{11.1}$$

Equation (11.1) may be equivalently represented using the matrix formula:

$$\mathbf{p}(t)^T = \mathbf{p}(t-1)^T \mathbf{P}. \tag{11.2}$$

Let $\mathbf{p}(0)$ be a probability vector specifying the initial probability distribution of $\tilde{\mathbf{x}}(0)$. Also, if $\tilde{\mathbf{x}}(t)$ takes on the $j$th state, $\mathbf{e}^j$, at time $t$ with probability one, then $\mathbf{p}(t)$ is the $j$th $m$-dimensional column vector of the $m$-dimensional identity matrix. Thus, using (11.2), the $k$th element of the row vector $\mathbf{p}(1)^T = \mathbf{p}(0)^T\mathbf{P}$ is the probability that the Markov chain will transition from the initial state $\mathbf{e}^j$ to the state $\mathbf{e}^k$. Thus, $\mathbf{p}(1)$ specifies the probability of reaching each possible state in the state space from the initial state $\mathbf{p}(0)$ after the *first step* of the Markov chain.

The probability that the Markov chain will transition from the initial state of the Markov Chain to a particular state $k$ in *two steps* is computed using (11.2) as follows. First, compute

$$\mathbf{p}(1)^T = \mathbf{p}(0)^T \mathbf{P} \tag{11.3}$$

and then compute

$$\mathbf{p}(2)^T = \mathbf{p}(1)^T \mathbf{P}. \tag{11.4}$$

Substituting, (11.3) into (11.4) then gives the two-step transition formula:

$$\mathbf{p}(2)^T = \mathbf{p}(0)^T \mathbf{P}^2.$$

| | Sunny Tomorrow | Rainy Tomorrow | Cloudy Tomorrow |
|---|---|---|---|
| Sunny Today | $P_{1,1} = 0.2$ | $P_{1,2} = 0.2$ | $P_{1,3} = 0.6$ |
| Rainy Today | $P_{2,1} = 0.1$ | $P_{2,2} = 0.2$ | $P_{2,3} = 0.7$ |
| Cloudy Today | $P_{3,1} = 0.0$ | $P_{3,2} = 0.8$ | $P_{3,3} = 0.2$ |

**FIGURE 11.1**
**An example of a transition matrix for a first-order Markov chain with three states.** Each random variable in the first-order Markov chain has three states: `SUNNY`, `RAINY`, and `CLOUDY`. The transition matrix in this example specifies the probability that tomorrow's weather will be `RAINY` given today's weather is `SUNNY`. So, for example, according to the transition matrix, if today is `SUNNY` then: (i) the probability that tomorrow's weather will remain `SUNNY` is 0.2, (ii) the probability that tomorrow's weather will be `CLOUDY` is 0.6, and (iii) the probability that tomorrow's weather will be `RAINY` will be 0.2. A graphical representation of this information is provided by the state transition graph which is depicted in Figure 11.2.

Proceeding in a similar manner,

$$\mathbf{p}(3)^T = \mathbf{p}(0)^T \mathbf{P}^3$$

yielding the general $m$-step transition formula:

$$\mathbf{p}(m) = \mathbf{p}(0)^T \mathbf{P}^m$$

where $\mathbf{p}(m)$ specifies the probability vector $\mathbf{p}(m)$ for $\tilde{\mathbf{x}}(m)$.

In practice, the transition matrix for a first-order Markov stochastic sequence of $d$-dimensional state vectors can be astronomically large. For example, suppose that a particular state of the Markov chain is a $d$-dimensional binary vector $\mathbf{x} \in \{0, 1\}^d$. If the configuration space consists of $2^d$ states and $d = 100$, this implies that the number of states, $M$, in $\Omega$ is $2^{100}$. Thus, the transition matrix will consist of $2^{100} \times 2^{100}$ elements.

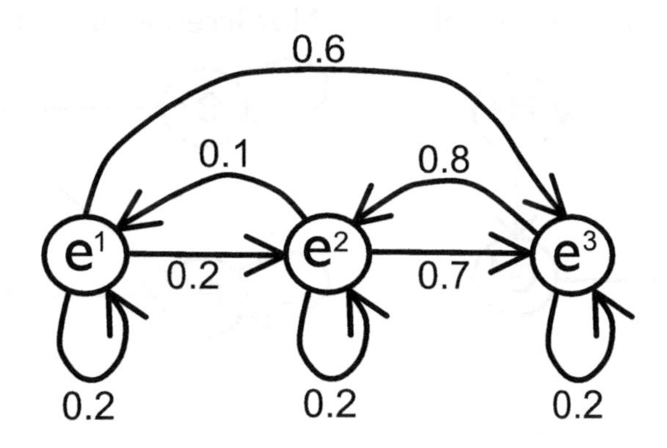


**FIGURE 11.2**
**An example of a state transition graph for a first-order Markov chain with three states.** The state transition graph provides a method for graphically illustrating the information in the transition matrix for the first-order Markov chain (see Figure 11.1).

### 11.1.2 Convergence Analysis of Monte Carlo Markov Chains

**Definition 11.1.4** (Irreducible Markov Chain). Let $\Omega \equiv \{\mathbf{e}^1, \ldots, \mathbf{e}^M\} \subset \mathcal{R}^d$. Let $\mathbf{P}$ be an $M$-dimensional transition matrix for a first-order Markov chain $\tilde{\mathbf{x}}(0), \tilde{\mathbf{x}}(1), \ldots$ on $\Omega$ such that the $ij$th element of $\mathbf{P}$ is the probability $p(\tilde{\mathbf{x}}(t+1) = \mathbf{e}^j | \tilde{\mathbf{x}}(t) = \mathbf{e}^i)$ for $i, j = 1, \ldots, M$. If all off-diagonal elements of $\mathbf{P}^m$ are strictly positive for some positive integer $m$ then the Markov chain is called *irreducible.* □

Less formally, if for every pair of states $\mathbf{e}^j$ and $\mathbf{e}^k$ in the state space $\Omega$ there is a positive probability that the Markov chain initiated at state $\mathbf{e}^j$ will eventually visit $\mathbf{e}^k$, then the Markov chain is called *irreducible.*

If $\mathbf{P}$ is the transition matrix for a Markov chain and all off-diagonal elements of $\mathbf{P}$ are strictly positive, this is a sufficient condition for ensuring that the Markov chain is irreducible.

Note that the property that a Markov chain is irreducible is important because such a Markov chain will eventually search all states in the state space. Examples of irreducible Markov chains are shown in Figure 11.3.

**Definition 11.1.5** (Aperiodic Markov Chain). Let $\Omega \equiv \{\mathbf{e}^1, \ldots, \mathbf{e}^M\} \subset \mathcal{R}^d$. Let $\mathbf{P}$ be an $M$-dimensional transition matrix for a first-order Markov chain $\tilde{\mathbf{x}}(0), \tilde{\mathbf{x}}(1), \ldots$ on $\Omega$. Let

$$S_k \equiv \left\{n \in \mathbb{N}^+ : \text{ the } k\text{th on-diagonal element of } \mathbf{P}^n \text{ is positive}\right\}.$$

The *period* of state $\mathbf{e}^k$ is the greatest common divisor (or common factor) of the elements of $S_k$ when $S_k$ is not empty. If $S_k$ is empty, then the period of state $\mathbf{e}^k$ is defined as zero. The state $\mathbf{e}^k$ is *aperiodic* if the period of $\mathbf{e}^k$ is equal to one. If all states in the Markov chain are aperiodic, then both the Markov chain and transition matrix $\mathbf{P}$ are also called *aperiodic.* □

Assume a particular Markov chain initiated in state $\mathbf{e}^k$ always returns to state $\mathbf{e}^k$ in $n$ iterations. It then follows that the Markov chain when initiated in state $\mathbf{e}^k$ will return to state $\mathbf{e}^k$ in $Mn$ iterations. The period of state $\mathbf{e}^k$ is the largest integer $n$ such that the

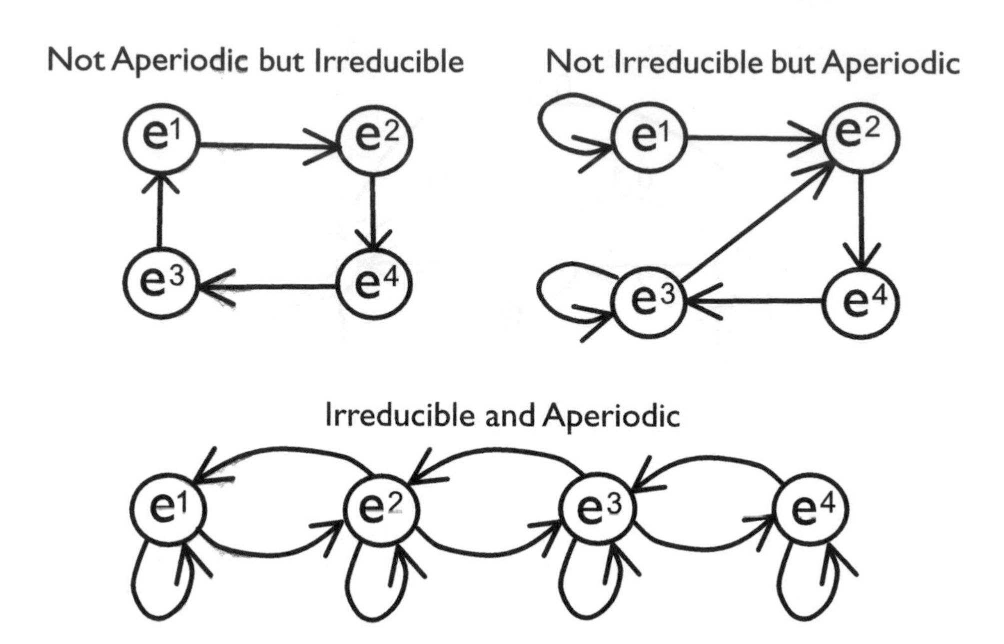


**FIGURE 11.3**
**Examples of first-order Markov chain state transition graphs which are either irreducible or aperiodic.** If the Markov chain is irreducible and has at least one self-loop, then the Markov chain is also aperiodic.

Markov chain returns to $\mathbf{e}^k$ in multiples of $n$. If the Markov chain initiated at state $\mathbf{e}^k$ always returns to state $\mathbf{e}^k$ in six iterations and always returns to state $\mathbf{e}^k$ in twenty-one iterations, then the period of the state is equal to three. When a state in the state space of a Markov chain has the property that the state is periodically reached by the Markov chain with some period greater than one, then that state is called a periodic state of the Markov chain. A state that is not a periodic state is called an aperiodic state. In summary, a state in a first-order Markov chain defined on a finite state space is called an aperiodic state when the chain is initiated at $\mathbf{e}^k$ but the chain does not periodically return to state $\mathbf{e}^k$.

A sufficient condition for a state $\mathbf{e}^k$ of a Markov chain specified by transition matrix $\mathbf{P}$ to be aperiodic is that there is some strictly positive probability that a Markov chain initiated at state $\mathbf{e}^k$ remains in state $\mathbf{e}^k$. In other words, if the $k$th on-diagonal element of the transition matrix is strictly positive, then the $k$th state of the Markov chain is aperiodic. Exercise 11.1-4 shows that if an irreducible Markov chain has one aperiodic state, then the entire Markov chain is aperiodic. Examples of aperiodic and irreducible Markov chains are shown in Figure 11.3.

The following theorem provides a method for constructing MCMC sampling algorithms that generate samples from a computationally intractable probability mass function $p$. The basic idea is to construct a computationally tractable Markov chain which generates a stochastic sequence that converges in distribution to a random vector whose distribution is the computationally intractable pmf $p$. Observations of the tail end of this stochastic sequence will then be approximately independent and identically distributed with pmf $p$ thus providing a computationally tractable method for sampling from $p$.

**Definition 11.1.6** (Markov Chain Stationary Distribution). Let $\mathbf{P}$ be an $M$-dimensional Markov transition matrix for a first-order Markov chain on $\Omega$. If there exists an

$M$-dimensional probability vector $\mathbf{p}^*$ such that $(\mathbf{p}^*)^T\mathbf{P} = (\mathbf{p}^*)^T$, then $\mathbf{p}^*$ is called the *stationary distribution* for $\mathbf{P}$. □

**Theorem 11.1.1** (MCMC Convergence Theorem). *Let $\Omega$ be a finite subset of $\mathcal{R}^d$. Let $\mathbf{P}$ be an $M$-dimensional Markov transition matrix for an irreducible and aperiodic first-order Markov chain $\tilde{\mathbf{x}}(0), \tilde{\mathbf{x}}(1), \ldots$ where the random vector $\tilde{\mathbf{x}}(t)$ has support $\Omega$. Let $\lambda_2$ be the second largest eigenvalue of $\mathbf{P}$. Let $\mathbf{p}(0)$ denote the initial probability vector specifying the probability distribution for $\tilde{\mathbf{x}}(0)$. Let $\mathbf{p}^*$ be a stationary distribution for $\mathbf{P}$. Then, as $m \to \infty$, $\tilde{\mathbf{x}}(m)$ converges in distribution to a unique random vector $\tilde{\mathbf{x}}^*$ whose stationary distribution is $\mathbf{p}^*$. In addition, as $m \to \infty$:*

$$\left|(\mathbf{p}^*)^T - (\mathbf{p}(0))^T\mathbf{P}^m\right| = O\left(|\lambda_2|^m\right). \tag{11.5}$$

*Proof.* Using the Jordan Matrix Decomposition Theorem (e.g., Theorem 10.2, Nobel and Daniel 1977, p. 362), it follows that one can write:

$$\mathbf{P} = \mathbf{RJR}^{-1} \tag{11.6}$$

where $\mathbf{R}$ is a non-singular matrix whose elements may be complex numbers and $\mathbf{J}$ may be either singular or non-singular.

The columns of $\mathbf{R}$ are denoted by $\mathbf{r}_1, \ldots, \mathbf{r}_M$. The rows of $\mathbf{R}^{-1}$ are called the eigenvectors of $\mathbf{P}^T$ and denoted by $\ddot{\mathbf{r}}_1^T, \ldots, \ddot{\mathbf{r}}_M^T$. The matrix $\mathbf{J}$ consists of $k$ on-diagonal submatrices called *Jordan Blocks* (Nobel and Daniel 1977) such that the $i$th $m_i$-dimensional submatrix corresponds to an eigenvalue $\lambda_i$.

Note that since:

$$\mathbf{P}^2 = \mathbf{PP} = \mathbf{RJR}^{-1}\mathbf{RJR}^{-1} = \mathbf{RJ}^2\mathbf{R}^{-1}$$

it follows that for every non-negative integer $m$:

$$\mathbf{P}^m = \mathbf{RJ}^m\mathbf{R}^{-1}. \tag{11.7}$$

The algebraic multiplicity of an eigenvalue $\lambda$, $m_\lambda$, of a square matrix $\mathbf{P}$ is the number of times $\lambda$ is a root of $\det(\mathbf{P} - \lambda\mathbf{I}) = 0$. The geometric multiplicity, $k_\lambda$, is defined as the number of Jordan Blocks generated from a Jordan Decomposition of $\mathbf{P}$ associated with eigenvalue $\lambda$ of $\mathbf{P}$.

Since the state transition matrix $\mathbf{P}$ is both irreducible and aperiodic then the Perron-Frobenius Theorem (Bremaud [1999] (2013) Theorem 1.1, pp. 197-198) may be applied to show that: (1) the eigenvalue with the largest magnitude in $\mathbf{P}$, $\lambda_1$, has geometric and algebraic multiplicity of one, (2) the largest eigenvalue in $\mathbf{P}$, $\lambda_1$, is unique and equal to one, and (3) $\mathbf{p}^*$ is the only vector that satisfies $(\mathbf{p}^*)^T\mathbf{P} = (\mathbf{p}^*)^T$ and $\mathbf{p}^*$ has strictly positive elements which sum to one.

Thus, using the definition of a Jordan Block, the matrix $\mathbf{J}^m$ may be rewritten as:

$$\mathbf{J}^m - \mathbf{1}_M\ddot{\mathbf{r}}_1^T + O(|\lambda_2|^m) \tag{11.8}$$

where $1 > |\lambda_2|$ and $\mathbf{1}_M$ and $\ddot{\mathbf{r}}_1$ are respectively the first column of $\mathbf{R}$ and the first row of $\mathbf{R}^{-1}$.

Thus, for every initial state probability vector $\mathbf{p}(0)$ whose elements are non-negative real numbers that sum to one:

$$\mathbf{p}(m)^T = \mathbf{p}(0)^T\mathbf{P}^m = \mathbf{p}(0)^T\mathbf{1}_M(\ddot{\mathbf{r}}_1)^T + O\left(|\lambda_2|^m\right) = (\ddot{\mathbf{r}}_1)^T + O\left(|\lambda_2|^m\right). \tag{11.9}$$

Equation (11.9) implies the probability vector $\mathbf{p}(m)^T \to (\ddot{\mathbf{r}}_1)^T$ as $m \to \infty$ for every initial probability vector $\mathbf{p}(0)$ with geometric convergence rate specified in (11.5). Since $\mathbf{p}^*$ is the unique stationary distribution of the chain by the Perron-Frobenius Theorem (see Winkler 2003, Theorem 4.3.2), it follows that $\mathbf{p}^*$ and $\ddot{\mathbf{r}}_1$ must be equal.

■

### 11.1.3 Hybrid MCMC Algorithms

In some applications, one might have several different MCMC algorithms that converge to a common stationary distribution $\mathbf{p}^*$ and one wishes to combine all of these MCMC algorithms to create a new MCMC algorithm with better or different convergence properties that also converges to $\mathbf{p}^*$. One possible way to do this is to make a probabilistic transition step using the stochastic transition step $T_1$ from one Markov chain and then follow that with a stochastic transition step $T_2$ from another Markov chain. This is an example of creating an entirely new hybrid MCMC chain using the principle of "composition". Alternatively, if one has two different MCMC chains with stochastic transitions $T_1$ and $T_2$, then the two chains may be combined to construct a new Markov chain where one selects a transition step $T_1$ from the first Markov chain with probability $\alpha$ and selects transition step $T_2$ from the second Markov chain with probability $1 - \alpha$. This is an example of constructing an entirely new hybrid Markov chain from a "mixture" of two other Markov chains.

Hybrid MCMC algorithms provide a useful tool for designing customized MCMC algorithms from other existing MCMC algorithms. An important practical application of such customized MCMC algorithms is the development of Monte Carlo Markov chains that implement customized strategies for exploring large state spaces. Such strategies can dramatically improve both algorithm speed and the quality of the results generated from the MCMC algorithm. Suppose one has prior heuristic knowledge that a few specific important regions of the state space should be visited with high probability but these regions are widely separated in the state space. Such heuristic prior knowledge can be embedded into a hybrid Markov chain algorithm by taking a large step into one of the important state space regions and then following this large step by a series of small steps intended to explore that desirable region. These local steps are then followed by a large step designed to explore a more distant, yet equally important, region of the state space. Thus, the dynamics of the hybrid MCMC algorithm can be visualized as the composition of two distinct Markov chains. One of the Markov chains takes large steps with the purpose of moving from one large desirable region of the state space to another distant large desirable region, while the other Markov chain takes small steps with the purpose of carefully exploring a particular local region of the state space located within a larger desirable region.

A second important practical application for hybrid MCMC algorithms is that it may be easier to mathematically derive or implement an MCMC algorithm for solving a specific task in a computationally efficient manner if one has the option of combining different MCMC algorithms in different ways.

**Definition 11.1.7** (Hybrid MCMC). Let $\mathbf{P}_1, \ldots, \mathbf{P}_K$ be a collection of $K$ (not necessarily unique) $M$-dimensional Markov transition matrices which respectively specify $K$ first-order Markov chains with a common stationary distribution $\mathbf{u}^*$ on a common finite state space $\Omega \subset \mathcal{R}^d$. Let $\alpha_1, \ldots, \alpha_K$ be non-negative real numbers which sum to one. A first-order Markov chain on $\Omega$ specified by the transition matrix

$$\mathbf{M} \equiv \sum_{k=1}^{K} \alpha_k \mathbf{P}_k$$

is called a *hybrid mixture MCMC* with respect to $\mathbf{P}_1, \ldots, \mathbf{P}_K$ and $\alpha_1, \ldots, \alpha_K$. A first-order Markov chain on $\Omega$ specified by the transition matrix

$$\mathbf{C} \equiv \prod_{k=1}^{K} \mathbf{P}_k$$

is called a *hybrid composition MCMC* with respect to $\mathbf{P}_1, \ldots, \mathbf{P}_K$. □

**Theorem 11.1.2** (Hybrid MCMC Convergence). *Let* $\mathbf{P}_1, \ldots, \mathbf{P}_K$ *be a collection of* $K$ *(not necessarily unique)* $M$*-dimensional Markov transition matrices which respectively specify* $K$ *first-order Markov chains with a common stationary distribution specified by the* $M$*-dimensional probability vector* $\mathbf{u}^*$. *Let* $\alpha_1, \ldots, \alpha_K$ *be non-negative real numbers which sum to one. An aperiodic and irreducible Markov chain on* $\Omega$ *specified by either the transition matrix*

$$\mathbf{M} \equiv \sum_{k=1}^{K} \alpha_k \mathbf{P}_k$$

*or the transition matrix*

$$\mathbf{C} \equiv \prod_{k=1}^{K} \mathbf{P}_k.$$

*converges in distribution to a unique random vector whose stationary distribution is the probability vector* $\mathbf{u}^*$.

*Proof.* The MCMC convergence theorem (Theorem 11.1.1) implies the aperiodic and irreducible Markov chain specified by the transition matrix $\mathbf{M}$ converges to a unique stationary probability distribution. Since $\mathbf{p}^*$ is a stationary distribution for $\mathbf{P}_1, \ldots, \mathbf{P}_K$ and $\sum \alpha_k = 1$ it follows that:

$$(\mathbf{p}^*)^T \mathbf{M} = (\mathbf{p}^*)^T \sum_{k=1}^{K} \alpha_k \mathbf{P}_k = \sum_{k=1}^{K} \alpha_k (\mathbf{p}^*)^T \mathbf{P}_k = \sum_{k=1}^{K} \alpha_k (\mathbf{p}^*)^T = (\mathbf{p}^*)^T.$$

Thus, the first-order Markov chain specified by the transition matrix $\mathbf{M}$ converges to the unique stationary distribution $\mathbf{p}^*$.

Using a similar argument, the MCMC convergence theorem (Theorem 11.1.1) implies that the aperiodic and irreducible Markov chain specified by the transition matrix $\mathbf{C}$ converges to a unique stationary probability distribution. Since $\mathbf{p}^*$ is a stationary distribution for $\mathbf{P}_1, \ldots, \mathbf{P}_K$ it follows that:

$$(\mathbf{p}^*)^T \mathbf{C} = (\mathbf{p}^*)^T \prod_{k=1}^{K} \mathbf{P}_k = (\mathbf{p}^*)^T.$$

Thus, the first-order Markov chain specified by the transition matrix $\mathbf{C}$ also converges to the unique stationary distribution $\mathbf{p}^*$. ■

### 11.1.4 Finding Global Minimizers and Computing Expectations

The MCMC Convergence Theorem provides a procedure for designing a Markov chain to generate samples from a desired Gibbs probability mass function. Thus, the samples which are generated most frequently will correspond to the global maximizers of the Gibbs probability mass function $p_{\mathcal{T}}(\mathbf{x}) = (1/Z_{\mathcal{T}}) \exp(-V(\mathbf{x})/\mathcal{T})$ where $V : \Omega \to \mathcal{R}$ and the positive temperature constant $\mathcal{T}$.

The following theorem (Gibbs Distribution Temperature Theorem 11.1.3) and Figure 11.4 show that when the temperature constant $\mathcal{T}$ in (11.10) is large, $p_{\mathcal{T}}$ is approximately a uniform distribution over the state space. When $\mathcal{T}$ is close to zero, then $p_{\mathcal{T}}$ is approximately a uniform distribution over the global minimizers of the energy function $V$. This suggests a useful heuristic strategy for searching for global minimizers. Begin with a larger value of the temperature constant $\mathcal{T}$, let the chain converge to a stationary distribution, and keep track of the states associated with the smallest values of $V$. Then, decrease the temperature constant $\mathcal{T}$, initiate multiple chains at the initial states with the smallest values of $V$ found in the previous step, and let each chain converge to a stationary distribution. Continue in this manner until the temperature constant $\mathcal{T}$ is very close to zero.

**Theorem 11.1.3** (Gibbs Distribution Temperature Dependence). *Let $\mathcal{T} \in (0,\infty)$. Let $\Omega \equiv \{\mathbf{x}_1,\ldots,\mathbf{x}_M\}$ be a finite subset of $\mathcal{R}^d$. Let $V : \Omega \rightarrow \mathcal{R}$. Let $G$ be a subset of $\Omega$ which contains all $K$ global minimizers of $V$. Let $p_{\mathcal{T}} : \Omega \rightarrow [0,1]$ be defined such that for all $\mathbf{x} \in \Omega$:*

$$p_{\mathcal{T}}(\mathbf{x}) = \frac{\exp(-V(\mathbf{x})/\mathcal{T})}{Z_{\mathcal{T}}} \tag{11.10}$$

*and*

$$Z_{\mathcal{T}} = \sum_{\mathbf{y}\in\Omega} \exp(-V(\mathbf{y})/\mathcal{T}).$$

*As $\mathcal{T} \rightarrow 0$, (i) $p_{\mathcal{T}}(\mathbf{x}) \rightarrow 1/K$ for all $\mathbf{x} \in G$, and (ii) $p_{\mathcal{T}}(\mathbf{x}) \rightarrow 0$ for all $\mathbf{x} \notin G$. In addition, (iii) as $\mathcal{T} \rightarrow \infty$, $p_{\mathcal{T}}(\mathbf{x}) \rightarrow 1/M$ for all $\mathbf{x} \in \Omega$,*

*Proof.* Note that $K \geq 1$ since $\Omega$ is a finite set. Let $V^* = V(\mathbf{x})$ for all $\mathbf{x} \in G$. Let

$$\delta_{\mathcal{T}}(\mathbf{y}) \equiv \exp[-(V(\mathbf{y}) - V^*)/\mathcal{T}]$$

which has the property that $\delta_{\mathcal{T}}(\mathbf{y}) \rightarrow 0$ as $\mathcal{T} \rightarrow 0$ if $\mathbf{y} \notin G$ and $\delta_{\mathcal{T}}(\mathbf{y}) = 1$ if $\mathbf{y} \in G$. Now rewrite $p_{\mathcal{T}}$ as:

$$p_{\mathcal{T}}(\mathbf{x}) = \frac{\delta_{\mathcal{T}}(\mathbf{x})}{\sum_{\mathbf{y}\in G}\delta_{\mathcal{T}}(\mathbf{y}) + \sum_{\mathbf{y}\notin G}\delta_{\mathcal{T}}(\mathbf{y})} = \frac{\delta_{\mathcal{T}}(\mathbf{x})}{K + \sum_{\mathbf{y}\notin G}\delta_{\mathcal{T}}(\mathbf{y})}. \tag{11.11}$$

Conclusions (i) and (ii) of the theorem follow directly from (11.11) by letting $\mathcal{T} \rightarrow 0$. Conclusion (iii) of the theorem follows directly from (11.11) by noting that as $\mathcal{T} \rightarrow \infty$, $\delta_{\mathcal{T}}(\mathbf{x}) \rightarrow 1$ which implies that $p_{\mathcal{T}}(\mathbf{x}) \rightarrow 1/M$. ∎

Another important application of the MCMC sampling algorithm involves computing expectations and useful marginal probabilities. The MCMC sampling algorithm generates a stochastic sequence $\{\tilde{\mathbf{x}}(t)\}$ that converges in distribution to a sequence of independent and identically distributed random vectors with common pmf $p : \Omega \rightarrow [0,1]$.

Let $\Omega$ be a finite subset of $\mathcal{R}^d$. Let $\boldsymbol{\phi} : \Omega \rightarrow \mathcal{R}^k$. Assume the computational goal of MCMC sampling is to compute the expectation

$$E\{\boldsymbol{\phi}(\tilde{\mathbf{x}})\} \equiv \sum_{\mathbf{x}\in\Omega} \boldsymbol{\phi}(\mathbf{x})p(\mathbf{x}), \tag{11.12}$$

yet evaluation of (11.12) is computationally impossible because the number of elements in $\Omega$ is very large.

A Monte Carlo approximation to (11.12) can be computed using the formula:

$$\bar{\boldsymbol{\phi}}_L(t) = (1/L)\sum_{s=0}^{L-1} \boldsymbol{\phi}(\tilde{\mathbf{x}}(t-s)) \tag{11.13}$$

where $L$ is called the *block size.* The block size $L$ specifies the length of the portion of the stochastic sequence which is used to estimate the expectation $E\{\boldsymbol{\phi}(t)\}$. Typically, this is the last $L$ observations in the stochastic sequence.

Unfortunately, the observations in a first-order MCMC are identically distributed but only asymptotically independent so the Strong Law of Large Numbers for i.i.d random vectors (see Theorem 9.3.1) is not technically applicable.

Fortunately, Theorem 11.1.4 avoids these issues by providing a Strong Law of Large Numbers that works for a stochastic sequence of random vectors generated by a Monte Carlo Markov Chain operating in a finite state space.

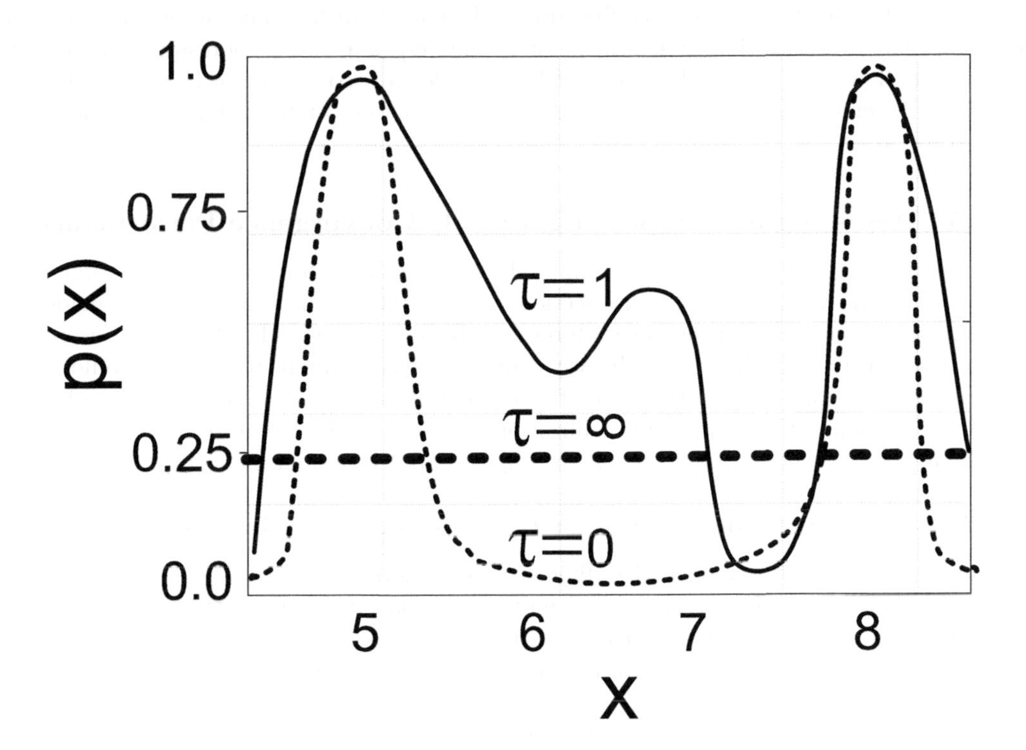


**FIGURE 11.4**
**Shape of Gibbs density as a function of the temperature parameter $\mathcal{T}$.** When the temperature parameter $\mathcal{T}$ is close to infinity, the Gibbs density is uniform over the sample space. When the temperature parameter $\mathcal{T}$ is close to zero, the Gibbs density is uniform over the set of most probable points $\{5, 8\}$ in the sample space. Typically, it is assumed that $\mathcal{T} = 1$ corresponds to the undistorted shape of the Gibbs density.

**Theorem 11.1.4** (MCMC Strong Law of Large Numbers). *Let the stochastic sequence of d-dimensional random vectors $\tilde{\mathbf{x}}(0), \tilde{\mathbf{x}}(1), \tilde{\mathbf{x}}(2), \ldots$ be an irreducible first-order Markov chain on a finite state space $\Omega$ which converges to some d-dimensional random vector $\tilde{\mathbf{x}}^*$ with stationary pmf $p : \Omega \rightarrow [0, 1]$. Let $\phi : \Omega \rightarrow \mathcal{R}^k$ and*

$$E\{\phi(\tilde{\mathbf{x}})\} \equiv \sum_{\mathbf{x} \in \Omega} \phi(\mathbf{x}) p(\mathbf{x}).$$

*Let*

$$\bar{\phi}_L \equiv (1/L) \sum_{t=0}^{L-1} \phi(\tilde{\mathbf{x}}(t)).$$

*Then, $\bar{\phi}_L \rightarrow E\{\phi(\tilde{\mathbf{x}})\}$ with probability one as $L \rightarrow \infty$. In addition, $\bar{\phi}_L \rightarrow E\{\phi(\tilde{\mathbf{x}})\}$ in mean-square as $L \rightarrow \infty$.*

*Proof.* Since $\Omega$ is finite state space, $\phi$ is bounded on $\Omega$. Convergence with probability one then follows immediately from the MCMC Strong Law of Large Numbers (see Theorem 4.1 of Bremaud 1999, p. 111; also see Theorem 3.3 of Bremaud 1999, p. 105). Since the state space is finite, convergence in mean-square follows from the Stochastic Convergence Relationships Theorem 9.3.8. ■

Note that even if an irreducible first-order Markov chain has not completely converged to a stationary distribution, it is still possible that $\bar{\phi}_L$ may be a good approximation for $E\{\phi(\tilde{\mathbf{x}})\}$ in Theorem 11.1.4. Theorem 11.1.4 provides a Strong Law of Large Numbers for correlated stochastic sequences generated by an irreducible first-order Markov chain on a finite state space.

### 11.1.5 Assessing and Improving MCMC Convergence Performance

There is not a wide consensus regarding effective methods for assessing convergence of Monte Carlo Markov chains. In this section, one simple example method is provided for assessing convergence but other procedures (see Robert and Casella 2004 for a useful discussion) should be considered as well. Any convergence assessment procedure should always be repeated multiple times with different initial conditions in order to obtain insights into variability associated with multiple realizations of exactly the same Monte Carlo Markov Chain. In practice, such repetitions may be implemented using parallel computing methods.

#### 11.1.5.1 Assessing Convergence for Expectation Approximation

Assume the stochastic sequence of $d$-dimensional random vectors $\tilde{\mathbf{x}}(0), \tilde{\mathbf{x}}(1), \tilde{\mathbf{x}}(2), \ldots$ is an irreducible first-order Markov chain on a finite state space $\Omega$ that converges in distribution to some $d$-dimensional random vector $\tilde{\mathbf{x}}^*$. Let $\phi : \mathcal{R}^d \to \mathcal{R}^k$.

In many MCMC applications, an important goal is to compute the expectation

$$\bar{\phi}^* \equiv E\{\phi(\tilde{\mathbf{x}}^*)\}$$

as in (11.12). Theorem 11.1.4 suggests that a useful approximation for $\bar{\phi}^*$ is to use the formula:

$$\bar{\phi}_L(t) \equiv \frac{1}{L} \sum_{i=0}^{L-1} \phi(\tilde{\mathbf{x}}(t-i)) \tag{11.14}$$

since Theorem 11.1.4 provides conditions for $\bar{\phi}_L(t)$ to converge in mean-square to $\bar{\phi}^*$ or equivalently that

$$E\{|\bar{\phi}_L(t) - \bar{\phi}^*|^2\} \to 0.$$

Therefore, if one had a method for estimating $(1/k)E\{|\bar{\phi}_L(t) - \bar{\phi}^*|^2\}$, this would be helpful for the purposes of assessing the effectiveness of using $\bar{\phi}_L(t)$ as an approximation for $\bar{\phi}^*$. Notice that the division by the dimensionality $k$ is included in order to measure the expected deviation magnitude for an individual element of $\bar{\phi}_L(t)$.

One possible estimator for $(1/k)E\{|\bar{\phi}_L(t) - \bar{\phi}^*|^2\}$ is

$$\hat{\sigma}_L^2(t) \equiv \frac{1}{kL} \sum_{i=0}^{L-1} \left|\bar{\phi}_L(t-i) - \hat{\mu}_L(t-i)\right|^2 \tag{11.15}$$

where $\hat{\mu}_L(t) \equiv (1/L) \sum_{i=0}^{L-1} \bar{\phi}_L(t-i)$.

The statistic $\hat{\sigma}_L(t)$ is called the *MCMC simulation error.* When the MCMC simulation error $\hat{\sigma}_L(t)$ is less than some positive number $\epsilon$, this is a useful indicator that the Monte Carlo sample average $\bar{\phi}_L(t)$ closely approximates its asymptotic value $\bar{\phi}^*$.

In some cases, because of computational constraints, it may not be possible to choose $t$ and $L$ large enough to ensure that $\hat{\sigma}_L(t)$ is close to zero. In such cases, one should report $\hat{\sigma}_L(t)$ as the MCMC simulation error associated with using $\bar{\phi}_L(t)$ to approximate $\bar{\phi}^*$. It is also important to choose $L$ large enough so that low-probability values of $\tilde{\mathbf{x}}^*$ are observed sufficiently frequently when $L$ observations of the Markov chain are sampled.

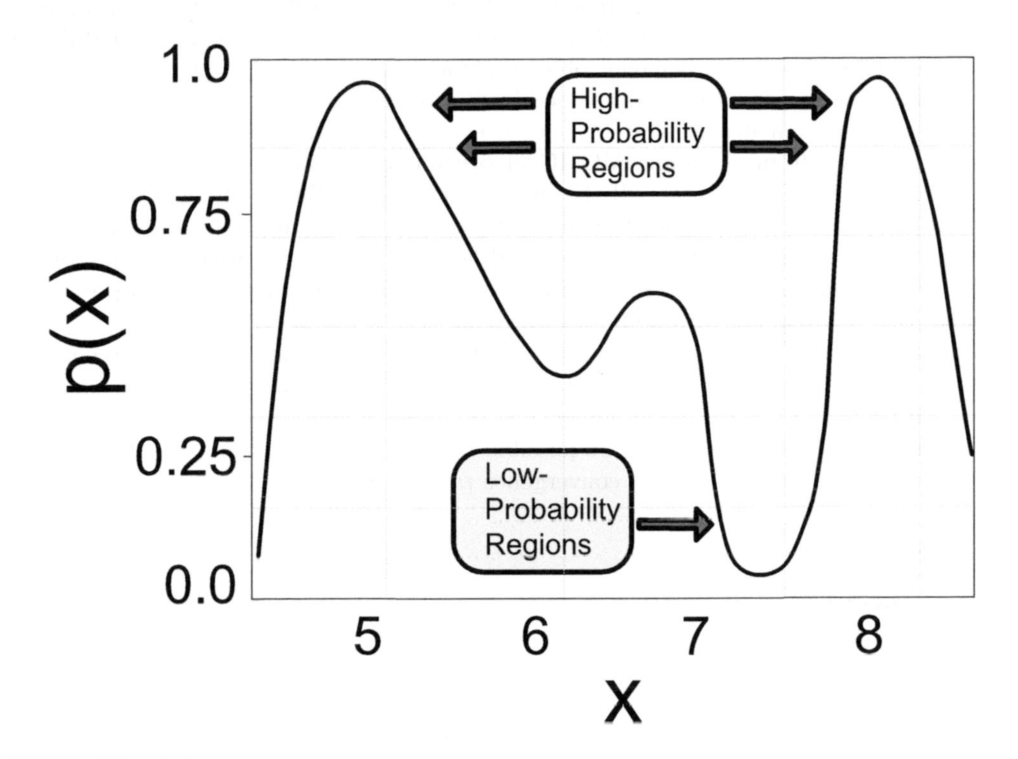


**FIGURE 11.5**
The presence of low-probability regions surrounding a high-probability region of interest can dramatically slow down stochastic search.

#### 11.1.5.2 MCMC Convergence Challenges and Heuristics

Although convergence in distribution to a unique random vector is guaranteed as the number of iterations of MCMC sampling algorithm becomes large, the system state may become trapped in regions of the state space surrounded by low-probability transitions. This situation is illustrated in Figure 11.5. Although convergence in distribution may still be ensured in such cases, the convergence rate can be so slow that the MCMC sampling algorithm becomes impractical. Assessing convergence for this general situation can be challenging because different properties of the approximating probability distribution for $\tilde{\mathbf{x}}(t)$ may possess different convergence rates.

To avoid the problem of becoming trapped near low-probability regions of the state space, one strategy would be to increase the temperature constant $\mathcal{T}$ in order to increase the probability that the Markov chain can traverse a low-probability region. Once the MCMC has converged to a stationary distribution, the temperature can then be reduced and this process repeated. At low temperatures, the Markov chain will visit global minimizers of the energy function of the stationary distribution very frequently. This method of progressively decreasing the temperature constant is useful in applications where MCMC sampling is used to solve optimization problems.

A second important heuristic strategy for traversing low-probability regions involves taking combinations of both larger and smaller steps. The larger steps allow one to jump from one region of the state space to another to avoid becoming trapped while the smaller steps are designed to zoom in to a solution. That is, one Markov chain visits different regions of the Markov field with different probabilities. Then, within that region, another MCMC explores state values of the random variables within the current region with different probabilities. The asymptotic behavior of such algorithms can be characterized using the hybrid MCMC convergence theory developed in Section 11.1.3.

A third important heuristic strategy is to generate multiple stochastic sequences generated by sampling from the same MCMC algorithm with different random seeds from different regions of the state space. If the different chains exhibit fundamentally different asymptotic properties, this implies that one or more of the chains have become trapped in a low-probability region of the state space. When this strategy is used, it is important to initiate the different chains at widely different points in the state space. As previously mentioned, this technique is also useful for assessing convergence.

A fourth important heuristic strategy for avoiding low-probability regions is to consider alternative MCMC sampling algorithms. Quite different MCMC sampling algorithms with different convergence rates can be designed such that they converge to the same stationary distribution. Thus, it may sometimes be possible to design a customized MCMC sampling algorithm which will possess a faster convergence rate for specific applications.

## Exercises

11.1-1. Referring to Figure 11.3, explain why the upper-left state-transition graph is an irreducible but not aperiodic chain, the upper-right state-transition graph is an aperiodic but not irreducible chain, and the bottom state-transition graph is both irreducible and aperiodic.

11.1-2. *Stochastic Search in a Model Space.* Suppose there is a collection of $d$ input variables and different subsets of these $d$ input variables correspond to different learning machine architectures. Let $\mathbf{x} \equiv [x_1, \ldots, x_d] \in S$ where $S \equiv \{0,1\}^d$ is a binary vector whose $j$th element is equal to 1 if the $j$th input variable is included in the input pattern for a particular learning machine. Thus, each element of $\{0,1\}^d$ corresponds to a different learning machine architecture. The finite state space $S$ is called a "model space".

Consider the following stochastic search algorithm through model space which generates a stochastic sequence $\tilde{\mathbf{x}}(0), \tilde{\mathbf{x}}(1), \ldots$ through model space. A particular realization of this stochastic sequence is generated using the following algorithm. Let the notation $y = \text{RANDOM}(Q)$ where $Q$ indicates that $y$ is a sample from a uniform distribution with support $Q$.

**Step 1.** $t \Leftarrow 0$. $\mathbf{x}(0) \Leftarrow \text{RANDOM}\left(\{0,1\}^d\right)$
**Step 2.** $k \Leftarrow \text{RANDOM}(\{1, \ldots, d\})$
**Step 3.** $\alpha \Leftarrow \text{RANDOM}([0,1])$
**Step 4.** $\mathbf{x}(t+1) \Leftarrow \mathbf{x}(t)$.
**Step 5.** **If** $\alpha > 0.8$ **Then** $x_k(t+1) \Leftarrow 1 - x_k(t)$ (i.e., "flip $k$th bit in $\mathbf{x}(t)$ )
**Step 6.** $t \Leftarrow t+1$ and then **Go To** Step 2.

Is the Markov chain $\tilde{\mathbf{x}}(0), \tilde{\mathbf{x}}(1), \ldots$ irreducible? Is the Markov chain aperiodic?

11.1-3. Modify the model search algorithm in Exercise 11.1-2 as follows.
First, replace **Step 1** with **Step 1'.** $t \Leftarrow 0$. $\mathbf{x}(0) \Leftarrow \mathbf{0}_d$
and second replace **Step 5** with **Step 5'.** $x_k(t+1) \Leftarrow 1$

Is the resulting new Markov chain $\tilde{\mathbf{x}}(0), \tilde{\mathbf{x}}(1), \ldots$ irreducible? Is the Markov chain aperiodic?

11.1-4. *Irreducible Chain with One Aperiodic State Is Irreducible.* Show that if an irreducible Markov chain has one aperiodic state, then the chain is aperiodic.

11.1-5. Consider the transition matrix $\mathbf{M}$ for the Markov chain in Figure 11.1. Is the Markov chain irreducible? Is the Markov chain aperiodic? Write a computer program to compute a numerical approximation for the stationary distribution $\mathbf{p}^*$ for the Markov chain in Figure 11.2 by computing $\mathbf{M}^{100}$. Verify that $\mathbf{p}^*$ is a stationary distribution by showing that $(\mathbf{p}^*)^T\mathbf{M} = (\mathbf{p}^*)^T$.

11.1-6. Let $S \equiv \{0,1\}^d$. Let $\mathbf{W}$ be a real symmetric matrix whose on-diagonal elements are equal to zero. Let $p(\mathbf{x}) = Z^{-1}\exp(-\mathbf{x}^T\mathbf{W}\mathbf{x}/\mathcal{T})$ where

$$Z = \sum_{\mathbf{y} \in S} \exp(-\mathbf{y}^T\mathbf{W}\mathbf{y}/\mathcal{T}).$$

Consider an MCMC algorithm which first picks an element of $\mathbf{x}(t)$ at random (e.g., the $i$th element) and then sets the value of that element, $x_i(t)$, equal to the number one with probability

$$p(\tilde{x}_i(t+1) = 1|\mathbf{x}(t)).$$

As $\mathcal{T}$ becomes close to zero or close to infinity, explain qualitatively how this changes the dynamics of the Markov chain.

11.1-7. Let $\mathbf{p}^*$ be an $M$-dimensional probability vector specifying the stationary distribution of an $M$-state Markov chain. Assume $M$ is very large (e.g., $M = 2^{10000}$). Explain how the MCMC Strong Law of Large Numbers can be used to approximately evaluate the expectation $\sum_{k=1}^{M} g_k p_k^*$ where $p_k^*$ is the $k$th element of $\mathbf{p}^*$ and $g_1, \ldots, g_M$ are known constants.

## 11.2 MCMC Metropolis-Hastings (MH) Algorithms

In this section, the important class of Metropolis-Hastings (MH) MCMC algorithms is introduced.

### 11.2.1 Metropolis-Hastings Algorithm Definition

The essential idea of the Metropolis-Hastings algorithm is that a candidate global state $\mathbf{c}$ is generated with some probability $q(\mathbf{c}|\mathbf{x}(t))$ from the current global state $\mathbf{x}(t)$ at iteration $t$. Then, the candidate state $\mathbf{c}$ is accepted with acceptance probability $\rho(\mathbf{x}(t), \mathbf{c})$ so that $\mathbf{x}(t+1) = \mathbf{c}$. If the candidate state $\mathbf{c}$ is not accepted, then $\mathbf{x}(t+1) = \mathbf{x}(t)$. This procedure generates a stochastic sequence $\tilde{\mathbf{x}}(1), \tilde{\mathbf{x}}(2), \ldots$ which converges in distribution at a geometric convergence rate to the desired stationary distribution $p : \Omega \to [0,1]$. The conditional probability mass function $q(\mathbf{c}|\mathbf{x})$ is called the "MH proposal distribution". The MH proposal distribution specifies the machine learning engineer's MH optimization algorithm choice. The desired stationary distribution $p$ is a Gibbs energy function with energy function $V : \Omega \rightarrow \mathcal{R}$. Typically, the goal is to either minimize $V$ or compute expectations with respect to $p(\mathbf{x}) = (1/Z)\exp(-V(\mathbf{x}))$.

**Definition 11.2.1** (MH Proposal Distribution). Let $\Omega$ be a finite subset of $\mathcal{R}^d$. The *MH proposal distribution* $q : \Omega \times \Omega \rightarrow [0,1]$ on $\Omega$ is defined such that:

- (i) $q(\cdot|\mathbf{x}) : \Omega \rightarrow [0,1]$ is a conditional pmf for all $\mathbf{x} \in \Omega$,
- (ii) $q(\mathbf{c}|\mathbf{x}) > 0$ if and only if $q(\mathbf{x}|\mathbf{c}) > 0$ for all $\mathbf{x}, \mathbf{c} \in \Omega$.

□

The proposal distribution is a specification of an algorithm that takes a current global state $\mathbf{x}(t)$ at iteration $t$ of the algorithm and transforms it into a new candidate global state $\mathbf{c}$ with probability $q(\mathbf{c}|\mathbf{x}(t))$.

**Definition 11.2.2** (Metropolis-Hastings Algorithm). Let $\Omega$ be a finite subset of $\mathcal{R}^d$. Let $p : \Omega \rightarrow (0,1]$ be a *Gibbs pmf* defined such that for all $\mathbf{x} \in \Omega$:

$$p(\mathbf{x}) = (1/Z)\exp[-V(\mathbf{x})] \text{ where } Z = \sum_{\mathbf{y}\in\Omega} \exp[-V(\mathbf{y})]. \tag{11.16}$$

Let $q : \Omega \times \Omega \rightarrow [0,1]$ be the proposal distribution defined in Definition 11.2.1. A *Metropolis-Hastings Markov Chain* is a first-order Markov chain that generates the stochastic sequence of random vectors $\tilde{\mathbf{x}}(1), \tilde{\mathbf{x}}(2), \ldots$ according to the following *Metropolis-Hastings algorithm.*

- *Step 1.* Choose an initial state $\mathbf{x}(0) \in \Omega$. Let $t = 0$.
- *Step 2.* Choose a candidate state $\mathbf{c}$ with proposal probability $q(\mathbf{c}|\mathbf{x}(t))$.
- *Step 3.* Accept the candidate state $\mathbf{c}$ with *acceptance probability* $\rho(\mathbf{x}(t), \mathbf{c})$ and reject the candidate state $\mathbf{c}$ with *rejection probability* $1 - \rho(\mathbf{x}(t), \mathbf{c})$ where

$$\rho(\mathbf{x}, \mathbf{c}) \equiv \min\left\{1, \left(\frac{q(\mathbf{x}|\mathbf{c})}{q(\mathbf{c}|\mathbf{x})}\right) \exp[-(V(\mathbf{c}) - V(\mathbf{x}))]\right\}. \tag{11.17}$$

- *Step 4.* Let $t = t + 1$, and then Go To Step 2.

□

The following algorithm may be useful for designing computationally efficient algorithms for sampling from a proposal probability distribution.

Let the notation $y \Leftarrow \text{RANDOM}(0,1)$ denote a function which randomly samples a number from a uniform distribution on the interval $(0,1)$ and assigns that number to the variable $y$.

The cut-off thresholds $L_1, L_2, \ldots, L_M$ in Algorithm (11.2.1) are then constructed such that the lengths of the intervals $(0, L_1)$, $(L_1, L_2)$, $\ldots$, $(L_{M-1}, L_M)$ are non-increasing. Next, the algorithm checks if a uniformly distributed random number takes on a value in the largest intervals before checking if it takes on a value in the smallest intervals. This means that on the average, the algorithm will terminate more rapidly relative to an algorithm which does not check the largest intervals first. On the other hand, Algorithm (11.2.1) may be less efficient in situations when the computing time associated with computing a global maximizer of $q(\cdot|\mathbf{x})$ on $C$ in Step 9 of Algorithm 11.2.1) is large and all $M$ intervals $(0, L_1), \ldots, (L_{M-1}, L_M)$ are approximately equal in length.

### Recipe Box 11.1 Expectations Using Metropolis-Hastings

- **Step 1: Design configuration space $\Omega$ and the energy function $V$.** Let the configuration space $\Omega$ be a finite subset of $\mathcal{R}^d$. The MCMC algorithm sampling algorithm will be designed such that it asymptotically visits $\mathbf{x}$ with probability

$$p(\mathbf{x}) = (1/Z)\exp(-V(\mathbf{x})) \text{ where } Z = \sum_{\mathbf{y}\in\Omega} \exp(-V(\mathbf{y}))$$

with energy function $V : \Omega \rightarrow \mathcal{R}$.

- **Step 2: Design the MH proposal distribution.** The MH proposal distribution $q(\mathbf{c}|\mathbf{x})$ specifies the probability of visiting candidate state $\mathbf{c}$ given the current state $\mathbf{x}$ for all $\mathbf{x}, \mathbf{c} \in \Omega$. It is assumed that $q(\mathbf{c}|\mathbf{x}) > 0$ if and only if $q(\mathbf{x}|\mathbf{c}) > 0$ for all $\mathbf{x}, \mathbf{c} \in \Omega$. Design MH proposal distribution $q$ so the Metropolis-Hastings Markov Chain in Definition 11.2.2 is irreducible and aperiodic. Then, let $t = 0$.

- **Step 3 : Generate a candidate state.** Given the current state of the chain $\mathbf{x}(t)$, choose a candidate state $\mathbf{c}$ with probability $q(\mathbf{c}|\mathbf{x}(t))$.

- **Step 4 : Decide whether or not to accept candidate state.** Set $\mathbf{x}(t+1)$ equal to $\mathbf{c}$ with acceptance probability $\rho(\mathbf{x}(t), \mathbf{c})$ and set $\mathbf{x}(t+1)$ equal to $\mathbf{x}(t)$ with rejection probability $1 - \rho(\mathbf{x}(t), \mathbf{c})$ where

$$\rho(\mathbf{x}, \mathbf{c}) \equiv \min\left\{1, \left(\frac{q(\mathbf{x}|\mathbf{c})}{q(\mathbf{c}|\mathbf{x})}\right) \exp[-(V(\mathbf{c}) - V(\mathbf{x}))]\right\}. \tag{11.18}$$

- **Step 5: Estimate expectation.** Let $\phi : \mathcal{R}^d \rightarrow \mathcal{R}^k$. Compute:

$$\bar{\phi}_L(t) \equiv \frac{1}{L} \sum_{i=0}^{L-1} \phi(\tilde{\mathbf{x}}(t-i)).$$

- **Step 6: Check for convergence.** Let

$$\hat{\sigma}_L^2(t) \equiv \frac{1}{kL} \sum_{i=0}^{L-1} \left|\bar{\phi}_L(t-i) - \hat{\boldsymbol{\mu}}_L(t-i)\right|^2$$

where $\hat{\boldsymbol{\mu}}_L(t) \equiv (1/L)\sum_{i=0}^{L-1} \bar{\phi}_L(t-i)$. Let $\epsilon$ be a small positive number specifying precision of estimation. If the MCMC simulation error $\hat{\sigma}_L(t)$ is less than $\epsilon$, then return $\bar{\phi}_L(t)$ as an estimator for

$$E\{\tilde{\mathbf{x}}\} = \sum_{\mathbf{x}\in\Omega} \phi(\mathbf{x})p(\mathbf{x});$$

otherwise let $t = t + 1$ and Go To Step 3.

**Algorithm 11.2.1** A Proposal Distribution Sampling Strategy

```
1: procedure PROPOSALSAMPLING( q(·|x), C ≡ {c^1,...,c^M} )
2:     Let y ⇐ RANDOM(0,1).
3:     Let c* be a global maximizer of q(·|x) on C.
4:     Let L_1 ⇐ q(c*|x).
5:     if y ≤ L_1 then return c*
6:     end if
7:     Remove c* from the set C.
8:     for k = 2 to M do
9:         Let c* be a global maximizer of q(·|x) on C.
10:        Let L_k ⇐ L_{k-1} + q(c*|x).
11:        if L_{k-1} < y ≤ L_k then return c*
12:        end if
13:        Remove c* from the set C.
14:    end for
15: end procedure
```

In some cases, the conditional probability mass function $q(\cdot|\mathbf{x})$ will have a particular functional form such as a Gibbs pmf that requires computation of a computationally intensive normalization constant $Z$ for a given $\mathbf{x}$. For example, define an MH proposal distribution $q : \Omega \times \Omega \rightarrow (0,1]$ such that for all $\mathbf{c} \in \Omega$:

$$q(\mathbf{c}|\mathbf{x}) = (1/Z)\exp\left(-U(\mathbf{c},\mathbf{x})\right), Z = \sum_{\mathbf{a}\in\Omega} \exp[-U(\mathbf{a},\mathbf{x})].$$

In such cases, the normalization constant $Z$ should be computed only once each time the Algorithm (11.2.1) is called. In addition, savings in computation can be further realized by using either limited precision or lookup-table strategies for implementing exponentiation operations.

## 11.2.2 Convergence Analysis of Metropolis-Hastings Algorithms

**Theorem 11.2.1** (Metropolis-Hastings Convergence Theorem). *Let $\Omega \equiv \{\mathbf{x}^1,\ldots,\mathbf{x}^M\}$ be a finite subset of $\mathcal{R}^d$. Let $\mathcal{A}$ be a Metropolis-Hastings algorithm defined with respect to MH proposal specification $q : \Omega \times \Omega \rightarrow [0,1]$ and a Gibbs pmf $p : \Omega \rightarrow [0,1]$ with energy function $V : \Omega \rightarrow \mathcal{R}$. Assume the Metropolis-Hastings algorithm generates an aperiodic and irreducible Markov chain of d-dimensional random vectors $\tilde{\mathbf{x}}(1), \tilde{\mathbf{x}}(2), \ldots$ such that the M-dimensional probability vector $\mathbf{p}(t)$ specifies the distribution of $\tilde{\mathbf{x}}(t)$. Let the M-dimensional probability vector $\mathbf{p}^*$ be defined such that the kth element of $\mathbf{p}^*$ is $p(\mathbf{x}^k)$ for $k = 1,\ldots,M$.*

*Then, $\tilde{\mathbf{x}}(t)$ converges to a random vector $\tilde{\mathbf{x}}^*$ in distribution as $t \rightarrow \infty$ such that $|\mathbf{p}(t) - \mathbf{p}^*| = O(|\lambda_2|^t)$ where $|\lambda_2|$ is the magnitude of the second largest eigenvalue of the transition matrix $\mathbf{M}$ of the Markov chain.*

*Proof.* Let $m_{jk}$ denote the probability that the MH Markov chain visits state $\mathbf{x}^k$ given the current state of the MH Markov chain is $\mathbf{x}^j$ for $j,k = 1,\ldots,M$.

The MH algorithm in Definition 11.2.2 can then be equivalently specified by defining

$$m_{jk} = q(\mathbf{x}^k|\mathbf{x}^j)\min\left\{1, \rho(\mathbf{x}^j,\mathbf{x}^k)\right\} \tag{11.19}$$

where

$$\rho(\mathbf{x}^j,\mathbf{x}^k) = \frac{p(\mathbf{x}^k)q(\mathbf{x}^j|\mathbf{x}^k)}{p(\mathbf{x}^j)q(\mathbf{x}^k|\mathbf{x}^j)}.$$

Using the definition of $m_{jk}$,

$$p(\mathbf{x}^j)m_{jk} = p(\mathbf{x}^j)q(\mathbf{x}^k|\mathbf{x}^j)\min\left\{1, \frac{p(\mathbf{x}^k)q(\mathbf{x}^j|\mathbf{x}^k)}{p(\mathbf{x}^j)q(\mathbf{x}^k|\mathbf{x}^j)}\right\}.$$

$$p(\mathbf{x}^j)m_{jk} = \min\left\{p(\mathbf{x}^j)q(\mathbf{x}^k|\mathbf{x}^j), p(\mathbf{x}^k)q(\mathbf{x}^j|\mathbf{x}^k)\right\}.$$

$$p(\mathbf{x}^j)m_{jk} = p(\mathbf{x}^k)q(\mathbf{x}^j|\mathbf{x}^k)\min\left\{\frac{p(\mathbf{x}^j)q(\mathbf{x}^k|\mathbf{x}^j)}{p(\mathbf{x}^k)q(\mathbf{x}^j|\mathbf{x}^k)}, 1\right\} = p(\mathbf{x}^k)m_{kj}.$$

Thus,

$$p(\mathbf{x}^j)m_{jk} = p(\mathbf{x}^k)m_{kj}. \tag{11.20}$$

Summing over both sides of (11.20) gives:

$$\sum_{j=1}^{M} p(\mathbf{x}^j)m_{jk} = \sum_{j=1}^{M} p(\mathbf{x}^k)m_{kj} = p(\mathbf{x}^k)\sum_{j=1}^{M} m_{kj} = p(\mathbf{x}^k)$$

which is equivalent to

$$(\mathbf{p}^*)^T\mathbf{M} = (\mathbf{p}^*)^T \tag{11.21}$$

where the $k$th element of $\mathbf{p}^*$ is equal to $p(\mathbf{x}^k)$ for $k = 1, \ldots, M$.

Therefore, since the Markov chain is aperiodic and irreducible and $\mathbf{p}^*$ is a stationary distribution for the chain by (11.21), it follows from the MCMC Convergence Theorem (Theorem 11.1.1) that the theorem's conclusion holds. ■

### 11.2.3 Important Special Cases of the Metropolis-Hastings Algorithm

In this section, a variety of important special cases of the Metropolis-Hastings algorithm are presented. Note that these may be combined in various ways to produce an even greater variety of hybrid MCMC-MH algorithms!

**Definition 11.2.3** (Independence Sampler). Let $q : \Omega \times \Omega \rightarrow [0, 1]$ be an MH proposal specification for a Metropolis-Hastings algorithm $\mathcal{A}$. If there exists a pmf $f : \Omega \rightarrow [0, 1]$ such that $q(\mathbf{c}|\mathbf{x}) = f(\mathbf{c})$ for all $\mathbf{x}, \mathbf{c} \in \Omega$, then $\mathcal{A}$ is called a *Metropolis-Hastings Independence Sampler Algorithm.* □

The MH Independence Sampler starts with the current state $\mathbf{x}$ and then selecting another state $\mathbf{c}$ in the state space $\Omega$ at random with probability $f(\mathbf{c})$ where $f$ is defined as in Definition 11.2.3. The probability of selecting the candidate state $\mathbf{c}$ is not functionally dependent upon $\mathbf{x}$. One can interpret the MH Independence Sampler as a type of random search algorithm that doesn't necessarily visit each state in the finite state space with equal probability.

**Definition 11.2.4** (Metropolis Algorithm). Let $q : \Omega \times \Omega \rightarrow [0, 1]$ be an MH proposal specification for a Metropolis-Hastings algorithm $\mathcal{A}$. If $q(\mathbf{c}|\mathbf{x}) = q(\mathbf{x}|\mathbf{c})$ for all $\mathbf{x}, \mathbf{c} \in \Omega$, then $\mathcal{A}$ is called a *Metropolis Algorithm.* □

The Metropolis Algorithm generates a candidate state $\mathbf{c}$ at random with some probability that is functionally dependent upon the current state $\mathbf{x}$. A critical requirement of the Metropolis Algorithm is that the probability that a candidate state $\mathbf{c}$ is visited, given the Markov chain's current state is $\mathbf{x}$, must be exactly equal to the probability that the candidate state $\mathbf{x}$ is visited given the current state is $\mathbf{c}$. Because $q(\mathbf{x}|\mathbf{c}) = q(\mathbf{c}|\mathbf{x})$ for the Metropolis algorithm, the candidate state $\mathbf{c}$ is always accepted if $V(\mathbf{c}) \leq V(\mathbf{x})$. If $V(\mathbf{c}) > V(\mathbf{x})$, then the candidate state $\mathbf{c}$ is accepted with probability $\exp(-V(\mathbf{c}) + V(\mathbf{x}))$ and otherwise rejected.

**Example 11.2.1** (Bit Flipping Metropolis Proposal Distribution). Let the finite state space $\Omega \equiv \{0,1\}^d$. Let $\mathbf{x} = [x_1, \ldots, x_d]^T$. Let $0 < \alpha < 1$. Assume that the probability of modifying the $k$th element of $\mathbf{x}$, $x_k$, to take on the value $1 - x_k$ is equal to $\alpha$ and the probability of not modifying $x_k$ is equal to $1 - \alpha$. Let $\delta_{c,x} = 1$ if $c = x$ and let $\delta_{c,x} = 0$ if $c \neq x$. Define a Metropolis proposal distribution $q$ such that:

$$q(\mathbf{c}|\mathbf{x}) = \prod_{k=1}^{d} \alpha^{1-\delta_{c_k,x_k}} (1-\alpha)^{\delta_{c_k,x_k}}.$$

△

**Definition 11.2.5** (Random Scan Gibbs Sampler). Let $\Omega$ be a finite subset of $\mathcal{R}^d$. Let $\tilde{\mathbf{x}} \equiv [\tilde{x}_1, \ldots, \tilde{x}_d]^T$ be a $d$-dimensional random vector whose probability distribution is specified by a Gibbs pmf $p : \Omega \to [0,1]$ with energy function $V : \Omega \to \mathcal{R}$. Let $q : \Omega \times \Omega \to [0,1]$ be a proposal distribution for a Metropolis-Hastings algorithm $\mathcal{A}$ with stationary distribution $p$ specified by $V$. Let $r : \{1, \ldots, d\} \to (0,1]$ be a positive-valued probability mass function. Let $\mathbf{c}^k = [x_1, \ldots, x_{k-1}, c_k, x_{k+1}, \ldots, x_d]$. Define the proposal distribution $q(\mathbf{c}^k|\mathbf{x}) = r(k)p(c_k|\mathcal{N}_k(\mathbf{x}))$ where the local conditional probability

$$p\left(c_k|\mathcal{N}_k(\mathbf{x})\right) = \exp\left(-V(\mathbf{c}^k) + V(\mathbf{x})\right) / Z(\mathcal{N}_k(\mathbf{x})), \tag{11.22}$$

where $\mathcal{N}_k(\mathbf{x})$ is a realization of the neighbors of $x_k$. The algorithm $\mathcal{A}$ is called a *Random Scan Gibbs Sampler.* □

The random scan Gibbs sampler works by selecting the $k$th element of the current state vector $\mathbf{x}$ with probability $r(k)$, and then changing the value of the $k$th element of $\mathbf{x}$, $x_k$, to $c_k$, with probability $p(c_k|\mathcal{N}_k(\mathbf{x}))$ where $p(\mathbf{x})$ is the desired stationary probability distribution.

To show that the random scan Gibbs sampler is a special case of the Metropolis-Hastings algorithm, note that since $\mathcal{N}_k(\mathbf{x}) = \mathcal{N}_k(\mathbf{c}^k)$:

$$\rho(\mathbf{x}, \mathbf{c}^k) = \left(\frac{q(\mathbf{x}|\mathbf{c}^k)}{q(\mathbf{c}^k|\mathbf{x})}\right) \exp[-V(\mathbf{c}^k) + V(\mathbf{x})] = \left(\frac{r(k)p(x_k|\mathcal{N}_k(\mathbf{x}))}{r(k)p(c_k|\mathcal{N}_k(\mathbf{c}^k))}\right) \exp[-V(\mathbf{c}^k) + V(\mathbf{x})]$$

$$= \frac{p(\mathbf{x})/p(\mathcal{N}_k(\mathbf{x}))}{p(\mathbf{c}^k)/p(\mathcal{N}_k(\mathbf{x}))} = \exp[-V(\mathbf{x}) + V(\mathbf{c}^k)] \exp[-V(\mathbf{c}^k) + V(\mathbf{x})] = 1.$$

Therefore, the probability that $\mathbf{x}(t+1) = \mathbf{c}^k$ is equal to the proposal probability for the random scan Gibbs sampler.

Note that, as noted in Recipe Box 10.1, the computation of the quantity $-V(\mathbf{c}^k) + V(\mathbf{x})$ can be simplified if $V$ can be expressed as the sum of local potential functions. In such cases, the local potential functions which are not functionally dependent on $x_k$ can be omitted from the summation without consequence.

**Definition 11.2.6** (Deterministic Scan Gibbs Sampler). Let $\Omega$ be a finite subset of $\mathcal{R}^d$. Let $\tilde{\mathbf{x}} \equiv [\tilde{x}_1, \ldots, \tilde{x}_d]^T$ be a $d$-dimensional random vector whose probability distribution is specified by a Gibbs pmf $p : \Omega \to [0,1]$ with energy function $V : \Omega \to \mathcal{R}$. Let $q : \Omega \times \Omega \to [0,1]$ be a proposal distribution for a Metropolis-Hastings algorithm $\mathcal{A}$ with stationary distribution $p$ specified by $V$. If $q(\mathbf{c}|\mathbf{x}) = \prod_{k=1}^{d} p\left(c_k|\mathcal{N}_k(\mathbf{x})\right)$ then $\mathcal{A}$ is called a *Deterministic Scan Gibbs Sampler.* □

The deterministic scan Gibbs sampler is almost exactly the same as the random scan Gibbs sampler. However, rather than randomly sampling the $d$ random variable in the Markov random field specified by the Gibbs pmf, the $d$ random variables are updated in some deterministic order and then this deterministic sequence of updates is repeated. The deterministic scan Gibbs sampler is not technically a Metropolis-Hastings algorithm but rather a composite hybrid MCMC whose component Markov chains are Metropolis-Hastings Markov chains defined by Gibbs sampler transitions of the form of (11.22).

**Definition 11.2.7** (Random Scan Block Gibbs Sampler). Let the $d$-dimensional random vector $\tilde{\mathbf{x}} \equiv [\tilde{\mathbf{x}}_1, \ldots, \tilde{\mathbf{x}}_K]^T$ where $\tilde{\mathbf{x}}_k$ is a $d/K$-dimensional random subvector for $k = 1, \ldots, K$. Let the probability distribution of $\tilde{\mathbf{x}}$ be a Gibbs pmf $p : \Omega \to [0, 1]$ with energy function $V : \Omega \to \mathcal{R}$. Let $q : \Omega \times \Omega \to [0, 1]$ be a proposal distribution for a Metropolis-Hastings algorithm $\mathcal{A}$ with stationary distribution $p$ specified by $V$. Let $r : \{1, \ldots, K\} \to (0, 1]$ be a positive-valued probability mass function. If $q(\mathbf{c}^k|\mathbf{x}) = r(k)p(\mathbf{c}_k|\mathcal{N}_k(\mathbf{x}))$ where the local conditional probability

$$p(\mathbf{c}_k|\mathcal{N}_k(\mathbf{x})) = \exp\left(-V(\mathbf{c}^k) + V(\mathbf{x})\right)/Z(\mathcal{N}_k(\mathbf{x})), \tag{11.23}$$

and $\mathbf{c}^k = [\mathbf{x}_1, \ldots, \mathbf{x}_{k-1}, \mathbf{c}_k, \mathbf{x}_{k+1}, \ldots, \mathbf{x}_K]$, then $\mathcal{A}$ is called a *Random Scan Block Gibbs Sampler.* □

The concept of a block sampler where the blocks have the same or different sizes can be used to generate useful variants of many types of Metropolis-Hastings algorithms. The key advantage of the block sampler method is that updating multiple state variables at each iteration has the potential to greatly improve the quality of each update. The key disadvantage of block methods is they require more computation per step. Proper selection of the block size can make block sampler methods very effective.

### 11.2.4 Metropolis-Hastings Machine Learning Applications

**Example 11.2.2** (MCMC Model Search Methods). Usually the best generalization performance for the learning machine is achieved using only a subset of the $d$ variables in the input pattern vector $\mathbf{s}$. The reason for the improvement in generalization performance is typically that some of the input variables in the $d$-dimensional input pattern vector $\mathbf{s}$ are either irrelevant or highly correlated. By eliminating irrelevant and redundant input variables, the number of free parameters in the model can often be reduced resulting in a reduction of the effects of overfitting and improved generalization performance (see Chapter 14).

Even if $d$ is only moderately large (e.g., $d = 100$), the number of possible subsets of input variables, $2^d - 1$, is astronomically large (e.g., $2^{100} \approx 10^{30}$). Let $\mathbf{m} \in \{0, 1\}^d$ be a mask vector defined such that the $k$th element of $\mathbf{m}$, $m_k$, is equal to one if the $k$th element of $\mathbf{s}$ is included as an input variable in the regression model. Set the $k$th element of $\mathbf{m}$, $m_k$, equal to zero if the $k$th element of $\mathbf{s}$ should not be included as an input variable in the model.

Let $\mathbf{x}(t) \equiv [\mathbf{s}(t), y(t)]$ where $y(t)$ is the desired response of the learning machine to input pattern $\mathbf{s}(t)$ for $t = 1, \ldots, n$. Let $c$ be a loss function which specifies the incurred loss $c(\mathbf{x}; \boldsymbol{\theta}, \mathbf{m})$ received by a learning machine with knowledge state parameter vector $\boldsymbol{\theta}$ when it experiences environmental event $\mathbf{x}$ using the input variables specified by $\mathbf{m}$. Let the empirical risk function $\hat{\ell}_n : \mathcal{R}^q \to \mathcal{R}$ be defined such that for all $\mathbf{m}$:

$$\hat{\ell}_n(\boldsymbol{\theta}; \mathbf{m}) = (1/n) \sum_{i=1}^{n} c(\tilde{\mathbf{x}}(i); \boldsymbol{\theta}, \mathbf{m})$$

and let $\hat{\boldsymbol{\theta}}_n(\mathbf{m})$ be a global minimizer of $\hat{\ell}_n(\cdot; \mathbf{m})$ on parameter space $\Theta$ which is a closed, bounded, convex subset of $\mathcal{R}^q$.

Using the methods of Section 16.1, the expected prediction error of the learning machine on a novel test data set can be estimated using the function $V : \{0,1\}^d \rightarrow \mathcal{R}$ defined such that:

$$V(\mathbf{m}) = \hat{\ell}_n(\hat{\boldsymbol{\theta}}_n(\mathbf{m}); \mathbf{m}) + (1/n)\,\mathrm{tr}\left(\hat{\mathbf{A}}_n^{-1}(\mathbf{m})\hat{\mathbf{B}}_n(\mathbf{m})\right) \tag{11.24}$$

where $\hat{\mathbf{A}}_n(\mathbf{m})$ is the Hessian of $\hat{\ell}_n(\cdot; \mathbf{m})$ evaluated at $\hat{\boldsymbol{\theta}}_n(\mathbf{m})$ and

$$\hat{\mathbf{B}}_n(\mathbf{m}) \equiv (1/n) \sum_{i=1}^{n} \hat{\mathbf{g}}(\tilde{\mathbf{x}}(i); \mathbf{m})[\hat{\mathbf{g}}(\tilde{\mathbf{x}}(i); \mathbf{m})]^T$$

where $\hat{\mathbf{g}}(\mathbf{x}; \mathbf{m}) \equiv (dc(\mathbf{x}; \boldsymbol{\theta}, \mathbf{m})/d\boldsymbol{\theta})^T$ evaluated at $\hat{\boldsymbol{\theta}}_n(\mathbf{m})$.

In order to find good choices for $\mathbf{m}$, one can construct an MCMC whose stationary distribution is a Gibbs pmf with energy function $V$ as defined in (11.24). Then, one samples from the MCMC to generate a stochastic sequence of states $\tilde{\mathbf{m}}(0), \tilde{\mathbf{m}}(1), \ldots$. The realizations of states which are observed most frequently will have smaller values of $V$ so the algorithm involves simply keeping track of the values of $V(\tilde{\mathbf{m}}(t))$ which are smallest in a list. △

**Example 11.2.3** (MCMC Model Averaging Methods). In Example 11.2.2, the key idea was that one has a large collection of models where each model has some measure of predictive performance denoted by $V(\mathbf{m})$ where $\mathbf{m} \in \{0,1\}^d$ is a binary vector whose $k$th element is equal to one if and only if the $k$th input variable is included in the model. The objective of Example 11.2.2 was to select one model out of this large collection of models for the purpose of inference and learning. In the real world, however, typically there will be many models (possibly hundreds or thousands) which exhibit excellent generalization performance in a model space consisting of $2^d - 1$ models where $d$ is moderately large (e.g., $d = 100$). Furthermore, some of these models may be great for some inference tasks but not others.

To address this difficulty, an alternative approach for improving predictive performance is not to select one out of many models for the purposes of prediction, but rather use all of the models for prediction. This may be accomplished by computing a weighted average of the predictions of all of the models in such a way that the predictions of more predictive models have greater weights than less predictive models.

Let $\ddot{y}(\mathbf{s}; \boldsymbol{\theta}, \mathbf{m})$ denote the predicted response of model $\mathbf{m}$ to input pattern $\mathbf{s}$ given parameter vector $\boldsymbol{\theta}$. Let $p_{\mathcal{M}}$ denote a probability mass function that assigns a probability $p_{\mathcal{M}}(\mathbf{m})$ to the model specified by $\mathbf{m}$. Let $\hat{\boldsymbol{\theta}}_n(\mathbf{m})$ be the parameter estimates for model $\mathbf{m}$. Bayesian Model Averaging (BMA) methods (see Example 12.2.5 and Exercise 16.2-6) use the formula

$$\bar{y}(\mathbf{s}) = \sum_{\mathbf{m} \in \{0,1\}^d} \ddot{y}(\mathbf{s}; \hat{\boldsymbol{\theta}}_n(\mathbf{m}), \mathbf{m}) p_{\mathcal{M}}(\mathbf{m}) \tag{11.25}$$

to generate an aggregate predicted response $\bar{y}(\mathbf{s})$ for a given input pattern vector $\mathbf{s}$ and all of the models in the model space.

The formula in (11.25), in practice, is computationally challenging to evaluate because when $d$ is even moderately large the summation in (11.25) is computationally intractable to evaluate. However, the expectation in (11.25) can be approximated by constructing a Monte Carlo Markov Chain whose stationary pmf is a Gibbs pmf $p_{\mathcal{M}}$ specified by the energy function $V(\mathbf{m})$. For example, for $T$ sufficiently large:

$$\bar{y}(\mathbf{s}) = \sum_{\mathbf{m} \in \{0,1\}^d} \ddot{y}(\mathbf{s}; \hat{\boldsymbol{\theta}}_n(\mathbf{m}), \mathbf{m}) p_{\mathcal{M}}(\mathbf{m}) \approx (1/T) \sum_{t=1}^{T} \ddot{y}(\mathbf{s}; \hat{\boldsymbol{\theta}}_n(\tilde{\mathbf{m}}(t)), \tilde{\mathbf{m}}(t))$$

where the stochastic sequence $\tilde{\mathbf{m}}(0), \tilde{\mathbf{m}}(1), \ldots$ is generated by an MCMC whose stationary pmf is a Gibbs pmf $p_{\mathcal{M}}$ with the energy function $V(\mathbf{m})$ in (11.24). △

**Example 11.2.4** (Probabilistic Inference in Markov Random Fields). Let $\tilde{\mathbf{x}}$ be a $d$-dimensional random vector whose $k$th element, $x_k$, is defined such that $x_k = 1$ means proposition $k$ holds and $x_k = 0$ means proposition $k$ does not hold. Assume that the probability distribution of $\tilde{\mathbf{x}}$ satisfies the positivity condition and is therefore a Markov random field. An important application of MCMC methods is to perform approximate inferences in Markov random fields in order to compute the probability that one proposition holds given knowledge that other propositions hold or do not hold. An example of this type of problem was discussed in the context of Markov logic nets (see Example 1.6.6). Solutions to this type of inference problem in Markov random fields are also relevant to inference in Bayesian knowledge networks since a Bayesian knowledge network that satisfies the positivity condition can be equivalently represented as a Markov random field.

Consider the MRF knowledge representation in Figure 10.7 where $p(x_1, x_2, x_3, x_4, x_5, x_6)$ is an MRF and thus can be represented as a Gibbs pmf. The probabilistic knowledge representation is queried by constructing queries such as:

What is the probability that $x_4$ : `LUNG CANCER` holds given $x_2$ : `SMOKING` holds?

Such a query is equivalent to evaluating the conditional probability

$$p(x_4 = 1 | x_2 = 1) = \frac{\sum_{x_1, x_3, x_5, x_6} p(x_1, x_2 = 1, x_3, x_4 = 1, x_5, x_6)}{\sum_{x_1, x_3, x_4, x_5, x_6} p(x_1, x_2 = 1, x_3, x_4, x_5, x_6)}. \tag{11.26}$$

In this example, the summations are easy to compute but in practical applications when hundreds or thousands of random variables in the MRF are present, then approximate inference methods based upon MCMC sampling are often considered. The summations in (11.26) can be evaluated by computing appropriate expectations by sampling from $p(\mathbf{x})$ where $\mathbf{x} = [x_1, \ldots, x_6]$. In particular, let $\phi_{4,2}(\mathbf{x}) = 1$ if $x_4 = 1$ and $x_2 = 1$ and let $\phi_{4,2}(\mathbf{x}) = 0$ otherwise. The summation in the numerator of (11.26) can then be approximately computed using the formula:

$$p(x_2, x_4) = \sum_{x_1, x_3, x_5, x_6} p(x_1, \ldots, x_6) \approx (1/T) \sum_{t=1}^{T} \phi_{4,2}(\tilde{\mathbf{x}}(t))$$

where $T$ is a sufficiently large number and $\tilde{\mathbf{x}}(0), \tilde{\mathbf{x}}(1), \ldots$ are samples from an MCMC which is designed so that its stationary distribution is $p(\mathbf{x})$. △

**Example 11.2.5** (Implementation of Genetic Algorithms as MCMC Algorithms). For many nonlinear optimization problems, selecting the correct network architecture involves many decisions which include not only the number of input units (e.g., Example 11.2.2) but many other decisions such as the types of hidden units, the number of hidden unit layers, the nonlinear optimization strategy, the particular choice of initial connections, the learning rate adjustment scheme, and the choice of regularization terms One can develop a *coding scheme* where all such decisions are encoded as a binary vector $\boldsymbol{\zeta}_k \in \{0,1\}^d$ for the $k$th individual in a population of $M$ individuals. The binary vector $\boldsymbol{\zeta}_k$ is an abstract model of the chromosome for the $k$th individual where each element of $\boldsymbol{\zeta}_k$ is interpreted as an abstract model of a gene. The population genotype vector $\boldsymbol{\zeta} \equiv [\boldsymbol{\zeta}_1, \ldots, \boldsymbol{\zeta}_M]^T \in \{0,1\}^{dM}$.

Now define the population fitness function $V : \{0,1\}^{dM} \rightarrow \mathcal{R}$ which specifies the fitness of the population of $M$ learning machines for a given choice of $\boldsymbol{\zeta}$. For example, for a given training data sample $\mathbf{x}_1, \ldots, \mathbf{x}_n$ generated from the statistical environment, let the prediction error of the $k$th individual in the population be denoted as $\hat{\ell}_n^k$. Then, the population fitness function $V$ is defined such that $V(\boldsymbol{\zeta}) = (1/M) \sum_{k=1}^{M} \hat{\ell}_n^k$ for a particular population of $M$ learning machines specified by $\boldsymbol{\zeta}$ (see Example 11.2.2).

The objective in the design of a genetic algorithm is to progressively transform a population of learning machines to generate a sequence of population genotype vectors $\boldsymbol{\zeta}(0), \boldsymbol{\zeta}(1)), \ldots$. This can be accomplished by taking the current population genotype $\boldsymbol{\zeta}(t)$ and applying random bit flipping *mutation operators* (see Example 11.2.1) to transform $\boldsymbol{\zeta}(t)$ into $\boldsymbol{\zeta}(t+1)$. In addition, *crossover operators* may be defined according to the following algorithm. First, select two individuals (parents) in the population corresponding to the subvectors $\boldsymbol{z}_k$ and $\boldsymbol{z}_j$ where $k \neq j$. Second, pick an integer $i$ at random from $\{2, \ldots, d-1\}$ and randomly replace the one of the "parent" subvectors $\boldsymbol{z}_k$ and $\boldsymbol{z}_j$ with a new (child) "subvector" $\boldsymbol{z}_u$ whose first $i$ elements correspond to the first $i$ elements of $\boldsymbol{z}_k$ and whose remaining elements correspond to the $i+1$st through $d$th elements of $\boldsymbol{z}_j$.

Assume that Nature has some strategy for either: (i) eliminating the entire new population of learning machines which is equivalent to discarding $\boldsymbol{\zeta}(t+1)$ if it doesn't like the new population genotype and keeping the old population genotype $\boldsymbol{\zeta}(t)$, or (ii) eliminating the entire old population resulting in discarding $\boldsymbol{\zeta}(t)$ if it doesn't like the old population genotype and keeping the new population genotype $\boldsymbol{\zeta}(t+1)$.

Show that the mutation and crossover operators specify the proposal distribution of an MCMC algorithm. Design a strategy for "Nature" to either kill the new or old population genotype generated from the proposal distribution such that the sequence $\boldsymbol{\zeta}(0), \boldsymbol{\zeta}(1)), \ldots$ can be viewed as a realization of a Metropolis-Hastings Monte-Carlo Markov chain. Then, apply the Metropolis-Hastings convergence theorem to conclude that the resulting genetic algorithm will frequently generate populations with good population fitness scores. △

## Exercises

11.2-1. *Deterministic Scan Gibbs Sampler Convergence Analysis.* Show the Deterministic Scan Gibbs Sampler is a hybrid MCMC-MH algorithm.

11.2-2. *Random Block Gibbs Convergence Analysis.* Show the Random Block Gibbs Sampler is an MH algorithm.

11.2-3. *MCMC Approximation of Gibbs Density Normalization Constant.* Show how to use an MCMC algorithm to approximate the expectation in the denominator of (11.26) in Example 11.2.4.

11.2-4. Consider a Gibbs Sampler algorithm for sampling from the joint probability mass function $p : \{0,1\}^d \rightarrow (0,1)$ is defined such that for all $\mathbf{x} \in \{0,1\}^d$:

$$p(\mathbf{x}) = Z^{-1} \exp[-V(\mathbf{x})]$$

where

$$V(\mathbf{x}) = \sum_{i=1}^{d} x_i b_i + \sum_{i=3}^{d} \sum_{j=2}^{i-1} \sum_{k=1}^{j-1} x_i x_j x_k b_{ijk}$$

and $Z$ is a normalization constant which guarantees that $\sum_{\mathbf{x}} p(\mathbf{x}) = 1$. Obtain an explicit formula for $p(x_i(t+1) = 1 | \mathcal{N}_i(\mathbf{x}(t)))$.

11.2-5. *MCMC Algorithm for Image Texture Generation.* Design an MCMC algorithm that generates samples of image texture by sampling from an MRF texture model as defined in Example 1.6.3.

11.2-6. *MCMC Algorithm for Markov Logic Net Inference.* Use the results of Exercise 10.4.6 to derive an MCMC sampling algorithm for a Markov logic net. Explain how to use the sampling algorithm to approximate the probability that one logical formula holds given other logical formulas in the field hold.

11.2-7. *MCMC Algorithms for Solving Clustering Problems.* Design an MCMC algorithm that solves clustering problems of the type described in Example 1.6.4 by using the probabilistic interpretation developed in Exercise 10.4-8.

## 11.3 Further Readings

### Historical Origins of MCMC Methods

The essential idea of the Metropolis-Hastings algorithm can be traced back to the seminal work of Metropolis et al. (1953) and Hastings (1970). In the mid-1980s, Geman and Geman (1984), Ackley, Hinton, and Sejnowski (1985), Besag (1986), Smolensky (1986), Cohen and Cooper (1987), and others demonstrated how these techniques could be applied to a variety of inference and image processing problems. A useful introductory review of Markov chain Monte Carlo methods which includes not only a useful historical perspective but specific applications to Bayesian data analysis in astronomy can be found in Sharma (2017).

### Probabilistic Programming Languages

Currently, "probabilistic programming languages" have been developed that allow users to specify customized probabilistic models and then use Monte Carlo Markov Chain methods to sample from those models and generate inferences (e.g., Davidson-Pilon 2015; Pfeffer 2016).

### Simulated Annealing for Finding Global Optimizers

Simulated annealing methods where $\mathcal{T}$ is decreased at an appropriate logarithmic rate as the stochastic sequence evolves in time have established convergence in distribution to a uniform pmf over the global minimizers of $V$. However, the logarithmic convergence rate of simulated annealing algorithms is much slower than the geometric convergence rate of MH sampling algorithms. Winkler (2003) provides an accessible comprehensive theoretical discussion of simulated annealing supported by some empirical examples.

### Additional Introductions to MCMC Methods

Winkler (2003) provides a general discussion of asymptotic behavior of Monte Carlo Markov Chains as well as an overall discussion of the Gibbs Sampler, Metropolis, and the more general Metropolis-Hastings algorithm from both a theoretical and empirical perspective. Gamerman and Lopes (2006) provides an introduction to MCMC methods with a greater focus on computational examples. For a more comprehensive introduction to MCMC theory, the book by Robert and Casella (2004) is also highly recommended.

### Extensions of MCMC Methods to Infinite State Spaces

In this chapter, the MCMC sampling methods described are restricted to finite state spaces. However, a very important application of MCMC sampling is to support mixed effects linear and nonlinear regression modeling as well as Bayesian models. For such applications, the state space is not finite and the goal is to sample from a Gibbs density which may not be a probability mass function. Although the extension of the MCMC theory to the more general case involving continuous and mixed random vectors is slightly more complicated, it has important similarities to the finite state space case described in this chapter (see Robert and Casella 2004).

# 12

# *Adaptive Learning Algorithm Convergence*

**Learning Objectives**

- Design adaptive learning algorithms.
- Design adaptive reinforcement learning algorithms.
- Design adaptive expectation-maximization learning algorithms.
- Design Markov field learning algorithms.

Let $\ell : \mathcal{R}^q \rightarrow \mathcal{R}$ be an objective function for learning. Chapter 7 analyzed the behavior of deterministic batch learning algorithms for minimizing $\ell$, which update the current parameter vector $\boldsymbol{\theta}(t)$, to obtain a revised parameter vector $\boldsymbol{\theta}(t+1)$, using the iterated map:

$$\boldsymbol{\theta}(t+1) = \boldsymbol{\theta}(t) + \gamma_t \mathbf{d}_t(\boldsymbol{\theta}(t))$$

where the stepsize $\gamma_t$ and the search direction function $\mathbf{d}_t$ is chosen such that

$$\ell(\boldsymbol{\theta}(t+1)) \leq \ell(\boldsymbol{\theta}(t)). \tag{12.1}$$

This batch learning algorithm uses all of the training data each time the parameters of the learning machine are updated.

Adaptive learning algorithms, on the other hand, update the parameter values each time a new feature vector is generated from the environment. In order to develop a general theory of adaptive learning, this chapter considers the analysis of stochastic discrete-time dynamical systems of the form

$$\tilde{\boldsymbol{\theta}}(t+1) = \tilde{\boldsymbol{\theta}}(t) + \gamma_t \tilde{\mathbf{d}}_t(\tilde{\boldsymbol{\theta}}(t)) \tag{12.2}$$

where $\gamma_t$ is called the *stepsize* at iteration $t$ and the *random search direction function* $\tilde{\mathbf{d}}_t$ is chosen so that

$$E\{\ell(\tilde{\boldsymbol{\theta}}(t+1))|\boldsymbol{\theta}(t)\} \leq \ell(\boldsymbol{\theta}(t)).$$

This latter condition, which only requires that the expected value of the objective function evaluated at the system state $\tilde{\boldsymbol{\theta}}(t+1)$ does not increase, can be interpreted as a weaker version of the deterministic downhill condition in (12.1).

## 12.1 Stochastic Approximation (SA) Theory

### 12.1.1 Passive versus Reactive Statistical Environments

#### 12.1.1.1 Passive Learning Environments

In a passive statistical learning environment (see Section 1.7), the probability distribution that generates the environmental inputs to the inference and learning machine is not influenced by the machine's behavior or internal state. For example, consider a learning machine which learns by watching examples of how an expert pilot lands a helicopter. The learning machine can be trained on many examples of helicopter landings before it actually controls the landing of a helicopter. An advantage to this type of learning procedure is that the learning machine can avoid real (or simulated) helicopter crashes during the learning process.

In a passive statistical learning environment, the observed data $\mathbf{x}(1), \mathbf{x}(2), \ldots$ is assumed to be a realization of a sequence of independent and identically distributed $d$-dimensional random vectors $\tilde{\mathbf{x}}(1), \tilde{\mathbf{x}}(2), \ldots$ with common Radon-Nikodým density $p_e : \mathcal{R}^d \rightarrow [0, \infty)$. Let $c(\mathbf{x}, \boldsymbol{\theta})$ be the loss incurred by the learning machine for observing event $\mathbf{x}$ given $q$-dimensional knowledge parameter vector $\boldsymbol{\theta}$ for all $\boldsymbol{\theta}$ in the parameter space $\Theta$. The risk function, $\ell$, minimized by a stochastic approximation adaptive learning machine operating in a passive statistical environment is explicitly given by the formula:

$$\ell(\boldsymbol{\theta}) = \int c(\mathbf{x}, \boldsymbol{\theta}) p_e(\mathbf{x}) d\nu(\mathbf{x}). \tag{12.3}$$

#### 12.1.1.2 Reactive Learning Environments

In contrast, in a reactive statistical learning environment, the learning machine is constantly interacting with the environment and the statistical characteristics of the environment will change as a direct consequence of the actions of the learning machine. Such situations correspond to the case where the learning machine learns while it is actually landing a real (or simulated) helicopter. That is, taking the helicopter into a nose dive will result in a different collection of learning experiences than a more routine helicopter landing.

Let $p_e(\cdot|\cdot) : \mathcal{R}^d \times \mathcal{R}^q \rightarrow [0, \infty)$. Let a learning machine's probabilistic model of its environment be defined as the set of probability densities $\mathcal{M} \equiv \{p_e(\cdot|\boldsymbol{\theta}) : \boldsymbol{\theta} \in \Theta\}$. In a reactive statistical learning environment, the observed data $\mathbf{x}(1), \mathbf{x}(2), \ldots$ is assumed to be a realization of a sequence of independent and identically distributed $d$-dimensional random vectors $\tilde{\mathbf{x}}(1), \tilde{\mathbf{x}}(2), \ldots$ with common Radon-Nikodým density $p_e(\mathbf{x}|\boldsymbol{\theta}) : \mathcal{R}^d \rightarrow [0, \infty)$ when the learning machine's state of knowledge is $\boldsymbol{\theta}$. In this case, the density $p_e(\mathbf{x}|\boldsymbol{\theta})$ specifying the likelihood an event $\mathbf{x}$ is generated by the environment can be altered by the behavior of the learning machine. Since the learning machine's behavior is functionally dependent upon the learning machine's current parameter value $\boldsymbol{\theta}$, this relationship is represented by allowing the environmental density $p_e$ to be conditioned upon the learning machine's parameter vector $\boldsymbol{\theta}$. The statistical environment "reacts" to the learning machine's behavior.

As in a passive learning environment, $c(\mathbf{x}, \boldsymbol{\theta})$ is the loss incurred by the learning machine for observing event $\mathbf{x}$ for a given state of knowledge parameter vector $\boldsymbol{\theta} \in \Theta$. However, the risk function, $\ell$, minimized by a stochastic approximation adaptive learning machine operating in a reactive statistical environment is explicitly given by the formula:

$$\ell(\boldsymbol{\theta}) = \int c(\mathbf{x}, \boldsymbol{\theta}) p_e(\mathbf{x}|\boldsymbol{\theta}) d\nu(\mathbf{x}). \tag{12.4}$$

Note that a learning machine that minimizes the risk function for learning in (12.4), seeks a parameter vector $\boldsymbol{\theta}$ by simultaneously attempting to reduce the magnitude of the incurred loss (i.e., $c(\mathbf{x}, \boldsymbol{\theta})$) as well as the likelihood of the incurred loss for a particular event $\mathbf{x}$ (i.e., $p_e(\mathbf{x}|\boldsymbol{\theta})$).

### 12.1.2 Average Downward Descent

One approach to the analysis of the learning dynamics in Equation (12.2) is to rewrite Equation (12.2) as:

$$\tilde{\boldsymbol{\theta}}(t+1) = \tilde{\boldsymbol{\theta}}(t) + \gamma_t \bar{\mathbf{d}}_t(\tilde{\boldsymbol{\theta}}(t)) + \gamma_t \tilde{\mathbf{n}}_t \tag{12.5}$$

where $\bar{\mathbf{d}}_t(\boldsymbol{\theta}) \equiv E\{\tilde{\mathbf{d}}_t|\boldsymbol{\theta}\}$ is called the *average search direction function* and

$$\tilde{\mathbf{n}}_t \equiv \tilde{\mathbf{d}}_t(\tilde{\boldsymbol{\theta}}(t)) - \bar{\mathbf{d}}_t(\tilde{\boldsymbol{\theta}}(t))$$

is called the *additive noise error*. By rewriting (12.2) in this way, (12.2) can be interpreted as a deterministic descent algorithm as discussed in Chapter 7 which is driven by the noise error term $\tilde{\mathbf{n}}_t$. Note that $E\{\tilde{\mathbf{n}}_t|\boldsymbol{\theta}\} = E\{\tilde{\mathbf{d}}_t|\boldsymbol{\theta}\} - \bar{\mathbf{d}}_t(\boldsymbol{\theta}) = \mathbf{0}_q$ (see Example 8.4.6).

The average search direction function $\bar{\mathbf{d}}_t$ is chosen such that $\bar{\mathbf{d}}_t$ points downhill in the opposite direction of the gradient of $\ell$, $\mathbf{g}$ (see Figure 7.1). That is, it is assumed that $\bar{\mathbf{d}}_t^T \mathbf{g} \leq 0$.

For a passive statistical learning environment, the gradient $\mathbf{g}$ (when it exists) of (12.3) is given by the formula:

$$\mathbf{g}(\boldsymbol{\theta}) = \left[\int \frac{dc(\mathbf{x}, \boldsymbol{\theta})}{d\boldsymbol{\theta}} p_e(\mathbf{x}) d\nu(\mathbf{x})\right]^T$$

and the Hessian $\mathbf{H}$ of (12.4) (when it exists) is given by the formula:

$$\mathbf{H}(\boldsymbol{\theta}) = \int \frac{d^2 c(\mathbf{x}, \boldsymbol{\theta})}{d\boldsymbol{\theta}^2} p_e(\mathbf{x}) d\nu(\mathbf{x}).$$

For a reactive statistical learning environment, the gradient $\mathbf{g}$ of (12.4) (when it exists) is given by the formula:

$$\mathbf{g}(\boldsymbol{\theta}) = \left[\int \frac{dc(\mathbf{x}, \boldsymbol{\theta})}{d\boldsymbol{\theta}} p_e(\mathbf{x}|\boldsymbol{\theta}) d\nu(\mathbf{x}) + \int c(\mathbf{x}, \boldsymbol{\theta}) \frac{dp_e(\mathbf{x}|\boldsymbol{\theta})}{d\boldsymbol{\theta}} d\nu(\mathbf{x})\right]^T \tag{12.6}$$

which explicitly shows how the gradient of the risk function is explicitly dependent upon $p_e(\mathbf{x}|\boldsymbol{\theta})$. It is convenient to rewrite (12.6) as:

$$\mathbf{g}(\boldsymbol{\theta}) = \left[\int \frac{dc(\mathbf{x}, \boldsymbol{\theta})}{d\boldsymbol{\theta}} p_e(\mathbf{x}|\boldsymbol{\theta}) d\nu(\mathbf{x}) + \int c(\mathbf{x}, \boldsymbol{\theta}) \frac{d \log p_e(\mathbf{x}|\boldsymbol{\theta})}{d\boldsymbol{\theta}} p_e(\mathbf{x}|\boldsymbol{\theta}) d\nu(\mathbf{x})\right]^T \tag{12.7}$$

by using the identity

$$\frac{d \log p_e(\mathbf{x}|\boldsymbol{\theta})}{d\boldsymbol{\theta}} = \frac{1}{p_e(\mathbf{x}|\boldsymbol{\theta})} \frac{dp_e(\mathbf{x}|\boldsymbol{\theta})}{d\boldsymbol{\theta}}.$$

Let the notation $\mathbf{g}(\mathbf{x}, \boldsymbol{\theta}) \equiv (dc(\mathbf{x}, \boldsymbol{\theta})/d\boldsymbol{\theta})^T$. The Hessian, $\mathbf{H}$, of (12.4) in a reactive statistical learning environment (when it exists) is given by the formula:

$$\mathbf{H}(\boldsymbol{\theta}) = \int \frac{dc^2(\mathbf{x}, \boldsymbol{\theta})}{d\boldsymbol{\theta}^2} p_e(\mathbf{x}|\boldsymbol{\theta}) d\nu(\mathbf{x}) + \int \mathbf{g}(\mathbf{x}, \boldsymbol{\theta}) \frac{d \log p_e(\mathbf{x}|\boldsymbol{\theta})}{d\boldsymbol{\theta}} p_e(\mathbf{x}|\boldsymbol{\theta}) d\nu(\mathbf{x}) \tag{12.8}$$

$$+ \int \left(\frac{d \log p_e(\mathbf{x}|\boldsymbol{\theta})}{d\boldsymbol{\theta}}\right)^T \mathbf{g}(\mathbf{x}, \boldsymbol{\theta})^T p_e(\mathbf{x}|\boldsymbol{\theta}) d\nu(\mathbf{x}) + \int c(\mathbf{x}, \boldsymbol{\theta}) \frac{d^2 \log p_e(\mathbf{x}|\boldsymbol{\theta})}{d\boldsymbol{\theta}^2} p_e(\mathbf{x}|\boldsymbol{\theta}) d\nu(\mathbf{x})$$

$$+ \int c(\mathbf{x}, \boldsymbol{\theta}) \left(\frac{d \log p_e(\mathbf{x}|\boldsymbol{\theta})}{d\boldsymbol{\theta}}\right)^T \frac{d \log p_e(\mathbf{x}|\boldsymbol{\theta})}{d\boldsymbol{\theta}} p_e(\mathbf{x}|\boldsymbol{\theta}) d\nu(\mathbf{x}).$$

### 12.1.3 Annealing Schedule

The variance of the additive noise error term $\gamma_t \tilde{\mathbf{n}}_t$ in (12.5) is governed by the magnitude of the step size $\gamma_t$. In many of the applications considered here, the step size $\gamma_t$ is often referred to as the *learning rate*. Thus, $\gamma_1, \gamma_2, \ldots$ must decrease toward zero to ensure $\tilde{\boldsymbol{\theta}}(1), \tilde{\boldsymbol{\theta}}(2), \ldots$ converges with probability one to a particular point $\boldsymbol{\theta}^*$ in the state space. On the other hand, convergence to a local or global minimizer will fail if $\gamma_1, \gamma_2, \ldots$ approaches zero too rapidly.

The technical condition to ensure that the *annealing schedule* $\gamma_1, \gamma_2, \ldots$ decreases toward zero is

$$\sum_{t=0}^{\infty} \gamma_t^2 < \infty. \tag{12.9}$$

The technical condition to ensure that the annealing schedule $\gamma_1, \gamma_2, \ldots$ does not decrease too rapidly is

$$\sum_{t=0}^{\infty} \gamma_t = \infty. \tag{12.10}$$

A typical annealing schedule satisfying the technical conditions (12.9) and (12.10) is illustrated in Figure 12.1 where the stepsize $\gamma_t = \gamma_0$ for $t < T_0$ and $\gamma_t = \gamma_0/(1 + (t/\tau))$ for $t \geq T_0$. This annealing schedule assumes an initially constant stepsize $\gamma_0$ for the time period until time $T_0$. The time period where the stepsize is constant is called the "search phase" which allows the algorithm to explore the state space with a noise variance which is relatively large. During this phase, the algorithm has the opportunity to sample its statistical environment, so it is important that $T_0$ be chosen to be sufficiently large in learning

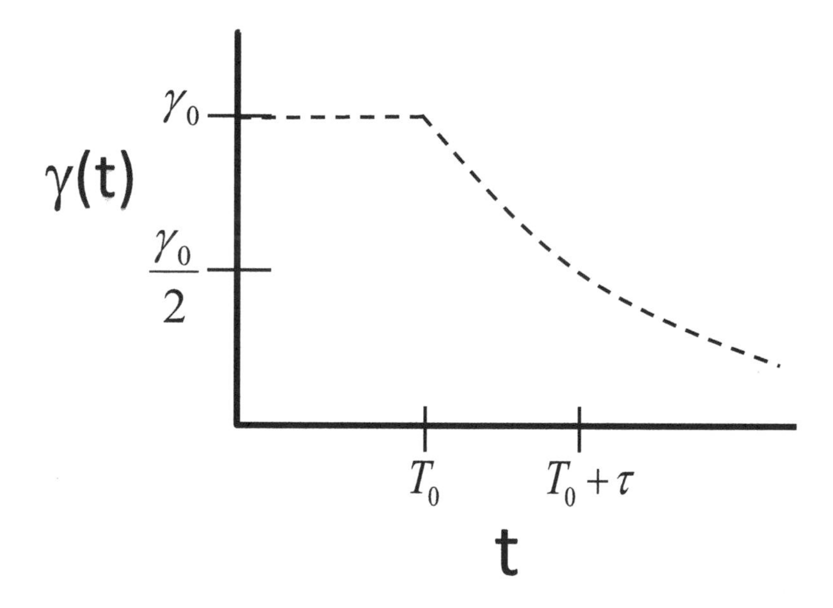


**FIGURE 12.1**
A typical stepsize annealing schedule for a stochastic approximation algorithm that satisfies Equation (12.9) and Equation (12.10). The stepsize is constant with value $\gamma_0$ until the time index reaches time $T_0$. At that point, the stepsize gradually decreases with a half-life of $\tau$. By selecting different choices for $\gamma_0$, $T_0$, and $\tau$, a variety of different annealing schedules can be generated.

machine applications so that the learning machine benefits from a wide range of learning experiences.

The time period following the search phase of the learning process is called the "converge phase" in which the sequence of stepsizes $\gamma_{T_0}, \gamma_{T_0+1}, \ldots$ are gradually decreased. Note that during this phase the stepsize decays with a "half-life" of $\tau$ (i.e., after approximately a time period of $\tau$ the stepsize is decreased by half). During this phase of learning, the goal is to converge to a final choice of learning parameter values.

In practice, different customized annealing schedules can be designed by comparing the sequence $\gamma_t = 1/(1+t)$ that satisfies (12.9) and (12.10) with a practical annealing schedule using the Limit Comparison Test (Theorem 5.1.3). For example, selecting the sequence of stepsizes $\gamma_1, \gamma_2, \ldots$ to be an increasing rather than decreasing sequence is also a useful strategy for supporting exploratory behaviors during the "search phase". The intuitive idea behind this heuristic is that the initial smaller learning rates allow the system to extract dominant statistical regularities and then the learning rate is gradually increased to accelerate the learning process during the initial search phase of the learning process. Eventually, however, the stepsize must gradually be decreased in order for the convergence analyses presented in this chapter to hold.

Exercise 12.1-2 provides an example of a smooth annealing schedule where stepsize is initially constant and then gradually decreases. Exercise 12.1-3 provides an example of a smooth annealing schedule where the stepsize is initially constant, then increases, and then decreases.

### 12.1.4 The Main Stochastic Approximation Theorem

The following stochastic convergence theorem has many important applications including the analysis of high-dimensional nonlinear systems that are minimizing either convex or non-convex smooth objective functions.

**Theorem 12.1.1** (Stochastic Approximation Theorem). *Let $\Theta$ be a closed, bounded, and convex subset of $\mathcal{R}^q$.*

- *Assume $\tilde{\mathbf{x}}_{\boldsymbol{\theta}}$ has Radon-Nikodým density $p_e(\cdot|\boldsymbol{\theta}) : \mathcal{R}^d \rightarrow [0, \infty)$ with respect to a sigma-finite measure $\nu$ for each $\boldsymbol{\theta} \in \Theta$.*
- *Assume $\ell : \mathcal{R}^q \rightarrow \mathcal{R}$ is a twice continuously differentiable function with a finite lower bound on $\mathcal{R}^q$. Let $\mathbf{g} \equiv (\nabla \ell)^T$. Let $\mathbf{H} \equiv \nabla^2 \ell$.*
- *Assume a positive number $x_{max}$ exists such that for all $\boldsymbol{\theta} \in \Theta$ the random vector $\tilde{\mathbf{x}}_{\boldsymbol{\theta}}$ with density $p_e(\cdot|\boldsymbol{\theta})$ satisfies $|\tilde{\mathbf{x}}_{\boldsymbol{\theta}}| \leq x_{max}$ with probability one.*
- *Let $\gamma_0, \gamma_1, \gamma_2, \ldots$ be a sequence of positive real numbers such that:*

$$\sum_{t=0}^{\infty} \gamma_t^2 < \infty \tag{12.11}$$

*and*

$$\sum_{t=0}^{\infty} \gamma_t = \infty. \tag{12.12}$$

- *Let $\mathbf{d}_t : \mathcal{R}^d \times \mathcal{R}^q \rightarrow \mathcal{R}^q$ be a piecewise continuous function on a finite partition of $\mathcal{R}^d \times \mathcal{R}^q$ for all $t \in \mathbb{N}$. When it exists, let*

$$\bar{\mathbf{d}}_t(\boldsymbol{\theta}) = \int \mathbf{d}_t(\mathbf{x}, \boldsymbol{\theta}) p_e(\mathbf{x}|\boldsymbol{\theta}) d\nu(\mathbf{x}).$$

- *Let $\tilde{\boldsymbol{\theta}}(0)$ be a q-dimensional random vector. Let $\tilde{\boldsymbol{\theta}}(1), \tilde{\boldsymbol{\theta}}(2), \ldots$ be a sequence of q-dimensional random vectors defined such that for $t = 0, 1, 2, \ldots$:*

$$\tilde{\boldsymbol{\theta}}(t+1) = \tilde{\boldsymbol{\theta}}(t) + \gamma_t \tilde{\mathbf{d}}_t \tag{12.13}$$

*where $\tilde{\mathbf{d}}_t \equiv \mathbf{d}_t(\tilde{\mathbf{x}}_{\boldsymbol{\theta}}(t), \tilde{\boldsymbol{\theta}}(t))$. Assume $\{\tilde{\mathbf{d}}_t\}$ is a bounded stochastic sequence when $\{\boldsymbol{\theta}(t)\}$ is a bounded stochastic sequence. Assume the distribution of $\tilde{\mathbf{x}}_{\boldsymbol{\theta}}(t)$ is specified by the conditional density $p_e(\cdot|\tilde{\boldsymbol{\theta}}(t))$.*

- *Assume there exists a positive number $K$ such that for all $\boldsymbol{\theta} \in \Theta$:*

$$\bar{\mathbf{d}}_t(\boldsymbol{\theta})^T \mathbf{g}(\boldsymbol{\theta}) \leq -K|\mathbf{g}(\boldsymbol{\theta})|^2. \tag{12.14}$$

*If there exists a positive integer $T$ such that $\tilde{\boldsymbol{\theta}}(t) \in \Theta$ for all $t \geq T$ with probability one, then $\tilde{\boldsymbol{\theta}}(1), \tilde{\boldsymbol{\theta}}(2), \ldots$ converges with probability one to the set of critical points of $\ell$ which are contained in $\Theta$.*

*Proof.* The statement and proof of this theorem is a variation of the theorem presented in Golden (2018) which extends earlier work by Blum (1954). Throughout the proof, assume there exists a positive integer $T$ such that $\tilde{\boldsymbol{\theta}}(t) \in \Theta$ for all $t \geq T$ with probability one.

Let $\tilde{\ell}_t \equiv \ell(\tilde{\boldsymbol{\theta}}(t))$ with realization $\ell_t \equiv \ell(\boldsymbol{\theta}(t))$. Let $\tilde{\mathbf{g}}_t \equiv \mathbf{g}(\tilde{\boldsymbol{\theta}}(t))$ with realization $\mathbf{g}_t \equiv \mathbf{g}(\boldsymbol{\theta}(t))$. Let $\tilde{\mathbf{H}}_t \equiv \mathbf{H}(\tilde{\boldsymbol{\theta}}(t))$ with realization $\mathbf{H}_t \equiv \mathbf{H}(\boldsymbol{\theta}(t))$.

*Step 1: Expand $\ell$ using a second-order mean value expansion.* Expand $\ell$ about $\tilde{\boldsymbol{\theta}}(t)$ and evaluate at $\tilde{\boldsymbol{\theta}}(t+1)$ using the mean value theorem to obtain:

$$\tilde{\ell}_{t+1} = \tilde{\ell}_t + \tilde{\mathbf{g}}_t^T \left(\tilde{\boldsymbol{\theta}}(t+1) - \tilde{\boldsymbol{\theta}}(t)\right) + \gamma_t^2 \tilde{R}_t \tag{12.15}$$

with

$$\tilde{R}_t \equiv (1/2)\tilde{\mathbf{d}}_t^T \mathbf{H}(\tilde{\boldsymbol{\zeta}}_t)\tilde{\mathbf{d}}_t \tag{12.16}$$

where the random variable $\tilde{\boldsymbol{\zeta}}_t$ can be defined as a point on the chord connecting $\tilde{\boldsymbol{\theta}}(t)$ and $\tilde{\boldsymbol{\theta}}(t+1)$. Substituting the relation

$$\gamma_t \tilde{\mathbf{d}}_t = \tilde{\boldsymbol{\theta}}(t+1) - \tilde{\boldsymbol{\theta}}(t)$$

into (12.15) gives:

$$\tilde{\ell}_{t+1} = \tilde{\ell}_t + \gamma_t \tilde{\mathbf{g}}_t^T \tilde{\mathbf{d}}_t + \gamma_t^2 \tilde{R}_t. \tag{12.17}$$

*Step 2: Identify conditions required for the remainder term of the expansion to be bounded.* Since, by assumption, $\{\tilde{\boldsymbol{\theta}}(t)\}$ is a bounded stochastic sequence and $\mathbf{H}$ is continuous, this implies that the stochastic sequence $\{\mathbf{H}(\tilde{\boldsymbol{\zeta}}_t)\}$ is bounded. In addition, by assumption $\{\tilde{\mathbf{d}}_t\}$ is a bounded stochastic sequence. This implies there exists a finite number $R_{max}$ such that for all $t = 0, 1, 2, \ldots$:

$$|\tilde{R}_t| \leq R_{max} \text{ with probability one.} \tag{12.18}$$

*Step 3: Compute bound on decrease of conditional expectation of objective function.* Taking the conditional expectation of both sides of (12.17) with respect to the conditional density $p_e$ and evaluating at $\boldsymbol{\theta}(t)$ and $\gamma_t$ yields:

$$E\left\{\tilde{\ell}_{t+1}|\boldsymbol{\theta}(t)\right\} = \ell_t + \gamma_t \mathbf{g}_t^T \bar{\mathbf{d}}_t + \gamma_t^2 E\{\tilde{R}_t|\boldsymbol{\theta}(t)\}. \tag{12.19}$$

Substituting the assumption $\bar{\mathbf{d}}_t(\boldsymbol{\theta})^T \mathbf{g}(\boldsymbol{\theta}) \leq -K|\mathbf{g}(\boldsymbol{\theta})|^2$ and the conclusion of Step 2 that $|\tilde{R}_t(\cdot)| \leq R_{max}$ with probability one into (12.19) gives:

$$E\left\{\tilde{\ell}_{t+1}|\boldsymbol{\theta}(t)\right\} \leq \ell_t - \gamma_t K|\mathbf{g}_t|^2 + \gamma_t^2 R_{max}. \tag{12.20}$$

*Step 4: Show a subsequence of* $\{|\tilde{\mathbf{g}}_t|^2\}$ *converges to zero wp1.* Since $\ell$ has a lower bound, $K$ is a finite positive number, and (12.11) hold by assumption, then the Almost Supermartingale Lemma (Theorem 9.3.4) can be applied to (12.20) on the set where $\{\tilde{\boldsymbol{\theta}}(t)\}$ and $\{\tilde{\mathbf{d}}_t\}$ are bounded with probability one to obtain the conclusion that:

$$\sum_{t=0}^{\infty} \gamma_t \left|\tilde{\mathbf{g}}_t\right|^2 < \infty \tag{12.21}$$

with probability one.

For every positive integer $t$, let

$$\tilde{a}_t^* \equiv \inf\left\{\left|\tilde{\mathbf{g}}_t\right|^2, \left|\tilde{\mathbf{g}}_{t+1}\right|^2, \ldots\right\}.$$

The sequence $\tilde{a}_1^*, \tilde{a}_2^*, \ldots$ is non-increasing with probability one and bounded from below by zero which implies this sequence is convergent with probability one to a random variable $\tilde{a}^*$ (see Theorem 5.1.1(vii); Rosenlicht, 1968, p. 50).

Assume that $\tilde{a}^*$ is positive and not equal to zero, using the assumption that $\sum_t \gamma_t = \infty$ in (12.12), then

$$\sum_{t=1}^{\infty} \gamma_t \tilde{a}_t \geq \tilde{a}^* \sum_{t=1}^{\infty} \gamma_t = \infty \tag{12.22}$$

with probability one.

Equation (12.21) holds for every subsequence of $\{|\tilde{\mathbf{g}}_t|^2\}$. Thus, (12.21) implies that:

$$\sum_{t=1}^{\infty} \gamma_t \tilde{a}_t < \infty \tag{12.23}$$

which directly contradicts the conclusion in (12.22). Thus, the sequence $\tilde{a}_1^*, \tilde{a}_2^*, \ldots$ is convergent with probability one to zero. Equivalently, a subsequence of $\{|\tilde{\mathbf{g}}_t|^2\}$ is convergent with probability one to zero.

*Step 5: Show the stochastic sequence* $\{\tilde{\boldsymbol{\theta}}(t)\}$ *converges to a random variable wp1.* Since $\ell$ has a lower bound, $K$ is a finite positive number, and (12.20) holds, the Almost Supermartingale Lemma ((Theorem 9.3.4) can be applied to (12.20) to show the stochastic sequence of $\ell(\tilde{\boldsymbol{\theta}}(1)), \ell(\tilde{\boldsymbol{\theta}}(2)), \ldots$ converges to some unknown random variable which will be denoted as $\tilde{\ell}^*$ with probability one. Since $\ell$ is continuous, this is equivalent to the assertion that $\tilde{\boldsymbol{\theta}}(1), \tilde{\boldsymbol{\theta}}(2), \ldots$ converges with probability one to some unknown random variable which will be denoted as $\tilde{\boldsymbol{\theta}}^*$ such that $\ell(\tilde{\boldsymbol{\theta}}^*) = \tilde{\ell}^*$ with probability one. By the assumption that with probability one every trajectory $\tilde{\boldsymbol{\theta}}(1), \tilde{\boldsymbol{\theta}}(2), \ldots$ is confined to the closed, bounded, and convex set $\Theta$, it follows that $\boldsymbol{\theta}^* \in \Theta$ with probability one.

*Step 6: Show the stochastic sequence* $\{|\tilde{\mathbf{g}}_t|^2\}$ *converges to zero wp1.* From Step 5, $\tilde{\boldsymbol{\theta}}(t) \to \tilde{\boldsymbol{\theta}}^*$ with probability one as $t \to \infty$. Since $\mathbf{g}$ is a continuous function, it then follows that $|\mathbf{g}(\tilde{\boldsymbol{\theta}}(1))|^2, |\mathbf{g}(\tilde{\boldsymbol{\theta}}(2))|^2, \ldots$ converges with probability one to $|\mathbf{g}(\tilde{\boldsymbol{\theta}}^*)|^2$. This is equivalent to the statement that every subsequence of $\{|\mathbf{g}(\tilde{\boldsymbol{\theta}}(t))|^2\}$ converges to $|\mathbf{g}(\tilde{\boldsymbol{\theta}}^*)|^2$ with probability one. That is, for every possible sequence of positive integers $t_1, t_2, \ldots$ the stochastic sequence $|\mathbf{g}(\tilde{\boldsymbol{\theta}}(t_1)|^2, |\mathbf{g}(\tilde{\boldsymbol{\theta}}(t_2)|^2, \ldots$ converges with probability one to $|\mathbf{g}(\tilde{\boldsymbol{\theta}}^*)|^2$.

From Step 4, there exists a sequence of positive integers, $k_1, k_2, \ldots$ such that the stochastic sequence $|\mathbf{g}(\tilde{\boldsymbol{\theta}}(k_1)|^2, |\mathbf{g}(\tilde{\boldsymbol{\theta}}(k_2)|^2, \ldots$ converges with probability one to zero. Thus, to avoid a contradiction, every subsequence of $\{|\mathbf{g}(\tilde{\boldsymbol{\theta}}(t))|^2\}$ converges to a random variable $|\mathbf{g}(\tilde{\boldsymbol{\theta}}^*)|^2$ with probability one and additionally with probability one: $|\mathbf{g}(\tilde{\boldsymbol{\theta}}^*)|^2 = 0$. Or, equivalently, $\{|\mathbf{g}(\tilde{\boldsymbol{\theta}}(t))|^2\}$ converges to 0 with probability one.

*Step 7: Show that* $\{\tilde{\boldsymbol{\theta}}(t)\}$ *converges to set of critical points.* From Step 6, $|\mathbf{g}(\tilde{\boldsymbol{\theta}}(t))|^2 \rightarrow 0$ with probability one. Using this result with the assumptions that $|\mathbf{g}|^2$ is a continuous function and $\tilde{\boldsymbol{\theta}}^* \in \Theta$, it follows that $\tilde{\boldsymbol{\theta}}(1), \tilde{\boldsymbol{\theta}}(2), \ldots$ converges with probability one to

$$\{\tilde{\boldsymbol{\theta}}^* \in \Theta : |\mathbf{g}(\tilde{\boldsymbol{\theta}}^*)|^2 = 0\}.$$

That is, $\tilde{\boldsymbol{\theta}}(1), \tilde{\boldsymbol{\theta}}(2), \ldots$ converges with probability one to the set of critical points of $\ell$ in $\Theta$. ∎

Note that a sufficient condition for the stochastic sequence $\{\tilde{\mathbf{x}}_{\boldsymbol{\theta}}\}$ to be bounded is that the range of $\tilde{\mathbf{x}}_{\boldsymbol{\theta}}$ is a bounded subset of $\mathcal{R}^d$. For example, if $\tilde{\mathbf{x}}_{\boldsymbol{\theta}}$ is a discrete random vector taking on values in a finite sample space, then this condition is automatically satisfied.

Assume $c : \mathcal{R}^d \times \Theta \rightarrow \mathcal{R}$ and $p_e : \mathcal{R}^d \times \Theta \rightarrow [0, \infty)$ are both twice continuously differentiable random functions. In addition, assume $c(\cdot; \boldsymbol{\theta})$ and $p_e(\cdot; \boldsymbol{\theta})$ are piecewise continuous on $\mathcal{R}^d$ for all $\boldsymbol{\theta} \in \Theta$. Then, with the assumption that $\{\tilde{\mathbf{x}}_{\boldsymbol{\theta}}\}$ is bounded and the assumption that $\{\tilde{\boldsymbol{\theta}}(t)\}$ is bounded, the Dominated Convergence Theorem implies that $\ell$, the gradient of $\ell$, and the Hessian of $\ell$ are continuous on $\Theta$.

The conclusion of the Stochastic Approximation Theorem essentially states that if the stochastic sequence of states $\tilde{\boldsymbol{\theta}}(1), \tilde{\boldsymbol{\theta}}(2), \ldots$ is confined to a closed and bounded region of the state space denoted by $\Theta$ with probability one, then $\tilde{\boldsymbol{\theta}}(t)$ converges to the set of critical points in $\Theta$ with probability one. Unless the dynamics of the descent algorithm's trajectories are artificially bounded, it is often difficult to show that the stochastic sequence of states $\tilde{\boldsymbol{\theta}}(1), \tilde{\boldsymbol{\theta}}(2), \ldots$ is confined to a closed and bounded region of the state space. Therefore, in practice, it is convenient to restate the conclusion of the Stochastic Approximation Theorem as either: (i) $\tilde{\boldsymbol{\theta}}(t)$ converges to the set of critical points in a closed and bounded set $\Theta$, or (ii) $\tilde{\boldsymbol{\theta}}(t)$ must eventually leave the set $\Theta$ with probability one. Although the conclusion of the theorem is relatively weak, the theorem is powerful because it is applicable to a large class of machine learning algorithms that minimize smooth objective functions which are not necessarily convex. For example, many missing data problems and deep learning neural network problems are non-convex learning problems characterized by situations where $\ell$ has multiple flats, saddlepoints, minimizers, and maximizers.

In the special case where the objective function for learning is convex (e.g., linear regression, logistic regression, and multinomial logit regression), then it is useful to choose $\Theta$ as a ball with an arbitrarily large radius. Another important special case occurs when $\Theta$ is chosen to contain exactly one critical point. If the conditions of the Stochastic Approximation Theorem hold and the stochastic sequence $\{\tilde{\boldsymbol{\theta}}(t)\}$ is convergent to a region of the state space which contains exactly one critical point with probability one, then it follows that $\{\tilde{\boldsymbol{\theta}}(t)\}$ must converge to that critical point with probability one.

**Example 12.1.1** (Convergence of Stochastic Gradient Descent). Assume a statistical environment where the observed training data $\mathbf{x}(1), \mathbf{x}(2), \ldots$ is assumed to be a realization of a bounded stochastic sequence of independent and identically distributed $d$-dimensional random vectors with common Radon-Nikodým density $p_e : \mathcal{R}^d \rightarrow [0, \infty)$ defined with respect to support specification measure $\nu$. Assume a machine learning algorithm operates by minimizing a twice continuously differentiable risk function $\ell : \mathcal{R}^q \rightarrow [0, \infty)$ defined such that for all $\boldsymbol{\theta} \in \mathcal{R}^q$:

$$\ell(\boldsymbol{\theta}) = \int c(\mathbf{x}, \boldsymbol{\theta}) p_e(\mathbf{x}) d\nu(\mathbf{x})$$

where $c$ is a continuously differentiable function with a finite lower bound.

**Recipe Box 12.1 Adaptive Learning Convergence Analysis (Theorem 12.1.1)**

- **Step 1: Identify the statistical environment.** A reactive statistical environment is modeled as a sequence of bounded, independent and identically distributed $d$-dimensional random vectors $\tilde{\mathbf{x}}(1), \tilde{\mathbf{x}}(2), \ldots$ with common density $p_e(\cdot|\boldsymbol{\theta})$ where $\boldsymbol{\theta} \in \mathcal{R}^q$. The density $p_e$ is not functionally dependent upon $\boldsymbol{\theta}$ for passive statistical environments.

- **Step 2: Check $\ell$ is twice continuously differentiable with a lower bound.** Let $\Theta$ be a convex, closed, and bounded subset of $\mathcal{R}^q$. Define a twice continuously differentiable function $\ell : \mathcal{R}^q \rightarrow \mathcal{R}$ such that for all $\boldsymbol{\theta} \in \mathcal{R}^q$:

$$\ell(\boldsymbol{\theta}) = \int c(\mathbf{x}, \boldsymbol{\theta}) p_e(\mathbf{x}|\boldsymbol{\theta}) d\nu(\mathbf{x}).$$

  Check that $\ell$ has a finite lower bound on $\mathcal{R}^q$.

- **Step 3: Check the annealing schedule.** Design a sequence of stepsizes $\gamma_1, \gamma_2, \ldots$ that satisfies (12.11) and (12.12) by comparing $\{\gamma_t\}$ to $\{1/t\}$ using the Limit Comparison Test for Adaptive Learning (Theorem 5.1.3). For example,

$$\gamma_t = \gamma_0(1 + (t/\tau_1))/(1 + (t/\tau_2)^2)$$

  is a step size sequence designed to have a search mode behavior when $t << \tau_2$ and a converge mode behavior when $t >> \tau_2$ for $0 < \tau_1 < \tau_2$ and $\gamma_0 > 0$.

- **Step 4: Identify the search direction function.** Let $\mathbf{d}_t : \mathcal{R}^d \times \mathcal{R}^q \rightarrow \mathcal{R}^q$ be a piecewise continuous function on $\mathcal{R}^d \times \mathcal{R}^q$ for each $t \in \mathbb{N}$. Rewrite the learning rule for updating parameter estimates using the formula:

$$\tilde{\boldsymbol{\theta}}(t+1) - \tilde{\boldsymbol{\theta}}(t) + \gamma_t \tilde{\mathbf{d}}_t$$

  where the *search direction* random vector $\tilde{\mathbf{d}}_t = \mathbf{d}_t(\tilde{\mathbf{x}}(t), \tilde{\boldsymbol{\theta}}(t))$ Check that $\{\tilde{\mathbf{d}}_t\}$ is a bounded stochastic sequence when $\{\tilde{\boldsymbol{\theta}}(t)\}$ is a bounded stochastic sequence.

- **Step 5: Show average search direction is downward.** Assume there exists a series of functions $\bar{\mathbf{d}}_1, \bar{\mathbf{d}}_2, \ldots$ such that:

$$\bar{\mathbf{d}}_t(\boldsymbol{\theta}) \equiv E\{\mathbf{d}_t(\tilde{\mathbf{x}}(t), \boldsymbol{\theta})|\boldsymbol{\theta}\} = \int \mathbf{d}_t(\mathbf{x}, \boldsymbol{\theta}) p_e(\mathbf{x}|\boldsymbol{\theta}) d\nu(\mathbf{x}).$$

  Show that there exists a positive number $K$ such that:

$$\bar{\mathbf{d}}_t(\boldsymbol{\theta})^T \mathbf{g}(\boldsymbol{\theta}) \leq -K|\mathbf{g}(\boldsymbol{\theta})|^2. \quad (12.24)$$

- **Step 6: Investigate asymptotic behavior.** Conclude that with probability one that either: (1) the stochastic sequence $\tilde{\boldsymbol{\theta}}(1), \tilde{\boldsymbol{\theta}}(2), \ldots, \tilde{\boldsymbol{\theta}}(t), \ldots$ does not remain in $\Theta$ for all $t > T$ for some positive integer $T$, or (2) $\tilde{\boldsymbol{\theta}}(1), \tilde{\boldsymbol{\theta}}(2), \ldots, \tilde{\boldsymbol{\theta}}(t), \ldots$ converges to the set of critical points in $\Theta$ as $t \rightarrow \infty$.

A gradient descent adaptive learning algorithm is now specified by the learning rule:

$$\tilde{\boldsymbol{\theta}}(t+1) = \tilde{\boldsymbol{\theta}}(t) - \gamma_t \mathbf{g}(\tilde{\mathbf{x}}(t), \tilde{\boldsymbol{\theta}}(t)) \tag{12.25}$$

where $\mathbf{g}(\boldsymbol{\theta}) \equiv [dc(\mathbf{x}, \boldsymbol{\theta})/d\boldsymbol{\theta}]^T$ for each $\mathbf{x} \in \mathcal{R}^d$. Assume the stepsize sequence is a sequence of positive numbers $\gamma_1, \gamma_2, \ldots$ such that $\gamma_t = 1$ for $t < T$ and $\gamma_t = 1/t$ for $t \geq T$ for some finite positive integer $T$. Show that the stochastic sequence $\tilde{\boldsymbol{\theta}}(0), \tilde{\boldsymbol{\theta}}(1), \ldots$ is either not bounded with probability one or converges to a set of critical points of $\ell$ with probability one.

**Solution.** The asymptotic behavior is analyzed following the steps in Recipe Box 12.1.

Step 1: The statistical environment is a stochastic sequence of bounded i.i.d. random vectors by assumption.

Step 2: The objective function $\ell$ is a twice-continuously differentiable function on $\mathcal{R}^q$. Define a subset of $\mathcal{R}^q$, $\Theta$, as a closed ball centered at the origin with a large finite radius $R$ (e.g., choose $R = 10^9$). The region $\Theta$ is closed, bounded, and convex. Since $c$ has a finite lower bound, $\ell$ has a finite lower bound.

Step 3: From Theorem 5.1.2, $\sum_{t=1}^{\infty} \gamma_t = \infty$ and $\sum_{t=1}^{\infty} \gamma_t^2 < \infty$.

Step 4: Let $\tilde{\mathbf{d}}_t = -\mathbf{g}(\tilde{\mathbf{x}}(t), \tilde{\boldsymbol{\theta}}(t))$. When $\tilde{\mathbf{x}}(0), \tilde{\mathbf{x}}(1), \ldots$ and $\tilde{\boldsymbol{\theta}}(0), \tilde{\boldsymbol{\theta}}(1), \ldots$ are bounded stochastic sequences and $\mathbf{g}$ is continuous, it follows that $\{|\tilde{\mathbf{d}}_t|\}$ is a bounded stochastic sequence.

Step 5: Let $\bar{\mathbf{g}}(\boldsymbol{\theta}) = E\{\mathbf{g}(\tilde{\mathbf{x}}(t), \boldsymbol{\theta})\}$. Define $\bar{\mathbf{d}}(\boldsymbol{\theta}) = -\bar{\mathbf{g}}(\boldsymbol{\theta})$, it follows that $\bar{\mathbf{d}}^T \bar{\mathbf{g}} \leq -|\bar{\mathbf{g}}|^2$ satisfies (12.14) for the case where $K = 1$.

Step 6: Conclude with probability one that either: (1) the stochastic sequence $\tilde{\boldsymbol{\theta}}(1), \tilde{\boldsymbol{\theta}}(2), \ldots, \tilde{\boldsymbol{\theta}}(t), \ldots$ generated by the learning algorithm does not remain in $\Theta$ with probability one for all $t > T$, or (ii) the stochastic sequence of parameter vectors $\tilde{\boldsymbol{\theta}}(1), \tilde{\boldsymbol{\theta}}(2), \ldots$ converges with probability one to the set of critical points in $\Theta$. △

**Example 12.1.2** (Convergence of Scaled Gradient Descent Adaptive Learning). Assume a statistical environment where the observed training data $\mathbf{x}(1), \mathbf{x}(2), \ldots$ is assumed to be a realization of a stochastic sequence of independent and identically distributed $d$-dimensional discrete random vectors with common probability mass function $p_e : \{\mathbf{x}^1, \ldots, \mathbf{x}^M\} \to [0, 1]$. Assume a machine learning algorithm operates by minimizing a twice continuously differentiable risk function $\ell : \mathcal{R}^q \to [0, \infty)$ defined such that for all $\boldsymbol{\theta} \in \mathcal{R}^q$:

$$\ell(\boldsymbol{\theta}) = \int c(\mathbf{x}, \boldsymbol{\theta}) p_e(\mathbf{x}) d\nu(\mathbf{x})$$

where $c$ is a continuously differentiable function with a finite lower bound. An adaptive learning algorithm is now specified by the learning rule:

$$\tilde{\boldsymbol{\theta}}(t+1) = \tilde{\boldsymbol{\theta}}(t) - \gamma_t \mathbf{M}_t \mathbf{g}(\tilde{\mathbf{x}}(t), \tilde{\boldsymbol{\theta}}(t)) \tag{12.26}$$

where $\mathbf{g}(\boldsymbol{\theta}) \equiv [dc(\mathbf{x}, \boldsymbol{\theta})/d\boldsymbol{\theta}]^T$ for each $\mathbf{x} \in \mathcal{R}^d$. Assume $\mathbf{M}_1, \mathbf{M}_2, \ldots$ is a sequence of symmetric positive definite matrices such that $\lambda_{min}$ is a positive number which is less than the smallest eigenvalue of $\mathbf{M}_t$ and $\lambda_{max}$ is a positive number which is greater than the largest eigenvalue of $\mathbf{M}_t$ for all $t \in \mathbb{N}$. Assume $\gamma_t = (1 + (t/1000))/(1 + (t/5000)^2)$ for all $t \in \mathbb{N}$. Analyze the asymptotic behavior of this adaptive learning algorithm.

**Solution.** The asymptotic behavior is analyzed following the steps in Recipe Box 12.1.

Step 1: The statistical environment is a stochastic sequence of discrete i.i.d. random vectors where each random vector takes on one of $M$ possible values.

Step 2: The objective function $\ell$ is a twice-continuously differentiable function on $\mathcal{R}^q$. Let $\Theta$ be defined as a closed, bounded, and convex subset of $\mathcal{R}^q$. Since $c$ is a continuous random function with a finite lower bound it follows that $\ell$ has a finite lower bound.

Step 3: From Theorem 5.1.2, $\sum_{t=1}^{\infty} \eta_t = \infty$ and $\sum_{t=1}^{\infty} \eta_t^2 < \infty$ when $\eta_t = 1/t$. Since as $t \to \infty$:

$$\frac{\gamma_t}{\eta_t} = \frac{1/t}{(1+(t/1000))/(1+(t/5000)^2)} \to K$$

where $K$ is a finite constant it immediately follows from Theorem 5.1.3 that $\sum_{t=1}^{\infty} \gamma_t = \infty$ and $\sum_{t=1}^{\infty} \gamma_t^2 < \infty$.

Step 4: The learning rule is:

$$\tilde{\boldsymbol{\theta}}(t+1) = \tilde{\boldsymbol{\theta}}(t) + \gamma_t \tilde{\mathbf{d}}_t$$

where $\tilde{\mathbf{d}}_t = -\mathbf{M}_t \mathbf{g}(\tilde{\mathbf{x}}(t)$.

Note that

$$|\tilde{\mathbf{d}}_t| = |\mathbf{M}_t \mathbf{g}(\tilde{\mathbf{x}}(t), \tilde{\boldsymbol{\theta}}(t))| = \sqrt{|\mathbf{M}_t \mathbf{g}(\tilde{\mathbf{x}}(t), \tilde{\boldsymbol{\theta}}(t))|^2}$$

$$= \sqrt{\mathbf{g}(\tilde{\mathbf{x}}(t), \tilde{\boldsymbol{\theta}}(t))^T \mathbf{M}_t^2 \mathbf{g}(\tilde{\mathbf{x}}(t), \tilde{\boldsymbol{\theta}}(t))} \leq \lambda_{max} |\mathbf{g}(\tilde{\mathbf{x}}(t), \tilde{\boldsymbol{\theta}}(t))|.$$

When $\tilde{\mathbf{x}}(0), \tilde{\mathbf{x}}(1), \ldots$ and $\tilde{\boldsymbol{\theta}}(0), \tilde{\boldsymbol{\theta}}(1), \ldots$ are bounded stochastic sequences and $\lambda_{max}|\mathbf{g}(\mathbf{x}, \boldsymbol{\theta})|$ is a continuous function, it follows that $|\tilde{\mathbf{d}}_t|$ is a bounded stochastic sequence.

Step 5: The expected search direction

$$\bar{\mathbf{d}}_t(\boldsymbol{\theta}) = E\{-\mathbf{M}_t \mathbf{g}(\tilde{\mathbf{x}}(t), \boldsymbol{\theta})\} = -\mathbf{M}_t \bar{\mathbf{g}}.$$

Since $\bar{\mathbf{d}}(\boldsymbol{\theta}) = -\mathbf{M}_t \bar{\mathbf{g}}(\boldsymbol{\theta})$, it follows that $\bar{\mathbf{d}}^T \bar{\mathbf{g}} \leq -\lambda_{min} |\bar{\mathbf{g}}|^2$ satisfies (12.14).

Step 6: Conclude with probability one that either: (1) the stochastic sequence $\tilde{\boldsymbol{\theta}}(1), \tilde{\boldsymbol{\theta}}(2), \ldots, \tilde{\boldsymbol{\theta}}(t), \ldots$ generated by the learning algorithm does not remain in $\Theta$ with probability one for all $t > T$, or (ii) the stochastic sequence of parameter vectors $\boldsymbol{\theta}(1), \boldsymbol{\theta}(2), \ldots$ converges with probability one to the set of critical points in $\Theta$. △

### 12.1.5 Assessing Stochastic Approximation Algorithm Convergence

**Example 12.1.3** (Polyak-Ruppert Averaging). Consider a generic stochastic approximation algorithm:

$$\tilde{\boldsymbol{\theta}}(t+1) = \tilde{\boldsymbol{\theta}}(t) + \gamma_t \tilde{\mathbf{d}}_t$$

with the property that $\tilde{\boldsymbol{\theta}}(t) \to \tilde{\boldsymbol{\theta}}^*$ as $t \to \infty$ with probability one. The goal of the generic stochastic approximation algorithm is to identify $\tilde{\boldsymbol{\theta}}^*$ which, in some special cases, could be a constant vector with zero variance. In general, however, $\tilde{\boldsymbol{\theta}}^*$ is a random vector with some covariance matrix.

In practice, the learning process stops after some finite period of time $T$. Although one could use $\tilde{\boldsymbol{\theta}}(T)$ as an estimator of $E\{\tilde{\boldsymbol{\theta}}^*\}$, an improved estimation strategy involves averaging the last $m$ observations in the stochastic sequence $\tilde{\boldsymbol{\theta}}(0), \tilde{\boldsymbol{\theta}}(1), \ldots, \tilde{\boldsymbol{\theta}}(T)$ to obtain the *Polyak-Ruppert estimator*

$$\bar{\boldsymbol{\theta}}_m(T) = (1/m) \sum_{k=0}^{m-1} \tilde{\boldsymbol{\theta}}(T-k). \tag{12.27}$$

The estimator $\bar{\boldsymbol{\theta}}_m(t)$ also converges to $\tilde{\boldsymbol{\theta}}^*$ as $T \to \infty$ since each term in the summand of (12.27) is converging to $\tilde{\boldsymbol{\theta}}^*$ with probability one (see Theorem 9.4.1). Therefore, $\bar{\boldsymbol{\theta}}_m(T)$ may also be used to estimate $\tilde{\boldsymbol{\theta}}^*$. In the special case where $m = 1$, then the Polyak-Ruppert estimator $\bar{\boldsymbol{\theta}}_m(T)$ and the estimator $\tilde{\boldsymbol{\theta}}(t)$ are equivalent. Increasing the block size $m$ results in a more robust estimator. See Exercise 12.2-1 for an adaptive version of this estimator with reduced computational requirements. △

Assume a stochastic approximation algorithm generates a stochastic sequence of $q$-dimensional random vectors $\tilde{\boldsymbol{\theta}}(0), \tilde{\boldsymbol{\theta}}(1), \ldots$ that converges to a $q$-dimensional random vector $\tilde{\boldsymbol{\theta}}^*$ with probability one. If the stochastic sequence $\tilde{\boldsymbol{\theta}}(0), \tilde{\boldsymbol{\theta}}(1), \ldots$ is bounded, the Stochastic Approximation Theorem 9.4.1 asserts $E\{|\tilde{\boldsymbol{\theta}}(t) - \tilde{\boldsymbol{\theta}}^*|^2\}$ converges to zero as $t \to \infty$. Therefore, an estimator of $(1/q)E\{|\tilde{\boldsymbol{\theta}}(t) - \tilde{\boldsymbol{\theta}}^*|^2\}$ for $t$ sufficiently large and a fixed positive integer $m$ is given by the formula:

$$\hat{\sigma}_m^2(t) = \frac{1}{mq} \sum_{k=0}^{m-1} |\tilde{\boldsymbol{\theta}}(t-k) - \bar{\boldsymbol{\theta}}_m(t)|^2 \tag{12.28}$$

where $\bar{\boldsymbol{\theta}}_m(t)$ is the Polyak-Ruppert estimator. If $\hat{\sigma}_m(t)$ is sufficiently small this provides evidence for convergence. Note that (12.28) is normalized by the dimensionality, $q$, of $\boldsymbol{\theta}$.

## Exercises

12.1-1. Let $\eta_1, \eta_2, \ldots$ be a sequence of positive constants. Suppose that $\eta_t = 1$ for $0 \leq t \leq 100$, $\eta_t = t/1000$ for $100 < t \leq 500$, and $\eta_t = 1/(200+t)$ for $t > 500$. Verify that as $t \to \infty$, the sequence of constants $\eta_1, \eta_2, \ldots$ satisfies (12.11) and (12.12).

12.1-2. *Constant and Then Decrease Smooth Annealing Schedule.* Consider the following stepsize annealing schedule (see Darken and Moody, 1992, for related discussions) for generating a sequence of positive stepsizes $\eta_1, \eta_2, \ldots$ given by:

$$\eta_t = \frac{\eta_0}{1 + (t/\tau)}. \tag{12.29}$$

Plot $\eta_t$ as a function of $t$ for $\tau = 100$ and $\eta_0 = 1$. Show that when $t << \tau$, $\eta_t \approx \eta_0$. Show that when $t \approx \tau$, $\eta_t \approx \eta_0/2$. Show that when $t >> \tau$, $\eta_t \approx (\eta_0 \tau)/t$. Then, use the Limit Comparison Test Theorem 5.1.3 to show that the sequence $\eta_1, \eta_2, \ldots$ satisfies the conditions (12.11) and (12.12).

12.1-3. *Constant, Increase, and Then Decrease Smooth Annealing Schedule.* Consider the following stepsize annealing schedule (see Darken and Moody, 1992, for related discussions) for generating a sequence of positive stepsizes $\eta_1, \eta_2, \ldots$ given by:

$$\eta_t = \frac{\eta_0 \left(1 + (t/\tau_1)\right)}{1 + (t/\tau_2)^2}. \tag{12.30}$$

Plot $\eta_t$ as a function of $t$ for $\tau_1 = 100$, $\tau_2 = 1000$, and $\eta_0 = 1$. Show that when $t << \tau_1$, $\eta_t \approx \eta_0$. Show that when $t >> \tau_1$ and $t << \tau_2$, $\eta_t \approx t\eta_0/\tau_1$. Show that when $t \approx \tau_2$, $\eta_t \approx \eta_0/2$. Show that when $t >> \tau_2$, $\eta_t \approx \eta_0(\tau_2)^2/t$. Then, use the Limit Comparison Test Theorem 5.1.3 to show that the sequence $\eta_1, \eta_2, \ldots$ satisfies the conditions (12.11) and (12.12).

12.1-4. Show that the assertion that the stochastic sequence $|\tilde{\boldsymbol{\theta}}(1)|, |\tilde{\boldsymbol{\theta}}(2)|, \ldots$ is not bounded with probability one is equivalent to the assertion that there exists a subsequence of $|\tilde{\boldsymbol{\theta}}(1)|, \tilde{\boldsymbol{\theta}}(2)|, \ldots$ which converges to $+\infty$ with probability one.

12.1-5. Using your knowledge of the relationships between the different types of stochastic convergence (see Theorem 9.3.8), show that the part of the conclusion of the Stochastic Approximation Theorem which states that $\tilde{\boldsymbol{\theta}}(1), \tilde{\boldsymbol{\theta}}(2), \ldots$ converges with probability one implies that $\tilde{\boldsymbol{\theta}}(1), \tilde{\boldsymbol{\theta}}(2), \ldots$ converges in distribution.

12.1-6.Show that the part of the conclusion of the Stochastic Approximation Theorem which states that $\tilde{\boldsymbol{\theta}}(1), \tilde{\boldsymbol{\theta}}(2), \ldots$ converges with probability one implies that $\tilde{\boldsymbol{\theta}}(1), \tilde{\boldsymbol{\theta}}(2), \ldots$ converges in mean-square. HINT: See Theorem 9.3.8.

12.1-7.Assume $\tilde{x}(0) = 0$. Consider the following stochastic dynamical system:

$$\tilde{x}(t+1) = \tilde{x}(t) - \eta_t[2\tilde{x}(t) - 10] - \eta_t \lambda \tilde{x}(t) + \eta_t \tilde{n}(t)$$

where $\eta_t = 1$ for $t \leq 1000$, $\eta_t = t/(1 + 10t + t^2)$ for $t > 1000$, $\lambda$ is a positive number, and $\{\tilde{n}(t)\}$ is a sequence of independent and identically distributed zero-mean truncated approximate Gaussian random variables constructed as in Example 9.3.2. Prove that $\tilde{x}(0), \tilde{x}(1), \ldots$ is not bounded or that $\tilde{x}(0), \tilde{x}(1), \ldots$ will converge with probability one to the number $10/(2+\lambda)$. HINT: Assume the stochastic dynamical system is a stochastic gradient descent algorithm that is minimizing some function $f(x)$ where $df/dx = 2x - 10 + \lambda x$.

## 12.2 Learning in Passive Statistical Environments Using SA

Stochastic approximation algorithms provide a methodology for the analysis and design of adaptive learning machines. Assume the statistical environment for an adaptive learning machine generates a sequence of training stimuli $\mathbf{x}(1), \mathbf{x}(2), \ldots$ which are presumed to be a realization of a sequence of independent and identically distributed random vectors $\tilde{\mathbf{x}}(1), \tilde{\mathbf{x}}(2), \ldots$ with common density $p_e : \mathcal{R}^d \rightarrow [0, \infty)$. It is assumed that the goal of the adaptive learning machine is to minimize the risk function $\ell : \mathcal{R}^q \rightarrow \mathcal{R}$ defined such that:

$$\ell(\boldsymbol{\theta}) = \int c(\mathbf{x}, \boldsymbol{\theta}) p_e(\mathbf{x}) d\nu(\mathbf{x}) \quad (12.31)$$

where $c(\mathbf{x}, \boldsymbol{\theta})$ is the penalty incurred by the learning machine for having the parameter vector $\boldsymbol{\theta}$ when the environmental event $\mathbf{x}$ is presented to the learning machine.

### 12.2.1 Implementing Different Optimization Strategies

**Example 12.2.1** (Minibatch Stochastic Gradient Descent Learning)**.** Let $c : \mathcal{R}^d \times \mathcal{R}^q \rightarrow \mathcal{R}$ be the loss function for the risk function in (12.31) for a passive statistical environment which generates the i.i.d observations $\tilde{\mathbf{x}}(1), \tilde{\mathbf{x}}(2), \ldots$. Let $\mathbf{g} : \mathcal{R}^d \times \mathcal{R}^q \rightarrow \mathcal{R}^q$ be defined such that $\mathbf{g}(\mathbf{x}, \cdot)$ is the gradient of $c(\mathbf{x}, \cdot)$ for each $\mathbf{x} \in \mathcal{R}^d$. Because $p_e$ is not functionally dependent upon the current state of the learning machine, this is an example of learning in a passive statistical environment.

A minibatch stochastic gradient descent algorithm is defined by beginning with an initial guess for the parameter values of the learning machine denoted by $\tilde{\boldsymbol{\theta}}(0)$, observing a *minibatch* $\tilde{\mathbf{x}}(1), \ldots, \tilde{\mathbf{x}}(m)$ of observations from the environment, and then updating that initial guess $\tilde{\boldsymbol{\theta}}(0)$ to obtain a refined estimate called $\tilde{\boldsymbol{\theta}}(k+1)$ using the iterative formula:

$$\tilde{\boldsymbol{\theta}}(k+1) = \tilde{\boldsymbol{\theta}}(k) - (\gamma_k/m) \sum_{j=1}^{m} \mathbf{g}\left(\tilde{\mathbf{x}}\left(j + (k-1)m\right), \boldsymbol{\theta}(k)\right). \quad (12.32)$$

After this update is completed, the next minibatch of $m$ observations $\tilde{\mathbf{x}}(m+1), \ldots, \tilde{\mathbf{x}}(2m)$ is selected and the update rule in (12.32) is applied again. The minibatches are chosen in this example to be non-overlapping in order to satisfy the i.i.d. assumption required by the version of the stochastic approximation theorem presented in this chapter.

The positive integer $m$ is called the *minibatch size.* Note that when $m = 1$, (12.32) corresponds to the standard stochastic gradient descent learning situation where the parameter vector $\tilde{\boldsymbol{\theta}}(k)$ is updated each time an observation $\tilde{\mathbf{x}}(k)$ is observed as in Example 12.1.1.

Use the Stochastic Approximation Theorem to provide sufficient conditions to ensure that the stochastic sequence $\tilde{\boldsymbol{\theta}}(1), \tilde{\boldsymbol{\theta}}(2), \tilde{\boldsymbol{\theta}}(3), \ldots$ converges with probability one to the set of critical points in $\Theta$ with probability one.

**Example 12.2.2** (Adaptive Learning for Logistic Regression). Let

$$S \equiv \{(\mathbf{s}^1, y^1), \ldots, (\mathbf{s}^M, y^M)\}$$

where $y^k \in \{0, 1\}$ is the desired binary response to input pattern vector $\mathbf{s}^k \in \mathcal{R}^d$ for $k = 1, \ldots, M$. Assume the statistical environment samples with replacement from the set $S$ and thus generates the stochastic sequence:

$$(\tilde{\mathbf{s}}(1), \tilde{y}(1)), (\tilde{\mathbf{s}}(2), \tilde{y}(2)), \ldots$$

where the discrete random vector $(\tilde{\mathbf{s}}(t), \tilde{y}(t))$ takes on one of the $M$ possible values in $S$ with equal probability for all $t \in \mathbb{N}$.

The objective function for learning, $\ell : \mathcal{R}^{d+1} \rightarrow \mathcal{R}$, is defined such that for all $\boldsymbol{\theta} \in \mathcal{R}^{d+1}$:

$$\ell(\boldsymbol{\theta}) = \lambda \left|\boldsymbol{\theta}\right|^2 - (1/M) \sum_{k=1}^{M} \left[y^k \log\left(\ddot{y}(\mathbf{s}^k, \boldsymbol{\theta})\right) + (1 - y^k) \log\left(1 - \ddot{y}(\mathbf{s}^k, \boldsymbol{\theta})\right)\right]$$

where $\lambda$ is a positive number and the predicted response

$$\ddot{y}(\mathbf{s}^k, \boldsymbol{\theta}) = \mathcal{S}\left(\psi_k(\boldsymbol{\theta})\right), \quad \psi_k(\boldsymbol{\theta}) = \boldsymbol{\theta}^T [(\mathbf{s}^k)^T, \ 1]^T$$

with $\mathcal{S}(\psi) = (1 + \exp(-\psi))^{-1}$.

Derive an Adaptive Minibatch Gradient Descent Learning Algorithm that will minimize $\ell$. The algorithm should operate by randomly sampling a minibatch of $B$ training vectors from the statistical environment and update the parameters after each minibatch. This strategy is useful because a minibatch of observations will provide higher-quality estimates of the true gradient (i.e., $\nabla \ell$) for each parameter update.

Specifically, show that your proposed algorithm has the property that if the stochastic sequence of parameter estimates $\tilde{\boldsymbol{\theta}}(1), \tilde{\boldsymbol{\theta}}(2), \ldots$ converges with probability one to some vector $\boldsymbol{\theta}^*$, then the vector $\boldsymbol{\theta}^*$ is a global minimizer of $\ell$.

**Solution.** The solution follows the approach presented in Recipe Box 12.1.

*Step 1.* The statistical environment is passive and consists of a sequence of independent and identically distributed discrete random vectors such that each random vector $(\tilde{\mathbf{s}}_t, \tilde{y}_t)$ in the stochastic sequence $\{(\tilde{\mathbf{s}}_t, \tilde{y}_t)\}$ takes on the value of $(\mathbf{s}^k, y^k)$ with probability $1/M$ for $k = 1, \ldots, M$.

*Step 2.* The loss function $c : \{(\mathbf{s}^1, y^1), \ldots, (\mathbf{s}^M, y^M)\} \times \Theta \rightarrow \mathcal{R}$ defined such that:

$$c\left((\mathbf{s}, y), \boldsymbol{\theta}\right) = -y \log\left(\ddot{y}(\mathbf{s}, \boldsymbol{\theta})\right) - (1 - y) \log\left(1 - \ddot{y}(\mathbf{s}, \boldsymbol{\theta})\right).$$

The expected loss function

$$\ell(\boldsymbol{\theta}) = \lambda |\boldsymbol{\theta}|^2 + (1/M) \sum_{k=1}^{M} c\left([\mathbf{s}^k, y^k], \boldsymbol{\theta}\right)$$

is twice continuously differentiable. Since $\ell$ is a cross-entropy function, it has a lower bound since the entropy of the data generating process is finite (see Theorem 13.2.1). Define $\Theta$ to be a large closed ball of radius $R$ (e.g., $R = 10^9$). The region $\Theta$ is closed, bounded, and convex.

*Step 3.* Choose $\gamma_t = 1$ for $t < T_0$ and $\gamma_t = 1/t$ for all $t \geq T_0$. This choice will satisfy (12.11) and (12.12).

*Step 4.* Let $\mathbf{x}(t) \equiv [\mathbf{s}(t), y(t)]$. Let the $t$th minibatch of $B$ observations be denoted by $\mathbf{X}_t \equiv [\mathbf{x}(1 + (t-1)B), \ldots, \mathbf{x}(tB)]$. An adaptive gradient descent algorithm corresponds to the case where the search direction function $\mathbf{d}$ is chosen such that:

$$\mathbf{d}(\mathbf{X}_t, \boldsymbol{\theta}) = -2\lambda\boldsymbol{\theta} - (1/B)\sum_{b=1}^{B} \ddot{\mathbf{g}}(\mathbf{x}(b + (t-1)B), \boldsymbol{\theta})$$

where

$$\ddot{\mathbf{g}}(\mathbf{x}(z), \boldsymbol{\theta}) = -(y(z) - \ddot{y}(\mathbf{s}(z), \boldsymbol{\theta}))\,[(\mathbf{s}(z)^T, \ 1]^T.$$

Let

$$\tilde{\mathbf{X}}_t \equiv [(\tilde{\mathbf{s}}(1 + (t-1)B), \tilde{y}(1 + (t-1)B)), \ldots, (\tilde{\mathbf{s}}(tB), \tilde{y}(tB))].$$

The stochastic sequence $\{|\mathbf{d}(\tilde{\mathbf{X}}_t, \tilde{\boldsymbol{\theta}})|\}$ is bounded since the search direction function $\mathbf{d}$ is continuous, it is assumed that $\{\tilde{\mathbf{x}}(t)\}$ is a bounded stochastic sequence, and it is additionally assumed that the case where $\{\tilde{\boldsymbol{\theta}}(t)\}$ is a bounded stochastic sequence holds.

*Step 5.* Let $\bar{\mathbf{g}}$ denote the gradient of $\ell$ which is a continuous function. The expected value of $\mathbf{d}\left(\tilde{\mathbf{X}}_t, \boldsymbol{\theta}\right)$ is

$$\bar{\mathbf{d}}(\boldsymbol{\theta}) = -2\lambda\boldsymbol{\theta} - (1/M)\sum_{k=1}^{M} \ddot{\mathbf{g}}([\mathbf{s}^k, y^k], \boldsymbol{\theta}) = -\bar{\mathbf{g}}(\boldsymbol{\theta}).$$

Therefore, $\bar{\mathbf{d}}(\boldsymbol{\theta})^T \bar{\mathbf{g}}(\boldsymbol{\theta}) \leq -|\bar{\mathbf{g}}(\boldsymbol{\theta})|^2$.

*Step 6.* The Hessian of $\ell$ is positive semidefinite everywhere because:

$$\nabla^2 \ell = 2\lambda \mathbf{I}_{d+1} + (1/M)\sum_{k=1}^{M} y^k(1 - y^k)[(\mathbf{s}^k)^T, \ 1]^T[(\mathbf{s}^k)^T, \ 1]$$

and therefore $\boldsymbol{\theta}^T\left(\nabla^2\ell\right)\boldsymbol{\theta} \geq 0$ for all $\boldsymbol{\theta} \in \mathcal{R}^{d+1}$. Thus, every critical point of $\ell$ is a global minimizer. When the Hessian of $\ell$ is positive semidefinite, then all of the critical points are located in a flat region and all of the critical points are global minimizers of $\ell$.

Using the Stochastic Approximation Theorem, it follows that the stochastic sequence $\tilde{\boldsymbol{\theta}}(1), \tilde{\boldsymbol{\theta}}(2), \ldots$ does not remain in $\Theta$ with probability one or the stochastic sequence $\tilde{\boldsymbol{\theta}}(1), \tilde{\boldsymbol{\theta}}(2), \ldots$ converges to a random vector taking on values in the set of global minimizers of $\ell$ in $\Theta$ with probability one. △

**Example 12.2.3** (Adaptive Momentum for Gradient Descent Deep Learning). Nonlinear regression models are widely used in a variety of supervised learning applications. For example, many deep supervised learning machines can be viewed as smooth nonlinear regression models provided that smooth basis functions are used. Assume that the scalar predicted response of a nonlinear regression model $\ddot{y}(\mathbf{s}; \boldsymbol{\theta})$ to an input pattern vector $\mathbf{s}$ is a twice continuously differentiable function of $\boldsymbol{\theta}$.

Let

$$S \equiv \{(\mathbf{s}^1, y^1), \ldots, (\mathbf{s}^M, y^M)\}$$

where $y^k \in \{0,1\}$ is the desired binary response to input pattern vector $\mathbf{y}^k \in \mathcal{R}^d$ for $k = 1, \ldots, M$. Assume

$$(\tilde{\mathbf{s}}(1), \tilde{y}(1)), (\tilde{\mathbf{s}}(2), \tilde{y}(2)), \ldots$$

is a stochastic sequence of i.i.d discrete random vectors with common probability mass function $p_e : S \to [0, 1)$.

Let the objective function for learning $\ell : \mathcal{R}^q \to \mathcal{R}$ be defined such that:

$$\ell(\boldsymbol{\theta}) = \sum_{k=1}^{M} c([\mathbf{s}^k, y^k], \boldsymbol{\theta}) p_e(\mathbf{s}^k, y^k)$$

where for some positive number $\lambda$

$$c([\mathbf{s}, y], \boldsymbol{\theta}) = \lambda \left|\boldsymbol{\theta}\right|^2 + (y - \ddot{y}(\mathbf{s}; \boldsymbol{\theta}))^2 . \tag{12.33}$$

Let

$$\ddot{\mathbf{g}}([\mathbf{s}, y], \boldsymbol{\theta})) \equiv \left(\frac{dc([\mathbf{s}, y], \boldsymbol{\theta})}{d\boldsymbol{\theta}}\right)^T .$$

Note that the Hessian of the regularization term $\lambda \left|\boldsymbol{\theta}\right|^2$ is equal to the identity matrix multiplied by $2\lambda$. If the smallest eigenvalue of the Hessian of the second term on the right-hand side of (12.33) is greater than $-2\lambda$ on some region of the parameter space $\Theta$, then the regularization term will ensure that $\ell$ is positive definite on that region. If a critical point of $\ell$, $\boldsymbol{\theta}^*$, exists in the interior of $\Theta$ where $\ell$ is positive definite, then that critical point $\boldsymbol{\theta}^*$ is a strict local minimizer.

In deep learning networks, a popular algorithm for minimizing $\ell$ is gradient descent with momentum defined as follows. Assume $\tilde{\boldsymbol{\theta}}(0)$ is a vector of zeros. Define the stochastic sequence $\tilde{\boldsymbol{\theta}}(0), \tilde{\boldsymbol{\theta}}(1), \ldots$ so that it satisfies for all $t \in \mathbb{N}$:

$$\tilde{\boldsymbol{\theta}}(t+1) = \tilde{\boldsymbol{\theta}}(t) + \gamma_t \tilde{\mathbf{d}}(t) \tag{12.34}$$

where

$$\tilde{\mathbf{d}}(t) = -\tilde{\mathbf{g}}(t) + \tilde{\mu}_t \tilde{\mathbf{d}}(t-1), \tag{12.35}$$

$$\tilde{\mathbf{g}}(t) \equiv \ddot{\mathbf{g}}\left([\mathbf{s}(t), y(t)], \tilde{\boldsymbol{\theta}}(t)\right),$$

$\gamma_t = (t/1000)$ for $t < 1000$ and $\gamma_t = 1/t$ for $t \geq 1000$, and $\tilde{\mu}_t \in [0, 1]$ is chosen to be equal to zero when $\tilde{\mathbf{d}}(t-1)^T \tilde{\mathbf{g}}(t) \leq 0$.

The case where $\tilde{\mu}_t = 0$ is called a "gradient descent step" and the case where $\tilde{\mu}_t > 0$ is called a "momentum step". Intuitively, the effect of a momentum step is to encourage the learning algorithm to move in a direction in the parameter space which is similar to the direction moved at the previous step.

First, show that if the stochastic sequence $\tilde{\boldsymbol{\theta}}(1), \tilde{\boldsymbol{\theta}}(2), \ldots$ remains in a sufficiently small open convex set containing a strict local minimizer $\boldsymbol{\theta}^*$ with probability one, then the stochastic sequence $\tilde{\boldsymbol{\theta}}(1), \tilde{\boldsymbol{\theta}}(2), \ldots$ converges to $\boldsymbol{\theta}^*$ with probability one.

Let $\Theta$ be an closed, bounded, and convex subset of $\mathcal{R}^q$. Now show that the stochastic sequence $\tilde{\boldsymbol{\theta}}(1), \tilde{\boldsymbol{\theta}}(2), \ldots$ will converge to the set of critical points in $\Theta$ or a subsequence of $\tilde{\boldsymbol{\theta}}(1), \tilde{\boldsymbol{\theta}}(2), \ldots$ will remain outside of $\Theta$ with probability one.

**Solution.** The solution follows the approach discussed in Recipe Box 12.1.

*Step 1.* The statistical environment is passive and consists of independent and identically distributed discrete random vectors such that each random vector takes on a finite number of values.

*Step 2.* The objective function $\ell : \mathcal{R}^q$ is twice continuously differentiable. Since $c$ is a continuous random function with a finite lower bound, $\ell$ has a finite lower bound. This problem considers two different choices for $\Theta$. Let $\Theta_1$ be a closed, bounded, and convex set that contains exactly one strict local minimizer $\boldsymbol{\theta}^*$ in its interior such that the Hessian of $\ell$ evaluated at $\boldsymbol{\theta}^*$ is positive definite. Let $\Theta_2$ be a closed, bounded, and convex subset of $\mathcal{R}^q$.

*Step 3.* Show that the choice of the sequence $\eta_1, \eta_2, \ldots$ satisfies (12.11) and (12.12).

*Step 4 and 5.* Under the assumption that $\tilde{\mathbf{d}}(t-1)^T\tilde{\mathbf{g}}(\mathbf{t})$ is strictly positive, rewrite the search direction (12.35) as

$$\tilde{\mathbf{d}}(t) = -\tilde{\mathbf{M}}_t\tilde{\mathbf{g}}(t) \tag{12.36}$$

where

$$\tilde{\mathbf{M}}_t = \mathbf{I} - \tilde{\mu}_t\tilde{\mathbf{d}}(t-1)\tilde{\mathbf{d}}(t-1)^T \tag{12.37}$$

and

$$\tilde{\mu}_t = \frac{\mu}{\tilde{\mathbf{d}}(t-1)^T\tilde{\mathbf{g}}(t)}. \tag{12.38}$$

Suppose that every realization of the stochastic sequence $\tilde{\mathbf{M}}_1, \tilde{\mathbf{M}}_2, \ldots$ has the property that the largest eigenvalue of every realization of $\tilde{\mathbf{M}}_t$ is less than or equal to $\lambda_{max}$ and the smallest eigenvalue of every realization of $\tilde{\mathbf{M}}_t$ is greater than or equal to $\lambda_{min}$. Then, Example 12.1.2 may be used to show that this satisfies the necessary downhill condition.

Every vector $\mathbf{e}$ in the $d-1$-dimensional linear subspace whose elements are orthogonal to $\mathbf{d}(t-1)$ has the property that $\mathbf{M}_t\mathbf{e} = \mathbf{e}$ where $\mathbf{M}_t$ is a realization of $\tilde{\mathbf{M}}_t$ as defined in (12.37). This means that the $d-1$ eigenvalues of $\mathbf{M}_t$ associated with the eigenvectors that span this $d-1$-dimensional subspace have the value of one. The condition that the largest eigenvalue of a realization of $\tilde{\mathbf{M}}_t$ is less than $\lambda_{max}$ is therefore achieved by choosing $\lambda_{max} = 1$ for these $q-1$ eigenvalues.

The remaining eigenvalue $\tilde{\lambda}_t$ associated with a realization of the eigenvector $\tilde{\mathbf{d}}(t-1)$ is given by the formula $\tilde{\lambda}_t = 1 - \tilde{\mu}_t|\tilde{\mathbf{d}}(t-1)|^2$. This eigenvalue formula is derived by showing that

$$\mathbf{M}_t\mathbf{d}(t-1) = (1 - \mu_t|\mathbf{d}(t-1)|^2)\mathbf{d}(t-1)$$

where $\mathbf{M}_t$ is a realization of $\tilde{\mathbf{M}}_t$ as defined in (12.37). Thus, this is the only condition which needs to be checked.

To check this constraint on $\tilde{\lambda}_t$, it is only necessary to show that $\lambda_{min} < \tilde{\lambda}_t < \lambda_{max}$ which yields the constraint:

$$\lambda_{min} \leq 1 - \tilde{\mu}_t|\tilde{\mathbf{d}}(t-1)|^2 \leq \lambda_{max}. \tag{12.39}$$

One possible algorithm for ensuring (12.39) holds is to simply check each step of the algorithm to determine if a particular realization of $\tilde{\mu}_t$ as defined in (12.38) satisfies the constraint in (12.39). If (12.39) is not satisfied, set $\tilde{\mu}_t = 0$ to realize a gradient descent step.

*Step 6.* Conclude that if $\tilde{\boldsymbol{\theta}}(1), \tilde{\boldsymbol{\theta}}(2), \ldots$ remains in $\Theta_1$ with probability one, then $\tilde{\boldsymbol{\theta}}(1), \tilde{\boldsymbol{\theta}}(2), \ldots$ will converge to $\boldsymbol{\theta}^*$ with probability one. In addition, conclude that if $\tilde{\boldsymbol{\theta}}(1), \tilde{\boldsymbol{\theta}}(2), \ldots$ remains in $\Theta_2$ with probability one, then $\tilde{\boldsymbol{\theta}}(1), \tilde{\boldsymbol{\theta}}(2), \ldots$ will converge to the set of critical points in $\Theta_2$ with probability one. △

**Example 12.2.4** (Random Block Gradient Coordinate Descent). Random block coordinate descent is a useful tool for dealing with non-convex optimization problems. The essential idea is that only a randomly chosen subset of parameters are updated at each iteration of the algorithm.

Let $\gamma_1, \gamma_2, \ldots$ be a sequence of positive stepsizes that satisfy the conditions of the Stochastic Approximation Theorem. Let

$$\ell(\boldsymbol{\theta}) \equiv \sum_S c(\mathbf{x}, \boldsymbol{\theta})p_e(\mathbf{x})$$

be an expected loss function defined with respect to loss function $c(\mathbf{x}, \boldsymbol{\theta})$ and probability mass function $p_e : \{\mathbf{x}^1, \ldots, \mathbf{x}^M\} \rightarrow [0, 1]$. Let $\mathbf{g}(\mathbf{x}, \boldsymbol{\theta}) = (dc/d\boldsymbol{\theta})^T$. Let $\tilde{\mathbf{x}}(1), \tilde{\mathbf{x}}(2), \ldots$ be a stochastic sequence of i.i.d. random vectors with common probability mass function $p_e$.

Let $Q \subseteq \{0, 1\}^q$ be a non-empty set of $q$-dimensional binary vectors such that $\mathbf{0}_q \notin Q$.

For random block coordinate descent, it is assumed that the data generating process works according to the following two-step process. First, sample from $p_e$ to obtain $(\tilde{\mathbf{s}}(t), \tilde{y}(t)) \in S$ at iteration $t$. Second, sample with replacement from $Q$ to generate a realization of the random vector $\tilde{\mathbf{m}}_t$ at iteration $t$. Third, update the parameter vector $\tilde{\boldsymbol{\theta}}(t)$ using the update rule:

$$\tilde{\boldsymbol{\theta}}(t+1) = \tilde{\boldsymbol{\theta}}(t) - \gamma_t \tilde{\mathbf{m}}_t \odot \mathbf{g}(\tilde{\mathbf{x}}(t), \tilde{\boldsymbol{\theta}}(t)). \tag{12.40}$$

Show that the average search direction, $\bar{\mathbf{d}}(\boldsymbol{\theta})$, is given by the formula:

$$\bar{\mathbf{d}}(\boldsymbol{\theta}) = -E\left\{\mathbf{d}\left([\tilde{\mathbf{s}}_t, \tilde{y}_t], \boldsymbol{\theta}\right)\right\} = -E\{\tilde{\mathbf{m}}(t)\} \odot \mathbf{g}.$$

This formula for the average search direction for block gradient descent shows that the $k$th parameter of the learning machine is updated using the learning rate $\gamma\delta_k$ where $\delta_k$ is the expected percentage of times that the $k$th element of $\tilde{\mathbf{m}}_t$ is observed to be equal to the number one.

Use the Stochastic Approximation Theorem to analyze the behavior of random block coordinate descent. △

### 12.2.2 Improving Generalization Performance

**Example 12.2.5** (Dropout Learning Convergence Analysis). In model search problems, the goal is to find the probability model in $\mathcal{M}$ which is the "best" and then estimate the parameters of that model and use that model with its estimated parameters for the purpose of inference and decision making.

Alternatively, rather than using just one of the models in $\mathcal{M}$, one could try to use all of the models in a model space $\mathcal{M}$ to support improved inference and decision-making performance. Inferences from multiple models are averaged together using frequentist model averaging (FMA) and Bayesian model averaging (BMA) methods (Burnham and Anderson 2010; Hjort and Claeskens 2003).

Dropout (Srivastava et al. 2014; Goodfellow et al. 2016) is a model averaging method which shares important similarities with FMA and BMA but differs from FMA/BMA methods because it allows for parameter sharing among the models in the model space. In this problem, a version of dropout learning is analyzed using the stochastic approximation analysis method introduced in this chapter.

Let $\Theta \subseteq \mathcal{R}^q$ be a closed, bounded and convex set. Let $p : \mathcal{R}^d \times \Theta \rightarrow [0, \infty)$ be defined such that $p(\cdot; \boldsymbol{\theta})$ is a Radon-Nikodým density with respect to support specification measure $\nu$ for each $\boldsymbol{\theta} \in \Theta$.

Let the *model space* $\mathcal{M} \equiv \{\mathcal{M}^1, \mathcal{M}^2, \ldots, \mathcal{M}^M\}$ be a finite set of $M$ probability models defined with respect to the same $q$-dimensional parameter space $\Theta$. In addition, assume that each $\mathcal{M}^k \in \mathcal{M}$ is defined such that

$$\mathcal{M}^k \equiv \left\{p\left(\mathbf{x}|\boldsymbol{\zeta}_k \odot \boldsymbol{\theta}\right) : \boldsymbol{\zeta}_k \in \{0, 1\}^q, \boldsymbol{\theta} \in \Theta\right\}, k = 1, \ldots, M.$$

The $q$-dimensional *parameter deletion specification mask* vector $\boldsymbol{\zeta}_k$ is defined such that the $j$th element of $\boldsymbol{\zeta}_k$ is equal to zero if and only if the $j$th parameter of the parameter vector $\boldsymbol{\theta}$ in model $\mathcal{M}^k$ is always constrained to be equal to zero.

Note that although all $M$ models share the same parameter space $\Theta$, this setup still permits specific models to introduce restrictions on the parameter space $\Theta$. So, for example, $\mathcal{M}_1$ may be a collection of probability densities that constrains 2 parameter values to be zero, while $\mathcal{M}_2$ may be a collection of densities that constrains 5 parameter values to be zero.

Next, it is assumed there exists a *model prior* probability mass function $p_{\mathcal{M}}$ that assigns a probability or preference for each model in the set of models. If all models are considered to be equally preferable, then one may choose $p_{\mathcal{M}}(\mathcal{M}^k) = 1/M$ for $k = 1, \ldots, M$. Drop-out deep learning methods can be interpreted as assuming a uniform distribution on a set of network architectures that have different numbers of hidden units.

It is also assumed that one can generate random samples from $p_{\mathcal{M}}$ and pick a model $\mathcal{M}^k$ at random with probability $p_{\mathcal{M}}(\mathcal{M}^k)$. If sampling from $p_{\mathcal{M}}$ is computationally intractable, then the Monte Carlo Markov Chain sampling methods of Chapter 11 may be used to approximately sample from $p_{\mathcal{M}}$ but the effects of this approximation will be ignored here.

The data observed by the learning machine is a sequence of vectors $\mathbf{x}_1, \mathbf{x}_2, \ldots$ which is assumed to be a realization of a stochastic sequence of independent and identically distributed random vectors $\tilde{\mathbf{x}}_1, \tilde{\mathbf{x}}_2, \ldots$ with common Radon-Nikodým density $p_e(\cdot)$ defined with respect to support specification measure $\nu$.

The objective function for learning is now defined by the formula:

$$\ell(\boldsymbol{\theta}) = \sum_{k=1}^{M} \int c(\mathbf{x}, \boldsymbol{\theta}, \mathcal{M}^k) p_{\mathcal{M}}(\mathcal{M}^k) p_e(\mathbf{x}) d\nu(\mathbf{x}) \tag{12.41}$$

which explicitly takes into account our uncertainty regarding which of the $M$ models in $\mathcal{M}$ actually generated the observed data. It should be noted that the model averaging loss function in (12.41) can be alternatively represented as:

$$\ell(\boldsymbol{\theta}) = \int \ddot{c}(\mathbf{x}, \boldsymbol{\theta}) p_e(\mathbf{x}) d\nu(\mathbf{x}), \quad \ddot{c}(\mathbf{x}, \boldsymbol{\theta}) = \sum_{k=1}^{M} c(\mathbf{x}, \boldsymbol{\theta}, \mathcal{M}^k) p_{\mathcal{M}}(\mathcal{M}^k).$$

For example, let $\ddot{y}(\mathbf{s}, \boldsymbol{\theta}, \mathcal{M}^k)$ be the predicted response of the learning machine given input pattern vector $\mathbf{s}$ when the learning machine assumes that model $\mathcal{M}^k$ should be used and that the parameter vector is $\boldsymbol{\theta}$. Assume the model prior $p_{\mathcal{M}}$ is known (e.g., $p_{\mathcal{M}}(\mathcal{M}^k) = 1/M$). Note that when generating predictions from this model it is necessary to average over the possible values of $\mathcal{M}^k$. So when the number of models in $\mathcal{M}$, $M$, is relatively small (e.g., $M = 10^5$) then the predicted response of the learning machine for a given input pattern vector $\mathbf{s}$ and parameter vector $\boldsymbol{\theta}$ is given by the formula:

$$\bar{\ddot{y}}(\mathbf{s}, \boldsymbol{\theta}) = \sum_{k=1}^{M} \ddot{y}(\mathbf{s}, \boldsymbol{\theta}, \mathcal{M}^k) p_{\mathcal{M}}(\mathcal{M}^k).$$

When $M$ is large, then an approximation based upon the law of large numbers can be used by generating random $T$ samples $\tilde{\mathcal{M}}_1, \ldots, \tilde{\mathcal{M}}_T$ from $p(\mathcal{M}^k)$ and using the formula:

$$\bar{\ddot{y}}(\mathbf{s}, \boldsymbol{\theta}) \approx (1/T) \sum_{t=1}^{T} \ddot{y}(\mathbf{s}, \boldsymbol{\theta}, \tilde{\mathcal{M}}_t).$$

Let the stepsize sequence $\gamma_1, \gamma_2, \ldots$ be a sequence of positive numbers that satisfies (12.11) and (12.12). A stochastic gradient descent algorithm that minimizes $\ell$ is given by the algorithm:

$$\tilde{\boldsymbol{\theta}}(t+1) = \tilde{\boldsymbol{\theta}}(t) - \gamma_t \ddot{\mathbf{g}}(\tilde{\mathbf{x}}(t), \tilde{\boldsymbol{\theta}}(t), \tilde{\mathcal{M}}_t) \tag{12.42}$$

where a realization of $\tilde{\mathcal{M}}_t$ is obtained by randomly sampling a model from $p_{\mathcal{M}}$ and a realization of $\tilde{\mathbf{x}}(t)$ is obtained by randomly sampling from $p_e$.

Explicitly identify a set of sufficient conditions that ensure the Stochastic Approximation Theorem can be used for analyzing the asymptotic behavior of (12.42). △

**Example 12.2.6** (Data Augmentation Methods for Improving Generalization). Data augmentation strategies are widely used in applications of deep learning to computer vision problems. In a typical application of this technique, the original training data consists of tens of thousands of photos of real-world situations. Then, this large collection of training data is augmented with hundreds of thousands of artificially generated photos. The artificially generated photos are often obtained by applying a variety of transformations to the original photos such as rotating the images at different angles, zooming in on different parts of the images, and chopping the original images into smaller pieces. The resulting augmented training data set is intended to provide a more realistic model of the richness and diversity of image data encountered in the real world. Such data augmentation strategies may be interpreted as extremely complicated nonlinear preprocessing transformations of the original data generating process (see Exercise 12.2-8).

The concept of data augmentation was previously discussed in the context of the Denoising Nonlinear Autoencoder (see Example 1.6.2). The Denoising Nonlinear Autoencoder works by injecting "feature noise" into the input pattern vector to improve generalization performance. Exercise 12.2-8 provides another data augmentation example which implements a data-dependent regularization strategy.

These ideas are now illustrated within the context of a simple, yet important, feature noise injection modeling strategy. Consider a medical application where each training vector is a five-dimensional random vector $\tilde{\mathbf{x}}$ consisting of the three real-valued random variables $\tilde{x}_1$, $\tilde{x}_2$, and $\tilde{x}_3$ which are used to respectively represent a patient's temperature, a patient's heart rate, and a patient's age as well as the two binary random variables $\tilde{x}_4$ and $\tilde{x}_5$ which respectively indicate the gender of the patient is female and whether the patient has hypertension. Suppose that a realization of $\tilde{\mathbf{x}}$ is observed in the training data set such as $\mathbf{x} = (98.6, 60, 40, 1, 0)$, then one might want to convey to the learning machine that, based upon observing this particular realization, it would not be surprising to observe vectors such as $(98.5, 59, 41, 1, 0)$, $99, 62, 39, 1, 0)$ or $(98.5, 60, 39, 0, 0)$. This type of information can be represented in the learning machine's design by feature noise injection. For example, add zero-mean Gaussian noise with some small variance to the first three elements of an observed training data vector and flip the fourth bit of the observed training data vector with some probability close to one-half. The training procedure in this case would involve selecting one of the $n$ training stimuli at random, then injecting feature noise, and then training the learning machine with the feature-noise injected training stimuli.

More formally, the feature noise injection process in this example is expressed by a $d$-dimensional *feature noise* random vector $\tilde{\mathbf{n}} \equiv [\tilde{n}_1, \ldots, \tilde{n}_5]$ defined such that $\tilde{n}_i$ has a zero-mean Gaussian density with variance $\sigma_n$ for $i = 1, \ldots 3$, $\tilde{n}_4$ takes on the value of 1 with probability 1/2 and takes on the value of 0 otherwise, and $\tilde{n}_5 = 1$ with probability $\alpha$ and $\tilde{n}_5 = 0$ with probability $1 - \alpha$. In addition, the data augmentation process is specified by the *data augmentation function* $\boldsymbol{\phi} : \mathcal{R}^d \times \mathcal{R}^d \rightarrow \mathcal{R}^d$ defined such that:

$$\boldsymbol{\phi}(\mathbf{x}) = [x_1 + \tilde{n}_1,\ x_2 + \tilde{n}_2,\ x_3 + \tilde{n}_3,\ \tilde{n}_4,\ \tilde{n}_5 x_5 + (1 - \tilde{n}_5)(1 - x_5)]^T.$$

If feature noise is not injected into the learning machine, then the expected loss function for learning is given by the formula:

$$\ell(\boldsymbol{\theta}) = \int c(\mathbf{x}, \boldsymbol{\theta}) p_e(\mathbf{x}) d\nu(\mathbf{x}) \tag{12.43}$$

where $c : \mathcal{R}^d \times \mathcal{R}^q \to \mathcal{R}$ is the loss function. Methods for designing adaptive learning algorithms for minimizing this expected loss function have been previously discussed.

The case where feature noise is injected (or equivalently the data generating process is transformed) is now considered. Let the $k$-dimensional feature noise vector $\tilde{\mathbf{n}}$ have Radon-Nikodým density $p_n : \mathcal{R}^k \to [0, \infty)$ with respect to support specification measure $\mu$. The data augmentation function is formally represented as the random function $\boldsymbol{\phi} : \mathcal{R}^d \times \mathcal{R}^k \to \mathcal{R}^d$ which specifies how the artificially generated $k$-dimensional noise vector $\tilde{\mathbf{n}}(t)$ is used to transform the original $d$-dimensional feature vector $\mathbf{x}(t)$. In this example, the augmented $d$-dimensional feature vector $\ddot{\tilde{\mathbf{x}}}(\mathbf{t})$ is generated using the formula: $\ddot{\tilde{\mathbf{x}}}(t) = \boldsymbol{\phi}(\tilde{\mathbf{x}}(t), \tilde{\mathbf{n}}(t))$. The goal of feature noise learning with respect to the no-feature-noise objective function $\ell$ in (12.43) is to minimize the feature-noise objective function:

$$\ddot{\ell}(\boldsymbol{\theta}) = \int c(\boldsymbol{\phi}(\mathbf{x}, \mathbf{n}), \boldsymbol{\theta})\, p_e(\mathbf{x}) p_n(\mathbf{n}) d\nu(\mathbf{x}) d\mu(\mathbf{n}). \tag{12.44}$$

A Stochastic Gradient Descent algorithm that minimizes the feature-noise objective function $\ddot{\ell}$ with respect to the no-feature-noise objective function $\ell$ may be constructed using an iterative formula of the form:

$$\tilde{\boldsymbol{\theta}}(t+1) = \tilde{\boldsymbol{\theta}}(t) - \gamma_t \mathbf{g}(\ddot{\tilde{\mathbf{x}}}(t), \tilde{\boldsymbol{\theta}}(t)) \tag{12.45}$$

where $\mathbf{g}$ is the gradient of $\ell$, $\ddot{\tilde{\mathbf{x}}}(t) = \boldsymbol{\phi}(\tilde{\mathbf{x}}(t), \tilde{\mathbf{n}}(t))$, $\tilde{\mathbf{x}}(t)$ is generated from the statistical environment, and $\tilde{\mathbf{n}}(t)$ is injected noise designed to improve generalization performance.

Show how to analyze the asymptotic behavior of the adaptive learning algorithm in (12.45) using the Stochastic Approximation Theorem. △

## Exercises

12.2-1. *Adaptive Averaged Stochastic Gradient Descent.* In averaged stochastic descent, one has a standard stochastic approximation algorithm which has updates of the form:

$$\tilde{\boldsymbol{\theta}}(t+1) = \tilde{\boldsymbol{\theta}}(t) + \gamma_t \tilde{\mathbf{d}}(t)$$

and suppose that $\tilde{\boldsymbol{\theta}}(t) \to \tilde{\boldsymbol{\theta}}^*$ with probability one as $t \to \infty$. In averaged stochastic descent, we define an additional update rule given by:

$$\bar{\boldsymbol{\theta}}(t+1) = \bar{\boldsymbol{\theta}}(t)(1-\delta) + \delta\tilde{\boldsymbol{\theta}}(t+1)$$

where $0 < \delta < 1$. The estimator $\bar{\boldsymbol{\theta}}(t+1)$ is used to estimate $\tilde{\boldsymbol{\theta}}^*$ rather than $\tilde{\boldsymbol{\theta}}(t+1)$. Prove that $\bar{\boldsymbol{\theta}}(t+1) \to \boldsymbol{\theta}^*$ with probability one as $t \to \infty$. Show that if $\delta = 1/(m+1)$ and $\bar{\boldsymbol{\theta}}(t) = (1/m) \sum_{j=1}^{m} \tilde{\boldsymbol{\theta}}(t-j+1)$, then

$$\bar{\boldsymbol{\theta}}(t+1) = \frac{1}{m+1} \sum_{j=1}^{m+1} \tilde{\boldsymbol{\theta}}(t-j+1).$$

12.2-2. *Adaptive Unsupervised Learning of High-Order Correlations.* Consider a statistical environment specified by a sequence of bounded, independent, and identically distributed random vectors $\tilde{\mathbf{x}}(1), \tilde{\mathbf{x}}(2), \ldots$ with common Radon-Nikodým density

$p_e : \mathcal{R}^d \rightarrow [0, \infty)$ with respect to some support specification measure $\nu$. Assume that the goal of an adaptive unsupervised learning machine is to learn high-order correlations from its statistical environment by minimizing the risk function

$$\ell(\boldsymbol{\theta}) = E\{\tilde{\mathbf{x}}(t)^T \mathbf{A} [\tilde{\mathbf{x}}(t) \otimes \tilde{\mathbf{x}}(t)] + \tilde{\mathbf{x}}(t)^T \mathbf{B} \tilde{\mathbf{x}}(t) + \mathbf{c}^T \tilde{\mathbf{x}}(t)\} + \lambda |\boldsymbol{\theta}|^2$$

where $\boldsymbol{\theta}^T \equiv [\mathbf{vec}(\mathbf{A})^T, \ \mathbf{vec}(\mathbf{B})^T, \ \mathbf{c}^T]$ and $\lambda > 0$. Provide explicit formulas for $\ell$ and $d\ell/d\boldsymbol{\theta}$ in terms of $p_e$. Design a stochastic approximation algorithm that minimizes $\ell$ which includes an explicit statement of the annealing schedule, the non-autonomous time-varying iterated map, and a step-by-step analysis of the asymptotic behavior of your proposed algorithm.

12.2-3. *Convergence of LMS Adaptive Learning Rule.* Consider a statistical environment defined by the sequence of independent and identically distributed random vectors:

$$(\tilde{\mathbf{s}}(0), \tilde{\mathbf{y}}(0)), (\tilde{\mathbf{s}}(1), \tilde{\mathbf{y}}(1)), \ldots, (\tilde{\mathbf{s}}(t), \tilde{\mathbf{y}}(t)), \ldots$$

where the probability distribution of the $t$th random vector is defined by the probability mass function which assigns a probability $p_k$ to the ordered pair $(\mathbf{s}^k, \mathbf{y}^k)$ where $k = 1 \ldots M$ and $M$ is finite. Assume $(\mathbf{s}^k, \mathbf{y}^k)$ is finite for $k = 1 \ldots M$ as well. An LMS (Least Means Squares) adaptive learning rule (Widrow and Hoff 1960) is defined as follows.

$$\tilde{\mathbf{W}}(t+1) = \tilde{\mathbf{W}}(t) + \gamma_t [\tilde{\mathbf{y}}(t) - \ddot{\mathbf{y}}(\tilde{\mathbf{s}}(t), \tilde{\mathbf{W}}(t))] \tilde{\mathbf{s}}(t)^T$$

where $\ddot{\mathbf{y}}(\tilde{\mathbf{s}}(t), \tilde{\mathbf{W}}(t)) = \mathbf{W}(t)\tilde{\mathbf{s}}(t)$. Let the parameter vector $\boldsymbol{\theta} \equiv \mathbf{vec}(\mathbf{W})$ contain the columns of $\mathbf{W}$. Provide sufficient conditions for ensuring that the LMS learning rule converges to the set of global minima of the objective function:

$$\ell(\boldsymbol{\theta}) = E\left\{|\tilde{\mathbf{y}} - \mathbf{W}\tilde{\mathbf{s}}|^2\right\} = \sum_{k=1}^{M} p_k |\mathbf{y}^k - \mathbf{W}\mathbf{s}^k|^2.$$

12.2-4. *Convergence of LMS Learning Rule with Regularization.* Let the function $\mathcal{J} : \mathcal{R}^q \rightarrow \mathcal{R}^q$ be defined such that the $k$th element of $\mathcal{J}([\theta_1, \ldots, \theta_q])$ is equal to $\log(1 + \exp(\theta_k))$ for $k = 1, \ldots, q$. Repeat Exercise 12.2-3 but use the objective function:

$$\ell(\boldsymbol{\theta}) = E\left\{|\tilde{\mathbf{y}} - \mathbf{W}\tilde{\mathbf{s}}|^2\right\} + \lambda(\mathbf{1}_q)^T \left(\mathcal{J}(\boldsymbol{\theta}) + \mathcal{J}(-\boldsymbol{\theta})\right).$$

12.2-5. *Convergence of Momentum Learning.* Repeat Example (12.2.3) but use the following specific choice for the response function $\ddot{y}$. Specifically, define $\ddot{y}$ such that:

$$\ddot{y}(\mathbf{s}, \boldsymbol{\theta}) = \sum_{j=1}^{M} w_j h(\mathbf{s}, \mathbf{v}_j)$$

where $\boldsymbol{\theta}^T \equiv [w_1, \ldots, w_M, \mathbf{v}_1^T, \ldots, \mathbf{v}_M^T]$, and $h(\mathbf{s}, \mathbf{v}_j) = \exp\left[-|\mathbf{s} - \mathbf{v}_j|^2\right]$. This network architecture is a multi-layer feedforward perceptron with one layer of radial basis function hidden units.

12.2-6. *Adaptive Minibatch Natural Gradient Descent Algorithm Learning.* Use the Stochastic Approximation Theorem to analyze the asymptotic behavior of an adaptive version of the Natural Gradient Descent Algorithm in Exercise 7.3-2.

12.2-7. *Adaptive L-BFGS Algorithm.* Design a minibatch adaptive learning version of the L-BFGS algorithm as described in Algorithm 7.3.3 by selecting a batch of observations, computing the L-BFGS search direction for that batch using Algorithm 7.3.3, then check if the angular separation between the gradient descent search direction and the L-BFGS search direction is less than 88 degrees. If the LBFGS descent search direction is not less than 88 degrees, then choose the gradient descent search direction.

12.2-8. *Eliminating Hidden Units Using Data-Dependent Regularization.* Following Chauvin (1989), add a data-dependent regularization term that attempts to minimize the number of active hidden units in Exercise 12.2-5 which has the form: $\lambda_2 E\left\{\sum_{j=1}^{M} |h(\tilde{\mathbf{s}}, \mathbf{v}_j)|^2\right\}$.

12.2-9. *Analysis of Hidden Units with Orthogonal Responses.* Repeat Exercise 12.2-5 but add a regularization term which attempts to adjust the parameters of the hidden units so that the parameter vector $\mathbf{v}_j$ for hidden unit $j$ is approximately orthogonal to a parameter vector $\mathbf{v}_k$ for hidden unit $k$. Using the notation in Exercise 12.2-5 this corresponds to a regularization term of the form:

$$\lambda_3 E\left\{\sum_{j=1}^{M} \sum_{k=(j+1)}^{M} h(\tilde{\mathbf{s}}, \mathbf{v}_j) h(\tilde{\mathbf{s}}, \mathbf{v}_k)\right\}$$

which is added on to the loss function $\ell$ where the expectation is taken with respect to $p_e$. Note that this regularization term is data dependent.

12.2-10. *Analysis of Data Augmentation for Computer Vision.* Show how the methods of Example 12.2.6 may be used to explicitly specify an empirical risk function that represents data augmentation strategies for transforming images to support deep learning applications. Then, explicitly specify a stochastic approximation learning algorithm which uses the data augmentation strategy you have proposed. The algorithm should be designed so that the Stochastic Approximation Theorem is applicable.

12.2-11. *RMSPROP Adaptive Learning Algorithm.* Specify an adaptive minibatch version of the RMSPROP algorithm in Exercise 7.3-3 which can be analyzed using the Stochastic Approximation Theorem.

12.2-12. *Adaptive Residual Gradient Reinforcement Learning.* Assume a statistical environment which generates a sequence of independent and identically distributed bounded observations $\tilde{\mathbf{x}}(1), \tilde{\mathbf{x}}_(2), \ldots$ with common density $p_e : \mathcal{R}^d \to [0, \infty)$ with respect to support specification measure $\nu$. Assume, in addition, that $\tilde{\mathbf{x}}(t) = [\mathbf{s}_0(t), \mathbf{s}_1(t), r_1(t)]$ where $\mathbf{s}_0(t)$ is the initial state of episode $t$, $\mathbf{s}_1(t)$ is the second state of episode $t$, and $r_{0,1}(t)$ is the reinforcement signal received when the learning machine transitions from the first state $\mathbf{s}_o(t)$ to the second state $\mathbf{s}_1$.

Let the reinforcement risk function, $\ell(\boldsymbol{\theta}) = \int c(\mathbf{x}, \boldsymbol{\theta}) p_e(\mathbf{x}) d\nu(\mathbf{x})$ with loss function

$$c(\mathbf{x}, \boldsymbol{\theta}) = \left(r_{0,1} - \left(\ddot{V}(\mathbf{s}_0; \boldsymbol{\theta}) - \mu \ddot{V}(\mathbf{s}_1; \boldsymbol{\theta})\right)\right)^2$$

where $\mathbf{x} = [\mathbf{s}_0, \mathbf{s}_1, r_{0,1}]$, $0 < \mu < 1$, and $\ddot{V}(\mathbf{s}; \boldsymbol{\theta}) = \boldsymbol{\theta}^T [\mathbf{s}^T \ \ 1]^T$. Let $\Theta$ be a closed and bounded set which contains the unique global minimizer of, $\boldsymbol{\theta}^*$, of $\ell$. Design a stochastic gradient descent algorithm which generates a stochastic sequence of states $\tilde{\boldsymbol{\theta}}(1), \tilde{\boldsymbol{\theta}}(2), \ldots$ that converges to $\boldsymbol{\theta}^*$ provided that $\tilde{\boldsymbol{\theta}}(1), \tilde{\boldsymbol{\theta}}(2), \ldots$ is bounded with probability one. HINT: Show that the objective function $\ell$ is a convex function on the parameter space.

12.2-13. *Recurrent Neural Network Learning Algorithm.* Show how to design an adaptive learning algorithm for the simple recurrent network (Example 1.5.7) or the GRU recurrent neural network (Example 1.5.8) where the parameters of the network are updated after each episode. Design the learning algorithm so that if the resulting learning algorithm generates a sequence of parameter estimates that converges to a large closed, bounded, and convex set $S$ with probability one, then it follows that the sequence of parameter estimates are converging with probability one to the critical points of the objective function which are located in $S$.

## 12.3 Learning in Reactive Statistical Environments Using SA

### 12.3.1 Policy Gradient Reinforcement Learning

Policy gradient reinforcement learning differs from off-policy value reinforcement learning because the actions of the learning machine alter the learning machine's statistical environment. Thus, the learning machine's statistical environment is specified in terms of the learning machine's parameter estimates which are continually evolving. Following the discussion in Section 1.7, assume the statistical environment of the learning machine may be modeled as a sequence of independent episodes where the probability density for episode

$$\tilde{\mathbf{x}}_i = [(\tilde{\mathbf{s}}_{i,1}, \tilde{\mathbf{a}}_{i,2}), \ldots, (\tilde{\mathbf{s}}_{i,T_i}, \tilde{\mathbf{a}}_{i,T_i}), T_i]$$

is specified by the probability density function

$$p(\mathbf{x}_i|\boldsymbol{\beta}) = p_e(\mathbf{s}_{i,1}) \prod_{k=1}^{T_i-1} [p(\mathbf{a}_{i,k}|\mathbf{s}_{i,k};\boldsymbol{\beta}) p_e(\mathbf{s}_{i,k+1}|\mathbf{s}_{i,k}, \mathbf{a}_{i,k})].$$

The density function $p_e(\mathbf{s}_{i,1})$ specifies the probability that the first state of the $i$th episode, $\mathbf{s}_{i,1}$, is generated by the environment. After this first state is generated, the learning machine, whose state of knowledge is denoted by $\boldsymbol{\beta}$, generates action $\mathbf{a}_{i,1}$ given environmental state $\mathbf{s}_{i,1}$ with probability $p(\mathbf{a}_{i,1}|\mathbf{s}_{i,1};\boldsymbol{\beta})$. The environment, in response to the learning machine's action, generates the next environmental state $\mathbf{s}_{i,2}$ with probability $p_e(\mathbf{s}_{i,2}|\mathbf{s}_{i,1}, \mathbf{a}_{i,1})$. The cycle of interactions between the learning machine and environment then continues until the end of the $i$th episode is reached. At the end of the episode, the learning process adjusts the parameter vector $\boldsymbol{\beta}$ which, in turn, influences future episodes which are generated from the learning machine's interactions with its environment.

Thus, this is an example of a learning problem in a reactive rather than a passive statistical environment. The *policy parameter vector* $\boldsymbol{\beta}$ is used to specify both the likelihood of episodes as well as the behavior of the learning machine. In this section, we introduce an additional *episode loss parameter vector* $\boldsymbol{\eta}$ which specifies the penalty incurred by the learning machine when it experiences episode $\mathbf{x}$ is equal to the number $c(\mathbf{x}, \boldsymbol{\eta})$ where $c$ is the *episode loss function*. Thus, the entire set of parameters of the learning machine are specified by a parameter vector $\boldsymbol{\theta}$ defined such that $\boldsymbol{\theta}^T = [\boldsymbol{\beta}^T, \boldsymbol{\eta}^T]$.

The reinforcement risk function $\ell$ is then defined such that:

$$\ell(\boldsymbol{\theta}) = \int c(\mathbf{x}, \boldsymbol{\eta}) p(\mathbf{x}|\boldsymbol{\beta}) d\nu(\mathbf{x}).$$

Interchanging integral and derivative operators, the derivative of $\ell$ is:

$$\frac{d\ell}{d\boldsymbol{\theta}} = \int \frac{dc}{d\boldsymbol{\eta}} p(\mathbf{x}|\boldsymbol{\beta}) d\nu(\mathbf{x}) + \int c(\mathbf{x}, \boldsymbol{\eta}) \frac{dp(\mathbf{x}|\boldsymbol{\beta})}{d\boldsymbol{\beta}} d\nu(\mathbf{x}). \tag{12.46}$$

Now substitute the identity

$$\frac{d \log p(\mathbf{x}|\boldsymbol{\beta})}{d\boldsymbol{\beta}} = \left( \frac{1}{p(\mathbf{x}|\boldsymbol{\beta})} \right) \frac{dp(\mathbf{x}|\boldsymbol{\beta})}{d\boldsymbol{\beta}}$$

into (12.46) to obtain:

$$\frac{d\ell}{d\boldsymbol{\theta}} = \int \frac{dc}{d\boldsymbol{\eta}} p(\mathbf{x}|\boldsymbol{\beta}) d\nu(\mathbf{x}) + \int c(\mathbf{x}, \boldsymbol{\eta}) \frac{d \log p(\mathbf{x}|\boldsymbol{\beta})}{d\boldsymbol{\beta}} p(\mathbf{x}|\boldsymbol{\beta}) d\nu(\mathbf{x}). \tag{12.47}$$

Now assume the reinforcement learning machine works as follows. The statistical environment generates an initial environmental state at random and then the environment and the learning machine interact to produce a random sample $\tilde{\mathbf{x}}_k$ using the learning machine's current parameters $\tilde{\boldsymbol{\theta}}_k$. The learning machine's current parameters are then updated using the formula:

$$\tilde{\boldsymbol{\theta}}_{k+1} = \tilde{\boldsymbol{\theta}}_k - \gamma_k \tilde{\mathbf{g}}_k$$

where

$$\tilde{\mathbf{g}}_k = \left[ \frac{dc(\tilde{\mathbf{x}}_k, \tilde{\boldsymbol{\eta}}_k)}{d\boldsymbol{\eta}} + c(\tilde{\mathbf{x}}, \tilde{\boldsymbol{\eta}}_k) \frac{d \log p(\tilde{\mathbf{x}}_k|\tilde{\boldsymbol{\beta}}_k)}{d\boldsymbol{\beta}} \right]^T. \tag{12.48}$$

and $\ddot{\mathbf{x}}_k$ is sampled from $p(\mathbf{x}|\boldsymbol{\beta}_k)$ where $\boldsymbol{\beta}_{k_0}$ is a realization of $\tilde{\boldsymbol{\beta}}_k$. This ensures that the conditional expectation of $\tilde{\mathbf{g}}_k$ given $\boldsymbol{\theta}_k$ is equal to (12.47) evaluated at $\boldsymbol{\theta}_k$. Or, in other words, $\tilde{\mathbf{g}}_k$ is an estimator of $d\ell/d\boldsymbol{\theta}$ in (12.47) evaluated at $\tilde{\boldsymbol{\theta}}_k$.

**Example 12.3.1** (Robot Control Problem)**.** The state of a robot at time $t$ is specified by the $d$-dimensional binary *situation state vector* $\mathbf{s}(t) \in \{0, 1\}^d$ which includes information such as measurements from the robot's visual and auditory sensors as well as measurements from the robot's current position of its arms and legs and the robot's current position as measured by GPS sensors. An action of the robot is specified by a $k$-dimensional *action vector* $\mathbf{a}(t)$ defined such that $\mathbf{a}(t) \in \mathcal{A}$ where $\mathcal{A} \equiv \{\mathbf{e}^1, \ldots, \mathbf{e}^M\}$ where $\mathbf{e}^k$ is the $k$th column of an $M$-dimensional identity matrix. When the robot generates an action $\mathbf{a}(t)$ in the context of an *initial situation state vector* $\mathbf{s}_I(t)$ this results in a new situation state vector $\mathbf{s}_F(t)$. The *episode vector* $\mathbf{x}(t) \equiv [\mathbf{s}_I(t), \mathbf{a}(t), \mathbf{s}_F(t))]$ is assumed to be a realization of the random vector $\tilde{\mathbf{x}}(t)$.

Assume the probability that the robot chooses action $\mathbf{a} \in \mathcal{A}$ given its current situation state vector $\mathbf{s}$ and its current parameter vector $\boldsymbol{\theta}$ is given by the formula

$$p(\mathbf{a}|\mathbf{s}, \boldsymbol{\theta}) = \mathbf{a}^T \mathbf{p}(\mathbf{s}, \boldsymbol{\theta})$$

where the $k$th element of the column vector $\mathbf{p}(\mathbf{s}, \boldsymbol{\theta})$ is the probability that the $k$th action, $\mathbf{e}^k$, is chosen by the robot for $k = 1, \ldots, M$. Assume that $\mathbf{log}\left(\mathbf{p}(\mathbf{s}, \boldsymbol{\theta})\right)$ is a continuously differentiable function of $\boldsymbol{\theta}$.

Let $p_e(\mathbf{s}_I)$ specify the probability that the environment generates $\mathbf{s}_I$. Let $p_e(\mathbf{s}_F|\mathbf{s}_I, \mathbf{a})$ specify the probability that the environment generates $\mathbf{s}_F$ given $\mathbf{s}_I$ and $\mathbf{a}$. The probability mass functions $p_e(\mathbf{s}_I)$ and $p_e(\mathbf{s}_F|\mathbf{s}_I, \mathbf{a})$ are not functionally dependent upon $\boldsymbol{\theta}$.

Define the probability that the environment generates an episode $\mathbf{x}$ as

$$p_e(\mathbf{x}|\boldsymbol{\theta}) = p_e(\mathbf{s}_I) p(\mathbf{a}|\mathbf{s}_I, \boldsymbol{\theta}) p_e(\mathbf{s}_F|\mathbf{s}_I, \mathbf{a}).$$

Then, for $\mathbf{x} = [\mathbf{s}_I, \mathbf{a}, \mathbf{s}_F]$:

$$\frac{dp_e(\mathbf{x}|\boldsymbol{\theta})}{d\boldsymbol{\theta}} = \frac{d\log p_e(\mathbf{x}|\boldsymbol{\theta})}{d\boldsymbol{\theta}} p_e(\mathbf{x}|\boldsymbol{\theta}) = \frac{d\log\left(\mathbf{a}^T\mathbf{p}(\mathbf{s}_I,\boldsymbol{\theta})\right)}{d\boldsymbol{\theta}} p_e(\mathbf{x}|\boldsymbol{\theta}).$$

Therefore, the reactive statistical environment of the robot learning machine is a sequence of independent episode vectors $\tilde{\mathbf{x}}_1, \tilde{\mathbf{x}}_2, \tilde{\mathbf{x}}_3, \ldots$ with common probability mass function $p_e(\mathbf{x}|\boldsymbol{\theta})$ for a given *knowledge state parameter vector* $\boldsymbol{\theta}$.

Let the value function $V(\mathbf{s}(t), \boldsymbol{\theta}) = \boldsymbol{\theta}[\mathbf{s}(t)^T, 1]$ be the estimated reinforcement the robot expects to receive over future episodes $\mathbf{x}(t+1), \mathbf{x}(t+2), \ldots$. The *predicted incremental reinforcement* $\ddot{r}\left(\mathbf{s}_I(t), \mathbf{s}_F(t)\right)$ is given by the formula:

$$\ddot{r}(\mathbf{s}_I(t), \mathbf{s}_F(t), \boldsymbol{\theta}) = V(\mathbf{s}_I(t), \boldsymbol{\theta}) - \lambda V(\mathbf{s}_F(t), \boldsymbol{\theta})$$

where $0 \leq \lambda \leq 1$ (see Section 1.7).

The expected loss $\ell$ is defined by the formula:

$$\ell(\boldsymbol{\theta}) = \sum_{\mathbf{x}} \left(r(\mathbf{s}_I, \mathbf{s}_F) - \ddot{r}(\mathbf{s}_I, \mathbf{s}_F, \boldsymbol{\theta})\right)^2 p_e(\mathbf{x}|\boldsymbol{\theta})$$

for all $\mathbf{x} \equiv [\mathbf{s}_I, \mathbf{a}, \mathbf{s}_F]$. The gradient of $\ell$ is

$$d\ell/d\boldsymbol{\theta} = \sum_{\mathbf{x}} -2\left(r(\mathbf{s}_I, \mathbf{s}_F) - \ddot{r}(\mathbf{s}_I, \mathbf{s}_F, \boldsymbol{\theta})\right) \frac{d\ddot{r}(\mathbf{s}_I, \mathbf{s}_F, \boldsymbol{\theta})}{d\boldsymbol{\theta}} p_e(\mathbf{x}|\boldsymbol{\theta})$$

$$+ \sum_{\mathbf{x}} \left(r(\mathbf{s}_I, \mathbf{s}_F) - \ddot{r}(\mathbf{s}_I, \mathbf{s}_F, \boldsymbol{\theta})\right)^2 \frac{d\log(p_e(\mathbf{x}|\boldsymbol{\theta}))}{d\boldsymbol{\theta}} p_e(\mathbf{x}|\boldsymbol{\theta}).$$

Show that the iterative algorithm

$$\tilde{\boldsymbol{\theta}}(t+1) = \tilde{\boldsymbol{\theta}}(t) + \gamma_t 2\left(r(\mathbf{s}_I, \mathbf{s}_F) - \ddot{r}(\mathbf{s}_I(t), \mathbf{s}_F(t), \tilde{\boldsymbol{\theta}}(t))\right) \frac{d\ddot{r}(\mathbf{s}_I(t), \mathbf{s}_F(t), \tilde{\boldsymbol{\theta}}(t))}{d\boldsymbol{\theta}}$$

$$-\gamma_t \left(r(\mathbf{s}_I, \mathbf{s}_F) - \ddot{r}(\mathbf{s}_I(t), \mathbf{s}_F(t), \tilde{\boldsymbol{\theta}}(t))\right)^2 \frac{d\log(p_e(\tilde{\mathbf{x}}(t)|\tilde{\boldsymbol{\theta}}(t)))}{d\boldsymbol{\theta}}$$

is a stochastic gradient descent algorithm that minimizes $\ell$ by applying the Stochastic Approximation Theorem. △

**Example 12.3.2** (Active Learning). In this problem, we discuss a learning machine that works in unsupervised learning mode but requests help when it observes an unfamiliar pattern.

Let $\boldsymbol{\theta} \in \Theta \subseteq \mathcal{R}^q$. Assume the data generating process is an i.i.d. bounded stochastic sequence of random vectors

$$(\tilde{\mathbf{s}}_1, \tilde{\mathbf{y}}_1, \tilde{m}_1), (\tilde{\mathbf{s}}_2, \tilde{\mathbf{y}}_2, \tilde{m}_2), \ldots$$

with common Radon-Nikodým density $p(\mathbf{s}, \mathbf{y}, m|\boldsymbol{\theta})$ for each $\boldsymbol{\theta} \in \Theta$. Assume, in addition, that $p(\mathbf{s}, \mathbf{y}, m|\boldsymbol{\theta})$ is factorizable such that:

$$p(\mathbf{s}, \mathbf{y}, m|\boldsymbol{\theta}) = p_e(\mathbf{s}, \mathbf{y}) p_m(m|\mathbf{s}, \boldsymbol{\theta}).$$

The density $p_e(\mathbf{s}, \mathbf{y})$ specifies the probability that an input pattern $\mathbf{s}$ and desired response $\mathbf{y}$ is generated from the statistical environment.

The conditional probability mass function $p_m$ specifies the probability that the learning machine asks for help. Define for each input pattern $\mathbf{s}_i$, a "help indicator" binary random variable $\tilde{m}_i$ which takes on the value of one when the learning machine is requesting the human to provide a desired response $\mathbf{y}_i$ for $\mathbf{s}_i$. When $\tilde{m}_i = 0$, the learning machine will ignore the desired response regardless of whether or not it is available. In particular, let the probability that the learning machine tells the human that a desired response $\mathbf{y}_i$ is required for a given input pattern $\mathbf{s}_i$ be equal to $p_m(\tilde{m}_i = 1|\mathbf{s}_i, \boldsymbol{\theta})$ and the probability that the learning machine does not request help is equal to $p_m(\tilde{m}_i = 0|\mathbf{s}_i, \boldsymbol{\theta})$.

Let $p_s(\mathbf{s}|\boldsymbol{\theta})$ specify the learning machine's model of the probability that $\mathbf{s}$ is observed in its environment given parameter vector $\boldsymbol{\theta}$ where $\boldsymbol{\theta} \in \Theta \subseteq \mathcal{R}^q$. Let the density $p(\mathbf{y}_i|\mathbf{s}_i, \boldsymbol{\theta})$ specify the learning machine's probabilistic model that $\mathbf{y}_i$ is observed for a given input pattern $\mathbf{s}_i$ for a given parameter vector $\boldsymbol{\theta}$.

Let the *supervised learning prediction error* $c^s([\mathbf{s}_i, \mathbf{y}_i], \boldsymbol{\theta})$ when the desired response $\mathbf{y}_i$ is observable be defined such that:

$$c^s([\mathbf{s}_i, \mathbf{y}_i], \boldsymbol{\theta}) = -\log p(\mathbf{y}_i|\mathbf{s}_i, \boldsymbol{\theta}) - \log p_s(\mathbf{s}_i|\boldsymbol{\theta})$$

Let the *unsupervised learning prediction error* $c^u(\mathbf{s}_i, \boldsymbol{\theta})$ when the desired response $\mathbf{y}_i$ is unobservable be defined such that:

$$c^u(\mathbf{s}_i, \boldsymbol{\theta}) = -\log p_s(s_i|\boldsymbol{\theta}).$$

The loss incurred by the learning machine for a given input pattern $\mathbf{s}_i$, the desired response vector $\mathbf{y}_i$ for $\mathbf{s}_i$, and the help indicator $\tilde{m}_i$ is defined such that:

$$c([\mathbf{s}_i, m_i, \mathbf{y}_i], \boldsymbol{\theta}) = m_i c^s([\mathbf{s}_i, \mathbf{y}_i], \boldsymbol{\theta}) + (1 - m_i) c^u(\mathbf{s}_i, \boldsymbol{\theta}).$$

Notice that when the help indicator $\tilde{m}_i = 1$, then the loss incurred by the learning machine is the supervised learning prediction error, $c^s([\mathbf{s}_i, \mathbf{y}_i], \boldsymbol{\theta})$ since the desired response component $\tilde{\mathbf{y}}_i$ is observable, while when $\tilde{m}_i = 0$ the incurred loss is $c^u(\mathbf{s}_i, \boldsymbol{\theta})$ since $\tilde{\mathbf{y}}_i$ is not observable.

The risk function for learning is specified by the formula:

$$\ell(\boldsymbol{\theta}) = \int c([\mathbf{s}, \mathbf{y}, m], \boldsymbol{\theta}) p_m(m|\mathbf{s}, \boldsymbol{\theta}) p_e(\mathbf{s}, \mathbf{y}) d\nu([\mathbf{s}, \mathbf{y}, m]).$$

Derive a stochastic gradient descent algorithm for teaching this learning machine to not only learn the data generating process, but also learn to ask for help. △

### 12.3.2 Stochastic Approximation Expectation Maximization

Many important problems involving "partially observable" data sets or "hidden variables" can be modeled as missing data processes (see Section 9.2.1, Section 10.1.4, and Section 13.2.5). In the following discussion it is assumed that the i.i.d. partially observable data generating process specified by the sequence of independent and identically distributed observations

$$(\tilde{\mathbf{x}}_1, \tilde{\mathbf{m}}_1), (\tilde{\mathbf{x}}_2, \tilde{\mathbf{m}}_2), \ldots \tag{12.49}$$

where the $k$th element of the complete-data record random vector $\tilde{\mathbf{x}}_t$ is observable by the learning machine if and only if the $k$th element of the binary mask random vector $\tilde{\mathbf{m}}_t$ takes on the value of one.

Let $\tilde{\mathbf{v}}_t$ denote the observable elements of $\tilde{\mathbf{x}}_t$ as specified by $\tilde{\mathbf{m}}_t$ and let $\tilde{\mathbf{h}}_t$ denote the unobservable elements of $\tilde{\mathbf{x}}_t$. Let $\tilde{\mathbf{S}}_t$ be the random observable-data selection matrix generated from $\tilde{\mathbf{m}}_t$ as discussed in Section 9.2.1 so that the observable component of $(\tilde{\mathbf{x}}_t, \tilde{\mathbf{m}}_t)$

is given by $(\tilde{\mathbf{v}}_t, \tilde{\mathbf{m}}_t)$ where $\tilde{\mathbf{v}}_t \equiv \tilde{\mathbf{S}}_t \tilde{\mathbf{x}}_t$. The collection of unobservable elements of $\tilde{\mathbf{x}}_t$ not selected by $\tilde{\mathbf{S}}_t$ given $\tilde{\mathbf{m}}_t$ will be referred to as the random vector $\tilde{\mathbf{h}}_t$.

Define a complete-data probability model $\mathcal{M} \equiv \{p(\mathbf{v}_m, \mathbf{h}_m|\boldsymbol{\theta}) : \boldsymbol{\theta} \in \Theta\}$ and define the missing-data mechanism model to be of type ignorable as discussed in Section 10.1.4. The observable data density

$$p(\mathbf{v}_m|\boldsymbol{\theta}) = \int p(\mathbf{v}_m, \mathbf{h}_m|\boldsymbol{\theta}) d\nu(\mathbf{h}_m)$$

where $\mathbf{v}_m$ is the set of observable elements in $\mathbf{x}$ specified by $\mathbf{m}$ and $\mathbf{h}_m$ is the set of unobservable elements in $\mathbf{x}$ specified by $\mathbf{m}$.

Now define the expected loss function

$$\ell(\boldsymbol{\theta}) = -\int \left(\log p(\mathbf{v}_m|\boldsymbol{\theta})\right) p_e(\mathbf{v}_m) d\nu(\mathbf{v}_m). \tag{12.50}$$

Equation (12.50) measures the similarity of the learning machine's model of the observed data components, $p(\mathbf{v}_m|\boldsymbol{\theta})$, with the distribution of the observable component of the data used by the environment, $p_e(\mathbf{v}_m)$, which actually generated the observable data components (see Section 13.2 for further discussion).

Now take the derivative of (12.50) with respect to $\boldsymbol{\theta}$ under the assumption that the interchange of derivative and integral operators is permissible to obtain:

$$\frac{d\ell}{d\boldsymbol{\theta}} = -\int \frac{d \log p(\mathbf{v}_m|\boldsymbol{\theta})}{d\boldsymbol{\theta}} p_e(\mathbf{v}_m) d\nu(\mathbf{v}_m) \tag{12.51}$$

where

$$\frac{d \log p(\mathbf{v}_m|\boldsymbol{\theta}))}{d\boldsymbol{\theta}} = \frac{1}{p(\mathbf{v}_m|\boldsymbol{\theta})} \int \frac{dp(\mathbf{v}_m, \mathbf{h}_m|\boldsymbol{\theta})}{d\boldsymbol{\theta}} d\nu(\mathbf{h}_m). \tag{12.52}$$

The derivative in the integrand of (12.52) is rewritten using the identity (e.g., see Louis, 1982; McLachlan and Krishnan, 2008):

$$\frac{dp(\mathbf{v}_m, \mathbf{h}_m|\boldsymbol{\theta})}{d\boldsymbol{\theta}} = \frac{d \log p(\mathbf{v}_m, \mathbf{h}_m|\boldsymbol{\theta})}{d\boldsymbol{\theta}} p(\mathbf{v}_m, \mathbf{h}_m|\boldsymbol{\theta}). \tag{12.53}$$

Substitution of (12.53) into (12.52) and then substituting the result into (12.51) gives:

$$\frac{d\ell}{d\boldsymbol{\theta}} = -\int\int \frac{d \log p(\mathbf{v}_m, \mathbf{h}_m|\boldsymbol{\theta})}{d\boldsymbol{\theta}} p(\mathbf{h}_m|\mathbf{v}_m, \boldsymbol{\theta}) p_e(\mathbf{v}_m) d\nu(\mathbf{h}_m) d\nu(\mathbf{v}_m) \tag{12.54}$$

where the density $p(\mathbf{h}_m|\mathbf{v}_m, \boldsymbol{\theta})$ is called the *stochastic imputation density.*

Now define a minibatch stochastic gradient descent where the expected search direction is chosen to be the negative gradient to obtain:

$$\tilde{\boldsymbol{\theta}}(k+1) = \tilde{\boldsymbol{\theta}}(k) - (\gamma_k/M) \sum_{j=1}^{M} \frac{d \log p(\tilde{\mathbf{v}}_k, \tilde{\mathbf{h}}^{jk}|\tilde{\boldsymbol{\theta}}(k))}{d\boldsymbol{\theta}} \tag{12.55}$$

where the minibatch $\tilde{\mathbf{h}}^{1k}, \ldots, \tilde{\mathbf{h}}^{Mk}$ at the $k$th learning trial is generated by first sampling a realization $\mathbf{v}_k$ from the environment and then sampling $M$ times from the stochastic imputation density $p(\mathbf{h}|\mathbf{v}_k, \boldsymbol{\theta}(k))$ using the sampled value $\mathbf{v}_k$ and the current parameter estimates $\boldsymbol{\theta}(k)$ at the $k$th learning trial. Note that even though (12.49) specifies a passive DGP, this is a reactive learning problem since the stochastic imputation density imputes missing data using the current parameter estimates.

This type of stochastic approximation algorithm is called a Stochastic Approximation Expectation Maximization (SAEM) algorithm. Note that for the SAEM algorithm, $M$ can be chosen equal to 1 or any positive integer. When the minibatch size $M$ is moderately large, an SAEM algorithm may be referred to as an MCEM (Monte Carlo Expectation Maximization) algorithm. When the minibatch size $M$ is very large (i.e., $M$ is close to infinity), then the SAEM algorithm closely approximates the deterministic Generalized Expectation Maximization (GEM) algorithm (see McLachlan and Krishnan 2008) in which the learning machine uses its current probabilistic model to compute the expected downhill search direction, takes a downhill step in that direction, updates its current probabilistic model, and then repeats this process in an iterative manner. The standard expectation-maximization (EM) algorithm can be interpreted as a special case of the GEM algorithm (see MacLachlan and Krishnan 2008).

**Example 12.3.3** (Logistic Regression with Missing Data). In this example, the adaptive learning algorithm in Example 12.2.2 is extended to handle situations where the predictors of the logistic regression are not always observable.

In a logistic regression model, the *conditional response complete data density*

$$p(y|\mathbf{s}, \boldsymbol{\theta}) = y p_y(\mathbf{s}; \boldsymbol{\theta}) + (1 - y)(1 - p_y(\mathbf{s}; \boldsymbol{\theta}))$$

where $p_y(\mathbf{s}; \boldsymbol{\theta}) \equiv \mathcal{S}(\psi(\mathbf{s}; \boldsymbol{\theta}))$ where $\mathcal{S}(\psi) = 1/(1 + \exp(-\psi))$ and $\psi(\mathbf{s}; \boldsymbol{\theta}) = \boldsymbol{\theta}^T [\mathbf{s}^T \quad 1]^T$. Since missing data is present, it is also necessary to explicitly define a model for the joint distribution of the predictors. For simplicity, assume the joint density of predictors and response variables may be factored so that:

$$p(y, \mathbf{s}|\boldsymbol{\theta}) = p(y|\mathbf{s}, \boldsymbol{\theta}) p_s(\mathbf{s}).$$

In this example, it is assumed that the DGP is of type MAR (see Section 9.2).

Partition the input pattern vector $\mathbf{s}$ into an observable component $\mathbf{s}^v$ and an unobservable component $\mathbf{s}^h$ so that $p_s(\mathbf{s})$ may be rewritten as $p_s([\mathbf{s}^v, \mathbf{s}^h])$. Now define the missing data risk function for learning as:

$$\ell(\boldsymbol{\theta}) = -\int \left( \log \sum_{\mathbf{s}^h} p(y, \mathbf{s}^v, \mathbf{s}^h | \boldsymbol{\theta}) \right) p_e(\mathbf{s}^v) d\nu(\mathbf{s}^v)$$

where the notation $\sum_{\mathbf{s}^h}$ denotes a summation over the possible values for $\mathbf{s}^h$.

The derivative of $-\log p(y|\mathbf{s}, \boldsymbol{\theta})$ is given by the formula:

$$-\frac{d \log p(y|\mathbf{s}, \boldsymbol{\theta})}{d\boldsymbol{\theta}} = -(y - p_y(\mathbf{s}; \boldsymbol{\theta})) \mathbf{s}^T.$$

The stochastic imputation density is computed from $p_s$ using the formula:

$$p_s(\mathbf{s}^h | \mathbf{s}^v) = \frac{p_s([\mathbf{s}^v, \mathbf{s}^h])}{\sum_{\mathbf{s}^h} p_s([\mathbf{s}^v, \mathbf{s}^h])}.$$

The final algorithm is then obtained using the MCEM missing data stochastic gradient descent algorithm specified in (12.55). The algorithm simply involves sampling from the missing data generating process to obtain $(y_t, \mathbf{s}_t^v)$, then "filling in" (also known as "imputing") the missing part of the input pattern by randomly generating $\mathbf{s}_t^h$ given $\mathbf{s}_t^v$ to obtain values for the unobservable components of $\mathbf{s}_t$. The observable component $\mathbf{s}_t^v$ and imputed component $\mathbf{s}_t^h$ generated from sampling from the stochastic imputation density are then concatenated together to obtain an input pattern vector $\ddot{\mathbf{s}}_t$ with no missing information.

The desired responses $\tilde{y}_t$ and $\ddot{\mathbf{s}}_t$ are then used to update the parameter vector

$$\tilde{\boldsymbol{\theta}}_{t+1} = \tilde{\boldsymbol{\theta}}_t + \gamma_t \left( \tilde{y}_t - \ddot{p}_y(\ddot{\mathbf{s}}_t; \tilde{\boldsymbol{\theta}}_t) \right) \ddot{\mathbf{s}}_t^T$$

where

$$\ddot{p}_y(\ddot{\mathbf{s}}_t;\boldsymbol{\theta}) \equiv \mathcal{S}\left(\boldsymbol{\theta}^T[\ddot{\mathbf{s}}_t^T \;\; 1]^T\right).$$

Use the Stochastic Approximation Theorem to show that if $\tilde{\boldsymbol{\theta}}_1, \tilde{\boldsymbol{\theta}}_2, \ldots$ remains with probability one in a convex, closed, and bounded region of the parameter space, then $\tilde{\boldsymbol{\theta}}_1, \tilde{\boldsymbol{\theta}}_2, \ldots$ converges with probability one to the set of critical points in that region. △

### 12.3.3 Markov Random Field Learning (Contrastive Divergence)

Given the positivity condition holds, a Markov random field can be represented as a Gibbs density. Let $V : \mathcal{R}^d \times \mathcal{R}^q \to \mathcal{R}$. Let $\Theta$ be a closed and bounded subset of $\mathcal{R}^q$. Assume for each $\boldsymbol{\theta} \in \Theta$ that the probability distribution of a $d$-dimensional random vector $\tilde{\mathbf{x}}$ is specified by a Gibbs density $p(\cdot|\boldsymbol{\theta}) : \mathcal{R}^d \to (0, \infty)$ defined such that

$$p(\mathbf{x}|\boldsymbol{\theta}) = [Z(\boldsymbol{\theta})]^{-1} \exp(-V(\mathbf{x};\boldsymbol{\theta})) \tag{12.56}$$

where the normalization constant $Z(\boldsymbol{\theta})$ is defined as:

$$Z(\boldsymbol{\theta}) = \int \exp(-V(\mathbf{y};\boldsymbol{\theta}))d\nu(\mathbf{y}). \tag{12.57}$$

A common method for estimating the parameters of a Gibbs density involves minimizing the maximum likelihood empirical risk function (see Section 13.2 for additional details). The maximum likelihood empirical risk function $\ell_n$ is defined such that:

$$\hat{\ell}_n(\boldsymbol{\theta}) \equiv -(1/n)\sum_{i=1}^{n} \log p(\mathbf{x}_i|\boldsymbol{\theta}) \tag{12.58}$$

on $\Theta$. The maximum likelihood empirical risk function $\hat{\ell}_n$ is an estimator of the risk function

$$\ell(\boldsymbol{\theta}) = -\int (\log p(\mathbf{x}|\boldsymbol{\theta}))\, p_e(\mathbf{x})d\nu(\mathbf{x}). \tag{12.59}$$

The derivative of (12.59) is:

$$\frac{d\ell}{d\boldsymbol{\theta}} = -\int \left(\frac{d\log p(\mathbf{x}|\boldsymbol{\theta})}{d\boldsymbol{\theta}}\right) p_e(\mathbf{x})d\nu(\mathbf{x}). \tag{12.60}$$

The derivative of $\log p(\mathbf{x}|\boldsymbol{\theta})$ using (12.56) is given by the formula:

$$\frac{d\log p(\mathbf{x}|\boldsymbol{\theta})}{d\boldsymbol{\theta}} = -\frac{dV(\mathbf{x};\boldsymbol{\theta})}{d\boldsymbol{\theta}} + \int \frac{dV(\mathbf{y};\boldsymbol{\theta})}{d\boldsymbol{\theta}} p(\mathbf{y}|\boldsymbol{\theta})d\nu(\mathbf{y}). \tag{12.61}$$

Substituting (12.61) into (12.60), then gives:

$$\frac{d\ell}{d\boldsymbol{\theta}} = \int \left(\frac{dV(\mathbf{x};\boldsymbol{\theta})}{d\boldsymbol{\theta}} - \frac{dV(\mathbf{y};\boldsymbol{\theta})}{d\boldsymbol{\theta}}\right) p_e(\mathbf{x})p(\mathbf{y}|\boldsymbol{\theta})d\nu(\mathbf{x})d\nu(\mathbf{y}). \tag{12.62}$$

An adaptive gradient descent learning algorithm which minimizes $\ell$ can be designed using the Stochastic Approximation Theorem as follows. Let

$$\tilde{\boldsymbol{\theta}}(t+1) = \tilde{\boldsymbol{\theta}}(t) - \gamma_t \left(\frac{dV(\tilde{\mathbf{x}}(t);\tilde{\boldsymbol{\theta}}(t))}{d\boldsymbol{\theta}} - \frac{dV(\tilde{\mathbf{y}}(t);\tilde{\boldsymbol{\theta}}(t))}{d\boldsymbol{\theta}}\right) \tag{12.63}$$

where at iteration $t$ $\tilde{\mathbf{x}}(t)$ is sampled from the actual statistical environment's density $p_e(\mathbf{x})$ and $\tilde{\mathbf{y}}(t)$ is sampled from the model's current theory about the statistical environment $p(\mathbf{y}|\tilde{\boldsymbol{\theta}}(t))$ using the current estimate $\tilde{\boldsymbol{\theta}}(t)$. Note that although $p_e$ is a passive DGP density, the density $p(\mathbf{y}|\boldsymbol{\theta}(t))$ is a reactive DGP density.

To obtain an improved sample average estimate of the gradient, one typically generates multiple samples $\tilde{\mathbf{y}}(t,1),\ldots,\tilde{\mathbf{y}}(t,M)$ from $p(\mathbf{y}|\tilde{\boldsymbol{\theta}}(t))$, computes the estimated gradient for each of the $M$ samples, and then averages the $M$ gradient estimates to obtain an improved estimate of the expected gradient. This strategy yields the revised learning rule:

$$\tilde{\boldsymbol{\theta}}(t+1) = \tilde{\boldsymbol{\theta}}(t) - \gamma_t \left( \frac{dV(\tilde{\mathbf{x}}(t);\tilde{\boldsymbol{\theta}}(t))}{d\boldsymbol{\theta}} - (1/M)\sum_{j=1}^{M} \frac{dV(\tilde{\mathbf{y}}(t,j);\tilde{\boldsymbol{\theta}}(t))}{d\boldsymbol{\theta}} \right). \tag{12.64}$$

Note that this stochastic approximation algorithm will converge for any choice of $M$ (including $M = 1$) since the expected value of the search direction in (12.64) is exactly equal to $-d\ell/d\boldsymbol{\theta}$ where $d\ell/d\boldsymbol{\theta}$ is defined in (12.62).

In practice, sampling from $p(\mathbf{y}|\tilde{\boldsymbol{\theta}}(t))$ is computationally challenging, so the Monte Carlo Markov Chain methods of Chapter 11 are used to develop approximate sampling schemes. When such methods are used, it is only required that the MCMC algorithm converge approximately to its asymptotic stationary distribution, provided that the MCMC algorithm generates a reasonably good approximation to the expected gradient. When an MCMC algorithm is used to sample from $p(\mathbf{y}|\tilde{\boldsymbol{\theta}}(t))$, then the stochastic approximation algorithm in (12.64) corresponds to a *contrastive divergence type* learning algorithm (Yuille 2005; Jiang et al. 2018). The minibatch size $m$ can be a fixed integer (e.g., $m = 3$ or $m = 100$) or $m$ can be varied (e.g., initially $m$ is chosen to be small and then gradually increased to some finite positive integer during the learning process).

### 12.3.4 Generative Adversarial Network (GAN) Learning

Assume that an environmental Data Generating Process (DGP) generates a sequence of $d$-dimensional feature vectors $\mathbf{x}_1, \mathbf{x}_2, \ldots$ corresponding to a particular realization of a stochastic sequence of independent and identically distributed $d$-dimensional discrete random vectors $\tilde{\mathbf{x}}_1, \tilde{\mathbf{x}}_2, \ldots$ with common probability mass function $p_e : S \to [0,1]$ where $S$ is a finite subset of $\mathcal{R}^d$.

A *discriminative unsupervised learning machine* expects that the probability that a particular $d$-dimensional training vector will occur in its statistical environment is given by the conditional probability mass function $p_D(\cdot|\boldsymbol{\eta}) : S \to [0,1]$ for a particular parameter vector $\boldsymbol{\eta}$. A *generative learning machine* generates a particular $d$-dimensional training vector for the discriminative learning machine using the conditional probability mass function $p_G(\cdot|\boldsymbol{\psi}) : S \to [0,1]$. This latter mass function is a reactive DGP density.

Define a GAN (Generative Adversarial Network) objective function $\ell(\boldsymbol{\eta},\boldsymbol{\psi})$ (see Goodfellow et al. 2014) such that:

$$\ell(\boldsymbol{\eta},\boldsymbol{\psi}) = \sum_{\mathbf{x}\in S}\sum_{\mathbf{z}\in S} c([\mathbf{x},\mathbf{z}];\boldsymbol{\eta}) p_e(\mathbf{x}) p_G(\mathbf{z}|\boldsymbol{\psi}) \tag{12.65}$$

where

$$c([\mathbf{x},\mathbf{z}];\boldsymbol{\eta}) = -\left(\log p_D(\mathbf{x}|\boldsymbol{\eta}) + \log[1 - p_D(\mathbf{z}|\boldsymbol{\eta})]\right).$$

The goal of the discriminative unsupervised learning machine is to find a probability mass function $p_D(\cdot|\boldsymbol{\eta}^*)$ which assigns high probability mass to observations $\mathbf{x}_1, \mathbf{x}_2, \ldots$ generated from the environmental DGP, while simultaneously assigning low probability mass to observations $\mathbf{z}_1, \mathbf{z}_2, \ldots$ generated from the generative learning machine $p_G(\cdot|\boldsymbol{\psi})$ for a given choice of $\boldsymbol{\psi}$.

On the other hand, the goal of the generative learning machine is to find a probability mass function $p_G(\cdot|\boldsymbol{\psi}^*)$ that increases the error rate of the discriminative learning machine by generating challenging training stimuli $\mathbf{z}_1, \mathbf{z}_2, \ldots$ that are assigned high probability mass by the discriminative learning machine.

A GAN consists of both the discriminative unsupervised learning machine and the generative learning machine which operate in an adversarial manner. The discriminative learning machine seeks to minimize $\ell(\cdot, \boldsymbol{\psi})$ for a fixed value of $\boldsymbol{\psi}$ while the generative learning machine seeks to maximize $\ell(\boldsymbol{\eta}, \cdot)$ for a fixed value of $\boldsymbol{\eta}$. Therefore, the goal of the GAN is to seek a saddlepoint $(\boldsymbol{\eta}^*, \boldsymbol{\psi}^*)$ such that both $|d\ell(\boldsymbol{\eta}^*, \boldsymbol{\psi}^*)/d\boldsymbol{\eta}| = 0$ and $|d\ell(\boldsymbol{\eta}^*, \boldsymbol{\psi}^*)/d\boldsymbol{\psi}| = 0$. A strategy for seeking such a saddlepoint using a stochastic gradient descent methodology may be implemented as follows.

- **Step 1: Discrimination Machine Learning Phase.** In the Discrimination Machine Learning Phase, each iteration of the stochastic gradient descent algorithm is a minibatch algorithm designed to minimize $\ell$ by adjusting the discrimination machine's parameter vector $\boldsymbol{\eta}$ and holding the generative learning machine's parameter vector $\boldsymbol{\psi}$ constant. With probability $p_e(\mathbf{x})$ a training vector $\mathbf{x}$ is generated from the environment and with probability $p_G(\mathbf{z}|\boldsymbol{\psi})$ a training vector $\mathbf{z}$ is generated from the *generative learning machine* at each iteration of the algorithm. Eventually, Phase 1 of the learning process is terminated. The Stochastic Approximation Theorem can be used to analyze and design the asymptotic behavior of the Phase 1 learning process since Phase 1 can be interpreted as minimizing $\ell(\cdot, \boldsymbol{\psi})$ for a fixed value of $\boldsymbol{\psi}$. Note that the learning process in Phase 1 corresponds to learning in a passive statistical environment.

- **Step 2: Generative Machine Learning Phase.** In the Generative Machine Learning Phase, each iteration of the stochastic gradient descent algorithm is a minibatch algorithm designed to maximize $\ell$ by adjusting the generative learning machine's parameter vector $\boldsymbol{\psi}$ and holding the discrimination machine's parameter vector $\boldsymbol{\eta}$ constant. With probability $p_G(\mathbf{z}|\boldsymbol{\psi})$ a training vector $\mathbf{z}$ is generated from the *generative learning machine* at each iteration of the algorithm. Eventually, Phase 2 of the learning process is terminated. The Stochastic Approximation Theorem can be used to analyze and design the asymptotic behavior of the Phase 2 learning process since Phase 2 can be interpreted as minimizing $-\ell(\boldsymbol{\eta}, \cdot)$ for a fixed value of $\boldsymbol{\eta}$. Note that the learning process in Phase 2 corresponds to learning in a reactive statistical environment.

- **Step 3: Go to Step 1 until Convergence to Saddlepoint.** The learning process then continues by following Phase 2 (Generative Machine Learning) with Phase 1 (Discriminative Machine Learning) and continuing this alternating block coordinate parameter update scheme until convergence to a saddlepoint is achieved.

## Exercises

12.3-1. *Robot Control with Multinomial Logistic Regression.* Derive a learning algorithm to control the robot in Example 12.3.1 where $\mathbf{p}(\mathbf{s}; \boldsymbol{\theta})$ is a four-dimensional column vector defined such that the $k$th element of $\mathbf{p}(\mathbf{s}; \boldsymbol{\theta})$, $p_k(\mathbf{s}; \boldsymbol{\theta})$, is defined such that for $k = 1, 2, 3$:

$$p_k(\mathbf{s}; \boldsymbol{\theta}) = \frac{\exp(\mathbf{w}_k^T \mathbf{s})}{\mathbf{1}_4^T \exp(\mathbf{W}\mathbf{s})}$$

and $p_4(\mathbf{s};\boldsymbol{\theta}) = 1/\left(\mathbf{1}_4^T\mathbf{exp}(\mathbf{Ws})\right)$. Derive an explicit learning rule to control this robot following the approach of Example 12.3.1.

12.3-2. *Lunar Lander Linear Control Law Design.* In Equation (1.60) of Example (1.7.4), a gradient descent algorithm is proposed for the purpose of estimating the parameters of a lunar lander control law. Derive an explicit formula for the gradient descent learning algorithm. Then, investigate the behavior of the learning rule using the Stochastic Approximation Theorem.

12.3-3. *Lunar Lander Nonlinear Control Law Design.* Enhance the lunar lander in Example 1.7.1 with an adaptive nonlinear control law using a function approximation method as described in Subsection 1.5.2. Then, use the Stochastic Approximation Theorem to prove that if the sequence of parameter estimates generated by the adaptive nonlinear control law converge to a neighborhood of a strict local minimizer then the parameter estimates converge to that strict local minimizer with probability one.

12.3-4. *Markov Logic Net Learning Algorithm.* Use the methods of Section 12.3.3 to derive a learning algorithm for the Markov Logic Net described in Example 11.2.4.

12.3-5. *Hessian of Expected Loss in Reactive Environments.* Equation 12.8 is the formula for the Hessian of Expected Loss in Reactive Environments. Derive that formula by taking the derivative of Equation 12.7.

## 12.4 Further Readings

Portions of Chapter 12 were originally published in Golden (2018) *Neural Computation, Vol., 30*, Issue 10, pp. 2805-2832 and are reprinted here courtesy of the MIT Press.

### Historical Notes

The terminology "stochastic approximation" most likely arose from the idea that stochastic observations could be used to approximate a function or its gradient for the purposes of optimization. Robbins and Monro (1951) is often attributed to the publication of the first algorithm in this area. Blum (1954) provided one of the first analyses of a stochastic approximation algorithm for an objective function defined on a multi-dimensional state space. The specific stochastic convergence theorem presented here is based upon the work of Blum (1954), White (1989a, 1989b), Golden (1996a), Benveniste et al. (1990, Part II Appendix), Sunehag et al. (2009), and Golden (2018).

### Stochastic Approximation Theorems

In the machine learning literature, a number of researchers have provided stochastic approximation theorems for passive learning environments (e.g., White 1989a, 1989b; Golden 1996a; Bottou 1998, 2004; Sunehag et al. 2009; Mohri et al. 2012; Toulis et al. 2014). Stochastic approximation theorems for reactive learning environments have also been published (e.g., Blum 1954; Benveniste et al. 1990; Gu and Kong 1998; Delyon et al. 1999; and Golden 2018). Reviews of the stochastic approximation literature can be found in Benveniste et al.

(1990), Bertsekas and Tsitsiklis (1996), Borkar (2008), Kushner and Yin (2003), Kushner (2010).

A unique feature of the theorem presented in this chapter is that the assumptions of the theorem are relatively easy to verify and the theorem is applicable to a wide range of important machine learning algorithms operating in both passive and reactive statistical environments. Useful mathematically rigorous introductions to the stochastic approximation literature may be found in Wasan ([1969] 2004), Benveniste et al. (1990), Bertsekas and Tsitsiklis (1996), Borkar (2008), and Douc et al. (2014).

## Expectation Maximization Methods

This chapter just briefly discussed expectation-maximization methods, but such methods are widely used in machine learning applications. Little and Rubin (2002, Chapter 8) provide a good mathematical introduction to the EM algorithm and its variants. A comprehensive but very accessible introduction to the EM algorithm is provided by McLachlan and Krishnan (2008). An introduction to the EM algorithm emphasizing machine learning mixture model applications can be found in Bishop (2006, Chapter 9).

## Policy Gradient Reinforcement

A discussion of policy gradient methods for reinforcement learning consistent with the discussion provided here can be found in Sugiyama (2015, Chapter 7). Such methods were first introduced into the machine learning literature by Williams (1992).

# Part IV

# Generalization Performance

# 13

## *Statistical Learning Objective Function Design*

**Learning Objectives**

- Design empirical risk functions for learning.
- Design maximum likelihood functions for learning.
- Design map estimation functions for learning.

Consider an inductive learning machine embedded within an environment that experiences a set of $n$ environmental events $\mathbf{x}_1, \ldots, \mathbf{x}_n$. For example, in a supervised learning paradigm, an environmental event $\mathbf{x}_i$ may be $(\mathbf{s}_i, y_i)$ where $y_i$ is the desired response for input pattern $\mathbf{s}_i$. During a "testing phase", novel environmental events are presented to the learning machine and, for each such novel event, the learning machine makes a prediction regarding its "preference" for that event.

For example, consider a linear prediction machine where the response $\ddot{y} : \mathcal{R}^d \times \mathcal{R}^q \rightarrow \mathcal{R}$ for a given input pattern $\mathbf{s}_i$ is specified as:

$$\ddot{y}(\mathbf{s}_i, \boldsymbol{\theta}) = \boldsymbol{\theta}^T \mathbf{s}_i$$

so that the machine's response $\ddot{y}(\mathbf{s}_i, \boldsymbol{\theta})$ is contingent not only upon the input $\mathbf{s}_i$ but also upon the machine's parameter values $\boldsymbol{\theta}$. During learning, the machine observes the $n$ environmental events $\{(\mathbf{s}_1, y_1), \ldots, (\mathbf{s}_n, y_n)\}$.

In particular, for each training stimulus input pattern $\mathbf{s}_i$, the machine makes a predicted response $\ddot{y}(\mathbf{s}_i, \boldsymbol{\theta})$ which is compared to the desired response $y_i$ by measuring the "prediction error" between predicted and desired response. For example, one such error measure would be the squared difference between the predicted and desired response $[y_i - \ddot{y}(\mathbf{s}_i, \boldsymbol{\theta})]^2$. The parameter estimate $\hat{\boldsymbol{\theta}}_n$ is computed by the machine's learning dynamics by minimizing the average prediction error

$$\hat{\ell}_n(\boldsymbol{\theta}) \equiv (1/n) \sum_{i=1}^{n} (y_i - \ddot{y}(\mathbf{s}_i, \boldsymbol{\theta}))^2$$

to obtain the empirical risk function global minimizer

$$\hat{\boldsymbol{\theta}}_n \equiv \arg\ \min \hat{\ell}_n(\boldsymbol{\theta}).$$

Two important questions regarding generalization performance of the learning machine immediately surface. First, if additional data was collected, what would be the behavior of the sequence of functions $\hat{\ell}_n, \hat{\ell}_{n+1}, \hat{\ell}_{n+2}, \ldots$? That is, how does the objective function for learning $\hat{\ell}_n$ evolve for different amounts of training data? A second question involves the global minimizer $\hat{\boldsymbol{\theta}}_n$. As additional data is collected, what is the behavior of the sequence of

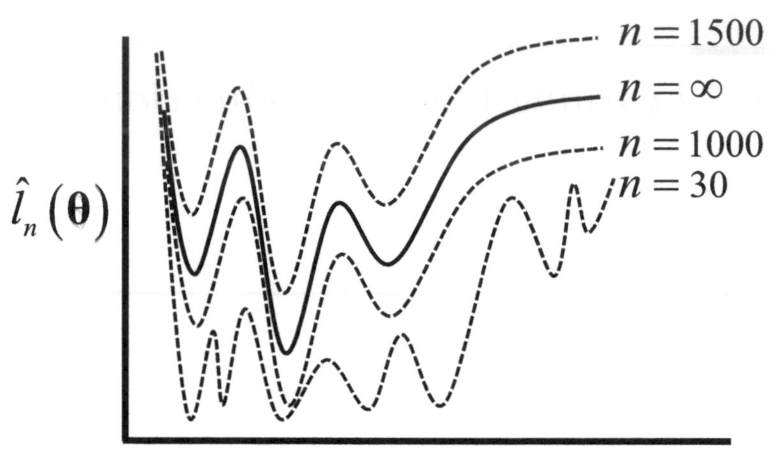


**FIGURE 13.1**
**Estimated prediction error as a function of the free parameters of a learning machine for different amounts of training data.** As more training data is collected, the parameter values that minimize the prediction error tend to converge to a particular asymptotic value.

global minimizers $\hat{\boldsymbol{\theta}}_n, \hat{\boldsymbol{\theta}}_{n+1}, \ldots$? These questions are fundamental to explicitly representing and then analyzing the computational goal of learning for the learning machine.

Figure 13.1 plots the shape of the objective function $\hat{\ell}_n$ as number of training stimuli, $n$, is varied. In particular, Figure 13.1 plots the shape of $\hat{\ell}_n$ for $n = 30$, $n = 1000$, $n = 1500$, and $n = \infty$. As the amount of training stimuli, $n$, is varied, the objective function shape changes erratically because the training samples are different. Similarly, the global minimizers of the five different objective functions will change their values erratically because of training sample differences.

Ideally, however, it would be desirable to show that when $n$ is sufficiently large that the objective functions $\hat{\ell}_n$, $\hat{\ell}_{n+1}$, $\hat{\ell}_{n+2}, \ldots$ are very similar and their respective global minimizers $\hat{\boldsymbol{\theta}}_n$, $\hat{\boldsymbol{\theta}}_{n+1}$, $\hat{\boldsymbol{\theta}}_{n+3}$, ... are very similar. Unfortunately however, without additional knowledge about the nature of the data generating process, such statements cannot be made.

Assume that the $n$ environmental events $\{(\mathbf{s}_1, y_1), \ldots, (\mathbf{s}_n, y_n)\}$ are a realization of a sequence of $n$ independent and identically distributed random vectors $\{(\tilde{\mathbf{s}}_1, \tilde{y}_1), \ldots, (\tilde{\mathbf{s}}_n, \tilde{y}_n)\}$ with a common DGP density $p_e(\mathbf{s}, y)$. Thus, $\hat{\ell}_n$ is recognized as a realization of the random function

$$\hat{\ell}_n(\boldsymbol{\theta}) \equiv (1/n) \sum_{i=1}^{n} (\tilde{y}_i - \ddot{y}(\tilde{\mathbf{s}}_i, \boldsymbol{\theta}))^2$$

with a random global minimizer

$$\hat{\boldsymbol{\theta}}_n \equiv \arg\ \min \hat{l}_n(\boldsymbol{\theta}).$$

Note that the "randomness" in $\hat{\ell}_n$ is generated from the "randomness" in the data set $\{(\tilde{\mathbf{s}}_1, \tilde{y}_1), \ldots, (\tilde{\mathbf{s}}_n, \tilde{y}_n)\}$.

## 13.1 Empirical Risk Function

The following definition provides a formal specification of the informal description of empirical risk functions in Chapter 1.

**Definition 13.1.1** (Empirical Risk Function (ERF)). Let $\tilde{\mathcal{D}}_n \equiv [\tilde{\mathbf{x}}_1, \ldots, \tilde{\mathbf{x}}_n]$ be a sequence of independent and identically distributed $d$-dimensional random vectors with common Radon-Nikodým density $p_e : \mathcal{R}^d \rightarrow [0, \infty)$. Let $\Theta \subseteq \mathcal{R}^q$. Let each *penalty term function* $k_n : \mathcal{R}^{d \times n} \times \Theta \rightarrow \mathcal{R}$ be a continuous random function, $n = 1, 2, 3, \ldots$. Let the *loss function* $c : \mathcal{R}^d \times \Theta \rightarrow \mathcal{R}$ be a continuous random function. The function $\ell(\cdot) \equiv E\{c(\tilde{\mathbf{x}}_i, \cdot)\}$ is called the *risk function.* Let the *penalized empirical risk function* $\ell_n : \mathcal{R}^{dn} \times \Theta \rightarrow \mathcal{R}$ be defined such that for all $\boldsymbol{\theta} \in \Theta$:

$$\hat{\ell}_n(\boldsymbol{\theta}) = k_n(\tilde{\mathbf{X}}, \boldsymbol{\theta}) + (1/n) \sum_{i=1}^{n} c(\tilde{\mathbf{x}}_i, \boldsymbol{\theta}).$$

The random function $\hat{\ell}_n$ is called an *empirical risk function* when $k_n = 0$ for $n = 1, 2, \ldots$. □

This definition of a loss function captures not only situations involving supervised learning but also situations involving unsupervised and reinforcement learning. For *unsupervised learning*, the loss function $c : \mathcal{R}^d \times \Theta \rightarrow \mathcal{R}$ is typically defined such that $c(\mathbf{x}, \boldsymbol{\theta})$ is the loss incurred by the learning machine when it experiences the event $\mathbf{x}$ in its statistical environment. For *supervised learning*, the $d$-dimensional $\mathbf{x}$ typically consists of an *input pattern record* subvector $\mathbf{s}$ and a *desired response* subvector $\mathbf{y}$. For *reinforcement learning*, the event $\mathbf{x}$ is a $d$-dimensional *episode vector* and is typically defined such that $\mathbf{x} \equiv [\boldsymbol{\xi}, \mathbf{m}]$ where the $d/2$-dimensional vector $\boldsymbol{\xi}$ is called the *fully observable episode vector* and a $d/2$-dimensional *unobservable data indicator vector* $\mathbf{m}$ is a vector consisting of ones and zeros. The fully observable $d/2$-dimensional episode vector $\boldsymbol{\xi}$ is defined such that

$$\boldsymbol{\xi} = [(\mathbf{s}_1, \mathbf{a}_1, r_1), \ldots, (\mathbf{s}_T, \mathbf{a}_T, r_T)]$$

where $T$ is the maximum length of episode $\boldsymbol{\xi}$, the $s$-dimensional state vector $\mathbf{s}_k$ is the $k$th state generated by the environment in episode $\boldsymbol{\xi}$, the $a$-dimensional state vector $\mathbf{a}_k$ is the $k$th action generated by the learning machine in episode $\boldsymbol{\xi}$, and $r_k$ is a scalar reinforcement signal received by the learning machine when the state of the environment changes from state $\mathbf{s}_k$ to $\mathbf{s}_{k+1}$ so that $(s + a + 1)T = d/2$.

The unobservable data indicator vector $\mathbf{m}$ is defined such that the $k$th element of $\mathbf{m}$ is nonzero if and only if the $k$th element of $\boldsymbol{\xi}$ is observable, $k = 1, \ldots, T$. By choosing the elements of $\mathbf{m}$ in an appropriate way episodes of different lengths can also be represented. For example, if all elements of $\mathbf{m}$ are equal to one except for the last $s + a + 1$ elements, this represents an episode consisting of only $T - 1$ actions.

The unobservable data indicator vector $\mathbf{m}$ may also be used to represent learning environments where the reinforcement signal $r_k$ is only observed some of the time. For example, when elements $(s + a + 1)$ and $2(s + a + 1)$ of the unobservable data indicator $\mathbf{m}$ are equal to zero, this indicates the reinforcement signals $r_1$ and $r_2$ for episodes 1 and 2 are not observable by the learning machine.

**Theorem 13.1.1** (Empirical Risk Function Convergence).

- *Let $\tilde{\mathcal{D}}_n \equiv [\tilde{\mathbf{x}}_1, \ldots, \tilde{\mathbf{x}}_n]$ be a sequence of independent and identically distributed $d$-dimensional random vectors with common Radon-Nikodým density $p_e : \mathcal{R}^d \rightarrow [0, \infty)$ defined with respect to support specification measure $\nu$.*

- *Let $\Theta$ be a closed, bounded, convex subset of $\mathcal{R}^q$. Let the continuous random function $c : \mathcal{R}^d \times \Theta \rightarrow \mathcal{R}$ be dominated by an integrable function on $\Theta$ with respect to $p$.*
- *Assume $k_n : \mathcal{R}^{d \times n} \times \Theta \rightarrow \mathcal{R}$ is a continuous random function on $\Theta$ such that $k_n(\tilde{\mathcal{D}}_n, \cdot) \rightarrow 0$ uniformly on $\Theta$ as $n \rightarrow \infty$ with probability one.*
- *For $n = 1, 2, 3, \ldots$, let the continuous random function $\ell_n : \mathcal{R}^{d \times n} \times \Theta \rightarrow \mathcal{R}$ be defined such that*

$$\ell_n(\tilde{\mathcal{D}}_n, \boldsymbol{\theta}) = k_n(\tilde{\mathcal{D}}_n, \boldsymbol{\theta}) + (1/n) \sum_{i=1}^{n} c(\tilde{\mathbf{x}}_i, \boldsymbol{\theta}). \quad \textit{Let } \hat{\ell}_n(\boldsymbol{\theta}) \equiv \ell(\tilde{\mathcal{D}}_n; \boldsymbol{\theta}).$$

- *Let*

$$\ell(\cdot) = \int c(\mathbf{x}, \cdot) p_e(\mathbf{x}) d\nu(\mathbf{x}).$$

*Then, $\hat{\ell}_n(\cdot) \rightarrow \ell(\cdot)$ uniformly on $\Theta$ as $n \rightarrow \infty$ with probability one where $\ell$ is continuous on $\Theta$.*

*Proof.* Using the Uniform Law of Large Numbers (Theorem 9.3.2),

$$\hat{\ell}'_n(\cdot) \equiv (1/n) \sum_{i=1}^{n} c(\tilde{\mathbf{x}}_i, \cdot)$$

converges uniformly on $\Theta$ with probability one to a continuous function $\ell(\cdot)$ as $n \rightarrow \infty$. Also, $k_n(\tilde{\mathcal{D}}_n, \cdot) \rightarrow 0$ uniformly with probability one on $\Theta$ as $n \rightarrow \infty$ by assumption. Therefore, $\hat{\ell}'_n + k_n(\tilde{\mathcal{D}}_n, \cdot) \rightarrow \ell$ uniformly on $\Theta$ with probability one as $n \rightarrow \infty$. ■

**Example 13.1.1** (Linear Regression Empirical Risk). Consider an objective function for learning which does not involve a penalty term:

$$\hat{\ell}_n(\boldsymbol{\theta}) = (1/n) \sum_{i=1}^{n} (\tilde{y}_i - \ddot{y}(\tilde{\mathbf{s}}_i, \boldsymbol{\theta}))^2 \tag{13.1}$$

where the response function $\ddot{y}$ is defined such that

$$\ddot{y}(\mathbf{s}, \boldsymbol{\theta}) = \boldsymbol{\theta}^T \mathbf{s}$$

and $(\tilde{\mathbf{s}}_1, \tilde{y}_1), \ldots, (\tilde{\mathbf{s}}_n, \tilde{y}_n)$ is a bounded stochastic sequence of $n$ independent and identically distributed $d$-dimensional random vectors with common density $p_e : \mathcal{R}^d \rightarrow [0, \infty)$ defined with respect to support specification measure $\nu$. The assumption that the sequence of random vectors is bounded is satisfied, for example, if each random vector $(\tilde{\mathbf{s}}_t, \tilde{y}_t)$ takes on a finite number of possible values.

This objective function for learning is an empirical risk function defined by:

$$\hat{\ell}'_n(\boldsymbol{\theta}) = (1/n) \sum_{i=1}^{n} c([y_i, \mathbf{s}_i], \boldsymbol{\theta}).$$

with loss function

$$c([y, \mathbf{s}], \boldsymbol{\theta}) = (y - \ddot{y}(\mathbf{s}, \boldsymbol{\theta}))^2 .$$

Assume the parameter space $\Theta$ is a closed, bounded, and convex set. The above assumptions satisfy the conditions of the Empirical Risk Function Convergence Theorem (Theorem 13.1.1). Thus, as $n \to \infty$:

$$\hat{\ell}_n(\boldsymbol{\theta}) \to \int (y - \ddot{y}(\mathbf{s}, \boldsymbol{\theta}))^2 p_e([y, \mathbf{s}]) d\nu([y, \mathbf{s}]) \tag{13.2}$$

uniformly on $\Theta$ with probability one. △

**Example 13.1.2** (Linear Regression with Regularization). A penalty term for minimizing network complexity (e.g., a ridge regression term) could be used to modify the empirical risk function in (13.1) of Example 13.1.1 to obtain a new *penalized* empirical risk function $\ell'$ defined by:

$$\hat{\ell}'_n(\boldsymbol{\theta}) = (1/n)|\boldsymbol{\theta}|^2 + (1/n)\sum_{i=1}^{n}(y - \ddot{y}(\mathbf{s}, \boldsymbol{\theta}))^2. \tag{13.3}$$

In this case, we again have the loss function

$$c([y, \mathbf{s}], \boldsymbol{\theta}) = (y - \ddot{y}(\mathbf{s}, \boldsymbol{\theta}))^2$$

but now the penalty function $k_n(\boldsymbol{\theta}) = (1/n)|\boldsymbol{\theta}|^2$. Since there exists a finite number $K$ such that $|\boldsymbol{\theta}|^2 \leq K$ on the closed, bounded, and convex parameter space $\Theta$, it follows that $K/n \to 0$ uniformly and hence $k_n \to 0$. Thus, $\{k_n\}$ is an acceptable sequence of penalty functions.

Using the Empirical Risk Function Convergence Theorem (Theorem 13.1.1), as $n \to \infty$:

$$\hat{\ell}'_n(\boldsymbol{\theta}) = \int (y - \ddot{y}(\mathbf{s}, \cdot))^2 p_e(|y, \mathbf{s}| d\nu(|y, \mathbf{s}|) \tag{13.4}$$

uniformly on $\Theta$ with probability one.

Note that constraint terms that do not converge to zero as a function of the sample size $n$ should be handled differently. For example, consider the objective function for learning:

$$\hat{\ell}''_n(\boldsymbol{\theta}) = \lambda|\boldsymbol{\theta}|^2 + (1/n)\sum_{i-1}^{n}(\tilde{y}_i - \ddot{y}(\tilde{\mathbf{s}}_i, \boldsymbol{\theta}))^2 \tag{13.5}$$

where $\lambda$ is a positive number. In order to apply the Empirical Risk Convergence Theorem in this case, choose:

$$c([y, \mathbf{s}], \boldsymbol{\theta}) = \lambda|\boldsymbol{\theta}|^2 + (y - \ddot{y}(\mathbf{s}, \boldsymbol{\theta}))^2$$

and set $k_n = 0$ for $n = 1, 2, 3, \ldots$.

Assume the parameter space $\Theta$ is a closed, bounded, and convex set. Using the Empirical Risk Convergence Theorem, as $n \to \infty$:

$$\hat{\ell}''_n(\boldsymbol{\theta}) \to \lambda|\boldsymbol{\theta}|^2 + \int (y - \ddot{y}(\mathbf{s}, \boldsymbol{\theta}))^2 p_e([y, \mathbf{s}]) d\nu([y, \mathbf{s}]) \tag{13.6}$$

uniformly on $\Theta$ with probability one.

The empirical risk function (13.3) therefore exhibits quite different asymptotic behavior from the empirical risk function in (13.5). In (13.3), the effects of the regularization term $|\boldsymbol{\theta}|^2$ become negligible for larger sample sizes. In (13.5), the effects of the regularization term $|\boldsymbol{\theta}|^2$ do not become negligible for larger sample sizes. The empirical risk function in (13.3) uses a regularization method appropriate for situations where the modeling problem is ill-posed for small samples but not ill-posed for sufficiently large samples. The empirical risk function constructed as in (13.5) uses a regularization method appropriate for situations where the modeling problem is intrinsically ill-posed and regularization is required for both small and very large sample sizes. △

Let $\boldsymbol{\theta}^*$ be a strict local minimizer of possibly a multimodal (non-convex) smooth function $\ell : \mathcal{R}^q \to \mathcal{R}$. Let the closed, bounded, convex subset $\Theta$ of $\mathcal{R}^q$ be defined such that $\boldsymbol{\theta}^*$ is the unique global minimizer of the restriction of $\ell$ to $\Theta$. Let $\hat{\ell}_n$ be an empirical risk function that converges uniformly to $\ell$ with probability one. Then, the following theorem provides conditions that ensure for each $n = 1, 2, \ldots$, there exists a (not necessarily unique) global minimizer of the empirical risk function $\hat{\ell}_n$ on $\Theta$ denoted by $\hat{\boldsymbol{\theta}}_n$. And, furthermore, the stochastic sequence $\hat{\boldsymbol{\theta}}_1, \hat{\boldsymbol{\theta}}_2, \ldots$ converges to $\boldsymbol{\theta}^*$ with probability one.

**Theorem 13.1.2** (Empirical Risk Minimizer Convergence).

- *Let $\tilde{\mathcal{D}}_n \equiv [\tilde{\mathbf{x}}_1, \ldots, \tilde{\mathbf{x}}_n]$ be a sequence of independent and identically distributed $d$-dimensional random vectors with common Radon-Nikodým density $p_e : \mathcal{R}^d \to [0, \infty)$ defined with respect to support specification measure $\nu$.*
- *Assume the continuous random function $c : \mathcal{R}^d \times \mathcal{R}^q \to \mathcal{R}$ is dominated by an integrable function on $\mathcal{R}^q$ with respect to $p_e$.*
- *Let $k_n : \mathcal{R}^{d \times n} \times \mathcal{R}^q \to \mathcal{R}$ be a continuous random function. Let $\hat{k}_n \equiv k_n(\tilde{\mathcal{D}}_n, \cdot)$. Assume $\hat{k}_n(\boldsymbol{\theta}) \to 0$ uniformly on $\mathcal{R}^q$ with probability one as $n \to \infty$.*
- *Assume the random function $\ell_n : \mathcal{R}^{d \times n} \times \mathcal{R}^q \to \mathcal{R}$ is defined such that for all $\boldsymbol{\theta} \in \mathcal{R}^q$:*

$$\ell_n(\tilde{\mathcal{D}}_n, \boldsymbol{\theta}) \equiv \hat{k}_n(\boldsymbol{\theta}) + (1/n) \sum_{i=1}^{n} c(\tilde{\mathbf{x}}_i; \boldsymbol{\theta}). \quad \text{Let } \hat{\ell}_n(\boldsymbol{\theta}) \equiv \ell(\tilde{\mathcal{D}}_n, \boldsymbol{\theta}).$$

- *Let $\Theta$ be a closed, bounded, and convex subset of $\mathcal{R}^q$ that contains a local minimizer, $\boldsymbol{\theta}^*$, of $\ell(\boldsymbol{\theta}) = \int c(\mathbf{x}; \boldsymbol{\theta}) p_e(\mathbf{x}) d\nu(\mathbf{x})$ such that $\boldsymbol{\theta}^*$ is also the unique global minimizer of the restriction of $\ell$ to $\Theta$.*

*(i) Then, $\hat{\boldsymbol{\theta}}_n \equiv \arg\min_{\boldsymbol{\theta} \in \Theta} \hat{\ell}_n(\boldsymbol{\theta})$ is a random vector for $n = 1, 2, \ldots$. (ii) Moreover, the stochastic sequence $\hat{\boldsymbol{\theta}}_1, \hat{\boldsymbol{\theta}}_2, \ldots$ converges to $\boldsymbol{\theta}^*$ with probability one.*

*Proof.* The proof is based upon the proof of Lemma 3 of Amemiya (1973).

*Proof of Part(i).* The proof of part (i) of the theorem's conclusion is established by using the Random Function Minimizers Are Borel-Measurable Theorem (Theorem 8.4.3) in conjunction with the assumptions that $\hat{k}_n$ and $c$ are continuous random functions to show that the global minimizer of $\hat{\ell}_n$ on $\Theta$, $\hat{\boldsymbol{\theta}}_n$, is a measurable function and hence a random vector for each $n = 1, 2, \ldots$.

*Proof of Part (ii).* Let $\mathcal{N}_{\boldsymbol{\theta}^*}$ be an open ball in $\Theta$ centered at $\boldsymbol{\theta}^*$ whose radius can be arbitrarily chosen. For example, the radius of the open ball $\mathcal{N}_{\boldsymbol{\theta}^*}$ can be chosen to be as small as desired. Part (ii) of the theorem's conclusion will be established by showing that given any arbitrary choice of the radius of the open ball $\mathcal{N}_{\boldsymbol{\theta}^*}$, that for $n$ sufficiently large, $\hat{\boldsymbol{\theta}}_n \in \mathcal{N}_{\boldsymbol{\theta}^*}$ with probability one.

Let $\neg\mathcal{N}_{\boldsymbol{\theta}^*} \equiv \{\boldsymbol{\theta} \in \Theta : \boldsymbol{\theta} \notin \mathcal{N}_{\boldsymbol{\theta}^*}\}$. The set $\neg\mathcal{N}_{\boldsymbol{\theta}^*}$ is a closed and bounded set since $\mathcal{N}_{\boldsymbol{\theta}^*}$ is an open set and $\Theta$ is a closed and bounded set. Since $\ell$ is a continuous function on the closed and bounded set $\neg\mathcal{N}_{\boldsymbol{\theta}^*}$, there exists a global minimizer of $\ell$ on $\Theta$, $\ddot{\boldsymbol{\theta}}^*$, which is an element of $\neg\mathcal{N}_{\boldsymbol{\theta}^*}$. Let $\epsilon = \ell(\ddot{\boldsymbol{\theta}}^*) - \ell(\boldsymbol{\theta}^*)$. Therefore, if

$$\ell(\hat{\boldsymbol{\theta}}_n) - \ell(\boldsymbol{\theta}^*) < \epsilon \tag{13.7}$$

this means that $\hat{\boldsymbol{\theta}}_n \in \mathcal{N}_{\boldsymbol{\theta}^*}$, which would establish Part(ii) of the theorem's conclusion.

To show this, note that it is possible to choose $n$ sufficiently large so that with probability one:

$$\ell(\hat{\boldsymbol{\theta}}_n) - \hat{\ell}_n(\hat{\boldsymbol{\theta}}_n) < \epsilon/2 \tag{13.8}$$

and

$$\hat{\ell}_n(\boldsymbol{\theta}^*) - \ell(\boldsymbol{\theta}^*) < \epsilon/2 \tag{13.9}$$

because $\hat{\ell}_n(\boldsymbol{\theta}) \rightarrow \ell(\boldsymbol{\theta})$ uniformly with probability one as $n \rightarrow \infty$ by the Empirical Risk Convergence Theorem (Theorem 13.1.1).

Substitute the assumption $\hat{\ell}_n(\hat{\boldsymbol{\theta}}_n) \leq \hat{\ell}_n(\boldsymbol{\theta}^*)$ into

$$\ell(\hat{\boldsymbol{\theta}}_n) - \ell(\boldsymbol{\theta}^*) = \ell(\hat{\boldsymbol{\theta}}_n) - \hat{\ell}_n(\hat{\boldsymbol{\theta}}_n) + \hat{\ell}_n(\hat{\boldsymbol{\theta}}_n) - \ell(\boldsymbol{\theta}^*)$$

to obtain:

$$\ell(\hat{\boldsymbol{\theta}}_n) - \ell(\boldsymbol{\theta}^*) \leq \left(\ell(\hat{\boldsymbol{\theta}}_n) - \hat{\ell}_n(\hat{\boldsymbol{\theta}}_n)\right) + \left(\hat{\ell}_n(\boldsymbol{\theta}^*) - \ell(\boldsymbol{\theta}^*)\right). \tag{13.10}$$

Now use (13.8) to bound the first difference on the right-hand side of (13.10) and use (13.9) to bound the second difference on the right-hand side of (13.10) to obtain:

$$\ell(\hat{\boldsymbol{\theta}}_n) - \ell(\boldsymbol{\theta}^*) \leq \left(\ell(\hat{\boldsymbol{\theta}}_n) - \hat{\ell}_n(\hat{\boldsymbol{\theta}}_n)\right) + \left(\hat{\ell}_n(\boldsymbol{\theta}^*) - \ell(\boldsymbol{\theta}^*)\right) < \epsilon/2 + \epsilon/2 = \epsilon$$

which establishes (13.7). ■

Note that the assumptions of Theorem 13.1.2 are satisfied provided that $\tilde{\mathbf{x}}_1, \ldots, \tilde{\mathbf{x}}_n$ is a bounded stochastic sequence and the continuous random function $c$ is piecewise continuous in its first argument since this implies that $c$ is dominated by an integrable function.

## Recipe Box 13.1 Convergence of Empirical Risk Minimizers (Theorem 13.1.1 and Theorem 13.1.2)

- **Step 1: Assume training data is bounded I.I.D.** Assume that the training data $\mathcal{D}_n \equiv [\mathbf{x}_1, \ldots, \mathbf{x}_n]$ is a realization of a bounded stochastic sequence of independent and identically distributed random vectors with common Radon Nikodým density $p_e : \mathcal{R}^d \rightarrow \mathcal{R}$ with respect to support specification measure $\nu$.

- **Step 2: Explicitly define empirical risk function.** Define the continuous random loss function $c(\mathbf{x}; \boldsymbol{\theta})$ and continuous random penalty function $k_n(\mathcal{D}_n, \boldsymbol{\theta})$. Let $\hat{k}_n(\cdot) \equiv k_n(\hat{\mathcal{D}}_n, \cdot)$ such that $\hat{k}_n \rightarrow 0$ as $n \rightarrow \infty$ with probability one. These definitions specify the empirical risk function:

$$\hat{\ell}_n(\boldsymbol{\theta}) = \hat{k}_n(\boldsymbol{\theta}) + (1/n)\sum_{i=1}^{n} c(\mathbf{x}_i; \boldsymbol{\theta}).$$

- **Step 3: Define restricted parameter space containing one minimizer.** Let $\ell : \mathcal{R}^q \rightarrow \mathcal{R}$ be defined such that for all $\boldsymbol{\theta} \in \mathcal{R}^q$:

$$\ell(\boldsymbol{\theta}) - \int c(\mathbf{x}; \boldsymbol{\theta}) p_e(\mathbf{x}) d\nu(\mathbf{x}).$$

  Let $\boldsymbol{\theta}^*$ be a local minimizer of $\ell$ on $\mathcal{R}^q$. Let $\Theta$ be a closed, bounded, and convex subset of $\mathcal{R}^q$ such that $\boldsymbol{\theta}^*$ is the unique global minimizer of the restriction of $\ell$ to $\Theta$.

- **Step 4: Conclude estimators are convergent.** Then, $\hat{\ell}_n \rightarrow \ell$ converges uniformly on $\Theta$ as $n \rightarrow \infty$ with probability one. In addition, for each $n = 1, 2, \ldots$ there exists a global minimizer, $\hat{\boldsymbol{\theta}}_n$, of the restriction of $\hat{\ell}_n$ to $\Theta$ which defines the stochastic sequence of minimizers $\hat{\boldsymbol{\theta}}_1, \hat{\boldsymbol{\theta}}_2, \ldots$ Moreover, this stochastic sequence of minimizers $\hat{\boldsymbol{\theta}}_1, \hat{\boldsymbol{\theta}}_2, \ldots$ converges to $\boldsymbol{\theta}^*$ with probability one.

**Example 13.1.3** (Convergence of Estimators for Multimodal (Non-convex) Risk Functions). The Empirical Risk Estimator Convergence Theorem (Theorem 13.1.2) provides a useful but weak insurance of convergence for smooth objective functions with multiple minima, maxima, and saddlepoints. Let $\tilde{\mathbf{x}}_1, \tilde{\mathbf{x}}_2, \ldots, \tilde{\mathbf{x}}_n$ be a bounded stochastic sequence of independent and identically distributed random vectors. Let $\hat{\ell}_n : \mathcal{R}^q \rightarrow \mathcal{R}$ be defined such that for all $\boldsymbol{\theta} \in \mathcal{R}^q$:

$$\hat{\ell}_n(\boldsymbol{\theta}) \equiv (1/n) \sum_{i=1}^{n} c(\tilde{\mathbf{x}}_i; \boldsymbol{\theta})$$

where $c$ is a continuous random function. Assume that the expected value of $\hat{\ell}_n$, $\ell : \mathcal{R}^q \rightarrow \mathcal{R}$, has multiple local minimizers. Assume, in addition, that $\boldsymbol{\theta}^*$ is a strict local minimizer of $\ell$ which implies there exists a closed, bounded, and convex set, $\Theta$, which contains $\boldsymbol{\theta}^*$ in its interior and has the property that $\boldsymbol{\theta}^*$ is the unique minimizer of $\ell$ on $\Theta$.

The Empirical Risk Function Convergence Theorem 13.1.1 ensures that $\hat{\ell}_n \rightarrow \ell$ as $n \rightarrow \infty$ with probability one. Let $\hat{\boldsymbol{\theta}}_n$ be a minimizer of $\hat{\ell}_n$ on $\Theta$ for $n = 1, 2, \ldots$. Then, the Empirical Risk Minimizer Convergence Theorem 13.1.2 ensures that the stochastic sequence of empirical risk minimizers $\hat{\boldsymbol{\theta}}_1, \hat{\boldsymbol{\theta}}_2, \ldots$ converges with probability one to the minimizer $\boldsymbol{\theta}^*$ pf the risk function $\ell$ as $n \rightarrow \infty$. Example 9.4.1 provides conditions ensuring that $\hat{\ell}_n(\hat{\boldsymbol{\theta}}_n) \rightarrow \ell(\boldsymbol{\theta}^*)$ as $n \rightarrow \infty$. △

## Exercises

13.1-1. *Risk Function for Linear Regression with Regularization.* Let $\mathbf{s}_i$ be a $d$-dimensional column vector specifying input pattern $i$ and let the real number $y_i$ be the desired response to $\mathbf{s}_i$ for $i = 1, \ldots, n$. Assume $(\mathbf{s}_1, y_1), \ldots, (\mathbf{s}_n, y_n)$ is a realization of a sequence of independent and identically distributed bounded random vectors. Consider the penalized least squares empirical risk function for linear regression

$$\hat{\ell}_n(\boldsymbol{\theta}) = \lambda|\boldsymbol{\theta}|^2 + (1/n) \sum_{i=1}^{n} (\tilde{y}_i - \ddot{y}(\tilde{\mathbf{s}}_i, \boldsymbol{\theta}))^2$$

where $\ddot{y}(\mathbf{s}, \boldsymbol{\theta}) = [\mathbf{s}^T \; 1]\boldsymbol{\theta}$. What additional assumptions (if any) are required to show, using the Empirical Risk Convergence Theorem, that $\hat{\ell}_n \rightarrow \ell$ uniformly on the parameter space $\Theta$ with probability one as $n \rightarrow \infty$? Provide an explicit expression for $\ell$ in terms of a Lebesgue integral and Radon-Nikodým density. Let $\hat{\boldsymbol{\theta}}_n$ be a local minimizer of $\hat{\ell}_n$ on $\Theta$. Assume $\boldsymbol{\theta}^*$ is a strict local minimizer of $\ell$ on $\Theta$. What additional assumptions (if any) are required to show, using the Empirical Risk Minimizer Convergence Theorem, that $\hat{\boldsymbol{\theta}}_n \rightarrow \boldsymbol{\theta}^*$ with probability one as $n \rightarrow \infty$?

13.1-2. *Risk Function for Logistic Regression with Regularization.* Let $\mathbf{s}_i$ be a $d$-dimensional column vector specifying input pattern $i$ and let the real number $y_i$ be the desired response to $\mathbf{s}_i$ for $i = 1, \ldots, n$. Assume $(\mathbf{s}_1, y_1), \ldots, (\mathbf{s}_n, y_n)$ is a realization of a sequence of independent and identically distributed random vectors. Assume that $(\mathbf{s}_i, y_i) \in \{0, 1\}^d \times \{0, 1\}$ for $i = 1, 2, \ldots$. Consider the penalized logistic regression empirical risk function

$$\hat{\ell}_n(\boldsymbol{\theta}) = (\lambda/n)\phi(\boldsymbol{\theta}) - (1/n) \sum_{i=1}^{n} [y_i \log \ddot{p}(\tilde{\mathbf{s}}_i, \boldsymbol{\theta}) + (1 - y_i) \log[1 - \ddot{p}(\tilde{\mathbf{s}}_i, \boldsymbol{\theta})]$$

where $\ddot{p}(\mathbf{s}, \boldsymbol{\theta}) = (1 + \exp(-[\mathbf{s}^T \; 1]\boldsymbol{\theta}))^{-1}$ and the smoothed L1 regularizer

$$\phi(\boldsymbol{\theta}) = \sum_{k=1}^{d+1} (\log(1 + \exp(-\theta_k)) + \log(1 + \exp(\theta_k))).$$

Repeat Exercise 13.1-1 using this risk function.

13.1-3. *Risk Function for Multinomial Logistic Regression.* Let $\mathbf{s}_i$ be a $d$-dimensional column vector specifying input pattern $i$ and let the $m$-dimensional column vector $\mathbf{y}_i$ be the desired response to $\mathbf{s}_i$ for $i = 1, \ldots, n$. Assume $(\mathbf{s}_1, \mathbf{y}_1), \ldots, (\mathbf{s}_n, \mathbf{y}_n)$ is a realization of a sequence of independent and identically distributed random vectors. Let $\mathbf{e}_1, \ldots, \mathbf{e}_m$ be the columns of an $m$-dimensional identity matrix. Assume that $(\mathbf{s}_i, \mathbf{y}_i) \in \{0,1\}^d \times \{\mathbf{e}_1, \ldots, \mathbf{e}_m\}$ for $i = 1, 2, \ldots$. In addition, define $\log : (0, \infty)^m \to \mathcal{R}^m$ such that $\log(\mathbf{x})$ is defined as a vector of the same dimension as $\mathbf{x}$ where the $k$th element of $\log(\mathbf{x})$ is the natural logarithm of the $k$th element of $\mathbf{x}$. Also define $\mathbf{exp}(\mathbf{x})$ as a vector of the same dimension as $\mathbf{x}$ where the $k$th element of $\mathbf{exp}(\mathbf{x})$ is $\exp(x_k)$ where $x_k$ is the $k$th element of $\mathbf{x}$. Consider the multinomial logistic regression empirical risk function

$$\hat{\ell}_n(\boldsymbol{\theta}) = -(1/n) \sum_{i=1}^{n} \mathbf{y}_i^T \log(\ddot{\mathbf{p}}(\mathbf{s}_i, \boldsymbol{\theta}))$$

where

$$\ddot{\mathbf{p}}(\mathbf{s}_i, \boldsymbol{\theta}) = \frac{\mathbf{exp}(\boldsymbol{\phi}_i)}{\mathbf{1}_m^T \mathbf{exp}(\boldsymbol{\phi}_i)}$$

and

$$\boldsymbol{\phi}_i = \mathbf{W}[\mathbf{s}_i^T \; 1]^T.$$

The last row of the matrix $\mathbf{W} \in \mathcal{R}^{m \times (d+1)}$ is a vector of zeros. The column vector $\boldsymbol{\theta} \equiv \mathbf{vec}([\mathbf{w}_1, \ldots, \mathbf{w}_{m-1}])$ where $\mathbf{w}_k^T$ is the $k$th row of $\mathbf{W}$. Repeat Exercise 13.1-1 using this risk function.

13.1-4. *Nonlinear Regression Risk Function with Large Sample Regularization.* Let the input pattern $\mathbf{s}_i$ be a $d$-dimensional column vector and let the scalar $y_i$ be the desired response to $\mathbf{s}_i$ for $i = 1, \ldots, n$. Assume $(\mathbf{s}_1, y_1, z_1), \ldots, (\mathbf{s}_n, y_n, z_n)$ is a realization of a sequence of independent and identically distributed random vectors with support $\{0,1\}^d \times \{0,1\} \times \{0\}$. Note that $\tilde{z}_i = 0$ for $i = 1, \ldots, n$.

Define the nonlinear regression empirical risk function

$$\hat{\ell}_n(\boldsymbol{\theta}) = \frac{1}{n} \sum_{i=1}^{n} (\tilde{z}_i - \lambda|\boldsymbol{\theta}_z|^2)^2 - [\tilde{y}_i \log \ddot{p}(\tilde{\mathbf{s}}_i, \boldsymbol{\theta}_y) + (1 - \tilde{y}_i) \log[1 - \ddot{p}(\tilde{\mathbf{s}}_i, \boldsymbol{\theta}_y)]] \quad (13.11)$$

where $\lambda$ is a positive number, $\ddot{p}(\mathbf{s}, \boldsymbol{\theta}_y) = (1 + \exp(-[\mathbf{s}^T \; 1]\boldsymbol{\theta}_y))^{-1}$, and $\boldsymbol{\theta} \equiv [\boldsymbol{\theta}_y \;\; \boldsymbol{\theta}_z]$. Provide a semantic interpretation of the first term on the right-hand side of (13.11) Repeat Exercise 13.1-1 using this risk function.

13.1-5. *Gaussian Model Risk Function.* Let $\mathbf{x}_i$ bea $d$-dimensional column vector specifying pattern $i$, $i = 1, \ldots, n$ where $d$ is a positive integer greater than 4. Assume $\mathbf{x}_1, \ldots, \mathbf{x}_n$ is a realization of a sequence of independent and identically distributed $d$-dimensional bounded random vectors $\tilde{\mathbf{x}}_1, \ldots, \tilde{\mathbf{x}}_n$ where for each $i = 1, \ldots, n$:

$\tilde{\mathbf{x}}_i \equiv [\tilde{x}_{i,1}, \ldots, \tilde{x}_{i,d}]$. Consider the empirical risk function for an unsupervised learning Gaussian model which is defined by:

$$\hat{\ell}_n(\boldsymbol{\theta}) = -(1/n) \sum_{i=1}^{n} \log \ddot{p}(\mathbf{x}_i, \boldsymbol{\theta})$$

where the notation $\ddot{p}_i(\boldsymbol{\theta})$ is defined such that:

$$\ddot{p}(\mathbf{x}_i, \boldsymbol{\theta}) = (2\pi)^{-d/2}(\det(\mathbf{C}))^{-1/2} \exp\left[-(1/2)(\mathbf{x} - \mathbf{m})^T \mathbf{C}^{-1}(\mathbf{x} - \mathbf{m})\right]$$

where $\boldsymbol{\theta} = [\mathbf{m}^T, \mathbf{vech}(\mathbf{C})^T]^T$. Repeat Exercise 13.1-1 using this risk function.

13.1-6. *Gaussian Mixture Model Risk.* Let $\mathbf{x}_i$ be a $d$-dimensional column vector specifying pattern $i$, $i = 1, \ldots, n$. Assume $\mathbf{x}_1, \ldots, \mathbf{x}_n$ is a realization of a sequence of independent and identically distributed $d$-dimensional bounded random vectors. Consider the empirical risk function for an unsupervised learning Gaussian mixture model which is defined by:

$$\hat{\ell}_n(\boldsymbol{\theta}) = -(1/n) \sum_{i=1}^{n} \log \ddot{p}(\mathbf{x}_i, \boldsymbol{\theta})$$

where $\ddot{p}_i(\boldsymbol{\theta})$ is defined such that:

$$\ddot{p}(\mathbf{x}_i, \boldsymbol{\theta}) = \sum_{k=1}^{m} \frac{\exp(\eta_k)}{\mathbf{1}_m^T \mathbf{exp}(\boldsymbol{\eta})} (\pi)^{-1/2} \exp(-|\mathbf{x}_i - \mathbf{m}_k|^2)$$

where $\boldsymbol{\theta} = [\boldsymbol{\eta}^T, \mathbf{m}_1^T, \ldots, \mathbf{m}_m^T]^T$ and $\boldsymbol{\eta} = [\eta_1, \ldots, \eta_m]^T$. Repeat Exercise 13.1-1 using this risk function.

13.1-7. *Perceptron Learning Empirical Risk Function.* Let $\hat{\ell}_n$ be an Empirical Risk Function as defined in Exercise 5.3-5 but include an L2 regularization term. Repeat Exercise 13.1-1 using this risk function.

## 13.2 Maximum Likelihood (ML) Estimation Methods

### 13.2.1 ML Estimation: Probability Theory Interpretation

Assume the observed data $\mathcal{D}_n = [\mathbf{x}_1, \ldots, \mathbf{x}_n]$ is a realization of a sequence of independent and identically distributed random vectors with a common environmental Radon-Nikodým density $p_e : \mathcal{R}^d \to [0, \infty)$ defined with respect to support specification measure $\nu$. In addition, assume the learning machine has a Bayesian probability model $\mathcal{M} \equiv \{p(\mathbf{x}|\boldsymbol{\theta}) : \boldsymbol{\theta} \in \Theta\}$ where each density in $\mathcal{M}$ is defined with respect to support specification measure $\nu$. It is assumed that $p(\mathbf{x}|\boldsymbol{\theta})$ has the semantic interpretation of the learning machine's expectation of observing $\mathbf{x}$ in its statistical environment when the learning machine's state of knowledge is $\boldsymbol{\theta}$.

Given this probability model $\mathcal{M}$, the likelihood of observing the first observation $\mathbf{x}_1$ is given by $p(\mathbf{x}_1|\boldsymbol{\theta})$ so choosing $\boldsymbol{\theta}^*$ to maximize $p(\mathbf{x}|\boldsymbol{\theta})$ on the parameter space yields the parameter vector $\boldsymbol{\theta}^*$ which makes the observation $\mathbf{x}_1$ most likely. The likelihood of observing the first two observations $\mathbf{x}_1$ and $\mathbf{x}_2$ is given by $p(\mathbf{x}_1|\boldsymbol{\theta})p(\mathbf{x}_2|\boldsymbol{\theta})$ since it is assumed that $\mathbf{x}_1$ and $\mathbf{x}_2$ are realizations of two statistically independent random vectors $\tilde{\mathbf{x}}_1$ and $\tilde{\mathbf{x}}_2$.

The likelihood of the observed data $\mathbf{x}_1, \ldots, \mathbf{x}_n$ is given by the

$$L_n(\mathcal{D}_n; \boldsymbol{\theta}) = p(\mathcal{D}_n|\boldsymbol{\theta}) = \prod_{i=1}^{n} p(\mathbf{x}_i|\boldsymbol{\theta})$$

where $\mathbf{x}_1, \ldots, \mathbf{x}_n$ is a presumed realization of a stochastic sequence of i.i.d. random vectors. The likelihood function $L_n : \mathcal{D}_n \times \Theta \rightarrow [0, \infty)$ is technically a random function of $\boldsymbol{\theta}$ defined with respect to the DGP density $p_e$. The random function $L_n$ is specified by the conditional density $p(\mathcal{D}_n|\boldsymbol{\theta})$.

Therefore, for a given $\boldsymbol{\theta}$, $L_n(\cdot; \boldsymbol{\theta})$ is a probability in the special case where $\mathcal{M}$ is a collection of probability mass functions. In the more general case where the data generating process is not a stochastic sequence of discrete random vectors, $L(\cdot; \boldsymbol{\theta})$ measures the likelihood of a particular data set $\mathcal{D}_n$ for a given $\boldsymbol{\theta}$. Finally, it is convenient for the purposes of the discussion here to compute the likelihood function with respect to a Bayesian probability model $\mathcal{M} \equiv \{p(\mathbf{x}|\boldsymbol{\theta}) : \boldsymbol{\theta} \in \Theta\}$. However, the likelihood function is also commonly defined with respect to the classical probability model $\mathcal{M} \equiv \{p(\mathbf{x}; \boldsymbol{\theta}) : \boldsymbol{\theta} \in \Theta\}$ as well.

The goal of maximum likelihood estimation is to find the parameter vector $\hat{\boldsymbol{\theta}}_n$ that maximizes the likelihood of the observed data $\hat{L}_n(\boldsymbol{\theta})$. If the probability model is possibly misspecified, then the goal of maximizing $\hat{L}_n(\boldsymbol{\theta})$ remains unchanged but the process is called *quasi-maximum likelihood estimation.* Since most machine learning algorithms use complex probabilistic models of their environments which are likely to be misspecified in many ways, quasi-maximum likelihood estimation is the rule rather than the exception in most machine learning analysis and design problems.

**Definition 13.2.1** (Maximum Likelihood Estimation). Let $\tilde{\mathcal{D}}_n \equiv [\tilde{\mathbf{x}}_1, \ldots, \tilde{\mathbf{x}}_n]$ be a sequence of independent and identically distributed $d$-dimensional random vectors with common Radon-Nikodým density $p_e : \mathcal{R}^d \rightarrow [0, \infty)$ with support specification measure $\nu$. Let

$$\mathcal{M} = \{p(\cdot|\boldsymbol{\theta}) : \boldsymbol{\theta} \in \Theta \subseteq \mathcal{R}^q\}$$

be a probability model where each element of $\mathcal{M}$ is in the support of $p_e$.

The random function $L_n : \mathcal{R}^{d \times n} \times \Theta \rightarrow [0, \infty)$ defined such that for all $\boldsymbol{\theta} \in \Theta$:

$$L_n(\mathcal{D}_n, \boldsymbol{\theta}) = \prod_{i=1}^{n} p(\mathbf{x}_i|\boldsymbol{\theta})$$

is called the *likelihood function.* The random function $\ell_n : \mathcal{R}^{d \times n} \times \Theta \rightarrow [0, \infty)$ defined such that for all $\boldsymbol{\theta} \in \Theta$:

$$\ell_n(\tilde{\mathcal{D}}_n, \boldsymbol{\theta}) \equiv -(1/n) \log L_n(\tilde{\mathcal{D}}_n, \boldsymbol{\theta}) \tag{13.12}$$

is called the *negative normalized log-likelihood function.* A global minimizer of $\ell_n(\tilde{\mathcal{D}}_n, \cdot)$ on $\Theta$ (when it exists) is called a *quasi-maximum likelihood estimator.* If, in addition, $p_e \in \mathcal{M}$, then a global minimizer of $\ell_n(\tilde{\mathcal{D}}_n, \cdot)$ on $\Theta$ (when it exists) is called a *maximum likelihood estimator.* □

The notation $\hat{\ell}_n \equiv \ell_n(\tilde{\mathcal{D}}_n, \cdot)$ is often used. Note that the negative normalized log-likelihood function

$$\hat{\ell}_n(\boldsymbol{\theta}) = -(1/n) \sum_{i=1}^{n} \log p(\tilde{\mathbf{x}}_i|\boldsymbol{\theta}) \tag{13.13}$$

is an empirical risk function. Thus, the Empirical Risk Function Convergence Theorem (see Theorem 13.1.1) may be used to provide conditions ensuring that $\hat{\ell}_n$ converges to

$$\ell(\boldsymbol{\theta}) = -\int p_e(\mathbf{x}) \log p(\mathbf{x}|\boldsymbol{\theta}) d\nu(\mathbf{x}). \tag{13.14}$$

uniformly with probability one.

Definition 13.2.1 is applicable to multimodal objective functions. To apply Definition 13.2.1 in such cases, a particular strict local minimizer, $\hat{\boldsymbol{\theta}}_n$, of $\hat{\ell}_n$ is chosen and then Definition 13.2.1 is applied to the restriction of $\hat{\ell}_n$ on a closed, bounded, and convex region, $\Theta$, chosen to contain $\hat{\boldsymbol{\theta}}_n$ for large $n$ with probability one and such that $\hat{\ell}_n$ is convex on $\Theta$. A local probability model is defined in such cases which restricts the parameter vector to take on values in $\Theta$.

**Example 13.2.1** (Likelihood Function for Coin Toss Experiment)**.** Let $\tilde{x}_i = 1$ if the $i$th coin flip in a series of coin flips takes on the value of "heads" and assume $\tilde{x}_i = 0$ if the outcome of the $i$th coin flip is "tails". Assume that $\tilde{x}_1, \tilde{x}_2, \ldots, \tilde{x}_n$ is a sequence of $n$ independent and identically distributed random variables. Let the free parameter $\theta$ denote the probability the coin takes on the value of heads when it is flipped. (i) Explicitly define a probability model $\mathcal{M}$ representation of this model. (ii) Construct an objective function $L_n : \mathcal{R}^n \times [0,1] \rightarrow \mathcal{R}$ such that a global maximizer of the random function $\hat{L}_n(\theta)$ makes a particular realization of $\tilde{x}_1, \ldots, \tilde{x}_n$ maximally likely with respect to $\mathcal{M}$.

**Solution.** (i) The coin probability model

$$\mathcal{M} \equiv \{p(x|\theta) \equiv \theta^x (1-\theta)^{1-x} : 0 \leq \theta \leq 1\}.$$

(ii) Choose

$$\hat{L}_n(\theta) = \prod_{i=1}^{n} \theta^{x_i} (1-\theta)^{(1-x_i)}.$$

△

**Example 13.2.2** (Likelihood Function (Discrete Random Vector))**.** Let $\Omega \equiv \{\mathbf{x}^1, \ldots, \mathbf{x}^M\}$ where $\mathbf{x}^k$ is a $d$-dimensional vector, $k = 1, \ldots, M$. Assume the data generating process samples with replacement from the set $\Omega$ to generate a stochastic sequence $\tilde{\mathbf{x}}_1, \tilde{\mathbf{x}}_2, \ldots$ of independent and identically distributed $d$-dimensional random vectors with common probability mass function $p_e(\mathbf{x}) = 1/M$. A researcher observed the outcomes of this data generating process and incorrectly assumes that the stochastic sequence $\tilde{\mathbf{x}}_1, \tilde{\mathbf{x}}_2, \ldots$ consists of independent and identically distributed observations with a common $d$-dimensional multivariate Gaussian density function whose mean $\boldsymbol{\mu}$ and covariance matrix $\mathbf{C}$ are free parameters. Let $\boldsymbol{\theta}$ be a $q$-dimensional parameter vector which contains the elements of $\boldsymbol{\mu}$ and the unique elements of $\mathbf{C}$. (i) Explicitly define a probability model $\mathcal{M}$ representation of this model. (ii) Construct an objective function $L_n : \mathcal{R}^{dn} \times \Theta \rightarrow \mathcal{R}$ such that a global maximizer of $L_n$ makes a particular realization of $\tilde{x}_1, \ldots, \tilde{x}_n$ maximally likely with respect to $\mathcal{M}$.

**Solution.** (i) Let $Q$ be the set of all $d$-dimensional positive definite symmetric matrices. Let the probability model $\mathcal{M} \equiv \left\{p(\cdot|\boldsymbol{\mu}, \mathbf{C}) : \boldsymbol{\mu} \in \mathcal{R}^d, \mathbf{C} \in Q\right\}$ where

$$p(\mathbf{x}|\mathbf{m}, \mathbf{C}) = (2\pi)^{-d/2} (\det(\mathbf{C}))^{-1/2} \exp\left[-(1/2)(\mathbf{x}-\mathbf{m})^T \mathbf{C}^{-1} (\mathbf{x}-\mathbf{m})\right]. \quad (13.15)$$

(ii) Construct objective function $\hat{L}_n$ such that:

$$\hat{L}_n(\boldsymbol{\theta}) = \prod_{i=1}^{n} p(\mathbf{x}_i|\mathbf{m}, \mathbf{C})$$

where $p(\mathbf{x}|\mathbf{m}, \mathbf{C})$ is defined as in (13.15). △

**Example 13.2.3** (Desired Responses Inform Risk Function Design)**.** The desired response of a supervised learning machine informs the design of the discrepancy function for the supervised learning machine. As noted in Section 1.2.1, the loss function $c([\mathbf{s}, \mathbf{y}], \boldsymbol{\theta})$ for a supervised learning machine generating a predicted response $\ddot{\mathbf{y}}(\mathbf{s}, \boldsymbol{\theta})$ for a given input pattern $\mathbf{s}$ is often conveniently defined in terms of a discrepancy function $D(\mathbf{y}, \ddot{\mathbf{y}}(\mathbf{s}, \boldsymbol{\theta}))$ which measures the deviation of the predicted response $\ddot{\mathbf{y}}(\mathbf{s}, \boldsymbol{\theta})$ from the desired response $\mathbf{y}$. A *maximum likelihood discrepancy function* is defined by the relationship:

$$c([\mathbf{s}, \mathbf{y}], \boldsymbol{\theta}) = D(\mathbf{y}, \ddot{\mathbf{y}}(\mathbf{s}, \boldsymbol{\theta})) = -\log p(\mathbf{y}|\mathbf{s}, \boldsymbol{\theta}) \tag{13.16}$$

where $p(\mathbf{y}|\mathbf{s}, \boldsymbol{\theta})$ specifies the learning machine's degree of belief that $\mathbf{y}$ is the desired response given input pattern $\mathbf{s}$ for a parameter vector $\boldsymbol{\theta}$. The support of $\tilde{\mathbf{y}}$ as specified by the conditional probability density $p(\mathbf{y}|\mathbf{s}, \boldsymbol{\theta})$ should be consistent with the observed range of values of the desired response $\mathbf{y}$. So, for example, if the desired response is a binary vector with $\mathbf{y} \in \{0, 1\}^m$ then defining a maximum likelihood discrepancy function assuming $\tilde{\mathbf{y}}$ is a conditional multivariate Gaussian with mean $\ddot{\mathbf{y}}(\mathbf{s}, \boldsymbol{\theta})$ would imply that the assumed probability model was intrinsically misspecified. △

**Example 13.2.4** (Maximum Likelihood Model for Real-Valued Targets). When the desired response $\mathbf{y}$ is an $m$-dimensional real vector, then a common choice in the statistics and machine learning literature is to choose a maximum likelihood discrepancy function with the assumption that the learning machine's probability model is specified by the multivariate Gaussian density function:

$$p(\mathbf{y}|\mathbf{s}, \boldsymbol{\theta}) = \pi^{-m/2} \exp\left(-|\mathbf{y} - \ddot{\mathbf{y}}(\mathbf{s}, \boldsymbol{\theta})|^2\right).$$

This corresponds to the assumption that the learning machine assumes the desired response $\mathbf{y}$ for a given input pattern $\mathbf{s}$ is equal to the learning machine's predicted response $\ddot{\mathbf{y}}(\mathbf{s}, \boldsymbol{\theta})$ plus zero-mean Gaussian noise with covariance matrix equal to $(1/2)\mathbf{I}_m$. Substituting this conditional multivariate Gaussian density assumption into (13.16) corresponds to the least-squares loss function

$$c([\mathbf{s}, \mathbf{y}], \boldsymbol{\theta}) = |\mathbf{y} - \ddot{\mathbf{y}}(\mathbf{s}, \boldsymbol{\theta})|^2 + (m/2) \log \pi.$$

△

**Example 13.2.5** (Maximum Likelihood Model for Binary Target Vectors). When the desired response $\mathbf{y}$ is an $m$-dimensional binary vector (i.e., $\mathbf{y} \in \{0, 1\}^m$), then a common choice in the statistics and machine learning literature is to choose a maximum likelihood discrepancy function with the assumption that the learning machine's probability model is specified by the Bernoulli-type conditional probability mass function:

$$p(\mathbf{y}|\mathbf{s}, \boldsymbol{\theta}) = \prod_{j=1}^{m} \ddot{p}_j(\mathbf{s}, \boldsymbol{\theta})^{y_j} (1 - \ddot{p}_j(\mathbf{s}, \boldsymbol{\theta}))^{1-y_j}$$

where $\ddot{p}_j(\mathbf{s}, \boldsymbol{\theta}) = (1 + \exp(-\phi_j(\mathbf{s}, \boldsymbol{\theta})))^{-1}$. The functions $\phi_1, \ldots, \phi_m$ can be linear or nonlinear smooth functions of the input pattern vector $\mathbf{s}$ and the learning machine's parameter vector $\boldsymbol{\theta}$.

This Bernoulli-type conditional probability mass function is then substituted into (13.16) to obtain the loss function:

$$c([\mathbf{s}, \mathbf{y}], \boldsymbol{\theta}) = -\sum_{j=1}^{m} (y_j \log \ddot{p}_j(\mathbf{s}, \boldsymbol{\theta}) + (1 - y_j) \log \ddot{p}_j(\mathbf{s}, \boldsymbol{\theta})). \tag{13.17}$$

The function $\ddot{p}_j(\mathbf{s}, \boldsymbol{\theta})$ is semantically interpreted as the probability that the $j$th element of the random desired response vector $\tilde{\mathbf{y}}$ is equal to one for a particular input pattern $\mathbf{s}$ and parameter vector $\boldsymbol{\theta}$. △

**Example 13.2.6** (Maximum Likelihood Model for Categorical Target Vectors)**.** When the desired response $\mathbf{y}$ is a categorical variable with $m$ possible values, a particular value of the categorical variable may be represented as a column of an $m$-dimensional identity matrix. A common choice in the statistics and machine learning literature is to choose a maximum likelihood discrepancy function with the assumption that the learning machine's probability model is specified by the multinomial logistic (or softmax) conditional probability mass function:

$$p(\mathbf{y}|\mathbf{s}, \boldsymbol{\theta}) = \mathbf{y}^T \mathbf{p}_y(\mathbf{s}, \boldsymbol{\theta})$$

where

$$\mathbf{p}_y(\mathbf{s}, \boldsymbol{\theta}) = \frac{\mathbf{exp}(\boldsymbol{\phi}(\mathbf{s}, \boldsymbol{\theta}))}{\mathbf{1}_m^T \mathbf{exp}(\boldsymbol{\phi}(\mathbf{s}, \boldsymbol{\theta}))}$$

where $\mathbf{exp}$ is a vector-valued function defined such that the $j$th element of $\mathbf{exp}(\boldsymbol{\psi})$ is equal to $\exp(-\psi_j)$ where $\psi_j$ is the $j$th element of $\boldsymbol{\psi}$.

The function $p(\mathbf{y}|\mathbf{s}, \boldsymbol{\theta})$ is semantically interpreted as the model's predicted probability that label $\mathbf{y}$ is appropriate given input pattern $\mathbf{s}$ and the model's parameter vector $\boldsymbol{\theta}$. The $m$-dimensional vector-valued function $\boldsymbol{\phi} : \mathcal{R}^d \times \mathcal{R}^q \rightarrow \mathcal{R}^m$ can be linear or nonlinear smooth functions of the input pattern vector $\mathbf{s}$ and the learning machine's parameter vector $\boldsymbol{\theta}$. The loss function is obtained by substituting the multinomial logistic conditional probability mass function into (13.16). △

**Example 13.2.7** (Maximum Likelihood Model for Nonlinear Least Squares)**.** Consider a supervised learning machine whose goal is to predict the numerical output $y_i$ given a particular $d$-dimensional input vector $\mathbf{s}_i$ for $i = 1, \ldots, n$. Assume that the functional form of the machine is given by a function $\ddot{y} : \mathcal{R}^d \times \mathcal{R}^q \rightarrow \mathcal{R}$ defined such that for a particular parameter vector $\boldsymbol{\theta}$ in the parameter space $\Theta$ and input pattern vector $\mathbf{s}$ that the machine's prediction of the output is $\ddot{y}(\mathbf{s}, \boldsymbol{\theta})$. It is proposed that the parameters of the learning machine should be chosen so that they minimize the sum-squared error function:

$$\hat{\ell}_n(\boldsymbol{\theta}) = (1/n) \sum_{i=1}^{n} (\tilde{y}_i - \ddot{y}(\tilde{\mathbf{s}}_i, \boldsymbol{\theta}))^2. \tag{13.18}$$

Assuming that the learning machine is solving a maximum likelihood estimation problem, provide an explicit probability model that specifies the learning machine's representation of its probabilistic environment. Such a probability model is useful because it provides a method for characterizing which types of statistical environments are "learnable" by the learning machine as well as which types of statistical environments can never be completely learned.

**Solution.** The first step is to make some general assumptions about how the data was generated by characterizing the learning machine's statistical environment. Assume that the input vectors $\mathbf{s}_1, \mathbf{s}_2, \ldots$ are a realization of a sequence of independent and identically distributed random vectors with common density $p_e(\mathbf{s})$. Assume that desired responses $y_1, y_2, \ldots$ are a realization of a sequence of random variables whose respective conditional densities are: $p_e(y|\mathbf{s}_1; \boldsymbol{\theta}), p_e(y|\mathbf{s}_2; \boldsymbol{\theta}), \ldots$.

The second step is to postulate a specific probability model for the learning machine. Let $\mathcal{M}$ be a probability model specification defined such that:

$$\mathcal{M} \equiv \{p_m(\cdot|\boldsymbol{\theta}) : \boldsymbol{\theta} \in \Theta\}$$

where $p_m$ is defined such that

$$p_m(y, \mathbf{s}|\boldsymbol{\theta}) = p(y|\mathbf{s}, \boldsymbol{\theta}) p(\mathbf{s})$$

where $p(\mathbf{s})$ specifies the model's expectation that $\mathbf{s}$ is observed in its statistical environment and

$$p(y|\mathbf{s},\boldsymbol{\theta}) = \sigma^{-1}(2\pi)^{-1/2}\exp\left(-\frac{(y-\ddot{y}(\mathbf{s};\boldsymbol{\theta}))^2}{2\sigma^2}\right)$$

specifies the model's expectation that $y$ is observed in its statistical environment given a particular value of $\mathbf{s}$ and a particular value of $\boldsymbol{\theta}$. Assume that $\sigma$ is a constant which has been provided to the learning machine. The goal of learning is to compute a maximum likelihood estimate of the parameter $\boldsymbol{\theta}$ with respect to the learning machine's probability model of the data generating process.

The negative normalized likelihood function is given by the formula:

$$\tilde{\ell}_n(\boldsymbol{\theta}) = -(1/n)\log\prod_{i=1}^{n} p_m(\tilde{y}_i,\tilde{\mathbf{s}}_i|\boldsymbol{\theta})$$

which can be rewritten after some algebra as:

$$\tilde{\ell}_n(\boldsymbol{\theta}) = (2n)^{-1}\log(2\pi\sigma^2) - (1/n)\sum_{i=1}^{n}\log p(\tilde{\mathbf{s}}_i) + (2n\sigma^2)^{-1}\sum_{i=1}^{n}(\tilde{y}_i - \ddot{y}(\tilde{\mathbf{s}}_i;\boldsymbol{\theta}))^2$$

or equivalently as:

$$\tilde{\ell}_n(\boldsymbol{\theta}) = K_1 + K_2\hat{\ell}_n(\boldsymbol{\theta})$$

where $K_1$ and $K_2$ are constants and $\hat{\ell}_n$ is defined in (13.18). △

**Example 13.2.8** (Large Sample Regularized Least Squares ML Estimation). Consider the general nonlinear regression modeling problem where $\ddot{y}(\mathbf{s};\boldsymbol{\theta})$ denotes the predicted response of a learning machine to input pattern vector $\mathbf{s}$ when the learning machine's parameter vector is $\boldsymbol{\theta}$. Consider the empirical risk function for learning:

$$\hat{\ell}_n(\boldsymbol{\theta}) = \lambda|\boldsymbol{\theta}|^2 + (1/n)\sum_{i=1}^{n}|\tilde{y}_i - \ddot{y}(\tilde{\mathbf{s}}_i;\boldsymbol{\theta})|^2 \tag{13.19}$$

where the term $\lambda|\boldsymbol{\theta}|^2$ does not converge to zero as $n \to \infty$. Such an objective function in (13.19) can be interpreted within a maximum likelihood estimation framework using the following argument.

Assume that the input vectors $\mathbf{s}_1, \mathbf{s}_2, \ldots$ are a realization of a sequence of independent and identically distributed random vectors with common density $p_e(\mathbf{s})$. The *augmented data trick* (e.g., Theil and Goldberger 1961; Montgomery and Peck 1982, p. 320) is then used to recode the original desired responses $y_1, y_2, \ldots$ as $(y_1, 0), (y_2, 0), \ldots$. It is assumed that the recoded desired responses $(y_1, 0), (y_2, 0), \ldots$ are a realization of a sequence of random variables whose respective conditional densities are: $p_e(y, \zeta|\mathbf{s}_1;\boldsymbol{\theta}), p_e(y, \zeta|\mathbf{s}_2;\boldsymbol{\theta}), \ldots$.

The second step is to postulate a specific probability model for the learning machine. Let $\mathcal{M}$ be a probability model specification defined such that:

$$\mathcal{M} \equiv \{p_m(\cdot|\boldsymbol{\theta}) : \boldsymbol{\theta} \in \Theta\}$$

where $p_m$ is defined such that

$$p_m(y,\zeta,\mathbf{s}|\boldsymbol{\theta}) = p(y,\zeta|\mathbf{s},\boldsymbol{\theta})p(\mathbf{s})$$

where $p(\mathbf{s})$ specifies the model's expectation that $\mathbf{s}$ is observed in its statistical environment. The density $p(y, \zeta|\mathbf{s}, \boldsymbol{\theta}) \equiv p(y|\mathbf{s}, \boldsymbol{\theta})p(\zeta|\boldsymbol{\theta})$. The density

$$p(y|\mathbf{s}, \boldsymbol{\theta}) = \sigma^{-1}(2\pi)^{-1/2} \exp\left(-\frac{(y - \ddot{y}(\mathbf{s}; \boldsymbol{\theta}))^2}{2\sigma^2}\right)$$

specifies the model's expectation that $y$ is observed in its statistical environment given a particular value of $\mathbf{s}$ and a particular value of $\boldsymbol{\theta}$ and constant positive number $\sigma$. The density

$$p(\zeta|\boldsymbol{\theta}) = (\sigma_\theta\sqrt{2\pi})^{-1} \exp\left(\frac{-(\zeta - |\boldsymbol{\theta}|)^2}{2\sigma_\theta^2}\right)$$

is a Gaussian density centered at $\boldsymbol{\theta}$ with constant variance $\sigma_\theta^2$.

The goal of learning is to compute a maximum likelihood estimate of the parameter $\boldsymbol{\theta}$ with respect to the learning machine's probability model of the data generating process.

The negative normalized likelihood function is given by the formula:

$$\tilde{\ell}_n(\boldsymbol{\theta}) = -(1/n) \log \prod_{i=1}^{n} p_m(\tilde{y}_i, \zeta_i, \tilde{\mathbf{s}}_i|\boldsymbol{\theta})$$

which can be rewritten after some algebra and substituting 0 for all values of $\zeta$ as:

$$\tilde{\ell}_n(\boldsymbol{\theta}) = K_1 + K_2\left(\lambda|\boldsymbol{\theta}|^2 + (1/n)\sum_{i=1}^{n} |\tilde{y}_i - \ddot{y}(\tilde{\mathbf{s}}_i; \boldsymbol{\theta})|^2\right) = K_1 + K_2\hat{\ell}_n(\boldsymbol{\theta})$$

where $K_1$ is a constant and $K_2$ is a positive constant. Thus, this particular probabilistic interpretation allows one to semantically interpret the regularization constant $\lambda$ as equal to $1/(2\sigma_\theta^2)$ for the case where $\lambda$ is a fixed constant. △

**Example 13.2.9** (Log-Likelihood for a Gibbs Density). Define a probability model specification $\mathcal{M} \equiv \{p(\mathbf{x}|\boldsymbol{\theta}) : \boldsymbol{\theta} \in \Theta\}$ such that

$$p(\mathbf{x}|\boldsymbol{\theta}) = \frac{\exp(-V(\mathbf{x}; \boldsymbol{\theta}))}{Z(\boldsymbol{\theta})}$$

is a Gibbs density. Then, the negative normalized log-likelihood function defined with respect to $\mathcal{M}$ and $\mathcal{D}_n$ is given by:

$$\hat{\ell}_n(\boldsymbol{\theta}) = -(1/n)\sum_{i=1}^{n} \log\left[\frac{\exp(-V(\tilde{\mathbf{x}}_i; \boldsymbol{\theta}))}{Z(\boldsymbol{\theta})}\right] = (1/n)\sum_{i=1}^{n} V(\tilde{\mathbf{x}}_i; \boldsymbol{\theta}) + \log Z(\boldsymbol{\theta}) \quad (13.20)$$

where the normalization constant $Z(\boldsymbol{\theta})$ is given by:

$$Z(\boldsymbol{\theta}) = \int \exp(-V(\mathbf{y}; \boldsymbol{\theta}))d\nu(\mathbf{y}).$$

Typically, the function $V(\mathbf{x}; \boldsymbol{\theta})$ is straightforward to evaluate. However, the normalization constant $Z(\boldsymbol{\theta})$ is a function of $\boldsymbol{\theta}$ and is computed by evaluating a high-dimensional integral. For example, when the Gibbs density $p(\cdot|\boldsymbol{\theta}) : \{0, 1\}^d \to [0, 1]$ is a probability mass function, the normalization constant $Z(\boldsymbol{\theta})$ is computed using the formula:

$$Z(\boldsymbol{\theta}) = \sum_{\mathbf{y} \in \{0,1\}^d} \exp(-V(\mathbf{y}; \boldsymbol{\theta}))$$

which is the sum of $2^d$ terms! △

### 13.2.2 ML Estimation: Information Theory Interpretation

Let $S$ be a finite set of objects. The *self-information* in *bits* received from observing the value $x$ of a discrete random variable $\tilde{x}$ with probability mass function $p : S \to [0, 1]$ is defined as $-\log_2 p(x)$ bits of information. Thus, more probable outcomes convey less information in bits. In the case where the natural log (i.e., $\log(\exp(x)) = x$) is used, then $-\log p(x)$ is the amount of self-information in *nats*.

**Example 13.2.10** (Self-Information in the Yes-No Question Game). Consider the following game where the first player secretly picks one of $M$ possible numbers with equal probability. The game also includes a second player whose goal is to guess the number by asking a minimal number of questions which can be answered either with a "yes" or "no" response.

- *First Player:* I am thinking of a number between 1 and 16.
- *Second Player:* Is the number greater than 8?
- *First Player:* No
- *Second Player:* Is the number greater than 4?
- *First Player:* No
- *Second Player:* Is the number greater than 2?
- *First Player:* No
- *Second Player:* Is the number equal to 1?
- *First Player:* Yes

The minimum number of yes-no questions required can be calculated using the formula for self-information. In particular, at most $\log_2 M$ yes-no questions are sufficient to ensure that the number is identified. For example, in the case where $M = 1$, only $\log_2 1 = 0$ yes-no question is required to identify the secret number. In the case where $M = 2$, only $\log_2 2 = 1$ yes-no questions are required to identify the secret number. In the example above where the goal of the Yes-No Question Game is to identify an unknown number between 1 and 16, it follows that $M = 16$, so at most $\log_2 16 = 4$ yes-no questions are sufficient for identifying the unknown number.

Suppose a third person, called "the cheater", is now involved in this game. The cheater's job is to secretly spy on the first player in order to determine the pattern vector chosen by the first player. Then, the cheater secretly tells the second player what number was chosen by the first player before the second player asks questions. If $M = 1$ the cheater conveys 0 bits (since $0 = \log_2 1$) of information to the second player by revealing the number. If $M = 2$ the cheater conveys 1 bit (since $1 = \log_2 2$) of information to the second player by revealing the secret number. If $M = 8$ the cheater conveys 3 bits of information (since $3 = \log_2 8$) to the second player by revealing the secret number. The number of bits of information conveyed by the cheater corresponds to the maximum number of yes-no questions that would have been required by the second player to identify the secret number without the cheater's help. $\triangle$

#### 13.2.2.1 Entropy

For discrete random variables, the expected value of the self-information of a random vector $\tilde{\mathbf{x}}$ with probability mass function $p : \Omega \to [0, 1]$ is given by the formula:

$$\mathcal{H}(\tilde{\mathbf{x}}) = E\{-\log p(\tilde{\mathbf{x}})\} = -\sum_{\mathbf{x} \in \Omega} p(\mathbf{x}) \log p(\mathbf{x})$$

where the summation is over the sample space of the discrete random vector $\tilde{\mathbf{x}}$. The number $\mathcal{H}(\tilde{\mathbf{x}})$ is called the *discrete entropy* for the discrete random vector $\tilde{\mathbf{x}}$.

The discrete entropy is the expected self-information per observation. Equivalently, the discrete entropy of $\tilde{\mathbf{x}}$ is the average number of yes-no questions required to identify a particular realization of $\tilde{\mathbf{x}}$ (see Example 13.2.10). Since $0 < p(\mathbf{x}) \leq 1$, the self-information for each realization of $\tilde{\mathbf{x}}$, $-\log p(\mathbf{x})$, is strictly non-negative which implies $\mathcal{H}(\tilde{\mathbf{x}}) \geq 0$.

Discrete entropy can also naturally be interpreted as specifying the amount of self-information provided by a"typical" data set for a given sample size $n$. Assume a discrete random vector $\tilde{\mathbf{x}}$ takes on the value of $\mathbf{x}^k$ with probability $p(\mathbf{x}^k)$ for $k = 1, \ldots, M$. A "typical" data set $\bar{\mathcal{D}}_n$ for a sample size $n$ would be expected to have approximately $np(\mathbf{x}^k)$ occurrences of the realization $\mathbf{x}^k$ for $k = 1, \ldots, M$.

The probability of a data set $\bar{\mathcal{D}}_n$ which contains exactly $np(\mathbf{x}^k)$ occurrences of $\mathbf{x}^k$ for $k = 1, \ldots, M$ is given explicitly by the formula:

$$p(\bar{\mathcal{D}}_n) = \prod_{k=1}^{M} p\left(\mathbf{x}^k\right)^{np(\mathbf{x}^k)} \tag{13.21}$$

where the self-information of the typical data set $\bar{\mathcal{D}}_n$ is given by the formula:

$$n\mathcal{H}(\tilde{\mathbf{x}}) = -\log p(\bar{\mathcal{D}}_n) = -n \sum_{k=1}^{M} p(\mathbf{x}^k) \log p(\mathbf{x}^k).$$

Since discrete entropy is the expected value of $-\log p(\tilde{\mathbf{x}})$ when $p$ is a probability mass function specifying the distribution of $\tilde{\mathbf{x}}$, it seems reasonable to define entropy in the more general case as the expected value of $-\log p(\tilde{\mathbf{x}})$ when $p$ is a Radon-Nikodým density function specifying the distribution of $\tilde{\mathbf{x}}$.

**Definition 13.2.2** (Entropy). Let $\tilde{\mathbf{x}}$ be a random vector with Radon-Nikodým density $p : \mathcal{R}^d \to [0, \infty)$ with respect to support specification measure $\nu$. Assume

$$\mathcal{H}(\tilde{\mathbf{x}}) = -\int p(\mathbf{x}) \log p(\mathbf{x}) d\nu(\mathbf{x})$$

is a finite number. The notation $\mathcal{H}(\tilde{\mathbf{x}})$ or the notation $\mathcal{H}(p)$ denotes the *entropy* of $\tilde{\mathbf{x}}$ with density $p$. If $\tilde{\mathbf{x}}$ is a discrete random vector, $\mathcal{H}(\tilde{\mathbf{x}})$ is called the *discrete entropy* of $\tilde{\mathbf{x}}$. If $\tilde{\mathbf{x}}$ is an absolutely continuous random vector, $\mathcal{H}(\tilde{\mathbf{x}})$ is called the *differential entropy* of $\tilde{\mathbf{x}}$. □

**Definition 13.2.3** (Empirical Entropy). Let $\tilde{\mathcal{D}}_n \equiv [\tilde{\mathbf{x}}_1, \ldots, \tilde{\mathbf{x}}_n]$ be a sequence of i.i.d. $d$-dimensional random vectors with common Radon-Nikodým density $p_e : \mathcal{R}^d \to [0, \infty)$ defined with respect to support specification measure $\nu$. The *empirical entropy*

$$\hat{\mathcal{H}}_n(\tilde{\mathcal{D}}_n) \equiv -(1/n) \sum_{i=1}^{n} \log p_e(\tilde{\mathbf{x}}_i).$$

□

Note that when $\mathcal{H}(\tilde{\mathbf{x}})$ is finite, then $E\left\{\hat{\mathcal{H}}_n(\tilde{\mathcal{D}}_n)\right\} = \mathcal{H}(\tilde{\mathbf{x}})$.

The semantic interpretation of entropy for the case of absolutely continuous and mixed random vectors, however, is not as straightforward as the discrete entropy case. For example, suppose $\tilde{\mathbf{x}}$ is an absolutely continuous random vector with a realization denoted by $\mathbf{x}$. First, $-\log p(\mathbf{x}) = \infty$ since $p(\mathbf{x}) = 0$. That is, the self-information of a realization of an absolutely continuous random vector conveys an infinite number of bits. Second, $\mathcal{H}(\tilde{\mathbf{x}})$ might be a negative number (see Example 13.2.11). That is, in some cases the "cheater" in Example 13.2.10 conveys a negative amount of information on the average which seems nonsensical.

**Example 13.2.11** (An Example where Entropy Is Negative). The entropy in bits of a uniformly distributed random variable $\tilde{x}$ on the interval $[a, b]$ is given by the formula:

$$\mathcal{H}(\tilde{x}) = -\int_a^b (1/(b-a)) \log_2(1/(b-a))dx = \log_2(b-a).$$

Choose $b = 1$ and $a = 0.5$, then $\mathcal{H}(\tilde{\mathbf{x}}) = \log_2(b-a) = -1$. △

Definition 13.2.4 provides a tool for constructing a more general semantic interpretation of entropy for random vectors which are discrete, absolutely continuous, or mixed. Definition 13.2.4 is based upon the discussion by Cover and Thomas (2006, p. 59).

**Definition 13.2.4** (Typical Set of Data Sets). Let $\tilde{\mathcal{D}}_n \equiv [\tilde{\mathbf{x}}_1, \ldots, \tilde{\mathbf{x}}_n]$ be a sequence of i.i.d. $d$-dimensional random vectors with common Radon-Nikodým density $p_e : \mathcal{R}^d \to [0, \infty)$ defined with respect to support specification measure $\nu$. For a given positive real number $\epsilon$ and positive integer $n$,

$$\left| -(1/n) \sum_{i=1}^{n} \log p_e(x_i) - \mathcal{H}(p_e) \right| \leq \epsilon \tag{13.22}$$

is called an $(\epsilon, n)$-*typical set of data sets* for $p_e$. An element of an $(\epsilon, n)$-*typical set of data sets* will be called an $(\epsilon, n)$-*typical data set* for a given positive $\epsilon$ and positive integer $n$. □

Let $p_e^n(\mathcal{D}_n) = \prod_{i=1}^{n} p_e(\mathbf{x}_i)$ for all $\mathcal{D}_n = [\mathbf{x}_1, \ldots, \mathbf{x}_n]$ in the support of $\tilde{\mathcal{D}}_n$. Then, rearrangement of the terms in (13.22) shows that an $(\epsilon, n)$-typical set of data sets may also be defined as:

$$\left\{ \mathcal{D}_n \in \mathcal{R}^{d \times n} : 2^{-(n/\log 2)(\mathcal{H}(p_e)+\epsilon)} \leq p_e^n(\mathcal{D}_n) \leq 2^{-(n/\log 2)(\mathcal{H}(p_e)-\epsilon)} \right\}. \tag{13.23}$$

If the entropy $\mathcal{H}(\tilde{\mathbf{x}})$ is finite, then the Strong Law of Large Numbers (Theorem 9.3.1) implies that $\hat{\mathcal{H}}_n(\tilde{\mathcal{D}}_n) \to \mathcal{H}(p_e)$ as $n \to \infty$ with probability one. Since convergence with probability one implies convergence in probability, it follows that for a given positive $\epsilon$, it is always possible to choose the sample size $n$ sufficiently large so that with sufficiently high probability:

$$\left| -(1/n) \sum_{i=1}^{n} \log p_e(\tilde{\mathbf{x}}_i) + \int p_e(\mathbf{x}) \log p_e(\mathbf{x}) d\nu(\mathbf{x}) \right| = \left| \hat{\mathcal{H}}_n(\tilde{\mathcal{D}}_n) - \mathcal{H}(p_e) \right| \leq \epsilon. \tag{13.24}$$

That is, the empirical entropy of each data set in the typical set of data sets is approximately equal to the entropy of the DGP. This semantic interpretation of the entropy holds not only for the discrete entropy case but for the more general definition of entropy which includes data generating processes consisting of absolutely continuous and mixed random vectors.

Finally, let $\mathcal{M} \equiv \{p(\mathbf{x}|\boldsymbol{\theta}) : \boldsymbol{\theta} \in \Theta\}$. Note that if $\hat{\ell}_n(\boldsymbol{\theta}^*) \equiv -(1/n) \log p(\mathbf{x}|\boldsymbol{\theta}^*)$, then $\hat{\mathcal{H}}_n(\tilde{\mathcal{D}}_n)$ in (13.24) may be interpreted as the negative normalized log-likelihood function evaluated at $\boldsymbol{\theta}^*$ provided $\boldsymbol{\theta}^*$ may be chosen such that $p(\mathbf{x}; \boldsymbol{\theta}^*) = p_e(\mathbf{x})$.

**Example 13.2.12** (Highly Probable Data Sets Are Not Typical Data Sets). A highly probable (or most probable) data set is not the same as a data set that is an element of the typical set of data sets. Consider a coin that is weighted to come up heads 60% of the time and come up tails 40% of the time. After 100 coin tosses, one would expect to see a sequence of observations where the outcomes of about 60 coin tosses were heads and the outcomes of about 40 of the coin tosses were tails. Such a sequence would be a typical data set. The probability of observing this sequence is $(0.6)^{60}(0.4)^{40}$. Now suppose that after 100 coin tosses, the outcome of all 100 coin tosses was heads. This latter observed sequence would not be a typical data set but would be a most probable data set whose probability is $(0.6)^{100}$.

#### 13.2.2.2 Cross-Entropy Minimization: ML Estimation

Assume that a data set $\mathcal{D}_n \equiv \{\mathbf{x}_1, \ldots, \mathbf{x}_n\}$ is a realization of a sequence of independent and identically distributed discrete random vectors $\tilde{\mathcal{D}}_n \equiv [\tilde{\mathbf{x}}_1, \ldots, \tilde{\mathbf{x}}_n]$, with common probability mass function $p_e : \Omega \rightarrow [0, 1]$.

Let

$$\mathcal{M} \equiv \{p(\mathbf{x}|\boldsymbol{\theta}) : \boldsymbol{\theta} \in \Theta\}$$

be a probability model where $p(\cdot|\boldsymbol{\theta}) : \Omega \rightarrow [0, 1]$ is a probability mass function on sample space $\Omega$ for each $\boldsymbol{\theta} \in \Theta$.

The negative normalized log-likelihood function, $\hat{\ell}_n$, defined with respect to $\mathcal{M}$ and $\tilde{\mathcal{D}}_n$ is defined such that

$$\hat{\ell}_n(\boldsymbol{\theta}) = -(1/n) \sum_{i=1}^{n} \log p(\tilde{\mathbf{x}}_i|\boldsymbol{\theta}). \tag{13.25}$$

When $p(\cdot|\boldsymbol{\theta})$ is a probability mass function for some fixed constant $\boldsymbol{\theta}$, then

$$\hat{\ell}_n(\boldsymbol{\theta}) = -(1/n) \sum_{i=1}^{n} \log p(\tilde{\mathbf{x}}_i|\boldsymbol{\theta}) = -(1/n) \log \prod_{i=1}^{n} p(\tilde{\mathbf{x}}_i|\boldsymbol{\theta}).$$

That is, $\hat{\ell}_n$ is the sample average self-information per observation $\tilde{\mathbf{x}}_i$ for a given $\boldsymbol{\theta}$.

The expected self-information per observation $\tilde{\mathbf{x}}_i$ for a given $\boldsymbol{\theta}$ is computed using the formula:

$$\ell(\boldsymbol{\theta}) = E\{\hat{\ell}_n(\boldsymbol{\theta})\} = E\left\{- \log p(\tilde{\mathbf{x}}_i|\boldsymbol{\theta})\right\}. \tag{13.26}$$

Equation (13.26) is the expected self-information per observation for a typical data set with $n$ records where the likelihood of the data set is computed using the approximating density $p(\cdot|\boldsymbol{\theta})$ rather than the DGP density $p_e$.

Assume a discrete random vector $\tilde{\mathbf{x}}$ takes on the value of $\mathbf{x}^k$ with probability $p_e(\mathbf{x}^k)$ for $k = 1, \ldots, M$. Assume the likelihood of an observation $\mathbf{x}$ is computed using the formula $p(\mathbf{x}; \boldsymbol{\theta})$ for a particular value of $\boldsymbol{\theta}$. That is, the likelihood of an observation is computed using the probability model $\mathcal{M}$. A typical data set $\bar{\mathcal{D}}_n$ would be expected to have approximately $np_e(\mathbf{x}^k)$ occurrences of the realization $\mathbf{x}^k$ which is assigned the likelihood $p(\mathbf{x}^k; \boldsymbol{\theta})$ by the probability model $\mathcal{M}$ for $k = 1, \ldots, M$.

A typical data set $\bar{\mathcal{D}}_n$ that contains exactly $np_e(\mathbf{x}^k)$ occurrences of $\mathbf{x}^k$ is assigned a probability by the model $\mathcal{M}$ so that the probability of $\mathcal{D}_n$ is given by:

$$p(\bar{\mathcal{D}}_n|\boldsymbol{\theta}) = \prod_{k=1}^{M} p\left(\mathbf{x}^k|\boldsymbol{\theta}\right)^{np_e(\mathbf{x}^k)}. \tag{13.27}$$

The self-information of $\bar{\mathcal{D}}_n$ per observation is given by the formula:

$$-(1/n) \log p(\bar{\mathcal{D}}_n|\boldsymbol{\theta}) = -(1/n)n \sum_{k=1}^{M} p_e(\mathbf{x}^k) \log p\left(\mathbf{x}^k|\boldsymbol{\theta}\right) = \ell(\boldsymbol{\theta}).$$

Thus, the expected negative normalized log-likelihood may be interpreted for the case of discrete random vectors as the self-information per observation of a typical data set where the likelihood of an observation is computed using an approximating density from the model. The expected negative normalized log-likelihood (or equivalently) the normalized self-information of a typical data set is called the "cross entropy" for the case of data generating processes consisting of discrete random vectors.

The previously presented semantic interpretation of cross entropy is only applicable to data generating processes consisting of discrete random vectors. The following definition of cross entropy is applicable to the more general situation where the data generating process is defined in terms of absolutely continuous and mixed random vectors as well as discrete random vectors.

**Definition 13.2.5** (Cross Entropy). Let $p_e : \mathcal{R}^d \rightarrow [0, \infty)$ and $p : \mathcal{R}^d \rightarrow [0, \infty)$ be two Radon-Nikodým density functions defined with respect to a common support specification measure $\nu$. Assume

$$\mathcal{H}(p_e, p) = -\int p_e(\mathbf{x}) \log p(\mathbf{x}) d\nu(\mathbf{x}) \tag{13.28}$$

is a finite number. Then, $\mathcal{H}(p_e, p)$ is called the *cross entropy* of $p$ with respect to $p_e$. □

Let $\mathcal{M} \equiv \{p(\cdot|\boldsymbol{\theta}) : \boldsymbol{\theta} \in \Theta\}$ be a parametric probability model. The expected value of the negative normalized log-likelihood evaluated at a particular $\boldsymbol{\theta}^* \in \Theta$ is simply the cross entropy of $p(\cdot; \boldsymbol{\theta}^*)$ with respect to the DGP density $p_e$.

**Definition 13.2.6** (Kullback-Leibler Information Criterion (KLIC)). Let $p : \mathcal{R}^d \rightarrow [0, \infty)$ and $p_e : \mathcal{R}^d \rightarrow [0, \infty)$ be two Radon-Nikodým density functions defined with respect to a common support specification measure $\nu$. Assume

$$D(p_e || p) \equiv -\int p_e(\mathbf{x}) \log \left[ \frac{p(\mathbf{x})}{p_e(\mathbf{x})} \right] d\nu(\mathbf{x}) \tag{13.29}$$

is a finite number. The number $D(p_e||p)$ is called the *Kullback-Leibler Information Criterion.* □

The Kullback-Leibler Information Criterion is also called the *relative entropy* or the *Kullback-Leibler divergence* of $p$ from $p_e$.

From the definition of the Kullback-Leibler Information Criterion,

$$D(p_e||p) = \mathcal{H}(p_e, p) - \mathcal{H}(p_e). \tag{13.30}$$

Note that (13.30) implies that a density $p$ that is a global minimizer of $D(p_e||p)$ is also a global minimizer of $\mathcal{H}(p_e, p)$ provided that $\mathcal{H}(p_e)$ is finite. In addition, inspection of (13.30) shows that $D(p_e||p)$ may be interpreted as the expected amount of information lost when $p$ is used for the purpose of representing $p_e$. In the special case where $p$ is a perfect representation of $p_e$ so that $p = p_e$ $\nu$-almost everywhere, then no information is lost and the expected information loss is zero bits.

The following definition (Definition 13.2.7) is helpful for providing a semantic interpretation of an information-theoretic interpretation of the negative normalized log-likelihood for situations where the probability model is possibly misspecified. This definition may be viewed as a natural extension of Definition 13.2.4.

**Definition 13.2.7** (Cross Entropy Typical Data Set). Let $\tilde{\mathcal{D}}_n \equiv [\tilde{\mathbf{x}}_1, \ldots, \tilde{\mathbf{x}}_n]$ be a sequence of i.i.d. $d$-dimensional random vectors with common Radon-Nikodým density $p_e : \mathcal{R}^d \rightarrow [0, \infty)$ defined with respect to support specification measure $\nu$. Let $p : \mathcal{R}^d \rightarrow [0, \infty)$ be a

Radon-Nikodým density defined with respect to $\nu$. For a given positive real number $\epsilon$ and positive integer $n$, the set

$$\left\{[\mathbf{x}_1, \ldots, \mathbf{x}_n] \in \mathcal{R}^{d \times n} : \left| -(1/n) \sum_{i=1}^{n} \log p(\mathbf{x}_i) - \mathcal{H}(p_e, p) \right| \leq \epsilon \right\} \tag{13.31}$$

is called the $(\epsilon, n)$-*cross entropy typical set of data sets* for $p$ relative to $p_e$. An element of an $(\epsilon, n)$ cross entropy typical set of data sets for $p$ relative to $p_e$ is called an $(\epsilon, n)$-cross entropy data set for $p$ relative to $p_e$. □

Let $\tilde{\mathcal{D}}_n$ be a sequence of $n$ independent and identically distributed random vectors with common density $p_e$ defined with respect to support specification measure $\nu$. Let the approximating model density $p$ be the probability distribution that the learning machine is using to approximately represent $p_e$. Then, the model-based likelihood of the observed data $\mathcal{D}_n \equiv [\mathbf{x}_1, \ldots, \mathbf{x}_n]$ computed using the approximating model density $p$ is given by the formula $p^n(\mathcal{D}_n) \equiv \prod_{i=1}^{n} p(\mathbf{x}_i)$.

Rearrangement of the terms in (13.31) show that an $(\epsilon, n)$-cross entropy typical set of data sets may also be defined as:

$$\left\{\mathcal{D}_n \in \mathcal{R}^{d \times n} : 2^{-(n/log2)(\mathcal{H}(p_e,p)+\epsilon)} \leq p^n(\mathcal{D}_n) \leq 2^{-(n/log2)(\mathcal{H}(p_e,p)-\epsilon)}\right\}. \tag{13.32}$$

If the cross entropy $\mathcal{H}(p_e, p)$ is finite, then the Strong Law of Large Numbers (Theorem 9.3.1) holds so that as $n \rightarrow \infty$:

$$-(1/n) \sum_{i=1}^{n} \log p(\mathbf{x}_i) \rightarrow \mathcal{H}(p_e, p) \tag{13.33}$$

with probability one.

Equations (13.31) and (13.32) therefore imply that the set of data sets generated from the DGP density $p_e$ have approximately the same likelihood $p^n(\mathcal{D}_n) = \prod_{i=1}^{n} p(\mathbf{x}_i)$ when $n$ is sufficiently large where the likelihood of the data set generated by the DGP density $p_e$ is computed using an approximating density $p$. These general semantic interpretations of cross entropy and relative entropy hold when the DGP density $p_e$ and approximating density $p$ are used to represent discrete, absolutely continuous, and mixed random vectors.

### 13.2.3 Properties of Cross-Entropy Global Minimizers

The following key theorem provides conditions ensuring a global minimizer, $\boldsymbol{\theta}^*$, of the expected negative normalized log-likelihood function

$$\ell(\boldsymbol{\theta}) = -\int p_e(\mathbf{x}) \log p(\mathbf{x}|\boldsymbol{\theta}) d\nu(\mathbf{x})$$

has the property that $p(\mathbf{x}|\boldsymbol{\theta}^*) = p_e(\mathbf{x})$ $\nu$-almost everywhere when the probability model

$$\mathcal{M} \equiv \{p(\mathbf{x}|\boldsymbol{\theta}) : \boldsymbol{\theta} \in \Theta\}$$

is correctly specified.

**Theorem 13.2.1** (Cross-Entropy Risk Function Lower Bound). *Let the DGP Radon-Nikodým density $p_e : \mathcal{R}^d \rightarrow [0, \infty)$ with respect to support specification measure $\nu$. Assume $\mathcal{H}(p_e)$ is a finite number. Let $\Theta$ be a subset of $\mathcal{R}^q$. Let the model*

$$\mathcal{M} \equiv \{p(\cdot|\boldsymbol{\theta}) : \boldsymbol{\theta} \in \Theta\}$$

*where the Radon-Nikodým density $p(\cdot|\boldsymbol{\theta}) : \mathcal{R}^d \rightarrow [0, \infty)$ is defined with respect to $\nu$ for each $\boldsymbol{\theta} \in \Theta$. Assume*

$$\ell(\boldsymbol{\theta}) \equiv -\int p_e(\mathbf{x}) \log p(\mathbf{x}|\boldsymbol{\theta}) d\nu(\mathbf{x})$$

*is finite for all $\boldsymbol{\theta} \in \Theta$. Then, $\ell(\boldsymbol{\theta}) \geq \mathcal{H}(p_e)$. In addition, if there exists a $\boldsymbol{\theta}^*$ such that $p(\mathbf{x}|\boldsymbol{\theta}^*) = p_e(\mathbf{x})$ $\nu$-almost everywhere, then $\ell(\boldsymbol{\theta}^*) = \mathcal{H}(p_e)$.*

*Proof.* The proof follows the approach of van der Vaart (1998, Lemma 5.35). Let

$$D\left(p_e || p(\cdot; \boldsymbol{\theta})\right) \equiv \ell(\boldsymbol{\theta}) + \int p_e(\mathbf{x}) \log p_e(\mathbf{x}) d\nu(\mathbf{x}).$$

$$D\left(p_e || p(\cdot; \boldsymbol{\theta})\right) = -\int \left[\log \frac{p(\mathbf{x}|\boldsymbol{\theta})}{p_e(\mathbf{x})}\right] p_e(\mathbf{x}) d\nu(\mathbf{x}).$$

Using the inequality $-\log R \geq -2(\sqrt{R} - 1)$,

$$D\left(p_e || p(\cdot; \boldsymbol{\theta})\right) \geq -\int 2\left(\sqrt{\frac{p(\mathbf{x}|\boldsymbol{\theta})}{p_e(\mathbf{x})}} - 1\right) p_e(\mathbf{x}) d\nu(\mathbf{x}) = -2\int \left[\sqrt{p(\mathbf{x}|\boldsymbol{\theta}) p_e(\mathbf{x})} - p_e(\mathbf{x})\right] d\nu(\mathbf{x}).$$

Since $\int p_e(\mathbf{x}) d\nu(\mathbf{x}) = \int p(\mathbf{x}|\boldsymbol{\theta}) d\nu(\mathbf{x}) = 1$ for all $\boldsymbol{\theta} \in \Theta$:

$$D\left(p_e || p(\cdot; \boldsymbol{\theta})\right) \geq \int \left[p_e(\mathbf{x}) - 2\sqrt{p(\mathbf{x}|\boldsymbol{\theta}) p_e(\mathbf{x})} + p(\mathbf{x}|\boldsymbol{\theta})\right] d\nu(\mathbf{x})$$

and it follows by rearranging terms in the integrand that

$$D\left(p_e || p(\cdot; \boldsymbol{\theta})\right) = \ell(\boldsymbol{\theta}) - \mathcal{H}(p_e) \geq \int \left(\sqrt{p_e(\mathbf{x})} - \sqrt{p(\mathbf{x}|\boldsymbol{\theta})}\right)^2 d\nu(\mathbf{x}) \geq 0. \quad (13.34)$$

Inspection of (13.34) shows that $\ell(\boldsymbol{\theta}) \geq \mathcal{H}(p_e)$. In addition, $p_e(\mathbf{x}) = p(\mathbf{x}|\boldsymbol{\theta})$ $\nu$-almost everywhere if and only if $\ell(\boldsymbol{\theta}) = \mathcal{H}(p_e)$. ■

The following theorem shows that if the parametric probability model is a canonical exponential family probability model, then the negative log-likelihood function $-\log p(\mathcal{D}_n|\cdot)$ is convex on the parameter space $\Theta$. Consequently, any strict local minimizer is the unique global minimizer on the parameter space (see Theorem 5.3.7).

**Theorem 13.2.2** (Canonical Exponential Family Likelihood Function Convexity). *Let $\Theta$ be a subset of $\mathcal{R}^q$. Let $p(\cdot|\boldsymbol{\theta})$ be a Radon-Nikodým density defined with respect to support specification measure $\nu$. Let $\boldsymbol{\Upsilon} : \mathcal{R}^d \rightarrow \mathcal{R}^q$ be a Borel measurable function. Let $\mathcal{M} \equiv \{p(\cdot|\boldsymbol{\theta}) : \boldsymbol{\theta} \in \Theta\}$ be a probability model specification defined with respect to support specification measure $\nu$ and for all $\boldsymbol{\theta} \in \Theta$: $p(\cdot|\boldsymbol{\theta})$ is a Gibbs density with energy function $V : \mathcal{R}^d \times \Theta \rightarrow \mathcal{R}$ defined such that*

$$V(\mathbf{x}; \boldsymbol{\theta}) = -\boldsymbol{\theta}^T \boldsymbol{\Upsilon}(\mathbf{x})$$

*and normalization constant $Z : \Theta \rightarrow \mathcal{R}$. Assume, in addition, that $\Theta$ has the property that $Z$ is bounded on $\Theta$. Then,*
*(i) $\Theta$ is a convex set,*
*(ii) $\log Z$ is convex on $\Theta$, and*
*(iii)$-\log p(\mathbf{x}|\cdot)$ is convex on $\Theta$ for all $\mathbf{x} \in \mathcal{R}^d$.*

*Proof.* Let $f(\cdot) \equiv (\exp[-\boldsymbol{\theta}_1^T \Upsilon(\cdot)])^{\alpha}$ and $g(\cdot) \equiv (\exp[-\boldsymbol{\theta}_2^T \Upsilon(\cdot)])^{(1-\alpha)}$ where $\alpha \in (0,1)$. Since

$$Z(\boldsymbol{\theta}) = \int \exp[-\boldsymbol{\theta}_1^T \Upsilon(\mathbf{s})] d\nu(\mathbf{s}) < \infty$$

on $\Theta$ and $\alpha \in (0,1)$:

$$\left| \int |f(\mathbf{s})|^{\frac{1}{\alpha}} d\mu(\mathbf{s}) \right|^{\alpha} = Z(\boldsymbol{\theta})^{\alpha} < \infty \tag{13.35}$$

and

$$\left| \int |g(\mathbf{s})|^{\frac{1}{1-\alpha}} d\nu(\mathbf{s}) \right|^{1-\alpha} = Z(\boldsymbol{\theta})^{1-\alpha} < \infty. \tag{13.36}$$

Since (13.35) and (13.36) hold for all $\alpha \in (0,1)$, Hölder's Inequality (Bartle 1966, p. 56) implies

$$\int f(\mathbf{s}) g(\mathbf{s}) d\nu(\mathbf{s}) \leq \left| \int |f(\mathbf{s})|^{\frac{1}{\alpha}} d\nu(\mathbf{s}) \right|^{\alpha} \left| \int |g(\mathbf{s})|^{\frac{1}{1-\alpha}} d\nu(\mathbf{s}) \right|^{(1-\alpha)}. \tag{13.37}$$

Substituting the definitions of $f(\cdot)$ and $g(\cdot)$ into (13.37) and rearranging terms gives for all $\boldsymbol{\theta}_1, \boldsymbol{\theta}_2 \in \Theta$:

$$Z\left(\alpha\boldsymbol{\theta}_1 + (1-\alpha)\boldsymbol{\theta}_2\right) = \int \exp\left[[(\alpha\boldsymbol{\theta}_1 + (1-\alpha)\boldsymbol{\theta}_2]^T \boldsymbol{\Upsilon}(\mathbf{s})\right] d\boldsymbol{\nu}(\mathbf{s}) < \infty \tag{13.38}$$

which implies $\alpha\boldsymbol{\theta}_1 + (1-\alpha)\boldsymbol{\theta}_2 \in \Theta$. Thus, $\Theta$ is a convex set. A similar substitution of $f$ and $g$ into (13.37) and taking the natural log gives:

$$\log Z\left(\alpha\boldsymbol{\theta}_1 + (1-\alpha)\boldsymbol{\theta}_2\right) \leq \alpha \log Z(\boldsymbol{\theta}_1) + (1-\alpha) \log Z(\boldsymbol{\theta}_2) \tag{13.39}$$

which implies that $\log Z(\cdot)$ is convex on $\Theta$. Since the Hessian of $V(\mathbf{x},\cdot)$ vanishes for all $\mathbf{x} \in \mathcal{R}^d$, this implies that $V(\mathbf{x},\cdot)$ is a convex function on $\Theta$ for all $\mathbf{x} \in \mathcal{R}^d$. Since $-\log p(\mathbf{x}|\cdot)$ is the sum of the convex functions $\log Z(\cdot)$ and $V(\mathbf{x},\cdot)$ with $\mathbf{x} \in \mathcal{R}^d$ on $\Theta$, it follows that $-\log p(\mathbf{x}|\cdot)$ is convex on $\Theta$ for all $\mathbf{x} \in \mathcal{R}^d$. ■

**Example 13.2.13** (Convexity Results for Nonlinear Regression Models)**.** The Exponential Family Likelihood Convexity Theorem is applicable to many important nonlinear regression modeling problems in machine learning. Consider a nonlinear regression probability model where the learning machine expects to observe an $m$-dimensional vector $\mathbf{y}$ given a $d$-dimensional input vector $\mathbf{s}$ and $q$-dimensional parameter vector $\boldsymbol{\theta}$ with a likelihood specified by the density $p(\mathbf{y}|\mathbf{s},\boldsymbol{\theta})$. For many nonlinear regression modeling problems, it is desirable to construct a probability model specification

$$\mathcal{M} \equiv \{p(\mathbf{y},\mathbf{s}|\boldsymbol{\theta}) : \boldsymbol{\theta} \in \Theta\}$$

which is a member of the exponential family. To achieve this, assume $p(\mathbf{y},\mathbf{s}|\boldsymbol{\theta})$ may be rewritten such that:

$$p(\mathbf{y},\mathbf{s}|\boldsymbol{\theta}) = p(\mathbf{y}|\mathbf{s},\boldsymbol{\theta}) p_s(\mathbf{s}).$$

Most standard nonlinear regression models including multilayer smooth perceptron architectures assume that $p(\mathbf{y}|\mathbf{s},\boldsymbol{\theta})$ is a member of the exponential family. The assumption $p_s(\mathbf{s})$ is not functionally dependent on $\boldsymbol{\theta}$ ensures the joint distribution $p(\mathbf{y},\mathbf{s}|\boldsymbol{\theta})$ is in the exponential family as well. Intuitively, this assumption simply means that the nonlinear regression model is not attempting to model the distribution of the input pattern $\tilde{\mathbf{s}}$ but is instead focusing upon the conditional density of the response pattern $\tilde{\mathbf{y}}$ for a given input pattern $\mathbf{s}$. △

### 13.2.4 Pseudolikelihood Empirical Risk Function

The following objective function for learning is useful for parameter estimation in Markov random fields.

**Definition 13.2.8** (Pseudolikelihood Function)**.** Let $\tilde{\mathcal{D}}^n \equiv [\tilde{\mathbf{x}}_1, \ldots, \tilde{\mathbf{x}}_n]$ be a sequence of $n$ independent and identically distributed $d$-dimensional random vectors with common support $S$. Let

$$\mathcal{M} = \{p(\cdot|\boldsymbol{\theta}) : \boldsymbol{\theta} \in \Theta \subseteq \mathcal{R}^q\}$$

be a probability model where each density in $\mathcal{M}$ is a homogeneous Markov random field with respect to some neighborhood system $\{\mathcal{N}_1, \ldots, \mathcal{N}_d\}$ with support $S$. Let the random function $\ell_n : \mathcal{R}^{dn} \times \Theta \to [0, \infty)$ be defined such that for all $\boldsymbol{\theta} \in \Theta$ and for all $\mathcal{D}_n$ in the support of $\tilde{\mathcal{D}}_n$:

$$\hat{\ell}_n(\boldsymbol{\theta}) = -(1/n) \log \left[ \prod_{i=1}^{n} \prod_{j=1}^{d} p\left(\tilde{x}_i(j) | \mathcal{N}_j(\tilde{\mathbf{x}}_i), \boldsymbol{\theta}\right) \right] \tag{13.40}$$

where $\tilde{x}_i(j)$ denotes the $j$th element of $\tilde{\mathbf{x}}_i$ for $i = 1, \ldots, n$ and $j = 1, \ldots, d$. The function $\hat{\ell}_n$ in (13.40) is not a true likelihood function because

$$(\tilde{x}_i(1), \mathcal{N}_1(\tilde{\mathbf{x}}_i)), \ldots, (\tilde{x}_i(d), \mathcal{N}_d(\tilde{\mathbf{x}}_i)) \tag{13.41}$$

are not independent and identically distributed for each $i \in \{1, \ldots, n\}$ even though $\tilde{\mathbf{x}}_1, \ldots, \tilde{\mathbf{x}}_n$ are i.i.d. For this reason, the random function $\hat{\ell}_n$ is called a *pseudolikelihood function.* A minimizer of $\hat{\ell}_n$ in (13.40) is called a *pseudo-maximum likelihood estimate.* □

The objective function $\hat{\ell}_n$ in (13.40) is an empirical risk function if one defines the loss function

$$c(\mathbf{x}_i, \boldsymbol{\theta}) \equiv -\sum_{j=1}^{d} \log p\left(x_i(j) | \mathcal{N}_j(\mathbf{x}_i), \boldsymbol{\theta}\right) \tag{13.42}$$

and specifies $\hat{\ell}_n$ such that $\hat{\ell}_n(\boldsymbol{\theta}) = (1/n) \sum_{i=1}^{n} c(\tilde{\mathbf{x}}_i, \boldsymbol{\theta})$ since $\tilde{\mathbf{x}}_1, \tilde{\mathbf{x}}_2, \ldots$ are independent and identically distributed.

By strategically reducing the amount of training data in such a way so that the assumption of statistical independence is not violated, one obtains a true likelihood function. This approach is called *Besag's Coding Assumption* (Besag, 1974, 1975). Assume that the $j$th element of the $i$th observation $\tilde{\mathbf{x}}_i$ is only included in the construction of the likelihood in (13.40) if the neighborhood of $\tilde{x}_i(j)$ does not overlap with the other neighborhoods of random variables included in the construction of (13.40). This non-overlap condition may be formally expressed in terms of the neighborhood graph $\mathcal{G} = (\mathcal{V}, \mathcal{E})$ for the MRF where $\mathcal{V} \equiv \{1, \ldots, d\}$. Let $S_j$ denote a subset of $\mathcal{V}$ which includes $j$ and its neighbors. A strong version of the non-overlap condition may be expressed as:

$$S_j \cap (\cup_{k \neq j} S_k) \; = \; \emptyset, j = 1, \ldots, d \tag{13.43}$$

so that (13.40) now becomes:

$$\hat{\ell}_n(\boldsymbol{\theta}) = -(1/n) \log \left[ \prod_{i=1}^{n} \prod_{j \in \Omega_i} p\left(\tilde{x}_i(j) | \mathcal{N}_j(\tilde{\mathbf{x}}_i), \boldsymbol{\theta}\right) \right] \tag{13.44}$$

where $\Omega_i$ is defined such that for all $j, k \in \Omega_i$: $S_j \cap S_k = \emptyset$.

Equation (13.44) is a now a true likelihood function because it selects out a subset of the $d$ random vectors in (13.41) which are independent. Besag's Coding Assumption is illustrated in Figure 13.2. However, the difficulty with this approach is that constraints imposed on the likelihood function effectively reduce the amount of training data.

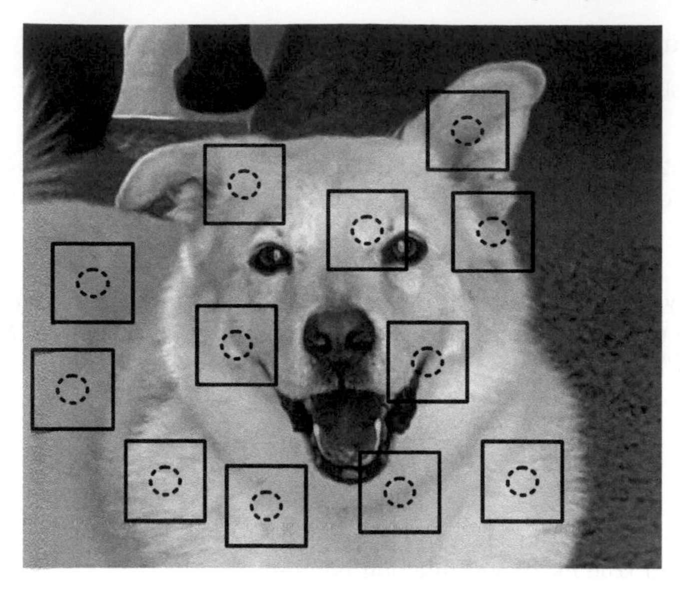

**FIGURE 13.2**
**Example of Besag's (1975) coding assumption.** Each pixel value in this image is interpreted as a realization of a random variable. The conditional density of the collection of random pixels in a dashed circle is functionally dependent only upon the values of the pixels located outside the dashed circle yet inside the square that contains the dashed circle. If two squares overlap, this indicates that the two groups of random pixels in the two dashed circles are not statistically independent. In this case, since none of the neighborhoods of the random pixel groups overlap, the likelihood of observing the circled pixel groups associated with these neighborhoods can be calculated by simply multiplying together the conditional densities for the random pixel groups in the dashed circles conditioned upon the pixels outside the dashed circles but inside the square surrounding the circle.

Finally, the terms pseudolikelihood function and quasi-likelihood function have different meanings. Every pseudolikelihood function is a quasi-likelihood function because a pseudolikelihood function is possibly misspecified. On the other hand, every quasi-likelihood function is not necessarily a pseudolikelihood function since misspecification of the quasi-likelihood can arise without violating conditional independence assumptions.

### 13.2.5 Missing-Data Likelihood Empirical Risk Function

Following the discussion of partially observable i.i.d. stochastic sequences in Chapter 9, define a partially observable data set of $n$ observations as a realization of the partially observable i.i.d. stochastic sequence $(\tilde{\mathbf{v}}_1, \tilde{\mathbf{h}}_1, \tilde{\mathbf{m}}_1), \ldots, (\tilde{\mathbf{v}}_n, \tilde{\mathbf{h}}_n, \tilde{\mathbf{m}}_n)$ where $\tilde{\mathbf{v}}_i$ specifies the visible or observable component of the $i$th data record, $\tilde{\mathbf{h}}_i$ specifies the hidden or unobservable component of the $i$th data record, and the binary random mask vector $\tilde{\mathbf{m}}_i$ specifies which components of the original data record are observable for $i = 1, \ldots, n$.

The observable data likelihood function $\hat{\ell}_n$ for an observable data sequence $(\tilde{\mathbf{v}}_1, \tilde{\mathbf{m}}_1), (\tilde{\mathbf{v}}_2, \tilde{\mathbf{m}}_2), \ldots$ given the assumption of an ignorable missing-data mechanism (see Chapter 10) is specified by the negative normalized likelihood function $\hat{\ell}_n$ defined such that:

$$\hat{\ell}_n(\boldsymbol{\theta}) = -(1/n) \sum_{i=1}^{n} \log p(\tilde{\mathbf{v}}_i | \boldsymbol{\theta}) \tag{13.45}$$

where

$$p(\mathbf{v}|\boldsymbol{\theta}) = \int p(\mathbf{v}, \mathbf{h}|\boldsymbol{\theta}) d\nu(\mathbf{h})$$

is the observable marginal density of the partially observable complete-data density $p(\mathbf{v}, \mathbf{h}|\boldsymbol{\theta})$ for the pattern of missingness $\mathbf{m}$ associated with the observable data record $(\mathbf{v}, \mathbf{m})$.

**Example 13.2.14** (Observable-Data Likelihood Notation and Calculations). Suppose that one has a data record with three binary random variables $x_1, x_2, x_3$ and assume the third variable's value is not observable. This data record is represented as the complete-data record $(\mathbf{x}, \mathbf{m})$ where $\mathbf{x} = (x_1, x_2, x_3)$ and the mask vector $\mathbf{m} = (m_1, m_2, m_3)$.

Let each element of a complete-data probability model be a probability mass function

$$p(\mathbf{x}|\boldsymbol{\theta}) \equiv \prod_{j=1}^{3} \theta_j^{x_j} (1 - \theta_j)^{1-x_j}$$

where $\mathbf{x} = [x_1, x_2, x_3]$ and $\boldsymbol{\theta} = (\theta_1, \theta_2, \theta_3)$ where $0 < \theta_1, \theta_2, \theta_3 < 1$.

Given the assumption that the missing-data probability model has an ignorable missing-data mechanism, the likelihood of the observable data record $\mathbf{v} = [x_1, x_2]$ associated with the mask $\mathbf{m} = [1, 1, 0]$ is then computed by adding

$$p(x_1, x_2, x_3 = 0|\boldsymbol{\theta}) = \theta_1^{x_1}(1 - \theta_1)^{1-x_1}\theta_2^{x_2}(1 - \theta_2)^{1-x_2}(1 - \theta_3)$$

and

$$p(x_1, x_2, x_3 = 1|\boldsymbol{\theta}) = \theta_1^{x_1}(1 - \theta_1)^{1-x_1}\theta_2^{x_2}(1 - \theta_2)^{1-x_2}\theta_3$$

to obtain the marginal probability

$$p(x_1, x_2|\boldsymbol{\theta}) = p(x_1, x_2, x_3 = 0|\boldsymbol{\theta}) + p(x_1, x_2, x_3 = 1|\boldsymbol{\theta}) = \theta_1^{x_1}(1 - \theta_1)^{1-x_1}\theta_2^{x_2}(1 - \theta_2)^{1-x_2}.$$

The marginal probability for each observable data record is then computed in a similar manner. The product of these marginal probabilities is defined as the missing-data likelihood function. △

**Example 13.2.15** (Missing Information Principle). When the missing-data probability model is postulated in terms of an ignorable missing-data mechanism, the observable-data likelihood function is often specified entirely in terms of a complete-data probability model. The observable-data likelihood function is given by the formula in (13.45). The Missing Information Principle shows how to express the first and second derivatives of the observable-data likelihood function using the complete-data probability model specified by the density $p(\mathbf{v}_i, \mathbf{h}_i|\boldsymbol{\theta})$ where $\mathbf{v}_i$ are the observable components of the $i$th data record $\mathbf{x}_i$ and $\mathbf{h}_i$ are the unobservable components of $\mathbf{x}_i$. Note that the dimensionality of $\mathbf{v}_i$ and $\mathbf{h}_i$ will vary as a function of the pattern of missingness $\mathbf{m}_i \in \{0, 1\}^d$ associated with the $i$th data record.

The first derivative of the observable-data likelihood function

$$\frac{d\hat{\ell}_n(\boldsymbol{\theta})}{d\boldsymbol{\theta}} = -(1/n)\sum_{i=1}^{n} \frac{d \log p(\mathbf{v}_i|\boldsymbol{\theta})}{d\boldsymbol{\theta}}. \tag{13.46}$$

Note that

$$\frac{d \log p(\mathbf{v}|\boldsymbol{\theta})}{d\boldsymbol{\theta}} = \frac{dp(\mathbf{v}|\boldsymbol{\theta})}{d\boldsymbol{\theta}} \left(\frac{1}{p(\mathbf{v}|\boldsymbol{\theta})}\right). \tag{13.47}$$

The derivative of $p(\mathbf{v}|\boldsymbol{\theta}) = \int p(\mathbf{v}, \mathbf{h}|\boldsymbol{\theta}) d\nu(\mathbf{h})$ with respect to $\boldsymbol{\theta}$ is:

$$\frac{dp(\mathbf{v}|\boldsymbol{\theta})}{d\boldsymbol{\theta}} = \int \frac{dp(\mathbf{v}, \mathbf{h}|\boldsymbol{\theta})}{d\boldsymbol{\theta}} d\nu(\mathbf{h}) \tag{13.48}$$

assuming interchange of derivative and integral operators is permissible.

Substituting the identity $dp/d\boldsymbol{\theta} = (d \log p/d\boldsymbol{\theta})\, p$ into (13.48) and dividing by $p(\mathbf{v}|\boldsymbol{\theta})$ then yields:

$$\frac{1}{p(\mathbf{v}|\boldsymbol{\theta})}\frac{dp(\mathbf{v}|\boldsymbol{\theta})}{d\boldsymbol{\theta}} = \int \frac{d \log p(\mathbf{v}, \mathbf{h}|\boldsymbol{\theta})}{d\boldsymbol{\theta}} \frac{p(\mathbf{v}, \mathbf{h}|\boldsymbol{\theta})}{p(\mathbf{v}|\boldsymbol{\theta})} d\nu(\mathbf{h}), \quad (13.49)$$

which is equivalent to:

$$\frac{1}{p(\mathbf{v}|\boldsymbol{\theta})}\frac{dp(\mathbf{v}|\boldsymbol{\theta})}{d\boldsymbol{\theta}} = \int \frac{d \log p(\mathbf{v}, \mathbf{h}|\boldsymbol{\theta})}{d\boldsymbol{\theta}} p(\mathbf{h}|\mathbf{v}, \boldsymbol{\theta}) d\nu(\mathbf{h}). \quad (13.50)$$

Then, combine (13.50) with (13.47) and substitute into (13.46) to obtain the final result:

$$\frac{d\hat{\ell}_n(\boldsymbol{\theta})}{d\boldsymbol{\theta}} = -(1/n)\sum_{i=1}^{n} \int \frac{d \log p(\tilde{\mathbf{v}}_i, \mathbf{h}|\boldsymbol{\theta})}{d\boldsymbol{\theta}} p(\mathbf{h}|\tilde{\mathbf{v}}_i, \boldsymbol{\theta}) d\nu(\mathbf{h}). \quad (13.51)$$

Equation (13.51) has an important semantic interpretation. The gradient of the negative log-likelihood for the $i$th partially observable data record is equal to the conditional expectation of the gradient of the negative log-likelihood for the $i$th complete-data gradient conditioned upon the observable components of the $i$th data record. In the special case where there is no missing data, then (13.51) reduces to the usual complete-data gradient of $\hat{\ell}_n$.

The Hessian of the observable-data negative normalized log-likelihood function $\hat{\ell}_n$, $\bar{\mathbf{H}}_n(\boldsymbol{\theta})$, may be obtained by taking the derivative of (13.51) (for details see Golden, Henley, White, and Kashner 2019; also see Louis 1982; Little and Rubin 2002, Section 8.4; also see Exercise 5.2-19). In particular, let the column vector $\boldsymbol{\delta}(\mathbf{v}, \mathbf{h})$ be defined such that:

$$\boldsymbol{\delta}(\mathbf{v}, \mathbf{h}; \boldsymbol{\theta})^T = \frac{d \log p(\mathbf{v}, \mathbf{h}|\boldsymbol{\theta})}{d\boldsymbol{\theta}} - \int \frac{d \log p(\mathbf{v}, \mathbf{h}|\boldsymbol{\theta})}{d\boldsymbol{\theta}} p(\mathbf{h}|\mathbf{v}, \boldsymbol{\theta}) d\nu(\mathbf{h}). \quad (13.52)$$

Let

$$\mathbf{H}(\mathbf{v}; \boldsymbol{\theta}) = \int \left( \frac{-d^2 \log p(\mathbf{v}, \mathbf{h}|\boldsymbol{\theta})}{d\boldsymbol{\theta}^2} - \boldsymbol{\delta}(\mathbf{v}, \mathbf{h}; \boldsymbol{\theta})\boldsymbol{\delta}(\mathbf{v}, \mathbf{h}; \boldsymbol{\theta})^T \right) p(\mathbf{h}|\mathbf{v}, \boldsymbol{\theta}) d\nu(\mathbf{h}). \quad (13.53)$$

Then, the Hessian of the observable-data negative normalized likelihood is given by the formula:

$$\bar{\mathbf{H}}_n(\boldsymbol{\theta}) = (1/n)\sum_{i=1}^{n} \mathbf{H}(\tilde{\mathbf{v}}_i; \boldsymbol{\theta}). \quad (13.54)$$

Note that the column vector $\boldsymbol{\delta}(\mathbf{v}, \mathbf{h}; \boldsymbol{\theta})$ (13.52) is equal to zero for the special case where there is no missing data (i.e., the mask vector $\mathbf{m}$ associated with the observable record $\mathbf{v}$ is a vector of ones). In this no-missing-data case, the second term on the right-hand side of (13.53) vanishes causing the Hessian of the observable-data negative normalized log-likelihood to reduce to the formula for the Hessian of the complete-data negative normalized log-likelihood.

Inspection of (13.53) can provide important insights into conditions when the Hessian of the observable-data likelihood function will be positive semidefinite. Assume the Hessian of the complete-data model likelihood function is positive semidefinite on the parameter space. That is, the first term on the right-hand side of (13.53) is positive semidefinite. Even in this case, (13.53) shows that the Hessian of the observable-data likelihood is not positive semidefinite on the parameter space because it is the difference of two positive semidefinite matrices. However, suppose there exists a convex region of the parameter space, $\Omega$, where the complete-data model likelihood Hessian is positive definite and the amount of missingness is sufficiently small so that the second term on the right-hand side of (13.53) is also sufficiently small on $\Omega$. In this case, it follows that the observable-data negative normalized log-likelihood Hessian will be positive definite on $\Omega$, which implies the observable-data negative normalized log-likelihood will be convex on $\Omega$.

△

## Exercises

13.2-1. *Construct Probability Models from Empirical Risk Functions under ML Assumption.* Interpret the Empirical Risk Functions in Exercise 13.1-1, Exercise 13.1-2, Exercise 13.1-3, Exercise 13.1-4, Exercise 13.1-5, Exercise 13.1-6, and Exercise 13.1-7 as negative normalized log-likelihood functions. Discuss what assumptions must hold for each of these Empirical Risk Functions for a quasi-maximum likelihood estimator of the Empirical Risk Function to converge to a strict local minimizer with probability one.

## 13.3 Maximum a Posteriori (MAP) Estimation Methods

Maximum likelihood estimation methods assume that the objective of the parameter estimation methodology is to choose a parameter value $\boldsymbol{\theta}$ that makes the observed data $\mathcal{D}_n$ most likely by maximizing $p(\mathcal{D}_n|\boldsymbol{\theta})$ with respect to $\boldsymbol{\theta}$. This computational objective is not conceptually satisfying. A more satisfactory computational objective would be to find the value of parameter value $\boldsymbol{\theta}$ that maximizes $p(\boldsymbol{\theta}|\mathcal{D}_n)$ for a given data set $\mathcal{D}_n$. That is, find the parameter value $\boldsymbol{\theta}$ that is most likely given the observed data $\mathcal{D}_n$. This strategy for estimating the parameter vector $\boldsymbol{\theta}$ is called *MAP estimation.*

Specifically, the *posterior density*

$$p(\boldsymbol{\theta}|\mathcal{D}_n) = \frac{p(\mathcal{D}_n|\boldsymbol{\theta})p_{\theta}(\boldsymbol{\theta})}{p(\mathcal{D}_n)} \tag{13.55}$$

where $p_{\theta}$ is called the *parameter prior.* The parameter prior density $p_{\theta}$ is required so that the joint distribution of $\tilde{\boldsymbol{\theta}}$ and $\tilde{\mathcal{D}}^n$ is properly defined. The parameter prior density $p_{\theta}$ specifies prior beliefs about the probability distribution of $\tilde{\boldsymbol{\theta}}$ (see Section 13.3.1 for further discussion).

The denominator on the right-hand side of (13.55) is called the *marginal likelihood.* Since the marginal likelihood $p(\mathcal{D}_n)$ is not functionally dependent upon the parameter vector $\boldsymbol{\theta}$, maximizing the left-hand side of (13.55) may be achieved by simply maximizing the numerator on the right-hand side of (13.55).

### 13.3.1 Parameter Priors and Hyperparameters

The parameter prior $p_{\theta}$ provides a convenient method for representing information about the likelihood of different choices of parameter values in order to accelerate the learning process and improve generalization performance. One can interpret the parameter prior

$p_\theta$ as a type of hint for the learning machine. If it is a good hint, then learning will be accelerated and generalization performance will be enhanced. If it is a bad hint, then the learning process might be disrupted and the generalization performance of the machine could be degraded. It is not recommended that $p_\theta$ be interpreted as the likelihood that $\boldsymbol{\theta}$ is the "true parameter value" because in the presence of model misspecification such a semantic interpretation is meaningless.

An important issue associated with selecting a prior $p_\theta$ within a Bayesian parameter estimation framework, is that the random variable $\tilde{\boldsymbol{\theta}}$ is not observable and is not functionally dependent upon the observed data. Thus, $p_\theta$ is a specification of the learning machine's prior beliefs regarding an appropriate choice of parameter vector $\boldsymbol{\theta}$. That is, $p_\theta(\boldsymbol{\theta})$ specifies the learning machine's belief $p_\theta(\boldsymbol{\theta})$ that $\boldsymbol{\theta}$ is a good choice for the parameter values of the learning machine. Two important strategies for constructing $p_\theta$ are now considered.

First, one may choose a *vague parameter prior* such as a uniform density that takes on the value of zero for impossible parameter values. The essential idea in this case is to place very weak constraints on the range of possible values of $\boldsymbol{\theta}$. For example, one might specify the $p_\theta(\theta) = 1/1000$ for $|\theta| < 500$ and $p_\theta(\theta) = 0$ for $|\theta| \geq 500$. Other choices might be a Gaussian distribution. The range of parameter values for a uniform density or the variance associated with a Gaussian density might be obtained purely through knowledge of the learning machine's environment or could actually be chosen based upon previously measured properties of the statistical environment. Although extremely weak prior knowledge constraints might seem relatively obvious to the modeler (i.e., you wouldn't want to use a model in which $\theta$ was some extremely small or an extremely large number even if this was the estimated value), such constraints are not obvious to the estimation algorithm. The concept of a "prior" allows the modeler to share vague a priori knowledge about the range of parameter values to be expected and as more data is collected, the effects of this prior knowledge will (generally) become negligible as explained in the following section.

A second strategy for selecting a parameter prior is to estimate the prior using some data. Ideally, the data set used to estimate the prior should be different from the training data set used to estimate the model parameters. A parameter prior which is estimated from some data is called an *empirical parameter prior*. Suppose that a parameter prior $p_\theta$ is used to compute the posterior density $p(\boldsymbol{\theta}|\mathcal{D}_n)$. The parameters of the parameter prior are called "hyperparameters".

**Definition 13.3.1** (Hyperparameter). Let $\mathcal{M} \equiv \{p(\mathbf{x}|\boldsymbol{\theta}) : \boldsymbol{\theta} \in \Theta\}$ be a probability model. Let the *parameter prior model* $\mathcal{M}_\theta \equiv \{p_\theta(\boldsymbol{\theta}|\boldsymbol{\eta}) : \boldsymbol{\eta} \in \zeta\}$ be a set of parameter prior densities for $\mathcal{M}$. Then, $\boldsymbol{\eta} \in \zeta$ is called a *hyperparameter* for $\mathcal{M}_\theta$. □

In some special cases, the parametric form of the posterior density $p(\boldsymbol{\theta}|\mathcal{D}_n)$ for a given data set $\mathcal{D}_n$ will have the same parametric form as $p_\theta$. Thus, if a new data set $\ddot{\mathcal{D}}_n$ is obtained, then density $p(\boldsymbol{\theta}|\mathcal{D}_n)$ may be used as the parameter prior for computing the posterior density $p(\boldsymbol{\theta}|\ddot{\mathcal{D}}_n)$ under the assumption that $\mathcal{D}_n$ is now treated as a known constant. In this special case, the parameter prior is called a "conjugate density" for the posterior density.

**Definition 13.3.2** (Conjugate Prior Density). Let $\mathcal{M}$ be a collection of probability density functions defined with respect to some support specification measure $\nu$. Assume $p_\theta \in \mathcal{M}_\theta$. For each $p(\cdot|\boldsymbol{\theta}) \in \mathcal{M}$, the density $p_\theta : \Theta \to [0, \infty)$ is called a *conjugate prior density* for $p(\cdot|\boldsymbol{\theta})$ with respect to $\mathcal{M}_\theta$ if

$$p(\boldsymbol{\theta}|\mathcal{D}_n) \equiv \frac{p(\mathcal{D}_n|\boldsymbol{\theta})p_\theta(\boldsymbol{\theta})}{\int p(\mathcal{D}_n|\boldsymbol{\theta})p_\theta(\boldsymbol{\theta})d\boldsymbol{\theta}}$$

is in $\mathcal{M}_\theta$. □

**Example 13.3.1** (Gaussian Conjugate Prior). Let $\sigma_e \in (0, \infty)$ and let

$$\boldsymbol{\theta}^T \equiv [\mathbf{m}^T, \mathbf{vech}(\mathbf{C})] \in \mathcal{R}^d \times \mathcal{R}^{d(d-1)/2}.$$

In this example, it is shown that the parameter prior

$$p_{\boldsymbol{\theta}}(\mathbf{m}) = (\sigma_o \sqrt{2\pi})^{-1} \exp\left(-\frac{|\mathbf{m}|^2}{2\sigma_o^2}\right)$$

is a conjugate prior with hyperparameter $\sigma_o$ for probability model

$$\mathcal{M} \equiv \{p(\cdot|\boldsymbol{\theta}) : \boldsymbol{\theta} \in \Theta\}$$

where for all $\mathbf{x}$ in the support of $p(\cdot|\boldsymbol{\theta})$:

$$p(\mathbf{x}|\boldsymbol{\theta}) = (2\pi)^{-d/2}(\det(\mathbf{C}))^{-1/2} \exp[-(1/2)(\mathbf{x}-\mathbf{m})^T \mathbf{C}^{-1}(\mathbf{x}-\mathbf{m})].$$

Let $p(\mathcal{D}_n|\boldsymbol{\theta}) = \prod_{i=1}^n p(\mathbf{x}_i|\boldsymbol{\theta})$ which implies that:

$$p(\boldsymbol{\theta}|\mathcal{D}_n) = (1/Z) \exp\left(-\frac{1}{2}\sum_{i=1}^n (\mathbf{x_i} - \mathbf{m}^T)\mathbf{C}^{-1}(\mathbf{x_i} - \mathbf{m}) - \frac{|\mathbf{m}|^2}{2\sigma_o^2}\right)$$

which is a multivariate Gaussian density conditioned upon $\mathcal{D}_n$ with the same functional form as $p_{\boldsymbol{\theta}}$. △

### 13.3.2 Maximum a Posteriori (MAP) Risk Function

**Definition 13.3.3** (MAP Risk Function). Let $\Theta \subseteq \mathcal{R}^q$ be a nonempty, closed, and bounded set. Let $\tilde{\theta}$ be a $q$-dimensional random vector with a Radon-Nikodým density *parameter prior* $p_{\theta} : \mathcal{R}^q \to [0, \infty)$. Let $\tilde{\mathcal{D}}_n \equiv [\tilde{\mathbf{x}}_1, \ldots, \tilde{\mathbf{x}}_n]$ be a sequence of independent and identically distributed $d$-dimensional random vectors with common Radon-Nikodým density $p_e(\cdot|\boldsymbol{\theta}) : \mathcal{R}^d \to [0, \infty)$ for each $\boldsymbol{\theta} \in \Theta$. Let

$$\mathcal{M} = \{p(\cdot|\boldsymbol{\theta}) : \boldsymbol{\theta} \in \Theta\}$$

be a probability model where each element of $\mathcal{M}$ is in the support of $p_e$. A *MAP Risk Function* is a function $p : \Theta \times \mathcal{R}^{d \times n} \to [0, \infty)$ defined such that for all $\boldsymbol{\theta} \in \Theta$:

$$p(\boldsymbol{\theta}|\mathcal{D}_n) = \frac{p(\mathcal{D}_n|\boldsymbol{\theta}) p_{\boldsymbol{\theta}}(\boldsymbol{\theta})}{p(\mathcal{D}_n)} \tag{13.56}$$

where $p(\mathcal{D}_n|\boldsymbol{\theta}) = \prod_{i=1}^n p(\mathbf{x}_i|\boldsymbol{\theta})$.

A *MAP estimate,*

$$\bar{\boldsymbol{\theta}}_n \equiv \arg\max_{\boldsymbol{\theta} \in \Theta} p(\boldsymbol{\theta}|\mathcal{D}_n)$$

is a global maximizer of the MAP Risk Function on the parameter space $\Theta$. The density $p(\cdot|\mathcal{D}_n)$ is called the *Bayes posterior density.* □

Since the denominator, $p(\mathcal{D}_n)$, in (13.56) is constant on the parameter space $\Theta$, it is computationally easier to simply maximize the numerator of (13.56), $p(\mathcal{D}_n|\boldsymbol{\theta})p_{\boldsymbol{\theta}}(\boldsymbol{\theta})$, to obtain the MAP estimate rather than directly maximizing (13.56). With this comment in mind, a MAP Risk Function is typically defined by choosing

$$c(\mathbf{x}, \boldsymbol{\theta}) \equiv -\log\left(p(\mathcal{D}_n|\boldsymbol{\theta})p_{\boldsymbol{\theta}}(\boldsymbol{\theta})\right) = -\log\left(p_{\boldsymbol{\theta}}(\boldsymbol{\theta}) \prod_{i=1}^n p(\mathbf{x}_i|\boldsymbol{\theta})\right)$$

yielding the MAP penalized empirical risk function

$$\bar{\ell}_n(\boldsymbol{\theta}) = -(1/n) \log p_{\boldsymbol{\theta}}(\boldsymbol{\theta}) - (1/n) \sum_{i=1}^{n} \log p(\tilde{\mathbf{x}}_i|\boldsymbol{\theta}). \tag{13.57}$$

By assuming that $p_{\boldsymbol{\theta}}$ is strictly positive and continuous on a closed and bounded parameter space $\Theta$ and assuming that $\hat{\boldsymbol{\theta}}_1, \hat{\boldsymbol{\theta}}_2, \ldots$ remains in $\Theta$ with probability one, then it immediately follows that $-(1/n) \log p_{\boldsymbol{\theta}}(\boldsymbol{\theta}) \to 0$ with probability one as $n \to \infty$. Thus, given appropriate regularity conditions, both the MAP empirical risk function $\bar{\ell}_n$ and ML empirical risk function $\hat{\ell}_n$ converge uniformly with probability one to the Kullback-Leibler Information Criterion plus a constant (see Theorem 13.1.1).

## Recipe Box 13.2 Risk Function Design Using Model

- **Step 1: Explicitly define the probability model.** Let

$$\mathcal{M} \equiv \{p(\mathbf{x}|\boldsymbol{\theta}) : \boldsymbol{\theta} \in \Theta\}$$

be a probability model. Note that each element of $\mathcal{M}$ assigns a likelihood to some environmental event $\mathbf{x}$ given the learning machine's knowledge state $\boldsymbol{\theta}$.

- **Step 2: Explicitly define the parameter prior for the model.** Let $p_{\boldsymbol{\theta}}(\boldsymbol{\theta})$ be the parameter prior for the model $\mathcal{M}$. Note that $p_{\boldsymbol{\theta}}$ is only functionally dependent upon the parameter vector $\boldsymbol{\theta}$ and not upon the data $\mathbf{x}$.

- **Step 3: Construct MAP risk function.**

$$\hat{\ell}_n(\boldsymbol{\theta}) = (1/n)k(\boldsymbol{\theta}) + (1/n) \sum_{i=1}^{n} c(\mathbf{x}_i; \boldsymbol{\theta})$$

where $k(\boldsymbol{\theta}) \equiv -\log p_{\boldsymbol{\theta}}(\boldsymbol{\theta})$ and $c(\mathbf{x}_i; \boldsymbol{\theta}) = -\log p(\mathbf{x}|\boldsymbol{\theta})$.

Note that the MAP risk function in (13.57) has an important semantic interpretation. In the initial stage of learning (when $n$ is small), the effects of prior knowledge provided to the learning machine as specified by the prior $p_{\boldsymbol{\theta}}$ are more influential while the negative normalized log-likelihood term

$$-(1/n) \sum_{i=1}^{n} \log p(\mathbf{x}_i|\boldsymbol{\theta})$$

(which represents the "effects of experience") in (13.57) is less influential. As $n \to \infty$, the effects of prior knowledge (i.e., the effects of the prior $p_{\boldsymbol{\theta}}$) become negligible and MAP estimation is asymptotically equivalent to Maximum Likelihood estimation.

Given a probability model and parameter prior, it is straightforward to construct a MAP empirical risk function for a learning machine (see Recipe Box 13.2). On the other hand, in many machine learning applications, an empirical risk function is given as the objective function for learning but a probability model is not provided. Golden (1988a, 1988c) suggested that if one assumes the learning machine is a MAP estimation algorithm, then one can in some cases deduce the learning machine's probability model from the empirical risk function. Recipe Box 13.2 demonstrates this construction which is also straightforward but requires the assumption that the learning machine's empirical risk function may be interpreted as a MAP empirical risk function.

## Recipe Box 13.3 Model Design Given Risk Function

- **Step 1: Explicitly define penalized empirical risk function.** Let $\Theta$ be a closed, bounded, and convex subset of $\mathcal{R}^q$. Let $k : \mathcal{R}^q \to \mathcal{R}$ and $c : \mathcal{R}^d \times \mathcal{R}^q \to \mathcal{R}$ be continuous random functions. Write the penalized empirical risk function in the following form:

$$\hat{\ell}_n(\boldsymbol{\theta}) = (1/n)k(\boldsymbol{\theta}) + (1/n)\sum_{i=1}^{n} c(\mathbf{x}_i; \boldsymbol{\theta})$$

  where $\boldsymbol{\theta}$ is a $q$-dimensional parameter vector and the $n$ $d$-dimensional vectors $\mathbf{x}_1, \ldots, \mathbf{x}_n$ specify the training data.

- **Step 2: Explicitly define statistical environment.** Assume that $\mathbf{x}_1, \ldots, \mathbf{x}_n$ is a realization of a stochastic sequence of $n$ independent and identically distributed $d$-dimensional random vectors $\tilde{\mathbf{x}}_1, \ldots, \tilde{\mathbf{x}}_n$ with common Radon-Nikodým density $p_e : \mathcal{R}^d \to [0, \infty)$ defined with respect to support specification measure $\nu$.

- **Step 3: Try to interpret loss function as a likelihood.** Specify the support, $S$, of $\tilde{\mathbf{x}}$ and then define a probability density $p(\mathbf{x}|\boldsymbol{\theta})$ with support $S$ such that the loss function $c(\mathbf{x}; \boldsymbol{\theta}) = -\log p(\mathbf{x}|\boldsymbol{\theta})$.

- **Step 4: Try to interpret penalty function as a prior.** Specify the support $S_\theta$ of $\tilde{\boldsymbol{\theta}}$ and then define a probability density $p_\theta$ with support $S_\theta$ such that the penalty term $k(\boldsymbol{\theta}) = -\log p_\theta(\boldsymbol{\theta})$. In addition, choose $p_\theta$ such that $p_\theta$ is continuous and positive on $\Theta$.

- **Step 5: Interpret estimation as MAP estimation.** The parameter estimation algorithm for the learning machine may now be interpreted as a MAP estimation algorithm with respect to the constructed probability model.

**Example 13.3.2** (Penalized Least Squares Estimation as MAP Estimation). Consider a supervised learning machine which has the objective of learning to generate a desired scalar response $y_i$ given a $d-1$-dimensional input pattern vector $\mathbf{s}_i$ for $i = 1, \ldots, n$. In particular, assume the learning machine seeks a minimizer of the penalized empirical risk function

$$\hat{\ell}_n(\boldsymbol{\theta}) = (1/n)k(\boldsymbol{\theta}) + (1/n)\sum_{i=1}^{n} c([y_i, \mathbf{s}_i], \boldsymbol{\theta}) \tag{13.58}$$

where $c$ and $k$ are continuously differentiable random functions. Given the assumption that the learning machine's goal is to maximize the likelihood of its parameters given the observed data, construct a probability model which is consistent with this goal and (13.58) (see Recipe Box 13.3).

**Solution.** First, assume that $(\mathbf{s}_1, y_1), \ldots, (\mathbf{s}_n, y_n)$ is a realization of a stochastic sequence of $n$ i.i.d. $d$-dimensional random vectors $(\tilde{\mathbf{s}}_1, \tilde{y}_1), \ldots, (\tilde{\mathbf{s}}_n, \tilde{y}_n)$ with some common density $p_e : \mathcal{R}^d \rightarrow [0, \infty)$ with respect to a support specification measure $\nu$.

Second, assume there exists a probability density function $p(y_i|\mathbf{s}_i; \boldsymbol{\theta})$ such that:

$$c([y_i, \mathbf{s}_i], \boldsymbol{\theta}) = -\log\left(p(y_i|\mathbf{s}_i; \boldsymbol{\theta}) p_s(\mathbf{s}_i)\right)$$

where $p_s(\mathbf{s})$ is the probability that input pattern vector $\mathbf{s}$ is presented to the learning machine. Therefore,

$$p(y|\mathbf{s}; \boldsymbol{\theta}) = \exp(-c([y, \mathbf{s}]; \boldsymbol{\theta}))/Z(\boldsymbol{\theta}, \mathbf{s})$$

where

$$Z(\boldsymbol{\theta}, \mathbf{s}) = \int \exp(-c([y, \mathbf{s}]; \boldsymbol{\theta})) d\nu(y)$$

provided that $Z(\boldsymbol{\theta}, \mathbf{s})$ is finite.

Third, interpret the penalty function as a prior by defining the parameter prior $p_\theta$ such that $-\log p_\theta(\boldsymbol{\theta}) = k(\boldsymbol{\theta})$ or equivalently $p_\theta(\boldsymbol{\theta}) = \exp(-k(\boldsymbol{\theta}))/Z(\boldsymbol{\theta})$ provided that the normalization constant $Z(\boldsymbol{\theta}) \equiv \int \exp(-k(\boldsymbol{\theta})) d\boldsymbol{\theta}$ is finite. △

### 13.3.3 Bayes Risk Interpretation of MAP Estimation

For expository reasons, assume the parameter space $\Theta$ has been discretized so that the random parameter vector can take on only a finite number of values. This corresponds to the case where each element of the parameter vector is represented by a finite precision number rather than a number with infinite precision. Each of the finite possible choices for the parameter estimate $\hat{\boldsymbol{\theta}}_n$ has probability

$$p(\boldsymbol{\theta}|\mathcal{D}_n) = \frac{p(\mathcal{D}_n|\boldsymbol{\theta}) p_\theta(\boldsymbol{\theta})}{p(\mathcal{D}_n)}$$

which may be interpreted as the probability of $\boldsymbol{\theta}$ given the observed data $\mathcal{D}_n$. Let $\boldsymbol{\theta}^*$ be the parameter vector which identifies the probability distribution in the probability model which best approximates the data generating process in some sense. The logic of MAP estimation would then be to select the *most probable* parameter vector $\hat{\boldsymbol{\theta}}_n$ as an estimator of $\boldsymbol{\theta}^*$ so that $p(\hat{\boldsymbol{\theta}}_n|\mathcal{D}_n) \geq p(\boldsymbol{\theta}|\mathcal{D}_n)$ for all $\boldsymbol{\theta}$ in the parameter space $\Theta$. Note that when the parameter space is a finite set, MAP estimation corresponds to a minimum probability of error decision rule.

Let $u : \Theta \rightarrow \mathcal{R}$ be a loss function defined such that $u(\boldsymbol{\theta}, \boldsymbol{\eta})$ is the penalty incurred by the learning machine for choosing $\boldsymbol{\theta}$ instead of $\eta$ for its parameter vector. When the parameter space has been discretized the *Bayes estimator* (discretized case) is defined as a global minimizer of:

$$\ell_n(\boldsymbol{\theta}) = \sum_{\boldsymbol{\eta} \in \Theta} u(\boldsymbol{\theta}, \boldsymbol{\eta}) p(\boldsymbol{\eta}|\mathcal{D}_n). \tag{13.59}$$

In the case of MAP estimation (i.e., a minimum probability of error decision rule),

$$u(\boldsymbol{\theta}, \boldsymbol{\eta}) = 1 \text{ if } \boldsymbol{\theta} \neq \boldsymbol{\eta} \text{ and } u(\boldsymbol{\theta}, \boldsymbol{\eta}) = 0 \text{ otherwise.}$$

Such a minimum probability of error selection strategy is potentially problematic because the selection strategy assumes the penalty loss received for choosing a parameter vector very similar to the optimal parameter vector is not appropriately adjusted for similarity. For example, suppose the correct parameter vector was $\boldsymbol{\eta} = [0, 0, 0]$. A MAP risk function assigns a penalty loss $u(\boldsymbol{\eta}, \boldsymbol{\theta}_n^1) = 1$ for choosing the parameter vector $\hat{\boldsymbol{\theta}}_n^1 = [10^{-5}, 0, 10^{-1}]$ which is numerically similar to $\boldsymbol{\theta} = \mathbf{0}$ but also assigns the same penalty loss $u(\boldsymbol{\eta}, \boldsymbol{\theta}_n^2) = 1$ for choosing the parameter vector $\hat{\boldsymbol{\theta}}_n^2 = [100, 1000, -500]$ which is numerically quite different from $\boldsymbol{\theta} = \mathbf{0}$.

Ideally, it would be more desirable if the penalty loss $u(\boldsymbol{\theta}_n^1, \boldsymbol{\eta})$ was less than $u(\boldsymbol{\theta}_n^2, \boldsymbol{\eta})$ so that the loss function would reflect the qualitative concept that the estimator $\boldsymbol{\theta}_n^1$ was "better" than the estimator $\boldsymbol{\theta}_n^2$ in the sense that $\boldsymbol{\theta}_n^1$ is closer to $\mathbf{0}$.

More generally, a *Bayes estimator* (general case) is defined as choosing parameter estimates that minimize the expected loss function

$$\ell_n(\boldsymbol{\eta}) = \int_{\boldsymbol{\theta} \in \Theta} u(\boldsymbol{\eta}, \boldsymbol{\theta}) p(\boldsymbol{\theta} | \mathcal{D}_n) d\boldsymbol{\theta}. \tag{13.60}$$

The relationship of the MAP estimator to the general form of the Bayes estimator in (13.60) can be explored in several ways. The first semantic interpretation of the MAP estimator as a Bayes estimator assumes a loss function $u$ defined such that $u(\boldsymbol{\eta}, \boldsymbol{\theta}) = 0$ if $|\boldsymbol{\eta} - \boldsymbol{\theta}| < \delta$ and $u(\boldsymbol{\eta}, \boldsymbol{\theta}) = 1$ otherwise. This case approximates the previous discretization analysis of MAP estimation when $\delta$ is chosen to be a very small positive number. The resulting Bayes estimator is not technically a MAP estimator, but in many cases may be a reasonable approximation if $\delta$ is sufficiently small and the density $p(\boldsymbol{\theta} | \mathcal{D}_n)$ is sufficiently smooth.

A second semantic interpretation of the MAP estimator as a Bayes estimator assumes a *squared error* loss $u(\boldsymbol{\eta}, \boldsymbol{\theta}) = |\boldsymbol{\eta} - \boldsymbol{\theta}|^2$. Unlike the minimum probability of error loss function, the squared error loss function has an intuitive appeal for the case where $\boldsymbol{\eta}$ and $\boldsymbol{\theta}$ are real vectors because the magnitude of the loss increases as the dissimilarity in a Euclidean distance sense increases between $\boldsymbol{\eta}$ and $\boldsymbol{\theta}$. The minimizer of a Bayes estimator with a squared loss function is called a *minimum mean-square estimator.*

When the squared loss function is substituted into (13.60) one obtains:

$$\ell_n(\boldsymbol{\theta}) = \int_{\boldsymbol{\eta} \in \Theta} |\boldsymbol{\theta} - \boldsymbol{\eta}|^2 p(\boldsymbol{\eta} | \mathcal{D}_n) d\boldsymbol{\eta}. \tag{13.61}$$

Now set

$$\frac{d\ell_n}{d\boldsymbol{\theta}} = \int_{\boldsymbol{\eta} \in \Theta} 2(\boldsymbol{\theta} - \boldsymbol{\eta}) p(\boldsymbol{\eta} | \mathcal{D}_n) d\boldsymbol{\eta} = 2\boldsymbol{\eta} - 2 \int_{\boldsymbol{\eta} \in \Theta} \boldsymbol{\eta} p(\boldsymbol{\eta} | \mathcal{D}_n) d\boldsymbol{\eta}$$

equal to a vector of zeros and it follows that $d\ell_n / d\boldsymbol{\theta}$ vanishes when $\boldsymbol{\theta}$ is equal to the mean of the posterior density $p(\boldsymbol{\eta} | \mathcal{D}_n)$, $\bar{\boldsymbol{\eta}}$),

$$\bar{\boldsymbol{\eta}} = \int_{\boldsymbol{\eta} \in \Theta} \boldsymbol{\eta} p(\boldsymbol{\eta} | \mathcal{D}_n) d\boldsymbol{\eta}.$$

Thus, choosing $\boldsymbol{\theta} = \bar{\boldsymbol{\eta}}$ is a critical point of $\ell_n(\boldsymbol{\theta})$. Since

$$\frac{d^2 \ell_n}{d\boldsymbol{\theta}^2} = 2\mathbf{I}$$

which is a positive definite matrix on the entire parameter space, it follows from the Strict Global Minimizer Test (Theorem 5.4.4) that $\bar{\boldsymbol{\eta}}$ is the unique strict global minimizer of $\ell_n$ on the parameter space.

Thus the analysis shows that the mean, $\bar{\boldsymbol{\eta}}$, of the posterior density $p(\boldsymbol{\theta} | \mathcal{D}_n)$ is a minimizer of the Bayes risk with a squared error loss function. If the posterior density has the property that the MAP estimator (i.e., the posterior density mode) and *posterior mean estimator*, $\bar{\boldsymbol{\eta}}$, are the same, then MAP estimation and minimization of the Bayes risk with squared error loss function are the same.

Theorem 15.2.1 provides sufficient conditions for ensuring the posterior density $p(\boldsymbol{\theta}|\mathcal{D}_n)$ is approximately Gaussian. Therefore, in the special case where the conditions of Theorem 15.2.1 hold, the MAP estimator and posterior mean estimator are approximately equal when the sample size $n$ is sufficiently large.

## Exercises

13.3-1. *Construct Probability Models from Risk Functions under MAP Assumption.* Construct MAP estimation objective functions using each of the empirical risk functions Exercise 13.1-1, Exercise 13.1-2, Exercise 13.1-3, Exercise 13.1-4, Exercise 13.1-5, Exercise 13.1-6, and Exercise 13.1-7 with a Gaussian prior, a Gamma prior, a uniform prior, and a hyperbolic secant prior.

13.3-2. *Identify Conditions for Estimator Convergence.* Provide explicit conditions ensuring convergence of the MAP estimators associated with each of the MAP estimation objective functions you constructed in Exercise 13.3-1 by using the Empirical Risk Minimizer Convergence Theorem.

## 13.4 Further Readings

### Model Misspecification Framework

It is important to emphasize that the theoretical framework presented in this chapter (as well as subsequent Chapters 14, 15, and 16) does not require that the probability model assumed by the learning machine about its environment is correctly specified. The framework proposed here follows the theoretical framework proposed by White (1982, 1994; also see Golden 1995, 2003).

### Consistent Estimation of Local Minimizers of Non-convex Functions

In many machine learning applications, one encounters risk functions which may be multimodal with many strict local minimizers. It is important for any comprehensive theory of estimation and inference to be able to handle such situations. For example, bootstrap and cross-validation methods are widely used in the machine learning literature to estimate sampling error. Bootstrap methods (see Chapter 14) implicitly assume convergence to a strict local minimizer and do not require convergence to a global minimizer of the empirical risk function over the parameter space. Like bootstrap methods, the mathematical theory presented in this chapter as well as Chapter 15 and Chapter 16, is applicable to a local probability model (see Definition 10.1.9) whose parameter space is a neighborhood of a strict local minimizer of the empirical risk function.

The concept of estimation and inference using only local rather than global identifiability assumptions is well known in the mathematical statistical literature. Essentially every classical treatment of $m$-estimation or maximum likelihood estimation uses assumptions that only require local identifiability (e.g., see Serfling 1980, pp. 144-145, 248-250 for a typical main-stream discussion of maximum likelihood estimation and $m$-estimation) yet the important benefits of requiring only a local identifiability assumption is often not sufficiently emphasized in the machine learning literature. The advantage of just requiring local identifiability is of great importance for work in the machine learning literature since the stricter requirement of global identifiability is frequently violated in the many machine learning applications involving non-convex empirical risk functions.

### Empirical Risk Convergence Theorems

The Empirical Risk Convergence Theorem is essentially a direct consequence of the well-known Uniform Law of Large Numbers (e.g., Jennrich, 1969; White, 1994, 2001; van der Vaart, 1998). White (1994), White (2001), and Davidson (2002) provide a useful review of more general versions of the Uniform Law of Large Numbers applicable to both correlated and nonstationary stochastic sequences. Finally, it should be emphasized that the conditions provided in the Empirical Risk Minimizer Convergence Theorem are stronger than necessary (e.g., see White, 1994, Theorem 3.4) but the conditions presented here are relatively straightforward to verify in many important engineering applications.

Golden (2003) discusses a version of the Empirical Risk Convergence Theorem for bounded $\tau$-dependent stochastic sequences. The Empirical Risk Minimizer Convergence Theorem was constructed for this text for both practical and pedagogical reasons. Most published versions of the Empirical Risk Minimizer Convergence Theorem assume weaker conditions and are applicable to a broader range of statistical environments (e.g., Serfling, 1980; White, 1982, 1994; Winkler, 2003; van der Vaart, 1998). Estimators as defined by the Empirical Risk Minimizer Convergence Theorem are called *M-estimators* in the statistics literature (e.g., Huber 1964, 1967; Serfling, 1980, van der Vaart, 1998). The "augmented data trick" has been referred to as "mixed estimation" in the statistics literature (e.g., Theil and Goldberger, 1961; Montgomery and Peck, 1982, p. 320).

### Maximum Likelihood Estimation

The discussion of maximum likelihood estimation is a standard and essential topic in mathematical statistics (e.g., Serfling 1980; van der Vaart 1998; White 1994, 2001; Davidson 2002) as well as machine learning (Hastie et al. 2001). Quasi-maximum likelihood estimation is discussed in the mathematical statistics literature often in the context of M-estimation (e.g., Huber 1967; Serfling 1980, van der Vaart 1998; Davidson 2002) but the philosophical emphasis upon the issue of model misspecification originates from the econometrics literature (e.g., White, 1982, 1994).

Discussions of pseudo-maximum likelihood estimation may be found in Winkler (2003; also see Besag, 1974). Pseudo-maximum likelihood estimation was introduced into the Markov random field literature by Besag (1975). A recent review of pseudo-maximum likelihood estimation may be found in Varin et al. (2011). Good introductions to information theory can be found in Cover and Thomas (2006) and MacKay (2003).

### Maximum Likelihood Estimation in the Presence of Missing Data

Discussions of maximum likelihood estimation in the presence of missing data can be found in Little and Rubin (2002), McLachlan and Krishnan (2008), and Golden, Henley, White, and Kashner (2019).

## MAP Estimation and Bayes Estimation

MAP and Bayes estimation are standard topics in mathematical statistics (Berger, 2006; Lehman and Casella, 1998) and machine learning (Hastie et al., 2001; Murphy, 2012; Duda et al., 2001). Golden (1988a, 1988c) proposed the idea of constructing probability models for machine learning algorithms under the assumption the classification algorithm for the machine learning algorithm is assumed to be a MAP estimation algorithm.

# 14

# *Simulation Methods for Evaluating Generalization*

**Learning Objectives**

- Estimate cross-validation generalization error.
- Apply bootstrap convergence theorem.
- Estimate nonparametric bootstrap generalization error.
- Estimate parametric bootstrap decision error.

A statistic is simply a function of a random sample that identifies particular features of that random sample. In practice, statistical inferences are based upon investigating the behavior of statistics.

Statistics computed from a particular realization of a random sample (i.e., a particular data set) almost inevitably will result in biased estimates and inferences. Figure 14.1 shows the results of fitting a correctly specified linear regression model to two distinct realizations of the same random sample which consists of $n$ observations (e.g., $n$ training stimuli). Even though the model is correctly specified, two distinct parameter estimates are obtained. The difference in parameter estimates is due to sampling error.

Fitting the regression model to multiple realizations of the same random sample has several advantages. First, the variation or sampling error in the parameter estimates for a given sample size $n$ can be better characterized. Second, by averaging the parameter estimates obtained from different samples of size $n$, it is sometimes possible to obtain a new estimate which is better than the parameter estimates obtained from any individual sample. On the other hand, it may be computationally expensive to collect multiple realizations of the same random sample.

One approach which avoids the computational cost of collecting additional data but attempts to benefit from the advantage of collecting multiple realizations of the same random sample is based upon dividing an existing data set in half. One portion of the data set called the "training data set" is used to estimate model parameters. The fit of the model to the training data set is called the "training data model fit". Then, the remaining portion of the data set, called the "test data set", is used to evaluate model performance using parameters estimated from the training data set. The fit of the model to the test data set using only parameters estimated from the training data is called the "test data model fit". The purpose of this procedure is to evaluate the model's generalization performance on novel data sets which were never experienced by the learning machine during the learning process.

Figure 14.2 illustrates a learning curve which plots performance on both a training data set and a test data set. In this case the model is not over-fitting the data and exhibits good generalization performance. As the number of training examples increases, the training data

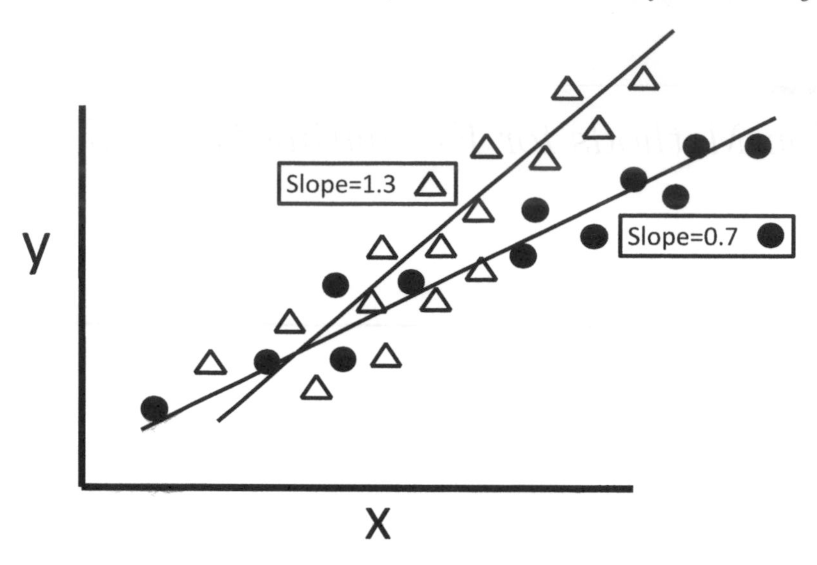


**FIGURE 14.1**
**Fit of one linear regression model to two data sets generated by the same DGP.** The estimated parameters for the two models are different even though both data sets were generated from exactly the same statistical environment. The differences in the parameter estimates are due to the fact that the data sets have finite sample size. The error induced as a result of variation in statistical characteristics between data sets generated from the same environment is called *sampling error.*

model fit systematically decreases indicating not only that the optimization algorithm is working correctly but also that the learning machine is capable of representing relevant statistical regularities in the training data set. In addition, as the number of training examples increases, the test data model fit tends to decrease, indicating that the statistical regularities extracted from the training data set are common to the test data set. Notice that the average performance on the test data set is usually slightly worse than average performance on the training data set since the test data set contains test stimuli never actually observed by the learning machine during the training process.

Figure 14.3 provides a second example in which over-fitting is present. In a manner similar to the example depicted in Figure 14.2, model fit to the training data improves as more training examples are presented to the learning machine. However, after a certain number of training examples have been presented, the model fit to the test data tends to *increase* while the model fit to the training data continues to decrease. One possible interpretation of this overfitting phenomenon is that in the initial stages of learning, the learning machine is extracting important statistical regularities which are common to both the training data and test data sets. In the later stages of learning, when the model fit to the training data and test data sets diverge, however, the learning machine is extracting statistical regularities which are present in the training data but absent in the test data. Such a situation can arise when the learning machine's ability to represent and learn an arbitrary data generating process's structure is *too* flexible and powerful.

The train-test method illustrated in Figure 14.2 and Figure 14.3 can thus provide critical insights regarding the presence or absence of sampling error (i.e., variations in statistical characteristics across data sets of a fixed sample size $n$). Still, although the train-test method is appealing because of its simplicity, it has a number of problems. First, only a portion

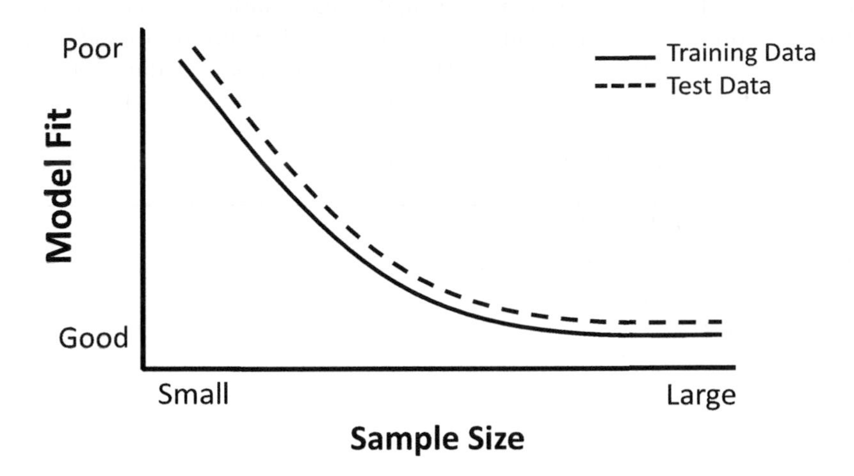


FIGURE 14.2
**A model with good generalization performance.** As sample size increases, the fit of the model to both the training and the test data will tend to decrease in situations where no over-fitting is present. Note that the fit of the model to the test data set is typically slightly worse than the fit of the model to the training data set.

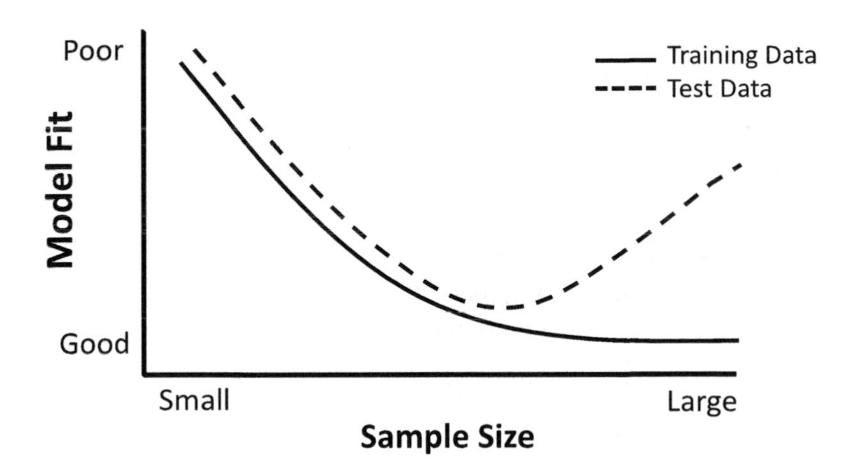


FIGURE 14.3
**A model which fits the training data but exhibits an overfitting phenomenon.** As the sample size increases, performance on the training and test data sets initially tends to decrease; however, at a certain point, the model fit on the training data will continue to decrease, while the model fit on the test data will begin to increase.

of the data sample is used for estimation and only a portion of the data sample is used for evaluation. Thus, the amount of available information in the full data set which is comprised of both the training and test data is not fully utilized. Second, if crucial statistical regularities are absent by chance in either the training or test data set but present in the other data set, then the evaluated performance of the model using the train-test method will be different relative to the case where those critical statistical regularities are simultaneously present (or simultaneously absent) in both the training and test data sets.

**Algorithm 14.0.1** Train-Test Method

1: **procedure** TRAINTEST($\ell_n, \{\mathcal{D}_n^1, \mathcal{D}_n^2\}$ )
2: $\quad \boldsymbol{\theta}_n^1 \Leftarrow \arg\min_{\boldsymbol{\theta}} \ell_n(\mathcal{D}_n^1, \boldsymbol{\theta})$
3: $\quad$ **return** $\boldsymbol{\theta}_n^1$ ▷ Training Data Parameter Estimates
4: $\quad$ **return** $\ell_n(\mathcal{D}_n^1, \boldsymbol{\theta}_n^1)$ ▷ Training Data Model Fit
5: $\quad$ **return** $\ell_n(\mathcal{D}_n^2, \boldsymbol{\theta}_n^1)$ ▷ Test Data Model Fit
6: **end procedure**

## 14.1 Sampling Distribution Concepts

### 14.1.1 K-Fold Cross-Validation

The K-Fold Cross-Validation method is an improvement on train-test cross-validation. To introduce this idea, consider the train-test cross-validation method previously described (see train-test algorithm 14.0.1). The data set is divided in half and the first half of the data set is used to estimate parameters and the second half is used to estimate out-of-sample prediction error using the parameter estimates from the first half of the data set. Next, the second half of the data set is used to estimate model parameters and thc first half of the data set is used to estimate out-of-sample prediction error using the parameter estimates from the second half of the data set. The final estimate of out-of-sample prediction error is obtained by computing the average of both out-of-sample prediction error estimates. This would be an example of a 2-Fold Cross-Validation methodology.

**Algorithm 14.1.1** K-Fold Cross-Validation

1: **procedure** KFOLDCROSSVALID($\ell_n, \mathcal{D}_n$ )
2: $\quad$ Let $\{\mathcal{D}_m^1, \ldots, \mathcal{D}_m^K\}$ be a finite partition of $n$ record data set $\mathcal{D}_n$, $m = n/K$.
3: $\quad$ **for** $j = 1$ to $K$ **do**
4: $\quad\quad \neg\mathcal{D}_{n-m}^j \equiv \{\mathcal{D}_m^1, \ldots, \mathcal{D}_m^{j-1}, \mathcal{D}_m^{j+1}, \ldots, \mathcal{D}_m^K\}$.
5: $\quad\quad \hat{\boldsymbol{\theta}}_{n-m}^j \equiv \arg\min \ell_{n-m}(\neg\mathcal{D}_{n-m}^j, \boldsymbol{\theta})$ ▷ Parameter estimates for $\neg\mathcal{D}_{n-m}^j$
6: $\quad$ **end for**
7: $\quad \bar{\ell}_{n-m} \Leftarrow (1/K)\sum_{j=1}^K \ell_{n-m}(\neg\mathcal{D}_{n-m}^j, \hat{\boldsymbol{\theta}}_{n-m}^j)$ ▷ Average In-Sample Fit
8: $\quad \bar{\ell}_m \Leftarrow (1/K)\sum_{j=1}^K \ell_m(\mathcal{D}_m^j, \hat{\boldsymbol{\theta}}_{n-m}^j)$ ▷ Average Out-of-Sample Fit
9: $\quad \hat{\sigma}_{n-m} \Leftarrow \sqrt{\frac{1}{K-1}\sum_{j=1}^K(\ell_{n-m}(\neg\mathcal{D}_{n-m}^j, \hat{\boldsymbol{\theta}}_{n-m}^j) - \bar{\ell}_{n-m})^2}$ ▷ In-Sample Sampling Error
10: $\quad \hat{\sigma}_m \Leftarrow \sqrt{\frac{1}{K-1}\sum_{j=1}^K(\ell_m(\mathcal{D}_m^j, \hat{\boldsymbol{\theta}}_{n-m}^j) - \bar{\ell}_m)^2}$ ▷ Out-of-Sample Sampling Error
11: **end procedure**

Algorithm 14.1.1 provides an explicit procedure for using K-fold cross-validation to estimate the in-sample model fit, average in-sample fit sampling error, average out-of-sample model fit, in-sample sampling error, and out-of-sample sampling error. The data set with $n$ observations is first divided into $K$ groups each consisting of $m = (n/K)$ observations. One of the $K$ groups is chosen to be the $j$th out-of-sample group and the parameters are estimated from the remaining $K-1$ groups for $j = 1, \ldots, K$. The average model fit of the $K$ out-of-sample groups is used to estimate the average out-of-sample model fit and the sample standard deviation of the $K$ out-of-sample model fit estimates is an estimator of the out-of-sample fit sampling error. The average in-sample model fit and in-sample sampling error are computed in a similar manner. If the data sample consists of $n$ records, then $n$-fold cross-validation (i.e., $K = n$) is particularly effective but it is also the most computationally demanding since it involves $n$ parameter estimations involving data sets consisting of $n-1$ observations. The $n$-fold cross-validation method is also called the "leave-one-out" cross-validation method.

### 14.1.2 Sampling Distribution Estimation with Unlimited Data

**Definition 14.1.1** (Statistic). Let $\tilde{\mathcal{D}}_n \equiv [\tilde{\mathbf{x}}_1, \ldots, \tilde{\mathbf{x}}_n]$ be a sequence of $n$ independent and identically distributed $d$-dimensional random vectors with common Radon-Nikodým density $p_e$ defined with respect to support specification measure $\nu$. Let $\boldsymbol{\Upsilon} : \mathcal{R}^{d \times n} \rightarrow \mathcal{R}^r$ be a Borel-measurable function. The *DGP sampling density* of $\hat{\boldsymbol{\Upsilon}}_n$ is defined as the probability density

$$\hat{p}_e^n(\mathcal{D}_n) = \prod_{i=1}^{n} \hat{p}_e(\mathbf{x}_i).$$

□

**Definition 14.1.2** (Sampling Error). Let $\hat{\boldsymbol{\Upsilon}}_n$ be a statistic. The square root of the trace of the covariance matrix of $\hat{\boldsymbol{\Upsilon}}_n$ is called the *sampling error* for $\hat{\boldsymbol{\Upsilon}}_n$. □

The in-sample model fit, in-sample fit sampling error, out-of-sample model fit, and out-of-sample fit sampling error in the K-Fold Cross-Validation Algorithm (Algorithm 14.1.1) are examples of statistics. The probability distribution of a statistic is called its sampling distribution. The standard deviation of the sampling distribution is called the sampling error.

Unfortunately, in practice, the DGP sampling density $p_e^n$ of $\hat{\Upsilon}_n$ is not directly observable. However, a frequency histogram for the sampling density of $\hat{\Upsilon}_n$, $p_e^n$, can be approximated by generating $m$ data sets from the data generating process such that each data set has exactly $n$ data records. Next, the statistic $\hat{\Upsilon}_n$ is computed for each of the $m$ data sets. Then, the percentage of times the statistic $\hat{\Upsilon}_n$ falls into one out of a finite number of bins is recorded. These results may then be plotted as a frequency histogram as illustrated in Figure 14.4.

As the number of data sets, $m$, becomes large, the frequency histogram more closely approximates the probability density function of $\hat{\boldsymbol{\Upsilon}}_n$ for a fixed sample size $n$. In order to characterize the sampling error associated with a data sample of size $n$, the number of data sets $m$ is chosen to be very large. Increasing the number of data sets $m$ improves the quality of the frequency histogram approximation. This averaged sampling error may be interpreted as a minimum mean square estimator as discussed in Section 13.3.3. Increasing the sample size $n$ will tend to decrease the sampling error of $\hat{\boldsymbol{\Upsilon}}_n^k$. The sampling error of $\hat{\boldsymbol{\Upsilon}}_n^k$ corresponds to the standard deviation of the frequency histogram in Figure 14.4.

Algorithm (14.1.2) shows how to use these ideas to estimate the minimum mean square estimator of the statistic $\hat{\boldsymbol{\Upsilon}}_n^k$ and the sampling error of $\hat{\boldsymbol{\Upsilon}}_n^k$.

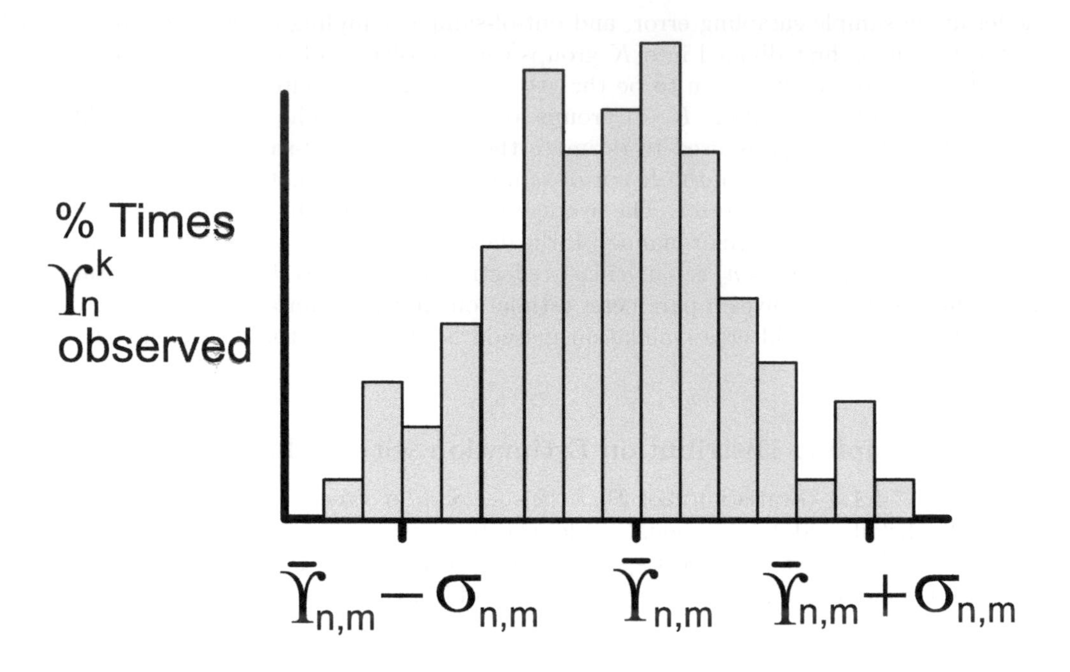


**FIGURE 14.4**
Frequency histogram of the set of $m$ observed values $\{\boldsymbol{\Upsilon}_n^1, \ldots, \boldsymbol{\Upsilon}_n^k, \ldots, \boldsymbol{\Upsilon}_n^m\}$ of the statistic $\tilde{\boldsymbol{\Upsilon}}_n^k$ for a fixed sample size $n$. The frequency histogram approximates the probability density of the statistic $\hat{\boldsymbol{\Upsilon}}_n^m$. As $m$ increases, the quality of the histogram approximation tends to improve. In addition, the mean $\bar{\boldsymbol{\Upsilon}}_{n,m}$ and standard deviation $\sigma_{n,m}$ of the $m$ observed values of the statistic $\hat{\boldsymbol{\Upsilon}}_n^k$ for a fixed sample size $n$ correspond respectively to an improved estimate of the expected value of $\hat{\boldsymbol{\Upsilon}}_n^k$ and its sampling error respectively.

**Algorithm 14.1.2** Estimating Sampling Distributions with Unlimited Data

1: **procedure** UNLIMITEDDATASAMPLING($\boldsymbol{\Upsilon}_n$, $\{\mathcal{D}_n^1, \mathcal{D}_n^2, \ldots\}$ )
2: $\quad \epsilon = 0.0001$ ▷ $\epsilon$ is a small positive number.
3: $\quad m \Leftarrow 0$
4: $\quad SE(\bar{\boldsymbol{\Upsilon}}_{n,m}) \Leftarrow \infty$
5: $\quad$ **repeat**
6: $\quad\quad m \Leftarrow m + 1$
7: $\quad\quad \boldsymbol{\Upsilon}_n^m \Leftarrow \boldsymbol{\Upsilon}_n(\mathcal{D}_n^m)$ ▷ $\boldsymbol{\Upsilon}_n^m$ is $r$-dimensional
8: $\quad\quad \bar{\boldsymbol{\Upsilon}}_{n,m} \Leftarrow (1/m)\sum_{k=1}^m \boldsymbol{\Upsilon}_n^k$
9: $\quad\quad \sigma(\bar{\boldsymbol{\Upsilon}}_{n,m}) \Leftarrow \sqrt{\frac{1}{m-1}\sum_{k=1}^m \left|\boldsymbol{\Upsilon}_n^k - \bar{\boldsymbol{\Upsilon}}_{n,m}\right|^2}$
10: $\quad\quad SE(\bar{\boldsymbol{\Upsilon}}_{n,m}) \Leftarrow \sigma(\bar{\boldsymbol{\Upsilon}}_{n,m})/\sqrt{m}$
11: $\quad$ **until** $SE(\bar{\boldsymbol{\Upsilon}}_{n,m}) < \epsilon$
12: $\quad$ **return** $\bar{\boldsymbol{\Upsilon}}_{n,m}$ ▷ Return Estimator
13: $\quad$ **return** $\sigma(\bar{\boldsymbol{\Upsilon}}_{n,m})$ ▷ Return Sampling Error with Approximation Error
14: $\quad$ **return** $SE(\bar{\boldsymbol{\Upsilon}}_{n,m})$ ▷ Return Approximation Error Magnitude
15: **end procedure**

In practice, the unlimited data sampling algorithm (Algorithm 14.1.2) is not commonly used because it requires that $m$ data sets each consisting of $n$ observations are collected. Typically, data collection is expensive in terms of time and money. In addition, if one does collect $m$ data sets consisting of $n$ observations each, it would be more desirable to use all $mn$ observations to estimate the model parameters to improve the quality of the parameter estimators. Nevertheless, Algorithm 14.1.2 is very useful for illustrating a number of key sampling distribution concepts and minor variants of Algorithm 14.1.2 yield practical and widely used parameter estimation simulation methods.

## 14.2 Bootstrap Methods for Sampling Distribution Simulation

In this section, practical bootstrap simulation methods for sampling error estimation are described. These methods may be interpreted as natural extensions of the Unlimited Data Sampling algorithm described in Algorithm 14.1.2. Assume that the observed data $\mathbf{x}_1, \mathbf{x}_2, \ldots$ is a realization of a sequence of *i.i.d.* observations $\tilde{\mathbf{x}}_1, \tilde{\mathbf{x}}_2, \ldots$ with common density $p_e$. The DGP sampling distribution is specified by the density function

$$p_e^n(\mathcal{D}_n) = \prod_{i=1}^{n} p_e(\mathbf{x}_i).$$

The Unlimited Data Sampling Method specified in Algorithm 14.1.2 and Figure 14.5 samples from $p_e^n$ to generate $m$ random data sets $\tilde{\mathcal{D}}_n^1, \ldots, \tilde{\mathcal{D}}_n^m$ where each random data set consists of $n$ data records.

The key idea of bootstrap simulation methods is to replace $p_e^n$ with an estimate of $p_e^n$ which will be denoted as $\hat{p}^n$. The density $\hat{p}^n$ specifies the bootstrap probability distribution which becomes a better approximation of the actual data sample distribution $p_e^n$ when the sample size $n$ is sufficiently large. Once the bootstrap distribution has been constructed, the sampling distribution of the statistic of interest is estimated. However, unlike the unlimited data sampling method which uses the true data generating density $p_e^n$, the bootstrap method uses an approximation $\hat{p}^n$ for generating $m$ random data sets. As in the unlimited data sampling method, as $m \to \infty$, the frequency histogram constructed from realizations of $\hat{\boldsymbol{\Upsilon}}_n^k$ (see Figure 14.4) more closely approximates the sampling distribution for $\hat{\boldsymbol{\Upsilon}}_n^k$ for $k = 1, \ldots, m$.

### 14.2.1 Bootstrap Approximation of Sampling Distribution

**Definition 14.2.1** (Parametric Bootstrap Sampling Density). Let $\tilde{\mathcal{D}}_n \equiv [\tilde{\mathbf{x}}_1, \tilde{\mathbf{x}}_2, \ldots, \tilde{\mathbf{x}}_n]$ be a sequence of $n$ independent and identically distributed $d$-dimensional random vectors with common Radon-Nikodým density $p_e : \mathcal{R}^d \to [0, \infty)$ with respect to support specification measure $\nu$. Let $\Theta \subseteq \mathcal{R}^q$. Let $p : \mathcal{R}^d \times \Theta \to [0, \infty)$ be defined such that $p(\mathbf{x}; \cdot)$ is continuous on $\Theta$ for each $\mathbf{x} \in \mathcal{R}^d$. Define a probability model

$$\mathcal{M} \equiv \{p(\cdot; \boldsymbol{\theta}) : \boldsymbol{\theta} \in \Theta\}.$$

Let $\boldsymbol{\theta}_n : \mathcal{R}^{d \times n} \to \mathcal{R}^q$ be a Borel-measurable function. Let $\hat{\boldsymbol{\theta}}_n \equiv \boldsymbol{\theta}_n(\tilde{\mathcal{D}}_n)$ for $n = 1, 2, \ldots$. Assume $\hat{\boldsymbol{\theta}}_n \to \boldsymbol{\theta}^*$ with probability one as $n \to \infty$. Now define the *parametric bootstrap sampling density* $\hat{p}_{\mathcal{M}}^n(\cdot; \hat{\boldsymbol{\theta}}_n)$ such that

$$\hat{p}_{\mathcal{M}}^n(\mathcal{D}_n; \hat{\boldsymbol{\theta}}_n) = \prod_{i=1}^{n} p(\mathbf{x}_i; \hat{\boldsymbol{\theta}}_n)$$

with respect to the probability model $\mathcal{M}$. □

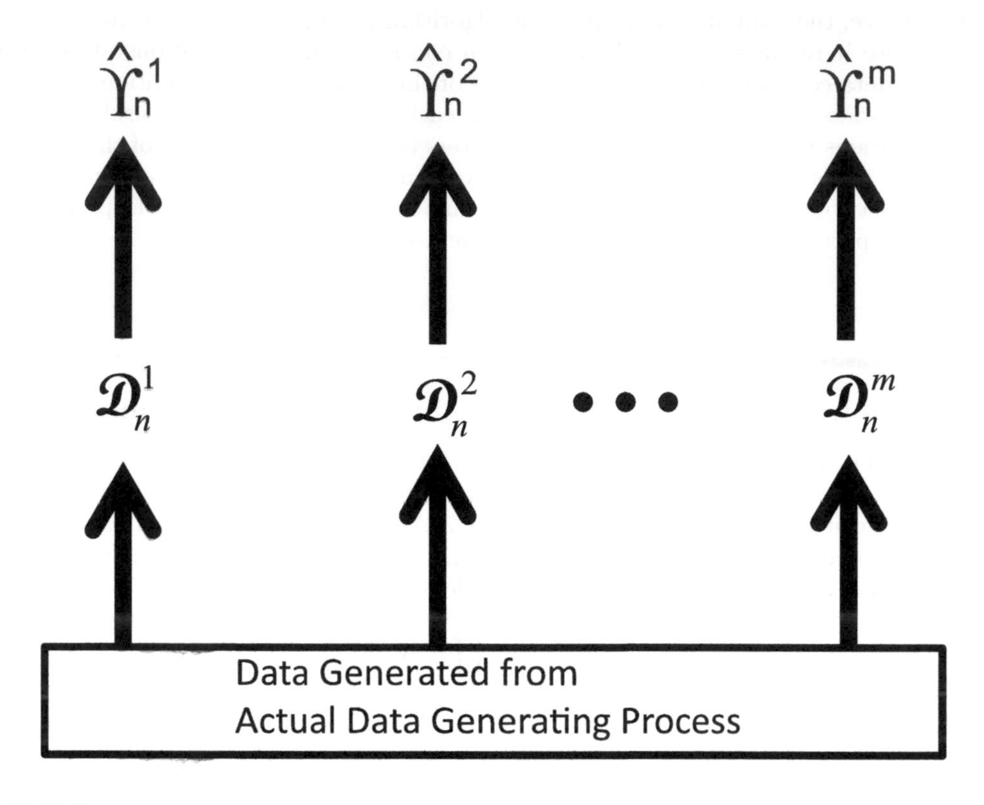


**FIGURE 14.5**
**Estimating the sampling error of a statistic by collecting additional data.** This figure depicts a methodology for estimating the sampling error for a statistic computed for a sample of size $n$. This is achieved by collecting data for $m$ data sets $\mathcal{D}_n^1, \ldots, \mathcal{D}_n^m$ such that each data set consists of exactly $n$ data records. Then, a realization of the statistic $\hat{\Upsilon}_n^k$ is computed for data set $\mathcal{D}_n^k$ for $k = 1, \ldots, m$. The mean and standard deviation of the $m$ statistics $\hat{\Upsilon}_n^1, \ldots, \hat{\Upsilon}_n^m$ are then used as estimators of the mean of $\hat{\Upsilon}_n^k$ and its standard deviation respectively.

Assume there exists a true parameter vector $\boldsymbol{\theta}^*$ such that the data generating process can be represented as sampling from the density $p(\cdot; \boldsymbol{\theta}^*)$. If $\tilde{\boldsymbol{\theta}}_n \to \boldsymbol{\theta}^*$ with probability one, then $p(\cdot; \tilde{\boldsymbol{\theta}}_n) \to p(\cdot; \boldsymbol{\theta}^*)$ with probability one because $p(\mathbf{x}; \cdot)$ is continuous on $\Theta$. However, a disadvantage of this methodology is that if model misspecification is present, the parametric bootstrap model approximation converges to a best approximating distribution rather than the distribution which actually generated the observed data.

Unlike the parametric bootstrap model approximation, the nonparametric bootstrap model approximation makes minimal assumptions about the probability distribution which generated the data. In addition, under general conditions, the nonparametric bootstrap sampling distribution will converge to the DGP sampling distribution. The essential idea is that the relative frequency of a data record in a data set is used as an estimator of the probability that the DGP density generates that data record. The sampling process for this nonparametric bootstrap density estimator can be implemented in a straightforward manner by simply sampling with replacement from a given data set.

**Definition 14.2.2** (Nonparametric Bootstrap Density). Let $\tilde{\mathcal{D}}_n \equiv [\tilde{\mathbf{x}}_1, \ldots, \tilde{\mathbf{x}}_n]$ be a sequence of $n$ independent and identically distributed $d$-dimensional random vectors with common Radon-Nikodým density $p_e$ with respect to support specification measure $\nu$. Let $\mathcal{D}_n$ be a realization of $\tilde{\mathcal{D}}_n \equiv [\tilde{\mathbf{x}}_1, \ldots, \tilde{\mathbf{x}}_n] \in \mathcal{R}^{d \times n}$. Let $\hat{p}_e : \mathcal{R}^d \to [0, \infty)$ be defined such that $\hat{p}_e(\mathbf{x}) = 1/n$ when $\mathbf{x} \in \mathcal{D}_n$, and $\hat{p}_e(\mathbf{x}) = 0$ when $\mathbf{x} \notin \mathcal{D}_n$. Let $\hat{p}_e^n : \mathcal{R}^{d \times n} \to [0, \infty)$ be a Radon-Nikodým density defined such that $\hat{p}_e^n(\mathcal{D}_n) = \prod_{i=1}^n \hat{p}(\mathbf{x}_i)$. The density $\hat{p}_e^n$ is called the *nonparametric bootstrap sampling density.* □

Let $\hat{P}_n$ and $P_e$ be the cumulative distribution functions specified by the densities $\hat{p}_e$ and $p_e$ respectively. By the Glivenko-Cantelli Uniform Law of Large Numbers (see Theorem 9.3.3), $\hat{P}_n$ converges uniformly to $P_e$ as $n \to \infty$.

### 14.2.2 Monte Carlo Bootstrap Sampling Distribution Estimation

**Definition 14.2.3** (Bootstrap Estimator). Let $\tilde{\mathcal{D}}_n \equiv [\tilde{\mathbf{x}}_1, \ldots, \tilde{\mathbf{x}}_n]$ be a bounded stochastic sequence of $n$ independent and identically distributed $d$-dimensional random vectors with common Radon-Nikodým density $p_e : \mathcal{R}^d \to [0, \infty)$ with respect to support specification measure $\nu$. Let $p : \mathcal{R}^d \to [0, \infty)$ and $\hat{p} : \mathcal{R}^d \to [0, \infty)$ be Radon-Nikodým densities defined with respect to $\nu$. Assume for each $n \in \mathbb{N}$, $\boldsymbol{\Upsilon}_n : \mathcal{R}^{d \times n} \to \mathcal{R}^r$ is a piecewise continuous function. For each $n \in \mathbb{N}$, define the $r$-dimensional vector:

$$\bar{\boldsymbol{\Upsilon}}_{n,\infty} = \int \boldsymbol{\Upsilon}_n(\mathcal{D}_n) \hat{p}^n(\mathcal{D}_n) d\nu^n(\mathcal{D}_n), \quad \hat{p}^n(\mathcal{D}_n) = \prod_{i=1}^n \hat{p}(\mathbf{x}_i). \tag{14.1}$$

Define the $r$-dimensional vector:

$$\boldsymbol{\Upsilon}^* = \int \boldsymbol{\Upsilon}_n(\mathcal{D}_n) p^n(\mathcal{D}_n) d\nu^n(\mathcal{D}_n), \quad p^n(\mathcal{D}_n) = \prod_{i=1}^n p(\mathbf{x}_i). \tag{14.2}$$

Then, $\bar{\boldsymbol{\Upsilon}}_{n,\infty}$ is called the *bootstrap estimator* for the *bootstrap estimand* $\boldsymbol{\Upsilon}^*$. If $\hat{p}^n$ is a parametric bootstrap sampling density, then $\bar{\boldsymbol{\Upsilon}}_{n,\infty}$ is called the *parametric bootstrap estimator* for $\boldsymbol{\Upsilon}^*$ where $\boldsymbol{\Upsilon}^*$ is defined with respect to the model sampling density $p^n$. If $\hat{p}^n$ is a nonparametric bootstrap sampling density, then $\bar{\boldsymbol{\Upsilon}}_{n,\infty}$ is called the *nonparametric bootstrap estimator* for $\boldsymbol{\Upsilon}^*$ where $\boldsymbol{\Upsilon}^*$ is defined with respect to the true sampling density $p_e^n$. □

Less formally, the bootstrap estimator for $\boldsymbol{\Upsilon}^*$ in (14.2) is obtained by replacing the sampling distribution $p^n(\mathcal{D}_n)$ in (14.2) with either the parametric bootstrap sampling distribution estimator $\hat{p}_{\mathcal{M}}^n(\mathcal{D}_n)$ or the nonparametric bootstrap sampling distribution estimator $\hat{p}_e^n(\mathcal{D}_n)$.

**Example 14.2.1** (Bootstrap Sample Average Estimator for Smooth Functions). Let $\tilde{\mathbf{x}}_1, \ldots, \tilde{\mathbf{x}}_n$ be a bounded stochastic sequence of independent and identically distributed $d$-dimensional random vectors. Let $\mathbf{x}_1, \ldots, \mathbf{x}_n$ be a realization of $\tilde{\mathbf{x}}_1, \ldots, \tilde{\mathbf{x}}_n$. Assume

$$\boldsymbol{\Upsilon}_n(\mathcal{D}_n) = (1/n) \sum_{i=1}^n \boldsymbol{\phi}(\mathbf{x}_i) \tag{14.3}$$

where $\boldsymbol{\phi} : \mathcal{R}^d \to \mathcal{R}^r$ is a piecewise continuous function. Develop a method for computing a bootstrap approximation for estimating

$$E\{\boldsymbol{\phi}(\tilde{\mathbf{x}})\} = \int \boldsymbol{\phi}(\mathbf{x}) p_e(\mathbf{x}) d\nu(\mathbf{x}) \tag{14.4}$$

which is a computationally intractable multidimensional integral.

**Solution.** Substitute assumption (14.3) into (14.1) to obtain the bootstrap estimator:

$$\bar{\Upsilon}_{n,\infty} = \int \left( (1/n) \sum_{i=1}^{n} \phi(\mathbf{x}_i) \right) \prod_{i=1}^{n} \hat{p}(\mathbf{x}_i) d\nu^n(\mathcal{D}_n) = \int \phi(\mathbf{x}) \hat{p}(\mathbf{x}) d\nu(\mathbf{x}). \quad (14.5)$$

△

Unfortunately, however, the multidimensional integral for the bootstrap estimator in (14.1) is often computationally intractable for most machine learning applications. On the other hand, there is a simple computational approximation based upon the unlimited data sampling algorithm (Algorithm 14.1.2) that results in a very computationally effective estimator of the bootstrap estimator. This bootstrap estimator is called the *Monte Carlo Bootstrap estimator*.

**Definition 14.2.4** (Monte Carlo Bootstrap Estimator). Let $\tilde{\mathcal{D}}_n \equiv [\tilde{\mathbf{x}}_1, \ldots, \tilde{\mathbf{x}}_n]$ be a sequence of independent and identically distributed $d$-dimensional random vectors with common DGP Radon-Nikodým density $p_e : \mathcal{R}^d \rightarrow [0, \infty)$ with respect to support specification measure $\nu$. Let $\mathbf{\Upsilon}_n : \mathcal{R}^{d \times n} \rightarrow \mathcal{R}^r$ be a Borel-measurable function. The $r$-dimensional *Monte Carlo bootstrap estimator*, $\bar{\mathbf{\Upsilon}}_{n,m}$, for a sample of size $n$ is defined by

$$\bar{\mathbf{\Upsilon}}_{n,m} \equiv (1/m) \sum_{j=1}^{m} \mathbf{\Upsilon}_n(\tilde{\mathcal{D}}_n^{*j}) \quad (14.6)$$

where the elements of the stochastic sequence $\tilde{\mathcal{D}}_n^{*1}, \ldots, \tilde{\mathcal{D}}_n^{*m}$ are independent and identically distributed with a common bootstrap Radon-Nikodým density $\hat{p}^n : \mathcal{R}^{d \times n} \rightarrow [0, \infty)$ defined with respect to support specification measure $\nu^n$. Let

$$\hat{\sigma}(\mathbf{\Upsilon}_{n,m}) \equiv \sqrt{\frac{1}{m-1} \sum_{j=1}^{m} |\mathbf{\Upsilon}_n(\tilde{\mathcal{D}}_n^{*j}) - \bar{\mathbf{\Upsilon}}_{n,m}|^2}$$

be called the *sampling error estimator* for $\bar{\mathbf{\Upsilon}}_{n,m}$. Let $\hat{\sigma}(\mathbf{\Upsilon}_{n,m})/\sqrt{m}$ be called the *simulation error estimator* for $\bar{\mathbf{\Upsilon}}_{n,m}$. □

**Example 14.2.2** (Monte Carlo Bootstrap Sample Average Estimator). In this example, a Monte Carlo estimator is proposed for approximating the Bootstrap Sample Average Estimator for smooth functions in Example 14.2.1. Substitute (14.3) into (14.6) to obtain:

$$\bar{\mathbf{\Upsilon}}_{n,m} \equiv (1/m) \sum_{j=1}^{m} \mathbf{\Upsilon}_n(\tilde{\mathcal{D}}_n^{*j}) = (1/m) \sum_{j=1}^{m} \left( (1/n) \sum_{i=1}^{n} \phi(\tilde{\mathbf{x}}_i^{*j}) \right)$$

where $\tilde{\mathbf{x}}_i^{*j}$ is the $i$th observation sampled from the $j$th bootstrap data set for $j = 1, \ldots, m$ and $i = 1, \ldots, n$. Then, by the Strong Law of Large Numbers, it immediately follows that $\bar{\mathbf{\Upsilon}}_{n,m} \rightarrow \bar{\mathbf{\Upsilon}}_{n,\infty}$ as $m \rightarrow \infty$ for a fixed sample size $n$. That is, the Monte Carlo Bootstrap Estimator converges with probability one as the number of bootstrap samples becomes large to the Bootstrap Estimator.

In addition, for this choice of $\Upsilon_n$, it also follows from the Strong Law of Large Numbers that $\bar{\mathbf{\Upsilon}}_{n,m} \rightarrow E\{\phi(\tilde{\mathbf{x}})\}$ as $n \rightarrow \infty$ for a fixed number of $m$ bootstrap samples. △

The Monte Carlo Bootstrap Estimation methodology is equally applicable to both the parametric and the nonparametric bootstrap estimators. Algorithm 14.2.1 provides an example algorithm for implementing Monte Carlo Bootstrap Estimation.

**Algorithm 14.2.1** Monte Carlo Bootstrap Estimation

1: **procedure** BOOTSTRAPESTIMATION($\boldsymbol{\Upsilon}_n, \hat{P}(\cdot)$ )
2: $\epsilon = 0.0001$ ▷ $\epsilon$ is a small positive number.
3: $m \Leftarrow 0$
4: $\hat{\sigma}(\bar{\boldsymbol{\Upsilon}}_{n,m}) \Leftarrow \infty$
5: **repeat**
6: $m \Leftarrow m + 1$
7: Sample with replacement $n$ times from $\hat{P}(\cdot)$ to generate $\mathcal{D}_n^m$
8: $\boldsymbol{\Upsilon}_n^m \Leftarrow \boldsymbol{\Upsilon}_n(\mathcal{D}_n^m)$
9: $\bar{\boldsymbol{\Upsilon}}_{n,m} \Leftarrow (1/m)\sum_{k=1}^m \boldsymbol{\Upsilon}_n^k$
10: $\hat{\sigma}(\bar{\boldsymbol{\Upsilon}}_{n,m}) \Leftarrow \sqrt{\frac{1}{m-1}\sum_{k=1}^m \left|\boldsymbol{\Upsilon}_n^k - \bar{\boldsymbol{\Upsilon}}_{n,m}\right|^2}$
11: **until** $\hat{\sigma}(\bar{\boldsymbol{\Upsilon}}_{n,m})/\sqrt{m} < \epsilon$
12: **return** $\bar{\boldsymbol{\Upsilon}}_{n,m}$ ▷ Bootstrap Estimate
13: **return** $\hat{\sigma}(\bar{\boldsymbol{\Upsilon}}_{n,m})$ ▷ Bootstrap Estimate Sampling Error
14: **return** $\hat{\sigma}(\bar{\boldsymbol{\Upsilon}}_{n,m})/\sqrt{m}$ ▷ Bootstrap Estimate Simulation Error
15: **end procedure**

Algorithm 14.2.1 can be used to specify a *nonparametric bootstrap algorithm* with respect to a data set $\mathcal{D}_n \equiv \{\mathbf{x}_1, \ldots, \mathbf{x}_n\}$ by defining $\hat{P}$ in Algorithm 14.2.1 such that $\hat{P}$ is a function with domain $\mathcal{D}_n$ and range $\{1/n\}$. Then, in Step 7 of Algorithm 14.2.1 one samples $n$ times with replacement from $\hat{P}$ by sampling $n$ times with replacement from $\{\mathbf{x}_1, \ldots, \mathbf{x}_n\}$.

Algorithm 14.2.1 can also be used to specify a *parametric bootstrap algorithm*. Now let

$$\mathcal{M} \equiv \{p(\cdot; \boldsymbol{\theta}) : \boldsymbol{\theta} \in \Theta\}$$

be a probability model which has been fitted to some data set $\mathcal{D}_n$ to obtain parameter estimates $\hat{\boldsymbol{\theta}}_n$. The fitted probability model is a particular density in $\mathcal{M}$ denoted by $p(\cdot; \hat{\boldsymbol{\theta}}_n)$. In order to implement a *parametric bootstrap algorithm*, one defines $\hat{P}$ in Algorithm 14.2.1 so that $\hat{P}$ is specified by $p(\cdot; \hat{\boldsymbol{\theta}}_n)$. Then, in Step 7 of Algorithm 14.2.1, one samples $n$ times with replacement from $\hat{P}$ by sampling $n$ times with replacement from $p(\cdot; \hat{\boldsymbol{\theta}}_n)$.

It is important to emphasize that this methodology does not magically create additional data but simply provides a mechanism for estimating characteristics of a sample of size $n$. The number of observations, $n$, in each "bootstrap" data sample is never allowed to exceed the number of observations in the original data sample $\mathbf{x}_1, \ldots, \mathbf{x}_n$. On the other hand, the number of bootstrap data samples of size $n$ is equal to $m$ and $m$ should be chosen to be as large as required. As $m \to \infty$ with a fixed sample size $n$, the sampling error of the relevant bootstrap statistic will typically converge to a positive number while the simulation error will typically converge to zero. It may be useful at this point to compare the Unlimited Data Sampling Algorithm 14.1.2, which is described in Figure 14.5, with the Bootstrap approach described in Figure 14.6.

An important problem is concerned with choosing the number of bootstrap data samples, $m$. If sufficient computational resources are available, $m$ should be chosen sufficiently large so that the Bootstrap Estimate Simulation Error is sufficiently small. In cases where this is not possible, then the Bootstrap Estimate Simulation Error should be explicitly reported.

**Example 14.2.3** (Minimum Mean Square Error Parameter Estimators). Rather than use MAP estimation methods to estimate parameter values, the posterior mean estimator discussed in Section 13.3.3 may be preferred. The posterior mean estimator is a more robust minimum mean-square error Bayes estimator which may be defined by $\boldsymbol{\Upsilon}_n : \mathcal{R}^{d \times n} \to \mathcal{R}^r$

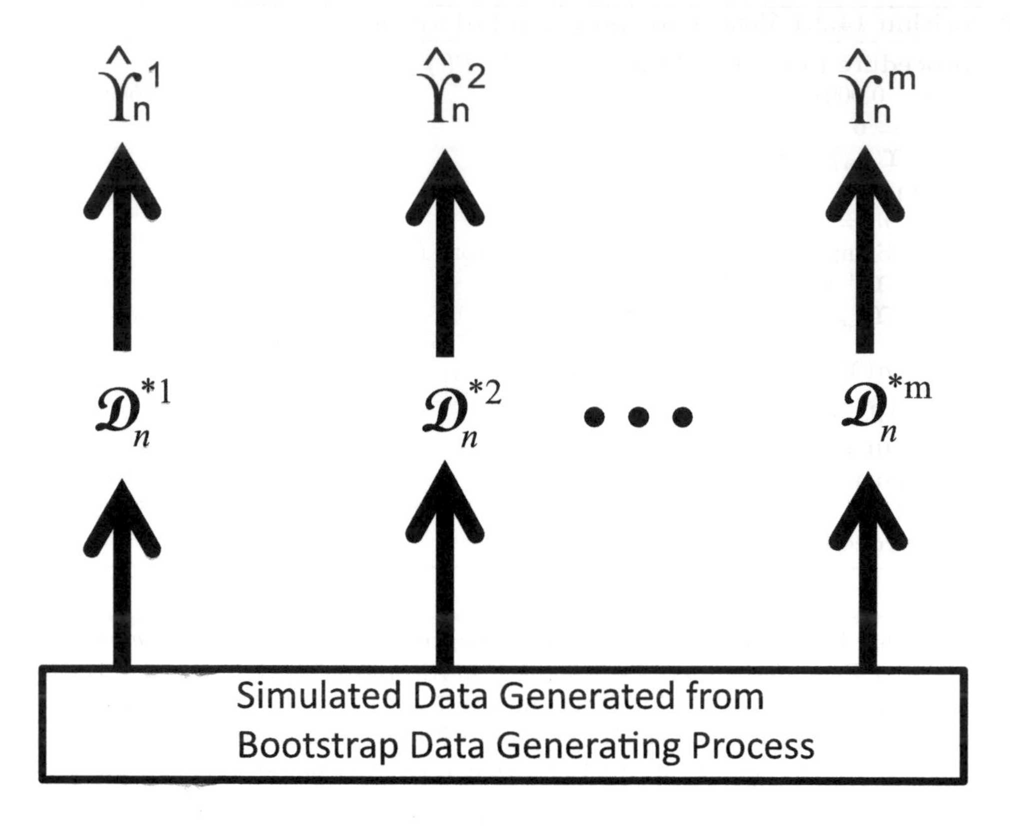


**FIGURE 14.6**
**Estimating the sampling error of a statistic using a bootstrap methodology.** This figure depicts a methodology for estimating the sampling error for a statistic computed for a sample of size $n$. This is achieved by $m$ data sets $\mathcal{D}_n^{*1}, \ldots, \mathcal{D}_n^{*m}$ from the bootstrap density such that each of the $m$ data sets consists of exactly $n$ data records. The mean and standard deviation of the $m$ statistics $\hat{\boldsymbol{\Upsilon}}_n^1, \ldots, \hat{\boldsymbol{\Upsilon}}_n^m$ computed respectively from the $m$ data sets correspond respectively to the bootstrap estimator and the bootstrap estimator sampling error. Compare this figure with Figure 14.5.

such that for all $\mathcal{D}_n \in \mathcal{R}^{d \times n}$:

$$\boldsymbol{\Upsilon}_n(\mathcal{D}_n) \equiv \boldsymbol{\theta}_n(\mathcal{D}_n).$$

Using this choice of $\boldsymbol{\Upsilon}_n$, the Monte Carlo Bootstrap Estimator

$$\bar{\boldsymbol{\Upsilon}}_{n,m} = (1/m) \sum_{j=1}^{m} \boldsymbol{\theta}_n(\tilde{\mathcal{D}}_n^j)$$

converges as $m \to \infty$ with probability one to the minimum mean-square error Bayes estimator

$$\bar{\boldsymbol{\theta}}_{n,\infty} = \int \boldsymbol{\theta}_n(\mathcal{D}_n) p_e^n(\mathcal{D}_n) d\nu^n(\mathcal{D}_n)$$

that minimizes the minimum mean-square error risk function

$$\Psi(\boldsymbol{\theta}) = \int \left|\boldsymbol{\theta}_n(\mathcal{D}_n) - \bar{\boldsymbol{\theta}}_{n,\infty}\right|^2 p_e^n(\mathcal{D}_n) d\nu^n(\mathcal{D}_n).$$

In practice, $\bar{\boldsymbol{\Upsilon}}_{n,m}$ is estimated by the following procedure. The procedure is implemented by direct application of the Monte Carlo Bootstrap Estimation Algorithm (14.2.1). Given a data sample of $n$ records, one samples with replacement $n$ times to create the first nonparametric bootstrap data set. Then, one estimates the parameters of the model using the first nonparametric bootstrap data set and denotes those parameter estimates as $\hat{\boldsymbol{\theta}}_n^{*1}$. In a similar manner, $m$ bootstrap data sets each consisting of $n$ records are constructed, and the model is fit to each of the $m$ bootstrap data sets to generate $m$ parameter estimates $\hat{\boldsymbol{\theta}}_n^{*1}, \ldots, \hat{\boldsymbol{\theta}}_n^{*m}$. An estimate of the minimum mean-square error Bayes estimator $\bar{\boldsymbol{\theta}}_{n,\infty}$ (see Section 13.3.3) is obtained using the posterior mean estimator:

$$\bar{\boldsymbol{\theta}}_{n,m} = (1/m) \sum_{j=1}^{m} \hat{\boldsymbol{\theta}}_n^{*j}.$$

The sampling error of $\hat{\boldsymbol{\theta}}_n$, $\hat{\sigma}_{n,m}$, is estimated using the formula:

$$\hat{\sigma}_{n,m}^2 = (1/m) \left[ \frac{1}{m-1} \sum_{j=1}^{m} \left| \hat{\boldsymbol{\theta}}_n^{*j} - \bar{\boldsymbol{\theta}}_{n,m} \right|^2 \right].$$

The number of bootstrap samples, $m$, is chosen to be sufficiently large so that the bootstrap simulation error $\hat{\sigma}_{n,m}/\sqrt{m}$ is sufficiently small. △

**Example 14.2.4** (Using Bootstrap Methods with Multimodal Objective Functions). Consider the problem of computing the minimum mean square estimator of a parameter estimate and its sampling error as in Example 14.2.3. The problem, however, is complicated by the fact that the parameter estimates are obtained by minimizing an empirical risk function which has multiple strict local minimizers. Assume the original data set consists of $n$ data records. In addition, assume a nonparametric bootstrap methodology is used to sample with replacement from the original data set of $n$ records for the purpose of generating $m$ bootstrap data sets each consisting of $n$ records.

An optimization algorithm is then used to minimize the empirical risk function for the first 99 bootstrap data sets and the algorithm generates parameter estimates for these first 99 bootstrap data sets which are strict local minimizers of their respective 99 different empirical risk functions. In addition, these 99 different strict local minimizers are very similar to each other. In fact, the average of these 99 bootstrap parameter estimates is found to be $\bar{\boldsymbol{\theta}}_{n,99} = [1, 2, -4, 7, 8, -2]$. The sampling error is found to be $\hat{\sigma}_{n,99} = 0.9$. In addition, it is empirically found that the 99 strict local minimizers are clustered together in parameter space and the radius of this cluster is $\hat{\sigma}_{n,99}$ as predicted by the asymptotic statistical theory.

However, the parameter estimate for the 100th bootstrap data set is equal to

$$\hat{\boldsymbol{\theta}}_n^{100} = [100, -1000, 1400, 12, 0.1, 0.2]$$

which is also a legitimate strict local minimizer of the empirical risk function. Should this bootstrap parameter estimate be averaged with the other 99 bootstrap parameter estimates to obtain the posterior mean estimator and the sampling error as defined in Example 14.2.3?

**Solution.** There is not a straightforward answer to this question. However, assume that the goal of the bootstrap sampling process is to characterize the sampling error associated with a parameter estimate which is a strict local minimizer of the empirical risk function. In addition, assume that: (1) the 99 parameter estimates from the first 99 bootstrap data samples are converging to a particular strict local minimizer, and (2) the parameter estimate from the 100th bootstrap data sample is converging to a different strict local minimizer. Given these assumptions, it would not be appropriate to combine the parameter estimate associated with the 100th bootstrap sample in bootstrap calculations of the type discussed in Example 14.2.3 with the first 99 parameter estimates.

If one did, in fact, combine all 100 parameter estimates, then this corresponds to a type of bootstrap estimator which is not considered here. The resulting bootstrap estimator would characterize not just variability in the different empirical risk minimizers due to sample size but variability in the parameter estimates due to properties of the optimization algorithm.

When the goal is to characterize the sampling error associated with estimating a particular strict local minimizer, $\boldsymbol{\theta}^*$, of the expected value of the empirical risk function, then minimizers of empirical risk functions constructed from bootstrap samples which are not converging to $\boldsymbol{\theta}^*$ should not be included in bootstrap calculations. △

**Example 14.2.5** (Bagging for Improving Predictive Accuracy). The standard empirical risk minimization method generates a predicted response $\ddot{y}(\mathbf{s};\hat{\boldsymbol{\theta}}_n)$ for a given input pattern $\mathbf{s}$ where $\hat{\boldsymbol{\theta}}_n$ is a strict global minimizer of an empirical risk function. This corresponds to the case where $\Upsilon_n(\tilde{\mathcal{D}}_n) = \ddot{y}(\mathbf{s};\hat{\boldsymbol{\theta}}_n)$ where $\mathcal{D}_n$ is a data set consisting of $n$ observations.

Given $\mathcal{D}_n$, one samples with replacement $n$ times to create the first non-parametric bootstrap data set. Then, one estimates the parameters of the model using the first non-parametric bootstrap data set and denotes that parameter estimate as $\hat{\boldsymbol{\theta}}_n^{*1}$. In a similar manner, $m$ bootstrap data sets each consisting of $n$ records are constructed, the model is fit to each of the $m$ bootstrap data sets to generate $m$ parameter estimates $\hat{\boldsymbol{\theta}}_n^{*1}, \ldots, \hat{\boldsymbol{\theta}}_n^{*m}$. The predicted response of the classifier is then computed using a weighted average of the responses of $m$ distinct classifiers using the formula:

$$\bar{y}_n(\mathbf{s}) = (1/m)\sum_{j=1}^{m} \ddot{y}\left(\mathbf{s}, \hat{\boldsymbol{\theta}}_n^{*j}\right).$$

This methodology of averaging a collection of regression models or classifiers that have been estimated using resampled data is referred to as *bagging* (bootstrap aggregation) in the machine learning literature.

The number of bootstrap data sets $m$ should be ideally chosen so the trace or determinant of the bootstrap estimation simulation error covariance matrix is sufficiently small. Provide an explicit formula for using the non-parametric bootstrap estimation method to estimate the sampling error of $\ddot{y}\left(\mathbf{s}, \hat{\boldsymbol{\theta}}_n^{j}\right)$ for a given input pattern $\mathbf{s}$.

**Solution.** The mean of $\ddot{y}\left(\mathbf{s}, \hat{\boldsymbol{\theta}}_n^{j}\right)$ is estimated using $\bar{y}_n(\mathbf{s})$. The sampling error of $\ddot{y}\left(\mathbf{s}, \hat{\boldsymbol{\theta}}_n^{j}\right)$ is estimated for each $\mathbf{s}$ using

$$\sigma(\hat{y}_n(\mathbf{s}, \hat{\boldsymbol{\theta}}_n)) \equiv \sqrt{\frac{1}{m-1}\sum_{j=1}^{m}\left(\ddot{y}\left(\mathbf{s}, \hat{\boldsymbol{\theta}}_n^{*j}\right) - \bar{y}_n(\mathbf{s})\right)^2}.$$

The simulation error for each $\mathbf{s}$ is $\sigma(\hat{y}_n(\mathbf{s}, \hat{\boldsymbol{\theta}}_n)/\sqrt{m}$. Ideally, $m$ should be chosen sufficiently large so that the simulation error is close to zero but if that is not possible, then the simulation error may be reported. △

**Example 14.2.6** (Model Selection Statistic Sampling Error). The model selection problem is concerned with deciding which of two models provides equivalent fits to the data generating process. Since model fit is computed using only a finite sample from the data generating process, estimates of model fit need to take into account the effects of sampling error.

Let $\mathbf{x}_1, \ldots, \mathbf{x}_n$ denote the observed data which is presumed to be a realization of a sequence of i.i.d. $d$-dimensional random vectors $\tilde{\mathbf{x}}_1, \ldots, \tilde{\mathbf{x}}_n$ with common density $p_e : \mathcal{R}^d \rightarrow [0, \infty)$.

Let the fit of model $\mathcal{M}_A$ to $\tilde{\mathbf{x}}_1, \ldots, \tilde{\mathbf{x}}_n$ be specified by the empirical risk function:

$$\hat{\ell}_n^A\left(\hat{\boldsymbol{\theta}}_n^A\right) = (1/n)\sum_{i=1}^{n} c^A\left(\tilde{\mathbf{x}}_i, \hat{\boldsymbol{\theta}}_n^A\right)$$

where $\hat{\boldsymbol{\theta}}_n^A$ is strict local minimizer of $\hat{l}_n^A \equiv (1/n)\sum_{i=1}^n c^A(\tilde{\mathbf{x}}_i, \cdot)$.

Let the fit of model $\mathcal{M}_B$ to $\tilde{\mathbf{x}}_1, \ldots, \tilde{\mathbf{x}}_n$ be specified by the empirical risk function:

$$\hat{\ell}_n^B\left(\hat{\boldsymbol{\theta}}_n^B\right) = (1/n)\sum_{i=1}^{n} c^B\left(\tilde{\mathbf{x}}_i, \hat{\boldsymbol{\theta}}_n^B\right)$$

where $\hat{\boldsymbol{\theta}}_n^B$ is a strict local minimizer of $\hat{l}_n^B \equiv (1/n)\sum_{i=1}^n c^B(\tilde{\mathbf{x}}_i, \cdot)$.

Show how to use a nonparametric bootstrap methodology to estimate the standard error of the difference in estimated model fits of two models to a common data generating process. That is, explain how to estimate the standard error of the model selection statistic:

$$\Upsilon_n(\tilde{\mathbf{x}}_1, \ldots, \tilde{\mathbf{x}}_n) = \hat{\ell}_n^A\left(\hat{\boldsymbol{\theta}}_n^A\right) - \hat{\ell}_n^B\left(\hat{\boldsymbol{\theta}}_n^B\right).$$

**Solution.** The following nonparametric bootstrap methodology may be used to estimate the standard error of a model statistic.

- *Step 1: Initialization.* Let $k = 0$.
- *Step 2 : Sample with replacement.* Let $k = k + 1$. Sample with replacement $n$ times from $\mathbf{x}_1, \ldots, \mathbf{x}_n$ to generate the $k$th nonparametric bootstrap sample $\mathbf{x}_1^{*k}, \ldots, \mathbf{x}_n^{*k}$.
- *Step 3: Estimate the expected value of the model selection statistic for $m$ bootstrap samples.* Let $\hat{\boldsymbol{\theta}}_n^{A,*k}$ and $\hat{\boldsymbol{\theta}}_n^{B,*k}$ be strict local minimizers of $\hat{\ell}_n^{A,*k}(\cdot) \equiv (1/n)\sum_{i=1}^n c^A(\mathbf{x}_i^{*k}, \cdot)$ and $\hat{\ell}_n^{B,*k}(\cdot) \equiv (1/n)\sum_{i=1}^n c^B(\mathbf{x}_i^{*k}, \cdot)$ over their respective parameter spaces. Let

  $$\delta_n^{*k} \equiv \ell_n^{A,*k}\left(\hat{\boldsymbol{\theta}}_n^{A,*k}\right) - \ell_n^{B,*k}\left(\hat{\boldsymbol{\theta}}_n^{B,*k}\right).$$

  Let

  $$\bar{\delta}_{n,m} \equiv \frac{1}{m}\sum_{k=1}^{m} \delta_n^{*k}.$$

- *Step 4: Estimate the sampling error for $m$ bootstrap samples.*

  $$\hat{\sigma}_{n,m} \equiv \sqrt{\frac{1}{m-1}\sum_{k=1}^{m}\left(\delta_n^{*k} - \bar{\delta}_{n,m}\right)^2}.$$

- *Step 5: Assess if the number of bootstrap data sets $m$ is sufficiently large.* The Monte Carlo nonparametric bootstrap simulation standard error of $\bar{\delta}_{n,m}$ is estimated using: $\hat{\sigma}_{n,m}/\sqrt{m}$. If $\hat{\sigma}_{n,m}/\sqrt{m}$ is sufficiently small, then the number of bootstrap data sets, $m$, is large enough to terminate the algorithm; otherwise go to Step 2.

△

## Exercises

14.2-1. *Detection of Misspecified Models.* A researcher has developed a procedure for the detection of model misspecification. The procedure will be evaluated with respect to a particular parametric probability model specification

$$\mathcal{M} \equiv \{p(\mathbf{x}|\boldsymbol{\theta}) : \boldsymbol{\theta} \in \Theta\}$$

and particular data set $\mathcal{D}_n \equiv \{\mathbf{x}_1, \ldots, \mathbf{x}_n\}$. Show how a parametric bootstrap methodology can be used to estimate with respect to $\mathcal{M}$ and $\mathcal{D}_n$ both: (1) the probability of incorrectly identifying a misspecified model as correctly specified, and (2) the probability of incorrectly identifying a correctly specified model as misspecified.

14.2-2. *Detection of Relevant Input Features.* The set $S$ of all connections originating from an input feature defines a vector-valued *input feature statistic* $\hat{\boldsymbol{\Upsilon}}_n$ which is a subset of the elements of $\hat{\boldsymbol{\theta}}_n$. The mean and standard deviation of $\hat{\boldsymbol{\Upsilon}}_n$ can then be used to estimate the magnitude of the connection strengths from the input feature and their respective sampling errors. If the sampling error is large relative to the magnitude of the connection strengths, this is an indication that the information provided by the input feature is not contributing in a systematic manner to the predictive performance of the learning given the current amount of training data. Provide specific details on how the sampling distribution of $\hat{\boldsymbol{\Upsilon}}_n$ may be used to calculate the minimum mean-square estimator and sampling error of $\hat{\boldsymbol{\Upsilon}}_n$ using the nonparametric bootstrap method.

14.2-3. *Model-Based Comparison of Statistical Environments.* Some parameter values of a learning machine might be equally effective for different statistical environments, while other parameter values might not be environment-invariant and must be re-estimated for different statistical environments. Let $\hat{\boldsymbol{\theta}}_1$ be the parameter estimates of a learning machine $\mathcal{M}$ which extracts statistical regularities from environment $E_1$. Let $\hat{\boldsymbol{\theta}}_2$ be the parameter estimates of a learning machine $\mathcal{M}$ which extracts statistical regularities from environment $E_2$. Let the *Between-Environments Statistic* $\hat{\Upsilon}_n \equiv \hat{\boldsymbol{\theta}}_1 - \hat{\boldsymbol{\theta}}_2$. Show how to use the nonparametric bootstrap method to estimate the minimum mean-square estimator and sampling error of $\hat{\Upsilon}_n$.

14.2-4. *Classification Error Performance Statistics.* The *accuracy* is defined as the number of times the classifier correctly classifies an input pattern divided by the total number of times an input pattern was presented to the classifier. If the accuracy for data set $\mathcal{D}_n^j$ is denoted as $a_n^j$, then a minimum mean-square estimate of classification accuracy and its respective sampling error may be obtained by computing respectively the sample mean and sample standard deviation of the bootstrap estimators $a_n^1, \ldots, a_n^m$ or their log-transformed counterparts $-\log(a_n^1), \ldots, -\log(a_n^m)$.

In addition, to computing accuracy, one often reports the *recall* (R) of the classifier which is the number of times the classifier correctly decides an input pattern should be assigned to the target category divided by the total number of times an input pattern from the target category was presented to the classifier. Another critical statistic is the *precision* (P) of the classifier, which is the number of times the classifier correctly decides an input pattern should be assigned to the target category divided by the total number of times the classifier decided the input pattern should be assigned to the target category. Show how to compute minimum mean-square estimates and their respective sampling errors for the accuracy (A), precision (P), and recall (R) of a classifier using the nonparametric bootstrap method.

14.2-5. *Empirical Investigation of Bootstrap Estimator Behavior.* Estimate the covariance matrix of the parameter estimates of a regression model using a bootstrap methodology. Then, systematically vary the number of data sets ($m$) and the sample size ($n$) in order to empirically study how effectively the large-sample conclusions of the Monte Carlo Bootstrap Estimator Consistency Theorem hold in an actual modeling problem.

## 14.3 Further Readings

### General Discussions of Bootstrap Methods

Discussions of bootstrap applications can be found in Efron (1982), Efron and Tibshirani (1993), and Davison and Hinkley (1997). More comprehensive discussions of mathematics underlying the bootstrap can be found in Bickel and Freedman (1981), Singh (1981), Manoukian (1986, pp. 50-54), van der Vaart (1998, Chapter 23), Shao and Tu (1995), Horowitz (2001), and Romano and Shaikh (2012). Davidson et al. (2003) reviews important developments in bootstrap theory not discussed in this chapter.

### Bootstrap Methods for Time-Series Data

Methods for using the bootstrap methodology for time-series data may be found in Politis and Romano (1994), Politis and White (2004), Patton et al. (2009), and Shao (2010).

### Subsampling Bootstrap Methods

The standard bootstrap method is based upon generating bootstrap data sets each consisting of the same number of data records as the original sample. This standard method was described in this chapter. Methods for using fewer than the original number of data records in the original data sample are called subsampling methods and are discussed in Politis and Romano (1994), Politis, Romano, and Wolf (2001), Bickel et al. (1997), Davison et al. (2003).

### Bootstrap Sampling without Replacement

The bootstrap methods presented here assume bootstrap sampling with replacement. Discussions of methods for bootstrap sampling without replacement may be found in Chao and Lo (1985), Bickel et al. (1997), and Davison et al. (2003).

# 15

# *Analytic Formulas for Evaluating Generalization*

**Learning Objectives**

- Estimate analytic formula generalization error.
- Derive model prediction confidence intervals.
- Derive statistical tests for model comparison decisions.

Different data sets corresponding to realizations of the same random sample will yield different empirical risk estimates for a common empirical risk function. The empirical risk estimate, $\hat{\boldsymbol{\theta}}_n$, is a strict local minimizer of the empirical risk function $\hat{\ell}_n$ which, in turn, is functionally dependent upon a realization, $\mathcal{D}_n$, of a data sample $\tilde{\mathcal{D}}_n$. Therefore, $\hat{\boldsymbol{\theta}}_n$ is a statistic which is functionally dependent upon the data $\tilde{\mathcal{D}}_n$ as well. The probability distribution of $\hat{\boldsymbol{\theta}}_n$ may then be characterized for large sample sizes using the nonparametric bootstrap methodology described in Chapter 14. The simulation methods introduced in Chapter 14 provided practical methods for evaluating model generalization performance by characterizing the sampling distribution of $\hat{\boldsymbol{\theta}}_n$.

In this chapter, methods for evaluating model generalization performance are introduced for characterizing the sampling distribution of $\hat{\boldsymbol{\theta}}_n$ which do not require the simulation methods used in Chapter 14. The key theoretical result shows that $\hat{\boldsymbol{\theta}}_n$ has a large sample multivariate Gaussian distribution with mean $\boldsymbol{\theta}^*$ and covariance matrix $(1/n)\mathbf{C}^*$ where the positive definite matrix $\mathbf{C}^*$ is approximated using the first and second derivatives of the empirical risk function. Notice that since $\mathbf{C}^*$ is divided by the sample size $n$, that the sampling error represented by the magnitude of $(1/n)\mathbf{C}^*$ becomes smaller and approaches zero as the sample size $n$ increases. As discussed in previous chapters, the sampling error characterizes how variations in the content of the data set for a fixed sample size $n$ are reflected in variations in the value of $\hat{\boldsymbol{\theta}}_n$.

## 15.1 Assumptions for Asymptotic Analysis

The following Empirical Risk Regularity Assumptions will be used frequently throughout this chapter and the next chapter.

**Definition 15.1.1** (Empirical Risk Regularity Assumptions).

- **A1 : I.I.D. Data Generating Process.** Let the data sample $\tilde{\mathcal{D}}_n \equiv [\tilde{\mathbf{x}}_1, \ldots, \tilde{\mathbf{x}}_n]$ be a sequence of independent and identically distributed $d$-dimensional random vectors with common Radon-Nikodým density $p_e : \mathcal{R}^d \to [0, \infty)$ defined with respect to support specification measure $\nu$.

- **A2: Smooth Loss Function.** Let $c : \mathcal{R}^d \times \mathcal{R}^q \to \mathcal{R}$ be a twice continuously differentiable random function on $\mathcal{R}^q$.
- **A3: Expectations Are Finite.** Let $\mathbf{g}(\mathbf{x}, \boldsymbol{\theta}) = [dc(\mathbf{x}, \boldsymbol{\theta})/d\boldsymbol{\theta}]^T$. Assume the functions $c$, $\mathbf{g}\mathbf{g}^T$, and $d^2c/d\boldsymbol{\theta}^2$ are dominated by integrable functions on $\mathcal{R}^q$ with respect to $p_e$.
- **A4: Smooth Penalty Function with Sufficiently Fast Vanishing Magnitude.** Let $k_n : \mathcal{R}^{d \times n} \times \mathcal{R}^q \to \mathcal{R}$ be a continuously differentiable random function for all $n \in \mathbb{N}$. Let $\hat{k}_n(\boldsymbol{\theta}) \equiv k_n(\tilde{\mathcal{D}}_n, \boldsymbol{\theta})$. Assume
$$\hat{k}_n \to 0 \text{ and } \sqrt{n}\left|\frac{d\hat{k}_n}{d\boldsymbol{\theta}}\right| \to 0$$
uniformly on $\mathcal{R}^q$ with probability one as $n \to \infty$.
- **A5: Construct a Local Parameter Space $\Theta$ Surrounding a Local Minimizer.** Let $\boldsymbol{\theta}^*$ be a strict local minimizer of
$$\ell(\boldsymbol{\theta}) = \int c(\mathbf{x}, \boldsymbol{\theta}) p_e(\mathbf{x}) d\nu(\mathbf{x})$$
on $\mathcal{R}^q$. Assume $\Theta$ is a closed, bounded, and convex subset of $\mathcal{R}^q$ such that $\boldsymbol{\theta}^*$ is the unique global minimizer of the restriction of $\ell$ to $\Theta$, and additionally $\boldsymbol{\theta}^*$ is in the interior of $\Theta$.
- **A6: Parameter Estimation Procedure.** Assume $\hat{\boldsymbol{\theta}}_n \equiv \arg\min \hat{\ell}(\boldsymbol{\theta})$ on $\Theta$ where
$$\hat{\ell}_n(\boldsymbol{\theta}) = \hat{k}_n(\boldsymbol{\theta}) + (1/n)\sum_{i=1}^{n} c(\tilde{\mathbf{x}}_i, \boldsymbol{\theta})$$
with probability one for all $n \in \mathbb{N}$.
- **A7: Hessian Positive Definite at Minimizer.** Let $\mathbf{A}$ be the Hessian of $\ell$. Assume $\mathbf{A}^* \equiv \mathbf{A}(\boldsymbol{\theta}^*)$ is positive definite.
- **A8: Outer Product Gradient Positive Definite at Minimizer.** Let the matrix-valued function $\mathbf{B} : \mathcal{R}^q \to \mathcal{R}^{q \times q}$ be defined such that for all $\boldsymbol{\theta} \in \Theta$:
$$\mathbf{B}(\boldsymbol{\theta}) = \int \mathbf{g}(\mathbf{x}, \boldsymbol{\theta})\mathbf{g}(\mathbf{x}, \boldsymbol{\theta})^T p_e(\mathbf{x}) d\nu(\mathbf{x}).$$
Assume $\mathbf{B}^* \equiv \mathbf{B}(\boldsymbol{\theta}^*)$ is positive definite.

□

Let $\tilde{\mathbf{g}}_i^T \equiv \mathbf{g}(\tilde{\mathbf{x}}_i, \hat{\boldsymbol{\theta}}_n)$. Let the notation

$$\hat{\mathbf{A}}_n \equiv (1/n)\sum_{i=1}^{n} \nabla_{\theta}^2 c(\tilde{\mathbf{x}}_i, \hat{\boldsymbol{\theta}}_n) \tag{15.1}$$

$$\hat{\mathbf{B}}_n \equiv (1/n)\sum_{i=1}^{n} \tilde{\mathbf{g}}_i \tilde{\mathbf{g}}_i^T, \tag{15.2}$$

and

$$\hat{\mathbf{C}}_n \equiv \hat{\mathbf{A}}_n^{-1}\hat{\mathbf{B}}_n\hat{\mathbf{A}}_n^{-1}. \tag{15.3}$$

Assumption **A1** states that the training data set $\mathcal{D}_n \equiv [\mathbf{x}_1, \ldots, \mathbf{x}_n]$ is assumed to be a particular realization of a sequence of independent and identically distributed $d$-dimensional random vectors $\tilde{\mathbf{x}}_1, \ldots, \tilde{\mathbf{x}}_{t-1}, \tilde{\mathbf{x}}_{t+1}, \ldots, \tilde{\mathbf{x}}_n$ with a common Radon-Nikodým density $p_e : \mathcal{R}^d \rightarrow [0, \infty)$.

This assumption must either be empirically checked or validated through knowledge of the data generating process. Consider the problem of verifying whether a sequence of coin flips of a fair coin are independent and identically distributed. Philosophically, one might argue that since it is the same coin which is flipped in the same manner and the outcome of one coin flip doesn't influence the outcome of the subsequent coin flip, it follows that the sequence of coin flips are independent and identically distributed. However, it is possible that the coin is not flipped in exactly the same manner each time corresponding to a violation of the identically distributed condition. And if the variation of coin flip is functionally dependent upon how the coin was previously flipped, this would correspond to a violation of the assumption of statistical independence. In practice, the assumption of stationarity can be checked by dividing a data set into multiple parts and examining if parameter values estimated for one part of the data set are comparable in value to parameter values estimated for another part of the data set. The assumption of independence can be checked by examining if the probability distribution of $\tilde{\mathbf{x}}_t$ is not functionally dependent upon $\tilde{\mathbf{x}}_1, \ldots, \tilde{\mathbf{x}}_{t-1}, \tilde{\mathbf{x}}_{t+1}, \ldots, \tilde{\mathbf{x}}_n$.

Assumption **A2** assumes the loss function $c : \mathcal{R}^d \times \Theta \rightarrow \mathcal{R}$ is twice continuously differentiable on the parameter space $\Theta$. In situations where a function $c$ is required that does not satisfy this condition, it is usually possible (in practice) to find a "smooth approximation". For example, the function $\phi : \mathcal{R} \rightarrow \{0, 1\}$ defined such that $\phi(x) = 1$ if $x > 0$ and $\phi(x) = 0$ otherwise is not a continuous function. However, a smoothed version of $\phi$, $\ddot{\phi}_\tau$, can be defined such that $\ddot{\phi}_\tau(x) = 1/(1 + \exp(-x/\tau))$ where $\tau$ is a positive number. As $\tau \rightarrow 0$, $\ddot{\phi}_\tau \rightarrow \phi$. Other examples of smooth approximations include the use of softplus transfer functions instead of rectified linear units as discussed in Section 1.3.3.

Assumption **A3** ensures that all relevant expectations are finite. A sufficient condition for **A3** to hold is that $c(\cdot, \boldsymbol{\theta})$ is piecewise continuous for all $\boldsymbol{\theta} \in \mathcal{R}^q$, $c(\mathbf{x}, \cdot)$ is twice continuously differentiable for all $\mathbf{x} \in \mathcal{R}^d$, and the data generating process $\tilde{\mathbf{x}}_1, \tilde{\mathbf{x}}_2, \ldots$ is a bounded stochastic sequence. The assumption of a bounded stochastic sequence is satisfied, for example, if $\tilde{\mathbf{x}}_t$ is a discrete random vector taking on at most a finite number of values.

Assumption **A4** ensures that the effects of the penalty term may be ignored when the sample size is sufficiently large. If no penalty term is present (i.e., $\hat{k}_n(\boldsymbol{\theta}) = 0$) then this assumption is automatically satisfied. Penalty terms such as $\hat{k}_n(\boldsymbol{\theta}) = 1/n$ and $\hat{k}_n(\boldsymbol{\theta}) = (1/n)\log n$ satisfy the conditions of **A4** since $\hat{k}_n \rightarrow 0$ and $\nabla_{\boldsymbol{\theta}} \hat{k}_n = \mathbf{0}^T$. If the parameter space is closed and bounded, then an $L_2$ regularization term such as $\hat{k}_n(\boldsymbol{\theta}) = (1/n)|\boldsymbol{\theta}|^2$ also satisfies **A4**. A penalty term such as $\hat{k}_n(\boldsymbol{\theta}) = n^{-1}|\boldsymbol{\theta}|_1$ does not satisfy **A4** because it is not continuously differentiable. However, when the parameter space is closed and bounded, then a smoothed $L_1$ penalty term such as $\hat{k}_n(\boldsymbol{\theta}) = n^{-1}\sqrt{|\boldsymbol{\theta}|^2 + \epsilon}$ satisfies **A4**.

Assumption **A5** is an assumption which allows the theory to be applied to multimodal nonconvex risk functions with multiple strict local minimizers. In practice, a particular strict local minimizer $\boldsymbol{\theta}^*$ of $\ell$ is obtained. Next, **A5** is satisfied by constructing a closed, bounded, and convex subset of $\mathcal{R}^q$, $\Theta$, such that: (1) $\boldsymbol{\theta}^*$ is the only local minimizer in $\Theta$, and (2) $\boldsymbol{\theta}^*$ is in the interior of $\Theta$.

Assumption **A6** states that the parameter estimate $\hat{\boldsymbol{\theta}}_n$ is a minimizer of a penalized empirical risk function over a restricted region of the parameter space. Note that, in practice, the parameter estimation is unconstrained and a strict local minimizer $\hat{\boldsymbol{\theta}}_n$ is obtained.

Assumption **A7** states that the Hessian of the risk function $\ell$ evaluated at $\boldsymbol{\theta}^*$, $\mathbf{A}^*$, is positive definite. In practice, this assumption cannot be directly verified since $\boldsymbol{\theta}^*$ and $p_e$ are

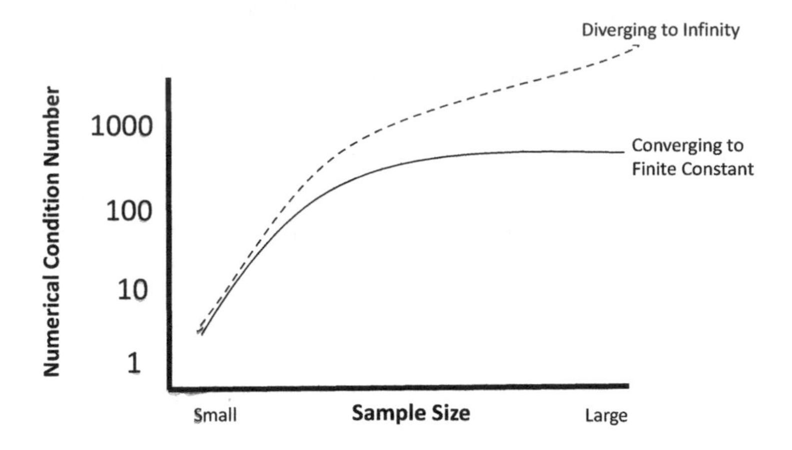


**FIGURE 15.1**
Two examples of how the condition number as a function of sample size may change. If the condition number is converging to a finite number, this is empirical evidence consistent with Assumption A7. If the condition number is diverging to infinity, this indicates a violation of Assumption A7.

not directly observable. However, $\mathbf{A}^*$ may be estimated by $\hat{\mathbf{A}}_n$ which is the Hessian of the empirical risk function evaluated at the parameter estimates $\hat{\boldsymbol{\theta}}_n$. Note that if the magnitude of the gradient of $\ell$ evaluated at $\boldsymbol{\theta}^*$ is zero and **A7** holds, then this is a necessary condition for $\boldsymbol{\theta}^*$ to be a strict local minimizer (see Theorem 5.3.3).

The methods of Example 5.3.6 can be used to investigate if $\hat{\mathbf{A}}_n$ is converging to a positive definite matrix. It is usually helpful to report the condition number (i.e., the ratio of the largest to smallest eigenvalue of $\hat{\mathbf{A}}_n$) as a rough measure of whether **A7** is satisfied (see Figure 15.1). A condition number close to unity is the best while very large condition numbers suggest **A7** may be violated. In actual applications, a good condition number will vary from one field of application to another. Acceptable condition numbers associated for good-quality data sets with small sampling error may be as low as 100 while acceptable condition numbers for poor-quality data sets with large sampling error could be very large (e.g., $10^{12}$).

Assumption **A8** is a requirement that the matrix $\mathbf{B}^*$ is positive definite. The matrix $\mathbf{B}^*$ is the expected value of an outer product constructed using the gradient of the random loss function evaluated at $\boldsymbol{\theta}^*$. The methods discussed in Example 5.3.6, which were used to empirically investigate if **A7** holds, are also used to investigate if **A8** holds.

Assumption **A8** is required for the Multivariate Central Limit Theorem to hold. A necessary but not sufficient condition for **A8** to hold is that the number of training stimuli is greater than or equal to the number of free parameters.

**Example 15.1.1** (Linear Regression Regularity Assumptions). Assume that the training data $\mathbf{x}_1, \ldots, \mathbf{x}_n$ is a realization of a bounded stochastic sequence of i.i.d. observations with common Radon-Nikodým density defined with respect to support specification measure $\nu$. In addition, assume that a training stimulus $\mathbf{x}_i \equiv (\mathbf{s}_i, y_i)$ specifies the desired response $y_i$ for a given input pattern vector $\mathbf{s}_i$ for $i = 1, \ldots, n$. Assume the goal of the parameter estimation process is to compute quasi-maximum likelihood estimates of a probability model which assumes that

$$p(y|\mathbf{s};\boldsymbol{\theta}) = (\sqrt{2\pi})^{-1}\exp\left(-(0.5)|y - \ddot{y}(\mathbf{s},\boldsymbol{\theta})|^2\right)$$

where $\ddot{y}(\mathbf{s},\boldsymbol{\theta}) = \boldsymbol{\theta}^T\mathbf{s}$. Assume the density of $\tilde{\mathbf{s}}$ is not functionally dependent upon the parameter vector $\boldsymbol{\theta}$ so that the density of $\tilde{\mathbf{s}}$ is specified by the probability density function $p_s$. Construct the Empirical Risk Function for this modeling problem and then discuss the conditions under which the Empirical Risk Regularity Conditions hold for this probability model.

**Solution.** Assumption **A1** holds by assumption. Since the goal is quasi-maximum likelihood estimation, the loss function

$$c(\mathbf{x};\boldsymbol{\theta}) = -\log p(y,\mathbf{s}|\boldsymbol{\theta}) = -\log p_s(\mathbf{s}) - \log p(y|\mathbf{s};\boldsymbol{\theta}).$$

Note that

$$\mathbf{g}(\mathbf{x};\boldsymbol{\theta}) = -(y - \ddot{y}(\mathbf{s},\boldsymbol{\theta}))\mathbf{s}^T$$

and the Hessian of $c$, $\mathbf{H}(\mathbf{x};\boldsymbol{\theta})$, is defined such that:

$$\mathbf{H}(\mathbf{x};\boldsymbol{\theta}) = \mathbf{s}\mathbf{s}^T.$$

Since $\mathbf{g}$ and $\mathbf{H}$ are continuous, it follows that **A2** holds.

Since $c$ is a continuous function in its first argument and $\{\tilde{\mathbf{x}}_t\}$ is a bounded stochastic sequence, it follows that **A3** holds.

Assumption **A4** holds because $\ddot{k}_n(\boldsymbol{\theta}) = 0$ in this case. Since the Hessian

$$\mathbf{A}(\boldsymbol{\theta}) = \int \mathbf{H}(\mathbf{x};\boldsymbol{\theta})p_e(\mathbf{x})d\nu(\mathbf{x}) - \int \mathbf{s}\mathbf{s}^T p_e(\mathbf{s})d\nu(\mathbf{s})$$

this implies that for any vector $\boldsymbol{\theta} \in \Theta$:

$$\boldsymbol{\theta}^T\mathbf{A}(\boldsymbol{\theta})\boldsymbol{\theta} = \int |\mathbf{s}^T\boldsymbol{\theta}|^2 p_e(\mathbf{s})d\nu(\mathbf{s}) \geq 0$$

and hence the Hessian $\mathbf{A}(\boldsymbol{\theta})$ is positive semidefinite on $\mathcal{R}^q$. By Theorem 5.3.5, this implies that the expected risk function $\ell$ is convex on the entire parameter space $\mathcal{R}^q$. Thus, a strict local minimizer of $\ell$ is the unique global minimizer of $\ell$ (see Theorem 5.3.3 and Theorem 5.3.7). Therefore, if a strict local minimizer of $\ell$ exists then **A5** is satisfied.

In order to empirically check **A5**, one typically checks if $\hat{\boldsymbol{\theta}}_n$ is a critical point of the empirical risk function. If the infinity norm of the gradient of $\hat{\ell}_n(\boldsymbol{\theta}) = (1/n)\sum_{i=1}^n c(\tilde{\mathbf{x}}_i,\boldsymbol{\theta})$ evaluated at $\hat{\boldsymbol{\theta}}_n$ is sufficiently small, then this is evidence that $\hat{\boldsymbol{\theta}}_n$ is a critical point of $\hat{\ell}_n$ and that a critical point of $\ell$ exists.

Assumption **A7** is empirically checked by examining if $\hat{\mathbf{A}}_n$ is positive definite at $\hat{\boldsymbol{\theta}}_n$. Assumption **A8** is empirically checked by examining if $\hat{\mathbf{B}}_n$ is positive definite at $\hat{\boldsymbol{\theta}}_n$. As discussed in Example 5.3.6, when $\hat{\mathbf{A}}_n$ and $\hat{\mathbf{B}}_n$ are respective estimators of $\mathbf{A}^*$ and $\mathbf{B}^*$ and the goal is to check that $\mathbf{A}^*$ and $\mathbf{B}^*$ are invertible, it is important to check that $\hat{\mathbf{A}}_n$ and $\hat{\mathbf{B}}_n$ are converging to invertible matrices. △

**Example 15.1.2** (Nonlinear Regression Regularity Assumptions). Assume that the training data $\mathbf{x}_1,\ldots,\mathbf{x}_n$ is a realization of a bounded stochastic sequence of i.i.d. observations with common Radon-Nikodým density defined with respect to support specification measure $\nu$. In addition, assume that a training stimulus $\mathbf{x}_i \equiv (\mathbf{s}_i, y_i, z_i)$ specifies the desired response $y_i$ for a given input pattern vector $\mathbf{s}_i$ for $i = 1,\ldots,n$. Assume, in addition, that $z_i = 0$ for $i = 1,\ldots,n$. The addition of the auxiliary variable $\tilde{z}_i$ is a data augmentation or recoding strategy which will be used to enforce a regularization term which does not vanish as the sample size becomes large. Assume a probability model

$$p(y|\mathbf{s};\boldsymbol{\theta}) = (\sqrt{2\pi})^{-1} \exp\left(-(0.5)|y - \ddot{y}(\mathbf{s},\boldsymbol{\theta})|^2\right)$$

where $\ddot{y}(\mathbf{s},\boldsymbol{\theta})$ is a nonlinear differentiable function of $\boldsymbol{\theta}$ and a piecewise continuous function of $\mathbf{s}$. For example, $\ddot{y}(\mathbf{s},\boldsymbol{\theta})$ could be the state of an output unit for a multi-layer feedforward perceptron network. Assume the density of $\tilde{\mathbf{s}}$ is not functionally dependent upon the parameter vector $\boldsymbol{\theta}$ so that the density of $\tilde{\mathbf{s}}$ is specified by the probability density function $p_s$. Assume

$$p(z|\boldsymbol{\theta}) = \frac{\exp(-(z - |\boldsymbol{\theta}|)^2/(2\sigma_z^2))}{\sigma_z\sqrt{2\pi}}$$

where the inverse of $2\sigma_z^2$ is called the regularization constant $\lambda$. Given these probabilistic modeling assumptions, construct an empirical risk function which can be interpreted as a negative normalized log-likelihood function. Then, discuss the conditions under which the Empirical Risk Regularity Conditions hold for this probability model and specifically explain what conditions are satisfied by definition and which conditions can be empirically examined.

**Solution.** Assumption **A1** holds by assumption. Since the goal is quasi-maximum likelihood estimation, the loss function

$$c(\mathbf{x};\boldsymbol{\theta}) = -\log p(y,\mathbf{s},z|\boldsymbol{\theta}) = -\log p(\mathbf{s}) - \log p(y|\mathbf{s};\boldsymbol{\theta}) - \log p(z|\boldsymbol{\theta}).$$

Let $\mathbf{g}_y$ and $\mathbf{H}_y$ denote the gradient and Hessian of $-\log p(y|\mathbf{s};\boldsymbol{\theta})$ respectively. Then, the gradient,

$$\mathbf{g}(\mathbf{x};\boldsymbol{\theta}) = \mathbf{g}_y + 2\lambda\boldsymbol{\theta},$$

and the Hessian of $c$, $\mathbf{H}(\mathbf{x};\boldsymbol{\theta})$, is

$$\mathbf{H}(\mathbf{x};\boldsymbol{\theta}) = \mathbf{H}_y + 2\lambda\mathbf{I}_q. \tag{15.4}$$

Since $\mathbf{g}$ and $\mathbf{H}$ are continuous, it follows that **A2** holds.

Since $c$ is continuous in its first argument and $\{\tilde{\mathbf{x}}_t\}$ is a bounded stochastic sequence, it follows that **A3** holds.

Assumption **A4** holds since $\tilde{k}_n(\boldsymbol{\theta}) = 0$.

In nonlinear machine learning applications such as multilayer feedforward perceptrons and mixture models, the Hessian $\mathbf{A}(\boldsymbol{\theta})$ is not positive semidefinite on the entire parameter space because the objective function is not convex on the parameter space. However, if the parameter space $\Theta$ is chosen to be a small neighborhood of a strict local minimizer $\hat{\boldsymbol{\theta}}_n$, then this would permit **A5** to be satisfied. Also note that if $\lambda$ is chosen to be a larger positive number, this can have the effect of forcing the Hessian $\mathbf{A}$ to be positive definite on $\Theta$ so that **A7** holds. However, increasing the value of $\lambda$ too much could sacrifice the quality of model fit for the benefit of a locally unique solution. Both Assumption **A7** and **A8** are empirically checked by examining the condition numbers of $\hat{\mathbf{A}}_n$ and $\hat{\mathbf{B}}_n$ respectively. △

## Exercises

15.1-1. Discuss the conditions under which the Empirical Risk Regularity Assumptions hold for a logistic regression probability model whose parameters are estimated using quasi-maximum likelihood estimation. Specifically explain what conditions are satisfied by definition and which conditions can be empirically examined.

15.1-2. Consider a feedforward perceptron architecture where the predicted response $\ddot{y}$ is a function of an input pattern vector $\mathbf{s}$ and a parameter vector $\boldsymbol{\theta}$ such that:

$$\ddot{y}(\mathbf{s}, \boldsymbol{\theta}) = \sum_{h=1}^{H} w_h \phi(\mathbf{v}_h^T \mathbf{s})$$

where $\boldsymbol{\theta} \equiv [w_1, \ldots, w_h, \mathbf{v}_1^T, \ldots, \mathbf{v}_h^T]^T$ and $\phi$ is defined such that $\phi(u) = 1/(1 + \exp(-u))$. Assume that the expected loss function, $\ell : \mathcal{R}^q \rightarrow \mathcal{R}$ is defined such that:

$$\ell(\boldsymbol{\theta}) = \lambda |\boldsymbol{\theta}|^2 + \sum_{k=1}^{M} p_k \left(y^k - \ddot{y}(\mathbf{s}^k, \boldsymbol{\theta})\right)^2 \tag{15.5}$$

where it is assumed the statistical environment presents the pattern vector $(\mathbf{s}^k, y^k)$ with probability $p_k$.

Note that if $[w_1^*, \ldots, w_H^*, (\mathbf{v}_1^*)^T, \ldots, (\mathbf{v}_H^*)^T]$ is a global minimizer of $\ell$ then it immediately follows that this is not a unique global minimizer since $[w_2^*, w_1^* \ldots, w_H^*, (\mathbf{v}_2^*)^T, (\mathbf{v}_1^*)^T, \ldots, (\mathbf{v}_H^*)^T]$ is also a global minimizer. Indeed, $H!$ such global minimizers can be identified. Explain why this assumption does not violate any of the Empirical Risk Function Regularity Assumptions when the parameter space is chosen appropriately.

Now consider a more serious problem where a global minimizer is found which has the property that $w_k^* = 0$ for some $k$ in $\{1, \ldots, H\}$. By examining different choices of $\mathbf{v}_k^*$ in this case, show that such a global minimizer cannot possibly be a strict local minimizer for the case where the $L_2$ regularization term is absent (i.e., $\lambda = 0$). Now show that for every positive $\lambda$ when $w_k^* = 0$ that the choice of $\mathbf{v}_k^*$ is uniquely determined.

Show that the Hessian of $\ell$ is equal to the sum of $2\lambda \mathbf{I}_q$ and the Hessian of the second term on the right-hand side of (15.5). Thus, the existence of the $L_2$ regularization term with a sufficiently large positive $\lambda$ ensures a critical point will be a strict local minimizer so that Assumption **A7** is satisfied.

What is the problem associated with making $\lambda$ too large?

## 15.2 Theoretical Sampling Distribution Analysis

Let $\boldsymbol{\theta} : \mathcal{R}^{d \times n} \rightarrow \mathcal{R}^q$ denote a parameter estimation procedure which maps a data set $\mathcal{D}_n$ into a $q$-dimensional parameter estimate. Chapter 14 showed how to use nonparametric simulation methods for the purpose of characterizing the asymptotic sampling distribution of the Monte Carlo Bootstrap Estimator $\Upsilon_n \equiv \boldsymbol{\theta}(\tilde{\mathcal{D}}_n)$. The following theorem shows how to obtain a similar result using a simple formula (see Figure 15.2). In particular, the following theorem provides conditions ensuring that $\boldsymbol{\theta}(\tilde{\mathcal{D}}_n)$ has an asymptotic multivariate Gaussian density centered at a strict local minimizer $\boldsymbol{\theta}^*$ of the expected value of the empirical risk function. In addition, an explicit formula for estimating the covariance matrix of this Gaussian density is provided.

**Theorem 15.2.1** (Empirical Risk Minimizer Asymptotic Distribution). *Assume the Empirical Risk Regularity Assumptions in Definition 15.1.1 hold with respect to a stochastic sequence of i.i.d. random vectors* $\tilde{\mathbf{x}}_1, \ldots, \tilde{\mathbf{x}}_n$ *with common DGP Radon-Nikodým density* $p_e : \mathcal{R}^d \rightarrow [0, \infty)$ *defined with respect to a support specification measure* $\nu$*, a loss function*

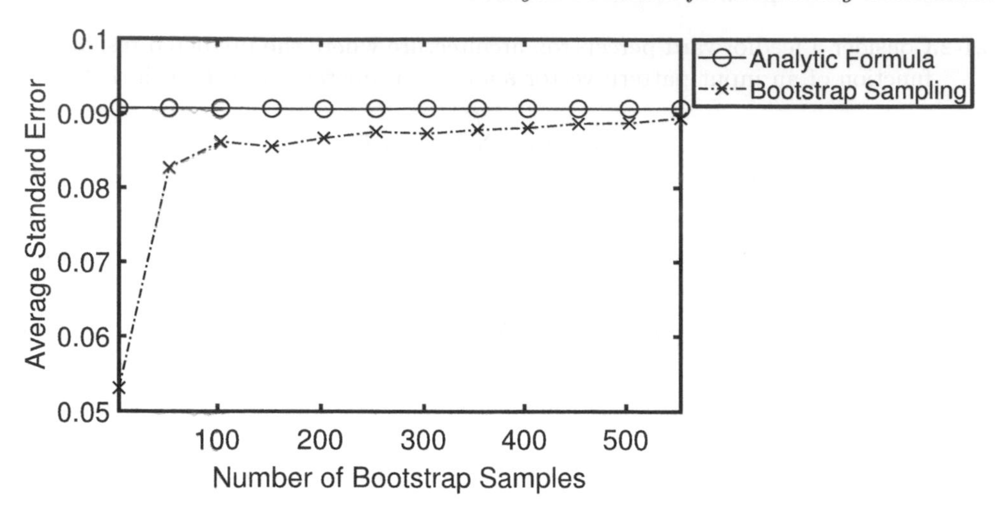


**FIGURE 15.2**
**Comparison of asymptotic formula for standard errors of the parameter estimates with a nonparametric bootstrap estimator.** Using the nonparametric bootstrap estimation algorithm described in Example 14.2.3, the sampling error of the parameter estimates of a logistic regression model for a fixed sample size $n$ was calculated for different numbers of bootstrap data sets. Each bootstrap data set consisted of $n$ data records. Using only the original training data, the low computation asymptotic formula summarized in Recipe Box 15.1 was also used to estimate the sampling error. As shown in the figure, the nonparametric bootstrap sampling error converges to the asymptotic formula sampling error (Recipe Box 15.1) as the number of bootstrap data set samples becomes large.

*$c : \mathcal{R}^d \times \mathcal{R}^q \rightarrow \mathcal{R}$, a penalty function $k_n : \mathcal{R}^{d \times n} \times \Theta \rightarrow \mathcal{R}$, restricted parameter space $\Theta \subseteq \mathcal{R}^q$, empirical risk function minimizer $\hat{\boldsymbol{\theta}}_n$ of*

$$\hat{\ell}_n(\boldsymbol{\theta}) \equiv \hat{k}_n(\boldsymbol{\theta}) + (1/n) \sum_{i=1}^{n} c(\tilde{\mathbf{x}}_i, \boldsymbol{\theta})$$

*on $\Theta$, and risk function minimizer $\boldsymbol{\theta}^*$ of*

$$\ell(\boldsymbol{\theta}) = \int c(\mathbf{x}, \boldsymbol{\theta}) p_e(\mathbf{x}) d\nu(\mathbf{x})$$

*on $\Theta$. Let $\mathbf{A}^*$ and $\mathbf{B}^*$ be defined as in **A7** and **A8** respectively in Definition 15.1.1. Let $\hat{\mathbf{A}}_n$, $\hat{\mathbf{B}}_n$, and $\hat{\mathbf{C}}_n$ be defined respectively as in (15.1), (15.2), and (15.3).*

*Then, as $n \rightarrow \infty$, $\sqrt{n}(\hat{\boldsymbol{\theta}}_n - \boldsymbol{\theta}^*)$ converges in distribution to a Gaussian random vector with mean $\mathbf{0}_q$ and covariance matrix $\mathbf{C}^* = [\mathbf{A}^*]^{-1}\mathbf{B}^*[\mathbf{A}^*]^{-1}$. In addition, as $n \rightarrow \infty$, $\sqrt{n}[\hat{\mathbf{C}}_n]^{-1/2}(\hat{\boldsymbol{\theta}}_n - \boldsymbol{\theta}^*)$ converges in distribution to a Gaussian random vector with mean $\mathbf{0}_q$ and covariance matrix $\mathbf{I}_q$.*

*Proof.* By Assumptions **A1**, **A2**, **A3**, **A4**, **A5**, **A6**, and Theorem 13.1.2, $\hat{\boldsymbol{\theta}}_n \rightarrow \boldsymbol{\theta}^*$ as $n \rightarrow \infty$ with probability one. Let $\tilde{c}_i(\boldsymbol{\theta}) \equiv c(\tilde{\mathbf{x}}_i, \boldsymbol{\theta})$. Let $\hat{\ell}_n \equiv (1/n) \sum_{i=1}^{n} \tilde{c}_i$. Let $\tilde{k}_n \equiv k_n(\mathcal{D}_n, \cdot)$. Let $\nabla \tilde{k}_n \equiv [dk_n(\tilde{\mathbf{x}}_i, \boldsymbol{\theta})/d\boldsymbol{\theta}]^T$. Using the Mean Value Theorem to expand $\nabla \tilde{c}_i(\boldsymbol{\theta})$ about $\boldsymbol{\theta}^*$ and evaluate at $\hat{\boldsymbol{\theta}}_n$ to obtain:

$$\nabla \hat{\ell}_n(\hat{\boldsymbol{\theta}}_n) = (1/n) \sum_{i=1}^{n} \nabla \tilde{c}_i(\boldsymbol{\theta}^*) + \left[ (1/n) \sum_{i=1}^{n} \nabla^2 \tilde{c}_{i,n}^* \right] (\hat{\boldsymbol{\theta}}_n - \boldsymbol{\theta}^*) + \nabla \tilde{k}_n \tag{15.6}$$

where the $j$th row of $\nabla^2 \ddot{c}^*_{i,n}$ is equal to the $j$th row of $\nabla^2 \tilde{c}_i$ evaluated at some point on the chord connecting $\hat{\boldsymbol{\theta}}_n$ and $\boldsymbol{\theta}^*$, $j = 1, \ldots, d$.

Let $\tilde{\mathbf{A}}^*_{i,n} \equiv (1/n) \sum_{i=1}^n \nabla^2 \ddot{c}^*_{i,n}$. By the definition of $\hat{\boldsymbol{\theta}}_n$, $\nabla \hat{\ell}_n(\hat{\boldsymbol{\theta}}_n) = \mathbf{0}_q$ with probability one. Thus, (15.6) becomes for $n$ sufficiently large with probability one:

$$\mathbf{0}_q = (1/n) \sum_{i=1}^n \nabla \tilde{c}_i(\boldsymbol{\theta}^*) + \mathbf{A}^*_{i,n}(\hat{\boldsymbol{\theta}}_n - \boldsymbol{\theta}^*) + \nabla \tilde{k}_n. \tag{15.7}$$

By the Random Function Existence Theorem (Theorem 8.4.1) and the assumption that $d^2 c / d\boldsymbol{\theta}^2$ is dominated by an integrable function, $\mathbf{A}$ is continuous on $\Theta$ which implies $\mathbf{A}$ is continuous on an open set containing $\boldsymbol{\theta}^*$. Since by Assumption **A7** $\mathbf{A}$ is positive definite at $\boldsymbol{\theta}^*$ and $\mathbf{A}$ is continuous by **A2**, if follows that $\mathbf{A}$ is positive definite on an open set containing $\boldsymbol{\theta}^*$. This, in turn, implies that there exists an open set containing $\boldsymbol{\theta}^*$ such that $\mathbf{A}^{-1}$ is continuous on that open set.

As $n \to \infty$, $\hat{\boldsymbol{\theta}}_n \to \boldsymbol{\theta}^*$ which implies since $\mathbf{A}^{-1}$ is continuous on an open set containing $\boldsymbol{\theta}^*$ that there exists an $n$ sufficiently large such that with probability one (15.6) can be rewritten as:

$$\sqrt{n}(\hat{\boldsymbol{\theta}}_n - \boldsymbol{\theta}^*) = -\sqrt{n}(\tilde{\mathbf{A}}^*_{i,n})^{-1}(1/n) \sum_{i=1}^n \nabla \tilde{c}_i(\boldsymbol{\theta}^*) + \sqrt{n}(\tilde{\mathbf{A}}^*_{i,n})^{-1} \nabla \tilde{k}_n. \tag{15.8}$$

Since $\boldsymbol{\theta}^*$ is the unique global minimizer of $\ell(\cdot)$, the gradient of $\ell(\cdot)$, $\nabla \ell(\cdot)$ evaluated at $\boldsymbol{\theta}^*$ is a vector of zeros. Since $dc/d\boldsymbol{\theta}$ is dominated by an integrable function this allows for the interchange of integral and gradient operators so that, $\nabla \ell(\boldsymbol{\theta}) = E\{\nabla \tilde{c}_i(\boldsymbol{\theta})\}$. Note that $\nabla \tilde{c}_1(\boldsymbol{\theta}^*), \nabla \tilde{c}_2(\boldsymbol{\theta}^*), \ldots$ is a sequence of *i.i.d.* $q$-dimensional random vectors with common mean $\nabla \ell(\boldsymbol{\theta}^*)$ and common covariance matrix

$$\mathbf{B}^* \equiv E\left\{ (1/n) \sum_{i=1}^n \nabla \tilde{c}_i(\boldsymbol{\theta}^*) \left[ \sum_{j=1}^n \nabla \tilde{c}_j(\boldsymbol{\theta}^*) \right]^T \right\} \tag{15.9}$$

which may be rewritten as:

$$\mathbf{B}^* \equiv E\left\{ (1/n) \sum_{i=1}^n \nabla \tilde{c}_i(\boldsymbol{\theta}^*)[\nabla \tilde{c}_i(\boldsymbol{\theta}^*)]^T \right\} + E\left\{ (1/n) \sum_{i=1}^n \sum_{j \neq i} \nabla \tilde{c}_i(\boldsymbol{\theta}^*)[\nabla \tilde{c}_j(\boldsymbol{\theta}^*)]^T \right\}. \tag{15.10}$$

Since $\nabla \tilde{c}_1(\boldsymbol{\theta}^*), \nabla c_2(\boldsymbol{\theta}^*), \ldots$ are independent, identically distributed, with common mean $\mathbf{0}$, it follows the second term on the right-hand side of (15.10) is equal to a matrix of zeros. By assumption **A8**, $\mathbf{B}^*$ is finite.

It then immediately follows from the Multivariate Central Limit Theorem (Chapter 9) that

$$\sqrt{n}(1/n) \sum_{i=1}^n \nabla \tilde{c}_i(\boldsymbol{\theta}^*)$$

converges in distribution to a $q$-dimensional Gaussian random vector with mean zero and covariance matrix $\mathbf{B}^*$. Since $[\tilde{\mathbf{A}}^*_{i,n}]^{-1}$ converges with probability one to $[\mathbf{A}^*]^{-1}$, it follows from Slutsky's Theorem (see Theorem 9.4.5) that the right-hand side of (15.8) given by

$$[\mathbf{A}^*_{i,n}]^{-1} \sqrt{n}(1/n) \sum_{i=1}^n \nabla \tilde{c}_i(\boldsymbol{\theta}^*)$$

converges in distribution to $[\mathbf{A}^*]^{-1} \tilde{\mathbf{n}}$ where $\tilde{\mathbf{n}}$ is a Gaussian random vector with zero mean and covariance matrix $\mathbf{B}^*$. By the Linear Transformation of Gaussian Random Vectors

Theorem (see Theorem 9.4.2), it follows that $[\mathbf{A}^*]^{-1}\tilde{\mathbf{n}}$ is a Gaussian random vector with mean $\mathbf{0}_q$ and covariance matrix $[\mathbf{A}^*]^{-1}\mathbf{B}^*[\mathbf{A}^*]^{-1}$.

Thus, $\sqrt{n}(\hat{\boldsymbol{\theta}}_n - \boldsymbol{\theta}^*)$ converges in distribution to a Gaussian random vector with mean $\mathbf{0}_q$ and covariance matrix $[\mathbf{A}^*]^{-1}\mathbf{B}^*[\mathbf{A}^*]^{-1}$. Moreover, since $[\hat{\mathbf{A}}_n)]^{-1} \rightarrow [\mathbf{A}^*]^{-1}$ and $\hat{\mathbf{B}}_n \rightarrow \mathbf{B}^*$ with probability one as $n \rightarrow \infty$, it follows from Slutsky's Theorem that $\sqrt{n}[\hat{\mathbf{C}}_n]^{-1/2}\left(\hat{\boldsymbol{\theta}}_n - \boldsymbol{\theta}^*\right)$ converges in distribution to a Gaussian random vector with mean $\mathbf{0}_q$ and covariance matrix $\mathbf{I}_q$. ■

**Example 15.2.1** (Parameter Estimate Standard Errors for Logistic Regression). The purpose of this example is to verify the assumptions of the Empirical Risk Function Regularity Conditions and then provide simple formulas for the standard errors of the parameter estimates. Assume the training data $\{(\mathbf{s}_1, y_1), \ldots, (\mathbf{s}_n, y_n)\}$ is a realization of a sequence of $n$ independent and identically distributed bounded random vectors with common density $p_e$. This satisfies Assumption **A1**. The empirical risk function $\hat{\ell}_n$ for a logistic regression model is specified by the formula:

$$\hat{\ell}_n(\boldsymbol{\theta}) = (1/n)\sum_{i=1}^{n} c([\mathbf{s}_i, y_i], \boldsymbol{\theta})$$

where

$$c([\mathbf{s}_i, y_i], \boldsymbol{\theta}) = -(1/n)\left(y_i \log p_i + (1 - y_i)\log(1 - p_i)\right),$$

$$p_i \equiv \mathcal{S}\left(\boldsymbol{\theta}^T \mathbf{u}_i\right),$$

$\mathbf{u}_i = [\mathbf{s}_i^T, 1]^T$, and $\mathcal{S}$ is defined such that $\mathcal{S}(\phi) = 1/(1 + \exp(-\phi))$.

The parameter space $\Theta$ is defined as a convex, closed, and bounded set which contains a strict local minimizer, $\hat{\boldsymbol{\theta}}_n$, of the objective function for logistic regression. In order to check if $\hat{\boldsymbol{\theta}}_n$ is a strict local minimizer, check that the infinity norm of the gradient of $\hat{\ell}_n(\hat{\boldsymbol{\theta}}_n)$, $\hat{\mathbf{g}}_n$, is sufficiently small. Also check that the condition number of the Hessian of $\hat{\ell}_n(\hat{\boldsymbol{\theta}}_n)$, $\hat{\mathbf{A}}_n$, is not excessively large. Also note that since $\hat{\ell}_n$ and $\ell$ are positive semidefinite on a convex parameter space, this implies that any strict local minimizer is the unique global minimizer on the parameter space by Theorem 5.3.7.

Show that formulas for $\hat{\mathbf{g}}_n$ and $\hat{\mathbf{A}}_n$ are given respectively by:

$$\hat{\mathbf{g}}_n = -(1/n)\sum_{i=1}^{n}(y_i - p_i)\mathbf{u}_i$$

and

$$\hat{\mathbf{A}}_n = (1/n)\sum_{i=1}^{n} p_i(1 - p_i)\mathbf{u}_i\mathbf{u}_i^T.$$

Since $\hat{\mathbf{A}}_n$ is positive definite at $\hat{\boldsymbol{\theta}}_n$, this is considered empirical evidence that **A7** holds since we have from Theorem 13.1.2 that $\hat{\boldsymbol{\theta}}_n \rightarrow \boldsymbol{\theta}^*$ with probability one as $n \rightarrow \infty$.

Now show that $\hat{\mathbf{B}}_n$ is computed using the formula:

$$\hat{\mathbf{B}}_n = (1/n)\sum_{i=1}^{n}(y_i - p_i)^2\mathbf{u}_i\mathbf{u}_i^T.$$

Let the square root of the $k$th on-diagonal element of the matrix

$$\hat{\mathbf{C}}_n = [\hat{\mathbf{A}}_n]^{-1}\hat{\mathbf{B}}_n\hat{\mathbf{A}}_n^{-1}.$$

be defined as $\hat{\sigma}_k$.

## Recipe Box 15.1 Derivation of Sampling Error Formula (Theorem 15.2.1)

- Assume the loss function $c : \mathcal{R}^d \times \mathcal{R}^q \to \mathcal{R}$, empirical risk function

$$\hat{\ell}_n(\boldsymbol{\theta}) = (1/n)\sum_{i=1}^{n} c(\tilde{\mathbf{x}}_i, \boldsymbol{\theta}),$$

restricted parameter space $\Theta$, empirical risk function minimizer $\hat{\boldsymbol{\theta}}_n$ of $\hat{\ell}_n$ on $\Theta$, risk function minimizer $\boldsymbol{\theta}^*$ of $\ell$ on $\Theta$, and matrices $\mathbf{A}^*$, $\mathbf{B}^*$ are defined as in the Empirical Risk Regularity Assumptions (see Definition 15.1.1) with penalty term $\hat{k}_n(\boldsymbol{\theta}) = 0$. The random function $\hat{\mathbf{g}}_i$ denotes the gradient of $c(\tilde{\mathbf{x}}_i, \cdot)$ with respect to $\boldsymbol{\theta}$. Let the Hessian of $\hat{\ell}_n$ be denoted by $\hat{\mathbf{A}}_n : \mathcal{R}^q \to \mathcal{R}^{q \times q}$.

- Expand $\hat{\mathbf{g}}_n$ in a first-order Taylor Series about $\boldsymbol{\theta}^*$ and evaluate at the critical point $\hat{\boldsymbol{\theta}}_n$ of $\hat{\mathbf{g}}_n$ to obtain:

$$\hat{\mathbf{g}}_n(\hat{\boldsymbol{\theta}}_n) \approx \hat{\mathbf{g}}_n(\boldsymbol{\theta}^*) + \hat{\mathbf{A}}_n(\boldsymbol{\theta}^*)\left(\hat{\boldsymbol{\theta}}_n - \boldsymbol{\theta}^*\right). \tag{15.11}$$

- Since $\hat{\boldsymbol{\theta}}_n$ is a critical point of $\hat{\ell}_n$ by assumption, the left-hand side of (15.11) is a vector of zeros so we have:

$$\mathbf{0}_q \approx \hat{\mathbf{g}}_n(\boldsymbol{\theta}^*) + \hat{\mathbf{A}}_n(\boldsymbol{\theta}^*)\left(\hat{\boldsymbol{\theta}}_n - \boldsymbol{\theta}^*\right). \tag{15.12}$$

- Use approximation $\hat{\mathbf{A}}_n(\boldsymbol{\theta}^*) \approx \mathbf{A}^*$ and rearrange terms in (15.12) to obtain:

$$\hat{\boldsymbol{\theta}}_n \approx \boldsymbol{\theta}^* - [\mathbf{A}^*]^{-1}\hat{\mathbf{g}}_n(\boldsymbol{\theta}^*). \tag{15.13}$$

- By the Multivariate Central Limit Theorem (Theorem 9.3.7), $\hat{\mathbf{g}}_n(\boldsymbol{\theta}^*)$ has an asymptotic Gaussian distribution with mean zero and covariance matrix $(1/n)\mathbf{B}^*$.

- By the Linear Transformation of Gaussian Random Variables Theorem (Theorem 9.4.2) and since $\hat{\mathbf{g}}_n(\boldsymbol{\theta}^*)$ is asymptotically Gaussian with covariance matrix $(1/n)\mathbf{B}^*$, $[\mathbf{A}^*]^{-1}\hat{\mathbf{g}}_n(\boldsymbol{\theta}^*)$ is an asymptotic Gaussian random vector with mean zero and covariance matrix

$$(1/n)\mathbf{C}^* \equiv (1/n)(\mathbf{A}^*)^{-1}\mathbf{B}^*(\mathbf{A}^*)^{-1}.$$

- Therefore, $\hat{\boldsymbol{\theta}}_n$ converges in distribution to a Gaussian vector with mean $\boldsymbol{\theta}^*$ and covariance matrix $(1/n)\mathbf{C}^*$ as $n \to \infty$.

The quantity $\hat{\sigma}_k/\sqrt{n}$ is called the standard error of $\hat{\theta}_n(k)$. The standard error is an estimator of the sampling error associated with estimating the $k$th element of $\hat{\boldsymbol{\theta}}_n$. △

Although the characterization of the asymptotic distribution of the empirical risk estimator is a critical result, in many applications one is interested in characterizing the asymptotic distribution of different types of functions of the empirical risk estimator. For example, if a

$q$-dimensional $\hat{\boldsymbol{\theta}}_n$ is the empirical risk estimator and $\mathbf{s}^T$ is the $k$th row of a $q$-dimensional identity matrix, then $\mathbf{s}^T\hat{\boldsymbol{\theta}}_n$ is the $k$th element of $\hat{\boldsymbol{\theta}}_n$. Thus, a linear function of the empirical risk estimator may be used to select out individual parameter estimators. As another example, let $\bar{y} : \mathcal{R}^u \times \Theta \rightarrow \mathcal{R}$ be a regression function for a nonlinear (or linear) regression model. The predicted response of the nonlinear regression model is $\bar{y}(\mathbf{s}, \hat{\boldsymbol{\theta}}_n)$ for a given input pattern $\mathbf{s}$ and given empirical risk estimator $\hat{\boldsymbol{\theta}}_n$. The sampling error in $\hat{\boldsymbol{\theta}}_n$ will generate sampling error in the predicted output $\bar{y}(\mathbf{s}, \hat{\boldsymbol{\theta}}_n)$.

The following theorem provides an alternative to using the nonparametric bootstrap methodology for characterizing the sampling error for a function $\phi$ of the empirical risk estimator $\hat{\boldsymbol{\theta}}_n$ given only a training data set.

**Theorem 15.2.2** (Function of Empirical Risk Minimizer Asymptotic Distribution). *Assume the Empirical Risk Regularity Assumptions in Definition 15.1.1 hold with respect to a stochastic sequence of i.i.d. random vectors $\tilde{\mathbf{x}}_1, \ldots, \tilde{\mathbf{x}}_n$ with common DGP Radon-Nikodým density $p_e : \mathcal{R}^d \rightarrow [0, \infty)$ defined with respect to a support specification measure $\nu$, a loss function $c : \mathcal{R}^d \times \mathcal{R}^q \rightarrow \mathcal{R}$, a penalty function $k_n : \mathcal{R}^{d \times n} \times \Theta \rightarrow \mathcal{R}$, restricted parameter space $\Theta \subseteq \mathcal{R}^q$, empirical risk function minimizer $\hat{\boldsymbol{\theta}}_n$ of*

$$\hat{\ell}_n(\boldsymbol{\theta}) \equiv \hat{k}_n(\boldsymbol{\theta}) + (1/n)\sum_{i=1}^{n} c(\tilde{\mathbf{x}}_i, \boldsymbol{\theta})$$

*on $\Theta$, and risk function minimizer $\boldsymbol{\theta}^*$ of*

$$\ell(\boldsymbol{\theta}) = \int c(\mathbf{x}, \boldsymbol{\theta}) p_e(\mathbf{x}) d\nu(\mathbf{x})$$

*on $\Theta$. Let $\mathbf{A}^*$ and $\mathbf{B}^*$ be defined as in* **A7** *and* **A8** *respectively in Definition 15.1.1. Let $\hat{\mathbf{A}}_n$, $\hat{\mathbf{B}}_n$, and $\hat{\mathbf{C}}_n$ be defined as in (15.1), (15.2), and (15.3) respectively.*

*Let $\phi : \mathcal{R}^q \rightarrow \mathcal{R}^r$ be a continuously differentiable function on $\Theta$. Assume $\mathbf{J} \equiv \nabla\phi$ is a continuous function on $\Theta$ such that $\mathbf{J}(\boldsymbol{\theta}^*)$ has full row rank $r$. Then, $\sqrt{n}\left(\phi(\hat{\boldsymbol{\theta}}_n) - \phi(\boldsymbol{\theta}^*)\right)$ converges in distribution to an $r$-dimensional Gaussian random vector with mean $\mathbf{0}_r$ and $r$-dimensional covariance matrix*

$$\mathbf{Q}^* = \mathbf{J}(\boldsymbol{\theta}^*)\mathbf{C}^*[\mathbf{J}(\boldsymbol{\theta}^*)]^T. \tag{15.14}$$

*In addition,*

$$\mathbf{Q}^* = \mathbf{J}(\hat{\boldsymbol{\theta}}_n)\hat{\mathbf{C}}_n[\mathbf{J}(\hat{\boldsymbol{\theta}}_n)]^T + o_{a.s.}(1). \tag{15.15}$$

*Proof.* Using the Mean Value Theorem,

$$\phi(\hat{\boldsymbol{\theta}}) = \phi(\boldsymbol{\theta}^*) + \mathbf{J}(\ddot{\boldsymbol{\theta}})(\hat{\boldsymbol{\theta}}_n - \boldsymbol{\theta}^*) \tag{15.16}$$

where each element of $\ddot{\boldsymbol{\theta}}$ is a possibly different point on the chord connecting $\boldsymbol{\theta}^*$ and $\hat{\boldsymbol{\theta}}_n$. Rearranging the terms in (15.16) and multiplying by $\sqrt{n}$ gives:

$$\sqrt{n}[\phi(\hat{\boldsymbol{\theta}}) - \phi(\boldsymbol{\theta}^*)] = \mathbf{J}(\ddot{\boldsymbol{\theta}})\sqrt{n}(\hat{\boldsymbol{\theta}}_n - \boldsymbol{\theta}^*). \tag{15.17}$$

As $n \rightarrow \infty$, $\hat{\boldsymbol{\theta}}_n \rightarrow \boldsymbol{\theta}^*$ with probability one as $n \rightarrow \infty$ which implies that the term $\mathbf{J}(\ddot{\boldsymbol{\theta}})$ on the right-hand side of (15.17) converges with probability one to $\mathbf{J}(\boldsymbol{\theta}^*)$ as $n \rightarrow \infty$. This term is multiplied by $\sqrt{n}(\hat{\boldsymbol{\theta}}_n - \boldsymbol{\theta}^*)$, which converges in distribution to a multivariate Gaussian

**Recipe Box 15.2 Formula for Estimating Sampling Error (Theorem 15.2.2)**

- **Step 1: Examine data generating process.**
Check the data sample $\mathbf{x}_1, \ldots, \mathbf{x}_n$ is a realization of a bounded sequence of $n$ i.i.d. random vectors.

- **Step 2: Check loss function is sufficiently smooth.**
Let $c : \mathcal{R}^d \times \Theta \rightarrow \mathcal{R}$ be a twice continuously differentiable random function. In addition, assume $c(\cdot; \boldsymbol{\theta})$ is piecewise continuous for each $\boldsymbol{\theta} \in \Theta$.

- **Step 3: Estimate parameters by minimizing empirical risk function.**
Use an optimization algorithm such as gradient descent to find a strict local minimizer, $\hat{\boldsymbol{\theta}}_n$, of the empirical risk function

$$\hat{\ell}_n(\boldsymbol{\theta}) \equiv (1/n) \sum_{i=1}^{n} c(\mathbf{x}_i, \boldsymbol{\theta}).$$

Numerically check $\hat{\boldsymbol{\theta}}_n$ is a critical point by verifying for large $n$ that the infinity norm of $\nabla \hat{\ell}_n(\hat{\boldsymbol{\theta}}_n)$ is sufficiently small.

- **Step 4: Check OPG and Hessian condition numbers.** Check the condition number of

$$\hat{\mathbf{B}}_n = (1/n) \sum_{i=1}^{n} \nabla c(\mathbf{x}_i, \boldsymbol{\theta})[\nabla c(\mathbf{x}_i, \boldsymbol{\theta})]^T$$

is converging to a finite positive number (see Example 5.3.6). Also check if $\hat{\boldsymbol{\theta}}_n$ is converging to a strict local minimizer by empirically checking if the condition number of $\hat{\mathbf{A}}_n \equiv \nabla^2 \hat{\ell}_n(\hat{\boldsymbol{\theta}}_n)$ is a finite positive number.

- **Step 5: Calculate the asymptotic distribution of estimates.** Calculate the asymptotic covariance matrix of the asymptotically Gaussian parameter estimates using the formula:

$$\hat{\mathbf{C}}_n = [\hat{\mathbf{A}}_n]^{-1} \hat{\mathbf{B}}_n [\hat{\mathbf{A}}_n].$$

The sampling error of the $k$th element of $\hat{\boldsymbol{\theta}}_n$ is equal to the square root of the $k$th on-diagonal element of $(1/n)\hat{\mathbf{C}}_n$.

- **Step 6: Compute asymptotic distribution of statistics.** Let $\phi : \mathcal{R}^q \rightarrow \mathcal{R}^r$ be a continuously differentiable function. Let $\mathbf{J}$ denote the derivative of $\phi$. Then, $\phi(\hat{\boldsymbol{\theta}}_n)$ is asymptotically Gaussian with mean $\phi(\boldsymbol{\theta}^*)$ whose covariance matrix, $(1/n)\mathbf{Q}^*$, can be estimated by the formula:

$$\hat{\mathbf{Q}}_n = \mathbf{J}(\hat{\boldsymbol{\theta}}_n) \hat{\mathbf{C}}_n \mathbf{J}(\hat{\boldsymbol{\theta}}_n)^T.$$

If the covariance matrix $\hat{\mathbf{Q}}_n$ is not invertible or appears to be converging to a non-invertible matrix, this indicates a violation of a key assumption of the asymptotic theory. The covariance matrix $\hat{\mathbf{Q}}_n$ may also be non-invertible due to a poor choice of the $\mathbf{J}$ matrix which must have full row rank when evaluated at the parameter estimates.

random vector with mean $\mathbf{0}_q$ and covariance matrix $\mathbf{C}^*$ by Theorem 15.2.1. The product of these two matrices converges by Slutsky's Theorem (Theorem 9.4.5) and the Linear Transformation of a Multivariate Gaussian Random Vector Theorem (Theorem 9.4.2) to a multivariate Gaussian random vector with mean $\mathbf{0}_r$ and covariance matrix $\mathbf{J}(\boldsymbol{\theta}^*)\mathbf{C}^*[\mathbf{J}(\boldsymbol{\theta}^*)]^T$.

Since: (i) $\hat{\mathbf{C}}_n \to \mathbf{C}^*$ and $\hat{\boldsymbol{\theta}}_n \to \boldsymbol{\theta}^*$ as $n \to \infty$ with probability one, and (ii) $\mathbf{C}$ and $\mathbf{J}$ are continuous functions, then (15.15) is implied by the Functions of Stochastic Sequences Theorem (Theorem 9.4.1). ∎

In order for the conditions of the Function of Empirical Risk Minimizer Asymptotic Distribution Theorem to hold, it is sufficient that the Empirical Risk Regularity Assumptions (Definition 15.1.1) hold and that in addition the derivative of $\phi$, $\mathbf{J} : \mathcal{R}^q \to \mathcal{R}^r$, is continuous and $\mathbf{J}^* \equiv \mathbf{J}(\boldsymbol{\theta}^*)$ has full row rank. The condition $\mathbf{J}$ is continuous can be checked analytically. If the condition number of

$$\hat{\mathbf{Q}}_n \equiv \mathbf{J}(\hat{\boldsymbol{\theta}}_n)\hat{\mathbf{C}}_n[\mathbf{J}(\hat{\boldsymbol{\theta}}_n)]^T$$

is excessively large or its largest eigenvalue is close to a numerical zero, this indicates a possible failure of Assumption **A7**, Assumption **A8**, or the assumption that $\mathbf{J}^*$ has full row rank.

**Example 15.2.2** (Gaussian Mixture Model for Unsupervised Learning). A simplified GMM (Gaussian Mixture Model) is analyzed. A GMM for unsupervised learning is a model specification defined such that each element is a weighted sum of Gaussian density functions such that the non-negative weights sum to one. In particular, the training data is assumed to be a realization, $\mathbf{x}_1, \ldots, \mathbf{x}_n$ of a bounded stochastic sequence of independent and identically distributed $d$-dimensional random vectors with common Radon-Nikodým density defined with respect to support specification measure $\nu$ such that the support of $\tilde{\mathbf{x}}_i$ is $\mathcal{R}^d$.

Let

$$p(\mathbf{x}|\mathbf{m}^k) = (\sqrt{2\pi})^{-d} \exp\left(-(1/2)|\mathbf{x} - \mathbf{m}^k|^2\right)$$

be the probability that training stimulus $\mathbf{x}$ is a member of the $k$th category. Let

$$q_k(\boldsymbol{\beta}) = \exp(\beta_k) / \sum_{j=1}^{K} \exp(\beta_j) \text{ with } \boldsymbol{\beta} \equiv [\beta_1, \ldots, \beta_K]$$

be the likelihood that a training stimulus is drawn from the $k$th category. Let the Gaussian Mixture Model be a collection of Radon-Nikodým densities

$$\mathcal{M} \equiv \{p(\cdot|\boldsymbol{\theta}) : \boldsymbol{\theta} \in \Theta\}$$

where for each $\boldsymbol{\theta} \in \Theta$:

$$p(\mathbf{x}|\boldsymbol{\theta}) = \sum_{k=1}^{K} q_k(\boldsymbol{\beta}) p(\mathbf{x}|\mathbf{m}^k)$$

is the probability assigned by the learning machine when it observes training stimulus $\mathbf{x}$ given the learning machine's parameter vector is $\boldsymbol{\theta} \equiv [\boldsymbol{\beta}^T, (\mathbf{m}^1)^T, \ldots, (\mathbf{m}^K)^T]^T$.

Derive a formula for the standard error of the maximum likelihood estimators $\hat{\beta}_n(k)$ and $\hat{\mathbf{m}}^k$. Discuss all assumptions that are necessary for these formulas to be valid.

**Solution.** The negative normalized log-likelihood function for this problem is defined by:

$$\hat{\ell}_n(\boldsymbol{\theta}) = (1/n) \sum_{i=1}^{n} c(\mathbf{x}_i, \boldsymbol{\theta})$$

where $c(\mathbf{x}_i, \boldsymbol{\theta}) = -\log p(\mathbf{x}_i|\boldsymbol{\theta})$. The maximum likelihood estimators $\hat{\beta}_n(k)$ and $\hat{\mathbf{m}}^k$ are obtained by computing a strict local minimizer of $\hat{\ell}_n$.

The gradient of $c(\mathbf{x}, \boldsymbol{\theta})$ is given by the formula:

$$\mathbf{g}(\mathbf{x}, \boldsymbol{\theta}) \equiv \left[ \frac{dc(\mathbf{x}_i, \boldsymbol{\theta})}{d\beta}, \frac{dc(\mathbf{x}_i, \boldsymbol{\theta})}{d\mathbf{m}^1}, \ldots, \frac{dc(\mathbf{x}_i, \boldsymbol{\theta})}{d\mathbf{m}^k} \right]^T$$

where

$$\frac{dc(\mathbf{x}_i, \boldsymbol{\theta})}{d\beta} = -[p(\mathbf{x}_i|\boldsymbol{\theta})]^{-1} \sum_{k=1}^{K} \left( \frac{dq_k}{d\beta} \right)^T p(\mathbf{x}|\mathbf{m}^k)$$

and

$$\frac{dc(\mathbf{x}_i, \boldsymbol{\theta})}{d\mathbf{m}^k} = -[p(\mathbf{x}_i|\boldsymbol{\theta})]^{-1} \sum_{k=1}^{K} q_k(\beta) p(\mathbf{x}|\mathbf{m}^k) \left( \mathbf{x} - \mathbf{m}^k \right).$$

Use the gradient of $c(\mathbf{x}, \boldsymbol{\theta})$ and the maximum likelihood estimate $\hat{\boldsymbol{\theta}}_n$ to compute $\hat{\mathbf{B}}_n$ in (15.2). Next, find the Hessian of $c(\mathbf{x}, \boldsymbol{\theta})$ by taking the derivative of $\mathbf{g}(\mathbf{x}, \boldsymbol{\theta})$ with respect to $\boldsymbol{\theta}$ and denote the result as $\mathbf{A}(\mathbf{x}; \boldsymbol{\theta})$. Use $\mathbf{A}(\mathbf{x}; \boldsymbol{\theta})$ to compute $\hat{\mathbf{A}}_n$ in (15.1).

Assumption **A1** is satisfied by assumption. In addition, $c$ is a twice continuously differentiable random function so **A2** is satisfied. Since the DGP is bounded and **A2** holds, it follows that **A3** holds. Assumption **A4** holds since the penalty term $\hat{k}_n(\boldsymbol{\theta}) = 0$.

Assumption **A5** will hold in a sufficiently small neighborhood surrounding a strict local minimizer of $\ell$. Note however, that Gaussian mixture models often contain complicated saddlepoint surfaces and flat regions as well as multiple strict local minimizers. Such parameter estimation challenges can be partially addressed by reducing or merging categories, introducing regularization terms, and placing strategic restrictions on the parameter space.

Assumptions **A5**, **A7**, and **A8** are empirically checked as follows. Check that $\hat{\boldsymbol{\theta}}_n$ is converging to a critical point by checking the infinity norm of the gradient evaluated at $\hat{\boldsymbol{\theta}}_n$ is sufficiently small. Check that $\hat{\mathbf{A}}_n$ and $\hat{\mathbf{B}}_n$ are converging to positive definite matrices by checking that their magnitudes are not too small and their condition numbers are not too large.

Next compute $\hat{\mathbf{C}}_n = \hat{\mathbf{A}}_n^{-1} \hat{\mathbf{B}}_n [\hat{\mathbf{A}}_n]^{-1}$. The standard error for the $k$th element of $\hat{\boldsymbol{\beta}}_n$ is the square root of the $k$th on-diagonal element of $\hat{\mathbf{C}}_n$ divided by the square root of $n$. The standard error for the $j$th element of the $k$th mean vector $\hat{\mathbf{m}}_k$ is the square root of the $z$th on-diagonal element of $\hat{\mathbf{C}}_n$ divided by the square root of $n$ where $z = kK + j$. △

## Exercises

15.2-1. *Linear Regression.* Let $\{(\tilde{\mathbf{s}}_1, \tilde{y}_1), \ldots, (\tilde{\mathbf{s}}_n, \tilde{y}_n)\}$ be a sequence of $n$ independent and identically distributed observations. Assume the probability density of the response scalar variable $\tilde{y}_i$ is a Gaussian random variable with mean $\boldsymbol{\theta}^T \mathbf{s}_i$ and variance one for a given input pattern vector $\mathbf{s}_i$, $i = 1, \ldots, n$. Derive a formula for the covariance matrix of the maximum likelihood estimate of $\boldsymbol{\theta}$. Discuss the appropriateness of all assumptions required for your analysis to be valid.

15.2-2. *Linear Regression with Smooth $L_1$ Regularizer.* Let $\{(\tilde{\mathbf{s}}_1, \tilde{y}_1), \ldots, (\tilde{\mathbf{s}}_n, \tilde{y}_n)\}$ be a sequence of $n$ independent and identically distributed random vectors. Assume the probability density of the response scalar variable $\tilde{y}_i$ is a Gaussian random variable with mean $\boldsymbol{\theta}^T \mathbf{s}_i$ and variance one for a given $d$-dimensional input pattern vector $\mathbf{s}_i$, $i = 1, \ldots, n$. Also assume a hyperbolic secant prior on $\boldsymbol{\theta}$ defined such that

$$p_{\boldsymbol{\theta}}(\boldsymbol{\theta}) = \prod_{j=1}^{d+1} \left[\exp\left((\pi/2)\theta_j\right) + \exp\left(-(\pi/2)\theta_j\right)\right]^{-1}.$$

Derive a formula for the covariance matrix of the MAP estimates of $\boldsymbol{\theta}$. Discuss the appropriateness of all assumptions required for your analysis to be valid.

15.2-3. *Sigmoidal Perceptron Model with $L_2$ Regularizer.* Let $\{(\tilde{\mathbf{s}}_1, \tilde{y}_1), \ldots, (\tilde{\mathbf{s}}_n, \tilde{y}_n)\}$ be a sequence of $n$ independent and identically distributed random vectors such that the response random variable $\tilde{y}_i \in \{0, 1\}$ for a given input pattern vector $\mathbf{s}_i$ for $i = 1, 2, \ldots$. Assume the predicted response $\ddot{y}(\mathbf{s}, \boldsymbol{\theta})$ is defined by the formula:

$$\ddot{y}(\mathbf{s}, \boldsymbol{\theta}) = \mathbf{w}^T \mathbf{h}(\mathbf{s}, \boldsymbol{\theta})$$

where the $k$th element of $\mathbf{h}(\mathbf{s}, \boldsymbol{\theta})$, $h_k$, is the function defined such that:

$$h_k(\mathbf{s}, \boldsymbol{\theta}) = \mathcal{S}\left(\mathbf{v}_k^T [\mathbf{s}^T, 1]^T\right)$$

where $\mathcal{S}(\phi) \equiv 1/(1 + \exp(-\phi))$. Let $\hat{\boldsymbol{\theta}}_n$ be a strict local minimizer of the penalized empirical risk function

$$\hat{\ell}_n(\boldsymbol{\theta}) = \lambda |\boldsymbol{\theta}|^2 + \sum_{i=1}^{n} \left(\tilde{y}_i - \ddot{y}(\tilde{\mathbf{s}}_i, \boldsymbol{\theta})\right)^2.$$

Provide an explicit formula for the asymptotic covariance matrix of $\hat{\boldsymbol{\theta}}$. Discuss what assumptions are required for your derivation to be valid with respect to the theorems introduced in this chapter.

15.2-4. *Comparison of Nonparametric Bootstrap and Analytic Formulas.* Conduct a simulation study for comparing the non-parametric bootstrap estimator of the asymptotic covariance matrix introduced in Chapter 14 with the analytic formula for the asymptotic covariance matrix introduced in this section.

## 15.3 Confidence Regions

Assume the data $x_1, \ldots, x_n$ is a realization of a stochastic sequence of i.i.d. 1-dimensional scalar random variables $\tilde{x}_1, \ldots, \tilde{x}_n$ with common probability density $p_e : \mathcal{R} \rightarrow [0, \infty)$. Let $c : \mathcal{R} \times \mathcal{R} \rightarrow \mathcal{R}$ be a loss function for the empirical risk function

$$\hat{\ell}_n(\theta) = (1/n) \sum_{i=1}^{n} c(\tilde{\mathbf{x}}_i, \theta).$$

The domain of $\hat{\ell}_n$ is the one-dimensional parameter space $\Theta \subseteq \mathcal{R}$. The scalar estimator $\hat{\theta}_n$ of $\theta^*$ is a strict local minimizer of $\hat{\ell}_n$. Let $\theta^*$ be a strict local minimizer of $E\{\hat{\ell}_n\}$. Assume $\hat{\theta}_n \rightarrow \theta^*$ as $n \rightarrow \infty$ with probability one.

Assume the conditions of the Empirical Risk Asymptotic Estimator Distribution Theorem (Theorem 15.2.1) hold so that $\hat{\theta}_n - \theta^*$ is asymptotically Gaussian with mean $\theta^*$ and variance $\hat{\sigma}_n/\sqrt{n}$. Let $\tilde{Z}_n$ denote a Gaussian random variable with mean zero and variance equal to 1. Let $Z_\alpha$ be defined such that the probability $|\tilde{Z}_n| \leq Z_\alpha$ is equal to $1-\alpha$. For example, if $\alpha = 0.05$ then $Z_\alpha = 1.96$. Then, it follows that the probability that $\theta^*$ lies in the random interval

$$\tilde{\Omega}_n(\alpha) \equiv \left\{ \theta \in \mathcal{R} : \hat{\theta}_n - Z_\alpha \frac{\hat{\sigma}_n}{\sqrt{n}} \leq \theta \leq \hat{\theta}_n + Z_\alpha \frac{\hat{\sigma}_n}{\sqrt{n}} \right\}$$

is equal to $1-\alpha$. The interval $\tilde{\Omega}_n(\alpha)$ is called a $(1-\alpha)100\%$ confidence interval for $\theta^*$.

The next theorem generalizes these ideas for the case where the parameter estimate $\hat{\boldsymbol{\theta}}_n$ is a vector rather than a scalar.

**Theorem 15.3.1** (Model Statistic Confidence Region). *Assume the conditions of the Model Statistic Distribution Theorem (Theorem 15.2.2) hold with respect to a DGP Radon-Nikodým density $p_e : \mathcal{R}^d \rightarrow [0,\infty)$, restricted parameter space $\Theta$, loss function $c : \mathcal{R}^d \times \Theta \rightarrow \mathcal{R}$, penalty function $k_n : \mathcal{R}^{d \times n} \times \Theta \rightarrow \mathcal{R}$, and continuously differentiable function $\boldsymbol{\phi} : \mathcal{R}^q \rightarrow \mathcal{R}^r$. Let $\mathbf{J} \equiv d\boldsymbol{\phi}/d\boldsymbol{\theta}$. Assume $\mathbf{J}(\boldsymbol{\theta}^*)$ has full row rank $r$. Let $\hat{\mathbf{Q}}_n$ be defined as in Equation 15.14 by the formula:*

$$\hat{\mathbf{Q}}_n \equiv \mathbf{J}(\hat{\boldsymbol{\theta}}_n)\hat{\mathbf{C}}_n[\mathbf{J}(\hat{\boldsymbol{\theta}}_n)]^T$$

*where $\hat{\mathbf{C}}_n \equiv \hat{\mathbf{A}}_n^{-1}\hat{\mathbf{B}}_n\hat{\mathbf{A}}_n^{-1}$ as in (15.3). Let*

$$\hat{\mathcal{W}}_n \equiv n\left(\boldsymbol{\phi}(\hat{\boldsymbol{\theta}}_n) - \boldsymbol{\phi}(\boldsymbol{\theta}^*)\right)^T [\hat{\mathbf{Q}}_n]^{-1}\left(\boldsymbol{\phi}(\hat{\boldsymbol{\theta}}_n) - \boldsymbol{\phi}(\boldsymbol{\theta}^*)\right). \tag{15.18}$$

*Let $K_\alpha$ be defined such that the probability a chi-squared random variable with $r$ degrees of freedom, $\tilde{\chi}^2(r)$, is less than $K_\alpha$ is equal to $1-\alpha$. Then, as $n \rightarrow \infty$, $\hat{\mathcal{W}}_n \rightarrow \tilde{\chi}^2(r)$ in distribution. In addition, the probability that $\boldsymbol{\phi}(\boldsymbol{\theta}^*)$ is in the region*

$$\tilde{\Omega}_n(\alpha) \equiv \left\{ \boldsymbol{\zeta} \in \mathcal{R}^r : n\left(\boldsymbol{\zeta} - \boldsymbol{\phi}(\hat{\boldsymbol{\theta}}_n)\right)^T [\hat{\mathbf{Q}}_n]^{-1}\left(\boldsymbol{\zeta} - \boldsymbol{\phi}(\hat{\boldsymbol{\theta}}_n))\right) \leq K_\alpha \right\}$$

*converges to $1-\alpha$ as $n \rightarrow \infty$.*

*Proof.* Let $\mathbf{Q}^* \equiv \nabla\boldsymbol{\phi}(\boldsymbol{\theta}^*)\mathbf{C}^*[\nabla\boldsymbol{\phi}(\boldsymbol{\theta}^*)]^T$. First, note that since $\nabla\boldsymbol{\phi}(\boldsymbol{\theta}^*)$ has full row rank $r$ and $\mathbf{C}^*$ is positive definite, then $\mathbf{Q}^*$ is positive definite. Therefore,

$$\tilde{\mathbf{z}}_n \equiv \sqrt{n}\,[\mathbf{Q}^*]^{-1/2}\,(\hat{\boldsymbol{\phi}}_n - \boldsymbol{\phi}^*)$$

converges in distribution to an $r$-dimensional multivariate Gaussian random vector with mean zero and covariance matrix equal to the identity matrix.

Second, let

$$\hat{\mathcal{W}}_n^* \equiv n\left(\hat{\boldsymbol{\phi}}_n - \boldsymbol{\phi}^*\right)[\mathbf{Q}^*]^{-1}\left(\hat{\boldsymbol{\phi}}_n - \boldsymbol{\phi}^*\right)^T. \tag{15.19}$$

Third, since the sum of squares of $r$ asymptotically distributed normalized Gaussian random variables, $|\tilde{\mathbf{z}}_n|^2$, has an asymptotic chi-squared random distribution with $r$ degrees of freedom and since $\mathbf{Q}^*$ is positive definite, this implies

$$\hat{\mathcal{W}}_n^* \equiv n\left|[\mathbf{Q}^*]^{-1/2}(\hat{\boldsymbol{\phi}}_n - \boldsymbol{\phi}^*)\right|^2 = n\,|\tilde{\mathbf{z}}_n|^2 \tag{15.20}$$

converges in distribution to a chi-squared random variable with $r$ degrees of freedom.

Fourth, since $\hat{\mathbf{Q}}_n \rightarrow \mathbf{Q}^*$ with probability one as $n \rightarrow \infty$ and $\sqrt{n}\left(\hat{\boldsymbol{\phi}}_n - \boldsymbol{\phi}^*\right) = O_p(1)$, it follows from subtracting (15.20) from (15.18) to obtain:

$$\hat{\mathcal{W}}_n - \hat{\mathcal{W}}_n^* = O_p(1)o_p(1)O_p(1) = o_p(1).$$

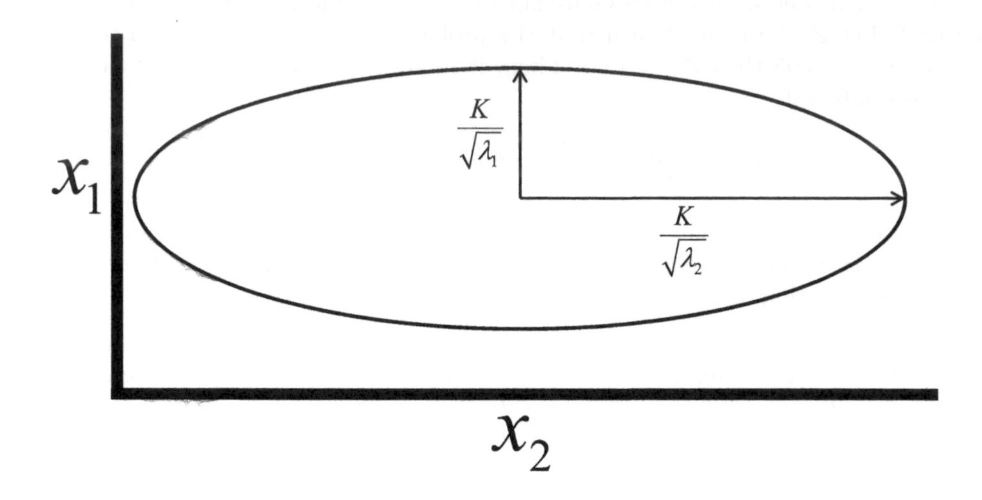


**FIGURE 15.3**
**Ellipsoidal confidence regions.** This figure depicts an ellipsoidal confidence region $\{\mathbf{x} \in \mathcal{R}^2 : \mathbf{x}^T \mathbf{C}^{-1} \mathbf{x} \leq K^2\}$ where $\mathbf{C}$ is a positive definite diagonal matrix with eigenvalues $\lambda_1$ and $\lambda_2$.

Thus,

$$\hat{\mathcal{W}}_n = \hat{\mathcal{W}}_n^* + (\hat{\mathcal{W}}_n - \hat{\mathcal{W}}_n^*) = \hat{\mathcal{W}}_n^* + o_p(1)$$

has a chi-squared distribution with $r$ degrees of freedom by Slutsky's Theorem.

Since $\hat{\mathcal{W}}_n$ has an asymptotic chi-squared probability distribution with $r$ degrees of freedom, it follows that the probability that $\boldsymbol{\phi}(\boldsymbol{\theta}^*) \in \tilde{\Omega}_n(\alpha)$ is equal to $1 - \alpha$ by the definition of $\tilde{\Omega}_n(\alpha)$. ■

**Definition 15.3.1** (Confidence Region). Assume the assumptions of Theorem 15.3.1 hold with respect to $\tilde{\Omega}_n(\alpha)$ and $\boldsymbol{\theta}^*$. Then, $\tilde{\Omega}_n(\alpha)$ is called a *confidence region* for $\boldsymbol{\theta}^*$.

The terminology $(1 - \alpha)100\%$ random confidence region $\tilde{\Omega}_n(\alpha)$ for $\boldsymbol{\theta}^*$ means that the probability the random confidence region $\tilde{\Omega}_n(\alpha)$ contains the strict local minimizer $\boldsymbol{\theta}^*$ is approximately equal to $1 - \alpha$ when the sample size $n$ is sufficiently large.

It is important to emphasize that the confidence region $\tilde{\Omega}_n(\alpha)$ is random but the strict local minimizer $\boldsymbol{\theta}^*$ in Theorem 15.3.1 is a constant. Therefore, except in the special case of credible confidence regions (see Example 15.3.3), it is misleading to make statements such as the probability $\boldsymbol{\theta}^*$ takes on values in some region with probability approximately $1 - \alpha$. However, the statement that the probability the event $\boldsymbol{\theta}^* \in \tilde{\Omega}_n(\alpha)$ is approximately $1 - \alpha$ is technically correct provided one keeps in mind that $\boldsymbol{\theta}^*$ is a constant and $\tilde{\Omega}_n(\alpha)$ is a random region.

The confidence region $\tilde{\Omega}_n(\alpha)$ has the shape of a hyperellipsoid whose principal axis lengths are proportional respectively to the reciprocal of the square roots of the eigenvalues of the matrix $\hat{\mathbf{C}}_n$. The directions of the principal axes of the hyperellipsoid are specified by the eigenvectors of $\hat{\mathbf{C}}_n$ (see Figure 15.3).

**Example 15.3.1** (Classical Confidence Regions for Maximum Likelihood Estimation). Assume the empirical risk function is a Maximum Likelihood empirical risk function and $\hat{\boldsymbol{\theta}}_n$ is a maximum likelihood estimate (see Section 13.2.1). Assume, in addition, that $\hat{\boldsymbol{\theta}}_n$ is converging with probability one to a true parameter $\boldsymbol{\theta}^*$ which is a strict local maximizer of $p(\mathcal{D}_n|\boldsymbol{\theta})$. In this case, $(1-\alpha)100\%$ confidence region for $\boldsymbol{\theta}^*$ , $\tilde{\Omega}_n(\alpha)$, contains the true parameter vector $\boldsymbol{\theta}^*$ with approximate probability $1-\alpha$ for sufficiently large sample size $n$. A major difficulty with this approach is that it assumes the true parameter value exists. In many real-world situations, confidence intervals for parameter values that minimize the expected negative log-likelihood function are desired even in situations where those parameter values are not exactly the true parameter values. If the true parameter values do not exist, then this concept of a confidence region is meaningless.

**Example 15.3.2** (Confidence Regions for Quasi-Maximum Likelihood Estimation). Assume the empirical risk function is a Maximum Likelihood empirical risk function and $\hat{\boldsymbol{\theta}}_n$ is a quasi-maximum likelihood estimate (see Section 13.2.1). Unlike Example 15.3.1 which imposes the requirement that the true parameter values exist, it is assumed only that the quasi-maximum likelihood estimator $\hat{\boldsymbol{\theta}}_n$ converges with probability one to a strict local minimizer $\boldsymbol{\theta}^*$. The strict local minimizer $\boldsymbol{\theta}^*$ minimizes the cross entropy between the approximating density $p(\cdot;\boldsymbol{\theta})$ and the DGP density $p_e$. In this case, the random $(1-\alpha)100\%$ confidence region for $\boldsymbol{\theta}^*$, $\tilde{\Omega}_n(\alpha)$, contains the local maximum, $\boldsymbol{\theta}^*$, of the likelihood function $p(\mathcal{D}_n|\boldsymbol{\theta})$ with probability approximately equal to $1-\alpha$ for sufficiently large sample size $n$.

**Example 15.3.3** (Credible Regions for MAP Estimation). Assume the empirical risk function is a MAP empirical risk function and $\hat{\boldsymbol{\theta}}_n$ is a MAP estimate (see Section 13.3). In this case, the $(1-\alpha)100\%$ confidence region for $\boldsymbol{\theta}^*$ is called an approximate *high posterior density credible region*. In this case, $\boldsymbol{\theta}^*$ is approximately the mode of the posterior density $p(\boldsymbol{\theta}|\mathcal{D}_n)$. For the special MAP empirical risk function case, the confidence region for $\boldsymbol{\theta}^*$ has an alternative semantic interpretation which is that the random vector $\tilde{\boldsymbol{\theta}}$ with density $p(\boldsymbol{\theta}|\mathcal{D}_n)$ is an element of $\tilde{\Omega}_n(\alpha)$ with probability $1-\alpha$ for sufficiently large sample size $n$.

**Example 15.3.4** (Numerical Evaluation of Chi-Squared Distribution Function). Most statistical and algorithm development software environments provide subroutines for computing the cumulative distribution function for a chi-squared random variable. Let *df* be the degrees of freedom for a chi-squared random variable. Let $p_\alpha$ denote the probability that a chi-squared random variable with *df* degrees of freedom takes on values greater than $\chi^2_\alpha(df)$.

The MATLAB®software development environment provides the function GAMMAINC : $(0,\infty)\times[0,\infty)\rightarrow[0,1]$ defined such that $p_\alpha = 1 - \texttt{GAMMAINC}(\chi^2_\alpha/2, df/2)$ where $p_\alpha$ is the probability that a chi-squared random variable with *df* degrees of freedom exceeds the number $\chi^2_\alpha(df)$. For example, $p_\alpha = 1 - \texttt{GAMMAINC}(1.96^2/2, 1/2) = 0.05$.

**Example 15.3.5** (Confidence Intervals for Logistic Regression Parameters). Let $\mathbf{x}_1,\ldots,\mathbf{x}_n$ be a realization of a sequence of $n$ independent and identically distributed random vectors $\tilde{\mathbf{x}}_1,\ldots,\tilde{\mathbf{x}}_n$. Assume that $\tilde{\mathbf{x}}_i \equiv (\tilde{\mathbf{s}}_i, \tilde{y}_i)$ where $\tilde{y}_i \in \{0,1\}$ is the desired binary response variable and $\tilde{\mathbf{s}}_i$ is the input pattern vector. Let the expected loss function for a logistic regression model

$$\hat{\ell}_n(\boldsymbol{\theta}) \equiv (1/n)\sum_{i=1}^{n} c(\mathbf{x}_i,\boldsymbol{\theta})$$

where the loss function $c$ is defined such that

$$c(\tilde{\mathbf{x}}_i,\boldsymbol{\theta}) = \lambda|\boldsymbol{\theta}|^2 - \tilde{y}_i \log p(\tilde{\mathbf{s}}_i,\boldsymbol{\theta}) - (1-\tilde{y}_i)\log\left(1 - p(\tilde{\mathbf{s}}_i,\boldsymbol{\theta})\right)$$

and $p(\mathbf{s},\boldsymbol{\theta}) \equiv \mathcal{S}\left(\boldsymbol{\theta}^T[\mathbf{s}^T,1]^T\right)$ is the predicted probability that $\tilde{y}_i = 1$ given input pattern vector $\mathbf{s}$ when the learning machine's parameters are set to the value of $\boldsymbol{\theta}$. Let $\hat{\boldsymbol{\theta}}_n$ be the strict global minimizer of $\hat{\ell}_n$. Assume $\boldsymbol{\theta}^*$ is the strict global minimizer of the expected value of $\hat{\ell}_n$ such that $\hat{\boldsymbol{\theta}}_n \rightarrow \boldsymbol{\theta}^*$ with probability one as $n \rightarrow \infty$. Let $\tilde{\mathbf{g}}_i \equiv \left(dc(\tilde{\mathbf{x}}_i, \hat{\boldsymbol{\theta}}_n)/d\boldsymbol{\theta}\right)^T$. Let $\hat{\mathbf{A}}_n$ be the Hessian of $\hat{\ell}_n$ evaluated at $\hat{\boldsymbol{\theta}}_n$. Provide a formula for a $(1-\alpha)100\%$ confidence interval for a parameter value.

**Solution.** The key empirical assumptions that need to be checked are that the infinity norm of $\hat{\mathbf{g}}_n = (1/n)\sum_{i=1}^{n} \tilde{\mathbf{g}}_i$ appears to be converging to zero, and the condition numbers of $\hat{\mathbf{B}}_n$ and $\hat{\mathbf{A}}_n$ are not diverging to infinity. A 95% confidence interval on the parameter estimate for the $k$th element in $\hat{\boldsymbol{\theta}}_n$, $\hat{\theta}_n(k)$, is obtained by first estimating the asymptotic covariance matrix of the parameter estimates $\hat{\mathbf{C}}_n = [\hat{\mathbf{A}}_n]^{-1}\hat{\mathbf{B}}_n[\hat{\mathbf{A}}_n]^{-1}$. Then, the $(1-\alpha)100\%$ confidence interval for $\hat{\theta}_n(k)$ is given by the formula: $\hat{\theta}_n(k) \pm 1.96\sqrt{\hat{c}_n(k,k)/n}$ where $\hat{c}_n(k,k)$ is the $k$th on-diagonal element of $\hat{\mathbf{C}}_n$. The number $Z_\alpha = 1.96$ is defined such that the probability that a chi-squared random variable with one degree of freedom exceeds $Z_\alpha^2$ is equal to $\alpha$. Show how the inverse of the incomplete gamma function could be used to calculate the number $Z_\alpha$ for any specified $(1-\alpha)100\%$ confidence interval. △

**Example 15.3.6** (Confidence Interval for Logistic Regression Predictions)**.** For the logistic regression modeling problem in Example 15.3.5, provide a large sample approximation for the probability that $|p(\mathbf{s}, \hat{\boldsymbol{\theta}}_n) - p(\mathbf{s}; \boldsymbol{\theta}^*) \leq \epsilon$ for a given input pattern $\mathbf{s}$. Comment upon empirical statistics that should be checked to investigate the validity of the proposed formula.

**Solution.** Let $\hat{p}_n(\mathbf{s}) \equiv p(\mathbf{s}, \hat{\boldsymbol{\theta}}_n)$. Let $p^*(\mathbf{s}) \equiv p(\mathbf{s}, \boldsymbol{\theta}^*)$. The variance of $\hat{p}_n(\mathbf{s})$ is estimated using Equation 15.14 by the formula:

$$\hat{\sigma}_p^2(\mathbf{s}) = (1/n)\left[\frac{dp(\mathbf{s}, \hat{\boldsymbol{\theta}}_n)}{d\boldsymbol{\theta}}\right] \hat{\mathbf{C}}_n \left[\frac{dp(\mathbf{s}, \hat{\boldsymbol{\theta}}_n)}{d\boldsymbol{\theta}}\right]^T$$

where the sampling error of $\hat{p}_n(\mathbf{s})$ is then estimated by $\hat{\sigma}_p(\mathbf{s})$. Note that $\hat{\sigma}_p^2(\mathbf{s})$ must converge to a finite positive number if $\hat{\mathbf{C}}_n$ converges to an invertible matrix and the rank of $dp(\mathbf{s}, \boldsymbol{\theta}^*)/d\boldsymbol{\theta}$ is equal to one. Thus, $\hat{\sigma}_p^2(\mathbf{s})$ should be checked to make sure that it is not converging to zero.

Since $\hat{p}_n(\mathbf{s})$ is asymptotically Gaussian with mean $p^*(\mathbf{s})$ and variance $\hat{\sigma}_p^2(\mathbf{s})$. The probability that $|\hat{p}_n(\mathbf{s}) - p^*(\mathbf{s})| < \epsilon$ is the same as the probability that

$$\tilde{Z}_n \equiv \frac{|\hat{p}_n(\mathbf{s}) - p^*(\mathbf{s})|}{\hat{\sigma}_p(\mathbf{s})} < \frac{\epsilon}{\hat{\sigma}_p(\mathbf{s})}.$$

The probability $p_\alpha = 1 - \texttt{GAMMAINC}(Z_\alpha^2/2, 1/2)$ where GAMMAINC is defined as in Example 15.3.4 computes the probability $p_\alpha$ that $|\tilde{Z}_n| > Z_\alpha$ when $Z_\alpha = \epsilon/\hat{\sigma}_p(\mathbf{s})$.

△

## Exercises

15.3-1. *Linear Regression Prediction Error.* Let $\{(\tilde{\mathbf{s}}_1, \tilde{y}_1), \ldots, (\tilde{\mathbf{s}}_n, \tilde{y}_n)\}$ be a sequence of $n$ independent and identically distributed observations. Assume the probability density of the response scalar variable $\tilde{y}_i$ is a Gaussian random variable with mean $\ddot{y}(\mathbf{s}, \boldsymbol{\theta}) = \boldsymbol{\theta}^T \mathbf{s}_i$ and variance one for a given input pattern vector $\mathbf{s}_i$, $i = 1, \ldots, n$.

Let $\hat{\boldsymbol{\theta}}_n$ be a maximum likelihood estimate. Derive a formula for a 95% confidence interval for the random variable $\ddot{y}(\mathbf{s}, \hat{\boldsymbol{\theta}}_n)$ given an input pattern vector $\mathbf{s}$. Discuss the appropriateness of all assumptions required for your analysis to be valid.

15.3-2. *Linear Regression Prediction Confidence Region with Smooth $L_1$ Regularizer.* Let $\{(\tilde{\mathbf{s}}_1, \tilde{y}_1), \ldots, (\tilde{\mathbf{s}}_n, \tilde{y}_n)\}$ be a sequence of $n$ independent and identically distributed random vectors. Assume the probability density of the response scalar variable $\tilde{y}_i$ is a Gaussian random variable with mean $\ddot{y}(\mathbf{s}, \boldsymbol{\theta}) = \boldsymbol{\theta}^T \mathbf{s}_i$ and variance one for a given $d$-dimensional input pattern vector $\mathbf{s}_i$, $i = 1, \ldots, n$. Also assume a hyperbolic secant prior on $\boldsymbol{\theta}$ defined such that

$$p_{\theta}(\boldsymbol{\theta}) = \prod_{j=1}^{d+1} \left[\exp\left((\pi/2)\theta_j\right) + \exp\left(-(\pi/2)\theta_j\right)\right]^{-1}.$$

Let $\hat{\boldsymbol{\theta}}_n$ be a MAP estimate. Derive a formula for a 95% confidence interval for the random variable $\ddot{y}(\mathbf{s}, \hat{\boldsymbol{\theta}}_n)$ given an input pattern vector $\mathbf{s}$. Discuss the appropriateness of all assumptions required for your analysis to be valid.

15.3-3. *Sigmoidal Perceptron Model Confidence Region with $L_2$ Regularizer.* Let $\{(\tilde{\mathbf{s}}_1, \tilde{y}_1), \ldots, (\tilde{\mathbf{s}}_n, \tilde{y}_n)\}$ be a sequence of $n$ independent and identically distributed random vectors such that the response random variable $\tilde{y}_i \in \{0, 1\}$ for a given input pattern vector $\mathbf{s}_i$ for $i = 1, 2, \ldots$. Assume the predicted response $\ddot{y}(\mathbf{s}, \boldsymbol{\theta})$ is defined by the formula:

$$\ddot{y}(\mathbf{s}, \boldsymbol{\theta}) = \mathbf{w}^T \mathbf{h}(\mathbf{s}, \boldsymbol{\theta})$$

where the $k$th element of $\mathbf{h}(\mathbf{s}, \boldsymbol{\theta})$, $h_k$, is the function defined such that:

$$h_k(\mathbf{s}, \boldsymbol{\theta}) = \mathcal{S}\left(\mathbf{v}_k^T [\mathbf{s}^T, 1]^T\right)$$

where $\mathcal{S}(\phi) \equiv 1/(1 + \exp(-\phi))$. Let $\hat{\boldsymbol{\theta}}_n$ be a strict local minimizer of the penalized empirical risk function

$$\hat{\ell}_n(\boldsymbol{\theta}) = \lambda |\boldsymbol{\theta}|^2 + \sum_{i=1}^{n} \left(\tilde{y}_i - \ddot{y}(\tilde{\mathbf{s}}_i, \boldsymbol{\theta})\right)^2.$$

Derive a formula for a 95% confidence interval for the random variable $\ddot{y}(\mathbf{s}, \hat{\boldsymbol{\theta}}_n)$ given an input pattern vector $\mathbf{s}$. Discuss the appropriateness of all assumptions required for your analysis to be valid.

## 15.4 Hypothesis Testing for Model Comparison Decisions

Confidence regions are useful for characterizing the sampling error of a particular statistic. In many situations, however, a statistical test is desirable for the purpose of making a specific decision. For example, suppose the connections from a particular input unit have values which are very small in magnitude. One might be tempted to assume that input unit is not contributing to the predictive performance of the learning machine but it is possible that as more data is collected, the connections from that input unit will converge to values which are not non-zero. A statistical test can be used to support a decision regarding whether or not the connections from that input unit are influencing the learning machine's predictive performance.

### 15.4.1 Classical Hypothesis Testing

**Definition 15.4.1** (Classical Statistical Test). The *null hypothesis*, $H_o$, of a statistical test is an assertion that can be either true or false. Let the random data set $\tilde{\mathcal{D}}_n$ be a $d$ by $n$ dimensional random matrix which has probability distribution $P_e$ if and only if $H_o$ holds.

- A *statistical test* is a Borel-measurable function $\Upsilon_n : \mathcal{R}^{d \times n} \rightarrow \{0, 1\}$ defined such that for all $\mathcal{D}_n \in \mathcal{R}^{d \times n}$: $\Upsilon_n(\mathcal{D}_n) = 1$ is interpreted as the decision to decide $H_o$ is false. and $\Upsilon_n(\mathcal{D}_n) = 0$ is interpreted as the decision to decide $H_o$ is true.
- The *test statistic* is defined as $\tilde{\Upsilon}_n \equiv \Upsilon_n(\tilde{\mathcal{D}}_n)$.
- A *Type 1 error* occurs when $\tilde{\Upsilon}_n = 1$ and $H_o$ is true.
- A *Type 2 error* occurs when $\tilde{\Upsilon}_n = 0$ and $H_o$ is false.
- The *Type 1 error probability* or *p-value* of the statistical test is $E\{\tilde{\Upsilon}_n | H_o\}$.
- The *power* of the statistical test is $E\{\tilde{\Upsilon}_n | \neg H_o\}$.
- The *Type 2 error probability* of the statistical test is defined as $1 - E\{\tilde{\Upsilon}_n | \neg H_o\}$.

□

Informally, a statistical test is a function whose input is a statistic and whose binary output indicates whether or not to reject a null hypothesis. Incorrectly rejecting the null hypothesis is called a Type 1 error, while incorrectly accepting the null hypothesis is called a Type 2 error. The power of a statistical test is the probability of correctly rejecting the null hypothesis. The probability of a Type 1 error is called a p-value. It is important to emphasize that the p-value is the probability of a Type 1 error and is not interpretable as the probability that the null hypothesis is false.

In practice, one checks if the estimated p-value is less than some critical value $\alpha$ which is called the *significance level*. This controls the magnitude of the Type 1 error. To control the magnitude of the Type 2 error, statistical tests are often designed so that the probability of the Type 2 error converges to zero as the sample size becomes large. If the null hypothesis is rejected at the $\alpha$ significance level, then the statistical test is said to be significant at the $\alpha$ significance level. A typical choice for the significance level is $\alpha = 0.05$.

**Theorem 15.4.1** (Model Statistic Wald Test). *Assume the conditions of the Model Statistic Distribution Theorem (Theorem 15.2.2) hold with respect to a DGP Radon-Nikodým density* $p_e : \mathcal{R}^d \rightarrow [0, \infty)$*, restricted parameter space* $\Theta \subseteq \mathcal{R}^q$*, loss function* $c : \mathcal{R}^d \times \Theta \rightarrow \mathcal{R}$*, penalty function* $k_n : \mathcal{R}^{d \times n} \times \Theta \rightarrow \mathcal{R}$*, and continuously differentiable function* $\boldsymbol{\phi} : \mathcal{R}^q \rightarrow \mathcal{R}^r$*. Let* $\mathbf{J} \equiv d\boldsymbol{\phi}/d\boldsymbol{\theta}$*. Assume* $\mathbf{J}(\boldsymbol{\theta}^*)$ *has full row rank* $r$*. Let* $\hat{\mathbf{Q}}_n$ *be defined as in Equation 15.14 by the formula:*

$$\hat{\mathbf{Q}}_n \equiv \mathbf{J}(\hat{\boldsymbol{\theta}}_n)\hat{\mathbf{C}}_n[\mathbf{J}(\hat{\boldsymbol{\theta}}_n)]^T$$

*where* $\hat{\mathbf{C}}_n \equiv \hat{\mathbf{A}}_n^{-1}\hat{\mathbf{B}}_n\hat{\mathbf{A}}_n^{-1}$*. Let*

$$\hat{\mathcal{W}}_n \equiv n\boldsymbol{\phi}(\hat{\boldsymbol{\theta}}_n)^T[\hat{\mathbf{Q}}_n]^{-1}\boldsymbol{\phi}(\hat{\boldsymbol{\theta}}_n).$$

*If the null hypothesis* $H_o : \boldsymbol{\phi}(\boldsymbol{\theta}^*) = \mathbf{0}_r$ *is true, then* $\hat{\mathcal{W}}_n$ *converges in distribution to a chi-squared random variable with* $r$ *degrees of freedom as* $n \rightarrow \infty$*. If the null hypothesis* $H_o : \boldsymbol{\phi}(\boldsymbol{\theta}^*) = \mathbf{0}_r$ *is false, then* $\hat{\mathcal{W}}_n \rightarrow \infty$ *with probability one as* $n \rightarrow \infty$*.*

*Proof.* If the null hypothesis $H_o : \boldsymbol{\phi}(\boldsymbol{\theta}^*) = \mathbf{0}_r$ is true, then by Theorem 15.3.1, $\hat{\mathcal{W}}_n$ converges in distribution to a chi-squared random variable with $r$ degrees of freedom as $n \rightarrow \infty$. If the null hypothesis $H_o : \boldsymbol{\phi}(\boldsymbol{\theta}^*) = \mathbf{0}_r$ is false, then as $n \to \infty$, $\boldsymbol{\phi}(\hat{\boldsymbol{\theta}}_n) \to \boldsymbol{\phi}(\boldsymbol{\theta}^*)$ with probability one where $\boldsymbol{\phi}(\boldsymbol{\theta}^*)$ is a non-zero $r$-dimensional vector. Thus, as $n \to \infty$

$$\hat{\mathcal{W}}_n = n(\boldsymbol{\phi}(\hat{\boldsymbol{\theta}}_n))^T [\hat{\mathbf{Q}}_n]^{-1} (\boldsymbol{\phi}(\hat{\boldsymbol{\theta}}_n))$$

approaches $n$ multiplied by a constant with probability one. That is, $\hat{\mathcal{W}}_n \to \infty$ with probability one as $n \to \infty$. ■

The Wald Test Theorem is used as follows. The Wald test statistic $\hat{\mathcal{W}}_n$ is computed and then one computes using off-the-shelf software (e.g., see Example 15.3.4) the probability $p_\alpha$, a chi-squared random variable with $r$ degrees of freedom, is greater than $\hat{\mathcal{W}}_n$. If $p_\alpha$ is less than a specified significance level $\alpha$ (e.g., $\alpha = 0.05$), then decide to reject $H_o : \boldsymbol{\phi}(\boldsymbol{\theta}^*) = \mathbf{0}_r$; otherwise conclude there is not sufficient evidence to reject $H_o$.

## Recipe Box 15.3 Using the Wald Test (Theorem 15.4.1)

- **Step 1: Specify null hypothesis.**
  Specify an empirical risk function

$$\hat{\ell}_n(\boldsymbol{\theta}) = (1/n) \sum_{i=1}^{n} c(\tilde{\mathbf{x}}_i, \boldsymbol{\theta})$$

  where $c : \mathcal{R}^d \times \mathcal{R}^q \to \mathcal{R}$ is a twice continuously differentiable random function. Let $\ell(\boldsymbol{\theta}) \equiv E\{c(\tilde{\mathbf{x}}_i, \boldsymbol{\theta})\}$. Let $\boldsymbol{\theta}^*$ be a strict local minimizer of $\ell$. Define a continuously differentiable function $\boldsymbol{\phi} : \mathcal{R}^q \to \mathcal{R}^r$ in order to specify the null hypothesis

$$H_o : \boldsymbol{\phi}(\boldsymbol{\theta}^*) = \mathbf{0}_r.$$

  The derivative of $\boldsymbol{\phi}$, $\mathbf{J} : \mathcal{R}^q \to \mathcal{R}^r$, evaluated at $\boldsymbol{\theta}^*$ must have full row rank $r$.

- **Step 2: Compute sampling error covariance matrix.**
  Compute $\hat{\mathbf{Q}}_n$ following the procedures in Recipe Box 15.2 using $c$, $\boldsymbol{\phi}$, and $\mathbf{J}$.

- **Step 3: Compute the Wald statistic.**
  Compute

$$\hat{\mathcal{W}}_n = n\boldsymbol{\phi}(\hat{\boldsymbol{\theta}}_n)^T [\hat{\mathbf{Q}}_n]^{-1} \boldsymbol{\phi}(\hat{\boldsymbol{\theta}}_n).$$

  The Wald statistic $\hat{\mathcal{W}}_n$ is approximately a realization of a chi-squared random variable with $r$ degrees of freedom.

- **Step 4: Estimate Type 2 error probability.**
  Compute the Type 2 error probability

$$\hat{p}_n = 1 - \texttt{GAMMAINC}(\hat{\mathcal{W}}_n/2, r/2)$$

  where `GAMMAINC` is a software function such as defined in Example 15.3.4 that computes the probability that a chi-squared random variable with $r$ degrees of freedom exceeds $\hat{\mathcal{W}}_n$.

- **Step 5: Test the null hypothesis using the Wald test.**
  Define the significance level $\alpha$ for the Wald test (e.g., $\alpha = 0.05$ is typical).
  If $\hat{p}_n < \alpha$, then REJECT $H_o$; else DO NOT REJECT $H_o$.

A significance level of $\alpha$ means that the Type 1 error of the statistical test is bounded from above by $\alpha$. Or, in other words, this means that on the average $(1-\alpha)100\%$ of the time the statistical test will be significant by chance. This means that for $\alpha = 0.05$, one out of twenty statistical tests will be significant. If one therefore does a sufficient number of statistical tests at a fixed significance level $\alpha$, one will almost certainly obtain a significant result! This is called the *multiple comparisons problem.* Therefore, in order to avoid obtaining misleading results from statistical tests, it is important to make a careful distinction between planned comparisons and post hoc comparisons. A *planned comparison* is a statistical test on a data set which is chosen *before* the data set is analyzed. All planned comparisons should be reported in an analysis of a data set. Planned comparisons should be interpreted as evidence about the nature of observed statistical regularities with respect to a specific probability model.

On the other hand, a *post hoc comparison* is a statistical test on a data set which may be chosen *after* the data set is analyzed. Some or all post hoc comparisons can be reported and the significance level for post hoc comparisons does not have to be adjusted. Post hoc comparisons should be used to interpret conclusions associated with planned comparisons. In order to control the experiment-wise error rate for the multiple comparisons problem in situations involving multiple planned comparisons, one can use the Bonferroni Inequality method.

Let $\alpha$ denote the significance level of a statistical test of the null hypothesis that all $K$ planned comparisons are true. Let $\alpha_k$ denote the significance level of a statistical test of the null hypothesis that the $k$th planned comparison is true. The Bonferroni Inequality (Manoukian 1986, p. 49) states that a sufficient but not necessary condition for $\alpha$ to be less than some desired significance level $\alpha_{critical}$ is to choose $\alpha_k = \alpha_{critical}/K$.

**Example 15.4.1** (Logistic Regression Predictor Relevance). Design a statistical test for the logistic regression model described in Example 15.3.5 for testing the null hypothesis that the $k$th parameter value is equal to zero.

**Solution.** Calculate $\hat{\mathcal{W}}_n$ using the formula

$$\hat{\mathcal{W}}_n = (\mathcal{Z}_n(\alpha))^2$$

where $\mathcal{Z}_n(k) = \hat{\theta}_n(k)/\sqrt{\hat{c}_n(k,k)/n}$ and $\hat{c}_n(k,k)$ is the $k$th on-diagonal element of the asymptotic covariance matrix estimate of the $k$th parameter value. Then, calculate $\hat{p}_n$ using the incomplete gamma function using the formula

$$\hat{p}_n = 1 - \texttt{GAMMAINC}(|\mathcal{Z}_n(\alpha)|^2/2, 1/2).$$

If $\hat{p}_n < \alpha$, then reject the null hypothesis that the $k$th parameter value is equal to zero at the $\alpha$ significance level. △

**Example 15.4.2** (Significance of Regression Test: Logistic Regression). A Significance of Regression Test is designed to show that the predictors in a regression model are predictive. Design a statistical test for the logistic regression model in Example 15.3.5 which tests the null hypothesis that all predictors except for the intercept parameter are equal to zero.

**Solution.** Let $\mathbf{S}$ be a matrix with $q-1$ rows defined such that the $k$th row is the $k$th row of a $q$-dimensional identity matrix. Note that $\mathbf{S}$ has full row rank $q-1$. To test the null hypothesis $H_o : \mathbf{S}\boldsymbol{\theta}^* = \mathbf{0}_{q-1}$, calculate $\hat{\mathcal{W}}_n$ using the formula

$$\hat{\mathcal{W}}_n = n\hat{\boldsymbol{\theta}}_n^T \mathbf{S}^T \left(\mathbf{S}\hat{\mathbf{C}}_n\mathbf{S}^T\right)^{-1} \mathbf{S}\hat{\boldsymbol{\theta}}_n$$

where $\hat{\mathbf{C}}_n = \hat{\mathbf{A}}_n^{-1}\hat{\mathbf{B}}_n\hat{\mathbf{A}}_n^{-1}$ where $\hat{\mathbf{A}}_n$ and $\hat{\mathbf{B}}_n$ are defined using Empirical Risk Regularity Assumptions **A7** and **A8** respectively. Then, calculate $\hat{p}_n$ using the incomplete gamma function using the formula

$$\hat{p}_n = 1 - \texttt{GAMMAINC}(0.5|\mathcal{Z}_n(\alpha)|^2, (q-1)/2).$$

If $\hat{p}_n < \alpha$, then reject the null hypothesis that the $k$th parameter value is equal to zero at the $\alpha$ significance level. △

**Example 15.4.3** (Between-Groups Wald Test for Examining Parameter Invariance)**.** An important practical problem in model development is determining which parameter values are invariant across different statistical environments and which parameter values not invariant. For example, suppose that a probability model for speech recognition is developed. One might imagine that a subset of the optimal parameter values will be invariant across different languages, other optimal parameter values might be invariant across speakers within a specific language, while a third group of parameter values might be speaker-dependent. Let $\hat{\ell}_n : \mathcal{R}^q \rightarrow [0, \infty)$ be an empirical risk function. Let the $q$-dimensional random vector $\hat{\boldsymbol{\theta}}_j$ be the parameter estimates of $\boldsymbol{\theta}_j^*$ obtained by finding a strict local minimizer of $\hat{\ell}_n$ using data set $\mathcal{D}_n^j$ for $j = 1, 2, 3$. Let $\hat{\mathbf{C}}_j$ be the asymptotic covariance matrix estimator for $\hat{\boldsymbol{\theta}}_j$ for $j = 1, 2, 3$. Assume the data sets $\mathcal{D}_n^1$, $\mathcal{D}_n^2$, and $\mathcal{D}_n^3$ are, respectively, realizations of the three random data sets $\tilde{\mathcal{D}}_n^1$, $\tilde{\mathcal{D}}_n^2$, and $\tilde{\mathcal{D}}_n^3$. Let $\mathbf{S} \in \mathcal{R}^{r \times q}$ be a selection matrix whose rows are row vectors from a $q$-dimensional identity matrix. Show how to construct a Wald test to test the between-groups null hypothesis: $H_o : \boldsymbol{\theta}_1^* = \boldsymbol{\theta}_2^* = \boldsymbol{\theta}_3^*$.

**Solution.** Since $\tilde{\mathcal{D}}_n^1$, $\tilde{\mathcal{D}}_n^2$, and $\tilde{\mathcal{D}}_n^3$ are independent data sets, the covariance matrix of

$$\ddot{\boldsymbol{\theta}}^T \equiv \left[(\ddot{\boldsymbol{\theta}}_1)^T, (\ddot{\boldsymbol{\theta}}_2)^T, (\ddot{\boldsymbol{\theta}}_3)^T\right]$$

may be specified by the formula:

$$\hat{\mathbf{C}} = \begin{bmatrix} \hat{\mathbf{C}}_1 & \mathbf{0} & \mathbf{0} \\ \mathbf{0} & \hat{\mathbf{C}}_2 & \mathbf{0} \\ \mathbf{0} & \mathbf{0} & \hat{\mathbf{C}}_3 \end{bmatrix} \qquad (15.21)$$

where $\mathbf{0}$ is a $q$-dimensional matrix of zeros.

Then, define a selection matrix $\mathbf{S}$ such that:

$$\mathbf{S} = \begin{bmatrix} \mathbf{I}_q & -\mathbf{I}_q & \mathbf{0} \\ \mathbf{0} & \mathbf{I}_q & -\mathbf{I}_q \end{bmatrix} \qquad (15.22)$$

where $\mathbf{0}$ is a $q$-dimensional matrix of zeros. Note that $\mathbf{S}$ has been defined such that it has full row rank $2q$.

Using the same approach as in Example 15.4.2, the Wald statistic is computed using the selection matrix $\mathbf{S}$ defined in (15.22) and the covariance matrix estimator $\hat{\mathbf{C}}$ defined in 15.21. Then, the standard procedure for applying the Wald Test is applied using the resulting Wald statistic which has $2q$ degrees of freedom. △

### 15.4.2 Bayesian Hypothesis Testing

The classical hypothesis testing method (see Definition 15.4.1) reviewed in the previous sections is based upon the idea of reducing decision errors by keeping the Type 1 error probability below the significance level threshold and showing that the Type 2 error probability converges to zero.

An alternative to the classical hypothesis testing method described is Bayesian Hypothesis Testing. Within the Bayesian Hypothesis Testing framework, the probability that the null hypothesis holds is directly compared with the probability that the alternative hypothesis holds. Let $H_o$ denote the null hypothesis which corresponds to model $\mathcal{M}_o$. Let $H_a$ denote the alternative hypothesis which corresponds to model $\mathcal{M}_a$.

Assume the observed data $\mathbf{x}_1, \ldots, \mathbf{x}_n$ is a realization of a stochastic sequence of *i.i.d.* random vectors with common density $p_e : \mathcal{R}^d \rightarrow [0, \infty)$ defined with respect to support specification measure $\nu$. As discussed in Section 13.2, the likelihood of the observed data is given by the formula

$$p(\mathcal{D}_n|\boldsymbol{\theta}, \mathcal{M}) = \prod_{i=1}^{n} p(\mathbf{x}_i|\boldsymbol{\theta}, \mathcal{M}).$$

The parameter prior for model $\mathcal{M}$ is specified by $p_\theta(\boldsymbol{\theta}|\mathcal{M})$ as discussed in Subsection 13.3.1. The likelihood function $p(\mathcal{D}_n|\boldsymbol{\theta}, \mathcal{M})$ and parameter prior for the model $p_\theta(\boldsymbol{\theta}|\mathcal{M})$ are then combined to compute the marginal likelihood

$$p(\mathcal{D}_n|\mathcal{M}) = \int p(\mathcal{D}_n|\boldsymbol{\theta}, \mathcal{M}) p_\theta(\boldsymbol{\theta}|\mathcal{M}) d\boldsymbol{\theta}. \tag{15.23}$$

The probability of the model $\mathcal{M}$ given a data set $\mathcal{D}_n$ is given by the formula

$$p(\mathcal{M}|\mathcal{D}_n) = \frac{p(\mathcal{D}_n|\mathcal{M}) p(\mathcal{M})}{p(\mathcal{D}_n)}$$

where the model prior probability $p(\mathcal{M})$ specifies the preference for model $\mathcal{M}$ before the data set $\mathcal{D}_n$ is observed.

In Bayesian hypothesis testing, the probability of model $\mathcal{M}_o$ associated with the null hypothesis $H_o$ given the data set $\mathcal{D}_n$ is computed to obtain $p(\mathcal{M}_o|\mathcal{D}_n)$. Next, the probability of the model associated with the alternative hypothesis $H_a$ given the data set $\mathcal{D}_n$ is computed to obtain $p(\mathcal{M}_a|\mathcal{D}_n)$. The ratio of these two quantities is then computed using the formula:

$$\frac{p(\mathcal{M}_a|\mathcal{D}_n)}{p(\mathcal{M}_o|\mathcal{D}_n)} = \left(\frac{p(\mathcal{D}_n|\mathcal{M}_a)}{p(\mathcal{D}_n|\mathcal{M}_o)}\right) \frac{p(\mathcal{M}_a)}{p(\mathcal{M}_o)}$$

where $p(\mathcal{M}_a)$ and $p(\mathcal{M}_o)$ are the model prior probabilties. The ratio

$$K = \frac{p(\mathcal{D}_n|\mathcal{M}_a)}{p(\mathcal{D}_n|\mathcal{M}_o)}$$

is called the *Bayes Factor*. Note that the data-dependent Bayes Factor influences the calculation of the relative probabilities of two competing models $\mathcal{M}_o$ and $\mathcal{M}_a$, yet is not dependent upon the data-independent prior beliefs specifying the model priors. If the Bayes Factor is greater than 5, this is usually interpreted as strong evidence that the data supports the alternative hypothesis $H_a$ over the null hypothesis $H_o$ (e.g., see Kass and Raftery 1995 for a review).

In order to implement Bayesian Hypothesis Testing, it is necessary to compute an approximation of the computationally intractable marginal likelihood in (15.23). This can be accomplished in two different ways. First, simulation methods such as the Monte Carlo Markov chain methods in Chapter 11, the stochastic approximation methods in Chapter 12, or the bootstrap methods in Chapter 14 can be developed to obtain approximations for the marginal likelihood in (15.23). Second, a simple computationally inexpensive formula based upon the Laplace approximation can be developed using the Laplace Approximation methodology described in Chapter 16 (see Example 16.2.1).

## Exercises

15.4-1.*Predictor Relevance: Multinomial Logistic (Softmax) Model.* Let $\mathcal{I}_m \equiv \{\mathbf{e}_1, \ldots, \mathbf{e}_m\}$ be the set of $m$-dimensional column vectors of an $m$-dimensional identity matrix which represent $m$ distinct response categories. Let $\{\mathbf{s}_1, \mathbf{y}_1), \ldots, (\mathbf{s}_n, \mathbf{y}_n)\}$ be a realization of a sequence of $n$ independent and identically distributed random vectors such that $(\mathbf{s}_i, \mathbf{y}_i) \in \{0,1\}^d \times \mathcal{I}_m$, $i = 1, \ldots, n$. For a given binary input pattern $\mathbf{s}_i$, the desired categorical response is $\mathbf{y}_i$ for $i = 1, 2, \ldots$. Let $\mathbf{Q} \in \mathcal{R}^{m \times d}$ be defined such that $\mathbf{Q} = [\mathbf{W}^T, \mathbf{0}_d]^T$ where $\mathbf{W} \in \mathcal{R}^{(m-1) \times d}$. A multinomial logistic (softmax) regression model is defined such that the predicted probability of category $k$ is given by the formula:

$$p(\tilde{\mathbf{y}}_k = \mathbf{e}_k | \mathbf{s}_k, \boldsymbol{\theta}) = \mathbf{e}_k^T \mathbf{p}_k(\mathbf{s}, \boldsymbol{\theta})$$

where

$$\mathbf{p}_k(\mathbf{s}, \boldsymbol{\theta}) = \frac{\mathbf{exp}(\mathbf{Q}\mathbf{s}_k)}{\mathbf{1}_m^T \mathbf{exp}(\mathbf{Q}\mathbf{s}_k)}$$

with $\boldsymbol{\theta} \equiv \mathbf{vec}(\mathbf{W}^T)$. Maximum likelihood estimation is used to estimate the parameters of the softmax probability model with respect to the data generating process.

Derive an explicit formula for a statistical test intended to test the null hypothesis that the $k$th feature has no predictive influence. That is, test the null hypothesis that: $H_o : \bar{\mathbf{w}}_k = \mathbf{0}_{m-1}$ where $\bar{\mathbf{w}}_k$ is the $k$th column of $\mathbf{W}$ where $k \in \{1, \ldots, d\}$. Discuss which assumptions of the analysis automatically hold for the analysis due to the softmax regression modeling assumption and the data generating process modeling assumption. Also discuss which assumptions of the analysis need to be empirically checked.

15.4-2.Repeat Exercise 15.4-1 but now derive an explicit formula for a statistical test intended to test the null hypothesis that the $j$th output probability unit can be eliminated from the softmax network without reducing predictive performance. That is, test the null hypothesis that $H_o : \mathbf{y}_j = \mathbf{0}_d$ where $\mathbf{y}_j$ is the $j$th row of $\mathbf{W}$ where $j \in \{1, \ldots, (m-1)\}$.

15.4-3.*Softmax Misspecification Detection Method.* Show that by selecting $\mathbf{W}$ in an appropriate way, the softmax model in Exercise 15.4-1 can represent any arbitrary conditional probability distribution of categorical response pattern $\mathbf{y}_k$ given input pattern vector $\mathbf{s}_k$ where $\mathbf{s}_k$ is a column of an identity matrix. Call the probability model which allows $\mathbf{W}$ to be arbitrarily selected as the "full" or "encompassing" model. Define a probability model which restricts elements of $\mathbf{W}$ to particular constant values such as setting as disconnecting some input units by restricting their connection weights to zero as the "reduced" model.

Explain why testing the null hypothesis that a subset of the connection weights of $\mathbf{W}$ as in Exercise 15.4-1 is equal to zero is equivalent to testing the null hypothesis that the reduced model defined by the restricted connectivity pattern is correctly specified. Thus, statistical tests such as described in Exercise 15.4-1 can be interpreted as model misspecification tests (see Section 16.3 for further discussion).

15.4-4.*Statistical Environment Parameter Invariance.* Consider two statistical environments which respectively generate the independent and identically distributed observations $\tilde{\mathbf{x}}_1^1, \ldots, \tilde{\mathbf{x}}_n^1$ from common density $p_e^1$ and the independent and identically distributed observations $\tilde{\mathbf{x}}_1^2, \ldots, \tilde{\mathbf{x}}_n^2$ from common density $p_e^2$. Let

$$\mathcal{M} \equiv \{p(\mathbf{x}; \boldsymbol{\theta}) : \boldsymbol{\theta} \in \Theta \subseteq \mathcal{R}^q\}$$

be a probability model. Let $\hat{\boldsymbol{\theta}}_n^j$ denote the quasi-maximum likelihood estimates estimated using data sampled from density $p_e^j$ with respect to model $\mathcal{M}$ for $j = 1, 2$. Assume $\hat{\boldsymbol{\theta}}_n^j \rightarrow \boldsymbol{\theta}^{*j}$ with probability one as $n \rightarrow \infty$. Design a statistical test intended to test the null hypothesis $H_o : \boldsymbol{\theta}^1 = \boldsymbol{\theta}^2$. Then, derive a second statistical test to test the null hypothesis that the first and last element of $\boldsymbol{\theta}^1$ are equal to the first and last elements of $\boldsymbol{\theta}^2$.

## 15.5 Further Readings

### Maximum Likelihood Estimation for Possibly Misspecified Models

Huber (1964, 1967) provided early discussions of the concept of maximum likelihood estimation in the presence of possible model misspecification. White (1982) extended this theoretical framework for supporting maximum likelihood estimation and inference in the presence of possible model misspecification. Extensions of the results obtained in this chapter can be obtained for situations where the assumption of independent and identically distributed observations is relaxed. Specifically, extensions of these results will hold under the weaker assumptions that the observations are asymptotically independent, stationary, and bounded (e.g., White and Domowitz 1984; White 1994; Golden 2003). Golden et al. (2019) discuss estimation and inference for possibly misspecified models in the presence of partially observable data. Henley et al. (2020) and Kashner et al. (2020) provide practitioner-oriented reviews of methods for estimation and inference in the presence of possible model misspecification.

### Asymptotic Distribution of Parameter Estimates

The general method for characterizing the asymptotic consistency and asymptotic distribution of a strict local minimizer of an empirical risk function is known as M-estimation in the statistics literature (e.g., Huber 1964; Huber 1967; Serfling 1980, Chapter 7; White 1982; van der Vaart 1998, Chapter 3; White 1994). The M-estimation method is often attributed to the seminal contribution of Huber (1964). White (1989a, 1989b) discusses this method in the context of training multilayer neural networks by minimizing empirical risk functions. White and Domowitz (1984) discuss this method in the context of parameter estimation for nonlinear regression models.

### Model Selection Tests for Misspecified and Non-Nested Models

Wilks (1938) Generalized Likelihood Ratio Test (GLRT) is very popular among practitioners. The GLRT assumes that: (i) one model $\mathcal{M}_\theta$ is fully nested within another model $\mathcal{M}_\psi$ (i.e., $\mathcal{M}_\theta \subseteq \mathcal{M}_\psi$), and (ii) the full model $\mathcal{M}_\psi$ is correctly specified with respect to the DGP density $p_e$ (i.e., $p_e \in \mathcal{M}_\psi$). Let $\hat{\ell}_n(\hat{\boldsymbol{\theta}}_n)$ and $\hat{\ell}_n(\hat{\boldsymbol{\psi}}_n)$ respectively denote the fit of models $\mathcal{M}_\theta$ and $\mathcal{M}_\psi$ to the observed data generated from $p_e$. Assume the dimensionality of $\hat{\boldsymbol{\psi}}_n$ is $r$ and the dimensionality of $\hat{\boldsymbol{\theta}}_n$ is $q$. In addition, assume the Empirical Risk Regularity

Assumptions (see Definition 15.1.1) hold for both models. Given these assumptions, the likelihood ratio test statistic

$$\hat{\delta}_n \equiv -2n\left(\hat{\ell}_n(\hat{\boldsymbol{\psi}}_n) - \hat{\ell}_n(\hat{\boldsymbol{\theta}}_n)\right).$$

has an asymptotic chi-squared distribution with $r - q$ degrees of freedom.

Vuong (1989) showed that the asymptotic distribution of the test statistic $\hat{\delta}_n$ could be computed without assuming that either of the two models was correctly specified and without assuming that one model was nested within another. Vuong (1989) showed the asymptotic distribution of the test statistic $\hat{\delta}_n$ had a weighted chi-squared distribution in this generalization of the result of Wilks (1938). Golden (2003) shows how to modify Vuong's (1989) results for developing statistical tests for testing the null hypothesis that two empirical risk function have empirically equivalent fits to a stationary, bounded, $\tau$-dependent data generating process.

## Confidence Intervals for Deep Learning Network Predictions

The methods described in this chapter are commonly used to estimate confidence intervals on the expected response of linear regression and logistic regression models for a given input pattern. In addition, the application of asymptotic statistical theory such as M-estimation for the purpose of estimating standard errors in artificial neural networks has been previously discussed in the literature (e.g., White 1989a, 1989b; Lee et al. 1993; Golden 1996). Tibshirani (1996b) compared different methods for estimating confidence intervals for deep learning network predictions.

## Pruning and Interpreting Deep Neural Networks

Using the Wald test described in this chapter to determine which connection weights may be safely pruned is a standard methodology in linear regression, logistic regression, and multinomial logistic regression. Furthermore, one could apply the techniques in this chapter for the purpose of analyzing and interpreting the final layers of a deep learning neural network given the assumption all previous layers of connection weights are fixed.

In addition, methods similar to the Wald test methodology proposed in this chapter have been proposed as strategies for pruning deep learning networks (e.g., Mozer and Smolensky 1989; White 1989a, 1989b; Hassibi and Stork 1993; Hassibi et al. 1993; Lee et al. 1993; Hassibi et al. 1994; Golden 1996). However, Hessian methods have only recently been used for pruning large-scale deep learning networks (e.g., Molchanov et al. 2017; Molchanov et al. 2019). Still, the application of the techniques in this chapter for the purpose of the design and interpretation of deep learning neural nets remains an active area of empirical research.

# 16

# *Model Selection and Evaluation*

**Learning Objectives**

- Define and use model selection criterion (MSC) method.
- Derive cross-validation risk MSC.
- Derive Bayesian risk MSC.
- Derive misspecification detection MSC.

Assume the observed data $\tilde{\mathcal{D}}_n \equiv [\tilde{\mathbf{x}}_1, \ldots, \tilde{\mathbf{x}}_n]$ consists of independent and identically distributed observations with common probability density $p_e$ with respect to some support specification measure $\nu$. The researcher has $M$ probability models $\mathcal{M}_1, \ldots, \mathcal{M}_M$ where the $k$th probability model

$$\mathcal{M}_k \equiv \{p_k(\mathbf{x}|\boldsymbol{\theta}_k) : \boldsymbol{\theta}_k \in \Theta_k\}$$

consists of probability densities defined with respect to a common support specification measure $\nu$.

The *model selection problem* is concerned with determining which of the models $\mathcal{M}_1, \ldots, \mathcal{M}_M$ is the most appropriate model of the data generating process that generated $\tilde{\mathcal{D}}_n$. To achieve this goal, the performance of a model with respect to a particular statistical environment is often mapped into a number called a Model Selection Criterion (MSC). Once the MSC for each of the $M$ models is computed, then the set of models can be rank ordered or the model with the smallest MSC can be chosen. In this chapter, three distinct categories of model selection criteria are discussed: (1) Cross-Validation Risk MSC, (2) Bayesian Risk MSC, and (3) Misspecification Detection MSC.

The *Cross-Validation Risk MSC* is designed to characterize the generalization performance of a fitted probability model. Specifically, given that the fitted probability model was trained on one data sample, how will its prediction error (or equivalently empirical risk) change when it is tested on a novel data sample? Referring to Figure 16.1, the Cross-Validation Risk MSC may be interpreted as identifying the "best-fitting" probability density in the probability model $\mathcal{M}$ by estimating a parameter vector $\hat{\boldsymbol{\theta}}_n$ using the observed training data. The similarity of the estimated density in $\mathcal{M}$ to the estimated data generating density $p_e$ is then reported as the random variable $\hat{\ell}_n(\hat{\boldsymbol{\theta}}_n)$. The expected value of the similarity measure $\hat{\ell}_n(\hat{\boldsymbol{\theta}}_n)$ is defined as the cross-validation risk.

The *Bayesian Risk MSC* is designed to estimate the probability of model $\mathcal{M}_k$ given the data $\mathcal{D}_n$ for $k = 1, \ldots, M$ which is denoted as $p(\mathcal{M}_k|\mathcal{D}_n)$. This probability is useful because it can be used to select the model which is most probable given the data if one assumes that the *model selection loss* incurred for incorrectly selecting model $\mathcal{M}_k$ is equal to one

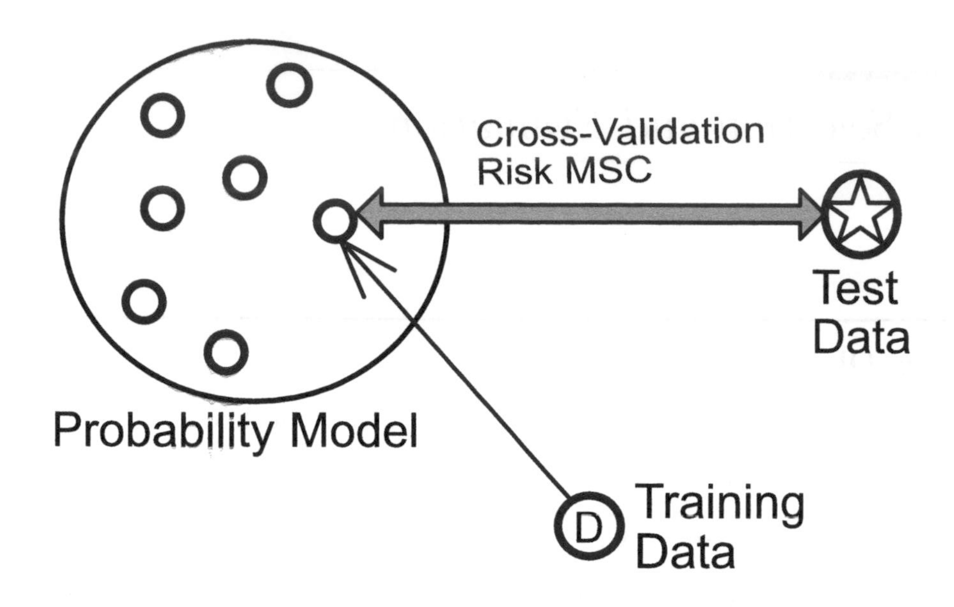


**FIGURE 16.1**
**The concept of a cross-validation risk model selection criterion.** A cross-validation risk MSC estimates the expected empirical risk on a test data set using parameter values estimated from a training data set. Equivalently, the cross-validation risk MSC estimates test data performance for the density in the model which best fits the training data.

and the *model selection loss* incurred for correctly selecting model $\mathcal{M}_k$ is equal to zero. For more general model selection loss functions, the Bayesian Risk MSC can be used to select the model which minimizes an expected model selection loss (see Exercise 16.2-1). Unlike the cross-validation risk MSC which is based upon a probability density in the model which "best-fits" the training data, the Bayesian Risk MSC integrates information from every probability density in the model (see Figure 16.2).

A *Misspecification Detection MSC* is used to assess if a given probability model can properly represent the data generating process by examining the model for evidence of misspecification (see Definition 10.1.7). The semantic interpretation of a misspecification detection MSC is that it measures evidence that supports the hypothesis that a particular probability model is misspecified with respect to a particular statistical environment. The hypothesis that a particular probability model is misspecified can be true or false as illustrated in Figure 16.3.

## 16.1 Cross-Validation Risk Model Selection Criteria

Consider the problem of evaluating the prediction error of a model as a function of sample size $n$. Let $\tilde{\mathbf{x}}_1, \tilde{\mathbf{x}}_2, \ldots$ be a stochastic sequence of independent and identically distributed $d$-dimensional random vectors. Let the first $n$ random vectors in the stochastic sequence be denoted as $\tilde{\mathcal{D}}_n^1$, the second $n$ random vectors in the stochastic sequence be denoted as $\tilde{\mathcal{D}}_n^2$, and so on.

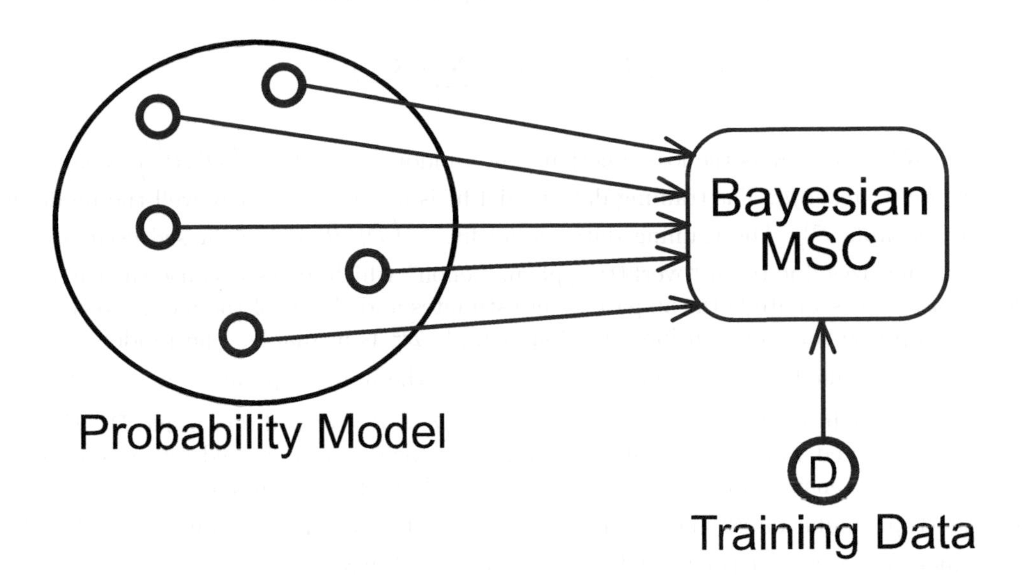


**FIGURE 16.2**
**The concept of a Bayesian model selection criterion.** A Bayesian risk MSC is computed using a weighted average of the likelihood of each density in the probability model given the observed data. The Bayesian model selection criteria uses all of the densities in the model since it summarizes information about the entire probability model.

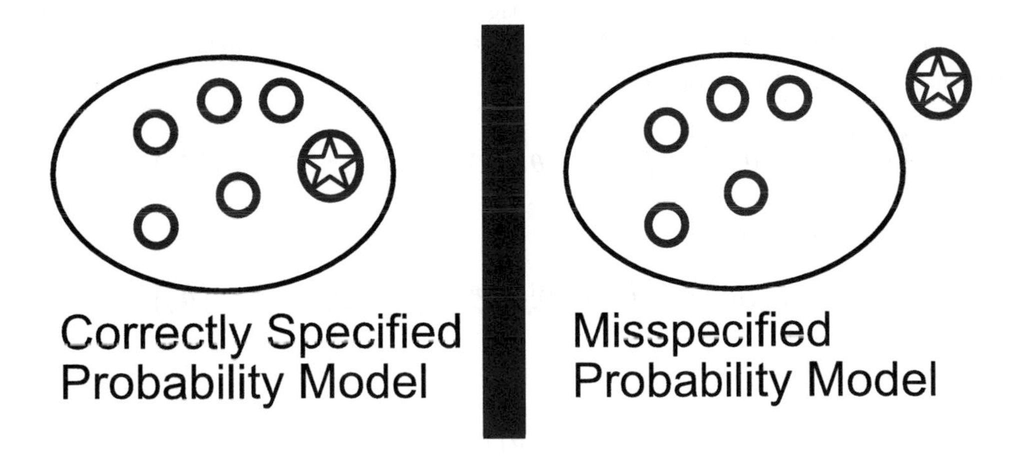


**FIGURE 16.3**
**The concept of a misspecification detection model selection criterion.** A misspecification detection model selection criterion measures evidence that model misspecification is present. A probability model is represented by a set of probability distributions which are represented as small circles. If the data generating process density (represented by a star in a circle) is not inside the probability model, this corresponds to the case of model misspecification.

Let $\hat{\boldsymbol{\theta}}_n^1$ be a strict global minimizer of the empirical risk function

$$\ell_n(\tilde{\mathcal{D}}_n^1, \cdot) = (1/n) \sum_{i=1}^{n} c(\tilde{\mathbf{x}}_i, \cdot)$$

where $c : \mathcal{R}^d \times \Theta \to \mathcal{R}$ is the loss function. The random variable $\ell_n\left(\tilde{\mathcal{D}}_n^1, \hat{\boldsymbol{\theta}}_n^1\right)$ is called the *training data model fit*. The training data model fit is a measure of how well the model fits the data sample $\tilde{\mathcal{D}}_n^1$. The training data model fit, $\ell_n\left(\tilde{\mathcal{D}}_n^1, \hat{\boldsymbol{\theta}}_n^1\right)$, is a biased estimator for finite sample sizes due to an "overfitting phenomenon" which arises because the same data set was used to estimate both the parameter estimates and the fit of the model to the data.

Now suppose that an additional random sample $\tilde{\mathcal{D}}_n^2$ is available. The random variable $\ell_n\left(\tilde{\mathcal{D}}_n^2, \hat{\boldsymbol{\theta}}_n^1\right)$ is called the *test data model fit* because the model's parameters are estimated using $\tilde{\mathcal{D}}_n^1$ yet the fit of the model is evaluated on the *out-of-sample* data sample $\tilde{\mathcal{D}}_n^2$. Furthermore, the expected value of the test data model fit provides an unbiased estimate of model fit. A natural extension of this methodology is to use K-fold cross-validation or nonparametric bootstrap simulation estimation methods to estimate the expected value of $\ell_n\left(\tilde{\mathcal{D}}_n^2, \hat{\boldsymbol{\theta}}_n^1\right)$. These methods were discussed in Chapter 14 (also see Exercise 16.1-3).

Figure 14.2 plots the expected values of $\ell_n\left(\tilde{\mathcal{D}}_n^1, \hat{\boldsymbol{\theta}}_n^1\right)$ and $\ell_n\left(\tilde{\mathcal{D}}_n^2, \hat{\boldsymbol{\theta}}_n^1\right)$ as a function of sample size using solid and dashed lines respectively. As the sample size increases, both measures of prediction error tend to decrease but the expected value of the test data model fit $\ell_n\left(\tilde{\mathcal{D}}_n^2, \hat{\boldsymbol{\theta}}_n^1\right)$ appears to be always greater than that of the expected value of the training data model fit $\ell_n\left(\tilde{\mathcal{D}}_n^1, \hat{\boldsymbol{\theta}}_n^1\right)$. Figure 14.3 illustrates an overfitting phenomenon which occurs when the probability model captures statistical regularities which are specific to the training data.

The following theorem theoretically investigates the empirical observations presented in Figure 14.2 and Figure 14.3. Since $\tilde{\mathcal{D}}_n^1$ and $\tilde{\mathcal{D}}_n^2$ are statistically independent and using the notation $\hat{\boldsymbol{\theta}}_n^1 = \boldsymbol{\theta}\left(\tilde{\mathcal{D}}_n^1\right)$ it follows that:

$$E\left\{\ell_n\left(\tilde{\mathcal{D}}_n^2, \hat{\boldsymbol{\theta}}_n^1\right)\right\} = \int \ell_n\left(\mathcal{D}_n^2, \boldsymbol{\theta}\left(\mathcal{D}_n^1\right)\right) p_e^n\left(\mathcal{D}_n^1\right) p_e^n\left(\mathcal{D}_n^2\right) d\nu\left(\mathcal{D}_n^1, \mathcal{D}_n^2\right)$$

which can be rewritten as:

$$E\left\{\ell_n\left(\tilde{\mathcal{D}}_n^2, \hat{\boldsymbol{\theta}}_n^1\right)\right\} = \int \ell\left(\boldsymbol{\theta}\left(\mathcal{D}_n^1\right)\right) p_e^n\left(\mathcal{D}_n^1\right) d\nu\left(\mathcal{D}_n^1\right) = E\left\{\ell\left(\hat{\boldsymbol{\theta}}_n^1\right)\right\} \tag{16.1}$$

where the expected loss

$$\ell(\boldsymbol{\theta}) = \int \ell_n\left(\mathcal{D}_n^2, \boldsymbol{\theta}\right) p_e^n\left(\mathcal{D}_n^2\right) d\nu\left(\mathcal{D}_n^2\right).$$

Equation (16.1) provides an explicit expression for the expected novel test data fit, $E\{\ell(\hat{\boldsymbol{\theta}}_n)\}$, in terms of only the training data parameter estimates $\hat{\boldsymbol{\theta}}_n$. The following theorem shows how to calculate the expected novel test data fit from the expected training data fit $E\{\hat{\ell}_n(\hat{\boldsymbol{\theta}}_n)\}$. Before presenting the theorem, however, a useful lemma, which is key to the theorem's proof, is presented.

**Lemma 16.1.1** (GAIC Lemma). *Assume the Empirical Risk Regularity Assumptions in Definition 15.1.1 hold with respect to a stochastic sequence of i.i.d. random vectors*

*$\tilde{\mathbf{x}}_1, \ldots, \tilde{\mathbf{x}}_n$ with common DGP Radon-Nikodým density $p_e : \mathcal{R}^d \rightarrow [0, \infty)$ defined with respect to a support specification measure $\nu$, a loss function $c : \mathcal{R}^d \times \mathcal{R}^q \rightarrow \mathcal{R}$, restricted parameter space $\Theta \subseteq \mathcal{R}^q$, empirical risk function minimizer $\hat{\boldsymbol{\theta}}_n$ of*

$$\hat{\ell}_n(\boldsymbol{\theta}) \equiv \hat{k}_n(\boldsymbol{\theta}) + (1/n) \sum_{i=1}^{n} c(\tilde{\mathbf{x}}_i, \boldsymbol{\theta})$$

*on $\Theta$, and risk function minimizer $\boldsymbol{\theta}^*$ of*

$$\ell(\boldsymbol{\theta}) = \int c(\mathbf{x}, \boldsymbol{\theta}) p_e(\mathbf{x}) d\nu(\mathbf{x})$$

*on $\Theta$. Let $\mathbf{A}^*$ and $\mathbf{B}^*$ be defined as in **A7** and **A8** respectively in Definition 15.1.1.*
*Then,*

$$E\{(\hat{\boldsymbol{\theta}}_n - \boldsymbol{\theta}^*)^T \mathbf{A}^* (\hat{\boldsymbol{\theta}}_n - \boldsymbol{\theta}^*)\} = (1/n) \operatorname{tr}\left([\mathbf{A}^*]^{-1} \mathbf{B}^*\right). \tag{16.2}$$

*Proof.*

$$E\{(\hat{\boldsymbol{\theta}}_n - \boldsymbol{\theta}^*)^T \mathbf{A}^* (\hat{\boldsymbol{\theta}}_n - \boldsymbol{\theta}^*)\} = \operatorname{tr}\left(E\left\{[\mathbf{A}^*]^{1/2}(\hat{\boldsymbol{\theta}}_n - \boldsymbol{\theta}^*)\left([\mathbf{A}^*]^{1/2}(\hat{\boldsymbol{\theta}}_n - \boldsymbol{\theta}^*)\right)^T\right\}\right)$$

$$= \operatorname{tr}\left([\mathbf{A}^*]^{1/2} E\left\{(\hat{\boldsymbol{\theta}}_n - \boldsymbol{\theta}^*)(\hat{\boldsymbol{\theta}}_n - \boldsymbol{\theta}^*)^T\right\} [\mathbf{A}^*]^{1/2}\right). \tag{16.3}$$

Now use the Asymptotic Estimator Distribution Theorem 15.2.1 to obtain:

$$E\left\{(\hat{\boldsymbol{\theta}}_n - \boldsymbol{\theta}^*)(\hat{\boldsymbol{\theta}}_n - \boldsymbol{\theta}^*)^T\right\} = (1/n)[\mathbf{A}^*]^{-1}\mathbf{B}^*[\mathbf{A}^*]^{-1}$$

and substitute this relation into (16.3) to obtain:

$$E\{(\hat{\boldsymbol{\theta}}_n - \boldsymbol{\theta}^*)^T \mathbf{A}^* (\hat{\boldsymbol{\theta}}_n - \boldsymbol{\theta}^*)\} = (1/n)[\mathbf{A}^*]^{1/2}[\mathbf{A}^*]^{-1}\mathbf{B}^*[\mathbf{A}^*]^{-1}[\mathbf{A}^*]^{1/2}. \tag{16.4}$$

Now use the result that the trace operator is invariant under cyclic permutations (Theorem 4.1.1) to obtain the relation:

$$\operatorname{tr}\left([\mathbf{A}^*]^{1/2}[\mathbf{A}^*]^{-1}\mathbf{B}^*[\mathbf{A}^*]^{-1}[\mathbf{A}^*]^{1/2}\right) = \operatorname{tr}\left([\mathbf{A}^*]^{-1}\mathbf{B}^*\right). \tag{16.5}$$

Substitute (16.5) into the right-hand side of (16.4) to complete the proof. ■

**Theorem 16.1.2** (Empirical Risk Overfitting Bias). *Assume the Empirical Risk Regularity Assumptions in Definition 15.1.1 hold with respect to a stochastic sequence of i.i.d. random vectors $\tilde{\mathbf{x}}_1, \ldots, \tilde{\mathbf{x}}_n$ with common DGP Radon-Nikodým density $p_e : \mathcal{R}^d \rightarrow [0, \infty)$ defined with respect to a support specification measure $\nu$, a loss function $c : \mathcal{R}^d \times \mathcal{R}^q \rightarrow \mathcal{R}$, restricted parameter space $\Theta \subseteq \mathcal{R}^q$, empirical risk function minimizer $\hat{\boldsymbol{\theta}}_n$ of*

$$\hat{\ell}_n(\boldsymbol{\theta}) \equiv (1/n) \sum_{i=1}^{n} c(\tilde{\mathbf{x}}_i, \boldsymbol{\theta})$$

*on $\Theta$, and risk function minimizer $\boldsymbol{\theta}^*$ of*

$$\ell(\boldsymbol{\theta}) = \int c(\mathbf{x}, \boldsymbol{\theta}) p_e(\mathbf{x}) d\nu(\mathbf{x})$$

*on $\Theta$. Let $\mathbf{A}^*$ and $\mathbf{B}^*$ be defined as in **A7** and **A8** respectively in Definition 15.1.1. Let $\hat{\mathbf{A}}_n$, $\hat{\mathbf{B}}_n$, and $\hat{\mathbf{C}}_n$ be defined as in (15.1), (15.2), and (15.3) respectively.*

*Then, as $n \to \infty$,*

$$E\{\ell(\hat{\boldsymbol{\theta}}_n)\} = E\{\hat{\ell}_n(\hat{\boldsymbol{\theta}}_n)\} + (1/n)\,\mathrm{tr}\left([\mathbf{A}^*]^{-1}\mathbf{B}^*\right) + o_p(1/n).$$

*In addition, as $n \to \infty$,*

$$E\{\ell(\hat{\boldsymbol{\theta}}_n)\} = E\{\hat{\ell}_n(\hat{\boldsymbol{\theta}}_n)\} + (1/n)\,\mathrm{tr}\left([\hat{\mathbf{A}}_n]^{-1}\hat{\mathbf{B}}_n\right) + o_p(1/n).$$

*Proof.* The proof follows the approach of Konishi and Kitagawa (2008; pp. 55-58), Linhart and Zucchini (1986, Appendix A), and Linhart and Volkers (1984, Proposition 2).

$$\ell(\hat{\boldsymbol{\theta}}_n) = \hat{\ell}_n(\hat{\boldsymbol{\theta}}_n) + \left(\ell(\hat{\boldsymbol{\theta}}_n) - \ell(\boldsymbol{\theta}^*)\right) + \left(\ell(\boldsymbol{\theta}^*) - \hat{\ell}_n(\boldsymbol{\theta}^*)\right) + \left(\hat{\ell}_n(\boldsymbol{\theta}^*) - \hat{\ell}_n(\hat{\boldsymbol{\theta}}_n)\right). \quad (16.6)$$

*Step 1: Calculate Estimation Error Using True Global Minimum*

Let $\delta_A(\lambda_n) \equiv \mathbf{A}(\ddot{\boldsymbol{\theta}}_n(\lambda_n)) - \mathbf{A}(\boldsymbol{\theta}^*)$. The second term, $\ell(\hat{\boldsymbol{\theta}}_n) - \ell(\boldsymbol{\theta}^*)$ on the right-hand side of (16.6) corresponds to the effects of parameter estimation error using the true risk function. To estimate the value of $\ell(\hat{\boldsymbol{\theta}}_n) - \ell(\boldsymbol{\theta}^*)$, expand $\ell$ about $\boldsymbol{\theta}^*$ and evaluate at $\hat{\boldsymbol{\theta}}_n$ to obtain:

$$\ell(\hat{\boldsymbol{\theta}}_n) = \ell(\boldsymbol{\theta}^*) + (\hat{\boldsymbol{\theta}}_n - \boldsymbol{\theta}^*)^T \nabla\ell(\boldsymbol{\theta}^*) + (1/2)(\hat{\boldsymbol{\theta}}_n - \boldsymbol{\theta}^*)\mathbf{A}^*(\hat{\boldsymbol{\theta}}_n - \boldsymbol{\theta}^*) + \tilde{R}_n \quad (16.7)$$

where

$$\tilde{R}_n \equiv (1/2)(\hat{\boldsymbol{\theta}}_n - \boldsymbol{\theta}^*)\delta_A(\lambda_n)(\hat{\boldsymbol{\theta}}_n - \boldsymbol{\theta}^*),$$

and $\ddot{\boldsymbol{\theta}}_n(\lambda_n) \equiv \boldsymbol{\theta}^* + \lambda_n(\hat{\boldsymbol{\theta}}_n - \boldsymbol{\theta}^*)$ with $\lambda_n \in [0, 1]$. By the Empirical Risk Minimizer Asymptotic Estimator Distribution Theorem (Theorem 15.2.2), $\sqrt{n}(\hat{\boldsymbol{\theta}}_n - \boldsymbol{\theta}^*)$ converges in distribution to a random vector such that $\sqrt{n}(\hat{\boldsymbol{\theta}}_n - \boldsymbol{\theta}^*) = O_p(1)$. And since $\mathbf{A}$ is continuous and $\hat{\boldsymbol{\theta}}_n \to \boldsymbol{\theta}^*$ with probability one by Theorem 13.1.2, Theorem 9.4.1 implies that $\mathbf{A}(\hat{\boldsymbol{\theta}}_n) \to \mathbf{A}^*$ with probability one as $n \to \infty$, it follows that $\delta_A(\tilde{\lambda}_n) = o_p(1)$ for $n$ sufficiently large. Thus,

$$\tilde{R}_n = O_p(n^{-1/2})o_p(1)O_p(n^{-1/2}) = o_p(1/n).$$

Now take the expectation of (16.7), using the GAIC Lemma Equation (16.2), and noting that $\nabla\ell(\boldsymbol{\theta}^*)$ vanishes one obtains:

$$E\{\ell(\hat{\boldsymbol{\theta}}_n)\} = \ell(\boldsymbol{\theta}^*) + (1/2)(1/n)\,\mathrm{tr}\left([\mathbf{A}^*]^{-1}\mathbf{B}^*\right) + o_p(1/n). \quad (16.8)$$

*Step 2: Calculate Empirical Risk Function Approximation Error*

The third term on the right-hand side of (16.6), $\ell(\boldsymbol{\theta}^*) - \hat{\ell}_n(\boldsymbol{\theta}^*)$, estimates the effects of estimating the true risk function $\ell$ using the empirical risk function $\hat{\ell}_n$ and evaluating at a true local risk minimizer $\boldsymbol{\theta}^*$. Since $\hat{\ell}_n$ is dominated by an integrable function

$$E\{\hat{\ell}_n(\boldsymbol{\theta}^*)\} = \ell(\boldsymbol{\theta}^*). \quad (16.9)$$

*Step 3: Calculate Estimation Error Using True Global Minimizer*

The fourth term on the right-hand side of (16.6), $\hat{\ell}_n(\boldsymbol{\theta}^*) - \hat{\ell}_n(\hat{\boldsymbol{\theta}}_n)$, corresponds to the effects of parameter estimation error with respect to the empirical risk function. To estimate the value of $\hat{\ell}_n(\boldsymbol{\theta}^*) - \hat{\ell}_n(\hat{\boldsymbol{\theta}}_n)$, expand $\hat{\ell}_n$ in a second-order Taylor expansion about $\boldsymbol{\theta}^*$ and evaluate at $\hat{\boldsymbol{\theta}}_n$ to obtain:

$$\hat{\ell}_n(\hat{\boldsymbol{\theta}}_n) = \hat{\ell}_n(\boldsymbol{\theta}^*) + (\hat{\boldsymbol{\theta}}_n - \boldsymbol{\theta}^*)^T \nabla\hat{l}_n(\boldsymbol{\theta}^*) + \frac{1}{2}(\hat{\boldsymbol{\theta}}_n - \boldsymbol{\theta}^*)^T \nabla^2\hat{\ell}_n(\ddot{\boldsymbol{\theta}}_n^{\lambda})(\hat{\boldsymbol{\theta}}_n - \boldsymbol{\theta}^*) \quad (16.10)$$

where for some $\lambda \in [0,1]$:

$$\ddot{\boldsymbol{\theta}}_n^{\lambda} \equiv \hat{\boldsymbol{\theta}}_n + (\boldsymbol{\theta}^* - \hat{\boldsymbol{\theta}}_n)\lambda.$$

Now expand $\nabla \hat{\ell}_n$ in a first-order Taylor expansion about $\boldsymbol{\theta}^*$ to obtain:

$$\nabla \hat{\ell}_n(\hat{\boldsymbol{\theta}}_n) = \nabla \hat{\ell}_n(\boldsymbol{\theta}^*) + \nabla^2 \hat{\ell}_n(\ddot{\boldsymbol{\theta}}_n^{\eta})(\hat{\boldsymbol{\theta}}_n - \boldsymbol{\theta}^*) \tag{16.11}$$

where $\ddot{\boldsymbol{\theta}}_n^{\eta}$ is a vector defined such that its $k$th element

$$\ddot{\theta}_{n,k}^{\eta_k} \equiv \hat{\theta}_{n,k} + (\theta_k^* - \hat{\theta}_{n,k})\eta_k$$

where $\eta_k \in [0,1]$, $\theta_k^*$ is the $k$th element of $\boldsymbol{\theta}^*$, and $\hat{\theta}_{n,k}$ is the $k$th element of $\hat{\boldsymbol{\theta}}_n$.

Since, by definition, $\nabla \hat{\ell}_n(\hat{\boldsymbol{\theta}}_n)$ is a vector of zeros, (16.11) becomes:

$$\nabla \hat{\ell}_n(\boldsymbol{\theta}^*) = -\nabla^2 \hat{\ell}_n\left(\ddot{\boldsymbol{\theta}}_n^{\boldsymbol{\eta}}\right)(\hat{\boldsymbol{\theta}}_n - \boldsymbol{\theta}^*). \tag{16.12}$$

Substitute (16.12) into (16.10) to obtain:

$$\hat{\ell}_n(\hat{\boldsymbol{\theta}}_n) = \hat{\ell}_n(\boldsymbol{\theta}^*) - (\hat{\boldsymbol{\theta}}_n - \boldsymbol{\theta}^*)^T \nabla^2 \hat{l}_n\left(\ddot{\boldsymbol{\theta}}_n^{\boldsymbol{\eta}}\right)(\hat{\boldsymbol{\theta}}_n - \boldsymbol{\theta}^*) +$$

$$(1/2)(\hat{\boldsymbol{\theta}}_n - \boldsymbol{\theta}^*)^T \nabla^2 \hat{\ell}_n\left(\ddot{\boldsymbol{\theta}}_n^{\boldsymbol{\lambda}}\right)(\hat{\boldsymbol{\theta}}_n - \boldsymbol{\theta}^*). \tag{16.13}$$

Since $\sqrt{n}(\hat{\boldsymbol{\theta}}_n - \boldsymbol{\theta}^*) = O_p(1)$ by Theorem 15.2.1 and Theorem 9.3.8, and since $\nabla^2 \hat{\ell}_n\left(\ddot{\boldsymbol{\theta}}_n^{\boldsymbol{\lambda}}\right) = \mathbf{A}^* + o_{a.s.}(1)$, these relations can be used in (16.13) to obtain:

$$\hat{\ell}_n(\boldsymbol{\theta}^*) - \hat{\ell}_n(\hat{\boldsymbol{\theta}}_n) = (1/2)(\hat{\boldsymbol{\theta}}_n - \boldsymbol{\theta}^*)^T \mathbf{A}^* (\hat{\boldsymbol{\theta}}_n - \boldsymbol{\theta}^*) + o_p(1/n). \tag{16.14}$$

Now take the expectation of (16.14), using the GAIC Lemma Equation (16.2), to obtain:

$$E\{\hat{\ell}_n(\boldsymbol{\theta}^*)\} - E\{\hat{\ell}_n(\hat{\boldsymbol{\theta}}_n)\} = (1/2)(1/n)\,\mathrm{tr}\left([\mathbf{A}^*]^{-1}\mathbf{B}^*\right) + o_p(1/n). \tag{16.15}$$

*Step 4: Calculate Empirical Risk Bias*

Now take the expectation of both sides of (16.6) to obtain:

$$E\left\{\ell(\hat{\boldsymbol{\theta}}_n)\right\} = E\left\{\hat{\ell}_n(\hat{\boldsymbol{\theta}}_n)\right\} + E\left\{\left(\ell(\hat{\boldsymbol{\theta}}_n)\right) - \ell(\boldsymbol{\theta}^*)\right\} +$$

$$E\left\{\left(\ell(\boldsymbol{\theta}^*) - \hat{\ell}_n(\boldsymbol{\theta}^*)\right)\right\} + E\left\{\left(\hat{\ell}_n(\boldsymbol{\theta}^*) - \hat{\ell}_n(\hat{\boldsymbol{\theta}}_n)\right)\right\}. \tag{16.16}$$

Now substitute the results of (16.8), (16.9), and (16.15) into (16.16) to obtain:

$$E\{\ell(\hat{\boldsymbol{\theta}}_n)\} = E\{\hat{\ell}_n(\hat{\boldsymbol{\theta}}_n)\} + \frac{\mathrm{tr}\left([\mathbf{A}^*]^{-1}\mathbf{B}^*\right)}{2n} + 0 + \frac{\mathrm{tr}\left([\mathbf{A}^*]^{-1}\mathbf{B}^*\right)}{2n} + o_p(1/n).$$

■

**Definition 16.1.1** (Cross-Validation Risk Criterion (CVRC)). Let $\hat{\ell}_n$, $\hat{\boldsymbol{\theta}}_n$, $\hat{\mathbf{A}}_n$, and $\hat{\mathbf{B}}_n$ be defined as in the Empirical Risk Overfitting Bias Theorem (Theorem 16.1.2). The *Cross-Validation Risk Criterion* (CVRC) is defined as:

$$\mathrm{CVRC} = \hat{\ell}_n(\hat{\boldsymbol{\theta}}_n) + (1/n)\,\mathrm{tr}\left([\hat{\mathbf{A}}_n]^{-1}\hat{\mathbf{B}}_n\right). \tag{16.17}$$

□

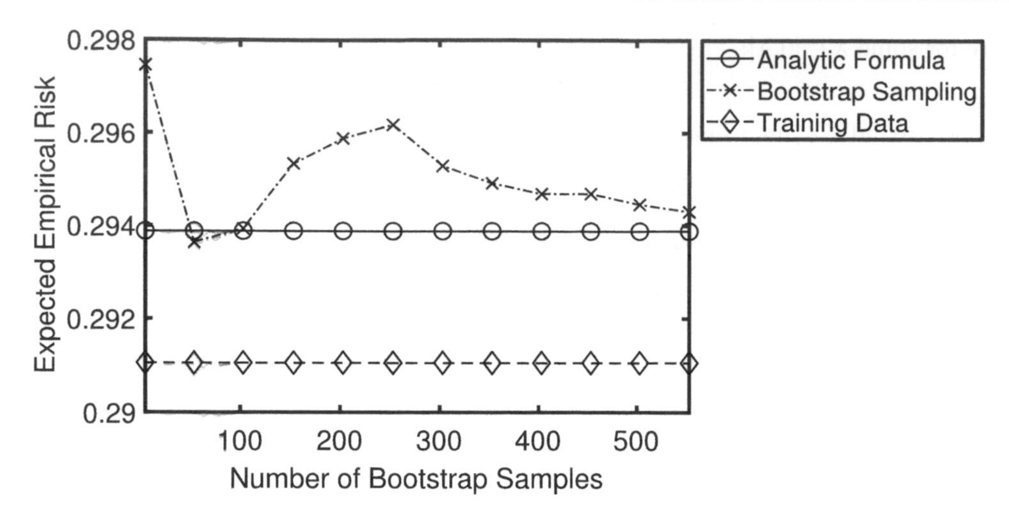


**FIGURE 16.4**
**Checking the asymptotic formula for estimating overfitting bias.** This figure uses crosses to depict a nonparametric bootstrap estimator of the estimated empirical risk on a test data set when logistic regression model parameters are estimated from a different training data set for different numbers of bootstrap samples (Bootstrap Sampling). In addition, the model fit of the estimated empirical risk function on the training data set is plotted using the diamond markers (Training Data). The training data was also used to calculate the cross-validation risk criterion which also estimates the expected fit of a model to novel test data (Analytic Formula). These latter results using the formulas provided in Recipe Box 16.1 are plotted using circle markers. For additional details of the simulation study, see Golden, Nandy, and Patel (2019).

In summary, the cross-validation risk criterion estimates the expected value of the model fit on a test data set using parameters estimated from a different training data set. In the special case where the empirical risk function is used to construct the Cross-Validation Risk Criterion is a negative normalized loglikelihood function, then the CVRC reduces to the important special case of the Generalization Akaike Information Criterion which is also known as the Takeuchi Information Criterion (Takeuchi 1976; see Bozdogan 2000 for a review). Figure 16.4 illustrates how overfitting bias may be estimated using either the CVRC analytic formula or the nonparametric bootstrap methodology introduced in Chapter 14.

**Definition 16.1.2** (Generalized Akaike Information Criterion (GAIC)). Let $\hat{\ell}_n$, $\hat{\boldsymbol{\theta}}_n$, $\hat{\mathbf{A}}_n$, and $\hat{\mathbf{B}}_n$ be defined as in the Empirical Risk Overfitting Bias Theorem. In addition, assume that $\hat{\ell}_n$ is a negative normalized log-likelihood function defined with respect to some probability model $\mathcal{M}$ and data generating density $p_e$. The *Generalized Akaike Information Criterion* (GAIC) is defined as:

$$\text{GAIC} = 2n\hat{\ell}_n(\hat{\boldsymbol{\theta}}_n) + 2\,\text{tr}\left([\hat{\mathbf{A}}_n]^{-1}\hat{\mathbf{B}}_n\right). \tag{16.18}$$

□

In the special case where the probability model $\mathcal{M}$ is correctly specified with respect to the data generating density $p_e$, the GAIC reduces as a special case to the popular Akaike Information Criterion (AIC)(see Exercise 16.1-4).

**Definition 16.1.3** (Akaike Information Criterion (AIC)). Let $\hat{\ell}_n$, $\hat{\boldsymbol{\theta}}_n$, and $q$ be defined as in the Empirical Risk Overfitting Bias Theorem (Theorem 16.1.2). In addition, assume that

$\hat{\ell}_n$ is a negative normalized log-likelihood function defined with respect to some probability model $\mathcal{M}$ and data generating density $p_e$. Assume, in addition, that $\mathcal{M}$ is correctly specified with respect to $p_e$. Then, the *Akaike Information Criterion* (AIC) is defined as:

$$\text{AIC} = 2n\hat{\ell}_n(\hat{\boldsymbol{\theta}}_n) + 2q. \tag{16.19}$$

□

**Recipe Box 16.1 Cross-Validation Model Selection Criterion (Theorem 16.1.2)**

- **Step 1: Examine data generating process.**
  Check the data sample $\mathbf{x}_1, \ldots, \mathbf{x}_n$ is a realization of a bounded sequence of $n$ i.i.d. random vectors.

- **Step 2: Check loss function is sufficiently smooth.**
  Check the loss function $c : \mathcal{R}^d \times \Theta \rightarrow \mathcal{R}$ is a twice continuously differentiable random function.

- **Step 3: Estimate parameters by minimizing empirical risk.** Use an optimization algorithm such as gradient descent to find a strict local minimizer, $\hat{\boldsymbol{\theta}}_n$, of the empirical risk function

$$\hat{\ell}_n(\boldsymbol{\theta}) = (1/n) \sum_{i=1}^{n} c(\mathbf{x}_i, \boldsymbol{\theta}).$$

  Check that $\hat{\boldsymbol{\theta}}_n$ is a critical point by verifying that the infinity norm of $\nabla \hat{\ell}_n(\hat{\boldsymbol{\theta}}_n)$ is sufficiently small. Define the parameter space as a neighborhood of a strict local minimizer of $E\{\hat{\ell}_n\}$ that always contains $\hat{\boldsymbol{\theta}}_n$.

- **Step 4: Empirically check OPG and Hessian condition numbers.**
  Compute $\hat{\mathbf{A}}_n \equiv \nabla^2 \hat{\ell}_n(\hat{\boldsymbol{\theta}}_n)$ and

$$\hat{\mathbf{B}}_n = (1/n) \sum_{i=1}^{n} \nabla c(\mathbf{x}_i, \boldsymbol{\theta})[\nabla c(\mathbf{x}_i, \boldsymbol{\theta})]^T.$$

  Check that $\hat{\mathbf{A}}_n$ and $\hat{\mathbf{B}}_n$ are converging to positive definite matrices by checking their condition numbers and their magnitudes.

- **Step 5: Compute CVRC model selection criterion.**

$$\text{CVRC} = \hat{\ell}_n(\hat{\boldsymbol{\theta}}_n) + (1/n)\,\text{tr}\left(\hat{\mathbf{A}}_n^{-1}\hat{\mathbf{B}}_n\right)$$

  which is an unbiased estimator of the expected performance of the learning machine on a novel test data set when the parameter vector $\hat{\boldsymbol{\theta}}_n$ which is estimated from the training data is used.

## Exercises

16.1-1. A researcher compares two models using the Akaike Information Criterion (AIC). Assume that either: (i) one or both models are misspecified, or (ii) the Hessian of the negative log-likelihood evaluated at the maximum likelihood estimates is converging to a singular matrix. Show that, under this assumption, the AIC cannot be interpreted as an estimator of the out-of-sample prediction error.

16.1-2. Derive a CVRC MSC for each model at the end of Section 13.1. Then, provide a discussion of the assumptions required for the conditions of the Empirical Risk Overfitting Bias (Theorem 16.1.2) to hold.

16.1-3. Show how to use the techniques in Chapter 14 to design nonparametric bootstrap simulation studies to evaluate the performance of the CVRC MSC.

16.1-4. Show how to derive the Akaike Information Criterion (AIC) from the Generalized Akaike Information Criterion (GAIC) for the special case where the probability model is correctly specified by using the results of Theorem 16.3.1.

16.1-5. Use the methods of Chapter 14 to design a nonparametric bootstrap methodology to estimate the sampling error of Cross-Validation Risk MSC and the simulation error associated with estimating that sampling error.

## 16.2 Bayesian Model Selection Criteria

### 16.2.1 Bayesian Model Selection Problem

Let $\mathcal{D}_n \equiv [\mathbf{x}_1, \ldots, \mathbf{x}_n]$ be a data set. The data set $\mathcal{D}_n$ is assumed to be a realization of a stochastic sequence of i.i.d. $d$-dimensional random vectors with common DGP density $p_e : \mathcal{R}^d \to [0, \infty)$ defined with respect to support specification measure $\nu$.

Let $\Omega_M \equiv \{\mathcal{M}_1, \ldots, \mathcal{M}_M\}$ be a set of probability models such that for $j = 1, \ldots, M$:

$$\mathcal{M}_j \equiv \{p_j(\mathbf{x}|\boldsymbol{\theta}) : \boldsymbol{\theta} \in \Theta_j\}.$$

The probability mass function $p_{\mathcal{M}} : \Omega_M \to [0, 1]$ is the *model prior* that specifies the probability $p_{\mathcal{M}}(\mathcal{M}_j)$ that model $\mathcal{M}_j$ is an appropriate probability model. Note that $p_{\mathcal{M}}$ is chosen based upon prior knowledge about the problem and is not estimated from $\mathcal{D}_n$. A common choice for $p_{\mathcal{M}}$ is to choose $p_{\mathcal{M}}(\mathcal{M}_j) = 1/M$ when the decision maker considers all $M$ probability models equally likely before observing the training data.

In the Bayesian framework presented here, the semantic interpretation of $p_{\mathcal{M}}(\mathcal{M}_j)$ does not denote the probability that model $\mathcal{M}_j$ generated the observed data since such an assumption requires that at least one of the $M$ models is correctly specified. Rather, $p_{\mathcal{M}}(\mathcal{M}_j)$ is interpreted as a probability that measures the decision maker's preference for model $\mathcal{M}_j$.

Let $p(\mathcal{D}_n|\boldsymbol{\theta}, \mathcal{M}) \equiv \prod_{i=1}^{n} p(\mathbf{x}_i|\boldsymbol{\theta})$. Let the negative normalized log-likelihood function

$$\hat{\ell}_n(\boldsymbol{\theta}) \equiv -(1/n) \log p(\mathcal{D}_n|\boldsymbol{\theta}, \mathcal{M}) = -(1/n) \log \prod_{i=1}^{n} p(\mathbf{x}_i|\boldsymbol{\theta}).$$

Then, $p(\mathcal{D}_n|\boldsymbol{\theta}, \mathcal{M}) = \exp(-n\hat{\ell}_n(\mathcal{D}_n, \boldsymbol{\theta}))$.

The *marginal likelihood* $p(\mathcal{D}_n|\mathcal{M}_j)$ may then be computed using the formula:

$$p(\mathcal{D}_n|\mathcal{M}_j) \equiv \int_{\Theta} p(\mathcal{D}_n|\boldsymbol{\theta}, \mathcal{M}_j) p_{\boldsymbol{\theta}}(\boldsymbol{\theta}|\mathcal{M}_j) d\boldsymbol{\theta} \tag{16.20}$$

where the density $p_{\boldsymbol{\theta}}(\cdot|\mathcal{M}_j)$ in (16.20) is the parameter prior for model $\mathcal{M}_j$. Typically, the computationally intractable integral in (16.20) must be approximated using numerical methods such as the Monte Carlo method discussed in Chapter 14 or using the Laplace approximation method which is discussed later on in this chapter.

Let $w_{j,k}$ denote the penalty a decision maker expects to receive if the decision maker chooses model $\mathcal{M}_j$ instead of model $\mathcal{M}_k$. For example, if $w_{j,k} = 0$ if $j = k$ and $w_{j,k} = 1$ if $j \neq k$, then this corresponds to the case where the decision maker expects to receive a penalty of 1 if a decision error is made and a penalty of 0 otherwise (i.e., all decision errors are equally costly).

The decision maker's Bayesian risk for choosing model $\mathcal{M}_j$ is then given by the Bayesian model risk function $R : \Omega_M \to \mathcal{R}$ defined such that:

$$R(\mathcal{M}_j) = \sum_{k=1}^{M} w_{j,k} p(\mathcal{M}_k|\mathcal{D}_n)$$

where

$$p(\mathcal{M}_k|\mathcal{D}_n) = \frac{p(\mathcal{D}_n|\mathcal{M}_k) p_{\mathcal{M}}(\mathcal{M}_k)}{p(\mathcal{D}_n)}. \tag{16.21}$$

Note that since

$$p(\mathcal{D}_n) = \sum_{z=1}^{M} p(\mathcal{D}_n|\mathcal{M}_z) p_{\mathcal{M}}(\mathcal{M}_z)$$

is a constant that is not a function of $\mathcal{M}_k$, it is often not necessary to explicitly compute $p(\mathcal{D}_n)$ for the purpose of determining if the risk, $R(\mathcal{M}_j)$, for model $\mathcal{M}_j$ is less than the risk, $R(\mathcal{M}_w)$, for selecting model $\mathcal{M}_w$.

Given the above Bayesian decision-making framework, the decision maker then chooses a model $\mathcal{M}^*$ that minimizes the Bayesian model risk function $R$.

### 16.2.2 Laplace Approximation for Multidimensional Integration

A major computational challenge associated with applying the Bayesian Model Selection Criterion in (16.21) is the computation of the marginal likelihood of the observed data in (16.20). For example, since

$$p(\mathcal{D}_n|\boldsymbol{\theta}, \mathcal{M}) = \exp(-n\ddot{\ell}_n(\boldsymbol{\theta})), \tag{16.22}$$

$$p(\mathcal{D}_n|\mathcal{M}) = \int_{\Theta_M} \exp(-n\hat{\ell}_n(\boldsymbol{\theta})) p_{\Theta}(\boldsymbol{\theta}|\mathcal{M}) d\boldsymbol{\theta}. \tag{16.23}$$

Monte Carlo simulation methods such as the bootstrap simulation methods described in Chapter 14 are often used to numerically evaluate (16.20).

The following theorem provides another important tool for evaluating multidimensional integrals of the form of (16.23).

**Theorem 16.2.1** (Laplace Approximation Theorem). *Let $\Theta \subseteq \mathcal{R}^q$ be an open set. Let $\ell : \Theta \to \mathcal{R}$ be a twice continuously differentiable function with a strict global minimizer, $\boldsymbol{\theta}^*$, in $\Theta$. Assume that $\nabla^2 \ell(\boldsymbol{\theta}^*)$ is positive definite. Let $\phi : \Theta \to [0, \infty)$ be continuous in a* *neighborhood of* $\boldsymbol{\theta}^*$ *such that* $\phi(\boldsymbol{\theta}^*) \neq 0$. *Assume that there exists a number* $n_0$ *such that for all* $n \geq n_0$:

$$\int_{\Theta} |\phi(\boldsymbol{\theta})| \exp(-n\ell(\boldsymbol{\theta}))d\boldsymbol{\theta} < \infty. \tag{16.24}$$

*Then,*

$$-(1/n)\log\left[\int_{\Theta} \phi(\boldsymbol{\theta}) \exp(-n\ell(\boldsymbol{\theta}))d\boldsymbol{\theta}\right] =$$

$$\ell(\boldsymbol{\theta}^*) - \frac{\log[\phi(\boldsymbol{\theta}^*)]}{n} + \frac{q}{2n}\log\left(\frac{n}{2\pi}\right) + \frac{\log(\det(\nabla^2\ell(\boldsymbol{\theta}^*)))}{2n} + o(1/n). \tag{16.25}$$

*Proof.* See Evans and Swartz (2005, Theorem 4.14, pp. 86-88; also see Hsu, 1948). ■

It is important to emphasize that this deterministic version of the Laplace Approximation assumes $\ell : \Theta \rightarrow \mathcal{R}$ is not functionally dependent on the $n$.

Although the technical details of the proof are beyond the scope of this text, the following heuristic proof is provided to obtain insights into the derivation of Theorem 16.2.1.

Let

$$Q_n(\boldsymbol{\theta}) \equiv -(1/n)\log\phi(\boldsymbol{\theta}) + \ell(\boldsymbol{\theta}).$$

so

$$\exp[-nQ_n(\boldsymbol{\theta})] = \phi(\boldsymbol{\theta})\exp[-n\ell(\boldsymbol{\theta})]. \tag{16.26}$$

Let $\boldsymbol{\theta}^* \equiv \arg\min \ell(\boldsymbol{\theta})$. If $n$ is sufficiently large, then

$$\boldsymbol{\theta}^* \approx \arg\min \ell(\boldsymbol{\theta}) - (1/n)\log\phi(\boldsymbol{\theta}) = \arg\min Q_n(\boldsymbol{\theta}). \tag{16.27}$$

Also if $n$ is sufficiently large, then

$$\nabla^2\ell(\boldsymbol{\theta}) \approx \nabla^2\left[\ell(\boldsymbol{\theta}) - (1/n)\log\phi(\boldsymbol{\theta})\right] = \nabla^2 Q_n(\boldsymbol{\theta}). \tag{16.28}$$

Now expand $Q_n$ in a second-order Taylor series about the strict global minimizer of $\ell$, $\boldsymbol{\theta}^*$, to obtain:

$$Q_n(\boldsymbol{\theta}) \approx Q_n(\boldsymbol{\theta}^*) + \nabla Q_n(\boldsymbol{\theta}^*)^T(\boldsymbol{\theta} - \boldsymbol{\theta}^*) + (1/2)(\boldsymbol{\theta} - \boldsymbol{\theta}^*)^T \nabla^2 Q_n(\boldsymbol{\theta}^*)(\boldsymbol{\theta} - \boldsymbol{\theta}^*). \tag{16.29}$$

Since $\boldsymbol{\theta}^*$ is approximately the strict local minimizer of $Q_n$ by (16.27), the term $\nabla Q_n(\boldsymbol{\theta}^*)$ in (16.29) vanishes. Substituting the relation (16.28) into (16.29) gives:

$$Q_n(\boldsymbol{\theta}) \approx Q_n(\boldsymbol{\theta}^*) + (1/2)(\boldsymbol{\theta} - \boldsymbol{\theta}^*)^T \nabla^2\ell(\boldsymbol{\theta}^*)(\boldsymbol{\theta} - \boldsymbol{\theta}^*). \tag{16.30}$$

Let $\mathbf{C}^* \equiv [\nabla^2\ell(\boldsymbol{\theta}^*)]^{-1}$. Then, multiply both sides of (16.30) by negative $n$ and use (16.26) to obtain:

$$\phi(\boldsymbol{\theta})\exp[-n\ell(\boldsymbol{\theta})] \approx$$

$$\phi(\boldsymbol{\theta}^*)\exp[-n\ell(\boldsymbol{\theta}^*)]\exp\left[-(n/2)(\boldsymbol{\theta} - \boldsymbol{\theta}^*)^T[\mathbf{C}^*]^{-1}(\boldsymbol{\theta} - \boldsymbol{\theta}^*)\right]. \tag{16.31}$$

Now integrate both sides of (16.31) to obtain:

$$\int \phi(\boldsymbol{\theta})\exp[-n\ell(\boldsymbol{\theta})]d\boldsymbol{\theta} \approx$$

$$\phi(\boldsymbol{\theta}^*)\exp[-n\ell(\boldsymbol{\theta}^*)]\int \exp\left[-(1/2)(\boldsymbol{\theta} - \boldsymbol{\theta}^*)^T[(1/n)\mathbf{C}^*]^{-1}(\boldsymbol{\theta} - \boldsymbol{\theta}^*)\right]d\boldsymbol{\theta}. \tag{16.32}$$

Using the definition of a multivariate Gaussian density with mean $\boldsymbol{\theta}^*$ and covariance matrix $(1/n)\mathbf{C}^*$ (see Definition 9.3.6) and the property that the integral of a probability density equals one, (16.32) may be rewritten as:

$$\int \phi(\boldsymbol{\theta}) \exp[-n\ell(\boldsymbol{\theta})]d\boldsymbol{\theta} \approx \phi(\boldsymbol{\theta}^*) \exp\left[-n\ell(\boldsymbol{\theta}^*)\right] (2\pi)^{q/2} \det\left((n^{-1}\mathbf{I}_q)\mathbf{C}^*\right)^{1/2}. \tag{16.33}$$

Now take the negative log of both sides of (16.33) and divide by $n$ to obtain:

$$-(1/n)\log\left[\int \phi(\boldsymbol{\theta}) \exp[-n\ell(\boldsymbol{\theta})]d\boldsymbol{\theta}\right] \approx$$

$$\ell(\boldsymbol{\theta}^*) - \frac{\log \phi(\boldsymbol{\theta}^*)}{n} + \frac{q}{2n}\log\left(\frac{n}{2\pi}\right) + \frac{\log(\det(\nabla^2 \ell(\boldsymbol{\theta}^*)))}{2n}.$$

### 16.2.3 Bayesian Information Criteria

In this section, the Laplace Approximation Theorem (Theorem 16.2.1) is applied for the purpose of deriving an approximation for the marginal likelihood of an $(\epsilon, n)$-cross entropy typical data set $\mathcal{D}_n \equiv [\mathbf{x}_1, \ldots, \mathbf{x}_n]$.

In classical Bayesian Model Selection, the negative normalized log-likelihood of the actual observed data set is computed with respect to the probability model $\mathcal{M}$. The likelihood of the actual observed data set $\mathcal{D}_n$ given $\mathcal{M}$ is given by the formula:

$$p(\mathcal{D}_n|\boldsymbol{\theta}, \mathcal{M}) = \exp\left(-n\ell_n(\boldsymbol{\theta})\right) \tag{16.34}$$

where $\ell_n(\boldsymbol{\theta})$ is a realization of the negative normalized log-likelihood

$$\hat{\ell}_n(\boldsymbol{\theta}) = -(1/n)\sum_{i=1}^{n} \log p(\tilde{\mathbf{x}}_i|\boldsymbol{\theta}).$$

However, to obtain a probability model with good generalization performance, it is desirable to consider an information-theoretic version of the classical Bayesian model selection approach where the likelihood of a typical data set rather than the observed data set is used.

Referring to Definition 13.2.7, define the likelihood of an $(\epsilon, n)$-cross entropy typical data set $\mathcal{D}_n$ using the formula:

$$\ddot{p}(\mathcal{D}_n|\boldsymbol{\theta}, \mathcal{M}) = \exp\left(-n\ell(\boldsymbol{\theta})\right) \tag{16.35}$$

where $\ell(\boldsymbol{\theta}) \equiv E\{\hat{\ell}_n(\boldsymbol{\theta})\}$. The notation $\ddot{p}(\mathcal{D}_n|\boldsymbol{\theta})$ indicates that $\ddot{p}(\mathcal{D}_n|\boldsymbol{\theta})$ is a large sample approximation of the marginal likelihood of an $(\epsilon, n)$-cross entropy typical data set defined with respect to density $p(\mathbf{x}|\boldsymbol{\theta})$ and DGP density $p_e$.

**Theorem 16.2.2** (Marginal Likelihood Approximation Theorem). *Assume the Empirical Risk Regularity Assumptions in Definition 15.1.1 hold with respect to a stochastic sequence of i.i.d. d-dimensional random vectors* $\tilde{\mathbf{x}}_1, \ldots, \tilde{\mathbf{x}}_n$ *with common DGP Radon-Nikodým density* $p_e$ *defined with respect to a support specification measure* $\nu$*, a loss function* $c : \mathcal{R}^d \times \mathcal{R}^q \rightarrow \mathcal{R}$ *defined such that* $c(\mathbf{x}, \boldsymbol{\theta}) = -\log p(\mathbf{x}|\boldsymbol{\theta})$*, a restricted parameter space* $\Theta \subseteq \mathcal{R}^q$*, empirical risk function minimizer* $\hat{\boldsymbol{\theta}}_n$ *of*

$$\hat{\ell}_n(\boldsymbol{\theta}) \equiv -(1/n)\sum_{i=1}^{n} \log p(\tilde{\mathbf{x}}_i|\boldsymbol{\theta})$$

*on $\Theta$, and risk function minimizer $\boldsymbol{\theta}^*$ of*

$$\ell(\boldsymbol{\theta}) \equiv -\int p_e(\mathbf{x}) \log p(\mathbf{x}|\boldsymbol{\theta}) d\nu(\mathbf{x})$$

*on $\Theta$. Let $\hat{\mathbf{A}}_n$ and $\hat{\mathbf{B}}_n$ be defined as in (15.1) and (15.2) respectively.*

*Let $\mathcal{M} \equiv \{p(\cdot|\boldsymbol{\theta}) : \boldsymbol{\theta} \in \Theta\}$ be a probability model. Assume the model parameter prior $p_\theta(\cdot|\mathcal{M})$ has the property that $p_\theta(\cdot|\mathcal{M})$ is continuous on $\Theta$.*

*Assume $\ell(\boldsymbol{\theta}) \geq 0$ for all $\boldsymbol{\theta} \in \Theta$. Let $\ddot{p}(\bar{\mathcal{D}}_n|\boldsymbol{\theta}, \mathcal{M}) \equiv \exp(-n\ell(\boldsymbol{\theta}))$ and*

$$\ddot{p}(\mathcal{D}_n|\mathcal{M}) \equiv \int_{\Theta_M} p_\theta(\boldsymbol{\theta}) \ddot{p}(\mathcal{D}_n|\boldsymbol{\theta}, \mathcal{M}) d\boldsymbol{\theta}. \tag{16.36}$$

*Then, for sufficiently large $n$,*

$$-(1/n) \log [\ddot{p}(\mathcal{D}_n|\mathcal{M})] = E\{\hat{\ell}_n(\hat{\boldsymbol{\theta}}_n)\} + \frac{q}{2n} \log n + \tilde{R}_n \tag{16.37}$$

*where*

$$\tilde{R}_n = \frac{\text{tr}(\hat{\mathbf{A}}_n^{-1}\hat{\mathbf{B}}_n)}{2n} - \frac{\log \left[p_\theta(\hat{\boldsymbol{\theta}}_n|\mathcal{M})\right]}{n} - \frac{q \log(2\pi)}{n} + \frac{\log(\det(\hat{\mathbf{A}}_n))}{2n} + o_p\left(\frac{1}{n}\right). \tag{16.38}$$

*Proof.* Since $\ell$ is a non-negative function, $p(\mathcal{D}_n|\boldsymbol{\theta}, \mathcal{M}) \leq 1$. Thus,

$$p(\mathcal{D}_n|\mathcal{M}) \equiv \int_{\Theta_M} p_\theta(\boldsymbol{\theta}) p(\mathcal{D}_n|\boldsymbol{\theta}, \mathcal{M}) d\boldsymbol{\theta} \leq \int_{\Theta_M} p_\theta(\boldsymbol{\theta}) d\boldsymbol{\theta} = 1 < \infty. \tag{16.39}$$

This ensures that the integral in (16.36) is finite so that (16.24) of the Laplace Approximation Theorem holds.

Let $\mathbf{A}^*$ be the Hessian of $\ell$ evaluated at $\theta^*$. Direct application of the Laplace Approximation Theorem gives:

$$-(1/n) \log p(\mathcal{D}_n|\mathcal{M}) = -(1/n) \log \left[\int_\Theta p_\theta(\boldsymbol{\theta}) \exp(-n\ell(\boldsymbol{\theta})) d\boldsymbol{\theta}\right] =$$

$$\ell(\boldsymbol{\theta}^*) - \frac{\log [p_\theta(\boldsymbol{\theta}^*)]}{n} + \frac{q}{2n} \log\left(\frac{n}{2\pi}\right) + \frac{\log(\det(\mathbf{A}^*))}{2n} + o\left(\frac{1}{n}\right). \tag{16.40}$$

Since $\hat{\mathbf{A}}_n = \mathbf{A}^* + o_p(1)$, $\log(\det(\cdot))$ is a continuous function of the set of positive definite matrices, and the Hessian of $\ell$ is positive definite in a neighborhood of $\boldsymbol{\theta}^*$, it follows that:

$$\log(\det(\hat{\mathbf{A}}_n) = \log(\det(\mathbf{A}^*)) + o_p(1). \tag{16.41}$$

In addition, since $\log p_\theta$ is a continuous function on $\Theta$ and $p_\theta(\cdot) > 0$ on $\Theta$,

$$\log p_\theta(\hat{\boldsymbol{\theta}}_n) = \log p_\theta(\boldsymbol{\theta}^*) + o_p(1) \tag{16.42}$$

for $n$ sufficiently large.

Substituting (16.41) and (16.42) into (16.40) gives:

$$-(1/n) \log \left[\int_\Theta p_\theta(\boldsymbol{\theta}) \exp(-n\ell(\boldsymbol{\theta})) d\boldsymbol{\theta}\right] =$$

$$\ell(\boldsymbol{\theta}^*) - \frac{\log \left[p_\theta(\hat{\boldsymbol{\theta}}_n)\right]}{n} + \frac{q}{2n} \log\left(\frac{n}{2\pi}\right) + \frac{\log(\det \hat{\mathbf{A}}_n))}{2n} + o_p\left(\frac{1}{n}\right). \tag{16.43}$$

Substituting (16.9) into (16.15) one obtains:

$$\ell(\boldsymbol{\theta}^*) = E\{\hat{\ell}_n(\hat{\boldsymbol{\theta}}_n)\} + \frac{\mathrm{tr}([\mathbf{A}^*]^{-1}\mathbf{B}^*)}{2n} + o_p(1/n). \tag{16.44}$$

In order to obtain (16.37) and (16.38), substitute (16.44) into (16.43) to complete the proof. ■

The assumption that $\ell$ is non-negative automatically holds for the case where the DGP is a stochastic sequence of discrete random vectors. In the more general case where the DGP is a stochastic sequence of absolutely continuous random vectors or mixed random vectors, $\ell$ can in some cases be explicitly computed and checked to see if it is non-negative. If $\ell$ is not a non-negative function or it is not possible to determine through mathematical analysis whether $\ell$ is a non-negative function, then $\hat{\ell}_n(\hat{\boldsymbol{\theta}}_n)$ can be computed and checked to see if this estimator is converging to a number which is greater than or equal to zero.

**Definition 16.2.1** (Bayesian Information Criterion (BIC)). Let $\hat{\ell}_n$, $\hat{\boldsymbol{\theta}}_n$, $n$, and $q$ be defined as in the Marginal Likelihood Approximation Theorem. The *Bayesian Information Criterion* is defined as:

$$\mathrm{BIC} = 2n\hat{\ell}_n(\hat{\boldsymbol{\theta}}_n) + q\log(n).$$

□

The Bayesian Information Criterion (BIC), which is also known as the Schwarz Information Criterion (SIC) (Schwarz 1978) provides a method for estimating the marginal likelihood of the training data set given a particular probability model.

The Marginal Likelihood Approximation Theorem 16.2.2 provides a second semantic interpretation of the BIC by showing that the BIC may also be semantically interpreted as the marginal likelihood of a cross entropy typical data set (see Definition 13.2.7) given a particular probability model. In particular, one uses the formula:

$$p(\mathcal{D}_n|\mathcal{M}_k) = \exp\left(-\frac{1}{2}\mathrm{BIC}\right) + O_p(1/n)$$

to calculate the marginal likelihood.

**Definition 16.2.2** (Cross-Entropy Bayesian Information Criterion (XBIC)). Let $\hat{\ell}_n$, $\hat{\mathbf{A}}^{-1}$, $\hat{\mathbf{B}}^{-1}$, and $p_\theta$ be defined as in the Marginal Likelihood Approximation Theorem. The *Cross-Entropy Bayesian Information Criterion* is defined as:

$$\mathrm{XBIC} = \mathrm{BIC} + \mathrm{tr}(\hat{\mathbf{A}}_n^{-1}\hat{\mathbf{B}}_n) - 2\log\left[p_\theta(\hat{\boldsymbol{\theta}}_n)\right] - q\log(2\pi) + \log(\det(\hat{\boldsymbol{A}}_n)) \tag{16.45}$$

□

The model selection criterion XBIC is a more accurate approximation of the marginal likelihood of a cross entropy typical data set since it explicitly incorporates terms of order $O_p(1/n)$ in (16.37) and (16.38). Thus,

$$p(\mathcal{D}_n|\mathcal{M}) \approx \exp\left(-\frac{1}{2}\mathrm{XBIC}\right)$$

may be used to approximate the marginal likelihood for a cross entropy typical data set when the sample size $n$ is sufficiently large.

## Recipe Box 16.2 BIC/XBIC Model Selection Criteria (Theorem 16.2.2)

- **Step 1: Specify parameter prior and model prior.**
Define a (possibly) local probability model $\mathcal{M} \equiv \{p(\cdot|\boldsymbol{\theta}) : \boldsymbol{\theta} \in \boldsymbol{\Theta}\}$ where $\boldsymbol{\Theta}$ is a closed and bounded parameter space. Select a parameter prior $p_{\boldsymbol{\theta}}(\boldsymbol{\theta}|\mathcal{M})$ such that $p_{\boldsymbol{\theta}}$ is continuous on $\boldsymbol{\Theta}$ and $p_{\boldsymbol{\theta}} > 0$ on $\boldsymbol{\Theta}$. Let $p_{\mathcal{M}}(\mathcal{M})$ denote the model prior probability of $\mathcal{M}$.

- **Step 2: Examine data generating process.**
Check the data sample $\mathbf{x}_1, \ldots, \mathbf{x}_n$ is a realization of a bounded sequence of $n$ i.i.d. random vectors.

- **Step 3: Check loss function is sufficiently smooth.**
Let $c : \mathcal{R}^d \times \Theta \to \mathcal{R}$ be a twice continuously differentiable random function defined such that: $c(\mathbf{x}; \boldsymbol{\theta}) = -\log p(\mathbf{x}|\boldsymbol{\theta})$.

- **Step 4: Compute maximum likelihood estimates.**
Use an optimization algorithm such as gradient descent to find a strict local minimizer, $\hat{\boldsymbol{\theta}}_n$, of the empirical risk function

$$\hat{\ell}_n(\boldsymbol{\theta}) \equiv -(1/n) \sum_{i=1}^{n} \log p(\mathbf{x}|\boldsymbol{\theta}).$$

- **Step 5: Check OPG and Hessian condition numbers.**
Check $|\nabla \hat{\ell}_n(\hat{\boldsymbol{\theta}}_n)|_\infty$ is sufficiently small. Check condition numbers of

$$\hat{\mathbf{B}}_n = (1/n) \sum_{i=1}^{n} \nabla \log p(\mathbf{x}_i|\boldsymbol{\theta})[\nabla \log p(\mathbf{x}_i|\boldsymbol{\theta})]^T$$

and $\hat{\mathbf{A}}_n \equiv \nabla^2 \hat{\ell}_n(\hat{\boldsymbol{\theta}}_n)$ are converging to finite positive numbers.

- **Step 6: Estimate cross-entropy typical data set marginal likelihood**
The marginal likelihood $\ddot{p}(\mathcal{D}_n|\mathcal{M})$ of a cross-entropy typical data set $\mathcal{D}_n$ is estimated by:

$$\ddot{p}(\mathcal{D}_n|\mathcal{M}) \approx \exp\left(-\frac{1}{2}\text{BIC}\right) \tag{16.46}$$

where $\text{BIC} \equiv -2n\hat{\ell}_n(\hat{\boldsymbol{\theta}}_n) + q \log n$. An improved estimator for $\ddot{p}(\mathcal{D}_n|\mathcal{M})$ may be obtained by substituting XBIC in (16.45) for BIC in (16.46).

- **Step 7: Estimate probability of model for a typical data set.**
Compute probability of model $\mathcal{M}_k$ given data set $\mathcal{D}_n$ using formula:

$$p(\mathcal{M}_k|\mathcal{D}_n) \approx \frac{\exp\left(-\frac{1}{2}\text{BIC}_k\right) p(\mathcal{M}_k)}{\sum_{j=1}^{M} \exp\left(-\frac{1}{2}\text{BIC}_j\right) p(\mathcal{M}_j)}$$

where $\text{BIC}_k$ is the BIC for model $\mathcal{M}_k$ for $k = 1, \ldots, M$. Replace BIC with XBIC to improve approximation quality for smaller sample sizes.

**Example 16.2.1** (Bayesian Hypothesis Testing Using a Laplace Approximation). Let $\hat{\ell}_n^a$ denote the negative normalized log-likelihood for model $\mathcal{M}_a$ and data set $\mathcal{D}_n$ evaluated at the $d_a$-dimensional strict local minimizer $\hat{\boldsymbol{\theta}}_n^a$. Let $\hat{\ell}_n^o$ denote the negative normalized log-likelihood for model $\mathcal{M}_o$ and data set $\mathcal{D}_n$ evaluated at the $d_o$-dimensional strict local minimizer $\hat{\boldsymbol{\theta}}_n^o$. Assume the conditions of Theorem 16.2.2 hold for models $\mathcal{M}_a$ and $\mathcal{M}_o$ with respect to $\mathcal{D}_n$. Using the formula for BIC as an approximation for the marginal likelihood, show the Bayes Factor $K$ described in Section 15.4.2 may be approximated by the formula:

$$K \approx \exp\left(n\hat{\ell}_n^o - n\hat{\ell}_n^a + (1/2)(d_o - d_a)\log n\right).$$

△

## Exercises

16.2-1. *Designing Bayesian Risk Decision Rules for Model Selection.* A consultant is asked to decide which of three possible probability models $\mathcal{M}_1$, $\mathcal{M}_2$, and $\mathcal{M}_3$ can be used to effectively solve a machine learning problem as inexpensively as possible. All three probability models exhibit reasonably good performance with respect to a given data set $\mathcal{D}_n$, but very different implementation costs. The consultant decides to estimate $p(\mathcal{M}_1|\mathcal{D}_n)$, $p(\mathcal{M}_2|\mathcal{D}_n)$, and $p(\mathcal{M}_3|\mathcal{D}_n)$. In addition, let the function $V$ which maps a model $\mathcal{M}$ into a number be used to calculate the cost in dollars for implementing each of the three models so that, for example, the cost in dollars for implementing model $\mathcal{M}_2$ is equal to $V(\mathcal{M}_2)$. Since the three models are each expected to exhibit reasonably good performance, assume the preference model prior probabilities $p(\mathcal{M}_1)$, $p(\mathcal{M}_2)$, and $p(\mathcal{M}_3)$ are equal to one another.

Show how to design a decision rule for choosing a model that minimizes expected costs by using the conditional probabilities $p(\mathcal{M}_1|\mathcal{D}_n)$, $p(\mathcal{M}_2|\mathcal{D}_n)$, and $p(\mathcal{M}_3|\mathcal{D}_n)$ as well as the associated implementation costs $V(\mathcal{M}_1)$, $V(\mathcal{M}_2)$, and $V(\mathcal{M}_3)$. Provide explicit yet approximate formulas for evaluating your decision rule by using the BIC approximation for the marginal likelihood $p(\mathcal{D}_n|\mathcal{M})$.

16.2-2. *Using BIC and XBIC to Support Bayesian Hypothesis Testing.* Use either the BIC or XBIC model selection criteria to derive useful approximations for computing the Bayes Factor in Section 15.4.2 to support Bayesian Hypothesis Testing.

16.2-3. *Checking Asymptotic Laplace Approximations Using Simulation Methods.* Use the methods of Chapter 14 to design a parametric bootstrap simulation methodology to evaluate performance of BIC and XBIC. Write a computer program to implement your methodology and use it to evaluate the relative performance of BIC and XBIC for a logistic regression model.

16.2-4. *Using Simulation Methods to Estimate Sampling Errors of Model Selection Criteria.* Use the methods of Chapter 14 to design a nonparametric bootstrap simulation methodology to estimate the sampling error and simulation error of the statistics BIC and XBIC for a logistic regression model.

16.2-5. *Bayesian Model Search Using BIC and XBIC.* Let $\mathcal{M}_1, \ldots, \mathcal{M}_m$ be a collection of a moderately but not excessively large number of models (e.g., $m = 100$ or

$m = 1000$). Let $\mathcal{D}_n \equiv \{(\mathbf{s}_1, y_1), \ldots, (\mathbf{s}_n, y_n)\}$ be a data set that is assumed to be a realization of a stochastic sequence of independent and identically distributed random vectors. Show how to use BIC to select the "most probable model given a cross entropy typical data set". Let $\ddot{y}(\mathbf{s}, \mathcal{M}_k, \mathcal{D}_n)$ be the predicted response of the learning machine given input pattern vector $\mathbf{s}$, the most probable model $\mathcal{M}_k$ and data set $\mathcal{D}_n$. See Example 11.2.2 for a related discussion.

16.2-6. *Bayesian Model Averaging Using BIC and XBIC.* Let $\mathcal{M}_1, \ldots, \mathcal{M}_m$ be a collection of a moderately but not excessively large number of models (e.g., $m = 100$ or $m = 1000$). Let $\mathcal{D}_n \equiv \{(\mathbf{s}_1, y_1), \ldots, (\mathbf{s}_n, y_n)\}$ be a data set that is assumed to be a realization of a stochastic sequence of independent and identically distributed random vectors. Let $\ddot{y}(\mathbf{s}, \mathcal{M}_k, \mathcal{D}_n)$ be the predicted response of the learning machine given input pattern vector $\mathbf{s}$, model $\mathcal{M}_k$ and data set $\mathcal{D}_n$. Rather than just picking one model to generate predictions as in Exercise 16.2-5, use the entire collection of models to generate a prediction using an ensemble average by computing the formula:

$$E\{\ddot{y}(\mathbf{s}, \tilde{\mathcal{M}}, \mathcal{D}_n)|\mathcal{D}_n\} = \sum_{k=1}^{m} \ddot{y}(\mathbf{s}, \mathcal{M}_k, \mathcal{D}_n) p(\mathcal{M}_k|\mathcal{D}_n).$$

Discuss how to obtain an explicit approximate formula for $p(\mathcal{M}_k|\mathcal{D}_n)$ using either the BIC or XBIC approximations. See Example 11.2.3 for a related discussion.

## 16.3 Model Misspecification Detection Model Selection Criteria

As previously noted, a misspecification detection model selection criterion is designed to assess how well a particular probability model represents its statistical environment. The Cross-Validation Risk Criterion (CVRC) is used to assess the predictive performance of a probability model for novel test data sets. The Bayesian Risk Model Selection Criterion (BRMSC) is used to assess the probability of a model for a given data set. However, it is possible for a probability model $\mathcal{M}$ to have both excellent generalization performance as assessed by the CVRC MSC and additionally the property that it is highly probable given the data as assessed by the BRMSC yet incorrectly represent its statistical environment.

### 16.3.1 Nested Models Method for Assessing Model Misspecification

One powerful and classical method relevant for the construction of misspecification detection model selection criteria is the well-known "nested models" method. Suppose the goal is to assess whether a particular model $\mathcal{M}$ with empirical risk model fit $\hat{\ell}_n(\hat{\boldsymbol{\theta}}_n)$ is misspecified. To apply the nested models method, one constructs a new correctly specified model $\mathcal{M}_e$ such that $\mathcal{M}_e \supseteq \mathcal{M}$. This correctly specified representation is typically achieved by including many additional parameters in $\mathcal{M}_e$. Let the empirical risk model fit for $\mathcal{M}_e$ be denoted as $\hat{\ell}_n^e\left(\hat{\boldsymbol{\theta}}_n^e\right)$. Now, using either the simulation methods of Chapter 14 or the analytic methods of Chapter 15, one can construct a test statistic $\mathcal{W}_n$ which has the property that large values of $\mathcal{W}_n$ indicate that observed differences in $\hat{\ell}_n(\hat{\boldsymbol{\theta}}_n)$ and $\hat{\ell}_n^e\left(\hat{\boldsymbol{\theta}}_n^e\right)$ are not likely due to chance which, in turn, implies model misspecification is present in $\mathcal{M}$. Vuong (1989; also see Golden, 2003) show how to relax the standard assumptions for model comparison which often require that: (i) the full model is correctly specified, and (ii) the models are fully nested.

### 16.3.2 Information Matrix Discrepancy Model Selection Criteria

In this section, a recently developed method for detecting the presence of model misspecification is described which is based upon the Information Matrix Equality. The exploitation of the Information Matrix Equality for the detection of model misspecification was originally proposed by White (1982, 1994) and has been further developed in recent work by Golden, Henley, White, and Kashner (2016, 2019).

**Theorem 16.3.1** (Information Matrix Equality). *Assume the data sample is a sequence of independent and identically distributed $d$-dimensional random vectors with common Radon-Nikodým density $p_e : \mathcal{R}^d \rightarrow [0, \infty)$ defined with respect to support specification measure $\nu$. Let $\Theta \subseteq \mathcal{R}^q$. Let the probability model specification*

$$\mathcal{M} \equiv \{p(\cdot; \boldsymbol{\theta}) : \boldsymbol{\theta} \in \Theta\}$$

*where $p(\cdot|\boldsymbol{\theta}) : \mathcal{R}^d \rightarrow [0, \infty)$ is a Radon-Nikodým density defined with respect to support specification measure $\nu$ for each $\boldsymbol{\theta} \in \Theta$. Let $\ell : \Theta \rightarrow \mathcal{R}$ be defined such that for all $\boldsymbol{\theta} \in \Theta$:*

$$\ell(\boldsymbol{\theta}) = -\int p_e(\mathbf{x}) \log p(\mathbf{x}; \boldsymbol{\theta}) d\nu(\mathbf{x})$$

*is finite. Assume $\log p$ is a twice continuously differentiable random function. Assume that the gradient*

$$\nabla \log p(\mathbf{x}; \boldsymbol{\theta}) \equiv \left[\frac{d \log p(\mathbf{x}; \boldsymbol{\theta})}{d\boldsymbol{\theta}}\right]^T, \tag{16.47}$$

*the outer product gradient matrix*

$$\nabla \log p(\mathbf{x}; \boldsymbol{\theta}) \nabla \log p(\mathbf{x}; \boldsymbol{\theta})^T, \tag{16.48}$$

*and the Hessian matrix*

$$\frac{d^2 \log p(\mathbf{x}; \boldsymbol{\theta})}{d\boldsymbol{\theta}^2} \tag{16.49}$$

*are dominated by integrable functions with respect to $p_e$. Assume, in addition, that there exists a finite number $K$ such that for all $\mathbf{x}$ in the support of $\tilde{\mathbf{x}}$:*

$$p(\mathbf{x}; \boldsymbol{\theta}) < K p_e(\mathbf{x}) \tag{16.50}$$

*on $\Theta$.*

*Let $\boldsymbol{\theta}^*$ be the unique global minimizer of $\ell$ on $\Theta$. Let*

$$\mathbf{A}(\boldsymbol{\theta}) = \int \nabla^2 \log p(\mathbf{x}; \boldsymbol{\theta}) p_e(\mathbf{x}) d\nu(\mathbf{x}).$$

*Let*

$$\mathbf{B}(\boldsymbol{\theta}) = \int \nabla \log p(\mathbf{x}; \boldsymbol{\theta}) \left[\nabla \log p(\mathbf{x}; \boldsymbol{\theta})\right]^T p_e(\mathbf{x}) d\nu(\mathbf{x}).$$

*Let $\mathbf{A}^* = \mathbf{A}(\boldsymbol{\theta}^*)$. Let $\mathbf{B}^* = \mathbf{B}(\boldsymbol{\theta}^*)$.*

- *If $\mathcal{M}$ is correctly specified with respect to $p_e$, then $\mathbf{A}^* = \mathbf{B}^*$.*
- *If $\mathbf{A}^* \neq \mathbf{B}^*$, then $\mathcal{M}$ is misspecified with respect to $p_e$.*

*Proof.* Note that:

$$\frac{d \log p(\mathbf{x};\boldsymbol{\theta})}{d\boldsymbol{\theta}} = (1/p(\mathbf{x};\boldsymbol{\theta}))\frac{dp(\mathbf{x};\boldsymbol{\theta})}{d\boldsymbol{\theta}}. \tag{16.51}$$

Take the derivative of

$$\int p(\mathbf{x};\boldsymbol{\theta})d\nu(\mathbf{x}) = 1 \tag{16.52}$$

and interchange the derivative and integral operators using (16.51), the assumption in (16.47), and assumption (16.50) with Theorem 8.4.4 to obtain

$$\int \frac{dp(\mathbf{x};\boldsymbol{\theta})}{d\boldsymbol{\theta}} d\nu(\mathbf{x}) = \mathbf{0}_q^T, \tag{16.53}$$

which may be rewritten as:

$$\int p(\mathbf{x};\boldsymbol{\theta}) \left[\frac{d \log p(\mathbf{x};\boldsymbol{\theta})}{d\boldsymbol{\theta}}\right] d\nu(\mathbf{x}) = \mathbf{0}_q^T. \tag{16.54}$$

Now take the derivative of (16.54), interchange the derivative and integral operators using Theorem 8.4.4, and use (16.51) with the assumptions in (16.48) and (16.49) with (16.50) to obtain

$$\int \left[\left(\frac{d \log p(\mathbf{x};\boldsymbol{\theta})}{d\boldsymbol{\theta}}\right)^T \frac{d \log p(\mathbf{x};\boldsymbol{\theta})}{d\boldsymbol{\theta}} + \frac{d^2 \log p(\mathbf{x};\boldsymbol{\theta})}{d^2\boldsymbol{\theta}}\right] p(\mathbf{x};\boldsymbol{\theta})d\nu(\mathbf{x}) = \mathbf{0}_{q\times q}. \tag{16.55}$$

Now evaluate (16.55) at the unique global minimizer of $\ell$, $\boldsymbol{\theta}^*$, and use the assumption $p_e = p(\cdot;\boldsymbol{\theta}^*)$ $\nu$-almost everywhere (i.e., the model is correctly specified) to obtain the formula:

$$\mathbf{B}^* - \mathbf{A}^* = \mathbf{0}_{q\times q}.$$

This establishes the first part of the theorem. The second part is the contrapositive statement of the first part. ∎

The Information Matrix Equality Theorem provides a useful method for detecting the presence of model misspecification. Assume $\hat{\mathbf{A}}_n$ and $\hat{\mathbf{B}}_n$ as defined in (15.1) and (15.2) converge respectively to $\mathbf{A}^*$ and $\mathbf{B}^*$ with probability one as $n \to \infty$. An *Information Matrix Discrepancy Measure* is a continuous function $\mathcal{I} : \mathcal{R}^{q\times q} \times \mathcal{R}^{q\times q} \to [0,\infty)$ defined such that $\mathcal{I}(\mathbf{A}^*, \mathbf{B}^*)$ measures the discrepancy between $\mathbf{A}^*$ and $\mathbf{B}^*$. $\mathcal{I}(\hat{\mathbf{A}}_n, \hat{\mathbf{B}}_n)$ is called an *Information Matrix Discrepancy Measure MSC.* Larger values of the Information Matrix Discrepancy MSC are interpreted as larger amounts of evidence for model misspecification.

The following two theorems provide important insights regarding good choices for examining the similarities and differences between $\hat{\mathbf{A}}_n$ and $\hat{\mathbf{B}}_n$ within a two-dimensional space.

**Theorem 16.3.2** (Trace-Trace Information Matrix Equality). *Let* $\mathbf{A}$ *and* $\mathbf{B}$ *be real positive definite symmetric* $q$*-dimensional matrices. Then,* $\mathbf{A} = \mathbf{B}$ *if and only if*

$$\text{tr}(\mathbf{A}^{-1}\mathbf{B}) = q \tag{16.56}$$

*and*

$$\text{tr}(\mathbf{B}^{-1}\mathbf{A}) = q. \tag{16.57}$$

*Proof.* The proof follows Lemma 1(i) of Cho and Phillips (2018). If $\mathbf{A} = \mathbf{B}$, then both (16.56) and (16.57) hold. Now show (16.56) and (16.57) imply $\mathbf{A} = \mathbf{B}$. Let $\mathbf{S} = \mathbf{A}^{-1/2}\mathbf{B}^{1/2}$. Then,

$$\mathbf{S}\mathbf{S}^T = \mathbf{A}^{-1/2}\mathbf{B}\mathbf{A}^{-1/2}$$

is a real symmetric matrix. Thus, $q$ orthonormal $q$-dimensional eigenvectors arranged in the columns of matrix $\mathbf{E}$ exist such that

$$\mathbf{E}^T\mathbf{A}^{-1/2}\mathbf{B}\mathbf{A}^{-1/2}\mathbf{E} = \mathbf{D} \tag{16.58}$$

where $\mathbf{D}$ is a $q$-dimensional diagonal matrix whose $k$th on-diagonal element, $\lambda_k$, is the eigenvalue associated with the $k$th column of $\mathbf{E}$.

Equation (16.56) implies that the arithmetic mean of the on-diagonal elements of $\mathbf{D}$, $(1/q)\sum_{k=1}^{q}\lambda_k$, is equal to one because by Theorem 4.1.1 and (16.58):

$$\mathrm{tr}\left(\mathbf{A}^{-1/2}\mathbf{B}\mathbf{A}^{-1/2}\right) = \mathrm{tr}\left(\mathbf{A}^{-1}\mathbf{B}\right) = \mathrm{tr}(\mathbf{D}) = q. \tag{16.59}$$

Equation (16.57) implies that the arithmetic means of the inverse of the on-diagonal elements of $\mathbf{D}$, $(1/q)\sum_{k=1}^{q}(\lambda_k)^{-1}$, is equal to one because by Theorem 4.1.1 and

$$\mathrm{tr}(\mathbf{B}^{-1}\mathbf{A}) = \mathrm{tr}\left(\mathbf{A}^{1/2}\mathbf{B}^{-1}\mathbf{A}^{1/2}\right) = \mathrm{tr}(\mathbf{D}^{-1}) = q. \tag{16.60}$$

By the geometric-arithmetic mean inequality (e.g., Uchida, 2008), it follows that the equality of the geometric and arithmetic means in this case imply that all of the eigenvalues of $\mathbf{D}$ are equal to the same value $\lambda$.

Then, (16.59) and (16.60) imply:

$$(1/q)\sum_{k=1}^{q}\lambda = 1 \tag{16.61}$$

which implies $\lambda = 1$. That is, $\mathbf{D} = \mathbf{I}_q$.

Using the result that $\mathbf{D}$ is the identity matrix and $\mathbf{E}\mathbf{E}^T = \mathbf{I}_q$, Equation (16.58) implies that

$$\mathbf{A}^{-1/2}\mathbf{B}\mathbf{A}^{-1/2} = \mathbf{E}\mathbf{D}\mathbf{E}^T = \mathbf{I} \tag{16.62}$$

which implies $\mathbf{A} = \mathbf{B}$. ■

**Theorem 16.3.3** (Trace-Determinant Information Matrix Equality). *Let* $\mathbf{A}$ *and* $\mathbf{B}$ *be real positive definite symmetric* $q$*-dimensional matrices. Then,* $\mathbf{A} = \mathbf{B}$ *if and only if*

$$\log\det\left(\mathbf{A}^{-1}\mathbf{B}\right) = 0 \tag{16.63}$$

*and*

$$\mathrm{tr}(\mathbf{A}^{-1}\mathbf{B}) = q. \tag{16.64}$$

*Proof.* The proof follows the proof of Lemma 1 of Cho and White (2014). If $\mathbf{A} = \mathbf{B}$, then both (16.63) and (16.64) hold. Now show (16.63) and (16.64) imply $\mathbf{A} = \mathbf{B}$.

As in the Proof of Theorem 16.3.2, construct the matrix $\mathbf{D}$ and eigenvalues $\lambda_1, \ldots, \lambda_q$.

Equation (16.63) implies that the geometric mean of the on-diagonal elements of $\mathbf{D}$, $\left(\prod_{k=1}^{q}\lambda_k\right)^{1/k}$, is equal to one since

$$\det\left(\mathbf{A}^{-1}\mathbf{B}\right) = \det\left(\mathbf{A}^{-1/2}\mathbf{B}\mathbf{A}^{-1/2}\right) = \det(\mathbf{D}).$$

The result then follows from arguments similar to the Proof of Theorem 16.3.2 (see Exercise 16.3-5). ■

**Recipe Box 16.3 IM Misspecification Discrepancy Measures (Theorem 16.3.1)**

- **Step 1: Check empirical risk regularity assumptions.** Check the assumptions of the Information Matrix Equality Theorem (Theorem 16.3.1) hold with respect to a data generating density $p_e$ and probability model specification:
$$\mathcal{M} \equiv \{p(\mathbf{x};\boldsymbol{\theta}) : \boldsymbol{\theta} \in \Theta \subseteq \mathcal{R}^q\}.$$
The parameter space may be a neighborhood of a strict local minimizer of the expected negative log likelihood function that specifies a local probability model.

- **Step 2: Compute maximum likelihood estimates.**
Let $c(\mathbf{x};\boldsymbol{\theta}) \equiv -\log p(\mathbf{x};\boldsymbol{\theta})$. Use an optimization algorithm such as gradient descent to find a strict local minimizer, $\hat{\boldsymbol{\theta}}_n$, of the empirical risk function
$$\hat{\ell}_n(\boldsymbol{\theta}) \equiv (1/n)\sum_{i=1}^{n} c(\mathbf{x}_i, \boldsymbol{\theta})$$
in the parameter space of a (possibly) local probability model. Check that $\hat{\boldsymbol{\theta}}_n$ is a strict local minimizer by verifying that the infinity norm of $\nabla\hat{\ell}_n(\hat{\boldsymbol{\theta}}_n)$ is sufficiently small. Check that condition numbers of $\hat{\mathbf{A}}_n \equiv \nabla^2\hat{\ell}_n(\hat{\boldsymbol{\theta}}_n)$ and
$$\hat{\mathbf{B}}_n = (1/n)\sum_{i=1}^{n} \nabla_{\boldsymbol{\theta}} c(\mathbf{x}_i, \hat{\boldsymbol{\theta}}_n)[\nabla_{\boldsymbol{\theta}} c(\mathbf{x}_i, \hat{\boldsymbol{\theta}}_n)]^T$$
are converging to finite positive numbers.

- **Step 3: Compute IM discrepancy measures.**
Larger values of any of the following three IM Discrepancy Measures indicates presence of model misspecification:
(*i*) $\mathcal{I}_{tr} \equiv \left|\log\left((1/q)\,\mathrm{tr}\left(\hat{\mathbf{A}}_n^{-1}\hat{\mathbf{B}}_n\right)\right)\right|$
(*ii*) $\ddot{\mathcal{I}}_{tr} \equiv \left|\log\left((1/q)\,\mathrm{tr}\left(\hat{\mathbf{B}}_n^{-1}\hat{\mathbf{A}}_n\right)\right)\right|$
(*iii*) $\mathcal{I}_{\det} \equiv \left|(1/q)\log\det\left(\hat{\mathbf{A}}_n^{-1}\hat{\mathbf{B}}_n\right)\right|$

**Definition 16.3.1** (IM Discrepancy Measures)**.** Assume the assumptions of Theorem 16.3.1 hold with respect to a probability model $\mathcal{M}$ and DGP density $p_e$. Let $\hat{\mathbf{A}}_n$ and $\hat{\mathbf{B}}_n$ be $q$-dimensional positive definite matrices defined as in (15.1) and (15.2) respectively with respect to $\mathcal{M}$ and $p_e$. The *Log Trace Discrepancy Measure*, $\mathcal{I}_{tr}$, is defined by the formula:
$$\mathcal{I}_{tr} = \left|\log\left((1/q)\,\mathrm{tr}\left(\hat{\mathbf{A}}_n^{-1}\hat{\mathbf{B}}_n\right)\right)\right|.$$
The *Inverted Log Trace Discrepancy Measure*, $\ddot{\mathcal{I}}_{tr}$, is defined by the formula:
$$\ddot{\mathcal{I}}_{tr} = \left|\log\left((1/q)\,\mathrm{tr}\left(\hat{\mathbf{B}}_n^{-1}\hat{\mathbf{A}}_n\right)\right)\right|.$$

The *Log Determinant IM Discrepancy Measure*, $\mathcal{I}_{det}$, is defined by the formula:

$$\mathcal{I}_{det} = \left|(1/q) \log \det \left(\hat{\mathbf{A}}_n^{-1} \hat{\mathbf{B}}_n\right)\right|.$$

□

## Exercises

16.3-1.Derive an Information Matrix Discrepancy MSC for each model at the end of Section 13.1. Discuss how to assess the assumptions for each of the derived Information Matrix Discrepancy MSCs to ensure the assumptions of the Information Matrix Equality Theorem (Theorem 16.3.1) holds.

16.3-2.Assume the assumptions of the Information Matrix Equality Theorem (Theorem 16.3.1) hold with respect to the $q$-dimensional matrices $\mathbf{A}^*$ and $\mathbf{B}^*$. Let $\hat{\mathbf{A}}_n$ and $\hat{\mathbf{B}}_n$ be defined as in (15.1) and (15.2). Let $\mathcal{I} : \mathcal{R}^{q \times q} \times \mathcal{R}^{q \times q} \rightarrow [0, \infty)$ be a continuous function. Show that as $n \rightarrow \infty$, $\mathcal{I}(\hat{\mathbf{A}}_n, \hat{\mathbf{B}}_n) \rightarrow \mathcal{I}(\mathbf{A}^*, \mathbf{B}^*)$ with probability one.

16.3-3.Use the methods of Chapter 14 to design a nonparametric bootstrap simulation study methodology to estimate the sampling error of $\mathcal{I}(\hat{\mathbf{A}}_n, \hat{\mathbf{B}}_n)$ and the simulation error associated with estimating that sampling error.

16.3-4.Use the methods of Chapter 14 to design a parametric bootstrap simulation study methodology intended to investigate how effectively the Information Matrix Discrepancy MSC detects the presence of model misspecification. Evaluate the effectiveness of the Information Matrix Discrepancy MSC by counting the percentage of times the Information Matrix Discrepancy MSC selects the correctly specified model from a group of competing models.

16.3-5.Complete the proof of Trace Determinant Information Matrix Equality Theorem 16.3.3 by using the result in (16.59) and then showing that (16.61) and (16.62) hold.

## 16.4 Further Readings

Henley et al. (2020) provides a comprehensive review of model selection, model misspecification, and model averaging methods relevant to the development and evaluation of sophisticated statistical models.

### Introduction to Model Selection

Good introductions to the model selection literature can be found in the special issue on Model Selection in the *Journal of Mathematical Psychology* (Myung et al., 2000). Konishi and Kitawaga (2008), Claeskens and Hort (2008), and Linhart and Zucchini (1986) provide more detailed discussions.

### Cross-Validation Risk Model Selection Criteria

Akaike (1973, 1974; also see Bozdogan 1987 for a review) originally proposed what is now called the Akaike Information Criterion (AIC) as an improved unbiased estimator of the likelihood of the observed data under the assumption of correct model specification. Stone (1977) showed that the AIC estimator was asymptotically equivalent to leave-one-out cross-validation simulation methodology as discussed in Chapter 14. The Generalized Akaike Information Criterion (GAIC), also known as the Takeuchi Information Criterion (TIC), is a generalization of the original AIC approach of Akaike (1973, 1974) but does not require the assumption that the probability model is correctly specified (see Takeuchi, 1976; for further discussion see Bozdogan 2000, and Konishi and Kitawaga 2008). Claeskens and Hjort (2008; Section 2.9) provide an analysis analogous to that of Stone (1977) showing that the GAIC estimator is asymptotically equivalent to the leave-one-out cross-validation estimator. The CVRC was originally derived by Linhard and Volkers (1984) and further discussed in Linhart and Zucchini (1986).

Golden, Nandy, and Patel (2019) provide simulation studies of the CVRC model selection criterion for logistic regression models. Some of the material presented in Section 16.1 was originally presented in Golden, Nandy, and Patel (2019).

## Bayesian Risk Model Selection Criteria

Although the AIC and BIC type model selection criteria look very similar from an algorithmic perspective, these criteria are derived with respect to fundamentally different objectives. While the AIC model selection criterion has the objective of providing an unbiased estimator of the "best-fitting" distribution in the probability model to the data using the likelihood function measure of model fit, the BIC (Bayesian Information Criterion) also known as the SIC (Schwarz Information Criterion) (Schwarz 1978) is designed to provide a large sample approximation to the marginal likelihood which corresponds to a weighted average over all possible distributions in the probability model to the data. Classical discussions of the use of the standard Laplace Approximation for deriving the Bayesian Information Criterion may be found in Claeskens and Hjort (2008, Chapter 3), Djuric (1998), Poskitt (1987, Corollary 2.2), Konishi and Kitagawa (2008, Chapter 9). Lv and Liu (2014) have recently provided new derivations of generalized versions of the classical Bayesian Information Criterion.

The new XBIC model selection criterion was originally introduced by Golden et al. (2015) at the NeurIPS 2015 Workshop on *Advances in Approximate Bayesian Inference.* The XBIC model selection criterion is closely related to information-theoretic motivated model selection criteria such as the CAIC (Consistent Akaike Information Criterion) (Bozdogan 2000), MDL (Minimum Description Length) model selection criteria (Barron et al., 1998; Grunwald, 2007; Hansen and Yu, 2001) and the MML (Minimum Message Length) model selection criteria (Wallace, 2005).

Kass and Raftery (1995) review many applications of BIC to problems of Bayesian Hypothesis Testing (see Section 15.4.2), Model Selection (see this Chapter; also see Section 15.4), and Model Averaging (see Example 11.2.3). Burnham and Anderson (2010) and Wasserman (2000) provide useful introductions to Bayesian model averaging.

## Model Misspecification Detection Model Selection Criteria

The Generalized Likelihood Ratio Test (Wilks, 1938) provides a mechanism for testing the null hypothesis that a proposed model provides a fit to the observed data which is equivalent to the fit of a more flexible model capable of representing any arbitrary data generating process. A limitation of the classical Generalized Likelihood Ratio Test is that it assumes that the more flexible model is correctly specified and contains the proposed model. Vuong (1989; also see Golden, 2003; also see Rivers and Vuong, 2002) show how the assumptions of correct specification and fully nested models can be relaxed to obtain more robust misspecification tests.

Golden, Henley, White, and Kashner (2013, 2016) have developed Generalized Information Matrix Tests (GIMTs) for the purpose of testing the null hypothesis that an Information Matrix Discrepancy Measure is zero following the approach of White (1982; also see White, 1994, for a discussion of the time-series case). The rejection of such a null hypothesis indicates the presence of model misspecification. Furthermore in a series of simulation studies, Golden, Henley, White, and Kashner (2013, 2016) evaluated the behavior of the Log Determinant IM Discrepancy Measure. In a related series of simulation studies, Golden, Henley, White, and Kashner (2016) evaluated the behavior of the Log Trace IM Discrepancy Measure.

Henley et al. (2020) and Kashner et al. (2020) provide practitioner-oriented reviews of a variety of methods for assessing model misspecification.

# References


Abadir, K. and J. Magnus 2002. Notation in econometrics: A proposal for a standard. *The Econometrics Journal 5*(1), 76–90.

Ackley, D. H., G. E. Hinton, and T. J. Sejnowski 1985. A learning algorithm for Boltzmann machines. *Cognitive Science 9*, 147–169.

Akaike, H. 1973. Information theory and an extension of the maximum likelihood principle. In *2nd International Symposium on Information Theory*, 267–281. Akadémiai Kiadó.

Akaike, H. 1974, December. A new look at the statistical model identification. *IEEE Transactions on Automatic Control 19*(6), 716–723.

Amari, S. 1967. A theory of adaptive pattern classifiers. *IEEE Transactions on Electronic Computers EC-16*(3), 299–307.

Amari, S.-I. 1998. Natural gradient works efficiently in learning. *Neural Computation 10*(2), 251–276.

Amemiya, T. 1973, November. Regression analysis when the dependent variable is truncated normal. *Econometrica 41*(6), 997–1016.

Anderson, J. A. and E. Rosenfeld 1998a. *Neurocomputing: Foundations of Research.* Bradford. Cambridge, MA: MIT Press.

Anderson, J. A. and E. Rosenfeld 1998b. *Neurocomputing 2: Directions for Research.* Bradford. Cambridge, MA: MIT Press.

Anderson, J. A. and E. Rosenfeld 2000. *Talking Nets: An Oral History of Neural Networks.* Bradford Book. Cambridge, MA: MIT Press.

Anderson, J. A., J. W. Silverstein, S. A. Ritz, and R. S. Jones 1977. Distinctive features, categorical perception, and probability learning: Some applications of a neural model. *Psychological Review 84*, 413–451.

Andrews, H. C. 1972. *Introduction to Mathematical Techniques in Pattern Recognition.* New York: Wiley-Interscience.

Bagul, Y. J. 2017. A smooth transcendental approximation to $|x|$. *International Journal of Mathematical Sciences and Engineering Applications 11*, 213–217.

Bahl, J. M. and C. Guyeux 2013. *Discrete Dynamical Systems and Chaotic Machines: Theory and Applications.* Numerical Analysis and Scientific Computing Series. Boca Raton, FL: CRC Press.

Baird, L. and A. Moore 1999. Gradient descent for general reinforcement learning. In M. Kearns, S. A. Solla, and D. A. Cohn (Eds.), *Advances in Neural Information Processing Systems*, Volume 11, Cambridge, MA: MIT Press.

Banerjee, S. and A. Roy 2014. *Linear Algebra and Matrix Analysis for Statistics.* Texts in Statistical Science. Boca Raton, FL: Chapman-Hall/CRC Press.

Barron, A., J. Rissanen, and B. Yu 1998, September. The minimum description length principle in coding and modeling. *IEEE Transactions on Information Theory 44*(6), 2743–2760.

Bartle, R. G. 1966. *The Elements of Integration.* New York: Wiley.

Bartlett, M. S. 1951, 03. An inverse matrix adjustment arising in discriminant analysis. *The Annals of Mathematical Statistics 22*(1), 107–111.

Bates, D. M. and D. G. Watts 2007. *Nonlinear Regression Analysis and Its Applications.*

Baydin, A. G., B. A. Pearlmutter, A. A. Radul, and J. M. Siskind 2017, January. Automatic differentiation in machine learning: A survey. *Journal of Machine Learning Research 18*(1), 5595–5637.

Beale, E., M. Kendall, and D. Mann 1967. The discarding of variables in multivariate analysis. *Biometrika 54*(3-4), 357–366.

Bearden, A. F. 1997. Utility representation of continuous preferences. *Economic Theory 10*, 369–372.

Bellman, R. 1961. *Adaptive Control Processes: A Guided Tour.* Princeton, NJ: Princeton University Press.

Bengio, Y., J. Louradour, R. Collobert, and J. Weston 2009. Curriculum learning. In *Proceedings of the 26th Annual International Conference on Machine Learning*, ICML '09, New York, NY, USA, 41–48. ACM.

Benveniste, A., M. Métivier, and P. Priouret 1990. *Adaptive Algorithms and Stochastic Approximations.* Applications of Mathematics Series. New York: Springer-Verlag.

Berger, J. O. 2006. *Statistical Decision Theory and Bayesian Analysis* (Second ed.). New York: Springer-Science.

Bertsekas, D. P. 1996. *Constrained Optimization and Lagrange Multiplier Methods.* Belmont, MA: Athena Scientific.

Bertsekas, D. P. and S. Shreve 2004. *Stochastic Optimal Control: The Discrete-time Case.* Athena Scientific.

Bertsekas, D. P. and J. N. Tsitsiklis 1996. *Neuro-dynamic Programming.* Belmont, MA: Athena Scientific.

Besag, J. 1974. Spatial interaction and the statistical analysis of lattice systems. *Journal of the Royal Statistical Society. Series B (Methodological) 36*, 192–236.

Besag, J. 1975. Statistical analysis of non-lattice data. *Journal of the Royal Statistical Society. Series D (The Statistician). 24*, 179–195.

Besag, J. 1986. On the statistical analysis of dirty pictures. *Journal of the Royal Statistical Society B48*, 259–302.

Bickel, P. J. and D. A. Freedman 1981. Some asymptotic theory for the bootstrap. *The Annals of Statistics 9*, 1196–1217.

Bickel, P. J., F. Gotze, and W. R. van Zwet 1997. Resampling fewer than $n$ observations: Gains, losses, and remedies for losses. *Statistica Sinica 7*, 1–31.

Billingsley, P. 2012. *Probability and Measure.* Wiley Series in Probability and Statistics. Hoboken, NJ: Wiley.

Bishop, C. M. 2006. *Pattern Recognition and Machine Learning.* Information Science. New York: Springer Verlag.

Blum, J. R. 1954, 12. Multidimensional stochastic approximation methods. *The Annals of Mathematical Statistics 25*(4), 737–744.

Borkar, V. S. 2008. *Stochastic Approximation: A Dynamical Systems Viewpoint.* India: Hindustan Book Agency.

Bottou, L. 1998. Online algorithms and stochastic approximations. In D. Saad (Ed.), *Online Learning and Neural Networks*, 146–168. Cambridge, UK: Cambridge University Press. revised, Oct 2012.

Bottou, L. 2004. Stochastic learning. In O. Bousquet and U. von Luxburg (Eds.), *Advanced Lectures on Machine Learning*, Lecture Notes in Artificial Intelligence, LNAI 3176, 146–168. Berlin: Springer Verlag.

Boucheron, S., G. Lugosi, and P. Massart 2016. *Concentration Inequalities: A Nonasymptotic Theory of Independence.* Oxford: Oxford University Press.

Bozdogan, H. 1987. Model selection and Akaike's information criterion (AIC): The general theory and its analytical extensions. *Psychometrika 52*, 345–370.

Bozdogan, H. 2000, March. Akaike's information criterion and recent developments in information complexity. *Journal of Mathematical Psychology 44*(1), 62–91.

Bremaud, P. 1999. *Markov Chains: Gibbs Fields, Monte Carlo Simulation, and Queues.* Texts in Applied Mathematics. Springer New York.

Burnham, K. P. and D. R. Anderson 2010. *Model Selection and Inference: A Practical Information-theoretic Approach.* New York: Springer.

Cabessa, J. and H. T. Siegelmann 2012. The computational power of interactive recurrent neural networks. *Neural Computation 24*(4), 996–1019.

Campolucci, P., A. Uncini, and F. Piazza 2000, August. A signal-flow-graph approach to on-line gradient calculation. *Neural Computation 12*(8), 1901–1927.

Cernuschi-Frias, B. 2007, 7. Mixed states Markov random fields with symbolic labels and multidimensional real values. Techreport arXiv:0707.3986, Institut National de Recherche en Enformatique et en Automatique (INRIA), France.

Chao, M.-T. and S.-H. Lo 1985. A bootstrap method for finite population. *Sankhya: The Indian Journal of Statistics 47*, 399–405.

Chauvin, Y. 1989. A back-propagation algorithm with optimal use of hidden units. In D. S. Touretzky (Ed.), *Advances in Neural Information Processing Systems 1*, 519–526. Morgan-Kaufmann.

Cho, J. S. and P. C. Phillips 2018. Pythagorean generalization of testing the equality of two symmetric positive definite matrices. *Journal of Econometrics 202*(2), 45–56.

Cho, J. S. and H. White 2014. Testing the equality of two positive-definite matrices with application to information matrix testing. In T. B. Fombay, Y. Chang, and J. Park (Eds.), *Essays in Honor of Peter C. B. Phillips*, Volume 33 of *Advances in Econometrics*, Bingley UK, 491–556. Emerald Group.

Cho, K., B. van Merrienboer, Ç. Gülçehre, F. Bougares, H. Schwenk, and Y. Bengio 2014. Learning phrase representations using RNN encoder-decoder for statistical machine translation. *CoRR abs/1406.1078*.

Claeskens, G. and N. L. Hjort 2008. *Model Selection and Model Averaging*. Cambridge Series in Statistics and Probabilistic Mathematics. New York: Cambridge University Press.

Cohen, F. S. and D. B. Cooper 1987. Simple parallel hierarchical and relaxation algorithms for segmenting noncausal Markovian random fields. *IEEE Transactions on Pattern Analysis and Machine Intelligence 9*(2), 195–219.

Cover, T. M. and J. A. Thomas 2006. *Elements of Information Theory*. New Jersey: Wiley.

Cowell, R. 1999a. Advanced inference in Bayesian networks. In M. I. Jordan (Ed.), *Learning in Graphical Models*, Cambridge, MA: MIT Press.

Cowell, R. 1999b. Introduction to inference for Bayesian networks. In M. I. Jordan (Ed.), *Learning in Graphical Models*, Cambridge, MA: MIT Press.

Cowell, R., P. Dawid, S. Lauritzen, and D. Spiegelhalter 2006. *Probabilistic Networks and Expert Systems: Exact Computational Methods for Bayesian Networks*. Information Science and Statistics. Springer Science & Business Media.

Cox, R. T. 1946. Probability, frequency and reasonable expectation. *American Journal of Physics 14*(1), 1–13.

Cristianini, N. and J. Shawe-Taylor 2000. *An Introduction to Support Vector Machines and Other Kernel-based Learning Methods*. Cambridge: Cambridge University Press.

Cross, G. R. and A. K. Jain 1983, January. Markov random field texture models. *IEEE Transactions on Pattern Analysis and Machine Intelligence 5*(1), 25–39.

Cybenko, G. 1989. Approximation by superpositions of a sigmoidal function. *Mathematics of Control, Signals, and Systems 2*, 303–314.

Darken, C. and J. Moody 1992. Towards faster stochastic gradient search. In J. E. Moody, S. J. Hanson, and R. P. Lippmann (Eds.), *Advances in Neural Information Processing Systems 4*, 1009–1016. Morgan-Kaufmann.

Davidson, J. 2002. *Stochastic Limit Theory: An Introduction for Econometricians*. Advanced Texts in Econometrics. Oxford: OUP Oxford.

Davidson-Pilon, C. 2015. *Bayesian Methods for Hackers: Probabilistic Programming and Bayesian Inference*. Addison Wesley Data and Analytics. New York: Addison Wesley Professional.

Davis, M. 2006. Why there is no such discipline as hypercomputation. *Applied Mathematics and Computation 178*(1), 4–7. Special Issue on Hypercomputation.

Davison, A. C. and D. V. Hinkley 1997. *Bootstrap Methods and Their Application*. Cambridge Series in Statistical and Probabilistic Mathematics. New York: Cambridge University Press.

Davison, A. C., D. V. Hinkley, and G. A. Young 2003. Recent developments in bootstrap methodology. *Statistical Science 18*, 141–157.

Delyon, B., M. Lavielle, and E. Moulines 1999, 03. Convergence of a stochastic approximation version of the EM algorithm. *The Annals of Statistics 27*(1), 94–128.

Dennis, J. E. and R. B. Schabel 1996. *Numerical Methods for Unconstrained Optimization and Nonlinear Equations.* Classics in Applied Mathematics. Englewoods, NJ: Society for Industrial and Applied Mathematics.

Djuric, P. M. 1998. Asymptotic MAP criteria for model selection. *IEEE Transactions on Signal Processing 46*, 2726–2735.

Domingos, P. and D. Lowd 2009. *Markov Logic: An Interface Layer for Artificial Intelligence*, Volume 3 of *Synthesis Lectures on Artificial Intelligence and Machine Learning.* Morgan & Claypool Publishers.

Dong, G. and H. Liu 2018. *Feature Engineering for Machine Learning and Data Analytics*, Volume First edition of *Chapman & Hall/CRC Data Mining and Knowledge Discovery Series.* Boca Raton, FL: CRC Press.

Doria, F. A. and J. F. Costa 2006. Introduction to the special issue on hypercomputation. *Applied Mathematics and Computation 178*(1), 1 – 3. Special Issue on Hypercomputation.

Douc, R., E. Moulines, and D. S. Stoffer 2014. *Nonlinear Time Series: Theory, Methods, and Applications with R Examples.* Texts in Statistical Science. Boca Raton, Florida: CRC Press.

Duchi, J., E. Hazan, and Y. Singer 2011. Adaptive subgradient methods for online learning and stochastic optimization. *Journal of Machine Learning Research 12*, 2121–2159.

Duda, R. O. and P. A. Hart 1973. *Pattern Classification and Scene Analysis.* New York: John Wiley and Sons.

Duda, R. O., P. E. Hart, and D. G. Stork 2001. *Pattern Classification* (Second ed.). New York: John Wiley & Sons.

Efron, B. 1982. *The Jackknife, the Bootstrap and Other Resampling Plans.* CA: Society for Industrial and Applied Mathematics.

Efron, B. and R. Tibshirani 1993. *An Introduction to the Bootstrap.* Number 57 in Monographs on Statistics and Applied Probability. 93004489 GB93-60388 Bradley Efron and Robert J. Tibshirani. Includes bibliographical references (p. [413]–425) and indexes.

Elman, J. L. 1990. Finding structure in time. *Cognitive Science 14*, 179–211.

Elman, J. L. 1991. Distributed representations, simple recurrent networks, and grammatical structure. *Machine Learning 7*, 195–225.

Evans, M. and T. Swartz 2005. *Approximating Integrals via Monte Carlo and Deterministic Methods.* Oxford Statistical Science Series. New York: Oxford.

Fourier, J. B. 1822. *Theorie Analytique de la Chaleur.* Paris: Chez Firmin Didot, Pere et Fils.

Franklin, J. N. 1968. *Matrix Theory.* Englewood Cliffs, NJ: Prentice-Hall.

Fukushima, K. 1980. Neocognitron: A self-organizing neural network model for a mechanism of pattern recognition unaffected by shift in position. *Biological Cybernetics 36*, 193–202.

Fukushima, K. and S. Miyake 1982. Neocognitron: A new algorithm for pattern recognition tolerant of deformations and shifts in position. *Pattern Recognition 15*(6), 455–469.

Gamerman, D. and H. F. Lopes 2006. *Markov Chain Monte Carlo: Stochastic Simulation for Bayesian Inference*. Texts in Statistical Science. Boca Raton, Florida: CRC Press.

Geman, S. and D. Geman 1984. Stochastic relaxation, Gibbs distributions, and the Bayesian restoration of images. *IEEE Transactions on Pattern Analysis and Machine Intelligence 6*, 721–741.

Géron, A. 2019. *Hands-on Machine Learning with Scikit-Learn and TensorFlow: Concepts, Tools, and Techniques to Build Intelligent Systems*. Sebastapol, CA: O'Reilly Media, Inc.

Getoor, L. and B. Taskar 2007. *Introduction to Statistical Relational Learning*, Volume 1. Cambridge, MA: MIT Press.

Goldberg, R. R. 1964. *Methods of Real Analysis*. MA: Xerox College Publishing.

Golden, R. M. 1986. The "brain-state-in-a-box" neural model is a gradient descent algorithm. *Journal of Mathematical Psychology 30*(1), 73–80 73 – 80.

Golden, R. M. 1988a. A unified framework for connectionist systems. *Biological Cybernetics 59*, 109–120.

Golden, R. M. 1988b. Relating neural networks to traditional engineering approaches. In *The Proceedings of the Artificial Intelligence Conference West*. Tower Conference Management Company.

Golden, R. M. 1988c. Probabilistic characterization of neural model computations. In D. Z. Anderson (Ed.), *Neural Networks and Information Processing*. AIP.

Golden, R. M. 1993. Stability and optimization analyses of the generalized brain-state-in-a-box neural network model. *Journal of Mathematical Psychology 37*, 282–298.

Golden, R. M. 1995. Making correct statistical inferences using a wrong probability model. *Journal of Mathematical Psychology 39*(1), 3–20.

Golden, R. M. 1996a. *Mathematical Methods for Neural Network Analysis and Design*. Cambridge, MA: MIT Press.

Golden, R. M. 1996b. Using Marr's framework to select appropriate mathematical methods for neural network analysis and design. In *Proceedings of the 1996 World Congress on Neural Networks*, NJ, 1007–1010. INNS Press, Erlbaum.

Golden, R. M. 1996c. Interpreting objective function minimization as intelligent inference. In *Intelligent Systems: A Semiotic Perspective, Proceedings of An International Multidisciplinary Conference. Volume 1: Theoretical Semiotics*. US Government Printing Office.

Golden, R. M. 2003. Discrepancy risk model selection test theory for comparing possibly misspecified or nonnested models. *Psychometrika 68*(2), 229–249.

Golden, R. M. 2018. Adaptive learning algorithm convergence in passive and reactive environments. *Neural Computation 30*, 2805–2832.

Golden, R. M., S. S. Henley, H. White, and T. M. Kashner 2013. New directions in information matrix testing: Eigenspectrum tests. In X. Chen and N. Swanson (Eds.), *Recent Advances and Future Directions in Causality, Prediction, and Specification Analysis*, 145–177. Springer.

Golden, R. M., S. S. Henley, H. White, and T. M. Kashner 2016. Generalized information matrix tests for detecting model misspecification. *Econometrics 4*(4), 1–24.

Golden, R. M., S. S. Henley, H. White, and T. M. Kashner 2019. Consequences of model misspecification for maximum likelihood estimation with missing data. *Econometrics 7*, 1–27.

Golden, R. M., S. Nandy, and V. Patel 2019. Cross-validation nonparametric bootstrap study of the Linhart-Volkers-Zucchini out-of-sample prediction error formula for logistic regression modeling. In *2019 Joint Statistical Meeting Proceedings.* Alexandria, VA: American Statistical Association.

Golden, R. M., S. Nandy, V. Patel, and P. Viraktamath 2015. *A Laplace approximation for approximate Bayesian model selection.* NIPS 2015 Workshop on Advances in Approximate Bayesian Inference, Palais des Congrès de Montréal, Montréal, Canada.

Goodfellow, I., Y. Bengio, and A. Courville 2016. *Deep Learning.* Cambridge, MA: MIT Press.

Goodfellow, I., J. Pouget-Abadie, M. Mirza, B. Xu, D. Warde-Farley, S. Ozair, A. Courville, and Y. Bengio 2014. Generative adversarial nets. In Z. Ghahramani, M. Welling, C. Cortes, N. D. Lawrence, and K. Q. Weinberger (Eds.), *Advances in Neural Information Processing Systems 27*, 2672–2680. Curran Associates, Inc.

Grandmont, J.-M. 1972. Continuity properties of a von Neumann-Morgenstern utility. *Journal of economic theory 4*(1), 45–57.

Grunwald, P. D. 2007. *The Minimum Description Length Principle.* Adaptive Computation and Machine Learning. Cambridge, MA: MIT Press.

Gu, M. G. and F. H. Kong 1998. A stochastic approximation algorithm with Markov chain Monte-Carlo method for incomplete data estimation problems. *Proceedings of the National Academy of Sciences of the United States of America*, 7270–7274.

Guliyev, N. J. and V. E. Ismailov 2018. Approximation capability of two hidden layer feedforward neural networks with fixed weights. *Neurocomputing 316*, 262–269.

Hansen, M. H. and B. Yu 2001. Model selection and the principle of minimum description length. *Journal of the American Statistical Association 96*(454), 746–774.

Hartigan, J. A. 1975. *Clustering Algorithms.* New York, NY, USA: John Wiley & Sons, Inc.

Harville, D. A. 2018. *Linear Models and the Relevant Distributions and Matrix Algebra.* Texts in Statistical Science. Boca Raton, FL: Chapman and Hall/CRC.

Hassibi, B. and D. G. Stork 1993. Second order derivatives for network pruning: Optimal brain surgeon. In S. J. Hanson, J. D. Cowan, and C. L. Giles (Eds.), *Advances in Neural Information Processing Systems 5*, San Mateo, CA, 164–171. Morgan Kaufmann.

Hassibi, B., D. G. Stork, and G. Wolff 1993. Optimal brain surgeon and general network pruning. In *IEEE International Conference on Neural Networks*, Volume 1, 239–299.

Hassibi, B., D. G. Stork, G. Wolff, and T. Watanabe 1994. Optimal brain surgeon: Extensions and performance comparisons. In J. D. Cowan, G. Tesauro, and J. Alspector (Eds.), *Advances in Neural Information Processing Systems 6*, 263–270. Morgan-Kaufmann.

Hastie, T., R. Tibshirani, and J. H. Friedman 2001. *The Elements of Statistical Learning: Data Mining, Inference, and Prediction.* Springer Series in Statistics. New York: Springer Science+Business Media.

Hastings, W. K. 1970, 04. Monte Carlo sampling methods using Markov chains and their application. *Biometrika 57*, 97–109.

Henley, S. S., R. M. Golden, and T. M. Kashner 2020. Statistical modeling methods: Challenges and strategies. *Biostatistics and Epidemiology*, *4*(1), 105–139.

Hinton, G. and S. Roweis 2002. Stochastic neighbor embedding. In *Proceedings of the 15th International Conference on Neural Information Processing Systems*, Volume C. Cortes and N.D. Lawrence and D.D. Lee and M. Sugiyama and R. Garnett. of *Neural Information Processing Systems 28*, Cambridge, MA, USA, 857–864. MIT Press.

Hiriart-Urruty, J. B. and C. Lemarechal 1993. *Convex Analysis and Minimization Algorithms 1: Fundamentals.* A Series of Comprehensive Studies in Mathematics. New York: Springer-Verlag.

Hirsch, M. W., S. Smale, and R. L. Devaney 2004. *Differential Equations, Dynamical Systems, and an Introduction to Chaos* (2nd ed.). Waltham, MA: Academic Press.

Hjort, N. L. and G. Claeskens 2003. Frequentist model average estimators. *Journal of the American Statistical Association 98*(464), 879–899.

Hoerl, A. E. and R. W. Kennard 1970. Ridge regression: Biased estimation for nonorthogonal problems. *Technometrics 12*(1), 55–67.

Hoffmann, P. H. W. 2016, Jul. A hitchhiker's guide to automatic differentiation. *Numerical Algorithms 72*(3), 775–811.

Hopcroft, J. E., R. Motwani, and J. D. Ullman 2001. *Introduction to Automata Theory, Languages, and Computation.* Boston: Addison-Wesley.

Hopfield, J. J. 1982. Neural networks and physical systems with emergent collective computational abilities. *Proceedings of the National Academy of Sciences 79*(8), 2554–2558.

Hornik, K. 1991. Approximation capabilities of multilayer feedforward networks. *Neural Networks 4*, 251–257.

Hornik, K., M. Stinchcombe, and H. White 1989. Multilayer feedforward networks are universal approximators. *Neural Networks 2*(5), 359–366.

Horowitz, J. L. 2001. The bootstrap. In J. J. Heckman and E. Leamer (Eds.), *Handbook of Econometrics*, The Netherlands. Elsevier Science B. V.

Hsu, L. C. 1948, 09. A theorem on the asymptotic behavior of a multiple integral. *Duke Mathematical Journal 15*(3), 623–632.

Huber, P. J. 1964, 03. Robust estimation of a location parameter. *The Annals of Mathematical Statistics 35*(1), 73–101.

Huber, P. J. 1967. The behavior of maximum likelihood estimates under nonstandard conditions. In *Proceedings of the Fifth Berkeley Symposium on Mathematical Statistics and Probability, Volume 1: Statistics*, Berkeley, Calif., 221–233. University of California Press.

Imaizumi, M. and K. Fukumizu 2019, 16–18 Apr. Deep neural networks learn non-smooth functions effectively. In *Proceedings of Machine Learning Research*, Volume 89, 869–878. PMLR.

Jaffray, J.-Y. 1975. Existence of a continuous utility function: An elementary proof. *Econometrica 43*, 981–983.

James, G., D. Witten, T. Hastie, and R. Tibshirani 2013. *An Introduction to Statistical Learning*, Volume 112 of *Springer Texts in Statistics*. New York: Springer.

Jennrich, R. I. 1969, 04. Asymptotic properties of non-linear least squares estimators. *The Annals of Mathematical Statistics 40*(2), 633–643.

Jiang, B., Tung-Yu, Y. Jin, and W. H. Wong 2018. Convergence of contrastive divergence algorithm in exponential family. *The Annals of Statistics 46*, 3067–3098.

Kaiser, M. S. and N. Cressie 2000. The construction of multivariate distributions from Markov random fields. *Journal of Multivariate Analysis 73*(2), 199–220.

Kalman, R. E. 1963. Mathematical description of linear dynamical systems. *Journal of the Society for Industrial and Applied Mathematics, Series A: Control 1*(2), 152–192.

Kalman, R. E., P. L. Falb, and M. A. Arbib 1969. *Topics in Mathematical Systems Theory*. International Series in Pure and Applied Mathematics. New York: McGraw-Hill.

Karr, A. F. 1993. *Probability*. Springer Texts in Statistics. New York: Springer Science Business Media.

Kashner, T. M., S. S. Henley, R. M. Golden, and X.-H. Zhou 2020. Making causal inferences about treatment effect sizes from observational datasets. *Biostatistics and Epidemiology 4*(1), 48–83.

Kass, R. E. and A. E. Raftery 1995. Bayes factors. *Journal of the American Statistical Association 90*(430), 773–795. Times Cited: 2571.

Klir, G. and B. Yuan 1995. *Fuzzy Sets and Fuzzy Logic*, Volume 4. New Jersey: Prentice Hall.

Klir, G. J. and T. A. Folger 1988. *Fuzzy Sets, Uncertainty and Information*. Englewood Cliffs, NJ: Prentice-Hall.

Knopp, K. 1956. *Infinite Sequences and Series*. New York: Dover.

Koller, D., N. Friedman, and F. Bach 2009. *Probabilistic Graphical Models: Principles and Techniques*. Cambridge, MA: MIT Press.

Kolmogorov, A. N. and S. V. Fomin 1970. *Introductory Real Analysis*. New York: Dover.

Konidaris, G. and A. Barto 2006. Autonomous shaping: Knowledge transfer in reinforcement learning. In *Proceedings of the 23rd International Conference on Machine Learning*, ICML '06, New York, NY, USA, 489–496. ACM.

Konishi, S. and G. Kitagawa 2008. *Information Criteria and Statistical Modeling*. Springer Series in Statistics. New York: Springer-Verlag.

Kopciuszewski, P. 2004. An extension of the factorization theorem to the non-positive case. *Journal of Multivariate Analysis 88*, 118–130.

Kushner, H. 2010. Stochastic approximation: A survey. *Computational Statistics 2*(1), 87–96.

Kushner, H. and G. Yin 2003. *Stochastic Approximation and Recursive Algorithms and Applications.* Stochastic Modelling and Applied Probability. New York: Springer-Verlag.

Landauer, T. K., D. S. McNamara, S. Dennis, and W. Kintsch 2014. *Handbook of Latent Semantic Analysis.* New York: Routledge.

Larson, H. J. and B. O. Shubert 1979. *Probabilistic Models in Engineering Sciences. Volume 1: Random Variables and Stochastic Processes.* New York: Wiley.

LaSalle, J. 1960, December. Some extensions of Liapunov's second method. *IRE Transactions on Circuit Theory 7*(4), 520–527.

LaSalle, J. P. 1976. *The Stability of Dynamical Systems.* Regional Conference Series in Applied Mathematics. Philadelphia: SIAM.

Lauritzen, S. L. 1996. *Graphical Models*, Volume 17 of *Oxford Statistical Science Series.* Oxford: Oxford University Press.

Le Cun, Y. 1985. Une procedure d'apprentissage ponr reseau a seuil asymetrique. In *Proceedings of Cognitiva*, Volume 85, 599–604.

Le Roux, N., P.-A. Manzagol, and Y. Bengio 2008. Topmoumoute online natural gradient algorithm. In J. C. Platt, D. Koller, Y. Singer, and S. T. Roweis (Eds.), *Advances in Neural Information Processing Systems 20*, 849–856, Curran Associates, Inc.

Lee, J. D. and T. J. Hastie 2015. Learning the structure of mixed graphical models. *Journal of Computational and Graphical Statistics 24*, 230–253.

Lee, T., H. White, and C. Granger 1993. Testing for neglected nonlinearity in time series models: A comparison of neural network methods and alternative tests. *Journal of Econometrics.*

Lehmann, E. L. and G. Casella 1998. *Theory of Point Estimation.* Springer Texts in Statistics. New York: Springer-Verlag.

Lehmann, E. L. and J. P. Roman 2005. *Testing Statistical Hypotheses.* Springer Texts in Statistics. New York: Springer.

Leshno, M., V. Lin, A. Pinkus, and S. Schocken 1993. Multilayer feedforward networks with a nonpolynomial activation function can approximate any function. *Neural Networks 6*, 861–867.

Lewis, F. L., D. Vrabie, and K. G. Vamvoudakis 2012, Dec. Reinforcement learning and feedback control: Using natural decision methods to design optimal adaptive controllers. *IEEE Control Systems Magazine 32*(6), 76–105.

Linhart, H. and P. Volkers 1984. Asymptotic criteria for model selection. *OR Spektrum 6*, 161–165.

Linhart, H. and W. Zucchini 1986. *Model Selection.* Wiley Series in Probability and Statistics. Hoboken, NJ: Wiley.

Little, R. J. A. and D. B. Rubin 2002. *Statistical Analysis with Missing Data.* Wiley Series in Probability and Statistics. Hoboken, NJ: Wiley.

Louis, T. A. 1982. Finding the observed information matrix when using the EM algorithm. *Journal of the Royal Statistical Society, Series B (Methodological) 44*, 226–233.

Luenberger, D. G. 1979. *Introduction to Dynamic Systems: Theory, Models, and Applications.* New York: Wiley.

Luenberger, D. G. 1984. *Linear and Nonlinear Programming*, Volume 67. Reading, MA: Addison-Wesley.

Lukacs, E. 1975. *Stochastic Convergence.* Probability and Mathematical Statistics Series. New York: Academic Press.

Lv, J. and J. S. Liu 2014. Model selection principles in misspecified models. *Journal of the Royal Statistical Society: Series B (Statistical Methodology) 76*(1), 141–167.

Mackay, D. 2003. *Information Theory, Inference, and Learning Algorithms.* Cambridge, UK: Cambridge University Press.

Maclennan, B. J. 2003. Transcending Turing computability. *Minds and Machines 13*, 3–22.

Magnus, J. R. 2010. On the concept of matrix derivative. *Journal of Multivariate Analysis 10*, 2200–2206.

Magnus, J. R. and H. Neudecker 2001. *Matrix Differential Calculus with Applications in Statistics and Econometrics* (Revised ed.). New York: Wiley.

Manoukian, E. B. 1986. *Modern Concepts and Theorems of Mathematical Statistics.* New York: Springer-Verlag.

Marlow, W. H. 2012. *Mathematics for Operations Research.* New York: Dover.

Marr, D. 1982. *Vision: A Computational Investigation into the Human Representation and Processing of Visual Information.* Cambridge, MA: MIT Press.

McCulloch, W. S. and W. Pitts 1943. A logical calculus of the ideas immanent in nervous activity. *The Bulletin of Mathematical Biophysics 5*(4), 115–133.

McLachlan, G. J. and T. Krishnan 2008. *The EM Algorithm and Extensions.* Wiley Series in Probability and Statistics. New York: Wiley.

McNeil, D. and P. Freiberger 1994. *Fuzzy Logic: The Revolutionary Computer Technology that Is Changing Our World.* New York: Touchstone.

Mehta, G. 1985. Continuous utility functions. *Economics Letters 18*, 113–115.

Metropolis, A. Rosenbluth, M. Rosenbluth, A. Teller, and E. J. Teller 1953, 01. Equation of state calculations by fast computing machines. *Journal of Chemical Physics 21*, 1087–1092.

Miller, A. J. 2002. *Subset Selection in Regression.* Monographs on Statistics and Applied Probability. Boca Raton, FL: Chapman Hall/CRC.

Minsky, M. and S. Papert 1969. *Perceptrons.* Cambridge, MA: MIT Press.

Mohri, M., A. Rostamizadeh, and A. Talwalkar 2012. *Foundations of Machine Learning.* Adaptive Computation and Machine Learning Series. Cambridge, MA: MIT Press.

Molchanov, P., A. Mallya, S. Tyreen, L. Frosio, and J. Kautz 2019, June. Importance estimation for neural network pruning. In *The IEEE Conference on Computer Vision and Pattern Recognition (CVPR).*

Molchanov, P., S. Tyreen, T. Karras, T. Aila, and J. Kautz 2017. Pruning convolutional neural networks for resource efficient transfer learning. In *The IEEE Conference on Computer Vision and Pattern Recognition (CVPR).*

Montgomery, D. C. and E. A. Peck 1982. *Introduction to Linear Regression Analysis.* Wiley Series in Probability and Mathematical Statistics. New York: Wiley.

Mozer, M. C. and P. Smolensky 1989. Skeletonization: A technique for trimming the fat from a network via relevance assessment. In D. S. Touretzky (Ed.), *Advances in Neural Information Processing Systems 1*, 107–115. Morgan-Kaufmann.

Muller, A. C. and S. Guido 2017. *Introduction to Machine Learning with Pylhon: A Guidc for Data Scientists.* Sebastopol, CA: O'Reilly Media.

Murphy, K. P. 2012. *Machine Learning: A Probabilistic Perspective.* Adaptive Computation and Machine Learning Series. Cambridge, MA: MIT Press.

Myung, I. J., M. R. Forster, and M. W. Browne 2000. Guest editors introduction: Special issue on model selection. *Journal of Mathematical Psychology 44*, 1–2.

Neudecker, H. 1969. Some theorems on matrix differentiation with special reference to Kronecker matrix products. *Journal of the American Statistical Association 64*, 953–963.

Nilsson, N. J. 1965. *Learning Machines: Foundations of Trainable Pattern-classifying Systems.* New York: McGraw-Hill.

Noble, B. and J. W. Daniel 1977. *Applied Linear Algebra* (Second ed.). NJ: Prentice-Hall.

Nocedal, J. and S. Wright 1999. *Numerical Optimization.* Springer Series in Operations Research and Financial Engineering. New York: Springer Science + Business Media.

Orhan, A. E. and Z. Pitkow 2018. Skip connections eliminate singularities. In *ICLR (International Conference on Learning Representations)*, Volume 6, https://iclr.cc/Conferences/2018.

Osowski, S. 1994, Feb. Signal flow graphs and neural networks. *Biological Cybernetics 70*(4), 387–395.

Pan, S. J. and Q. Yang 2010. A survey on transfer learning. *IEEE Transactions on Knowledge and Data Engineering 22*(10), 1345–1359.

Parker, D. B. 1985. *Learning Logic Report 47*, Volume TR-47. MIT Sloan School of Management, Cambridge, MA.

Patton, A., D. N. Politis, and H. White 2009. Correction to "automatic block-length selection for the dependent bootstrap". *Econometric Reviews 28*, 372–375.

Petzold, C. 2008. *The Annotated Turing: A Guided Tour through Turing's Historic Paper on Computability and the Turing Machine.* Indianapolis, IN: Wiley.

Pfeffer, A. 2016. *Practical Probabilistic Programming.* Manning Publications.

Pinkus, A. 1999. Approximation theory of the MLP model in neural networks. *Acta Numerica 8*, 143–195.

Politis, D. N. and J. P. Romano 1994. The stationary bootstrap. *Journal of the American Statistical Association 89*, 1303–1313.

Politis, D. N., J. P. Romano, and M. Wolf 2001. On the asymptotic theory of subsampling. *Statistica Sinica 11*, 1105–1124.

Politis, D. N. and H. White 2004. Automatic block-length selection for the dependent bootstrap. *Econometric Reviews 23*, 53–70.

Polyak, B. 1964, 12. Some methods of speeding up the convergence of iteration methods. *USSR Computational Mathematics and Mathematical Physics 4*, 1 –17.

Poskitt, D. S. 1987. Precision, complexity, and Bayesian model determination. *Journal of the Royal Statistical Society, Series B 49*, 199–208.

Ramirez, C., R. Sanchez, V. Kreinovich, and M. Argaez 2014. $\sqrt{x^2 + \mu}$ is the most computationally efficient smooth approximation to $|x|$: A proof. *Journal of Uncertain Systems 8*, 205–210.

Raschka, S. and V. Mirjalili 2019. *Python Machine Learning: Machine Learning and Deep Learning with Python, scikit-learn, and TensorFlow 2.* Packt Publishing.

Robbins, H. and S. Monro 1951, 09. A stochastic approximation method. *The Annals of Mathematical Statistics 22*(3), 400–407.

Robbins, H. and D. Siegmund 1971. A convergence theorem for nonnegative almost supermartingales and some applications. In J. S. Rustagi (Ed.), *Optimizing Methods in Statistics*, New York, 233–257. Academic Press.

Robert, C. P. and G. Casella 2004. *Monte Carlo Statistical Methods.* Springer Texts in Statistics. New York: Springer Science and Business Media.

Romano, J. P. and A. M. Shaikh 2012. On the uniform asymptotic validity of subsampling and the bootstrap. *The Annals of Statistics 40*, 2798–2822.

Rosenblatt, F. 1962. *Principles of Neurodynamics: Perceptrons and the Theory of Brain Mechanisms.* New York: Spartan Book.

Rosenlicht, M. 1968. *Introduction to Analysis.* Dover Books on Mathematics. Dover.

Rosenthal, J. S. 2016. *A First Look at Rigorous Probability Theory.* Singapore: World Scientific Publishing Company.

Rumelhart, D. E., R. Durbin, R. M. Golden, and Y. Chauvin 1996. Backpropagation: The basic theory. In Y. Chauvin and D. E. Rumelhart (Eds.), *Backpropagation: Theory, Architectures, and Applications*, Hillsdale, NJ, 1–34. Erlbaum.

Rumelhart, D. E., G. Hinton, and J. L. McClelland 1986. A general framework for parallel distributed processing. In D. E. Rumelhart and J. L. McClelland (Eds.), *Parallel Distributed Processing: Explorations in the Microstructure of Cognition. Foundations. 1*, Volume 1, Cambridge, MA, 45–76. MIT Press.

Rumelhart, D. E., G. E. Hinton, and R. J. Williams 1986. Learning representations by back-propagating errors. *Nature: International Journal of Science 323*(6088), 533–536.

Savage, L. J. (1954) 1972. *The Foundations of Statistics.* Wiley [Reprinted by Dover in 1972].

Schmidhuber, J. 2015. Deep learning in neural networks: An overview. *Neural Networks 61*, 85–117.

Schmidt, M., G. Fung, and R. Rosales 2007. Fast optimization methods for $L_1$ regularization: A comparative study and two new approaches. In J. N. Kok, J. Koronacki, R. L. d. Mantaras, S. Matwin, D. Mladenič, and A. Skowron (Eds.), *Machine Learning: ECML 2007*, Berlin, Heidelberg, 286–297. Springer Berlin Heidelberg.

Scholkopf, B. and A. J. Smola 2002. *Learning with Kernels: Support Vector Machines, Regularization, Optimization, and Beyond.* Adaptive Computation and Machine Learning. Cambridge, MA: MIT Press.

Schott, J. R. 2005. *Matrix Analysis for Statistics* (Second ed.). Wiley Series in Probability and Statistics. New Jersey: Wiley.

Schwarz, G. 1978, 03. Estimating the dimension of a model. *The Annals of Statistics 6*(2), 461–464.

Serfling, R. J. 1980. *Approximation Theorems of Mathematical Statistics.* Wiley Series in Probability and Statistics. New York: John Wiley & Sons.

Shalev-Shwartz, S. and S. Ben-David 2014. *Understanding Machine Learning: From Theory to Algorithms.* New York: Cambridge University Press.

Shao, J. and D. Tu 1995. *The Jackknife and the Bootstrap.* Springer Series in Statistics. New York: Springer-Verlag.

Shao, X. 2010. The dependent wild bootstrap. *Journal of the American Statistical Association 105*, 218–235.

Sharma, S. 2017. Markov chain Monte Carlo methods for Bayesian data analysis in astronomy. *Annual Review of Astronomy and Astrophysics 55*, 213–259.

Simon, H. A. 1969. *The Sciences of the Artificial.* Cambridge, MA: MIT Press.

Singh, K. 1981. On the asymptotic accuracy of Efron's bootstrap. *The Annals of Statistics 9*, 1187–1195.

Smolensky, P. 1986. Information processing in dynamical systems: Foundations of harmony theory. In *Parallel Distributed Processing: Explorations in the Microstructure of Cognition*, 194–281. Cambridge, MA: MIT Press.

Srivastava, N., G. Hinton, A. Krizhevsky, I. Sutskever, and R. Salakhutdinov 2014. Dropout: A simple way to prevent neural networks from overfitting. *Journal of Machine Learning Research 15*, 1929–1958.

Stone, M. 1977. An asymptotic equivalence of choice of model by cross-validation and Akaike's criterion. *Journal of the Royal Statistical Society. Series B (Methodological) 39*(1), 44–47.

Strang, G. 2016. *Introduction to Linear Algebra.* Wellesley, MA: Wellesley-Cambridge Press.

Sugiyama, M. 2015. *Statistical Reinforcement Learning: Modern Machine Learning Approaches*. Machine Learning and Pattern Recognition. Boca Raton, FL: Chapman & Hall/CRC.

Sunehag, P., J. Trumpf, S. Vishwanathan, and N. N. Schraudolph 2009. Variable metric stochastic approximation theory. In D. V. Dyk and M. Welling (Eds.), *Proceedings of the Twelfth International Conference on Artificial Intelligence and Statistics (AISTATS-09)*, Volume 5, 560–566. Journal of Machine Learning Research - Proceedings Track.

Sutskever, I., J. Marten, G. Dahl, and G. Hinton 2013. On the importance of initialization and momentum in deep learning. In S. Dasgupta and D. McAllester (Eds.), *Proceedings of the 30th International Conference on Machine Learning*, Volume 28 of *Proceedings of the 30th International Conference on Machine Learning*, 1139–1147. JMLR: Workshop and Conference Proceedings.

Sutton, R. S. and A. G. Barto 2018. *Reinforcement Learning: An Introduction* (second ed.). Adaptive Computation and Machine Learning Series. Cambridge, MA: MIT Press.

Takeuchi, K. 1976. Distribution of information statistics and a criterion of model fitting for adequacy of models. *Mathematical Sciences 153*, 12–18.

Taylor, M. E. and P. Stone 2009. Transfer learning for reinforcement learning domains: A survey. *Journal of Machine Learning Research 10*(Jul), 1633–1685.

Theil, H. and A. S. Goldberger 1961. On pure and mixed statistical estimation in economics. *Journal of Economic Review 2*, 65–78.

Tibshirani, R. 1996a. Regression shrinkage and selection via the lasso. *Journal of the Royal Statistical Society. Series B (Methodological) 58*(1), 267–288.

Tibshirani, R. 1996b. A comparison of some error estimates for neural network models. *Neural Computation 8*, 152–163.

Toulis, P., E. Airoldi, and J. Rennie 2014, 22–24 Jun. Statistical analysis of stochastic gradient methods for generalized linear models. In E. P. Xing and T. Jebara (Eds.), *Proceedings of the 31st International Conference on Machine Learning*, Volume 32 of *Proceedings of Machine Learning Research*, Beijing, China, 667–675. PMLR.

Uchida, Y. 2008. A simple proof of the geometric-arithmetic mean inequality. *Journal of Inequalities in Pure and Applied Mathematics 9*, 1–2.

van der Maaten, L. and G. Hinton 2008. Visualizing data using t-SNE. *Journal of Machine Learning Research 9*, 2579–2605.

van der Vaart, A. W. 1998. *Asymptotic Statistics*. Cambridge Series in Statistical and Probabilistic Mathematics. Cambridge: Cambridge University Press.

Vapnik, V. 2000. *The Nature of Statistical Learning Theory*. New York: Springer Science & Business Media.

Varin, C., N. Reid, and D. Firth 2011. An overview of composite likelihood methods. *Statistica Sinica 21*(1), 5–42.

Vidyasagar, M. 1993. *Nonlinear Systems Analysis* (2nd ed.). NJ: Prentice-Hall.

Vincent, P., H. Larochelle, Y. Bengio, and P.-A. Manzagol 2008. Extracting and composing robust features with denoising autoencoders. In *Proceedings of the 25th International Conference on Machine Learning*, ICML '08, New York, NY, USA, 1096–1103. ACM.

Von Neumann, J. and O. Morgenstern (1947) 1953. *Theory of Games and Economic Behavior.* Princeton University Press.

Vuong, Q. H. 1989. Likelihood ratio tests for model selection and non-nested hypotheses. *Econometrica 57*(307-333), 307–333.

Wade, W. R. 1995. *An Introduction to Analysis.* NJ: Prentice-Hall.

Wallace, C. 2005. *Statistical and Inductive Inference by Minimum Message Length (Information Science and Statistics).* Berlin, Heidelberg: Springer-Verlag.

Wapner, L. 2005. *The Pea and the Sun: A Mathematical Paradox.* Taylor & Francis.

Wasan, M. T. (1969) 2004. *Stochastic Approximation.* Cambridge Tracts in Mathematics and Mathematical Physics. New York: Cambridge University Press.

Wasserman, L. 2000. Bayesian model selection and model averaging. *Journal of Mathematical Psychology 44*, 92–107.

Watanabe, S. 2010. *Algebraic Geometry and Statistical Learning Theory.* Cambridge Monographs on Applied and Computational Mathematics. Cambridge: Cambridge University Press.

Werbos, P. 1974. *New Tools for Prediction and Analysis in the Behavioral Sciences.* Ph. D. dissertation. Harvard University.

Werbos, P. J. 1994. *The Roots of Backpropagation: From Ordered Derivatives to Neural Networks and Political Forecasting.* Adaptive and Learning Systems for Signal Processing, Communications, and Control. New York: Wiley.

White, H. 1982. Maximum likelihood estimation of misspecified models. *Econometrica 51*, 1–25.

White, H. 1989a. Learning in artificial neural networks: A statistical perspective. *Neural Computation 1*, 425–464.

White, H. 1989b. Some asymptotic results for learning in single hidden-layer feedforward network models. *Journal of the American Statistical Association 84*(408), 1003–1013.

White, H. 1994. *Estimation, Inference and Specification Analysis.* Number 22 in Econometric Society Monographs. New York: Cambridge university press.

White, H. 2001. *Asymptotic Theory for Econometricians.* UK: Emerald Publishing.

White, H. and I. Domowitz 1984. Nonlinear regression with dependent observations. *Econometrica 52*, 143–161.

White, H. L. 2004. *New Perspectives in Econometric Theory.* Economists of the Twentieth Century Series. Cheltenham, UK: Edward Elgar.

Widrow, B. and M. Hoff 1960. Adaptive switching circuits. In *IRE Wescon Convention Record*, Volume 54, 96–104. IRE Wescon Convention.

Wiering, M. and M. Van Otterlo 2012. *Reinforcement Learning*, Volume 12 of *Adaptation, Learning, and Optimization Series*. New York: Springer.

Wiggins, S. 2003. *Introduction to Applied Dynamical Systems and Chaos*. Texts in Applied Mathematics. New York: Springer-Verlag.

Wilks, S. S. 1938, 03. The large-sample distribution of the likelihood ratio for testing composite hypotheses. *The Annals of Mathematical Statistics 9*(1), 60–62.

Williams, R. J. 1992. Simple statistical gradient-following algorithms for connectionist reinforcement learning. *Machine Learning 8*, 229–256.

Winkler, G. 2003. *Image Analysis, Random Fields, and Dynamic Monte Carlo Methods: A Mathematical Introduction*. Stochastic Modeling and Applied Probability. New York: Springer.

Wolfe, P. 1969. Convergence conditions for ascent methods. *SIAM Review 11*(2), 226–235.

Wolfe, P. 1971. Convergence conditions for ascent methods. ii: Some corrections. *SIAM Review 13*(2), 185–188.

Yang, E., Y. Baker, P. Ravikumar, G. Allen, and Z. Liu 2014, 22–25 Apr. Mixed graphical models via exponential families. In S. Kaski and J. Corander (Eds.), *Proceedings of the Seventeenth International Conference on Artificial Intelligence and Statistics*, Volume 33 of *Proceedings of Machine Learning Research*, Reykjavik, Iceland, 1042–1050. PMLR.

Yuille, A. L. 2005. The convergence of contrastive divergences. In L. K. Saul, Y. Weiss, and L. Bottou (Eds.), *Advances in Neural Information Processing Systems 17*, 1593–1600. MIT Press.

Zheng, A. and A. Casari 2018. *Feature Engineering for Machine Learning*. Sebastopol, California: O'Reilly Media, Inc.

Zou, H. and T. Hastie 2005. Regularization and variable selection via the elastic net. *Journal of the Royal Statistical Society: Series B (Statistical Methodology) 67*(2), 301–320.

Zoutendijk, G. 1970. Nonlinear programming, computational methods. In J. Abadie (Ed.), *Integer and Nonlinear Programming*, Amsterdam, The Netherlands, 37–86. North Holland.

# *Algorithm Index*

# Subject Index